Lecture Notes in Computer Science 7282

Commenced Publication in 1973
Founding and Former Series Editors:
Gerhard Goos, Juris Hartmanis, and Jan van Leeuwen

Lecture Notes in Computer Science

Poika Isokoski   Jukka Springare (Eds.)

Haptics: Perception, Devices, Mobility, and Communication

International Conference, EuroHaptics 2012
Tampere, Finland, June 13-15, 2012
Proceedings, Part I

 Springer

Volume Editors

Poika Isokoski
Jukka Springare
School of Information Sciences
33014 University of Tampere, Finland
E-mail: {poika.isokoski, jukka-sakari.springare}@uta.fi

ISSN 0302-9743   e-ISSN 1611-3349
ISBN 978-3-642-31400-1   e-ISBN 978-3-642-31401-8
DOI 10.1007/978-3-642-31401-8
Springer Heidelberg Dordrecht London New York

Library of Congress Control Number: 2012940368

CR Subject Classification (1998): H.5.2, H.5, H.4, H.3, K.4, K.4.2

LNCS Sublibrary: SL 3 – Information Systems and Application, incl. Internet/Web
and HCI

Typesetting: Camera-ready by author, data conversion by Scientific Publishing Services, Chennai, India

Printed on acid-free paper

Springer is part of Springer Science+Business Media (www.springer.com)

Preface

Welcome to the proceedings of the EuroHaptics 2012 conference. EuroHaptics is the main meeting for European researchers in the field of haptics but it is also highly international, attracting researchers from all parts of the world. EuroHaptics 2012 took place in Tampere, Finland, during June 13–15. We received a total of 153 submissions in three categories (full papers, short papers, and demo papers). The review process led to 99 of these being accepted for publication (56 full papers, 32 short papers, and 11 demo papers). The material is divided into two volumes. The first volume contains the long papers and the second volume contains the short papers and demo papers. Owing to schedule restrictions some of the long and short papers were presented at the conference as oral presentations and others as posters. In the proceedings, however, all papers are equal.

Conferences cannot function without the challenging work that the referees do. They read and consider each submitted paper, often under a tight schedule, to help the Program Committee choose the best work to be presented. On behalf of the whole EuroHaptics 2012 Organizing Committee I thank the reviewers for their effort, including those that we may have unintentionally omitted from the listings here.

This collection of papers shows, once again, that the field of haptics is of interest in many areas of science and technology all over the world. I am happy to serve this community of researchers as the editor of these proceedings. Hopefully, they will serve us well as a reference and offer many interesting reading sessions.

June 2012   Poika Isokoski

Organization

EuroHaptics 2012 was organized in cooperation between the Tampere Unit for Human–Computer Interaction (TAUCHI) at the School of Information Sciences at the University of Tampere (UTA), Unit for Human Centered Technology (IHTE) at the Tampere University of Technology (TUT), and Nokia Research Center with many volunteers from other organizations.

Executive Committee

General Co-chairs

Roope Raisamo (UTA)
Kaisa Väänänen-Vainio-Mattila (TUT)
Jyri Huopaniemi (Nokia)

Program Co-chairs

Veikko Surakka (UTA)
Vuokko Lantz (Nokia)
Stephen Brewster (University of Glasgow)

Demo Chair

Markku Turunen (UTA)

Poster Chair

Johan Kildal (Nokia)

Workshop Chair

Viljakaisa Aaltonen (Nokia)

Publication Chair

Poika Isokoski (UTA)

Sponsor Chair

Arto Hippula (UTA)

Exhibit Chair

Topi Kaaresoja (Nokia)

Local Arrangements

Päivi Majaranta (UTA)
Teija Vainio (TUT)

Student Volunteer Chair

Thomas Olsson (TUT)

Program Committee

The Program Committee was led by the Program Chairs. The review process relied heavily on meta reviewers who synthesized a recommendation based on the reviews written for each paper by the primary referees. The meta reviewers, in alphabetical order, were:

Margarita Anastassova	Monica Bordegoni
Christos Giachritsis	Matthias Harders
Topi Kaaresoja	Kyung Ki-Uk
Stephen Laycock	Vincent Levesque
Karljohan Lundin Palmerius	David McGookin
Haruo Noma	Ian Oakley
Miguel Otaduy	Jee-Hwan Ryu
Ian Summers	Mark Wright

Referees

Marco Agus	Fawaz Alsulaiman
Hideyuki Ando	Hichem Arioui
Carlo Alberto Avizzano	Mehmet Ayyildiz
Jose M. Azorin	Soledad Ballesteros
Thorsten Behles	Gianni Borghesan
Diego Borro	Andrea Brogni
Ozkan Celik	Nienke Debats
Massimiliano Di Luca	Knut Drewing
Christian Duriez	Marc Ernst
Ildar Farkhatdinov	Irene Fasiello
Antonio Frisoli	Ignacio Galiana
Marcos Garcia	Carlos Garre
Daniel Gooch	Florian Gosselin
Burak Guclu	Blake Hannaford
Christian Hatzfeld	Vincent Hayward
Sandra Hirche	Joe Huegel
Barry Hughe	Rosa Iglesias
Miriam Ittyerah	Seokhee Jeon
Li Jiang	Lynette Jones
Christophe Jouffrais	Sang-Youn Kim
Yoshihiro Kuroda	Tomohiro Kuroda
Hoi Fei Kwok	Piet Lammertse
Anatole Lecuyer	Claudio Loconsole

Table of Contents – Part I

Long Papers

Erratum

Table of Contents – Part II

Short Papers

Demonstration Papers

A Novel Approach for Pseudo-haptic Textures Based on Curvature Information

Ferran Argelaguet, David Antonio Gómez Jáuregui,
Maud Marchal, and Anatole Lécuyer

INRIA Rennes
{fernando.argelaguet_sanz,david.gomez_jauregui,
maud.marchal,anatole.lecuyer}@inria.fr

Abstract. Pseudo-haptic textures allow to optically-induce relief in textures without a haptic device by adjusting the speed of the mouse pointer according to the depth information encoded in the texture. In this work, we present a novel approach for using curvature information instead of relying on depth information. The curvature of the texture is encoded in a normal map which allows the computation of the curvature and local changes of orientation, according to the mouse position and direction. A user evaluation was conducted to compare the optically-induced haptic feedback of the curvature-based approach versus the original depth-based approach based on depth maps. Results showed that users, in addition to being able to efficiently recognize simulated bumps and holes with the curvature-based approach, were also able to discriminate shapes with lower frequency and amplitude.

1 Introduction

Pseudo-haptic feedback is a well-known interaction technique that allows to create an illusion of haptic properties with the combination of visual stimuli and passive input devices [7]. When introducing a decoupling between the visual stimuli and the users' actions, the visual stimuli is dominant and can successfully create something of haptic illusion. Several studies proved that pseudo-haptic feedback can be used to simulate various haptic sensations such as friction [7], the degree of hardness or softness of an object [5], the mass of a virtual object [2] or the relief of a 2D image [6]. Its main advantage is that it does not require any dedicated hardware device to simulate haptic stimuli. For example, to simulate pseudo-haptic feedback in a desktop computer, a standard mouse is enough. Thus, the pseudo-haptic feedback can be used in a wide range of applications (e.g. 2D GUIs, video games, virtual reality, tactile images).

In this paper, we focus on the improvement of pseudo-haptic textures [6]. Pseudo-haptic textures allow to simulate the relief of a 2D texture by adjusting the control-display ratio of the mouse cursor. As the cursor advances along the texture, the CD ratio is adjusted in a per-pixel basis according to the information stored in a depth map. Unfortunately, according to haptic studies [11,12], using only depth information does not provide a strong cue for shape discrimination.

P. Isokoski and J. Springare (Eds.): EuroHaptics 2012, Part I, LNCS 7282, pp. 1–12, 2012.

Fig. 1. Comparison of a generated depth map (center) and normal map (right) of a particular viewpoint of the armadillo model (left). High frequency details cannot be stored in the depth map due to limited spatial resolution. In contrast, high frequency details are better preserved in the normal map.

In addition, it is difficult to balance the effect of the depth-based approach for high depth ranges.

In this work, we introduce the use of curvature information to enhance the illusion of the pseudo-haptic textures. The curvature is encoded into a normal map, which provides information about the curvature at each pixel and determines changes of the local orientation of the surface (see Figure 1). Normal maps are commonly used in computer graphics and haptics to store high frequency data, they can be computed procedurally or using a 3D graphic editing software. Although the curvature information of a surface is an important discriminant for haptic perception, as far as we know, there is no study exploring its integration with pseudo-haptic textures. We believe that curvature information can provide a stronger cue about the shape of textures than depth information.

The remaining of the paper is structured as follows. In Section 2, we discuss previous work on pseudo-haptic textures and haptic perception using curvature information. In Section 3, we present the proposed algorithm based on curvature information. Next, in Section 4, we detail and discuss the results of the user evaluation comparing our approach with the depth-based approach. Finally, Section 5 provides concluding remarks and future work.

2   Related Work

2.1   Pseudo-haptic Textures Based on Depth Maps

Several works have studied the efficiency of pseudo-haptic effects in order to simulate haptic properties. As one of the first experiments, Lécuyer et al. [7] simulated friction with a virtual cube moved by the user using a 2D mouse and a Spaceball. In this experiment, the virtual cube was decelerated when crossing a gray area by altering artificially the Control/Display ratio (Control refers to the speed of hand movement while Display refers to the speed of cursor movement). Here, users were able to perceive *"friction"* with an increased C/D ratio without using a haptic device.

In a posterior study, Lécuyer et al. [6] investigated the simulation of tactile sensations on 2D textures using a pseudo-haptic approach. The main idea consisted in controlling the speed of the mouse cursor as a function of the depth information of the texture over which the mouse cursor was traveling. For example, when the mouse cursor was moving along a surface with a positive slope, the speed of the cursor was decreased; on the opposite case, a negative slope produced an acceleration of the mouse cursor. In these experiments, the users were able to efficiently identify simulated bumps and holes in the texture. Hachisu et al. [4] also used the depth information of the textures to combine the pseudo-haptic effect with visual and tactile vibrations in order to strengthen the tactile perception.

Mensvoort et al. [8] compared the usability of mechanically simulated haptic textures with optically simulated haptic textures. The authors used the slope vector of the texture to increase or decrease the speed of the mouse cursor. In order to compute the slope, they used the depth map of the texture as source of information. Later, the same authors [9] compared the perceptual differences of recognizing bumps and holes using pseudo-haptic textures with the simulation generated by a mechanical force feedback device. The results showed that in some cases, for example, for subtle forces, optically simulated haptic feedback can be even more expressive than mechanical simulations of force feedback.

2.2   Haptic Perception Using Curvature Information

Although previous work present important improvements and applications of pseudo-haptic textures, there is no study that investigates the curvature information of the texture in order to enhance the pseudo-haptic effect. Several prior studies using haptic devices have shown that the orientation of the surface of contact is a dominant source of information for perceiving shape [3]. Bernard et al. [10] presented an experiment in which users had to distinguish an elliptical cylinder from a circular cylinder using a haptic stimuli. In the elliptical cylinder, the local curvature varies over the surface, whereas it is constant for the circular cylinder. The results showed that an ellipse can be distinguished from a circle when divergences or changes in curvature can be perceived. In contrast, when curvature information was lacking, shape recognition performance decreased.

Wijntjes et al. [11] investigated the specific contribution between the curvature (local orientation) and depth cues in order to perceive the shape of curved surfaces. Their results demonstrated that discrimination performances depend largely on the availability of local orientation. They also found that a curved shape that is defined solely by a height profile is hard to discriminate. In a more recent study, Zeng et al. [12] constructed a shape-simulating haptic device that depends totally on the curvature information. The haptic device was able to move and rotate a flat plate. The rotation angle was dependent on the position of the contact point and it varied following the position of fingertip in order to satisfy the local orientation of the curved surface. The results verified that users were able to efficiently perceive distinct curved shapes.

3   Pseudo-haptic Textures Based on Curvature

According to the results showed in haptic studies for curvature perception, we believe that the curvature and local changes of orientation can be exploited in order to enhance the exploration of a pseudo-haptic texture. While the original approach [6] only uses depth information to modify the Control/Display ratio, our proposed method relies on the curvature information of the surface encoded into a normal map. When the cursor moves from one pixel to another, the CD ratio is computed according to the curvature at the midpoint of both pixels and to the changes of the local orientation. Changes in the CD ratio can be seen as the effect of lateral forces [1]. In the following, we describe the details of our proposed approach that integrates the curvature information with the local changes of orientation as important cues for simulating pseudo-haptic textures. The algorithm differs from [6] mainly for the computation of the CD ratio.

3.1   Algorithm Description

Input. Assuming that the interactive area is restricted to the pseudo-haptic texture, the algorithm has as inputs (see Figure 2 Left): (a) the normal map (M_n), (b) the current mouse cursor position $p_m = (x_m, y_m)$, (c) the direction of the mouse **m** and (d) the physical distance covered by the mouse d_m.

Overview. According to the input variables, the algorithm updates the position of mouse cursor (starting at p_m) by traversing the texture along the direction defined by **m**. Each time the mouse cursor advances one pixel, a new CD ratio is computed according to the current and previous pixel. The traversal finishes when the accumulated distance is equal or greater than d_m. Additional details can be found in [6].

Curvature Computation. In order to compute the CD ratio, we need to retrieve the curvature information between pixels. Let consider two neighboring pixels $p_1 = (x_1, y_1, 0)$ and $p_2 = (x_2, y_2, 0)$ and their respective normals **n₁** and **n₂** (see Figure 2 Middle). We define the plane π considering the direction of the

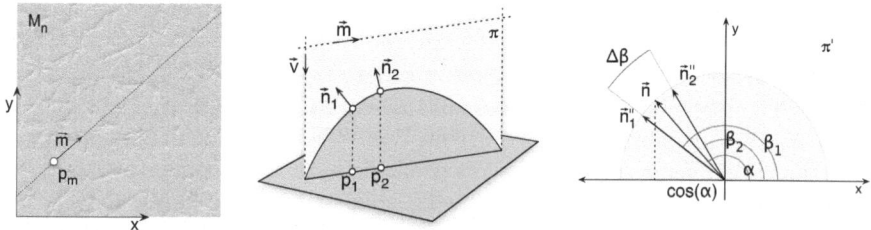

Fig. 2. Curvature computation summary. Left, top view of the pseudo-haptic texture. Middle, the computation of the CD ratio only takes into account the projection of the normals into the plane π. Right, relationship of the normals of the two pixels with α and β.

mouse $\mathbf{m}$, the viewing direction $\mathbf{v} = (0, 0, -1)$ and p_1. Then, we apply an affine transformation to π, transforming it into xy plane (referred as π'); the same transformation is applied to vectors $\mathbf{n_1}$ and $\mathbf{n_2}$ obtaining $\mathbf{n'_1}$ and $\mathbf{n'_2}$. Finally, we project the normals $\mathbf{n'_1}$ and $\mathbf{n'_2}$ into the plane π' obtaining $\mathbf{n''_1}$ and $\mathbf{n''_2}$ (see Figure 2 Right). Notice that the x-axis relates to the direction of movement, while the negative y-axis relates to the viewing direction. From $\mathbf{n''_1}$ and $\mathbf{n''_2}$, and using Equation 1, we can obtain the curvature represented by α and the local change in curvature ($\Delta\beta$).

$$\cos(\alpha) = (n''_{1x} + n''_{2x})/2$$
$$\Delta\beta = \arccos(n''_{1x}) - \arccos(n''_{2x})$$

$$(1)$$

The cosine of α determines the magnitude of the slope and its sign and $\Delta\beta$ provides information about the change of the local orientation. If $\Delta\beta > 0$, the slope is locally increasing and vice-versa.

CD Ratio Computation. The computation of the CD ratio has two components, the influence of the $\cos(\alpha)$ and the influence of $\Delta\beta$. First, the $\cos(\alpha)$ is considered as a lateral force which models the friction of the surface. If it is negative, it applies a force against the movement and vice-versa. As the cosine is normalized between -1 and 1 we need to apply two scaling constants determining the maximum and minimum CD ratio. Considering that a CD ratio of 1 refers to a flat surface ($\cos(\alpha) = 0$), the CD ratio is modeled as shown in Equation 2. k_{min} and k_{max} are user defined constants determining the maximum and minimum CD ratio.

$$\text{CD ratio} = \begin{cases} 1 - \cos(\alpha) \cdot (1 - 1/k_{min}) & \text{if } \cos(\alpha) > 0, k_{min} \geq 1 \\ 1 - \cos(\alpha) \cdot (k_{max} - 1) & \text{if } \cos(\alpha) < 0, k_{max} \geq 1 \end{cases}$$

$$(2)$$

However, this formulation accounts for the local curvature but not for changes in curvature. The same value of $\cos(\alpha)$ may result from different values of $\mathbf{n_1}$ and $\mathbf{n_2}$. As the computation of the CD ratio is done for each pair of pixels, without considering previous pixels, it would produce unnoticeable effects for sharp surfaces and edges. For that purpose, the factors k_{max} and k_{min} were adjusted according to $\Delta\beta$ (see Equation 3). $\Delta\beta$ has its units in degrees. If $\Delta\beta \approx 0$, the CD ratio only accounts for α. Now, k'_{max} and k'_{min} determine the maximum and minimum CD ratio when $\Delta\beta = 0$. In order to avoid extreme values of k_{min} and k_{max}, a lower and upper limit can be defined.

$$k_{max} = k'_{max} \cdot (1 + \Delta\beta)$$
$$k_{min} = k'_{min} \cdot (1 - \Delta\beta)$$

$$(3)$$

Figure 3 shows the behavior of the CD ratio according to α and $\Delta\beta$. The plot only considers smooth variations of curvature ($\Delta\beta = \pm 2$).

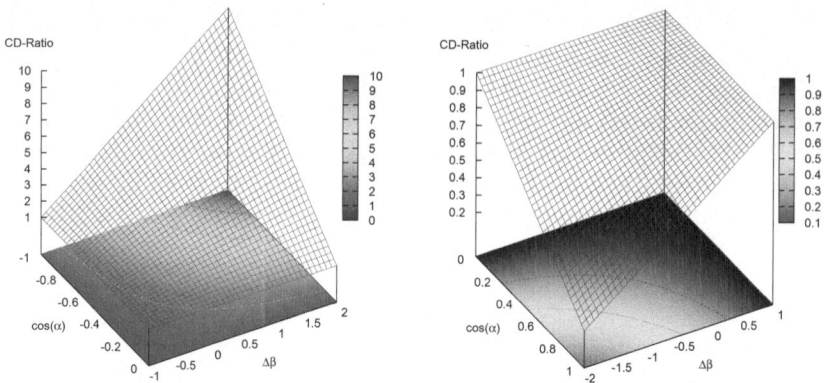

Fig. 3. CD-ratio variation as a function of $\Delta\beta$ and $\cos(\alpha)$ for (left) positive slopes, (right) negative slopes. The upper and lower limit for k_{max} and k_{min} were 30 and 5, and 4 and 1.5 for k'_{max} and k'_{min}.

3.2 Comparison with the Depth-Based Approach

Both approaches exhibit a similar behavior, crossing two pixels with a positive slope will require a CD ratio greater than 1, and vice-versa. However, the CD ratio for the depth-based approach is obtained taking into account depth changes between pixels (Δh). Changes in CD ratio exhibit a linear behavior according to Δh. Similar to our approach, it also requires to determine the maximum and minimum CD ratio, but as the depth map is typically normalized between 0 and 1, we also need to provide a scaling factor for the depth map.

In contrast to the linear behavior of Δh, the behavior of $\cos(\alpha)$ is not linear. It provides higher variation for values of $\alpha > 3\pi/4$ and $\alpha < \pi/4$ ($\alpha \in [0..\pi]$). This provides higher precision for surfaces with smaller slopes, thus being easier for the user to identify them.

Another difference is that our approach is able to simulate C^0 discontinuities. One of the results detailed in [6] showed that users tend to recognize sharp surfaces as smooth surfaces. As we consider local variations of curvature ($\Delta\beta$), non-continuous changes in curvature result in stronger variations of the CD ratio, which now will be noticeable for the user. All these differences are expected to provide better shape recognition.

4 User Evaluation

The analysis of the depth-based approach has revealed existing limitations when willing to recognize pseudo-haptic textures with varying depth ranges. In this user evaluation we want to explore the limits of the depth-based approach when trying to recognize shapes with different depth ranges and whether the curvature-based approach is able to provide better results. For this purpose we have conducted a similar experiment to [6], based on the recognition of simple targets with different shapes and profiles.

4.1   Procedure

The visual stimuli was a rectangular area shaded with a constant dark blue of 500px, placed at the center of the screen. Users were presented with a simple task: identify if the shape placed inside was a bump or a hole. They were instructed to perform horizontal mouse movements along the boundaries of the rectangular area. The only feedback provided was the speed of the mouse cursor, which was adjusted according to the pseudo-haptic technique being evaluated. For each shape, after ten seconds of exploration an answer screen was displayed, asking the user to classify the shape. A typical 2AFC protocol was employed, users could only choose between bump or hole. In order to explore if users were able to recognize the shape in less time, they could press the left mouse button to display the answer screen, answering before the time limit. Once the user chooses an option, the next target is "displayed". This process was repeated until the end of the experiment.

4.2   Experimental Design and Hypotheses

Twenty four different targets were considered in the experiment (see Figure 4). They were built according to the Shape (Bump or Hole), the simulated Profile (Linear, Gaussian and Polynomial), the Height (0.25 and 1, the depth map was normalized between -1 and 1) and two different Radiuses (100px and 200px).

According to the targets' configuration and the techniques evaluated, the independent variables were the pseudo-haptic technique (Depth and Curvature), the Shape, the Profile, the Heights (High, Low) and the Radius (Large, Small) of the targets, all considered as whithin-subject. Thus resulting in a 2×2×3×2×2 factorial design (48 combinations). The dependent variables were the number of misclassified shapes and the time to complete the task. For the task-completion-time, we only considered the time spent during the active exploration of the shape.

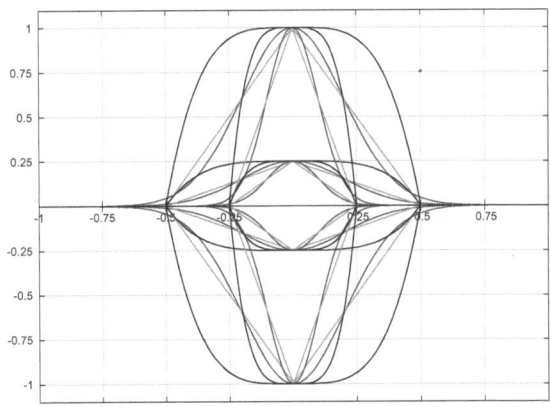

Fig. 4. Shapes used during the experiment. Linear (green), Polynomial (blue) and Gaussian (red). The Equation for the polynomial profile was $f(x) = 1 - x^4$.

The experiment was subdivided into two phases, one for each Technique. Each phase was also subdivided into three blocks. For each block, users had to classify the 24 targets which were ordered randomly. For each phase, the first block was considered as training/adaptation and was not considered in the analysis. Participants required on average twenty minutes to complete the experiment and they could take a break at the end of the first phase.

Our main hypotheses were that users would significantly misclassify less shapes when using the curvature-based approach (H1.1), with higher differences for targets with a smaller depth range (H1.2). Regarding recognition time, we also expected lower mean of recognition time for the curvature-based approach (H2.1). In addition, we also expected a lower mean for the recognition time for targets with higher depth range (H2.2).

4.3   Experimental Apparatus and Population

The experiments were conducted with a 24" inch monitor with a resolution of 1920×1200 pixels, using as input device a standard mouse. The mouse acceleration provided by the operative system was disabled. Participants were placed at 60 cm from the screen and operated the mouse with their dominant hand.

Twelve volunteers users (undergraduate and graduate students) took part in the experiment, aged from 23 to 31, 2 females and 11 males. All users had not experienced pseudo-haptic feedback before.

4.4   Results

For the statistical analysis we used ANOVA analysis. For all post-hoc comparisons, Bonferroni adjustments for $\alpha = 95\%$ were applied; only significant post-hoc comparisons are mentioned ($p < 0.05$).

Error Rate. Regarding the number of misclassified targets versus the Technique, Radius, Height, Profile and Shape, there were two main significant effects for Technique ($F_{1,11} = 8.11, p \leq 0.01$) and for Height ($F_{1,11} = 18.78, p < 0.001$). Post-hoc tests showed that users did more mistakes when employing the depth-based approach (7.47%) in comparison with the curvature-based approach (3.47%), thus accepting H1.1. Post-hoc tests also showed that users had more difficulties to classify targets with lower height, which also supports H1.2. No main effects were found for shape, radius and profile.

The two-way interactions found significant were Radius and Height ($F_{1,11} = 6.76, p \leq 0.01$), Technique and Height ($F_{1,11} = 5.42, p \leq 0.05$), Technique and Radius ($F_{1,11} = 6.64, p \leq 0.01$). The ANOVA also showed a three-way significant interaction between Technique, Radius and Height ($F_{1,11} = 9.58, p \leq 0.005$). Post-hoc tests showed that users performed worst when combining large radius with lower height, however a deeper analysis, showed that it only happened when the users were recognizing the shapes with the depth-based approach. The effect was non-significant for the curvature-based approach.

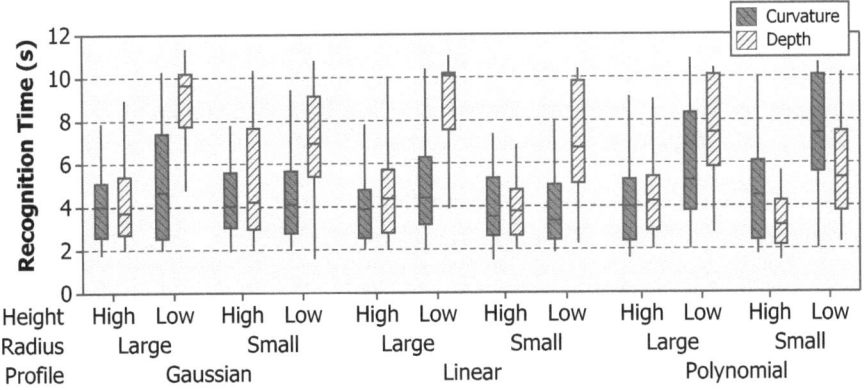

Fig. 5. Boxplot of the mean recognition time for each Technique, grouped by Profile, Radius and Height

Recognition Time. The ANOVA showed main effects for Technique ($F_{1,11} = 101.13, p < 0.001$), Radius ($F_{1,11} = 14.99, p < 0.001$) and Height ($F_1 = 276.88$, $p < 0.001$). Post-hoc tests showed that users were able to recognize the targets faster with the curvature-based approach ($6.03s$) than with the depth approach ($4.88s$), thus accepting H2.1. In addition, users required significantly more time to recognize targets with lower height and larger radius, which also supports H2.2.

The ANOVA also showed two-way interactions and three-way interaction for Technique, Radius and Height: Radius and Height ($F_{1,11} = 18.52, p < 0.001$), Technique and Radius ($F_{1,11} = 16.15, p < 0.001$), Technique and Height ($F_{1,11} = 68.53, p < 0.001$), and Technique, Radius and Height ($F_{1,11} = 11.64, p < 0.001$). Post-hoc tests showed that the combination of low height and large radius resulted in the targets which required more exploration time (see Figure 5), specially for the depth approach ($7.9s$) than for the curvature-based approach ($5.4s$).

There was also a two-way interaction effect between Technique and Profile ($F_{2,11} = 28.23, p < 0.001$) and a three-way interaction between Technique, Profile and Height ($F_{2,11} = 13.68, p < 0.001$). Post-hoc tests showed that users significantly needed more time to recognize the shape using the Depth-based approach for Gaussian and Linear profiles, specially for targets with lower height. In contrast, for the curvature-based approach users required significantly more time to recognize Polynomial profiles (see Figure 5).

4.5 Discussion

From the results, we can state that both techniques presented a similar behavior in terms of shape recognition except for the combination of large radius and low height. For these targets, users had a significant better recognition performance and required less time to recognize them with the curvature-based approach.

At the end of the experiment, most users reported that they focused more in the deceleration of the mouse cursor rather than in the acceleration to determine the shape. Several users explained the following simple strategy in order to locate the shapes: "if the mouse slows down in the first half of the shape it is classified as a bump and if it slows down on the second half of the shape it is classified as a hole".

Figure 6 depicts the variation of the CD-ratio when the mouse cursor moves horizontally from the left side of the target to the right size. Targets with large

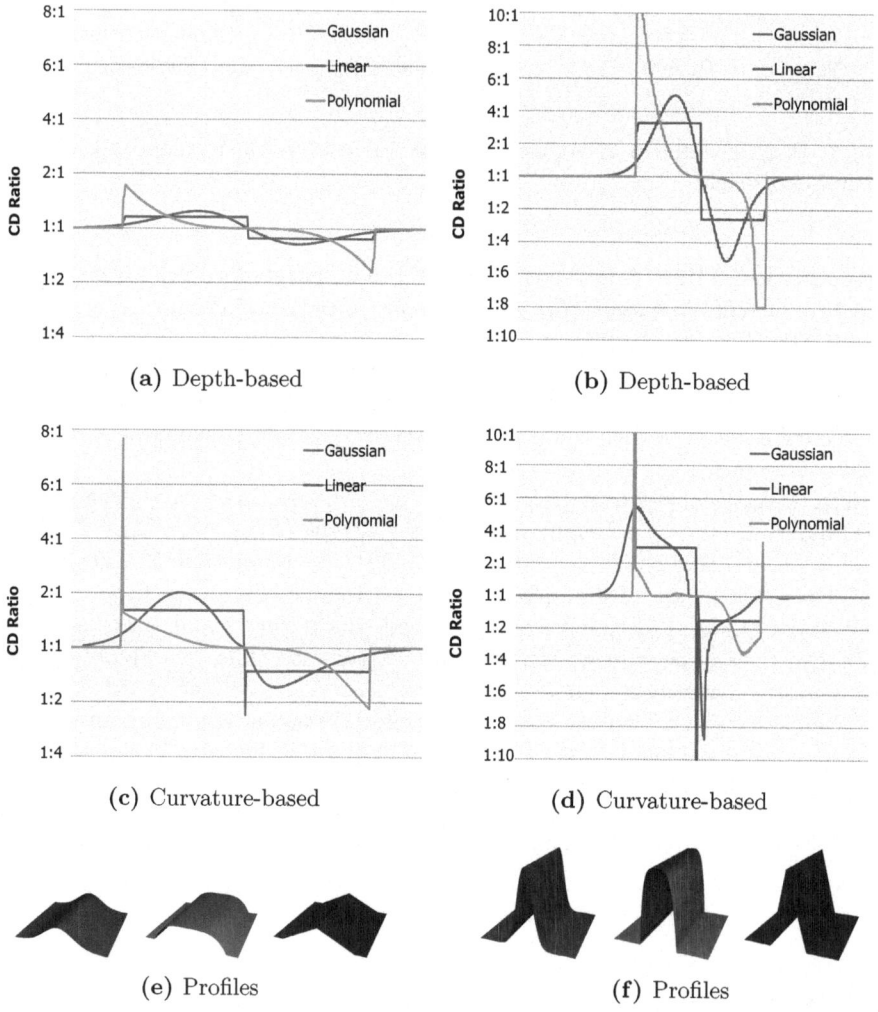

Fig. 6. Evolution of the CD ratio while moving the mouse from the leftmost of the target to the rightmost (bump shapes). (a,c), targets with large radius and lower height. (b,d), targets with small radius and higher height. (e,f), 3D representation of the simulated profiles.

radius and lower height presented the minimum CD ratio range, being the CD ratio range for the depth-based approach the smallest. This low CD ratio range resulted in less perceptible mouse speed changes which made the recognition of the targets more difficult. In contrast, for the curvature-based approach the CD ratio ranges were higher (see Figure 6, Left). That is our main explanation for users being able to easily recognize targets with the curvature-approach. Although we could have increased the acceleration and deceleration effect for the depth-based approach it would have increased too much the CD-ratio range for the other profiles. Acceleration and deceleration factors were adjusted taking into account the targets having higher depth ranges. Figures 6b and 6d depict the CD ratio range for targets with higher slope.

However, the curvature-based approach poses two main limitations. The first limitation is related to non-continuous surfaces. Our method is not able to recognize a discontinuity and will proceed as if the surface is continuous, which might result in an unexpected behavior for the user. A possible solution would be to detect the discontinuities using depth information and apply the depth-based approach for this situation. The second limitation is high frequency surfaces. While the depth-based approach will not be able to detect them, the curvature in this surfaces will change abruptly, resulting in rapid changes of the CD ratio. This effect can also result as an unexpected behavior for the user. A possible course of action can be focused on using depth information to smooth the normal map locally, for instance keep high frequency surfaces close to the viewer and smooth those further away.

On the other hand, the normal map provides additional information than using only a depth map; for continuous surfaces depth information can be inferred from the normal map. In addition, it does not require the usage of a scaling factor. Furthermore, our approach is also able to account for abrupt changes of curvature. For example, in the Linear and Polynomial profiles, when the cursor crosses a discontinuity (C^0 surface), due to the abrupt change of curvature a strong change of the CD ratio is provided.

5 Conclusion

Focusing on the limitation of depth for pseudo-haptic textures and the fact that the curvature information of the surface is considered as an important haptic cue, we have proposed a novel pseudo-haptic texture approach that takes into account the local orientation of the surface and how it varies. The Control/Display ratio is computed as a function of the slope of the surface and the changes of the local orientation.

An experimental study was conducted in order to compare the proposed curvature-based approach versus the original approach based on depth. The study consisted in evaluating the recognition of bumps and holes with different heights, radius and profiles. The only feedback provided was the variation of the speed of the mouse cursor. The results showed that participants made less number of mistakes when employing the curvature-based approach in comparison with the depth-based approach. Particularly, a significantly less number of errors was obtained for shapes of large radius and low height. Participants also required more time to recognize the shape of the targets with the depth-based approach.

As a future work, we plan to explore whether the curvature-based approach allows to provide better shape recognition for more complex shapes and surfaces and we also plan to extend the proposed approach to computer generated 3D environments as well as some other potential applications like surface analysis, medical simulators, navigation in 3D websites or video games.

References

1. de-la Torre, G.R., Hayward, V.: Force can overcome object geometry in the perception of shape through active touch. Nature 412, 445–448 (2001)
2. Dominjon, L., Lécuyer, A., Burkhardt, J.M., Richard, P., Richir, S.: Influence of control/display ratio on the perception of mass of manipulated objects in virtual environments. In: Proceedings of the IEEE International Conference on Virtual Reality, pp. 19–25 (2005)
3. Gordon, I., Morison, V.: The haptic perception of curvature. Perception and Psychophysics 31, 446–450 (1982)
4. Hachisu, T., Cirio, G., Marchal, M., Lécuyer, A., Kajimoto, H.: Pseudo-haptic feedback augmented with visual and tactile vibrations. In: Proceedings of International Symposium on VR Innovations (ISVRI 2011), pp. 331–332 (2011)
5. Lécuyer, A.: Simulating haptic feedback usinfg vision: A survey of research and applications of pseudo-haptic feedback. In: Presence: Teleoperators and Virtual Environments, vol. 18, pp. 39–53. MIT Press, Cambridge (2009)
6. Lécuyer, A., Burkhardt, J.-M., Etienne, L.: Feeling bumps and holes without a haptic interface: the perception of pseudo-haptic textures. In: Proceedings of the SIGCHI Conference on Human Factors in Computing Systems, CHI 2004, pp. 239–246. ACM (2004)
7. Lécuyer, A., Coquillard, S., Kheddar, A., Richard, P., Coiffet, P.: Pseudohaptic feedback: can isometric input devices simulate force feedback? In: IEEE International Conference on Virtual Reality (IEEE VR), pp. 83–90 (2000)
8. Mensvoort, K.v., Hermes, D.J., Monfort, M.V.: Usability of visually simulated force feedback. International Journal of Human-Computer Studies 66(6), 438–451 (2008)
9. van Mensvoort, K., Vos, P., Hermes, D.J., van Liere, R.: Perception of mechanically and optically simulated bumps and holes. ACM Transactions on Applied Perception 7(2), 1–24 (2010)
10. van der Horst, B.J., Kappers, A.M.L.: Using curvature information in haptic shape perception of 3D objects. Experimental Brain Research 190(3), 361–367 (2008)
11. Wijntjes, M.W.A., Sato, A., Hayward, V., Kappers, A.M.L.: Local surface orientation dominates haptic curvature discrimination. IEEE Transactions on Haptics 2(2), 94–102 (2009)
12. Zeng, T., Giraud, F., Lemaire-Semail, B., Amberg, M.: Haptic perception of curvature through active touch. In: IEEE World Haptics Conference (WHC), Istanbul, pp. 533–538 (2011)

Cursor Navigation Using Haptics for Motion-Impaired Computer Users

Christopher T. Asque, Andy M. Day, and Stephen D. Laycock

School of Computing Sciences, University of East Anglia, Norwich, NR4 7TJ, UK
c.asque@uea.ac.uk, {amd,sdl}@cmp.uea.ac.uk

Abstract. In recent years, typical desktop computer screen sizes and resolutions have increased significantly. The result of this is that a pointing device has to travel a much greater distance to navigate the whole of a computer screen. For motion-impaired operators that suffer from fatigue or have a limited range of movement this can make a computer inaccessible. This paper introduces a new method for cursor navigation using the Phantom Omni force-feedback device. The newly proposed workbox is designed to aid the operator with coarse navigation of the cursor and improve target selection. The proposed method can significantly reduce the effect of target distracters, which have been a major hindrance to the development of haptic assistance in graphical user interfaces (GUI). The workbox has shown to significantly improve computer access for operators with a limited range of movement by giving them the ability to navigate all of a computer screen.

Keywords: Haptics, haptic assistance, cursor navigation.

1 Introduction

Computer access for a person with disabilities can significantly improve their quality of life and provide them with much greater independence. Resources such as the Internet and Office applications are a major asset in both educational and working environments but without a suitable interface they cannot be exploited easily. A computer is a highly versatile tool where both software and hardware techniques can be developed to help overcome many obstacles that a disabled person may encounter.

One of the primary tasks when using a computer is to navigate the on-screen cursor using a pointing device. According to Dennerlein et al. the use of a pointing device accounts for 30-80% of all time spent working at a computer [7]. The use of a pointing device can be a challenge for many motion-impaired computer users. According to Hwang et al. symptoms such as tremor, spasm, muscle weakness, partial paralysis, or poor coordination can make standard pointing devices difficult, if not impossible, to use [12]. In recent years typical desktop computer screen sizes and resolutions have increased significantly. The result of this is that the pointing device has to move a much greater distance to navigate the

P. Isokoski and J. Springare (Eds.): EuroHaptics 2012, Part I, LNCS 7282, pp. 13–24, 2012.

whole computer screen, which is a major difficulty for people who suffer from fatigue or have a limited range of movement. Standard pointing devices often do not provide motion-impaired operators with sufficient access to a computer. Each disability is unique to each individual and although two people may have the same diagnosis their level of impairment may differ significantly. As a result there is not one single assistive input device that will meet the needs of all motion-impaired people, since each person's needs and abilities are different.

Alternative cursor control techniques have developed significantly in recent years such as head tracking and eye gaze. These offer the ability to control the cursor through processing video camera images to determine the direction of the cursor on screen. These have shown to be especially effective in providing computer access for people that are quadriplegic. However, previous studies have shown that it is often difficult for motion-impaired operators to perform precise manipulations using these techniques. Many difficulties have been highlighted with calibrating the equipment [18]. Modern image processing and eye-tracking systems have resolved many of these shortcomings but tracking systems are still comparatively expensive and require great user concentration and efforts to achieve precise cursor control.

Haptic technology utilises the sense of touch to enable the operator to physically interact with the environment in which they are working. The inclusion of haptic feedback in point-and-click tasks has been shown, in several cases, to improve interaction for able-bodied and motion-impaired computer users. These improvements have been observed both in terms of cursor navigation and target selection [13] [15]. There are many haptic devices available with varying degrees of freedom (DOF). According to Langdon et al. increasing the degrees-of-freedom can improve interaction rates if implemented carefully so that the extra freedom does not over complicate the interface or increase the cognitive workload [16]. There are valid reasons for choosing the 3DOF Phantom Omni such as the ability to pass over target distracters more easily. This is discussed in greater depth in Section 2.1.

The aim of this study is to explore ways to make a computer more accessible for motion-impaired users through haptic feedback. To permit the development of haptic assistance we introduce the concept of the workbox in which the user will interact. The technique utilises a rate / position hybrid method to permit rapid cursor navigation whilst allowing accurate target selection. The technique will benefit users that suffer from fatigue or have a limited range of movement as they will be able to navigate the whole of a computer screen from within the confines of the workbox. The implementation of this new technique is discussed in greater depth in Section 3.1.

The remainder of the paper is organised as follows: Section 2 gives an overview of work related to the development of haptic assistance for motion-impaired computer users. Section 3 describes the methods used to conduct the experiment. The results of the study are presented in Section 4. Finally, the conclusions and discussion are drawn in Section 5.

2   Related Work

A number of studies have been undertaken that have attempted to assist motion-impaired computer users in human-computer-interaction (HCI). For haptic assistance to truly benefit motion-impaired operators it must be able to be applied to existing interfaces that they wish to use. The following subsections identify a number of difficulties that have hampered the development of haptic technology in GUI's.

2.1   Target Distracters

Under normal conditions the majority of motion-impaired computer users do not have difficulty with the navigation phase of a point-and-click task [15]. However, when haptic cues are placed around icons to help with target selection this can disrupt the navigation phase due to the introduction of target distracters. A target distraction occurs when the cursor has to pass through an undesired haptic cue before reaching the destination. The target distracter can disrupt the position of the cursor due to the force imposed on the operator. Hwang et al. state that target arrangements requiring the cursor to pass through other haptically enabled items can be detrimental to user performance and should be avoided [12].

Studies specific to motion-impaired operators have identified distracters as an issue for haptic development [3] [11]. Gravity wells are commonly used as a method for assisting target selection [13] [15]. When exiting a gravity well distracter the cursor will often overshoot which can impede the next task. This problem is amplified with icons in close proximity of each other because the overshoot can land the cursor into an undesired neighbouring target. This increases user fatigue because the operator has to physically oppose the force of the gravity well before exiting the target.

Gunn et al. suggest that there may be valid reasons for a skilled user to want to ignore the advice provided by a computer system [10]. They go on to argue that force feedback might limit this ability to ignore the advice and therefore be less effective as an aid. This has been observed especially with a 2DOF device where target distracters make an interface frustrating to use due to the operator continually having to oppose forces from distracters to reach the destination. A 3DOF device offers the potential to reduce the effects of target distracters by allowing the operator to lift the stylus off the page, pass over the target distracters and then resume the target selection. However, this can disjoint interaction as the operator has to continuously lift the device off the page and then re-apply it for each operation. Asque et al. produced a haptic cone approach that does not impose a force on the operator and therefore distracters can be exited more easily [3]. Ideally the operator would not have to pass through any type of distracter on course to the target.

A possible solution to target distracters is to use target prediction techniques to only enable the haptic cues that the operator requires. A number of studies have investigated target prediction by analysing the cursor trajectory [2] [17].

These techniques monitor the current path of the cursor to predict the path and distance to the target. The difficulty with this technique is that operators do not always follow predictable paths towards a target. Holburt et al. have attempted to use target prediction to reduce the effects of target distracters for people with motion-impairments [11]. The targeting performance of the participants improved significantly when the haptic effect was applied to the correct target. Unfortunately within the study the rate of correctly predicted targets was only 23% which meant that the overall improvement was unclear. This low performance rate was due to the much lower predictability of data produced from motion-impaired operators and the sensitivity of the Logitech Wingman in a limited workspace.

2.2   Factors Affecting Fitts' Law

Fitts' Law is a mathematical model of human motor performance which predicts the movement time (**MT**) from one position to another as a function of the distance to a target (A) and its size (W). Fitts' Law is given below in Equation 1. The variables a and b are empirically-determined constants where a represents the start/stop time of the device and b is the speed of the device.

$$\mathbf{MT} = (a + b)\mathbf{ID} \text{ where } \mathbf{ID} = \log((A/W) + 1) \tag{1}$$

What can be deduced from Fitts' Law is that targets that are larger and closer together will be easier to select, whereas smaller targets that are further away will be more difficult. These factors have a large influence on the design of GUI's. Previous studies have shown that Fitts' Law is appropriate for motion-impaired operators [14]. Therefore, if it is possible to increase the size of the targets or reduce the distance between them then this could significantly improve interaction rates for motion-impaired users.

As the screen resolution of modern monitors increases the distance in cursor displacement between icons will often increase. The definition of Fitts' law indicates that mouse efficiency has decreased with the increase of screen resolution. For example, Microsoft Word 1.0 was designed for a screen resolution of 640x480 with toolbar button dimensions of 20x20. However, the same button in Microsoft Word 2010 may be displayed on a screen with resolutions in excess of 1920x1080. The button size will have remained unchanged but it is likely that the cursor will be much further away than it could have been on a 640x480 display. This can cause difficulties for motion-impaired operators because they have to physically move the pointing device a much greater distance to reach the destination. The increase in distance to the target will lead to an increase in movement time (MT) and user fatigue. The increase in resolution also means that the targets on screen are visually much smaller. Screen magnifiers are useful for increasing the visual size of the target but they do not increase its size in terms of device displacement. Previous studies have used techniques such as reducing the gain of the device when the operator passes over a target, therefore increasing its effective width [19]. The workspace of a pointing device and the cursor gain are closely related to the display resolution on a computer screen.

The gain or sensitivity of a pointing device determines how far the cursor moves on the screen for a given input movement. Koester et al. investigated the gain settings of pointing devices for users with physical impairments [14]. The results of the calibration did not provide a significant improvement in performance when compared to the Windows XP default. Fitts' law provides the most logical reason for this. An increase in gain will reduce the target distance (A) by reducing the amount of movement required by the device to translate the cursor a given distance on the screen. It will also reduce the effective target width (W) of the target and therefore increase the index of difficulty. The simultaneous changes in target distance and width tend to cancel each other out resulting in little performance change [14]. The limited workspace of haptic devices has been identified as a concern for the development of haptic assisted interfaces [8] [11] [21]. For example, the workspace of the Logitech Wingman is only 4cm x 4cm. The Phantom Omni has a larger workspace of 16cm x 12cm x 7cm. What this means is that the whole of the computer screen has to be mapped within the confinements of the workspace of that device to allow the operator to have direct positional control of the cursor. As a result the cursor gain often has to be quite high for haptic devices which increases their sensitivity. A large cursor gain will reduce the effective width of the targets and make them more difficult to select. Devices such as the mouse do not suffer as significantly from having to move larger distances because they have an unlimited workspace. When reaching the limits of the useable workspace, the operator can lift the mouse, and then put it down on a new location. This is often referred to as declutching.

Previous studies have investigated a hybrid position/rate control system to enable both accurate interaction and coarse positioning of the cursor in a large virtual environment (VE). Dominjon et al. proposed a bubble technique for interacting with large virtual environments using haptic devices with a limited workspace [8]. The operator is able to navigate the cursor by pressing against the semi-transparent sphere in the direction they wish to move. Force-feedback is provided for interactions between the probe and the sphere. When the cursor lies within the sphere it is positionally controlled. Casiez et al. propose a 2D passive haptic feedback system through an elastic ring on top of a touchpad to allow the user to switch from position to rate control without clutching. Results showed performance benefits when reaching distant targets, whilst maintaining position control for precise movements [5]. Stocks et al. state that the spherical navigation volume in the bubble approach does not correspond well to the workspace of the haptic device and so they propose a navigation cube that is automatically scaled to fit the workspace [21]. The navigation cube is used to translate a protein within the haptic workspace by moving the probe outside of the walls of the cube in the desired direction. The speed of translation is dependent on the distance the haptic probe is from the side of the cube it penetrated. This allows the user to move slightly outside the cube for fine navigation control and to move further for faster translation. In this approach no forces of interaction between the navigation cube and the probe are included to avoid confusion with forces with the biomolecule.

2.3 Joystick Control

Joystick control is especially useful for people with severe motion-impairments and a limited range of movement. A joystick can allow the operator to navigate the whole of the computer screen with small movements. An isometric joystick is a pointing device that is able to sense the applied force and translate that into a proportional velocity of the cursor on the screen. Many motion-impaired operators are experienced joystick users because they are often used on electric wheelchairs to give proportional control over speed and direction. The neutral position of the joystick is useful because it acts as a brake.

However, previous research has consistently shown joystick control to be slower and have a higher error rate than that of the mouse [4] [9]. A study by Mithal et al. found that participants complained that the isometric joystick "was hard to control" [20] and that they had a lot of trouble getting the joystick to stop in small targets. The reason for this is that tremor causes involuntary changes in the velocity at which the cursor moves. This makes it difficult for users to achieve fine control to stop the cursor at a desired point on the screen and explains why isometric joysticks are hard to control. The harder that a user pushes, the faster the cursor moves. This is referred to as first order control, while the mouse's mapping of mouse displacement to cursor displacement is called zero order control [20].

The workbox aims to overcome the difficulties of selecting small targets by allowing direct positional control. One of the major aims of the workbox is to assist those with severe motor handicaps that have a limited range of movement in their extremities. The workbox approach has been developed to overcome many of these shortcomings so that motion-impaired operators can use haptic assistance with existing GUI's. The workbox aims to reduce the effect of distracters, increase the effective width of targets and allows operators to navigate all of the computer screen. The approach and implementation is discussed in the following section.

3 Methods

This section introduces the workbox concept and the measures used to evaluate its performance. The haptic techniques presented in this paper have been implemented using an Open Source API, named CHAI3D [6]. The CHAI3D API uses Zilles and Salisbury's God-Object haptic rendering algorithm [22]. The algorithm tracks a history of contact with a surface. The position of the God-Object (proxy) is chosen to be the point which locally minimizes the distance to the Haptic Interface Point (HIP) along a surface and a restoring spring force is calculated between the two. The implementation of the workbox is discussed in the following section.

3.1 Workbox Implementation

The workbox approach is a rate / position hybrid system where coarse navigation is rate controlled and fine navigation is position controlled. The workbox can be considered as a workspace in which the user will interact. The operator will be able to navigate the whole of the computer screen whilst only moving within the confinements of the workbox. The coarse navigation of the cursor is rate controlled. This is achieved by pressing the proxy against the wall(s) of the workbox according to which direction the operator wishes the cursor to move. For example, if the operator wishes to move the cursor to the right hand side of the screen then they must press the proxy against the right hand wall of the workbox, as shown in Figure 1(a). The cursor speed is proportional to the force that the operator is applying to the wall. Unlike the approach by Stocks et al. [21] it was decided that force-feedback was required on the walls of the workbox because it was observed that the device felt too free without feedback and the cursor would often overshoot the target region. The feedback of the walls provides essential stability to the hand for motion-impaired operators to more accurately position the cursor. When the proxy is in contact with the wall the haptic cues surrounding the targets are disabled. This reduces the effect of target distracters when the operator is coarsely navigating the cursor. When the device switch is pressed the position of the workbox is locked to ensure that the cursor does not slip off the target should the operator accidentally press against a wall of the workbox. The square surrounding the cursor in Figure 1(b) gives a representation of the area of the screen that will be magnified in the workbox window. Once the square is placed around the desired target region then the operator will be able to begin the target selection phase.

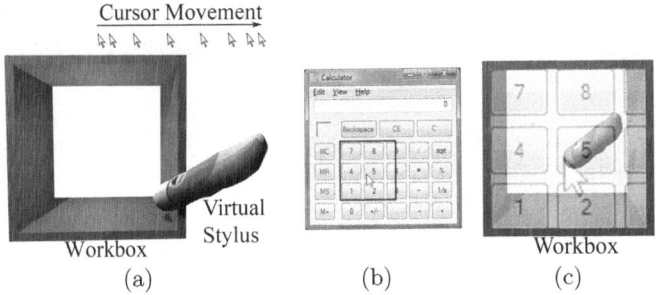

Fig. 1. An example of how to navigate the cursor to the right hand side of the screen by pressing the tool against the corresponding wall of the workbox (a). The original window with the black square indicating the area of the screen that will be magnified (b). The magnified semi-transparent window with the workbox behind (c).

When the proxy is not in contact with the walls of the workbox the operator has direct positional control of the cursor within the workbox region. This allows the precise manipulation of the cursor that is not possible with a joystick interface. During this phase the haptic cues surrounding the targets are re-enabled to aid target selection. The operator is also presented with a semi-transparent window which is overlaid on top of the workbox. This repeats a section of the screen surrounding the cursor, which is depicted in Figure 1(b) by the black square.

The cursor within the window will map directly to the tip of the virtual tool. The interface and haptic assistance can be scaled up to help people that have difficulty selecting small targets. The scaling increases the effective width of the targets which will make them easier to select. The interface within the workbox in Figure 1(c) has been scaled by a factor of three. The magnified window is not essential for interaction but is useful for giving a visual representation of the haptic assistance and the scaling of the interface.

User comfort is essential for motion-impaired operators especially when using a pointing device for long periods of time. When using the stylus grip it is necessary to provide a comfortable leaning position at the back of the workbox. A diagonal plane has been chosen to produce the parallelepiped shape shown in Figure 2(a). One of the key features of the technique is that the operator can navigate the whole screen within the confines of the workbox and so its position in the haptic workspace is crucial to user comfort. The workbox has been placed at the lower region of the y-axis to ensure that the operator can reach the four walls whilst still using the wrist rest provided with the Phantom Omni. An example of this is shown in Figure 2(b).

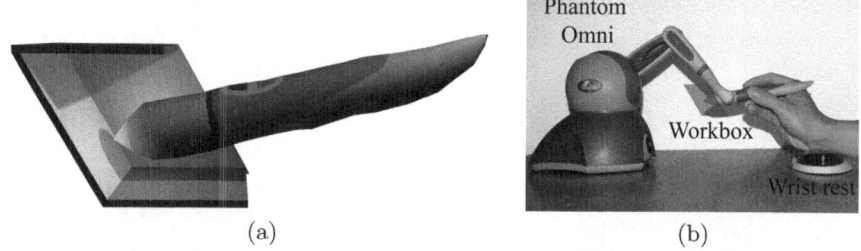

(a)   (b)

Fig. 2. A side view of the parallelepiped workbox with the Phantom Omni stylus (a) A scale view of the workbox and its position within the physical workspace of the Phantom Omni using the stylus grip (b)

3.2   Point-and-Click Task

The experiment often conducted with cursor analysis techniques is based on the ISO 9241-9 standard for pointing device evaluation [1]. The experiment consists of 15 circular targets arranged in a circular layout. This is often criticised for its use with haptics because it does not take into consideration the effect of target distracters and the circular layout is unrealistic for GUI's. To evaluate a more realistic interface the Windows On-Screen-Keyboard (OSK) has been chosen. The task required fifty successful selections, using the Phantom Omni, to produce a predefined sentence. Six participants were included in the study and each person was asked to repeat the experiment three times for each haptic condition over a twelve week period. The sessions were two hours long and the order of presentation of the haptic techniques was randomised. The target key is highlighted in red and data collection begins once the first target is selected. Any selections of surrounding keys were recorded by the cursor analysis but ignored in the textbox sentence. i.e. the operator was not required to delete undesired key selections. The test group consists of six participants, with varying degrees of motion-impairment, from the Norfolk and Norwich Scope Association (NANSA). The participants all had at least three years experience with the Phantom Omni and were familiar with the haptic feedback. The control condition was the experiment conducted with gravity wells only and the second condition was the workbox with gravity wells.

4 Results

In this section Condition 1 refers to the experiment conducted with direct positional control using gravity wells and Condition 2 refers to the workbox with gravity wells. All but one participant was able to complete the predefined sentence using direct positioning of the Phantom Omni and using the workbox approach. To provide statistical significance a paired t-test was performed to compare the mean for each condition.

4.1 Missed-Clicks and Movement time

For both conditions a mean of 0.867 and 0.667 missed-clicks were recorded respectively. There were no statistically significant differences between the two conditions (t = 0.557, p = 0.607). Previous studies have reported increased error rates when using rate controlled devices and so it is important to allow direct positional control for accurate targeting [4] [9] [20].

The mean task completion time for each condition was 175.164 and 223.542 seconds respectively. A statistically significant increase in movement time was recorded between the two conditions (t = -7.593, p = 0.002). This was expected and will be caused by the increased time of the rate control phase similar to that observed with isometric joysticks. However, the movement time when using the workbox only increased on average by ($\mu = 0.968$, $\sigma = 0.708$) seconds per selection. This increase in time is undesirable but may not be an issue if it allows an operator access to a computer that they may not otherwise have.

4.2 Distracters along Task Axis

Difficulties are often encountered with distracters especially if more than one distracter lies along the axis of approach. Within the predefined sentence was a transition from the P key to E. The cursor trajectories of each participant have been analysed for this segment. When the operator is pushing against the walls of the workbox the target distracters are disabled. Figure 3 shows the cursor trajectories of a participant using gravity wells with and without the workbox. It is clear that there is considerably less positional disruption of the cursor through other targets along the task axis when the workbox is used.

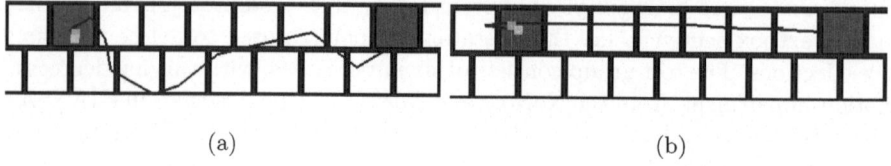

(a)   (b)

Fig. 3. A cursor trace between keys P and E for a participant using gravity wells (a) A cursor trace between keys P and E for a participant using gravity wells and the workbox (b)

4.3   Cursor Control for Severe Impairments

One of the participants has a severe motion-impairment and a very limited range of motion. They are able to navigate an electric wheelchair using a joystick. The participant is unable to operate the device switch and so it was not possible to include them in the point-and-click task on this occasion. However, to demonstrate the potential benefits of the workbox for someone with a very limited range of movement we asked the participant to position the cursor within four squares at the extremities of the screen using direct mapping and then using the workbox. The cursor trajectories of the two attempts are shown in Figures 4(a) and 4(b) respectively. The range of movement the participant was able to produce in the unassisted experiment was approximately 8cm x 8cm in device displacement. Cursor control at the extremities of the operator's movement were less controlled and so a workbox was chosen of size 4cm x 4cm. The workbox was positioned so that its origin was placed around the centre of the operator's movement range. Figure 4 illustrates the potential benefits for operators with a limited but controlled movement range.

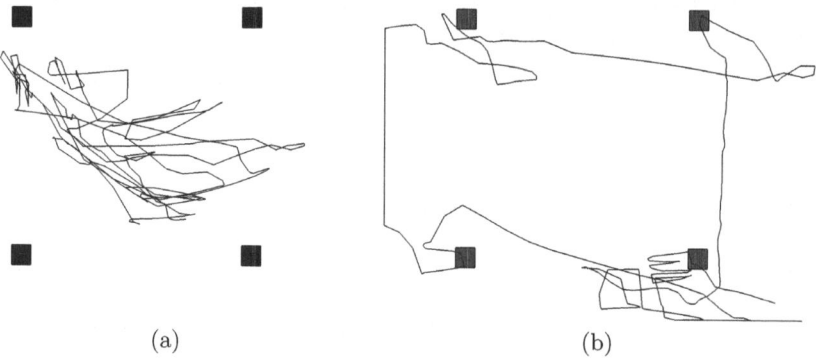

(a)   (b)

Fig. 4. The cursor trajectory of a motion-impaired operator with a limited range of movement using the direct mapping of the Phantom Omni (a) The cursor trajectory of a motion-impaired operator with a limited range of movement using a 4cm x 4cm workbox with the Phantom Omni (b)

5   Conclusions and Discussion

The workbox approach allows the operator to rapidly navigate large and complex GUI's with a haptic feedback device, whilst still permitting accurate and fast selection of icons. As screen size and resolutions continue to increase this will become more important especially for devices with a limited workspace.

Five of the six participants were able to perform a point-and-click task on the OSK using the workbox. Future work will aim to provide an external switching method for people more severely impaired such as the sixth participant. No detrimental effects on the clicking accuracy were recorded when using the workbox technique. A mean increase in 0.968s was recorded per selection but this is a small penalty if it allows the operator access to a computer. The technique has many benefits for people that suffer from fatigue or have a limited range of movement. An operator with only a limited range of movement (8cm x 8cm) was able to navigate the cursor accurately across the whole screen using a 4cm x 4cm workbox. People that suffer from fatigue at their extremities are able to navigate the screen whilst only moving within the confines of the parallelepiped volume.

Using the workbox approach the gain can be adjusted and it will not affect the distance that the device has to move as it is a function of force applied to the wall. This allows a lower gain to be permitted inside the workbox to increase the effective width of a target and aid icon selection. This gain would not be permissible on the whole display because the cursor would not be able to reach the extremities of the screen due to the limited workspace of a haptic device. By using the workbox approach it is possible to increase the size of the targets both visually in the magnified workbox window and physically in device displacement.

The majority of toolbars are designed in rows or columns. The workbox has shown to be effective when navigating along a task axis. When the proxy is in contact with the wall of the workbox the cursor is allowed to scroll freely as all haptic cues are disabled. The only distracters that can affect interaction lie within the confines of the workbox. The four walled approach allows the operator to accurately navigate the cursor using one or both axes at a time. It is anticipated that the results produced in this study will be useful in providing assistance that could significantly improve access to computer software.

Acknowledgments. The authors would like to thank the volunteers and staff at the Norfolk and Norwich Scope Association (NANSA).

References

1. ISO ergonomic requirements for office work with visual display terminals (VDTs). Part 9 - Requirements for non-keyboard input devices, International Organisation for Standardisation (1998)
2. Asano, T., Sharlin, E., Kitamura, Y., Takashima, K., Kishino, F.: Predictive interaction using the delphian desktop. In: Proc. UIST, pp. 133–141 (2005)
3. Asque, C.T., Day, A.M., Laycock, S.D.: Haptic assisted target acquisition in a visual point-and-click task for computer users with motion-impairments. IEEE Transactions on Haptics (2011)

4. Card, S.K., English, W.K., Burr, B.J.: Evaluation of mouse, rate-controlled isometric joystick, step keys, and text keys, for text selection on a crt. In: HCI, pp.
   386–392 (1987)
5. Casiez, G., Vogel, D., Pan, Q., Chaillou, C.: Rubberedge: reducing clutching by
   combining position and rate control with elastic feedback. In: Proc. of UIST,
   pp. 129–138 (2007)
6. Conti, F., Barbagli, F., Morris, D., Sewell, C., Grange, S.: CHAI3D open source
   haptic api, http://www.chai3d.org/
7. Dennerlein, J., Johnson, P.: Changes in upper extremity biomechanics across different mouse positions in a computer workstation. In: Ergonomics (2006)
8. Dominjon, L., Lécuyer, A., Burkhardt, J., Andrade-barroso, G., Richir, S.: The
   "bubble" technique: Interacting with large virtual environments using haptic devices with limited workspace. In: World Haptics, pp. 639–640 (2005)
9. Epps, B.W.: Comparison of six cursor control devices based on fitts law models.
   In: The Human Factors Society 30th Annual Meeting, vol. 30, pp. 327–331 (1986)
10. Gunn, C., Muller, W., Datta, A.: Performance improvement with haptic assistance:
    A quantitative assessment. In: World Haptics, USA, pp. 511–516 (2009)
11. Holbert, B., Huber, M.: Design and evaluation of haptic effects for use in a com-
    puter desktop for the physically disabled. In: Proc. of the PErvasive Technologies
    Related to Assistive Environments, Greece, pp. 9:1–9:8 (2008)
12. Hwang, F., Langdon, P., Keates, S., Clarkson, J.: The effect of multiple haptic dis-
    tractors on the performance of motion-impaired users. In: 6th ERCIM Workshop,
    Italy, pp. 14–25 (2003)
13. Keates, S., Hwang, F., Langdon, P., Clarkson, P.J., Robinson, P.: The use of cursor
    measures for motion-impaired computer users, vol. 2(1), pp. 18–29 (2002)
14. Koester, H., LoPresti, E., Simpson, R.: Toward goldilocks' pointing device: Deter-
    mining a "just right" gain setting for users with physical impairments. In: Proc. of
    ACM SIGACCESS, Baltimore, MD, USA, pp. 84–89 (2005)
15. Langdon, P., Hwang, F., Keates, S., Clarkson, P.J., Robinson, P.: Investigating
    haptic assistive interfaces for motion-impaired users: Force-channels and competi-
    tive attractive-basins. In: Proc. of Eurohaptics, UK, pp. 122–127 (2002)
16. Langdon, P., Keates, S., Clarkson, J., Robinson, P.: Using haptic feedback to en-
    hance computer interaction for motion-impaired users. In: Proc. of ICDVRAT 2000,
    Alghero, Italy, pp. 25–32 (September 2000)
17. Lank, E., Cheng, Y.C.N., Ruiz, J.: Endpoint prediction using motion kinematics.
    In: Proc. of the SIGCHI Conference on Human Factors in Computing Systems,
    New York, NY, USA, pp. 637–646 (2007)
18. Majaranta, P., Räihä, K.J.: Twenty years of eye typing: systems and design issues.
    In: Proc. of the 2002 Sym. on Eye Tracking Research & Applications, pp. 15–22
    (2002)
19. Mandryk, R.L., Gutwin, C.: Perceptibility and utility of sticky targets. In: Pro-
    ceedings of Graphics Interface 2008, pp. 65–72 (2008)
20. Mithal, A.K., Douglas, S.: Differences in movement microstructure of the mouse
    and the finger-controlled isometric joystick. In: Proceedings of the SIGCHI Con-
    ference on Human Factors in Computing Systems, pp. 300–307 (1996)
21. Stocks, M., Hayward, S., Laycock, S.: Interacting with the biomolecular solvent
    accessible surface via a haptic feedback device. BMC Structural Biology 9, 69
    (2009)
22. Zilles, C.B., Salisbury, J.K.: A constraint based God-Object method for haptic
    display. In: Proc. of the IEEE Conference on Intelligent Robots and Systems, pp.
    146–151 (1995)

Modifying an Identified Angle of Edged Shapes Using Pseudo-haptic Effects

Yuki Ban*, Takashi Kajinami, Takuji Narumi,
Tomohiro Tanikawa, and Michitaka Hirose

The University of Tokyo, Information Science and Technology,
7-3-1 Hongo Tokyo, Japan
{ban,kaji,narumi,tani,hirose}@cyber.t.u-tokyo.ac.jp
http://www.cyber.t.u-tokyo.ac.jp

Abstract. In this paper, we focus on modifying the identification of an angle of edges when touching it with a pointing finger, by displacing the visual representation of the user's hand in order to construct a novel visuo-haptic system. We compose a video see-through system, which enables us to change the perception of the shape of an object a user is visually touching, by displacing the visual representation of the user's hand as if s/he was touching the visual shape, when in actuality s/he is touching another shape.

We had experiments and showed participants perceived angles of edges that was the same as the one they were visually touching, even though the angles of edges they were actually touching was different. These results prove that the perceived angles of edges could be modified if the difference of angles between edges is in the range of $-35°$ to $30°$.

Keywords: Pseudo-haptics, Visuo-haptic interaction, Identified Shape Modification.

1 Introduction

Haptics have become an important modality in recent virtual reality (VR) systems, and several haptic devices have been recently developed [1,2,3]. However, because it is difficult to perfectly reproduce the force that we perceive when touching an object, most haptic devices exhibit very complicated problems. As a result, it is difficult to apply haptic devices to widely used systems, because a large amount of preparation work, such as installation and calibration, must be performed for each person.

While research on haptic presentation in VR systems often concerns active haptics, an increasing number of works focus on alternative approaches such as passive haptics, which include pseudo-haptics and sensory substitution. Pseudo-haptics represent a kind of cross modal phenomenon between our visual and

---

* Graduate School of Information Science and Technology the University of Tokyo.

P. Isokoski and J. Springare (Eds.): EuroHaptics 2012, Part I, LNCS 7282, pp. 25–36, 2012.

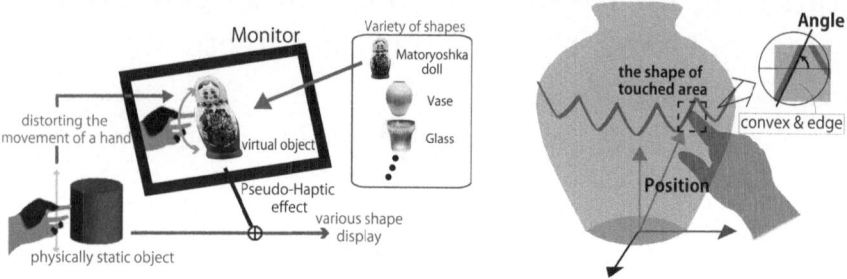

Fig. 1. Displaying the shape of an object using the pseudo-haptic effect

Fig. 2. The requirements for shape display

haptic senses [4]. The pseudo-haptic approach is a potential solution for exploiting boundaries and capabilities of the human sensory system to simulate haptic information without using active haptic systems. For example, when we are working on a computer, the slowdown of the cursor evokes a virtual frictional force on our hand holding the mouse. This phenomenon can potentially generate haptic sensations using only visual feedback, without the need of apply any physical devices.

In our research, we use this cross-modal effect to "change" the shape of an object, and construct a simple system that can display a variety of shapes, while the user touches only a simple static physical object (Fig.1). In other words, using a visual display and physical device, we aim to exploit visual feedback to widen the range of what can be physically presented by the device. This system evokes a pseudo-haptic effect by controlling the displacement of a userfs hand in the image showing the user touching the static object. By exploiting this effect in simple devices, we can change the user's perception of the shape.

In our system, we aim to display complicated shapes that are composed of primitives. Our system concept is presented in Fig.2. To realize this concept, we must develop two capabilities. First, we need to display primitives, i.e. convex, edge, concave, and so on, near the point of contact. Second, we need to set the relative postures of these primitives in an object. By combining these two capabilities, we can exploit the pseudo-haptic effect and display various shapes without applying any physical devices.

First, we used our simple system to confirm the possibility of displaying primitives of areas touched by the user. In particular, we proved that using the pseudo-haptic effect, users can perceive a variety of curved shapes, while touching only a physically static cylinder [5].

The next step in our research is to use our system to examine the possibility of exploiting the pseudo-haptic effect to modify an identified placement, in particular angles of primitives.

2   Related Work

While much research exists about cross modal effects between our haptic sensation and other sensations, here we mainly focus on effects between our haptic sensation and vision, which we aim to use in our system.

Haptic illusion which combines the presentation of forces with manipulated visual stimuli has a long history, dating back to Charpentier's size-weight illusion [6], which showed that subjects estimated the weights of objects with equal mass, based on their apparent visual size. In his work, it was revealed that subjects feel the object lighter when it appeared larger in their vision.

Pseudo-haptics, which is an illusional phenomena triggered by this characteristic of our senses, was first introduced by Lecuyer [7]. Lecuyer et al. had subjects push their thumbs against a piston, which in turn pushed against an isometric Spaceball device. Simultaneously, subjects were visually presented with a compressed virtual spring. Even though the Spaceball device was not compressed, the virtual spring influenced the perception of stiffness of the subjects. We can easily reproduce this effect using PowerCursor [8], a Flash toolkit used to create interfaces that a user can touch. Mensvoort developed this toolkit as part of his work on the feedback of simulated haptic [9].Pusch et al. proposed a pseudo-haptic approach, called hand-displacement-based pseudo-haptics (HEMP), which provides haptic-like sensations by displacing the visual representation of a userfs hand [10]. Specifically, their subjects wore a video see-through head-mounted display (HMD). When they placed their hands in the hole of a pipe, the HMD presented a virtual image of their hand moving to the right. As a result, subjects felt a force on their hands, even though they did not exert any physical force.

In addition, some research results show the potential of pseudo-haptic effects not only on our perception of force but also on our perception of texture and shape [11,12]. Research has shown that when we are presented with conflicting sensory stimuli, our vision usually dominates in our perception of a shape. Gibson's work [13] is an example of this type of research, demonstrating that subjects moving their hands along a straight surface while wearing distorting glasses, feels the straight surface as if it was curved. The work of Rock and Victor can be considered as another example of this type of research [14]. They asked their subjects to hold an object through a cloth while viewing the same object through a distorting lens. In this experiment, subjects matched the shape of the test object to the one most similar to the distorted visual image they saw, rather to the shape they actually touched. Kohli et al. proved that distorting a pointer showing the position of a device along a flat surface of a desk can change the perception of the shape of the surface [15]. Their work revealed that when subjects traced the device on the flat surface while being presented with the visual presentation as if they were tracing it on a curved surface, they visually perceived the curved one.

Other research work revealed that in some cases, even though visual stimuli are not given complete priority over haptic stimuli, cross-modal effects between the two sensations can influence our perception to some extent. Nakahara et al. found that in a mixed-realty system, when users are presented with haptic and

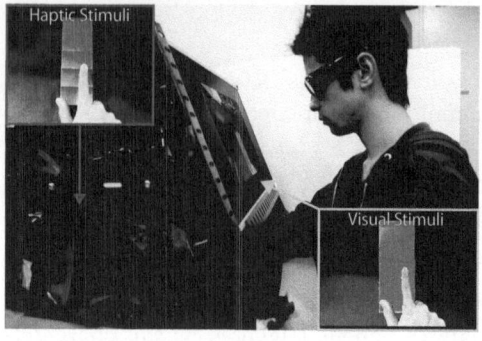
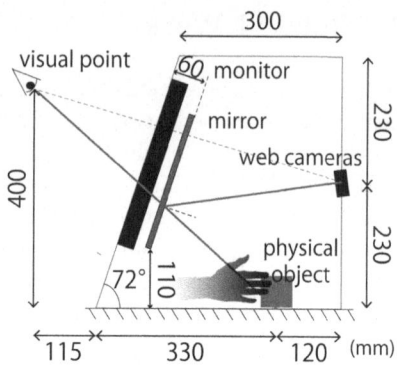

Fig. 3. Shape display system using effect of a visuo-haptic interaction

Fig. 4. Measurements of our shape display system

visual cube-shaped objects with discrepant edge curvatures, they perceive the curvatures to be somewhere in between the two objects [16].

3   Modification of an Identified Angle of Edges

In designing the display system for sensing a shape, we conducted an experiment on the pseudo-haptic effects on our perception of shape using a simple system we composed. In this experiment we confirmed the possibility that pseudo-haptic effects can assist in generating a perception of an angle of a shape. We focused on touching an object with a pointing finger, and experimented on the shape subjects felt when presented with visual and haptic stimuli independently.

3.1   Composition of Video See-through System

We constructed the simple video see-through system shown in Fig.3. In this system, users are shown that they are touching virtual objects whose shapes are different from the physical objects they are actually touching. They touch a physical object placed behind a visual monitor and view it through the 3D monitor.

We placed two web cameras at locations corresponding to a user's eyes using a mirror, and captured images around his hand. In this regard, the binocular parallax is specified by setting the distance between the cameras to 65mm. Then, the users sat on a chair and we set their heads to the position we established. Using the images captured by these cameras, the system realizes a stereoscopic video see-through display(Fig.4).

Touching the physically static object, whose shape is defined as $S_{physical}$, serves as the haptic stimuli. Watching the image of an object as if the user was touching another one, whose shape is defined as S_{visual}, serves as the visual stimuli. The shape of the object reported by the user as the one perceived as touching is defined as $S_{perceived}$. In this system, we aim to change $S_{perceived}$

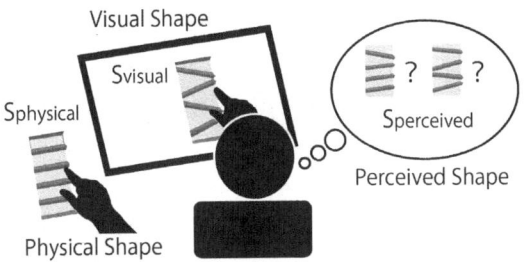

Fig. 5. Experiment for modifying the identified angle of edges on a board

without changing the physical shape of the object, by changing S_{visual} to a variety of $S_{physical}$ shapes and by provoking the pseudo-haptic effect.

The proposed system is implemented by postulating that we can combine primitives to display various complex shapes. Thus, we must be able to modify not only an identified primitive shape, but also an identified position and angle of a primitive shape. Using the video see-through system we composed, we first confirmed that an identified primitive curved surface shape can be modified. Next, we conducted an experiment on the effects of pseudo-haptics on our perception of angles of shapes. The shape of the object chosen to be touched by the subjects was selected to contain edges with various angles, so that we could easily measure the differences between $S_{physical}$, S_{visual} and $S_{perceived}$.

3.2   Algorithm for Visual Feedback Composition

In this paper, we construct an algorithm used to generate the necessary visual feedback to provoke the pseudo-haptic effect, and enable us to perceive a variety of angles of edges with only parallel edges.

We compose an image for visual feedback (I_{vf}) from the images taken by the two cameras attached(I_c), according to the following procedure(see Fig.6).

Calculation of the Distortion between $S_{physical}$ and S_{visual}. First, we compare $S_{physical}$ to S_{visual}, and decide how to modify the captured image.

Extraction of the shape of the object. First we set the background image(I in Fig.6), which was taken in advance. Then, on this background image we overlaid the virtual object S_{visual}, and extracted both ends of the angles of S_{visual}(II in Fig.6). In a similar way, we extracted both ends of the angles of the physical object $S_{physical}$, from the image taken by the web cameras(IV in Fig.6). Hereinafter, we refer to this image as the "captured image"(III in Fig.6).

Calculation of the warping space for the displacement of the hand. Using the positions of the ends of the edges we obtained, we warped the space for the displacement of the user's hand as if it was touching $S_{physical}$(V in Fig.6). To

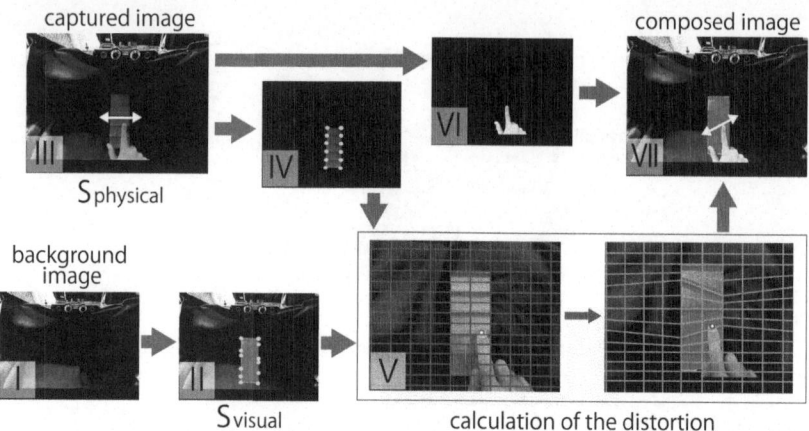

Fig. 6. Procedure for manipulating a visual stimulus

obtain this warping, we displaced point P in I_c to point P' in I_{vf} by the follow process.

Recognition of the Positional Relation between the Ends of Edges of $S_{physical}$ and Point P

In what follows, the x,y-coordinates of point P are denoted by (x_P, y_P). The point of origin and the x and y axis are set as shown in Fig.7. In this coordinate system, the x,y-coordinates of the ends of edges for both of $S_{physical}$ and $S_{virtual}$ are computed by the formula presented below. We name the ends of edges of the physical object $L_i(i = 0, \cdots, 4), R_i(i = 0, \cdots, 4)$, and name the ends of edges of the virtual object $L_i'(i = 0, \cdots, 4), R_i'(i = 0, \cdots, 4)$. The x coordinates of all $L_i(i = 0, \cdots, 4)$ are equal to each other. The same also holds for R, R' and L'.

First, we investigated the domain surrounded by the ends of the edges($A_i(i = 0, \cdots, 5)$) in which the point P belongs. Next, we recognize the positional relation between point P and its upper and lower edges. In Fig.7, the positions of points T and B, and the lengths of w_0, w_1, h_0, h_1 are calculated as follows (for the case of $P \in A_i$).

$$w_0 : w_1 = (x_P - x_{L_i}) : (x_{R_i} - x_P)$$
$$x_T = x_B = x_P$$
$$y_T = y_{L_i} + \frac{w_0}{w_0 + w_1} \cdot (y_{R_i} - y_{L_i})$$
$$y_B = y_{L_{i+1}} + \frac{w_0}{w_0 + w_1} \cdot (y_{R_{i+1}} - y_{L_{i+1}})$$
$$h_0 : h_1 = (y_P - y_T) : (y_B - y_P)$$

Displacement of Point P to Point P'

Using the ratios $w_0 : w_1$ and $h_0 : h_1$, we can determine the position of point

$P'(P'_x, P'_y)$ in I_{vf} that satisfies the positional relationship between the ends of the edges of $S_{physical}$ and P.

The following relation is used to determine T' and B' in Fig.7.

$$w'_0 : w'_1 = w_0 : w_1$$
$$h'_0 : h'_1 = h_0 : h_1$$
$$x_{T'} = x_{B'} = x_{L'_i} + \frac{w_0}{w_0 + w_1} \cdot (x_{R'_i} - x_{L'_i})$$
$$y_{T'} = y_{L'_i} + \frac{w_0}{w_0 + w_1} \cdot (y_{R_i} - y_{L_i})$$
$$y_{B'} = y_{L'_{i+1}} + \frac{w_0}{w_0 + w_1} \cdot (y_{R_{i+1}} - y_{L_{i+1}})$$

Using the positions of T' and B', we determine the position of point P'.

$$x_{P'} = x_{T'}$$
$$y_{P'} = y_{T'} + \frac{h'_0}{h'_0 + h'_1} \cdot (y_{B'} - y_{T'})$$

Replacement of the User's Hand. From the captured image, using color extraction we identify the area of the user's hand and the position of his fingertip(VI in Fig.6). Next, we calculate the distortion of the hand's movement by comparing the contour of the physical shape, with that of the visual shape(V in Fig.6). Based on this distortion, we place the image of the user's hand at the corresponding position as if s/he was touching the virtual object(VII in Fig.6). We calculate the distortion according to the procedure described above. If we assign the position of the fingertip in I_c to P, we can get the position of the displaced fingertip in I_{vf} as point P'.

Finally, from the two images captured from the right and left cameras, using the procedure described above we create the stereoscopic image.

4   Experiment

We investigated the ability of our shape-display system based on the pseudo-haptic effect to control the perception of angles of edges. We conducted an experiment to examine how the effect of S_{visual} changed $S_{physical}$ to $S_{perceived}$, and compared these three shapes. If $S_{perceived}$ resembled S_{visual} rather than $S_{physical}$, then we concluded that the visuo-haptic interaction was effectively provoked, meaning that the system worked as designed. Conversely, if $S_{perceived}$ resembled $S_{physical}$ rather than S_{visual}, then we concluded that the haptic stimuli was more influential for the perceived shape rather than the visual one. Thus, when the latter result was obtained, it indicated that it was difficult for our system to modify an identified an angle of shapes.

We chose five types of boards with edges of various angles shown in Fig.8, denoted as S_{visual}. For virtual bodies B_1-B_5, we presented subjects with the $S_{physical}$

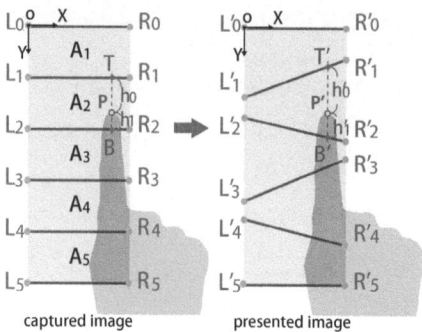

Fig. 7. Computing the position of a fingertip

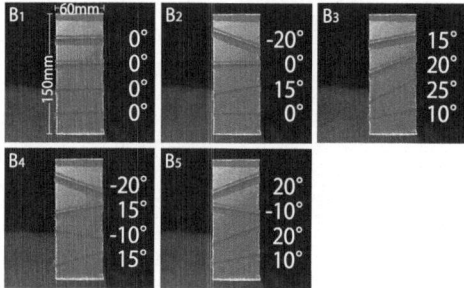

Fig. 8. S_{visual} as the visual stimuli

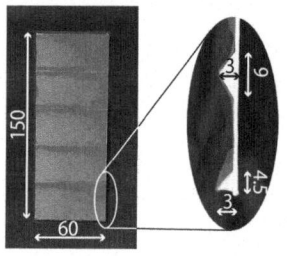

Fig. 9. Shape A_1 as haptic stimulus (mm)

shape A_1 (Fig.9). Virtual bodies were presented to the users in a random order. The reason why we presented users with boards with various angles of edges, such as B_1-B_5, not with only one edge, was because we predicted that the amount of modification of the perception of the angle was influenced not by the difference between the angle of the edge of $S_{physical}$ and S_{visual} ($\theta(= \theta_1 - \theta_0)$ in Fig:10) but by the relative difference between two successive angles ($\phi(= \phi_1 - \phi_0)$ in Fig:10).

We did not set a time limit for the subjects to answer what shape they felt, and we gave them the following four instructions.

- Subjects must watch the image presented on the monitor when touching the object.
- Subjects must touch the object from top to bottom as evenly as possible.
- Subjects must touch the object with one finger.
- Subjects can touch the object repeatedly.

We did not hold the heads of the subjects with any equipment. Instead, we instructed them not to move their heads from the position that we initially set in realizing the video see-through system. We measured the position of the heads of the subjects, and confirmed that the subjects watched the correct video see-through image on the monitor.

In the experiment, subjects were not aware that they actually touched only one kind of $S_{physical}$. In each trial the experimenter only went through the motions of changing $S_{physical}$. In addition, the experimenter did not indicate whether the hand position in the monitor was actually distorted or not, and thus the subjects did not know whether S_{visual} was the same as $S_{physical}$. Eight men and two women in their twenties participated in this experiment, and two trials were conducted for each S_{visual}.

The subjects rotated the edges on the device(Fig.11) and answered about the angle of the edges they felt touching. This device has four edges which can be rotated freely, and the size of the board and its edges is the same as $S_{physical}$. The subjects touched and rotated the edges, and the experimenter read the scale and measured the angles of $S_{perceived}$. The positional relationship of this device and the eyes of a subject was arranged with the positional relationship of $S_{physical}$ and a subject's eyes in a trial. The experimenter told the subjects in advance that they could touch and rotate edges repeatedly, and the experimenter set the angle of all edges to zero degree when one trail finished.

For this experiment, we conducted two types of control experiments. In the first, we asked subjects to perform almost the same task, but we only showed them S_{visual}, and did not show them their hands in the monitor. In the second control experiment, the subjects were shown the shapes they were actually touching as S_{visual}. Thus, in the second control experiment, $S_{physical}$ was the same as S_{visual}. We call $Ex_{w/oHand}$ the type of experiment in which we do not show the image of subject's hand on a monitor. The type of experiment that shows the displaced the image of subject's hand is called $Ex_{composed}$. Finally, the type of experiment that shows S_{visual} as the shape that subjects are actually touching, is called Ex_{real}. @ During Ex_{real}, we used A_i as $S_{physical}$ which has same angles as B_i ($i = 1, \cdots, 5$). Two trials were conducted for each S_{visual} in $Ex_{w/oHand}$ and two trials were conducted for each $S_{physical}$ in Ex_{real}. We computed the visuo-haptic effect by distorting the position of the hand touching the object and comparing the results of these experiments. By comparing the results of $Ex_{composed}$ and the result of Ex_{real}, we measured the effectiveness of the system, defined as the ability to modify the identified angle of edges.

When the results of $Ex_{composed}$ were similar to those of Ex_{real}, subjects perceived S_{visual} differently from the shape they were actually touching. It appears that if the results of $Ex_{w/oHand}$ indicate the subjects perceived S_{visual} without being shown the hand displacement, then the effect of modifying the identified angle was not provoked by the pseudo-haptic effect of displacing the hand.

5   Results and Discussion

First, during the experiment we ensured that the head of the subject did not move from the position we chose when implementing the video see-through system (error 5 mm). Thus, subjects watched the correct video-see-through image during the experiment.

We arrange the results according to the order of the angles of S_{visual}, in Fig.10 (θ_1 in Fig.12). The comparison of the results of $Ex_{composed}$ and Ex_{real}

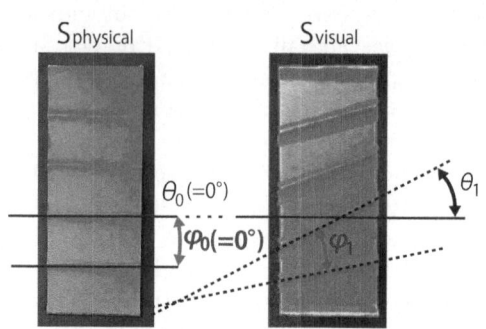

Fig. 10. Computing the position of a fingertip

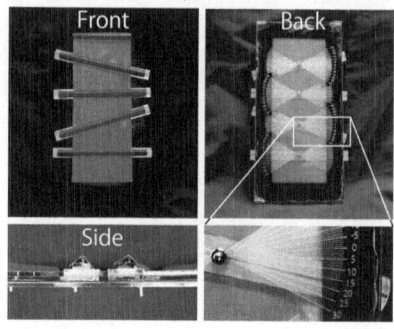

Fig. 11. Subjects answer the shape they feel by adjusting the angle of this device

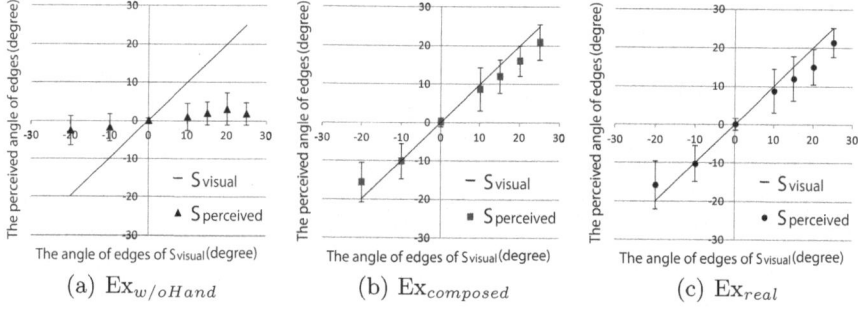

(a) $\mathrm{Ex}_{w/oHand}$ (b) $\mathrm{Ex}_{composed}$ (c) Ex_{real}

Fig. 12. The perceived angle of edges by the angle of edges of S_{visual} (means and SDs)

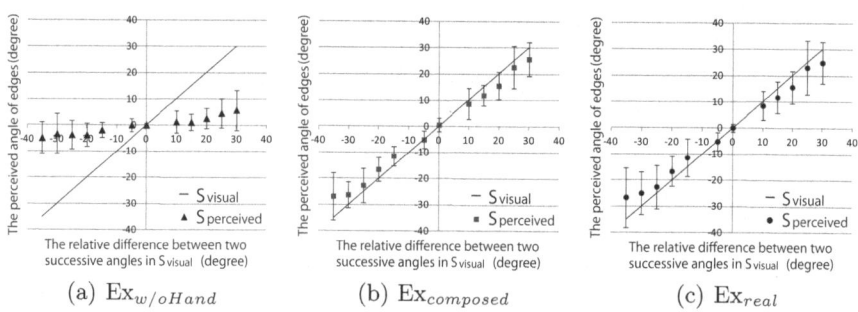

(a) $\mathrm{Ex}_{w/oHand}$ (b) $\mathrm{Ex}_{composed}$ (c) Ex_{real}

Fig. 13. The perceived relative difference between two successive angles (means and SDs)

shows that even though subjects were actually touching A_1 containing parallel edges, when we displayed the image of the displaced hand, they perceived a shape similar to that they perceived when touching A_2-A_5 as $S_{physical}$. These results indicate that even though subjects were touching parallel angles, the angles of edges identified by the subjects could be modified if they were in the range of $-20°$ to $25°$. Conversely, in $Ex_{w/oHand}$ were subjects actually touched shape A_1, Fig.12(a) shows the identified angle of each edge was in the neighborhood of $0°$. This result indicates that when we did not display the displaced image of the hand, subjects were strongly conscious of $S_{physical}$ as the haptic stimulus. The explanation of this behavior is that when a displaced image of the hand was not shown, subjects could not recognize the position they were touching, and thus they could not use the visual stimuli to perceive the angle. This result indicates that simply showing the shape of a virtual object that is different from the real one is insufficient, while showing the image of the displaced hand is critical.

Moreover, we arranged these results in the order of the relative difference between two successive angles of S_{visual} (ϕ_1 in Fig.13). This ordering of results shows that the results of $Ex_{composed}$ and Ex_{real} were almost in agreement, and the result of $Ex_{w/oHand}$ shows that the identified angle difference settled in the neighborhood of $0°$ as A_1. Thus, our system can modify the perception of the relative difference between two successive angles, if the angle differences are in the range of $-35°$ to $30°$.

These results differ from the results obtained by Johansson et al., stating that the direction of the pressure placed on a fingertip can be estimated from responses of afferent nerves [17]. In our system, it appears that the sense of direction of the pressure on the fingertip was modified by displacing the visual representation of the user's hand. In future work, we will analyze in more detail whether the inclination of a subject's finger changed during the experiment.

6   Conclusion and Future Work

This paper evaluate the effectiveness of modifying an identified angle of shapes using a visuo-haptic shape display system. This system uses the pseudo-haptic effect based on the visual displacement of a user's hand touching an object.

We postulated that we can display various shapes by combining primitives. Thus, we needed to modify not only identified primitives, but also their relative positions and angles identified in an object. In this study, we focused on modifying an identified angle of primitives, edges. We conducted an experiment to evaluate the effectiveness of the proposed system, and a large portion of the subjects felt that they were not touching the shape of the haptic stimuli, but rather the shape of the object visually presented. This experiment proved that our system can modify the perception of the relative difference between two successive angles, if the angle differences are in the range of $-35°$ to $30°$. Thus, we conclude that it is possible to modify an identified angle of primitives, such as an edge, by using visual feedback similar to the one used in our system. To derive more detailed specifications for the system, we should measure the range of perception of the angle that we can generate using pseudo-haptic effects, and examine whether it is possible to modify the identified position of a shape. Based on these measurements, we can decide on the range and accuracy of shapes we should generate with physical devices. Then, we can compose the physical devices and visual display to generate the visual feedback, and construct a system, which give users the sense of touching a variety of shapes.

References

1. 'phantom', http://www.sensable.com
2. Hoshino, H., Tachi, S.: Realization and utillization of 3d cyber space. a method to represent an arbitrary surface in encounter type shape representation system. Transactions 4, 445–454 (1999)
3. Iwata, H.: Project feelex: adding haptic surface to graphics. In: SIGGRAPH 2001 Conference Proceedings, pp. 469–475 (2001)
4. Lecuyer, A., Coquillart, S., Kheddar, A., Richard, P., Coiffet, P.: Pseudo-haptic feedback: Can isometric input devices simulate force feedback. In: Proc. IEEE VR, pp. 83–90 (2000)
5. Ban, Y., Kajinami, T., Narumi, T., Tanikawa, T., Hirose, M.: Modifying an Identified Angle of Edged Shapes Using Pseudo-haptic Effects. In: Isokoski, P., Springare, J. (eds.) EuroHaptics 2012, Part I. LNCS, vol. 7282, pp. 25–36. Springer, Heidelberg (2012)
6. Mrrray, D.J., Ellis, R.R., Bandomir, C.A., Ross, H.E.: Charpentier (1891) on the size-weight illusion. Perception and Psychophysics, 61 (8) (1999)
7. Lecuyer, A.: Simulting haptic feedback using vision: A survey of research and applications of pseudo-haptic feedback. Teleoperators and Virtual Environments 18(1), 39–53 (2009)
8. 'powercursor', http://www.powercursor.com/
9. van Mensvoort, K.: What you see is what you feel: exploiting the dominance of the visual over the haptic domain to simulate force-feedback with cursor displacements. In: Proceedings of the 4th Conference on Designing Interactive Systems: Processes, Practices, Methods and Techniques, pp. 345–348 (2002)
10. Pusch, A., Martin, O., Coquillart, S.: Hemp–hand-displacement-based pseudo-haptics: A study of a force field application and a behavioural analysis. International Journal of Human-Computer Studies 67(3), 256–268 (2009)
11. Iesaki, A., Somada, A., Kimura, A., Shibata, F., Tamura, H.: Psychophysical influence on tactual impression by mixedreality visual stimulation. Transactions of the Virtual Reality Society of Japan 13(2), 129–140 (2008)
12. Lecuyer, A., Burkhardt, J.M., Tan, C.H.: A study of the modification of the speed and size of the cursor for simulating pseudo-haptic bumps and holes. ACM Transaction on Applied Perception (ACM TAP) 5 (2008)
13. Gibson, J.J.: Adaptation, after-effect, and contrast in the perception of curved lines. Journal of Experimental Psychology 16, 1–31 (1933)
14. Rock, I., Victor, J.: Vision and touch: An experimentally created conflict the two senses. Science 143, 594–596 (1964)
15. Kohli, L.: Exploiting perceptual illusions to enhance passive haptics. In: IEEE VR Workshop on Perceptual Illusions in Virtual Environments, pp. 22–24 (2009)
16. Nakahara, M., Kitahara, I., Ohta, Y.: Sensory propety in fusion of visual/haptic cues by using mixed reality. In: Second Joint Eurohaptics Conference and Symposium on Haptic Interfaces For Virtual Environment and Teleoperator System (2007)
17. Johansson, R.S., Birznieks, I.: First spikes in ensembles of human tactile afferents code complex spatial fingertip events. Nature Neuroscience 7(2), 170–177 (2004)

Transparency Improvement in Haptic Devices with a Torque Compensator Using Motor Current

Ozgur Baser[1], E. Ilhan Konukseven[2], and Hakan Gurocak[1]

[1] School of Engineering and Computer Science, Washington State University Vancouver, USA
[2] Mechanical Engineering Department, Middle East Technical University Ankara, Turkey
ozgrbasr@gmail.com, konuk@metu.edu.tr, hgurocak@vancouver.wsu.edu

Abstract. Transparency of a haptic interface can be improved by minimizing the effects of inertia and friction through the use of model based compensators. However, the performance with these algorithms is limited due to the estimation errors in the system model and in the velocity and acceleration from quantized encoder data. This paper contributes a new torque compensator based on motor current to improve transparency. The proposed method was tested experimentally in time and frequency domains by means of an excitation motor attached at the user side of the device. The excitation motor enabled evaluation of the algorithms with smooth trajectories and high frequencies, which cannot be generated by user hand. Experimental results showed that the algorithm significantly improves transparency and doubles the transparency bandwidth.

Keywords: Haptic transparency, haptic stability, current feedback, transparency bandwidth, inertia/friction compensation.

1 Introduction

Performance of a haptic device is often evaluated by its force capacity, precision and transparency. Ideally the haptic device should transmit impedance of the virtual environment to the user without any distortions. In other words, the ratio of the transmitted impedance (Z_M) and the environment impedance (Z_E) should be unity for a desired bandwidth [1]. However, the dynamics of the haptic device disturbs the transmitted impedance. Parasitic torques/forces such as mass/inertia, friction and gravity in haptic interaction need to be reduced or eliminated to improve the transparency.

There are two ways to overcome this problem: (1) Re-design the haptic interface to minimize its dynamic effects, or (2) Improve the performance of the controller to meet the requirements for the transparent simulation [2-4]. Most of the techniques used for improving the transparency involve model based compensators. Linear lead-lag compensators have been shown to extend transparency bandwidth in simulation [5]. Adaptive control laws were used to increase performance criteria such as transparency or stability [6, 7]. In recent studies, either closed loop or open loop feedback with model based compensators were used for improving transparency and stability characteristics [8-10]. To evaluate the transparency characteristics McJunkin

P. Isokoski and J. Springare (Eds.): EuroHaptics 2012, Part I, LNCS 7282, pp. 37–46, 2012.
© Springer-Verlag Berlin Heidelberg 2012

classified haptic interaction into two types; active user interaction (AUI) and passive user interaction (PUI) [11, 12]. In an AUI application, whenever the user touches a virtual object, the device senses the motion and provides a force based on the virtual model. In the PUI case, the user does not generate motion. The passive user comes into contact with an object such as a rubber ball in a virtual tennis game. Maximization of the transparency, hence the haptic performance in both cases can only be performed by minimizing the disturbance torques/forces resulting from mass/inertia, gravity and friction.

Model-based compensators are employed in haptic interfaces to mask the inertia and friction of the system [8]. These compensators require accurate model identification. In addition, accurate velocity and acceleration estimations are needed which is quite challenging due to the quantized encoder positions. Furthermore, control gains cannot be too high due to instability issues.

In this research, we explored a torque compensator based on motor current (TCBMC) to improve the performance of the model-based compensators.

2   Torque Compensator Based on Motor Current (TCBMC)

There are two main control algorithms in haptic interfaces: impedance and admittance control [8, 13]. In impedance control, user motion is sensed and a reference force is computed based on the virtual environment model. This type of control strategy can be improved with a force-feedback loop leading to the so called closed loop impedance control (CLIC). CLIC algorithm cannot employ high controller gains due to the inherent device dynamics. To decrease the dynamic effects of the haptic device felt by the user, a model-based compensator can also be added to the closed loop impedance control (CLIC + MBC) [14]. However, the feedback loop and model-based compensator loop are coupled in this algorithm and may give worse results than CLIC.

The torque compensator based on motor current (TCBMC) creates an additional inner control loop (Figure 1c) to estimate exact interaction torques in actuator joint for effective transparency. It compares the torque output of the CLIC controller (Controller-I) to the actual motor torque based on instantaneous motor current. If any error exists between them, Controller-II compensates for it. A low-pass filter (LPF) with 40 Hz cut-off frequency is also employed to reduce the oscillatory effects of the noisy feedback. The motor current measurement was provided by the amplifier.

The proposed control algorithm is appropriate for digital applications and does not require any analog processing. The symbols Z_v, Z_u, K_p, K_c, x_u, x_h, θ_h, $\dot{\theta}_h$, J_v, J_a, F_u, τ_d, τ_c, τ_m, τ_u, v_a, i_m, R_m, L, K_{emf}, K_m symbols in Fig.1 correspond to impedance of virtual environment, impedance of user hand, gain of controller-I, gain of controller-II, motion of the user hand, motion of the haptic handle, joint motion of the device, joint velocity of the device, virtual Jacobian matrix, actual Jacobian matrix, user force, desired torque command, compensator torque command, joint torque applied by motor, joint torque applied by user, motor voltage, motor current, motor resistance, motor inductance, motor back EMF constant, motor torque constant, respectively.

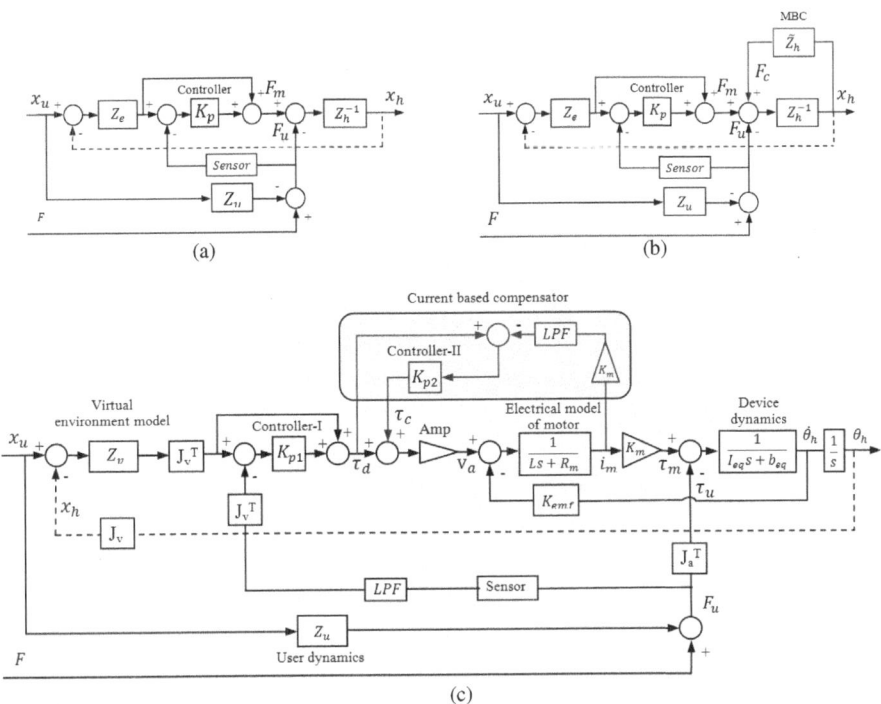

Fig. 1. Detailed Block Diagrams for (a) CLIC, (b) CLIC+MBC and (c) CLIC+TCMBC

3 Experiments and Results

3.1 Experimental Setup

We designed a 1-DOF haptic device with a rotary handle as shown in Figure 2. The back end of the setup is the haptic device where a brushless DC servomotor (*Parker BE232FJ*) is connected to a servo amplifier (*AMC-DPRANIE-015A400*). The actuator torque is increased using sprockets and chains with 96:20 gear ratio. The servomotor has a built-in 2000-line encoder. A torque transducer (*Futek FSH01987*) with ±10Nm range is attached to the handle. The front end of the setup has a second motor which is used to provide excitation inputs for the transparency experiments. The control algorithms are implemented using Quarc v2.2 and SIMULINK software. The computer is an Intel® Core™ i7-2600 (3.40 GHz, 8 MB cache, 4 cores), 4 GB Ram, AMD Radeon HD 6350 (512 MB) graphics card. The computer also has a Quanser Q4 interface card.

3.2 System Parameter Identification and Velocity/Acceleration Estimation

Total parasitic torque (τ) sensed by user comes from the equivalent inertial and frictional effects on the handle:

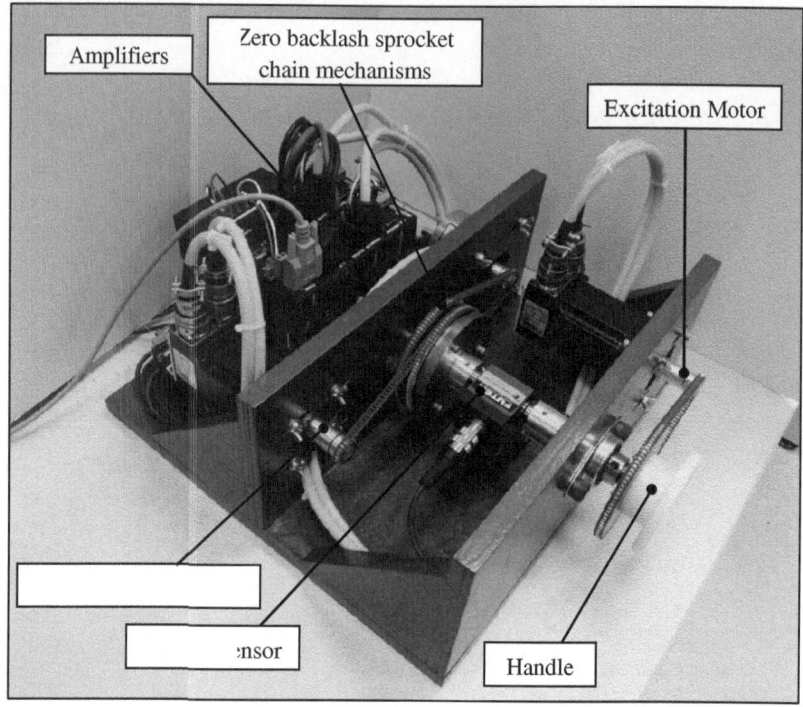

Fig. 2. Experimental setup with a 1-DOF haptic device

$$\tau = I_{eq}\alpha_o + B_{eq}\omega_o + \text{sgn}(\omega_o)T_{c_{eq}} \qquad (1)$$

I_{eq}, B_{eq}, T_{Ceq} in Eq.(1) are equivalent inertia, viscous friction and Coulomb friction induced in moving the handle, respectively. To find these equivalent parameters, a simple experiment was carried out. Torque measurements and estimations of velocity and acceleration of the haptic handle from the encoder data were stored while the handle was rotated at various speeds and accelerations. The Gauss-Newton nonlinear least-square estimation technique in [15] (MATLAB® "*lsqnonlin*") was applied to estimate the parameters using the parasitic torque equations based on the torque, velocity and acceleration measurements. As a result, equivalent static friction, viscous friction and inertia parameters were found as (I_{eq}= 0.001 kgm², B_{eq}=0.005 Nms, T_{Ceq}=0.15 Nm). RMS error of this estimation is 0.1629 Nm.

Velocity/acceleration estimation - Optical encoders are used in haptic devices. However, the encoders cause quantization error in position measurements. This makes it impossible to calculate the velocity and acceleration directly by numerical differentiation, especially at low velocities. As a result, the performance of the model-based algorithms is adversely affected. Therefore, it is necessary to use signal processing and estimation techniques.

There are various methods available in the literature for estimating velocity and acceleration including filtering techniques [16], observers [17, 18] and a first-order adaptive filtering method [19]. In this study, we used our second order enhanced adaptive windowing method to estimate both velocity and acceleration for model-based compensation [20, 21]. In this algorithm, coefficients of a second-order curve passing through all intermediate samples are dynamically determined using least squares method within an adaptive "moving" data window length. The size of the discrete data window is increased until the difference between the exact encoder position and the estimated position decreases to the maximum quantization error. Fig.3 shows the flowchart of the enhanced second order adaptive windowing method for velocity and acceleration estimation. y_k is last value of encoder position while Y_k denotes last calculated position using a model whose parameters (a, b, c) are dynamically determined via least squares method within an adaptive "moving" data window with length "i". Discrete data window is increased until the difference between exact encoder position (y_k) and estimated position (Y_k) decreases to maximum quantization error "d".

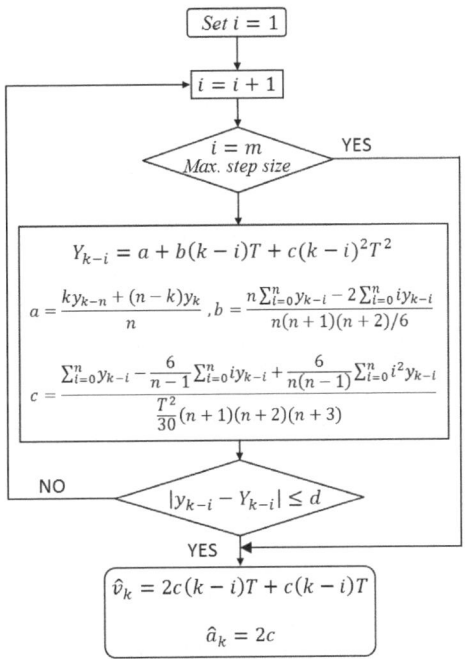

Fig. 3. Second order adaptive windowing algorithm

3.3   Free motion Transparency Experiment

In this experiment, we evaluated the performance of the controllers as the handle of the haptic device was rotated back-and-forth in free motion. The device did not simulate any virtual object interaction hence the virtual stiffness (K) and damping (B) elements were set to zero. It was desired that the device did not produce any frictional and inertial effect in the experiments.

The excitation motor rotated the handle with varying velocity and acceleration using a chirp signal ($\delta = X_{amp}sin\,(2\pi f_{max}t^2/t_{dur})$) with 1.1 radian amplitude, 10 seconds duration and 3 Hz maximum frequency. The interaction torque was recorded. As a measure of the free motion transparency, we plotted the interaction torque versus angular rotation. The interaction torque measurements are expected to be close to horizontal axis for maximum transparency.

Fig.4 presents the free-motion transparency measures for all three algorithms. As it can be seen, the closed loop algorithms give good performance. However, the proposed (CLIC+TCBMC) algorithm is superior compared to the others. In order to quantify the experimental results, RMS errors of the experiments were calculated as (0.0751 Nm, 0.0835 Nm, 0.0450 Nm) for CLIC, CLIC+MBC and CLIC+TCBMC respectively.

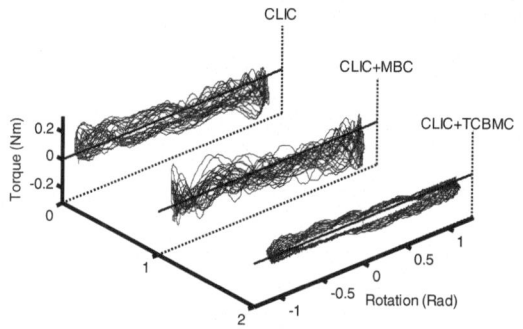

Fig. 4. Time domain transparency performance measures for free-motion

3.4   Virtual Load Simulation

In this experiment, torsion stiffness (K) and torsion damping (B) are assigned to the virtual environment model for the simulation.   The same excitation trajectory is used in the second experiments. The experiments were conducted for two different virtual models: (1) a virtual spring ($K=0.5$ Nm/rad), and (2) a virtual damping ($B=0.025$ Nms/rad).   For high transparency performance, the haptic device is expected to produce the torque required only for the simulation of the virtual environment.

Figures 5a-b and 5c-d show the experimental results along with the desired torque values (straight lines) and error plots. The closeness of the experimental performance to the desired torque specifies the performance of the algorithms in terms of transparency.   As it can be seen, the CLIC provides stable interaction torques. However, it is not enough for full transparency since the proportional gain cannot be increased to high values due to the resulting instability of the haptic device. The CLIC+MBC does not improve the haptic interaction effectively due to the imprecision in the estimation of velocity/acceleration based on encoder data and inaccuracy in the model identification. The proposed CLIC+TCBMC controller gives the highest transparency performance for the haptic interactions since it follows the desired torque line with minimal deviation. RMS errors of the virtual spring and virtual damping experimental results were calculated as (0.0832 Nm, 0.0756 Nm, 0.0426 Nm) and (0.0825 Nm, 0.0690 Nm, 0.0376 Nm) for CLIC, CLIC+MBC and CLIC+TCBMC, respectively.

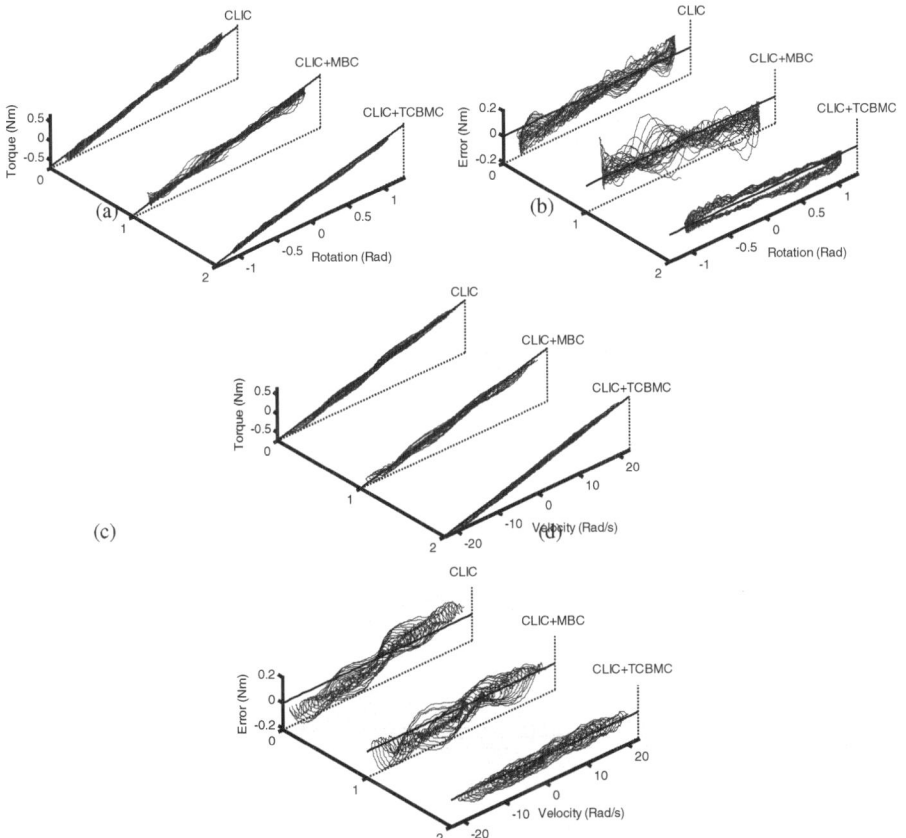

Fig. 5. Time domain transparency performance measures for (a-b) virtual spring with *K=0.5 Nm/rad*, and (c-d) virtual damping with *B=0.025 Nms/rad*

3.5   Transparency Bandwidth

Experimental results given above present the transparency performance measures in time domain. The transparency can also be evaluated in the frequency domain by means of the transparency bandwidth. It is a measure of the range of force/torque frequencies that can be displayed with the haptic interface.

In this experiment, the handle was rotated by the excitation motor using a chirp signal with 0.3 radian amplitude, 30 seconds duration and 10 Hz maximum frequency for a virtual spring simulation with *K=1 Nm/rad*. The transmitted impedance, Z_T was estimated as a function of frequency by dividing the cross power spectral density between the motion input and force output ($\phi_{VF}(j\omega)$) by the power spectral density of the motion input ($\phi_{VV}(j\omega)$) [22, 23].

$$Z_T(j\omega) = \frac{\phi_{VF}(j\omega)}{\phi_{VV}(j\omega)} \tag{2}$$

The transparency transfer function is computed by dividing the transmitted impedance Z_T by the desired impedance Z_D. The desired impedance in our experiments are $K=1$ Nm/rad for whole frequency range. The transparency bandwidth values are given according to the transparency transfer function limits +/-3 dB [23].

As shown in Figures 6a, 6b and 6c, the transparency bandwidth for the CLIC, CLIC+MBC and CLIC+TCBMC algorithms were 5.25 Hz, 4.51 Hz and 9.55 Hz, respectively. The proposed algorithm increases the transparency bandwidth almost twice as much. Besides, phase difference remains almost zero within the increased transparency bandwidth.

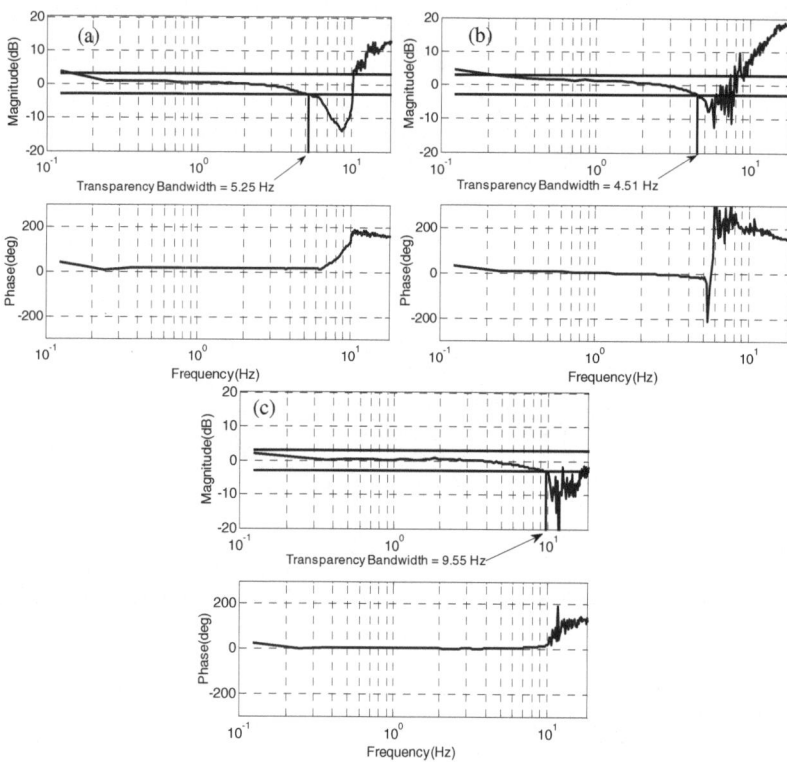

Fig. 6. Transparency performance measures for a virtual spring with $K=1$ Nm/rad in frequency domain.   (a) CLIC, (b) CLIC + MBC, (c) CLIC + TCBMC.

4   Conclusions

Ideally a haptic device should transmit impedance of the virtual environment to the user without any distortions. However, the dynamics of the haptic device disturbs the transmitted impedance reducing its transparency. Advanced force control algorithms were developed to improve the transparency of haptic devices. Most of the techniques used for improving the transparency involve model based compensators. They require accurate model identification and velocity/acceleration estimations. Furthermore, control gains cannot be too high due to instability issues.

The contribution of this research is a torque compensator based on motor current (TCBMC) to improve the performance of the model-based compensators. The performance of the proposed controller was compared to two common approaches found in the literature. The algorithms were tested experimentally by using an excitation motor attached to the user side of the device. This motor enabled us to conduct the experiments at high frequencies which could not be done accurately with manual input from the user. It is shown that the proposed algorithm leads to much better transparency.

Acknowledgement. The first author would like to thank TUBITAK (The Scientific and Technological Research Council of Turkey) for the research scholarship grant supporting his work at Washington State University Vancouver as a visiting scholar.

References

1. Lawrence, D.A.: Stability and transparency in bilateral teleoperation. IEEE Transactions on Robotics and Automation 9(3), 624–637 (1993)
2. McJunkin, S.T.: Transparency Improvement for Haptic Interfaces. PhD. Thesis (May 2007)
3. Vlachos, K., Papadopoulos, E.: Transparency maximization methodology for haptic devices. IEEE/ASME Transactions on Mechatronics 11(3), 249–255 (2006)
4. Colgate, J.E.: Robust impedance shaping telemanipulation. IEEE Transactions on Robotics and Automation 9(4), 374–384 (1993)
5. Fite, K.B., Speich, J.E., Goldfarb, M.: Transparency and Stability Robustness in Two-Channel Bilateral Teleoperation. Journal of Dynamic Systems, Measurement, and Control 123, 400–407 (2001)
6. Lee, H.K., Chung, M.J.: Adaptive Controller of a Master-Slave System for Transparent Teleoperation. Journal of Robotic Systems 15(8), 465–475 (1998)
7. Hashtrudi-Zaad, K., Salcudean, S.E.: Analysis of Control Architectures for Teleoperation Systems with Impedance/Admittance Master and Slave Manipulators. The International Journal of Robotic Research 20(6), 419–445 (2001)
8. Carignan, C.R., Cleary, K.R.: Closed-Loop Force Control for Haptic Simulation of Virtual Environments. The Electronic Journal of Haptics Research 1(2), 1–14 (2000)
9. Frisoli, A., Sotgiu, E., Avizzano, C.A., Checcacci, D., Bergamasco, M.: Force-based Impedance Control of a Haptic Master System for Teleoperation. Sensor Review 24(1), 42–50 (2004)
10. Bernstein, N.L., Lawrence, D.A., Pao, L.Y.: Friction Modeling and Compensation for Haptic Interfaces. In: Proceedings of the First Joint Eurohaptics Conference and Symposium on Haptic Interfaces for Virtual Environment and Teleoperator Systems, pp. 290–295 (2005)
11. McJunkin, S.T., O'Malley, M.K., Speich, J.E.: Transparency of a Phantom Premium Haptic Interface for Active and Passive Human Interaction. In: American Control Conference, pp. 3060–3065 (2005)

12. McJunkin, S.T., Li, Y., O'Malley, M.K.: Human-Machine Admittance and Transparency Adaptation in Passive User Interaction with a Haptic Interface. In: Proceedings of the First Joint Eurohaptics Conference and Symposium on Haptic Interfaces for Virtual Environment and Teleoperator Systems, pp. 283–289 (2005)
13. Gil, J.J., Sanchez, E.: Control Algorithms for Haptic Interaction and Modifying the Dynamical Behavior of the Interface. In: 2nd International Conference on Enactive Interfaces (2005)
14. Bernstein, N.L., Lawrence, D.A., Pao, L.Y.: Friction Modeling and Compensation for Haptic Interfaces. In: Proceedings of the First Joint Eurohaptics Conference and Symposium on Haptic Interfaces for Virtual Environment and Teleoperator Systems, pp. 290–295 (2005)
15. Hollerbach, J., Khalil, W., Gautier, M.: Model Identification. In: Siciliano, B., Khatib, O. (eds.) Springer Handbook of Robotics, pp. 321–344. Springer (2008)
16. Liu, G.: On velocity estimation using position measurements. In: American Control Conference, vol. 2, pp. 1115–1120 (2002)
17. Chan, S.P.: Velocity estimation for robot manipulators using neural network. Journal of Intelligent and Robotic Systems 23, 147–163 (1998)
18. Yusivar, F., Hamada, D., Uchida, K., Wakao, S., et al.: A new method of motor speed estimation using fuzzy logic algorithm. In: International Conference of Electric Machine Drives, pp. 278–280 (1999)
19. Janabi-Sharifi, F., Hayward, V., Chen, C.-S.J.: Discrete-Time Adaptive Windowing for Velocity Estimation. IEEE Transaction on Control System Technology 8(6) (2000)
20. Kilic, E., Baser, O., Dolen, M., Konukseven, E.I.: Enhanced adaptive windowing technique for velocity and acceleration estimation using incremental position encoders. In: IEEE International Conference on Signal and Electronic Systems, Gliwice, Poland (2010)
21. Baser, O., Kilic, E., Konukseven, E.I., Dolen, M.: A hybrid technique to estimate velocity and acceleration using low-resolution optical incremental encoders. In: IEEE International Conference on Signal and Electronic Systems, Gliwice, Poland (2010)
22. Fite, K., Speich, J., Goldfarb, M.: Loop Shaping for Transparency and Stability Robustness in Bilateral Telemanipulation. IEEE Trans. Rob. Autom. 20(3), 620–624 (2004)
23. Gupta, A., O'Malley, M.K.: Disturbance-Observer-Based Force Estimation for Haptic Feedback. Journal of Dynamic Systems, Measurement, and Control 133(1), 014505 (4 pages) (2011)

Finite Element Modeling of a Vibrating Touch Screen Actuated by Piezo Patches for Haptic Feedback

Buket Baylan, Ugur Aridogan, and Cagatay Basdogan

College of Engineering, Koc University, Istanbul, 34450, Turkey
{bbaylan,uaridogan,cbasdogan}@ku.edu.tr

Abstract. The aim of our work is to design a touch screen for displaying vibro-tactile haptic feedback to the user via piezo patches attached to its surface. One of the challenges in the design is the selection of appropriate boundary conditions and the piezo configurations (location and orientation) on the screen for achieving optimum performance within the limits of human haptic perception. To investigate the trade-offs in our design, we developed a finite element model of the screen and four piezo actuators attached to its surface in ABAQUS. The model utilizes the well-known Hooke's law between stress and strain extended by piezoelectric coupling. After selecting the appropriate boundary condition for the screen based on the range of vibration frequencies detectable by a human finger, the optimum configuration for the piezo patches is determined by maximizing the vibration amplitude of the screen for a unit micro Coulomb charge applied to each piezo patch. The results of our study suggest that the piezo patches should be placed close to the clamped sides of the screen where the boundary conditions are applied.

Keywords: touch screen, vibrotactile haptic feedback, finite element modeling, piezo patch actuators.

1   Introduction

The touch screens replace the mechanical buttons on mobile devices, touch pads, tablet PCs and other displays. While the screens available in the market today are sensitive to touch inputs and gestures, they do not enable the user to feel any programmable resistive forces as her/his finger moves on its surface. However, it is desirable to display some of the information through haptic channel in mobile devices, touch pads, tablet PCs and other interactive displays in order to alleviate the perceptual and cognitive load of the user since our visual and auditory channels are already highly overloaded. Moreover, haptic feedback is more personal and intimate than visual and auditory feedback and hence can enrich the user experience and perception of the interaction. We anticipate that the use of haptic feedback as an additional information channel in interactive displays will result in a new interaction paradigm, and enable novel applications in games, entertainment, education, internet-based business, and many more.

P. Isokoski and J. Springare (Eds.): EuroHaptics 2012, Part I, LNCS 7282, pp. 47–57, 2012.
© Springer-Verlag Berlin Heidelberg 2012

So far, various approaches have been followed to display haptic feedback on touch surfaces. Almost one and a half decade ago, Kaczmarek et al. [1] developed a touch surface for electrocutaneous stimulation of user's finger pad by applying current to it via the electrodes placed on the surface. They reported the difficulty of altering the tactile perception of the user by adjusting the applied current. Later, Kaczmarek et al. [2] developed another touch surface based on electrostatic actuation using a matrix of 7x7 pin electrodes, which are covered with an insulator layer to prevent direct contact of the finger pad with the electrodes. They conducted a user study and investigated the tactile perception of subjects by displaying four different biphasic waveforms through the touch surface. The results of the study showed that the sensitivity of the subjects to the positive pulses was less than that of the negative or biphasic pulses. Yamamato et al. [3] developed a telepresentation system for tactile exploration of remote surface textures. This system was made of two parts: a tactile sensor on the slave site and a tactile display utilizing an electrostatic actuator on the master site. As the user moves her/his finger on the display, the tactile sensor simultaneously scans the texture surface and the surface roughness recorded by the sensor is displayed to the user through the tactile display by applying two-phase cyclic voltage patterns to the electrodes. They conducted a user study and reported that the subjects correctly matched the textures at the remote site to the local ones with a success rate of 79%. Bau et al. [4] from Disney Research presented TeslaTouch, a touch screen providing haptic feedback to the user based on electrostatic actuation. The device controls the frictional force between the user finger and the screen by modulating the frequency and the amplitude of alternating electrostatic force. The results of the psychophysical studies performed with 10 subjects showed that the average frequency JND varied from 11% at 400 Hz to 25% at 120 Hz and the average amplitude JND was 1.16 dB and constant across all frequencies. Maaski and Toshiaki [5] integrated electric motors into a PDA to develop a vibrotactile haptic interface, which they called it "active click". They suggested that active click can improve the usability of touch panels, especially in noisy environments. Poupyrev and Maruyama [6] utilized piezo film actuators to design a PDA with haptic feedback to the user. They argued that further research is required to improve the quality of haptic interactions by developing formal design guidelines. Biet et al. [7] developed a tactile display using an array of piezo actuators attached to the back side of a metal plate. The plate was vibrated at an ultrasonic resonance frequency of 30.5 kHz, reaching to peak-to-peak amplitude of 2.3 μm and causing a squeeze film to form between the surfaces of the user's finger and the plate. By controlling the thickness of the squeeze film, square gratings were simulated. They conducted a user study with 12 subjects and investigated the slipperiness thresholds of the square gratings for the vibration amplitudes of 0, 0.5, 1.2, and 2.3 μm. Winfield et al. [8] developed TPaD by attaching a piezo disk to a glass plate. The plate was actuated at ultrasonic frequencies to modulate the friction coefficient between the finger and the plate surface based on the squeeze film effect. Chubb et al. [9] further extended this idea in ShiverPaD [9] by oscillating the plate in-plane at a frequency of 854 Hz using a voice coil. Hence, the ShiverPaD is capable of applying and controlling shear force on a finger regardless of its direction of motion.

Our group designs a touch screen actuated by piezo patches for displaying vibro-tactile haptic feedback to the user. One of the challenges in the design is the selection of appropriate boundary conditions and the piezo configurations on the screen for achieving optimum performance within the limits of human haptic perception. To investigate the design trade-offs, we developed a finite element model of the screen and four piezo actuators attached to its surface in ABAQUS. The model utilizes the well-known Hooke's law between stress and strain extended by piezoelectric coupling. After selecting the appropriate boundary condition for the screen based on the range of vibration frequencies that are detectable by a human user, the optimum configuration for the piezo patches is determined by maximizing the vibration amplitude of the screen while minimizing the power consumption of the piezo actuators.

2   Our Approach

To demonstrate our idea, we attach an array of thin-film piezo actuators at the back surface of a glass plate where computer-generated images are projected onto it through an LCD display (Figure 1). These piezo films are actuated by a signal genera-tor and an amplifier to generate vibrations on the front surface of the plate with vary-ing amplitudes and frequencies. The position of the user finger on the glass plate is sensed by an IR frame.

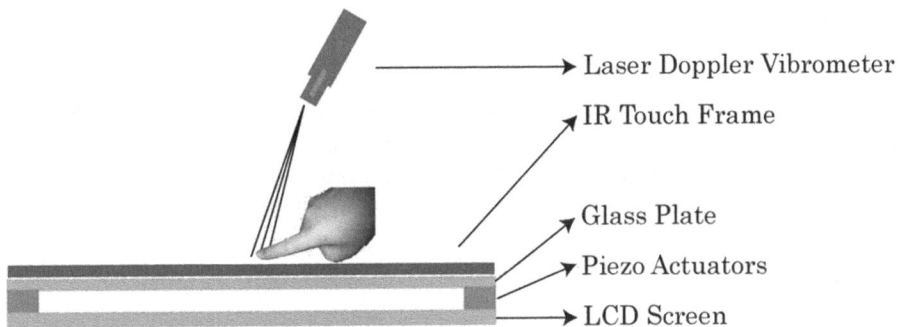

Fig. 1. The proposed vibro-tactile display: The user feels the vibrations generated by an array of piezo actuators glued to the back surface of a glass plate. The magnitude, frequency, and direction of the vibrations are tailored to induce application-specific tactile sensations on the user.

In order to construct the system shown in Figure 1, we first decided on the type of thin-film piezo actuators and then purchased them from the manufacturer (PI Dura-act P-876.SP1). In addition to being small and thin (16x13x0.5 mm), this type of piezoe-lectric patches are light-weighted (0.3 gram), hence the additional weight due to coupling with the glass plate is negligible compared to the own weight of the plate. Besides these advantages, the insulation layer enables it to be attached to the glass plate easily. We glued one of the piezo patches on the glass plate (230x180x3 mm) and conducted some initial experiments by applying alternating voltage to the patch through a signal generator and an amplifier to vibrate the plate (Figure 2). We observed that the vibration amplitudes generated by a single patch are not sufficient to be sensed by a human finger.

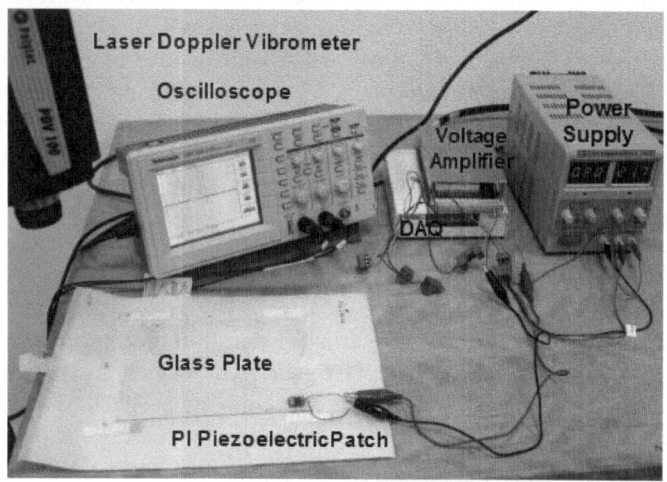

Fig. 2. Our experimental set-up

Hence, we decided to investigate the optimum number and placement of piezo patches on the glass plate. Since the piezo patches cannot be easily detached from the glass plate once they are glued to its surface, it is more convenient to make this analysis in simulation environment using ABAQUS finite element package. For this purpose, we developed the models of the glass plate, the piezo patch actuators, and the interactions between them in ABAQUS and then investigated the design trade-offs based on the amplitude and frequency of the vibrations of the glass plate and the power consumption of the patches.

For the glass plate and the piezoelectric actuator, element types C3D20 and C3D20E are used respectively. The material properties of the glass plate are taken from the literature. The material and electrical properties of the piezo patches are provided by the manufacturer. Our initial design consists of four piezoelectric actuator patches attached to a glass plate in 8 different configurations of the patches (Figure 3). By running simulations in ABAQUS, we first investigated the effect of 3 different boundary conditions on the resonance frequency of the plate and then the effect of piezo locations and orientations on the amplitude of the vibrations of the plate per unit charge applied to the piezo patches.

3   Modeling

The electro-elastic response of the glass plate is governed by the following finite element equations [10] [11]:

$$M\ddot{U} + K_{uu}U + K_{u\Phi}\Phi = F \tag{1}$$

$$K_{\Phi u}U + K_{\Phi\Phi}\Phi = G \tag{2}$$

where, M is the mass matrix, K_{uu} is the stiffness matrix, $K_{u\Phi}$ and $K_{\Phi u}$ are the piezoelectric coupling matrices, $K_{\Phi\Phi}$ is the capacitance matrix, U is the ment, Φ is the electrical potential, F is the externally applied force and G is the applied charge. In our case only charge is applied and the externally applied force is zero. Substituting zero for F into Eq. 1,

$$M\ddot{U} + K_{uu}U + K_{u\Phi}\Phi = 0 \tag{3}$$

$$M\ddot{U} + K_{uu}U = -K_{u\Phi}\Phi \tag{4}$$

and solving for Φ using Eq. 2, we obtain

$$\Phi = K_{\Phi\Phi}^{-1}[G - K_{u\Phi}^{T}U] \tag{5}$$

Then, substituting Eq. 5 into Eq. 4, the following relation is obtained,

$$M\ddot{U} + K_{uu}U = -K_{u\Phi}K_{\Phi\Phi}^{-1}G + K_{u\Phi}K_{\Phi\Phi}^{-1}K_{u\Phi}^{T}U \tag{6}$$

$$M\ddot{U} + [K_{uu} - K_{u\Phi}K_{\Phi\Phi}^{-1}K_{u\Phi}^{T}]U = -K_{u\Phi}K_{\Phi\Phi}^{-1}G \tag{7}$$

which, can be written as

$$M\ddot{U} + KU = T_{G\Phi}G \tag{8}$$

where

$$K = K_{uu} - K_{u\Phi}K_{\Phi\Phi}^{-1}K_{u\Phi}^{T} \tag{9}$$

$$T_{G\Phi} = -K_{u\Phi}K_{\Phi\Phi}^{-1} \tag{10}$$

Eq. 8 shows the relation between the applied charge (G) to the piezo actuators and the resulting displacement (U) in the glass plate. The difference in electric charge creates a potential difference in the piezo actuators, which causes the actuator to bend and the glass plate to deform. If the charge is applied to the piezo patches in the form of a sinusoidal signal, $G = |G|e^{j\omega t}$, the resulting displacements in the glass plate will be also a sinusoidal signal with a phase delay of Ψ, $U = |U|e^{j\omega t + \Psi}$. Then, we can obtain the frequency response function (FRF) as $H(\omega) = U(\omega)/G(\omega)$.

4   Results

Modal analysis was performed on the glass plate for 3 different boundary conditions; a) the glass was fully clamped from all edges, b) the long edges of the plate were

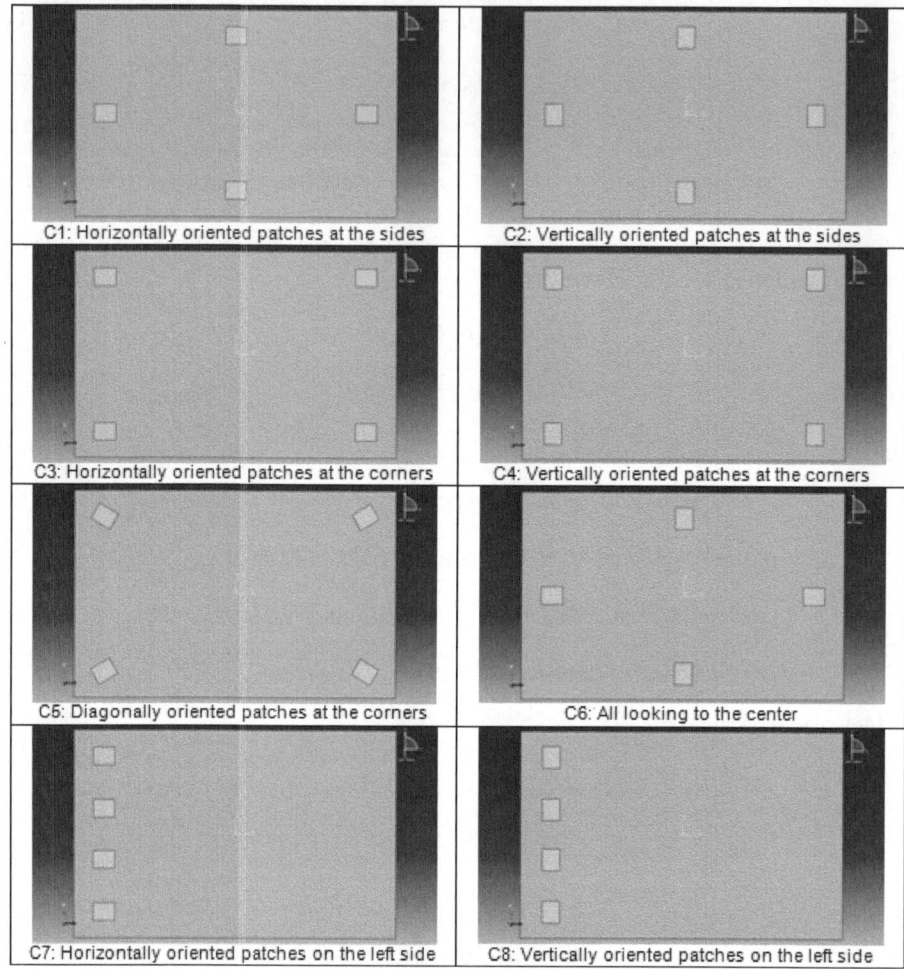

Fig. 3. The selected locations and orientations of the piezoelectric patches for the FE analysis

clamped while the short edges were left free, and c) the short edges of the plate were clamped while the long edges were left free. The first four mode shapes of the glass plate are also shown in Figure 4. The results obtained from ABAQUS are compared to the theoretical values [12] in Table 1. Since we aim to vibrate the glass plate at frequencies that are within the limits of human vibrotactile perception of 0.1-500 Hz [13], the third boundary condition was chosen for the further analysis.

Then, the out of plane displacements of the glass plate were calculated at 15 different locations on the surface of glass plate (Figure 5), covering an area of 153 mm by 77 mm. The frequency response function (FRF) of each point labeled as 1 to 15 in Figure 5 was calculated for the 8 different piezo configurations. As an exemplar, we show the FRFs of point 8 (center point) in Figure 6 for the x, y, z axes.

Table 1. The first four resonance frequencies of the glass plate for the 3 different boundary conditions

Boundary Conditions	First Resonance (Hz)	Second Resonance (Hz)	Third Resonance (Hz)	Fourth Resonance (Hz)
a)	ABAQUS: 690.29	ABAQUS: 1194.00	ABAQUS: 1605.30	ABAQUS: 2022.40
	Theoretical: 706.67	Theoretical: 1204.00	Theoretical: 1637.50	Theoretical: 1865.10
b)	ABAQUS: 523.02	ABAQUS: 588.56	ABAQUS: 830.60	ABAQUS: 1326.50
	Theoretical: 526.13	Theoretical: 591.87	Theoretical: 846.17	Theoretical: 1225.00
c)	ABAQUS: 318.46	ABAQUS: 417.19	ABAQUS: 824.88	ABAQUS: 881.28
	Theoretical: 321.38	Theoretical: 418.52	Theoretical: 772.66	Theoretical: 964.60

5   Discussion

The goal of our project is to produce controlled vibrations on the surface of a touch screen via piezo actuators attached to its surface. To achieve higher vibration amplitudes, higher voltages must be applied to the actuators, which results in higher energy consumption. However, most of the devices utilizing interactive touch screens, especially the mobile ones, are limited by power. Currently, there are no established methods on a) how many piezo patches must be used and b) how they must be attached to a touch screen to generate the desired haptic effects with minimum power. Since the piezo patches cannot be easily detached from a touch screen once they are glued to its surface, it is more convenient to make this analysis in simulation environment using a finite element package. In fact, this is the approach followed in this paper.

As shown in Figure 7, the displacement amplitudes of the center point (point 8 in Figure 5) at each resonance frequency is different for the 8 different piezo configurations. To further investigate the effect of the piezo configurations on the vibration amplitude of each measurement point for a unit micro Coulomb charge applied to

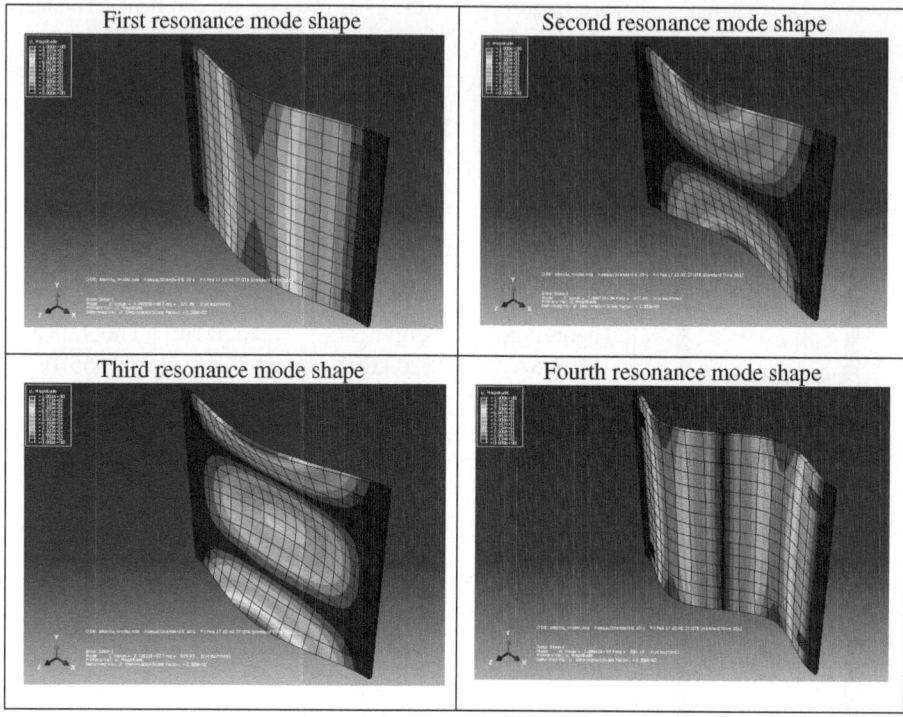

Fig. 4. The first four mode shapes of the glass plate

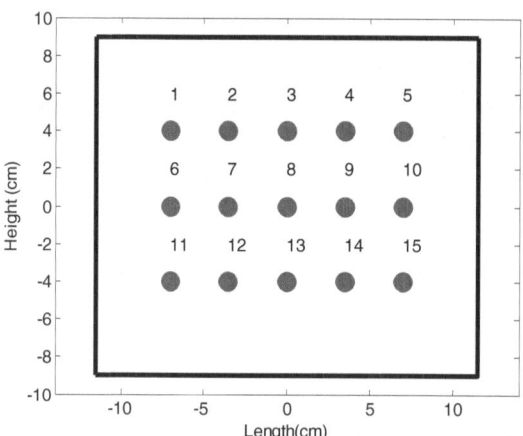

Fig. 5. Selected measurement points on the glass plate for the FE analysis

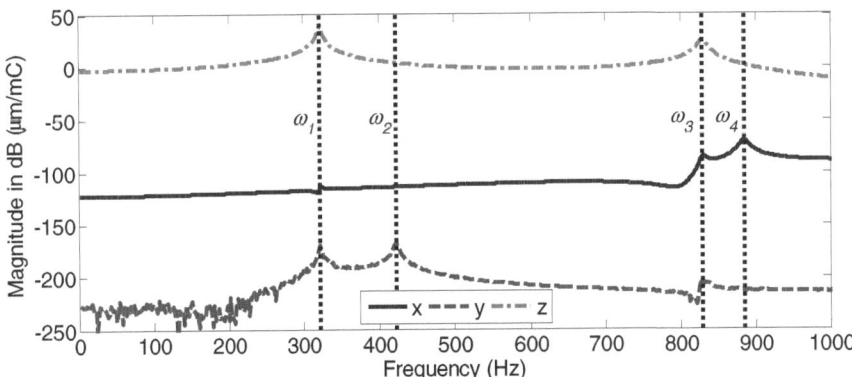

Fig. 6. The frequency response function (FRF) of the point 8 (the center point on the glass plate) for x, y and z axes. The first four resonance frequencies of the plate are marked on the figure.

each piezo (approximately equivalent to 100 V), the plot shown in Figure 8 was constructed from the FRFs. It is obvious from Figure 8 that some of the piezo configurations (C3, C4, C5, C7 and C8) lead to more displacement at the measurement points than the others. If the locations of the patches are inspected carefully for these configurations (see Figure 3), one can conclude it is preferable to place the patches close to the fixed boundaries of the glass plate and not close to its free boundaries. In fact, if the third and fourth resonances of the plate are also considered for the analysis (Figure 7), the piezo configurations C3, C4, and C5 are even more favorable.

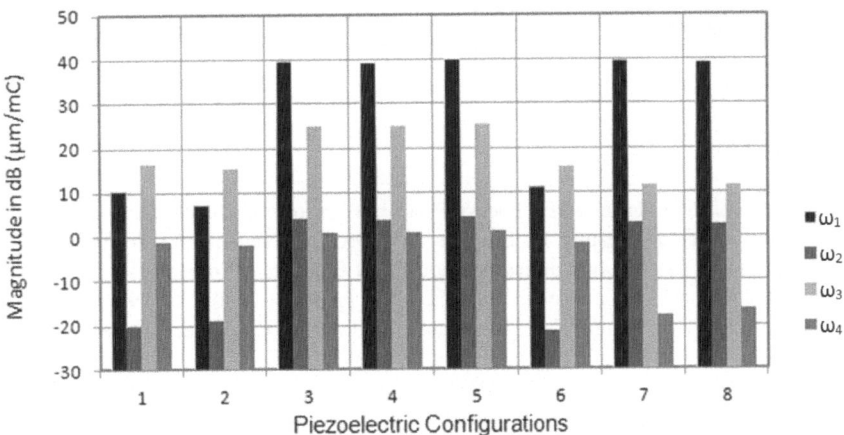

Fig. 7. The out-of-plane (z-axis) displacement amplitudes of the point 8 (the center point on the glass plate) for the first 4 resonance frequencies

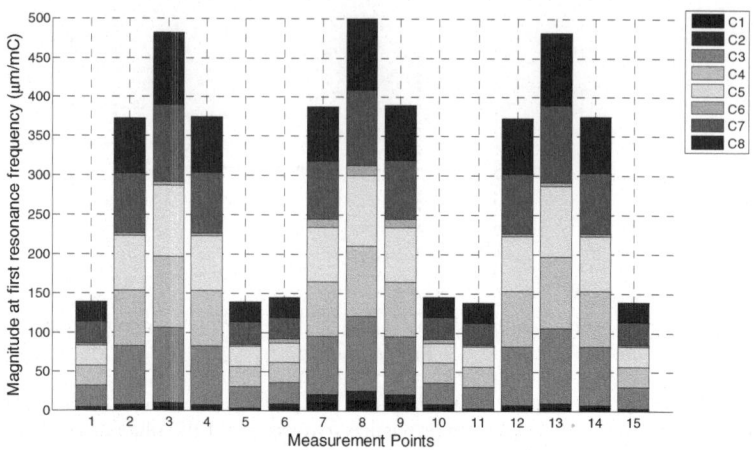

Fig. 8. The bar chart shows the vibration amplitude of each measurement point at the first resonance frequency for the 8 different piezo configurations

6 Future Work

To validate the results obtained through the finite element model, experimental modal analysis will be performed with four piezo patches attached to the glass plate. Then, user studies will be conducted to investigate the human haptic perception of various textures displayed on the glass plate by altering the frequency and the amplitude of the vibrations.

Acknowledgement. This work was supported by Turk Telekom under Grant Number 11315-01.

References

1. Kaczmarek, K.A., Tyler, M.E., Bach-y-Rita, P.: Electrotactile Haptic Display on the Fingertips: Preliminary Results. In: Engineering in Medicine and Biology Society, Engineering Advances: New Opportunities for Biomedical Engineers, Proceedings of the 16th Annual International Conference of the IEEE, vol. 2, pp. 940–941. IEEE (1994)
2. Kaczmarek, K.A., Nammi, K.K., Agarwal, A.K., Tyler, M.E., Haase, S.J., Beebe, D.J.: Polarity Effect in Electro-vibration for Tactile Display. IEEE Transactions on Biomedical Engineering 53(10), 2047–2054 (2006)
3. Yamamoto, A., Nagasawa, S., Yamamoto, H., Higuchi, T.: Electrostatic Tactile Display with Thin Film Slider and Its Application to Tactile Telepresentation Systems. IEEE Transactions on Visualization and Computer Graphics 12(2), 168–177 (2006)
4. Bau, O., Poupyrev, I., Israr, A., Harrison, C.: TeslaTouch: Electrovibration for Touch Surfaces. In: 23rd ACM Symposium on User Interface Software and Technology, pp. 283–292. ACM (2010)

5. Fukumoto, M., Sugimura, T.: Active Click: Tactile Feedback for Touch Panels. In: Conference on Human Factors in Computing Systems, pp. 121–122. ACM, New York (2001)
6. Poupyrev, I., Maruyama, S.: Tactile Interfaces for Small Touch Screens. In: 16th ACM Symposium on User Interface Software and Technology 2, pp. 217–220. ACM, New York (2003)
7. Biet, M., Casiez, G., Giraud, F., Lemaire-Semail, B.: Discrimination of Virtual Square Gratings by Dynamic Touch on Friction Based Tactile Displays. In: Symposium on Haptic Interfaces for Virtual Environment and Teleoperator Systems, pp. 41–48 (2008)
8. Winfield, L., Glassmire, J., Colgate, J.E., Peshkin, M.: T-PaD: Tactile Pattern Display through Variable Friction Reduction. In: 2nd Joint EuroHaptics Conference and Symposium on Haptic Interfaces for Virtual Environment and Teleoperator Systems, pp. 421–426. IEEE (2007)
9. Chubb, E.C., Colgate, J.E., Peshkin, M.A.: ShiverPad: A Device Capable of Controlling Shear Force on a Bare Finger. In: 3rd Joint EuroHaptics Conference and Symposium on Haptic Interfaces for Virtual Environment and Teleoperator Systems, pp. 18–23 (2009)
10. Piefort, V.: Finite Element Modeling of Piezoelectric Active Structures. Ph.D. thesis, Faculty of Applied Sciences, Universit'e Libre de Bruxelles, Belgium, p. 87 (2001)
11. IEEE Standard on Piezoelectricity. ANSI/IEEE Std. 176-1987. IEEE (1988)
12. Blevins, R.D.: Formulas for Natural Frequency and Mode Shape. Krieger, Florida (1984)
13. Jones, L.A., Sarter, N.B.: Tactile Displays: Guidance for Their Design and Application. Human Factors: The Journal of the Human Factors and Ergonomics Society 50, 90–111 (2008)

Evidence for 'Visual Enhancement of Touch' Mediated by Visual Displays and Its Relationship with Body Ownership

Valeria Bellan, Carlo Reverberi, and Alberto Gallace

Department of Psychology, University of Milan-Bicocca, Milan, Italy

Abstract. Several studies have shown that watching one's own body part improves tactile acuity and discrimination abilities for stimuli presented on that location. In our experiment we asked the participants to localize tactile stimuli presented on the left or right arm. During the task the participants were not allowed to watch their body, but they could see another person's left arm via a LCD display. This arm could be touched or not during the presentation of the tactile stimuli. We found that when the participants saw a finger touching the arm on the screen, their responses to the tactile stimuli presented on the left and on the right arm were faster than when the arm on the screen was approached but not touched. Critically, we did not find any illusion of ownership related to the hand seen on the screen. We concluded that the effects found might be mediated by higher order multisensory mechanisms related to the allocation of attentional resources to the body.

Keywords: Cross-modal, Multisensory, Perception, Touch; Vision.

1   Introduction

Tactile perception plays a key role in building boundaries between the self and the external world [10]. In fact, when we touch an external object we can feel both the incoming perception from the object itself and the presence of our body well differentiated from it. Several data suggest that vision of a person's own body part, together with proprioception or alone, can enhance the processing of tactile stimuli delivered to that location [17]. In particular, Kennett and colleagues [12] have shown that their participants could detect a tactile target delivered on their arms more effectively when they could directly see that part of the body, as compared to when its vision was prevented (an effect named 'Visual Enhancement of Touch' –VET–). Interestingly, research findings support the idea that there might be similar patterns of neural activation when the body is touched and when one sees another person being touched. In particular, Schaefer et al. [15] investigated whether the tactile perception of their participants could be modulated by watching a stranger's body part being touched on a screen, rather than by watching their own body. The authors measured the sensory thresholds of the participants' index finger, after viewing a right hand in a screen being touched by a stick, or after viewing the stick just touching the space beneath the

P. Isokoski and J. Springare (Eds.): EuroHaptics 2012, Part I, LNCS 7282, pp. 58–66, 2012.
© Springer-Verlag Berlin Heidelberg 2012

hand. They found a reduction of the sensory threshold specific for the right index finger, only after showing the video where the hand was touched by the stick. The authors concluded that the VET is linked to the observation of touch, rather than to the depiction of the body part per se. However, in Schaefer et al's study the participants performance was measured by means of a tactile threshold task. Considering that tactile thresholds are more related to relatively lower level neurocognitive mechanisms [3], one might wonder, whether similar results might be obtained also by means of tasks that involve higher order functions, such as the deployment of attention towards different body districts.

It is also worth considering here the question of whether the effect of watching another person's body on tactile information processing can be somehow related to an extension of body ownership towards the body part seen on the display. In fact, in the 'rubber hand illusion' (RHI) [2] the hidden participant's hand is brushed synchronously with a visible rubber hand. Using this procedure, the majority of participants, after a few seconds, starts to perceive touch as if it is coming from the position of the fake hand rather than from the real hand. Longo et al. [13] measured the participants' tactile acuity during the onset of the RHI. They found a significant enhancement of acuity when the illusion was induced compared to a control condition where the illusion was not generated. The authors suggested that the VET effect depends on seeing 'one's own' hand, rather than seeing 'a' hand and that there is a functional relation between the bodily self and tactile perception. However, also in this case, the task performed by the participants relies mainly on lower order neurocognitive mechanisms. Therefore, it remains unclear whether using a tactile task that involves higher order mechanisms, such as those responsible for the deployment of spatial attention, would lead to similar effects. As far as this point is concerned, it should be noted here that Moseley et al's [14] found a slow down of tactile information processing due to alterations of body ownership when a spatial discrimination task was used.

The aim of our research, is to verify whether seeing another person's hand being touched (on a pc screen) can affect the processing of tactile information presented on the participant's body when the spatial discrimination of the stimuli (rather than a threshold assessment task) is required. We will also analyze if an illusion of embodiment of the hand on the screen would occur and whether this would influence the processing of tactile stimuli.

2 Materials and Methods

Thirty-two (22 female) right-handed volunteers (age = 24.5 ± 3.7 years; education = 16.7 ± 1.8 years of school) took part in the experiment. They sat with both their forearms resting on a shelf 10 cm under a desk (see Fig.1). A 17' LCD screen showing the picture of a left arm and hand was placed on the desk. The gender of the arm on the screen was matched with the participant's gender. Four vibrotactile stimulators (Audiological Engineering Corporation) were applied on the dorsum of both hands and on the forearms, just below the elbow. The image of the arm was aligned with the participant's real left hand. The experiment was composed of two different parts. In the first part the participants were instructed to look at the screen displaying an index finger approaching the hand and touching it, while a 100 msec tactile stimulus was delivered to their real left hand. No response was required in this part of the experiment. The participants were randomly assigned to either the Illusion or NonIllusion group. In the Illusion group the presentation of the tactile stimulus was synchronized with the touch seen on the screen (a finger touching the hand), while in the Non-Illusion group the touch delivered to the participants' left hand was presented asynchronously (three seconds delayed) with respect to the touch viewed on the screen. After this first part of the experiment, all the participants were required to fill a questionnaire regarding their sense of ownership about the hand seen on the screen [2]. The questionnaire was composed of 9 statements. Each statement was followed by a 15cm long line with the endpoints indicating the degree of the participant's agreement (not at all vs. completely) towards that statement. The participants were required to mark their agreement along the line. In the second part of the experiment, both groups performed the same task. They were asked to look at the screen where a left arm and hand were displayed. An index finger approached the hand or the forearm, randomly touching one of them (Touch condition) or just approaching without touching (No-Touch condition) these body parts. A 100ms vibrotactile stimulus was synchronously delivered to the participant's right or left hand or forearm. A number of catch trials where the stimuli were delivered to both sides of the body at the same time was randomly interleaved to the left and right stimuli. The order of the stimulated locations was randomly varied within the experiment. The participants were asked to press either the right or left or both buttons of a pc mouse with the index and middle finger of their right hand, according to the side (or sides) where they perceived the tactile stimulation. A total of 240 trials balanced for each side of the body for each location and for each experimental condition were presented.

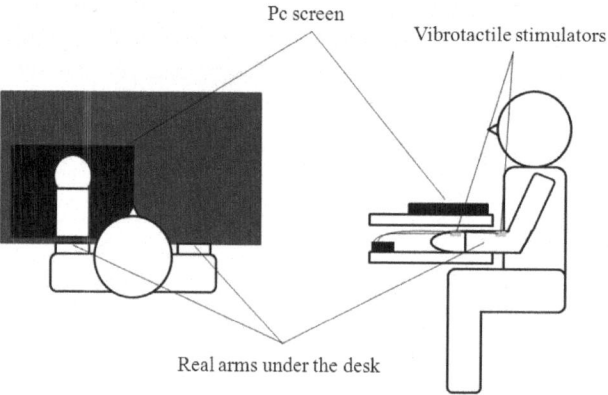

Fig. 1. Experimental setup used in the present experiment

3   Results

We analysed the participants' reaction times for each experimental condition. The accuracy of participants' responses were not analysed, because the amount of errors made in each condition approached zero for all of the participants. We also analysed the answers at the questionnaire for the two groups.

3.1   Reaction Times

A repeated-measures ANOVA with four within subjects factors (Touch: Touch Vs. NoTouch; Position Felt: left, right; Position Seen: hand Vs. arm; Position Touched: hand Vs. arm) and one between subject factor (Group: Illusion Vs. NonIllusion) was performed on the reaction times. The results of this analysis revealed the presence of a main effect of Touch [$F(1,30)=26.475$; p<0.001]. In particular, the participants were faster when the finger actually touched the hand presented on the screen (Touch condition) as compared to the condition where the hand on the screen was approached but not touching (Fig. 2, on the left). We also found a main effect of the Position Felt [$F(1,30)=9.0220$; p<0.005], suggesting faster responses for the stimuli administered on the left side of the participants' body as compared to stimuli presented on the right (Fig. 2, on the right).

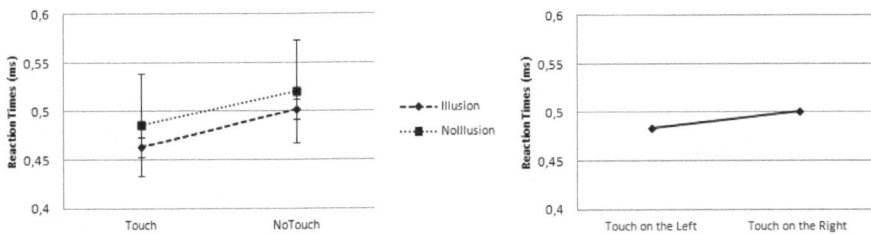

Fig. 2. Effect of Touch (graph on the left): the participants gave faster responses in Touch Condition (finger touching the hand on the screen) than in NoTouch Condition (finger just approaching). Effect of Position Felt (graph on the right): the participants gave faster responses for the tactile perception on the left side of the body than on the right. The error bars represent the standard error of the means of the reaction times.

We also found a main effect of the Position Seen [$F(1,30)=23.47$; $p<0.0001$], with faster responses when the participants saw the arm being touched as compared to when they saw the hand being touched. The analysis also revealed a significant interaction between Touch Seen and Position Felt [$F(1,30)=7.35$; $p=0.01$]. A post hoc LSD analysis on this interaction revealed that participants were faster in the Touch condition than in NoTouch condition both when the stimulus was delivered on the right (p<0.001) and on the left side (p<0.001) of the body. By contrast, in the NoTouch condition they were faster when the stimulus was delivered on the left (p<0.001) of the body compared to when it was delivered on the right. Another significant interaction was found for the factors of Touch Seen and Position Seen. A LSD post hoc test on this interaction revealed that in the Touch condition participants responded faster than in the NoTouch condition when the participants saw the hand being touched compared to the arm being touched (p<0.001). By contrast, in the NoTouch condition they were faster when they saw the arm touched compared to the hand (p<0.001). Finally, the post hoc analysis of the interaction between Position Felt and Position Touched also revealed that participants responded faster when their left arm was touched as compared to when their right arm was touched (p=0.02). Faster responses were also found when the left participant's arm was touched as compared to the condition where the left hand was touched (p<0.001). The analysis also revealed the presence of a significant interaction among all the four factors considered. In order to further explore this interaction, two separate ANOVAs one for each Position Felt (i.e. one for the condition Hand Touched and one for the condition Arm Touched) with the between subject factor of group (Illusion Vs. NonIllusion), and the within subject factors of Touch Seen, Position Felt and Position Seen were performed. The ANOVA on the Position Felt regarding the arm revealed the presence of a main effect of Touch [$F(1,30)=11.55$; $p=0.002$], with faster responses for Touch than NoTouch condition. A main effect of Position Felt [$F(1,30)=14.72$; $p=0.0006$], with faster responses when the stimulus was delivered on the left side of the body as compared to the right side of the body was also found. The ANOVA on the Position Felt regarding the hand revealed the presence of a main effect of Touch [$F(1,30)=18.61$; $p=0.0001$], with faster responses for Touch than NoTouch. A main effect of Position Seen [$F(1,30)=5.93$; $p=0.02$], highlighting faster responses when the participants saw the arm being touched compared to when they saw the hand being touched, was also found. The analysis also revealed the presence of significant interactions between the following factors: Position Seen and Group [$F(1,30)=4.59$; $p=0.04$], Touch and Position Felt [$F(1,30)=7.27$; $p=0.011$], and Touch and Position Seen [$F(1,30)=7.95$; $p=0.008$]. An LSD post hoc test on the Position Seen and Group interaction showed that the NonIllusion group was faster than the Illusion group only when the arm was seen (p=0.003). The LSD post hoc test on the Touch and Position Felt interaction showed that participants were faster in Touch as compared to NoTouch condition only when the stimulus was delivered on the right side of the body (p<0.0001). Finally, the post hoc test on the Touch and Position Seen interaction showed that in the NoTouch condition the participants gave faster responses when the touch on arm was seen (p<0.001) as compared to when the touch on the hand was seen. Moreover when the touch on hand was seen (but not when on the arm) the participants were faster for Touch condition (p<0.0001) than for the NoTouch condition. This analysis also highlighted the presence of an interaction among Touch, Position Seen and Group, that was further analysed by means of two further ANOVAs, one for the Illusion group and one for the NonIllusion group. In these ANOVAs we considered the within subject factors of Touch and of the Position Seen. In the ANOVA performed on the Illusion group we found a main effect of Touch [$F(1,31)=9.70$; $p=0.004$], with the participants being faster in the Touch than NoTouch condition. No other significant main effects or interactions were found. As far as the ANOVA on the NonIllusion group is concerned the main effect of Touch was also found [$F(1,31)=6.92$; $p=0.013$]. This ANOVA also revealed the presence of a main effect of Position Seen [$F(1,31)=10.06$; $p=0.003$], indicating faster reaction times when the touch was viewed on the arm as compared to the hand. A significant interaction between the two factors [$F(1,30)=12.18$; $p=0.001$] was also found in this analysis. An LSD post hoc test revealed that in NoTouch condition the participants' responses were faster when the touch on arm was viewed (p<0.0001) as compared to when the touch on hand was viewed and that, when the touch on hand was seen, the participants were faster in the Touch condition (p<0.0001) than in NoTouch condition.

3.2   Questionnaires

We performed a repeated-measures ANOVA on the responses given to the nine-question-Likert questionnaire with one within subject factor (question) and one between subject factor (Group: Illusion Vs. NonIllusion). We found a main effect of the question [$F(8,240)=11.80$; $p<0.0001$], revealing significantly different answers among the questionnaire. More importantly, this analysis did not reveal any significant difference between the two groups of participants [$F(8,240)=1.00$; $p=0.43$]. No significant interactions were found.

4   Conclusions and Discussion

The first aim of our research was to verify whether watching a bi-dimensional image of a hand on a screen being touched could affect the spatial processing of tactile information presented on the participant's hands or arms [cf. 15]. We hypothesized a facilitation in tactile processing when the participants could see a finger touching a limb on the screen, compared to viewing a finger just approaching but not touching the limb. We also wanted to assess if an embodiment of the hand seen on the screen occurs under these conditions of stimulus presentation and if this can affect tactile perception, in line with Longo et al.'s findings [13]. Our findings clearly suggest that the presentation of a bidimensional image of a body part can affect tactile processing. Indeed, we found that participants responded faster to the tactile stimuli presented on their own arms when they could see on a pc screen the limb of a stranger being touched, as compared to a condition where they could see the same limb being approached but not touched. It is important to note here that the effect that we found cannot be fully classifiable as a VET effect (i.e., a body-location specific enhancement of tactile processing caused by the vision of the location of the body where the stimuli are presented). In fact, the participants in our experiment responded faster to tactile stimuli when they could see a 'picture' of a hand being touched rather than when seeing their own body touched (like in the classic VET paradigm, [15]). Moreover, the enhancement of tactile processing following vision of the body in our experiment regarded both upper limbs rather than being specific to the limb shown on the display. This result would appear to contrast with those reported by Schaefer and colleagues [15], where showing the picture of a left index finger being touched resulted into an enhanced of the participants' tactile thresholds for stimuli presented on their own left index finger, but not on their right index finger. Though, an important difference between our study and the previous studies on this topic is the fact that we tested tactile processing using a spatial discrimination task, rather than a tactile threshold task [15] or a tactile acuity task [13]. The two latter tasks, in fact, involve lower order cognitive mechanisms, while the task performed by the participants in our experiment relies on higher level attentional mechanisms. Therefore, we suggest that seeing a person's body part being touched in our study might have directed attentional resources towards the tactile modality in a quite generalized fashion. This generalized focusing of attention is also confirmed by the fact that the responses were faster when the participants saw the finger on the screen touching the arm than when touching the hand, irrespective of feeling the touch on their own hand or arm [17]. One can argue that the tactile acuity decreases from the finger tips to the elbow [18], such as that the participants should be less sensitive for the touch on the arm (thus resulting in faster reaction times on the hand rather than on the arm). However, the fact that we set the intensity of the stimuli in order to be perceived by the participants as equally intense in all the four locations would seem to rule out this possibility. Alternatively, one might hypothesize that the touch on the arm was perceived by the participants as more salient because this part is less frequently touched than a hand is, leading to faster responses when that body part was touched. Our results also show that when the left part of the body was touched, the reaction times were shorter. In fact, the arm displayed was a left one, so it is reasonable that the participants could be more reactive to stimuli delivered on their left part of the body, as if they had a greater arousal for this side [16]. In our experiment we also examined if an illusion of embodiment of a stranger's hand occurs when the participants observe that hand being touched on a screen and when tactile stimuli are simultaneously presented on the participant's actual limb. The results from the questionnaire did not reveal any significant difference between the groups. This results clearly suggests that no illusion was elicited under these conditions of stimulus presentation. Nevertheless, we also found that the participants in the NonIllusion group provided faster responses when the touch on the screen was directed to the arm, as compared to when it was directed to the hand. Here it should be considered that the illusion procedure adopted in our experiment involved just a touch on the hand and the Illusion group felt this touch synchronously to the touch on the hand seen on the screen. By contrast, the NonIllusion group saw the same stimulation but in an asynchronous fashion. It might be that the Illusion condition may have induced a sort of hebbian association between the touch felt and the touch seen on the hand. As a consequence, the participants belonging to this group might have started to consider the whole arm as one thing. On the other hand, the NonIllusion group kept on considering hand and arm as two separate sites of touch, giving that the asynchronous stimulation condition did not create an equally strong association between the tactile and the visual stimuli. That is, the synchronous stimulation might have led to a more homogenous perception of the limb depicted on the screen than the asynchronous stimulation.

Taken together, the results reported in the present experiment support the idea that vision of a stranger's body part being touched, can affect the processing of tactile information presented on another person's body. This effect is present even when bidimensional simple stimuli displayed on a computer LCD screen are adopted. Our findings would seem to confirm that tactile receptive fields can be reorganized according to the visual fields, also when these refer to another person's body [13]. We can rule out the possibility that the enhancement in tactile detection that we found in our experiment was mediated by a variation of the participants' sense of body ownership. In fact, we did not find any significant differences between the Illusion and Non Illusion group in the participants' self-reported sense of ownership over the hand seen on the screen. By contrast, the effect that we found would seem to be more related to a general shift in the deployment of attentional resources towards the tactile modality, determined by the vision of a body part being touched.

To date the role of touch in new technologies is becoming more and more important, in particular for the development of graphic interfaces that can be used in everyday life, such as touch screens in mobile phones, tablet PC or learning aids (such as the Interactive Whiteboard; see [9]). Its importance also relates to more advanced devices, such as those used in 'teleoperation', a technology that allows trained surgeons to operate on patients located miles away from them [8]. A better understanding of the mechanisms at the basis of the interaction between vision and touch can certainly contribute to ameliorate these technologies by making them more suitable for the users. The fact that tactile processing improves on a certain body part, not only when that body part is directly seen, but also when the participants can observe touch on someone else limb displayed on a LCD screen should certainly be taken into account in the development of tele-robotics and tele-surgery devices [7], where people feel sensations on their own body, but watch it operating on a different body or object by means of a visual display.

References

1. Blakemore, S.-J., Bristow, D., Bird, G., Frith, C., Ward, J.: Somatosensory activations during the observation of touch and a case of vision-touch synaesthesia. Brain: a Journal of Neurology 128, 1571–1583 (2005)
2. Botvinick, M., Cohen, J.: Rubber hands "feel" touch that eyes see. Nature 391(6669), 756 (1998)
3. Connor, E.: Neural Coding of Tactile Texture: Comparison of Spatial Temporal Mechanisms for Roughness Perception. Stimulus 12, 3414–3426 (1992)
4. Dionne, J.K., Meehan, S.K., Legon, W., Staines, W.R.: Crossmodal influences in somatosensory cortex: Interaction of vision and touch. Human Brain Mapping 31(1), 14–25 (2010)
5. Driver, J., Grossenbacher, P.G.: Multimodal spatial constraints on tactile selective attention. In: Inui, T., McClelland, J.L. (eds.) Information Integration in Perception and Communication (Attention and Performance XVI), pp. 209–235. MIT Press, Cambridge (1996)
6. Ehrsson, H.H., Wiech, K., Weiskopf, N., Dolan, R.J., Passingham, R.E.: Threatening a rubber hand that you feel is yours elicits a cortical anxiety response. Proceedings of the National Academy of Sciences of the United States of America 104(23), 9828–9833 (2007)
7. Gallace, A., Spence, C.: The science of interpersonal touch: an overview. Neuroscience and Biobehavioral Reviews 34(2), 246–259 (2010)
8. Gallace, A., Tan, H.Z., Spence, C.: The body surface as a communication system: the state of art after 50 years of research. Presence: Teleoperators and Virtual Environments 16, 655–676 (2007)

9. Glover, D., Miller, D., Averis, D., Door, V.: The interactive whiteboard: a literature survey. Technology, Pedagogy and Education 14(2), 155–170 (2005)
10. Haggard, P., Taylor-Clarke, M., Kennett, S.: Tactile perception, cortical representation and the bodily self. Current Biology 13(5), R170–R173 (2003)
11. Honore, J., Bourdeaud'hui, M., Sparrow, L.: Reduction of cutaneous reaction time by directing eyes towards the source of stimulation. Neuropsychologia 27(3), 367–371 (1989)
12. Kennett, S., Taylor-Clarke, M., Haggard, P.: Noninformative vision improves the spatial resolution of touch in humans. Current Biology: CB 11(15), 1188–1191 (2001a)
13. Longo, M.R., Cardozo, S., Haggard, P.: Visual enhancement of touch and the bodily self. Consciousness and Cognition 17(4), 1181–1191 (2008)
14. Moseley, G.L., Olthof, N., Venema, A., Don, S., Wijers, M., Gallace, A., Spence, C.: Psychologically induced cooling of a specific body part caused by the illusory ownership of an artificial counterpart. Proceedings of the National Academy of Sciences of the United States of America 105(35), 13169–13173 (2008)
15. Schaefer, M., Heinze, H.-J., Rotte, M.: Viewing touch improves tactile sensory threshold. Neuroreport 16(4), 367–370 (2005)
16. Serino, A., Pizzoferrato, F., Làdavas, E.: Viewing a face (especially one's own face) being touched enhances tactile perception on the face. Psychological Science 19(5), 434–438 (2008)
17. Tipper, S.P., Lloyd, D., Shorland, B., Dancer, C., Howard, L.A., McGlone, F.: Vision influences tactile perception without proprioceptive orienting. Neuroreport 9(8), 1741–1744 (1998)
18. Weinstein, S.: Intensive and extensive aspects of tactile sensitivity as a function of body part, sex, and laterality. In: Kenshalo, D.R. (ed.) The Skin Senses, pp. 195–222. Thomas, Springfield, Ill (1968)

On the Perceptual Artifacts Introduced by Packet Losses on the Forward Channel of Haptic Telemanipulation Sessions

Fernanda Brandi, Burak Cizmeci, and Eckehard Steinbach

Institute for Media Technology
Technische Universität München
Munich, Germany
{fernanda.brandi,burak.cizmeci,eckehard.steinbach}@tum.de

Abstract. In this work we study the position and velocity signal reconstructions using predictive coding when packets are lost during telemanipulation sessions and classify the high-level haptic artifacts perceived by the users. The usage of packet-switched networks for bilateral telemanipulation systems is challenging due to several adversities such as low transmission rates, packet delays, jitter and losses. The previously proposed deadband-based haptic data reduction approaches selectively decrease the high transmission rate of the force-feedback and position/velocity samples on account of human perception limitations. Recently, an error-resilient perceptual haptic data reduction approach was proposed to address the packet losses in the feedback channel. However, the impact of faltered transmission on the forward channel and its subjective influence on the user are still an open issue and thus are treated in this paper.

Keywords: [Robotics and automation]: Teleoperators, [H.5.2.g]: Haptic I/O, [E.4.a]: Data compaction and compression, [C.2.1.g]: Network communications , [Signal Processing]: Error Correction, [Information Theory]: Error Compensation.

1 Introduction

1.1 Real-Time Haptic Applications

The frameworks that allow realistic real-time interactive communication and remote task realization are typically referred to as telepresence and telemanipulation (TPTM) systems. The degree to which the human is unable to distinguish between direct and teleinteraction with a (remote) environment is called *transparency* [1] and can be used to subjectively evaluate these frameworks. The more transparent the interaction is (or similarly, the more immersed the user is), the better is the TPTM system. Thus, transparency is highly desirable since it makes the subject reach out and experience the remote environment as if it was local.

TPTM systems are essentially composed of four basic elements, namely: 1) the *human operator* (OP); 2) the *human-system interface* (HSI) which consists of the

P. Isokoski and J. Springare (Eds.): EuroHaptics 2012, Part I, LNCS 7282, pp. 67–78, 2012.

devices that perform the transduction from the OP's actions into haptic signals
(e.g. acquiring position/velocity data) and also that provide haptic, visual and/or
auditory feedback to the user (e.g. displaying forces, video and/or sounds); 3)
the *teleoperator* (TOP) which can be a robot or an end-effector in a virtual
environment that follows the OP's actions; and, 4) a *communication link* through
which all the data is transmitted and which closes the loop between the OP and
the TOP.

Visual and auditory signals are unidirectionally transmitted from the TOP to
the OP while the haptic signal is being transmitted in both directions providing
an authentic interactive experience. Naturally, due to the closed-loop properties
of this real-time bidirectional communication, the actions performed by the OP
have a direct impact on the force-feedback that the user receives. Analogously,
the force-feedback exposed to the operator strongly interferes with his/her future actions. In this sense, to cope with the high temporal human perception
resolution [2] and to better maintain the stability of the control loop, haptic samples are typically acquired and transmitted at rates of up to 1 kHz. Moreover,
transmission delays and packet losses on the communication network challenge
the TPTM system's operability since they potentially destabilize the control
structure and impair the interaction [3,4]. Thus, since haptic samples need to be
sent at such high rates and with minimum latency, the haptic communication
fundamentally differs from typical video and audio data streaming. It is important to note that the low-latency constraint has a direct impact on suitable data
reduction schemes as described in [5].

1.2   Haptic Data Reduction

Block-based data compression schemes - typically used for video and audio data
compression - are not suitable for haptic signals. Due to hard low-latency constraints in bidirectional haptic communication, adding encoding delay is not
recommended hence block-based compression schemes should be avoided.

There are data compression schemes designed for real-time haptic communication [5,6,7]. The most efficient of them was proposed by Hinterseer *et al.* [5]
and this approach, in essence, exploits the human haptic perception limitations.
This psychophysically motivated scheme employs Weber's Law of Just Noticeable
Differences (JND) [8] to selectively eliminate samples and reduce the amount of
data to be transmitted. Weber's Law attests that the perceivability of changes in
a pairwise stimulus comparison experiment is proportional to the magnitude of a
reference stimulus I. This relation can be mathematically written as:

$$\frac{\Delta I}{I} = k \tag{1}$$

where ΔI is the smallest amount of change in the stimulus I that can be perceived by at least 50% of the users (or, correspondingly, the JND) and k is a
constant that describes the linear relation between these two intensities.

Weber's Law can be utilized to establish perceptual thresholds wherein stimuli
intensity variations cannot be detected. These regions can be referred to as the

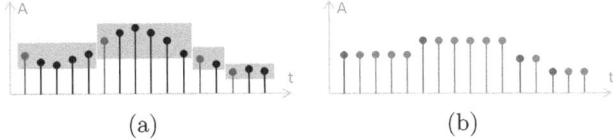

Fig. 1. (a) At the encoder side, samples are acquired. The deadband regions (gray) are calculated using the predictions of the encoded samples (red). The incoming samples that fall within the deadband thresholds (black) are not encoded. (b) At the decoder side, the signal is reconstructed using the decoded samples (red) and the predicted samples (green). In this example, the HLS approach is employed for prediction.

deadbands [5]. To achieve the overall goal of reducing the haptic data rate while maintaining good subjective experience, the signal can be analyzed on a sample basis deciding whether samples would cause a perceivable change for the user and, ultimately, if they should be encoded. Figure 1 illustrates the use of Weber's Law applied to a one-dimensional signal.

It can be observed in Figure 1 that whenever the incoming sample falls within the deadband thresholds, the difference between this sample and its prediction is said to be too small to be detected and therefore the acquired sample can be omitted and its prediction displayed instead. On the other hand, if the incoming sample provokes a perceivable change in relation to the predicted one (i.e. it violates the deadband), then that incoming sample should be encoded. Previous results for haptic signal compression show that prediction methods such as *hold last sample* (HLS) and *first-order linear prediction* (FOLP) can be successfully employed in combination with Weber's Law to achieve strong data reduction [5].

The remainder of this paper is organized as follows. In Section 2, the artifacts caused by packet loss in the force-feedback channel are revisited. In Sections 3 and 4 we present the contributions of this work. In Section 3 we describe the commanding modes of TPTM systems and the effects of packet losses on the reconstructed position/velocity signals when predictive coding is used. In Section 4 we illustrate the experimental procedure and present the observed high-level artifacts due to position/velocity packet losses. Lastly, in Section 5 we make some final comments about this work and draw some ideas for future work.

2   Packet Loss in the Feedback Channel

As mentioned in Section 1.2, the state-of-the-art in haptic data reduction is the direct use of Weber's Law to evaluate whether or not haptic samples should be transmitted. In the presence of such data reduction schemes where not every sample is sent, the loss of any packet in the communication network can be significantly impairing. A direct consequence of packet losses is the resulting asynchrony between both sender and receiver since they would have distinct reference samples and, thus, would generate different estimates. Employing a first-order linear predictor, the predicted force sample $\mathbf{f}_i^P$ can be calculated as:

$$\mathbf{f}_i^P = \frac{\mathbf{f}_j^U - \mathbf{f}_{j-1}^U}{t_j^U - t_{j-1}^U}(t_i^P - t_j^U) + \mathbf{f}_j^U \tag{2}$$

where $\mathbf{f}_j^U$ and $\mathbf{f}_{j-1}^U$ are, respectively, the current and previous update force samples, t_j^U and t_{j-1}^U their corresponding acquiring times, and t_i^P is the current time (corresponding to the moment of the prediction $\mathbf{f}_i^P$). In this work, the upper indices $(.)^U$ and $(.)^P$ represent, respectively, *update* and *prediction* samples. If no upper index is marked, it indicates the original signal sample. Moreover, $(.)_j$ always index the *updates* while $(.)_i$ refer to *acquired* or *predicted* samples.

The expression above is calculated at the sender side so the predicted sample $\mathbf{f}_i^P$ and the acquired force sample $\mathbf{f}_i$ can be compared. This comparison checks whether there is a deadband violation such as shown in the following evaluation and decides over sending or not sending a new update.

$$\begin{cases} ||\mathbf{f}_i - \mathbf{f}_i^P|| \leq ||\mathbf{f}_i^P|| \cdot k_f & \longrightarrow & \text{non-violation (use predicted sample)} \\ ||\mathbf{f}_i - \mathbf{f}_i^P|| > ||\mathbf{f}_i^P|| \cdot k_f & \longrightarrow & \text{violation (transmit force sample)} \end{cases} \tag{3}$$

In the expression above, k_f is the deadband parameter for the force signal as generally presented in Equation 1.

Furthermore, the receiver employs Equation 2 to calculate its predicted force samples - what we will call from now on $\hat{\mathbf{f}}_i^P$ to distinguish the sender's estimate from the receiver's reconstruction. In a lossless communication channel, $\hat{\mathbf{f}}_i^P = \mathbf{f}_i^P$ since all transmitted samples reach their destination and both sender and receiver are then in synchrony. In this case, if no updates are received it is exclusively due to the sender's decision of not transmitting anything. However, in an error-prone network, samples can be lost and since the transmission rate is not uniform, the receiver does not know when a sample is accidentally missing and hence it continues to estimate $\hat{\mathbf{f}}_i^P$ and in this case $\hat{\mathbf{f}}_i^P \neq \mathbf{f}_i^P$.

Previous works by Brandi *et al.* [9,10] investigated the effect of losing such packets in the force-feedback channel (i.e. from the TOP to the OP). In the cases described in [9,10], lost force update samples lead to three main artifacts, namely, *bouncing, roughness* and the *glue effect*. The *bouncing* generates a repelling force that pushes the end-effector away from the object that is being touched; the *roughness* displays an irregular texture sensation when the user runs the end-effector along the object's surface; and the *glue effect* is an attraction force that pulls the end-effector towards the object when the user ceases the contact with its surface. To minimize these artifacts, binary tree-based error-resilient encoding approaches based on [11] were successfully proposed yielding good subjective experience while maintaining excellent compression ratios.

3   Packet Loss in the Forward Channel

A general study for passifying TPTM systems in the presence of packet losses was performed in [12], however, for the simpler case where a hold last sample reconstruction is employed. Moreover, the artifacts induced by the packet loss in the forward channel (from the OP to the TOP) were never particularly discussed at a higher level, that is, how the loss of information during transmission is perceived by the operator and how it affects the teleoperator. Thus, the main goal of this work is to present the observed subjective artifacts resulting from position/velocity packet losses in telemanipulation sessions when the state-of-the-art deadband-based haptic data reduction approach [5] is employed in combination with a first-order linear predictor.

3.1 Velocity Commanding of TPTM Systems

In most TPTM system architectures, the position signal is directly acquired by the sensors in the HSI. Due to stability constraints [3,13] in the standard architecture of the telemanipulation systems, the velocity of the operator is transmitted instead of the position signal. The velocity $\mathbf{v}_i$ at the current time i can be obtained by differentiating the position signal as seen in Equation 4 below.

$$\mathbf{v}_i = \frac{\mathbf{p}_i - \mathbf{p}_{i-1}}{t_i - t_{i-1}} \tag{4}$$

where $\mathbf{p}_i$ and $\mathbf{p}_{i-1}$ are, respectively, the current and previous acquired position samples, and t_i and t_{i-1} their corresponding times. Since $t_i - t_{i-1}$ is always unitary (due to uniform sampling), this term could be omitted without loss of generality.

To decrease the transmission rate and the amount of data to be transmitted from the OP to the TOP, a perceptual data reduction scheme such as described in Section 1.2 can also be applied to the velocity signal. Nonetheless, when the state-of-the-art deadband compression scheme is employed [5], a slight position deviation is induced between the OP's acquired signal and what is displayed at the TOP. In order to minimize position deviation under variable time delay conditions, Chopra et al. [14] proposed an architecture including a position feedforward channel in addition to the velocity commanding. Using the deadband approach, the position updates are sent together with the current velocity samples in order to keep the position tracking accurate [13]. Therefore, depending on the application and the stability constraints, different types of commanding can be employed, this means that each transmitted data packet could contain only position samples, only velocity samples or both position and velocity samples as in [14] and this present work.

3.2 Impact of Packet Losses on the Signal Prediction

When predictive coding methods are applied together with deadband-based compression, packet losses can potentially degrade the transparency of the system since the sender and the receiver run out of synchrony. As explained in Section 2, both ends would use different reference samples (i.e. updates) to estimate future samples hence they would be no longer mutually consistent.

Also, there are two major differences between the packet loss effects on the forward and feedback channels. 1) In the case of force packet loss induced errors, the OP acts as an impedance factor on the HSI and consequently on the end-effector. This means that users can mildly compensate the force-feedback artifacts and still keep the system under control. 2) The prediction error in the forward channel propagates to two signal levels (velocity and position) unlike the errors in the feedback channel that only affect the force signal.

As mentioned in Section 3.1, velocity samples are transmitted. To calculate them at the sender side, an expression similar to Equation 2 is utilized.

$$\mathbf{v}_i^P = \frac{\mathbf{v}_j^U - \mathbf{v}_{j-1}^U}{t_j^U - t_{j-1}^U}(t_i^P - t_j^U) + \mathbf{v}_j^U \tag{5}$$

where the velocity update samples $\mathbf{v}_j^U$ and $\mathbf{v}_{j-1}^U$ can be calculated as follows.

$$\mathbf{v}_j^U = \frac{\mathbf{p}_j^U - \mathbf{p}_{j-1}^U}{t_j^U - t_{j-1}^U} \quad \text{and} \quad \mathbf{v}_{j-1}^U = \frac{\mathbf{p}_{j-1}^U - \mathbf{p}_{j-2}^U}{t_{j-1}^U - t_{j-2}^U} \tag{6}$$

Observe that in these last two expressions, only the latest update position samples (i.e. $\mathbf{p}_j^U$, $\mathbf{p}_{j-1}^U$ and $\mathbf{p}_{j-2}^U$) are employed and not the currently acquired position samples (i.e. $\mathbf{p}_i$, $\mathbf{p}_{i-1}$ and $\mathbf{p}_{i-2}$).

After calculating the predicted velocity sample, both $\mathbf{v}_i^P$ and the acquired velocity sample $\mathbf{v}_i$ are compared to decide over sending a new position/velocity sample. This evaluation can be written as below.

$$\begin{cases} ||\mathbf{v}_i - \mathbf{v}_i^P|| \leq ||\mathbf{v}_i^P|| \cdot k_v \longrightarrow \text{non-violation (use predicted sample)} \\ ||\mathbf{v}_i - \mathbf{v}_i^P|| > ||\mathbf{v}_i^P|| \cdot k_v \longrightarrow \text{violation (transmit position/velocity sample)} \end{cases} \tag{7}$$

where k_v is the deadband parameter for the velocity signal.

At the receiver side, two cases may occur: 1) the receiver does receive a new packet containing the update velocity $\mathbf{v}_j^U$, the position $\mathbf{p}_j^U$ and their time stamp t_j^U. In this case, both received velocity and position can be then directly displayed by the TOP. 2) the receiver does not receive any packets and therefore must use Equation 5 to predict the velocity $\hat{\mathbf{v}}_i^P$. Moreover, the estimated position $\hat{\mathbf{p}}_i^P$ to be displayed can be calculated as follows.

$$\hat{\mathbf{p}}_i^P = \hat{\mathbf{p}}_{i-1}^P + \hat{\mathbf{v}}_i^P(t_i - t_{i-1}) \tag{8}$$

where $\mathbf{p}_{i-1}^P$ is the last displayed position sample at the TOP. Since $t_i - t_{i-1}$ is always unitary (due to uniform displaying rate), this term could be omitted without loss of generality.

As mentioned before, if the communication channel is lossless, both ends are consistent thus the difference between $\mathbf{v}_i$ and $\hat{\mathbf{v}}_i^P$ ($= \mathbf{v}_i^P$) is already expected and said to be tolerable under human perception limitations. However, in the presence of communication unreliabilities, some packets may be lost and the estimated velocity at the receiver will diverge from the calculated velocity at the sender (i.e. $\hat{\mathbf{v}}_i^P \neq \mathbf{v}_i^P$). The direct impact of this event is clearly the end-effector displaying an incorrect velocity. However, there is another critical impact at the TOP side. As observed in Equation 8, whenever the velocity component $\hat{\mathbf{v}}_i^P$ is wrong, this error also influences the displayed position $\hat{\mathbf{p}}_i^P$. Thus, even if velocity packets arrive correctly after a first loss, this position deviation stays in the system. This error cumulation could be corrected when a position packet arrives (since it carries the correct current position $\mathbf{p}_i$) however, this might not be possible without special measures since the deviation can be too large and a sudden position correction could cause impairing events (e.g. jumps, ripples, unforeseen impacts) and, ultimately, the overall instability of the telemanipulation system.

4 Artifacts due to Position/Velocity Packet Losses

4.1 Experimental Setup

The experimental phase was motivated by our goal of observing and describing the high-level artifacts caused by packet losses in the forward channel. Although the theoretical description of the asynchrony between OP and TOP was described in the previous sections, a subjective analysis is still necessary to evaluate how the users perceive such inconsistencies.

The experimental setup employed a SensAble Phantom Omni as the human-system interface at the operator side. A CHAI3D [15] interactive virtual environment was utilized to emulate the telemanipulation ambient. In this environment, a small spherical end-effector was controled by the user. Additionally, there was a static toroidal object in the center of the workspace with which the user could interact. This experimental setup and the flow of the bidirectional data are depicted in Figure 2. The acquiring rate is set to 1 kHz and the deadband parameter k_v was kept at 15% since this value has previously shown a good compromise between data reduction and subjective quality [16].

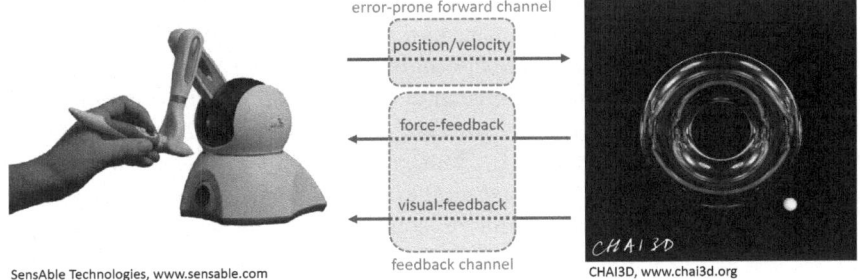

Fig. 2. Experimental setup utilized in this work. On the left hand side, the operator manipulates the HSI and receives the force-feedback through it. Moreover, the teleoperator (virtual end-effector) is depicted as the white sphere on the lower right hand side. The forward and feedback communication channels connect both ends.

4.2   Experimental Observations: Haptic Artifacts

End-effector drifting: this artifact consists of the end-effector arbitrarily continuing its motion even when the users have already stopped, changed their velocity or trajectory. This phenomenon is related to what is described in Section 3.2 concerning velocity errors. When a packet is lost, the velocity $\hat{\mathbf{v}}_i^P$ at the receiver continues to be predicted instead of updated (since the update was lost) thus it immediately differs from the intended $\mathbf{v}_i^P$ calculated at the sender. The user loses control of the TOP until a new update successfully arrives and the mismatch between $\hat{\mathbf{v}}_i^P$ and $\mathbf{v}_i^P$ is eliminated. The *drifting* artifact can be observed in Figure 3(c) where deviating spikes can be seen in the predicted velocity (red curve) at the receiver. The acquired velocity and its correct prediction are shown respectively in Figures 3(a) and 3(b). In the experiments we have observed that the deviation of the predicted velocity $\hat{\mathbf{v}}_i^P$ at the receiver was considerably higher than the one encountered in the velocity prediction $\mathbf{v}_i^P$ at the sender. Naturally, this difference depends on the packet loss probability q thus in this work we tested scenarios with $q = 5\%$, 10%, 20%, 30% and 40%, gauging average distortions (solid blue curve on the left side of Figure 4) from 6 to 21 times higher than what was expected by the sender. Also, it can be observed in this left plot of Figure 4 that the maximum deviation at the receiver (dashed blue curve) in the best case ($q = 5\%$) and in the worst case ($q = 40\%$) were, respectively 5 and 22 times greater than what was calculated at the sender. These results show that even with low packet loss probability, the *drifting* can be a significant issue.

End-effector "jump": this artifact is characterized by a sudden correction in the end-effector's location in case position commanding [3,13] is utilized. This correction is usually useful to avoid mismatches between the OP and the TOP positions however, due to the packet losses, it may have occurred *end-effector drifting* and the difference between these two position signals can be significant. When this deviation is large, the repositioning of the end-effector can mainly cause two phenomena depending on the kind of environment the telemanipulation is being run. 1) In a virtual environment, the end-effector can be instantaneously readjusted to the correct location but the user clearly notices this flickering since the motion is not continuous (i.e. it disappears from one place and appears somewhere else). 2) In a real environment, if the TOP tries to correct itself in a short time span the robot might jolt or ripple, hit something in the environment while in the correction trajectory and could even ultimately become unstable. These *"jumps"* are depicted by the spikes in the predicted position (red curve) in Figure 5(b). Moreover, it can be observed on the right hand side of Figure 4 that the maximum position deviation (dashed blue curve) is two to ten-fold higher than what is computed at the sender (dashed black curve). This shows quantitatively that the position correction, even in the best case scenarios (low q) can still lead the system into repositioning problems.

Workspace displacement: this artifact is the direct result of the aforementioned *drifting* whenever the position is not corrected to avoid the *"jump"* artifact. When a packet is lost and the velocity $\hat{\mathbf{v}}_i^P$ is different from the intended velocity $\mathbf{v}_i^P$, the error is propagated to the predicted position $\hat{\mathbf{p}}_i^P$ as seen in Equation 8.

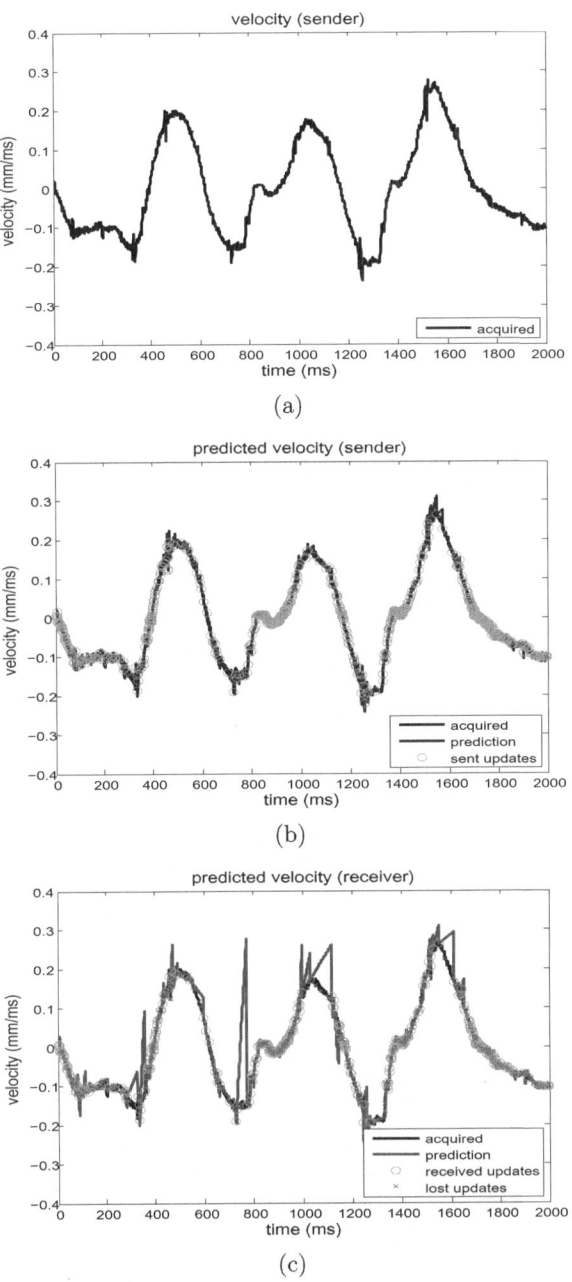

Fig. 3. (a) Velocity signal (black) derived from the acquired position signal at the sender. (b) Velocity signal (black), predicted velocity signal at the sender (blue) and sent update samples (green circles). (c) Velocity signal (black), predicted velocity signal at the receiver (red), received update samples (green circles) and lost update samples (pink crosses). It shows the *end-effector drifting* artifact. All curves depict the z-axis.

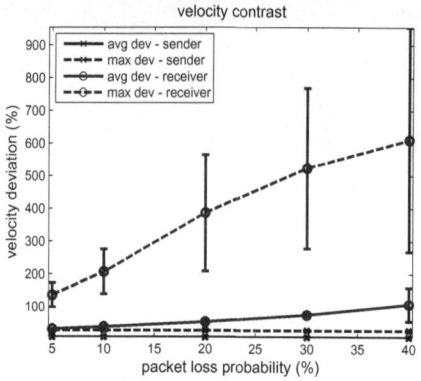 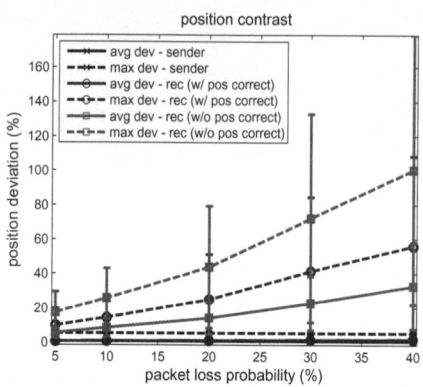

Fig. 4. (left) Average and maximum velocity deviations at sender (black) and receiver (blue). (right) Average and maximum position deviations at sender (black) and receiver with (blue) and without (red) position correction.

When no position commanding is employed, the position acquired at the HSI is never input to the TOP and all position prediction error is accumulated. What could be observed during the experimental sessions is the progressively non-corresponding workspace. This means that the TOP's three-dimensional space no longer matches the one utilized by the user hence seriously impairing the task performance. This artifact can be observed in Figure 5(c) where the predicted position (red curve) at the receiver significantly differs from the acquired position (black curve). Furthermore, it can be seen on the right hand side of Figure 4 that the average position at the receiver (solid red curve) is differing from 10 to 60 times comparing to the sensed position at the HSI (solid black curve). This shows that when no position correction is applied, there is a considerably large *displacement* even when the packet loss probability is low.

5   Conclusion

This work demonstrates the theoretical influence of losing position/velocity packets on the forward channel of TPTM systems when predictive and perceptual coding are employed. Moreover, we describe the high-level artifacts that are perceived by the users. We show that the losses can considerably affect the teleoperator position and velocity degrading the transparency and stability of the system. When packets are lost, the teleoperator continues to predict the position/velocity samples using different reference samples as utilized at the sender therefore creating an asynchrony between both ends. These cases can provoke three distinct artifacts, namely, the *end-effector drifting*, the *end-effector "jump"* and, lastly, the *workspace displacement*.

We plan to propose a joint error-resilient velocity data-reduction approach and position compensation scheme in the future to address the packet losses in the forward channel and ultimately mitigate the artifacts presented in this work.

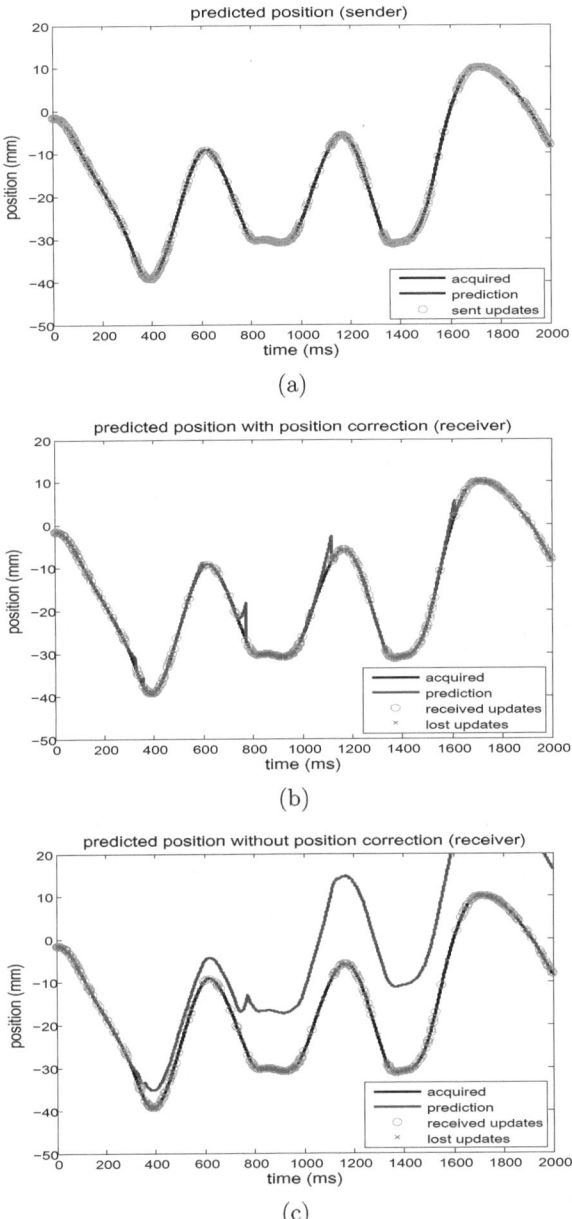

(a)

(b)

(c)

Fig. 5. (a) Position signal (black), predicted position signal at the sender (blue) and sent update samples (green circles). (b) Position signal (black), predicted position signal at the receiver (red) with position commanding, received update samples (green circles) and lost update samples (pink crosses). It shows the *end-effector "jump"* artifact. (c) The same as (b) but without position commanding. It shows the *workspace displacement* artifact. All curves depict the z-axis.

Acknowledgment. This work has been supported by the European Research Council under the European Union's Seventh Framework Programme (FP7/2007-2013) / ERC Grant agreement no. 258941.

References

1. Hirche, S.: Haptic Telepresence in Packet Switched Communication. PhD thesis. VDI-Verlag, Düsseldorf, Germany (2005)
2. Goldstein, E.B.: Sensation and Perception. 6th edn. Wadsworth (2002)
3. Anderson, R., Spong, M.: Bilateral control of teleoperators with time delay. IEEE Trans. Automatic Control 34, 494–501 (1989)
4. Shi, Z., Zou, H., Rank, M., Chen, L., Hirche, S., Müller, H.J.: Effects of packet loss and latency on the temporal discrimination of visual-haptic events. IEEE Trans. on Haptics 3(1), 28–36 (2010)
5. Hinterseer, P., Hirche, S., Chaudhuri, S., Steinbach, E., Buss, M.: Perception-based data reduction and transmission of haptic data in telepresence and teleaction systems. IEEE Trans. on Signal Processing 56(2), 588–597 (2008)
6. Ortega, C.S.A., Kolahdouzan, M.R.: A comparison of different haptic compression techniques. In: Proc. of Int. Conf. on Multimedia and Expo (ICME), Lausanne, Switzerland (August 2002)
7. Ortega, A., Liu, Y.: Lossy compression of haptic data. Prentice-Hall (2002)
8. Burdea, G.C.: Force and Touch Feedback for Virtual Reality. Wiley, New York (1996)
9. Brandi, F., Kammerl, J., Steinbach, E.: Error-resilient perceptual coding for networked haptic interaction. In: Proc. of ACM Multimedia, Firenze, Italy, pp. 351–360 (October 2010)
10. Brandi, F., Steinbach, E.: Low-complexity error-resilient data reduction approach for networked haptic sessions. In: Proc. of IEEE Int. Symposium on Haptic Audio-Visual Environments and Games (HAVE), Qinhuangdao, China (October 2011)
11. Chou, P.A., Miao, Z.: Rate-distortion optimized streaming of packetized media. IEEE Trans. on Multimedia 8(2), 390–404 (2006)
12. Hirche, S., Buss, M.: Packet loss effects in passive telepresence systems. In: Proc. of the IEEE Conference on Decision and Control (CDC), Atlantis, Paradise Island, Bahamas, vol. 4, pp. 4010–4015 (December 2004)
13. Hirche, S., Buss, M., Hinterseer, P., Steinbach, E.: Towards deadband control in networked teleoperation systems. In: Proc. of International Federation of Automatic Control World Congress (IFAC), Prague (July 2005)
14. Chopra, N., Spong, M.W., Hirche, S., Buss, M.: Bilateral teleoperation over the internet: the time varying delay problem. In: Proc. of American Control Conference, vol. 1, pp. 155–160 (June 2003)
15. Conti, F., Barbagli, F., Morris, D., Sewell, C.: CHAI 3D: An open-source library for the rapid development of haptic scenes. In: Proc. of IEEE World Haptics, Pisa, Italy (March 2005)
16. Kammerl, J., Hinterseer, P., Steinbach, E.: A novel signal reconstruction algorithm for perception based data reduction in haptic signal communication. In: Proc. of Int. Workshop on Networking Technology for Robotics and Applications, with Int. Conf. on Computer Communications and Networks, Honolulu, Hawaii (August 2007)

Tactile Emotions: A Vibrotactile Tactile Gamepad for Transmitting Emotional Messages to Children with Autism

Gwénaël Changeon, Delphine Graeff, Margarita Anastassova, and José Lozada

CEA, LIST, Sensorial and Ambient Interfaces Laboratory,
91191 – Gif-sur-Yvette CEDEX, France
{gwenael.changeon,margarita.anastassova,jose.lozada}@cea.fr

Abstract. This paper presents an innovative vibrotactile gamepad used as complementary therapeutic aids for transmitting emotional messages to children with autism. The gamepad includes eight electromagnetic actuators embedded in the gamepad shell to provide distributed tactile feedback. The location of the actuators was defined by measuring the contact zones of the hand and the gamepad with a number of users. Actuators are designed to cover the 0-500 Hz bandwidth with a 0.5 N maximum force. The integrated gamepad was submitted to two user evaluations: one (13 users) to define the tactile patterns needed for control and the second (9 users) to test the acceptability and the usefulness of the device for the enhancement of emotional competences of children with autism. Encouraging results are shown.

Keywords: Haptics, Tactile stimulation, Vibrotactile gamepad, Electromagnetic actuators, Autism, Sensory and Touch therapy, Serious game.

1   Introduction

Autism is a neurodevelopmental disorder characterized by impaired social interaction and communication, and by restricted and repetitive behaviour. Sensory distortions are also observed in people with autism. For example, tactile perception is often impaired resulting in hypo- or hypersensitivity. This impairment in tactile perception affects, in turn, social relations because touch is of crucial importance for communication and emotional understanding [1]. First-person accounts of individuals with autism have often described unusual or excessive emotional reactions to tactile stimulations. In an autobiographic report, Grandin [2], an American researcher with autism, writes: *"I pulled away when people tried to hug me, because being touched sent overwhelming wave stimulation through my body...Small itches and scratches that most people ignore were torture."* She also noted that certain clothing textures could make her extremely anxious, distracted, and nervous [3]. Another person with autism states: *"To be just lightly touched appeared to make my nervous system whimper, as if the nerve ends were curling up. If anyone hit on the terrible idea of tickling me, I died."* [4]. Similar experiences have also been described by Willey [5] and Williams

P. Isokoski and J. Springare (Eds.): EuroHaptics 2012, Part I, LNCS 7282, pp. 79–90, 2012.
© Springer-Verlag Berlin Heidelberg 2012

[6]. They have been confirmed by experimental studies showing that adults with autism have significantly lower tactile perceptual threshold, but only for vibration at 200 Hz [7], which suggests that these people would be hypersensitive to high frequency vibrations (i.e. "buzzes"), but not to low frequency vibrations (i.e. "flutters"). Parents and caregivers report similar observations. Thus, a community interview-based study involving parents and caregivers found that in a sample of 75 children with autism, 52% were hypersensitive to touch [8]. Some aspects of touch such as hugging firmly with a parent, having a back rub, engaging in a rough play with an adult, feeling substances in one's hands or feeling the wind in one's face were reported as positive experiences. Other aspects such as having one's ears cleaned, one's face touched or one's hair cut were reported as negative experiences. These unusual reactions, if repetitive, may provoke stress and anxiety-related behaviour, as well as a negative impact on self-concept and social interactions [9-11].

However, despite this impact on self-concept, communication, social relations and daily activities, there are only few technological endeavours to convey emotions to individuals with autism using touch. A first example in this direction is a group of haptic interfaces developed at the MIT [12]. The devices called *Touch Me*, *Squeeze Me*, and *Hurt Me* rely on vibrotactile, pneumatic and heat pump actuation (Fig. 1).

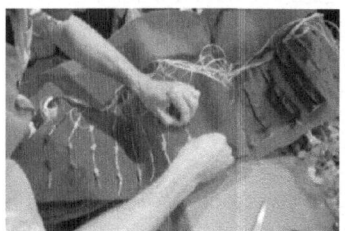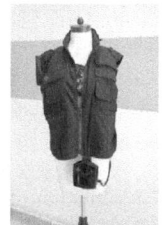

Fig. 1. Some existing haptic interfaces for conveying emotions for autism [12]

Touch Me consists of a flexible vibrotactile motor array in a soft enclosure that can be applied to large areas of the body such as arms, legs and the chest. This array can be remotely actuated by a caregiver controlling the location and the intensity of the vibrotactile actuation.

Squeeze Me is a vest that simulates hugs. Its design has been based on the assumption that people with autism tend to sometimes better tolerate electronic compared to human touch. Pneumatic chambers around the shoulders, chest and back are inflated by a portable air compressor to momentarily provide distributed pressure. A physical safety system prevents over-inflation or over-compression through pressure-release valves and fasteners with engineered strain releases. Users have to press a button to initiate the inflation and subsequent deflation of the vest, which is fixed in pressure and duration.

Hurt Me is a pneumatic bracelet which inflates a pressure bladder studded with plastic teeth. It generates controlled pain and is intended to be used by persons with autism having tendencies towards self-harm. *Hurt Me* aims to substitute the need to self-mutilate by supplying a painful stimulus controlled by the patient and designed with built-in safety limits.

Other similar haptic interfaces for emotional communication have been developed, though they have not been applied to the therapy of people with autism. For example, the *Hug-Over-A-Distance jacket* contains air compartments that inflate quickly all around the torso to simulate a real hug [13]. *Heart2Heart* [14] is a wearable electronic garment that physically and emotionally links two remote individuals.

These developments, as well as the sensory impairments in autism show that there is real need for vibrotactile devices providing distributed stimulations at different parts of the body. Such devices can be very useful as complementary therapeutic aids because they can partially substitute human touch which is not always well-tolerated by people with autism. They can thus help these individuals get used to being touched. Furthermore, touch is an emotional sense and emotions involve perceptual, sensory and motor experiences (referred to as "embodiment"). In this sense, such devices can be a way of embodying emotions for persons with autism. The device presented below is a contribution in this direction.

2   Actuation Concept

A tactile gamepad was developed to give emotional feedback to children with autism. It will be used as a therapeutic aid. It is composed of 8 tactile actuators distributed over the gamepad body. In order to provide an efficient tactile stimulation, the actuators' position was defined by measuring the contact zones between the hand and the gamepad during the game. The interface used for this measurement, as well as the tests done to measure the contact zones are described below.

Fig. 2 shows the "measuring" gamepad redesigned to embed 16 tactile sensors on its right half. As the gamepad is symmetrical, the contact zone is identical on both sides. The two buttons at locations 3 and 4 were replaced by a hard surface to let the users position their hands as they wish on the overall surface of the shell.

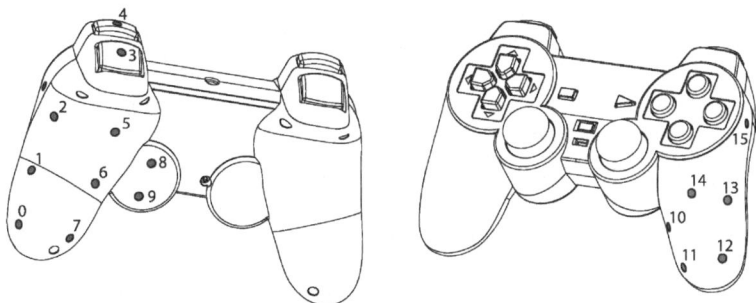

Fig. 2. Gamepad equipped with 16 contact sensors

Fig. 3 presents the measurement process. Each sensor is made of one conductive inset placed on the gamepad surface. The user wears a conductive bracelet. The contact of the hand on an inset closes the electrical loop and switches the associated

MOSFET transistor. The signal of each sensor is collected using a digital input of a National Instruments data acquisition Board (DAQ) and processed via a *Labview* application.

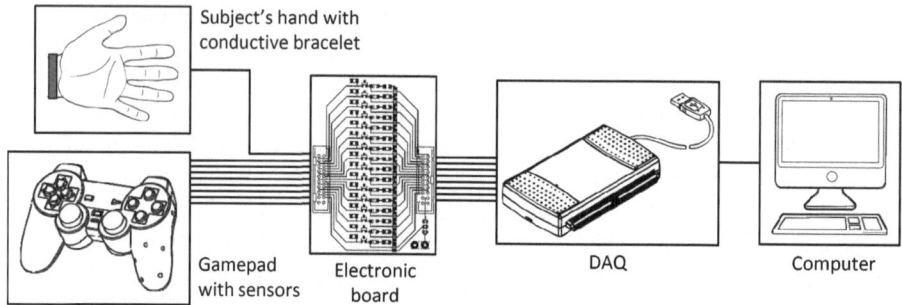

Fig. 3. Experimental equipment diagram: when the user touches a sensor on the gamepad the electrical loop is closed provoking a digital input on the acquisition board. Data is recorded and processed using *Labview*.

20 subjects participated in the experiments done to measure the contact zones. There were 7 women and 13 men with different hand sizes. Several video game backgrounds were also used. The experiments consisted in playing 3 distinct types of games: a racing one, a one sports and an action one. The subjects had to play every game for 3 minutes. Also, their hand size was measured. The user experience with the game was evaluated by a questionnaire.

Fig. 4 shows the median percentage of time each sensor was in contact with the user's hand. Sensors *0, 1* and *7* were almost permanently used. Sensors *5, 12* and *13* were also widely used. On the contrary, sensors *6* and *15* were used only occasionally. All the other sensors (2, 3, 4, 6, 8, 9, 10, 11, 14 and 15) were almost never used.

Regarding the influence of the game type on the use of the sensors, there was a difference between the racing game and the other types of games for sensors *5, 12, 13* and *15*. In the racing game, sensors *5* and *15* were used more often than in the other game types. There was the opposite effect for sensors *12* and *13*. This may be due to the buttons used to interact with the different types of games. The racing game requires pushing the same button (the throttle) with the thumb for long periods of time. Therefore, the hand stays immobile on the gamepad. Since the thumb is bended, the hand position does not allow touching the upper part of the side of the gamepad handle. On the contrary, in the sport or action game, the user needs to push the buttons sporadically. The thumb is less bended and the hand is covering a bigger part of the handle side.

The video game background has also an influence on grasping. In fact, expert players have a more stationary grasp than beginners. They get a percentage close to 100% for the sensors they use and 0% for the others. Because they have an unsteady grasp, beginners have less distinctive results. They obtain percentages between 20% and 80%.

The hand size does not seem to be an important factor for determining the contact zones between the hand and the gamepad. Distant sensors such as *3, 4, 8, 9, 10* or *11*, were used both by users with small hands and by users with big hands. The usage patterns of the other sensors are also similar for users with the different hand sizes.

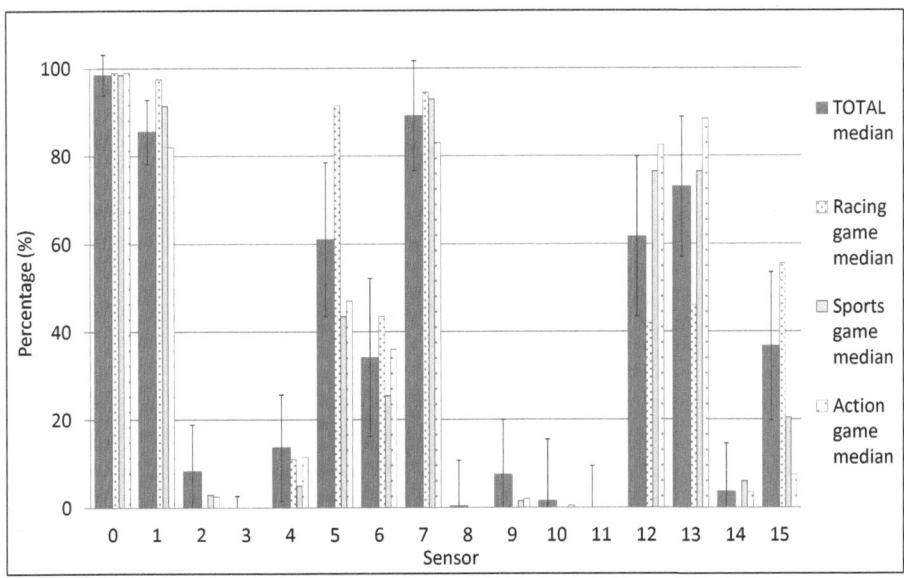

Fig. 4. Experimental results: median percentage of time each sensor is in contact with the hand for each type of game

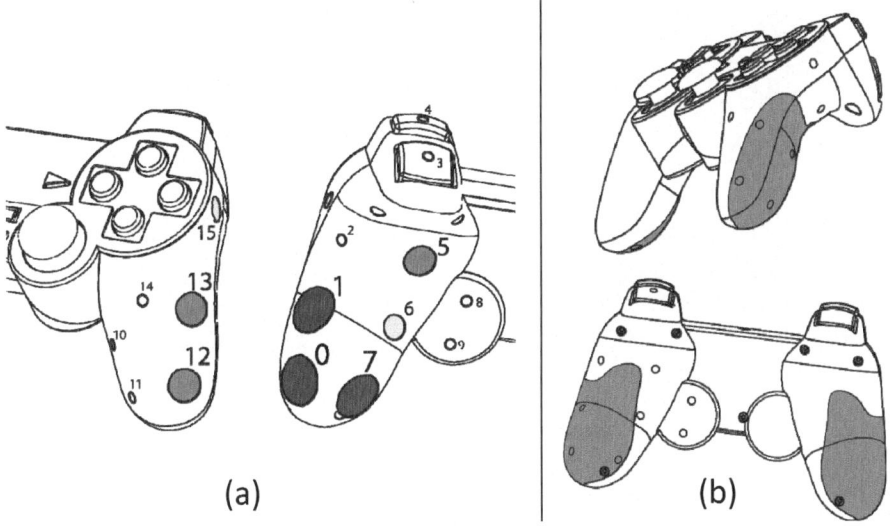

Fig. 5. (a) Schematic view of the contact percentage of time for each sensor. (b) Contact area for actuators embedding

Fig. 5a shows a view of the gamepad with the sensors displayed according to their percentage of use. These results helped us define the areas in which to embed the actuators to ensure a maximum probability of achieving stable contact between the user's hand and the gamepad when playing different types of games. (See Fig. 5b)

3   Actuation Design and Implementation

Gamepads are usually equipped with rumble motors which have a high response time. Furthermore, the vibration frequency is linked to its amplitude and the minimum frequency is often up to 20 Hz with low vibration amplitude.

The actuator we propose is designed to provide vibrations which have independent frequency and amplitude. It also allows reaching very low frequencies (i.e. lower than 1 Hz).

Fig. 6 shows a CAD view of the complete actuator. It is composed of a permanent magnet (1), a coil (2), a ferromagnetic core (3) and a surrounding armature (4) to guide the magnetic flux and avoid energy losses in the air. When a current is applied to the coil, it produces a magnetic field that interacts with the magnet's permanent field. The magnetic interaction generates a controlled attraction/repulsion force. A flexible brass membrane (5) is used to guide the magnet along the vertical axis. A silicone spacer (6) is used to set a constant distance between the magnetic core and the magnet and to filter vibration from the membrane to the actuator body. Finally, a plastic block (7) with a diameter of 3 mm is glued on the top side of the membrane to provide a sufficient pressure on the hand. The complete actuator has a diameter of 9 mm. It is 9 mm high without the plastic block.

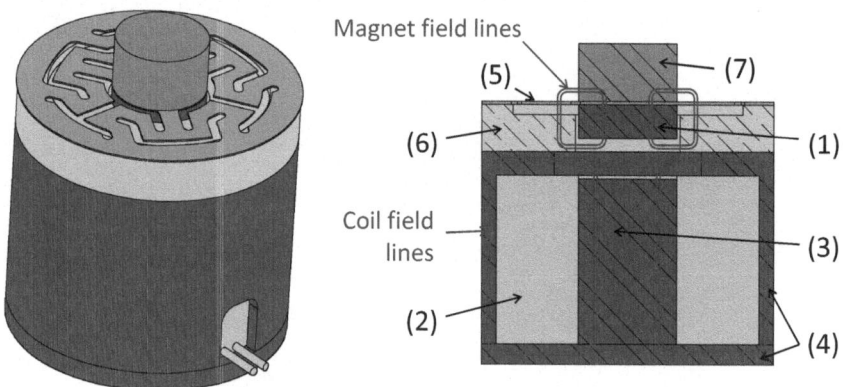

Fig. 6. Cross-sectional view of the complete actuator

The disc magnet (diameter 3 mm, thickness 1 mm) made of Neodymium-Iron-Boron (NdFeB) has a magnetization of 1.4 T. The coil (OD 8 mm, ID 3 mm, and thickness 5.5 mm) is made of 0.15 mm copper wire with 350 turns. The surrounding iron armature and the ferromagnetic core (iron) have a relative permeability of 1000.

Even with no energy supply, the magnet's field interacts with the ferromagnetic parts and the magnet is attracted by the core. The flexible membrane behaves like a spring and stores this offset force. As a consequence, the force provided by the actuator is the difference between the case where no current is applied to the coil and the case with current. Fig. 7 shows the results of a 2D axisymmetric finite element simulation carried out using *FEMM* (Finite Element Method Magnetics) with and without current. The ferromagnetic core and armature are designed to guide the magnetic field avoiding magnetic saturation. These simulation results are also used to calculate the actuation force. At 0.5 A, the calculated force is 0.55 N.

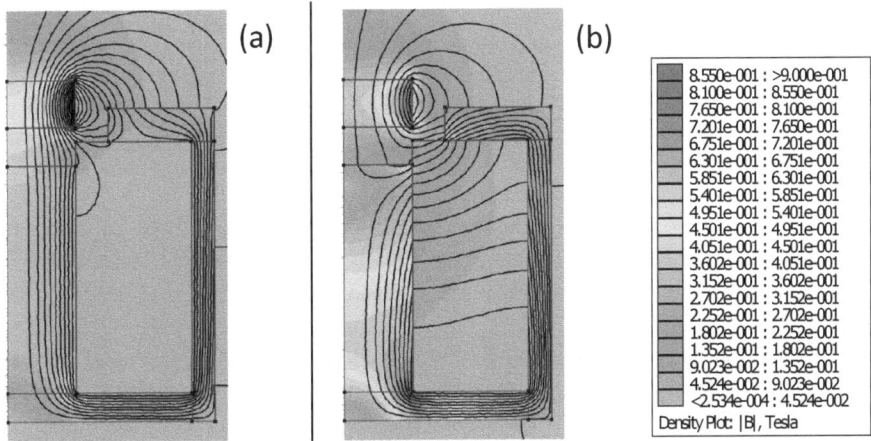

Fig. 7. Finite element simulation of the actuator: magnetic field at 0 A (a) and at 0.5 A (b)

The prototype of the actuator has a resistance of 6.2 Ω and an inductance of 1.9 mH. Fig. 8 presents the frequency response of the actuator.

In order to determine the stiffness of the membrane the modification of the resonance frequency of the system was measured with and without an additional mass. Equation (1) gives ω_0 and $\omega_{0_{\Delta m}}$ the resonant angular frequencies respectively for a system without and with additional mass as a function of m the effective mass, Δm the additional mass and K the stiffness. From (1) we obtain Equation (2) that gives the effective mass.

$$\begin{cases} \omega_0 = \sqrt{\dfrac{K}{m}} \\ \omega_{0_{\Delta m}} = \sqrt{\dfrac{K}{m+\Delta m}} \end{cases} \tag{1}$$

$$m = \dfrac{\dfrac{\omega_{0_{\Delta m}}^2}{\omega_0^2} \cdot \Delta m}{1 - \dfrac{\omega_{0_{\Delta m}}^2}{\omega_0^2}} \tag{2}$$

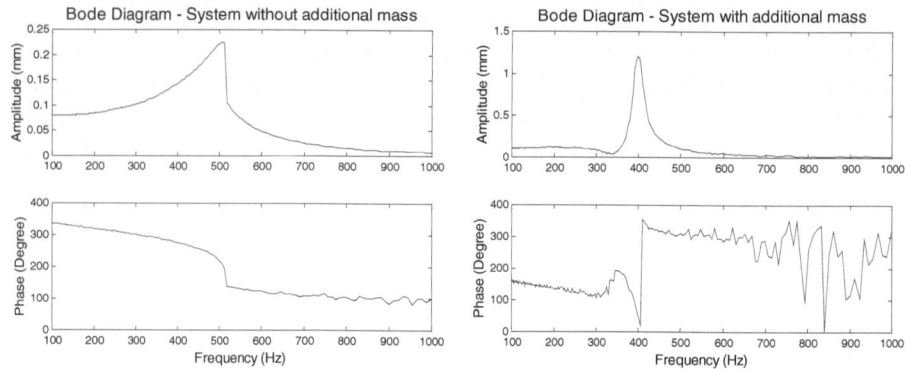

Fig. 8. Experimental results: Bode diagrams with and without additional mass used to determine the actuator characteristics. A sinusoidal signal of amplitude 0.5 V and a frequency from 100 to 1000 Hz was applied to the actuator. The displacement is measured with a Keyence LC-2220 Laser sensor.

Fig. 8 present the Bode diagrams for the system with and without an additional mass. The resonant frequencies are $f_{0_{\Delta m}}$ = 405 Hz and f_0 = 515 Hz respectively for the system with and without added mass. Then, it appears that m = 0.0814 g and K = 852 N.m^{-1}.

The actuation force F is given by $F = -K.x$ where x is the displacement from the equilibrium position. At 0.5 A, the measured displacement is 0.6 mm, what is equivalent to a force F = 0.51 N. This value is very close to simulation results (0.55 N).

According to [15], the bandwidth over which the human tactile system can detect vibration ranges from 0.4 Hz to 700 Hz; with an optimal detection at 200 Hz. The actuator is well adapted for tactile stimulation since it can provide stimulations up to 500 Hz.

Eight actuators were embedded into the gamepad positioned as shown on Fig. 9 according to the experimental results. The gamepad shell was rebuilt to integrate new slots for the actuators and the electronic board. Each actuator is wedged between the shell and a cap. The caps are mainly made of a central silicone membrane. This membrane is useful for filtering the vibrations and isolates the shell ensuring a local tactile stimulation.

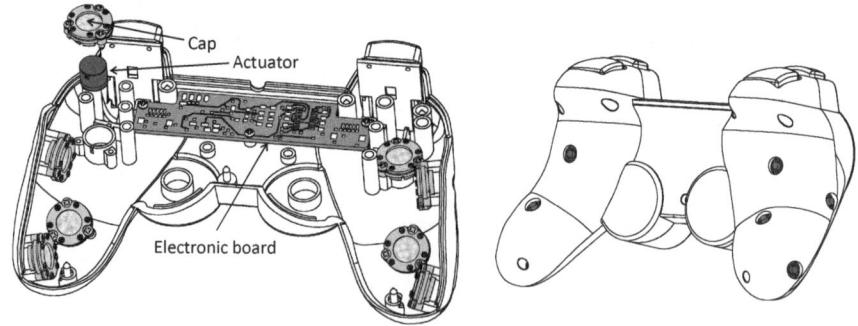

Fig. 9. Gamepad interior view: 8 actuators and the electronic board are embedded in new special slots

Fig. 10 shows a schematic view of the control principle. The actuators behavior is controlled by a microcontroller that drives 8 MOSFET transistors and a programmable power supply. Each transistor is connected to an actuator coil. The vibration frequency is controlled by the transistor switching frequency. The amplitude of the vibration is given by the programmable power supply that can deliver 4 different energy levels.

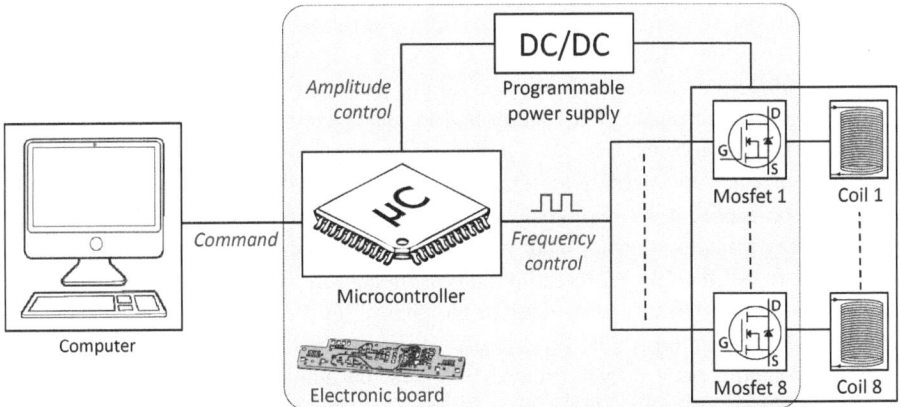

Fig. 10. Electronics and command functional diagram

The device described above allows the creation of compound vibrotactile stimuli called "vibrotactile patterns", which can be linked in order to form a vibrotactile "language". A vibrotactile language is thus a structured set of vibrotactile patterns, which convey a certain sense and represent several more or less abstract concepts (e.g. emotions, navigational concepts). If well-designed, these vibrotactile patterns and the respective vibrotactile language should be as easy to understand as visual icons.

In the case of the gamepad described in this paper, the created vibrotactile patterns are stored in the microcontroller. They are called by a command coming from the computer.

A user study focusing on the creation and validation of such vibrotactile patterns for conveying emotions has been done. After this first study, the most intuitive vibrotactile emotional patterns have been selected, integrated in a specifically designed game for the enhancement of the emotional competences of children with autism and presented to such children. The understanding and use of emotional vibrotactile patterns by children with autism has been evaluated in this second user study. These two studies are briefly described below.

4   User Studies

The first user study had two major objectives. First, we wanted to see how vibrotactile emotional patterns could be produced and how this production can be assisted by external aids (audio or verbal). Second, we wanted to test the recognition of tactile patterns by users who had not created them.

In order to achieve these objectives, we applied a two-phases procedure. In a first phase, 9 users (3F/6M), aged from 22 to 25 (M = 23, SD = 0.9), created a number of vibrotactile patterns corresponding to six basic emotions. The basic emotions which simulated in vibrotactile patterns were anger, disgust, fear, happiness, sadness and surprise. The users could manipulate the following parameters: amplitude, frequency, duration of the activation of the actuators and the spatial distribution in order to convey an emotional meaning. They could also use 2 types of different aids: 1) verbal aids in the form of words and idioms describing different sensations and phenomena conveying emotions, and 2) audio aids in the form of extracts of soundtracks again conveying different emotions. In the second phase, 4 users (2F/2M) who had not participated in the first phase of the study had to match the created vibrotactile patterns with the 6 basic emotions.

A total of 144 vibrotactile patterns were created. Both the verbal and the audio aid proved to be useful. The vibrotactile pattern creators relied equally on both of these aids (i.e. they relied on the audio aid in 51% of the cases and on the verbal aid on 49% of the cases). All the 144 vibrotactile patterns created in the first phase of the experiment were matched to basic emotions in the phase 2 of the test. One user managed to work on all 144 patterns. The other users worked on 96 patterns each. In this way, every vibrotactile patterns was treated by 3 test participants. All the patterns were matched to basic emotions. The easiest "correct" associations were the ones with the emotions of surprise and anger. The most difficult "correct" association was the one with the disgust. These results show that the vibrotactile gamepad allows the creation of a large variety of vibrotactile patterns conveying different emotional messages. These patterns can also been recognized by people who have not created them.

The second user study was based on the first one. Thus, the most intuitive emotional patterns corresponding to each of the 6 basic emotions had been selected and integrated in a specifically designed video game for the enhancement of the emotional competences of children with autism. Nine children, all of them boys aged from 10 to 16 participated in the study. The typical test setting is presented in Fig. 11.

Fig. 11. Test settings. The child uses the tactile gamepad with a serious game assisted by a therapist.

All the children participating in the study had problems identifying basic emotions. As far as tactile perception was concerned, 3 of them were hypersensitive, 2 of them were hyposensitive and the others had no particular distortions. Despite these perceptual differences, all of them easily accepted the use of the vibrotactile gamepad.

No rejection of tactile stimulation was noted. Furthermore, our observations showed that 50% of the participants relied on vibrotactile stimulations in order to identify, recognize and memorize the different emotions presented in the game. Another result showed that the vibrotactile stimulations were used by the children to help focusing their attention on the events in the game. Vibrotactile stimulations also seemed to help reducing stereotypical behaviours.

5   Conclusion

This paper presents a tactile gamepad capable of conveying emotions via a tactile stimulation. The device is composed of eight actuators distributed on the shell. In order to identify the locations for the actuators, another gamepad was equipped with 16 sensors. 20 subjects played three different kinds of video games using this "measuring" gamepad. Results shows the zones were the hand is in quasi-permanent contact with the gamepad. Actuators were specifically designed for providing haptic feedback. Each actuator vibration can be modulated in frequency (0-500 Hz) and amplitude (0-0.55 N) independently.

The integrated gamepad was used in two tests. First, tactile patterns were created by a group of users and tested in order to define the most adequate stimulation for each emotion. Then, the second test was done with 9 children. The gamepad and the tactile patterns were integrated with a serious game. The preliminary evaluation shows a good acceptance for tactile stimulation. Tactile stimulation seems to be useful for emotion recognition and to help reducing stereotypical behaviours. These results, though very preliminary, are encouraging. Specific tests on the advantages of using the vibrotactile gamepad for the enhancement of the emotional competences of children with autism are currently being done.

References

1. Dickie, V.A., Baranek, G.T., Schultz, B., Watson, L.R., McComish, C.S.: Parent reports of sensory experiences of preschool children with and without autism: A qualitative study. American Journal of Occupational Therapy 63, 172–181 (2009)
2. Grandin, T.: My experiences with visual thinking, sensory problems and communication difficulties. Web Page Article from the Center for the Study of Autism (1996)
3. Grandin, T.: Thinking in pictures: My life with autism. Doubleday, New York (1995)
4. Gerland, G.: A real person: Life on the outside. Souvenir Press Ltd, London (1997)
5. Willey, L.H.: Pretending to be normal: Living with Asperger syndrome. Jessica Kingsley, London (1999)
6. Williams, D.: Nobody nowhere: The extraordinary biography of an autistic girl. Avon, New York (1994)
7. Blakemore, S.J., Tavassoli, T., Calo, S., Thomas, R.M., Catmur, C., Frith, U., Haggard, P.: Tactile sensitivity in Asperger syndrome. Brain and Cognition 61, 5–13 (2006)
8. Bromley, J., Hare, D.J., Davison, K., Emerson, E.: Mothers supporting children with autistic spectrum disorders: social support, mental health status and satisfaction with services. Autism 8, 409–423 (2004)

9. Ben-Sasson, A., Carter, A.S., Briggs-Gowan, M.J.: Sensory over-responsivity in elementary school: prevalence and social-emotional correlates. Journal of Abnormal Child Psychology 37, 705–716 (2009)
10. Goldsmith, H.H., Van Hulle, C.A., Arneson, C.L., Schreiber, J.E., Gernsbacher, M.A.: A population-based twin study of parentally reported tactile and auditory defensiveness in young children. Journal of Abnormal Child Psychology 34, 393–407 (2006)
11. Hilton, C.L., Harper, J.D., Kueker, R.H., Lang, A.R., Abbacchi, A.M., Todorov, A., La-Vesser, P.D.: Sensory responsiveness as a predictor of social severity in children with high-functioning autism spectrum disorders. Journal of Autism and Developmental Disorders 40, 937–945 (2010)
12. Vaucelle, C., Bonanni, L., Ishii, H.: Design of haptic interfaces for therapy. In: Proceedings of ACM CHI 2009, Boston, MA, USA, April 4-9 (2009)
13. Mueller, F., Vetere, F., Gibbs, M.R., Kjeldskov, J., Pedell, S., Howard, S.: Hug over a distance. In: Proceedings of CHI 2005 (2005)
14. Grimmer, N.: Heart2Heart (2001),
    http://www.baychi.org/calendar/20010508/#1
15. Johansson, R.S., Vallbo, A.B.: Tactile sensibility in the human hand: relative and absolute densities of four types of mechanoreceptive units in glabrous skin. The Journal of Physiology, 283–300 (1979)

Cognitive Load Can Explain Differences in Active and Passive Touch

George H. Van Doorn[1], Vladimir Dubaj[2], Dianne B. Wuillemin[1],
Barry L. Richardson[1], and Mark A. Symmons[1]

[1] Psychological Studies, Monash University, Churchill, 3842, Australia
[2] Department of Physiology, Monash University, Clayton, 3800, Australia
george.vandoorn@monash.edu

Abstract. Active touch is often described as yielding "better-quality" information than passive touch. However, some authors have argued that passive-
guided movements generate superior percepts due to a reduction in demands on
the haptic sensory system. We consider the possibility that passive-guided conditions, as used in most active-passive comparisons, are relatively free from
cognitive decision-making, while active conditions involve cognitive loads that
are quite high and uncharacteristic of normal sensory processes. Thus studies
that purport to show differences in active and passive touch may instead be revealing differences in the amount of cognition involved in active and passive
tasks. We hypothesized that passive-guided conditions reduce not the *sensory*
load but the *cognitive* load that active explorers must bear. To test this hypothesis Blood Oxygen Level Dependent (BOLD) activity was measured using functional Magnetic Resonance Imaging (fMRI) during active and passive-guided
fingertip exploration of 2D raised line drawings. Active movements resulted in
greater activation (compared with passive movements) in areas implicated in
higher order processes such as monitoring and controlling of goal-directed behavior, attention, execution of movements, and error detection. Passive movements, in contrast, produced greater BOLD activity in areas associated with
touch perception, length discrimination, tactile object recognition, and efference
copy. The activation of a greater number of higher-order processing areas during active relative to passive-guided exploration suggests that instances of passive-guided superiority may not be due to the haptic system's limited ability to
cope with *sensory* inputs, but rather the restriction imposed by the use of a single finger such that active exploration may require cognitive strategies not de-
manded in the passive condition. Our findings suggest that previous attempts to
compare active and passive touch have, in order to simplify tasks, inadvertently
introduced cognitive load at the expense of normal sensory inputs.

Keywords: Active, passive-guided, fMRI, raised lines, cognitive load.

1 Introduction

Gibson's [1] work distinguishing touching from being touched provided insight into
how perceptual awareness may be modulated by self-generated actions (see also [2]).

P. Isokoski and J. Springare (Eds.): EuroHaptics 2012, Part I, LNCS 7282, pp. 91–102, 2012.
© Springer-Verlag Berlin Heidelberg 2012

In Gibson's experiments passive touch consisted of having a stimulus pressed upon the skin with no (intended) subsequent movement. However, the term "passive touch" has more recently been applied to situations where the stimulus is moved across the skin by an experimenter, or where the hand of the participant is guided in such a way that the input can resemble that of active exploration, but with movement being externally-generated [3].

Self-generated movements involve the intent to, the planning of, the preparation for, and execution of, movement [4]. None of these are thought to be involved in passive exploration. Active movements may also contain a copy of the efferent signal used to predict the sensory consequences of motor commands [5]. The efference copy is used to make a comparison between the expected and actual movement, allowing updating to fine-tune action. Christensen et al. [6] suggest that the comparison process is likely mediated by brain structures associated with motor preparation and sensory-motor integration, such as the intraparietal [7] and ventral premotor cortices [8]. Since passive movements do not generate an equivalent motor signal, the consequences of these movements cannot be compared with a prediction. This may be the mechanism by which people distinguish their active from their passive movements [5].

Although active touch is often described as yielding "better-quality" information, some authors have argued that passive movements generate superior percepts. For example, Magee and Kennedy [9] have suggested that the *act of planning* (see Grezes & Decety's [4] features of self-generated movements outlined above) associated with active touch overtaxes "the limited resources of the haptic system" (p. 288). Passive touch is not so taxed, they claim – an idea that has remained speculative for 30 years.

Richardson and Wuillemin [10] were able to partly replicate Magee and Kennedy's [9] finding of passive superiority but suggested that their explanation of active inferiority (i.e. overtaxing the haptic system) was implausible. Richardson, Wuillemin and MacKintosh [11] argued that the "act of planning" is more cognitive than sensory and that the responsibility for the production of movement is not a potent determinant of, or a genuine component in, the accuracy of tactual pattern recognition. Thus, Magee and Kennedy's [9] task may not have been one that revealed limitations in the haptic sensory system, but a task that instead revealed how their active subjects were obliged to recruit more cognitive strategies than were their passive counterparts. Symmons, Richardson and Wuillemin [12] critically reviewed many active-passive comparisons and concluded that task-type was the significant variable determining whether active or passive touch was superior, or whether they were equivalent. They proposed a theory of "cognitive complexity" to explain their findings. A means of further exploring the cognitive differences between active and passive touch could include the observation of brain areas associated with cognition that are hypothesized to be recruited during active, but not passive, exploration.

Some brain structures have been shown to produce differential levels of activation during active and passive movements. Simões-Franklin, Whitaker and Newell [13] had participants perform roughness estimates of sandpaper in active (i.e. subjects freely explored the surface) and passive (i.e. sandpaper was moved under each person's finger) conditions. Using fMRI they found greater activation during active exploration, relative to passive exploration, in the anterior cingulate cortex

(ACC)/pre-supplementary motor area (pre-SMA). Mima et al. [14] compared active and passive flexion-extension movements of a finger joint. Using Positron Emission Tomography they showed that brain structures activated to a greater extent by active movement included the left primary sensorimotor complex localized to the antero-medial part, the SMA, the left lateral premotor cortex, the left inferior parietal lobe, the right superior temporal lobule extending to the inferior parietal lobe, the right cerebellum, and the bilateral basal ganglia. Weiller et al. [15] provided further empirical support for active and passive differences in brain activity, finding that when active was compared with passive movement of the right elbow, significant regional Cerebral Blood Flow increases appeared in the inferior SMA, the dorsal ACC, the precuneus, and bilaterally in the posterior part of the putamen.

Activation of these areas is associated with a variety of functions. For example, activation of the primary sensorimotor complex, basal ganglia, dorsal ACC, SMA and cerebellum are all associated with the execution of motor tasks [16-19]. The putamen plays a role in motor tasks and movement sequences [20], while the inferior parietal lobe is thought to be important in the integration of sensorimotor signals [21].

Consistent with the ideas expressed by other researchers (e.g. [9]), some of these areas are associated with "planning". The ACC has been associated with executive function [22] while the pre-SMA has been linked to functions such as movement selection. Passingham (1993, cited in [4]) found that the lateral premotor cortex is "associated with planning, programming, initiation, guidance, and execution of simple and skilled motor tasks" (p. 10). The SMA has a role, according to Shima and Tanji [23], in planning complex movements and coordinating movements involving both hands. Further, the SMA projects directly to the primary motor cortex and is thought to be functionally involved in simple tasks, motor execution and automatic performance [4].

The question here is whether or not active inputs are different from those that arise from passive-guided movements. Most of the studies mentioned above did not directly address what we consider to be the crux of the active-passive debate – they did not explore active-passive differences in haptic information gathering. Here we are interested in how the possible "purposefulness and responsibility" of active exploration may enhance perception for active subjects or, conversely, how the "freedom from decision-making" in passive-guided conditions may favor passive explorers. This experiment attempts to explore what "active" sensing entails in terms of the recruitment of neural systems. Specifically, this experiment was designed to investigate how the brain receives and processes sensory signals during active and passive-guided exploration of 2D, raised line stimuli.

Our hypotheses were that active and passive-guided exploration would result in similar sensory processes, but that active exploration would result in more cognitive activity as demonstrated by greater mean signal amplitude and cluster size in specific brain areas (e.g. ACC). Further, if participants in the passive-guided conditions are "freed from decision-making", and thus can devote resources to mentally image the stimuli and contemplate the identifying names, greater activation should be evident in some visual and auditory-related areas, relative to active conditions.

2   Method

2.1   Participants

Six men and six women (M = 27.4 years, SD = 8.9 years) participated. All were right-handed according to the Edinburgh Handedness test [24], with no history of neurological illness.

2.2   Stimuli and Apparatus

Stimuli consisted of 13 continuously traceable figures (see Figure 1 for examples) produced using raised line paper. The tangible lines were approximately 1 mm in height. Ten of these figures were used as experimental stimuli, two were used for practice purposes, and one figure was used in a pre-experiment training session. The stimuli were randomly allocated to active/passive conditions between subjects.

Fig. 1. Examples of 2D raised line stimuli used in the experiment

2.3   Procedure

Experimental Task. fMRI sessions were configured in a block design. The duration of a block was 28.1 seconds and each block comprised 5 consecutive scans (TRs), each lasting 5.62 seconds. The 5 consecutive scans gave participants adequate time to build a mental representation of a 2D stimulus, while a 28.1 second "rest" condition allowed the hemodynamic response to return to baseline before the next experimental condition. Each block contained one of three conditions: active exploration, passive exploration, or a baseline "rest" condition.

Ten exploratory blocks were used (5 active and 5 passive). Blocks were presented to half of the participants as: active, rest, passive, rest, active, rest, etc. To avoid order effects, the other half completed the trials in the reverse order (i.e. passive, rest, active, rest, etc.). The start and end of active and passive blocks were signaled to the experimenter via headphones – only the experimenter heard the auditory cues. The experimenter indicated the start of each block to participants by tapping the thigh contralateral to the hand used to explore the stimuli; one tap indicated an active block, two taps a passive block. The start of the resting phase was also signaled by a thigh tap. During "rest" participants lay quietly without any intentional movement. Participants were asked to keep their eyes closed for the duration of their session.

Controls. To hold constant the effects of tactile stimulation and make the active and passive conditions as similar as possible, the experimenter's hand was placed on each subject's hand during every trial, including rest. In the passive condition the experimenter's hand guided the subject's hand around the stimuli, while in the active condition the experimenter's hand lightly gripped each participant's exploratory hand while they moved themselves around each stimulus.

Practice Phase. Participants needed to learn to circumnavigate the stimuli within 5.62 seconds during the experimental task. A practice phase occurred before the experiment proper (and before entering the fMRI machine) in which participants were given instructions advising them that they were required to navigate around each stimulus in approximately 5 seconds. They were shown how to explore the stimuli in a clockwise manner, starting at the bottom left-hand corner of each stimulus, and using the index finger of their right hand. Exploration was to be in a smooth and continuous fashion for the duration of each trial; it was explicitly stated that in the active condition participants were not to repeatedly examine a small section of the stimulus and ignore the rest but to continuously move around the stimulus in the manner just described. During the practice phase participants were given two stimuli that did not appear in the experimental set.

Pre-experiment Training. When participants entered the fMRI machine they were given one more stimulus (that was distinct from the experiment and practice stimuli) so that they could practice exploring stimuli under the limitations imposed by this machine (e.g. variation in body position). They were monitored visually while they performed this task to ensure that it was performed correctly.

Participants were also told that at no time were they to talk while the fMRI machine was operational. Further, they were not required, at any stage, to identify (verbally or otherwise) the figure they were exploring. Identification of each stimulus was not considered important. Post-scanning questioning provided participants with the opportunity to verbalize whether or not they felt they had correctly identified the stimuli – feedback was given at this point.

2.4 Design and Analysis

Functional images were acquired on a 1.5 Tesla MR Siemens scanner using a standard head coil. Spin-echo echoplanar imaging was used to acquire functional images (TR = 5620 ms, TE = 62 ms, flip angle = 90°, matrix = 128 x 128, voxel size = 5.0 mm x 5.0 mm x 5.0 mm) of the whole head. Structural scans were also acquired for each participant (matrix = 256 x 256, voxel size = 1 x 1 x 1, TR = 1940 ms, TE = 3.08 ms, TI = 1100 ms, flip angle = 15°), consisting of 176 coronal slices.

Image processing and statistical analysis of fMRI data were performed with SPM5 (http://www.fil.ion.ucl.ac.uk/spm/software) running on Matlab 7 (The MathWorks, Natick, Massachusetts, United States). Images were converted from the raw DICOM format to the ANALYZE format adopted in the SPM package. The functional images were realigned to correct for head movements, normalized to standard brain coordinates (Montreal Neurological Institute, MNI, standard space) and spatially smoothed with an isotropic 8 mm full-width at half maximum Gaussian kernel.

Individual subject-level statistical analyses were performed using the General Linear Model. One active condition, one passive condition and a rest condition (baseline) were modeled using a canonical hemodynamic response function. Contrast maps of active vs. passive, passive vs. active, passive vs. rest, and active vs. rest were obtained. Group analysis, comparing the means of various measures, was done using a random effects inference with a one-sample t-test.

Cluster detection was performed on all voxels above the $p = .05$ threshold to determine clusters significantly activated between conditions (active > rest, passive > rest, active > passive, passive > active). Significant differences were assumed at $p = .05$, corrected for multiple comparisons using False Discovery Rate. The averaged structural brain of the 12 participants was used to overlay functional maps using MRIcro (http://www.cabiatl.com/mricro/mricro/index.html). Functional anatomical regions corresponding to the MNI coordinates were obtained using probabilistic atlases (Harvard-Oxford Cortical, Harvard-Oxford Subcortical Atlases and the Cerebellar Atlas in MNI152 space after Normalization with FLIRT) integrated with FSLView version 4.1 (FMRIB Software Library, www.fmrib.ox.ac.uk/fsl).

3   Results

To assess differences in activation between active and passive-guided exploration tasks, these conditions were contrasted directly. Active exploration was found to produce a greater number of activated regions (see Figure 2). Clusters of activation in this task appear to fall into 4 general areas: parietal regions (superior parietal lobule and supramarginal gyrus), occipital regions (right lateral occipital cortex and left occipital cortex), cerebellum (right vermis VI) and the frontal area (frontal orbital cortex and ACC). This last area, the ACC, produced the largest area of activation ($K_E = 1163$).

During the passive-guided exploration task, relative to active, significantly greater activation was observed in four regions: right middle temporal gyrus, right parietal operculum cortex, the right lateral occipital cortex, and the left middle temporal gyrus (see Figure 3).

Compared to the rest condition participants' passive exploration of raised line drawings via experimenter-guided hand movements resulted in clusters of activation in both ipsilateral and contralateral postcentral gyri. These areas correspond to the domain of the somatosensory cortex and are expected as they are the primary regions receiving sensory afferents. Activation was also evident within the cerebellum which, although speculative, may be related to proprioceptive afferents from the arm during the passive movement.

The active exploration task, relative to rest, shares similar regions of activation to the passive exploration task. That is, activation of both ipsilateral and contralateral somatosensory areas as well as numerous sub-regions of both cerebellar hemispheres. Activation of the precentral gyri was also detected, which corresponds to motor function.

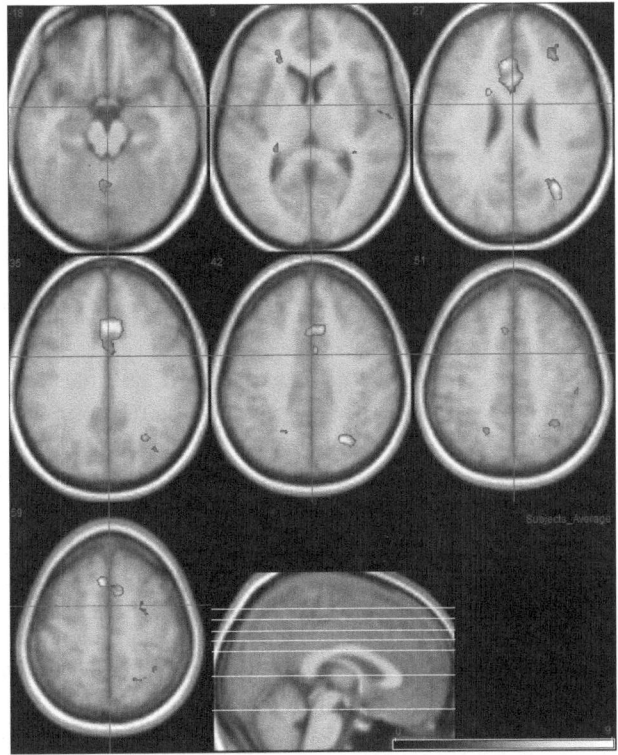

Fig. 2. Areas of significant increase of BOLD (red) when active movements were contrasted with passive-guided movements. Pixels significant at $p = .05$ are displayed. The statistical parametric map was overlaid onto the averaged brain of the 12 participants.

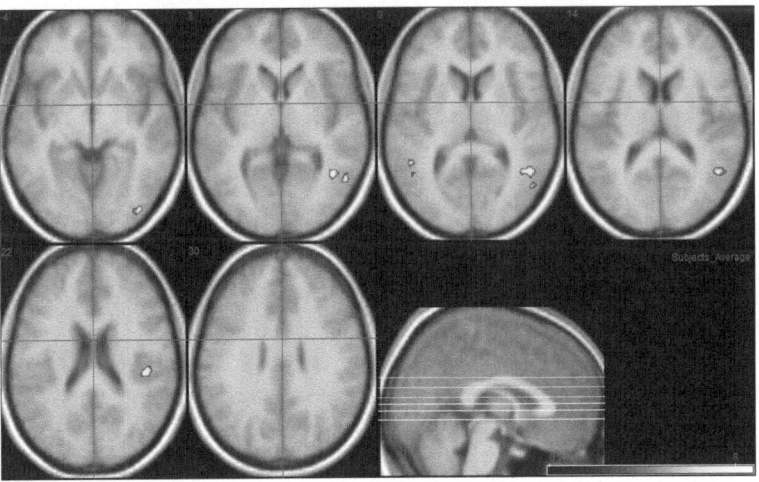

Fig. 3. Areas of significant increase of BOLD (red) when passive movements were contrasted against active movements. Pixels significant at $p = .05$ are displayed. The statistical parametric map was overlaid onto the averaged brain of the 12 participants.

4   Discussion

Consistent with Weiller et al. [15] and Simões-Franklin et al. [13] we found that when active movement was compared with passive there was a significant BOLD increase in the ACC. Stephan et al. [19] suggest that the dorsal ACC is associated with the execution of movement, while Bush et al. [22] proposed that this area is coupled with executive functions such as monitoring and controlling goal-directed behavior. The most popular interpretation of ACC activation, however, is that the magnitude of the ACC's response to a given action is proportional to the perceived likelihood of an error occurring during that action [25]. This seems to serve as "an early-warning system that recruits cognitive control" [25, p. 1120] and adjusts motor acts in response to incorrect execution of those acts [26]. Increased activity in the ACC during active movements relative to passive is, perhaps, explained by the possibility of committing an error in this condition, e.g. falling off the raised line, ending exploration too soon, or continuing to explore "incorrectly". The experimenter kept each participant's finger "on the line" in the passive-guided condition, and thus errors were not possible. Alternatively, the ACC may have been less responsive during passive movements because being "passive" is incongruous with being "goal-directed", and thus passive movements may not take movement errors into account. These differences in the activation levels of brain regions associated with planning are of relevance to findings reported by Magee and Kennedy [9] and others suggesting that active perception in tactile recognition tasks is hampered by planning.

Activation of the frontal orbital cortex suggests that higher order processing was recruited for active but not passive conditions. Although the exact role of this region remains unclear, Bechara, Damasio and Damasio [27] have put forward a theory that the frontal orbital cortex is part of a large-scale system that mediates decision making. The activation of the frontal orbital cortex, and the ACC for that matter, provide preliminary evidence in support of Symmons et al's [12] cognitive load hypothesis (i.e. performance during active exploration may be impeded because of the cognitive strategies an active explorer is obliged to recruit).

An increase in activity in the parietal lobe in active relative to passive-guided conditions may reflect a transient increase in attentional processes [28]. Pardo, Fox and Raichle [29] and Rosen et al. [30] also found activation in the superior parietal lobe during attentionally demanding tasks.

Much like Reddy, Floyer, Donaghy and Matthews [31] we found that the cerebellum was activated more during the active task than the passive. Reddy et al. argue that the cerebellum may generate the prediction that is compared with actual sensory feedback used to coordinate on-going movement. Mima et al. [14] and Gao et al. [32] have made similar attributions of function during active movements.

The lateral occipital cortex was, overall, more active during the active movement task, while the middle temporal gyrus was more active in the passive-guided condition (relative to one another). Both of these areas are part of the ventral ("object") pathway [33], which is specialized for form discrimination and object identification [34-35]. Interestingly, James, Servos, Kilgour, Huh and Lederman [36] and Kitada, Johnsrude, Kochiyama and Lederman [37] have shown that the occipital-temporal pathway is involved in haptic shape perception, particularly of three-dimensional objects. The late visual processing regions (i.e. middle temporal gyrus) contribute to object naming and recognition memory [38].

Due to an absence of evidence for active/passive differences in perceptual judgments (i.e. actual ability to identify stimuli) we can do little more than speculate as to why this area (i.e. the middle temporal gyrus) was more responsive during passive tasks. It may have been that freedom from movement planning and execution leaves the brain with additional resources which it can then devote to other tasks, in this case object identification. We provide this interpretation of our results cautiously in light of past work regarding "intentions" [39]. Specifically, even if active and passive movement patterns were equivalent, the resulting patterns of brain activity and the perceptual consequences may not have been depending on the intention (e.g. "action" or "perception") caused by these movement types. As an example, Zelano et al. [40] found that odorants were processed differently (i.e. gated) depending on whether a person's intention was to breathe or to smell.

Understanding the differences in activation of various brain areas during active and passive exploration is a matter of some interpretation. In brain areas where blood flow during active trials significantly exceeds that of the passive trials, some indication is provided of those aspects of performance that may either give a relative advantage to active performance (as reported in Philips [41], for example) or impede performance compared with that of the passive condition (see [9], for example). For the active task, these areas include the right lateral occipital cortex, cingulate gyrus, right supramarginal gyrus and vermis VI in the cerebellum.

Comparatively greater activation in brain areas during passive performance may mean that these areas become relatively active when no longer inhibited by other areas involved in active performance, or may compensate in some way for the absence of usual active input. Alternatively, such differences could also signal attenuation of activity in those areas during active task performance. If we are considering an argument consistent with that proffered by Blakemore et al. [5], it may be that one of these areas is the site of the hypothesized "forward model". Voss, Ingram, Haggard and Wolpert [42] suggest that the likely site of attenuation of afferent sensation during voluntary actions arises upstream of the primary motor cortex. Thus, although greater activity in areas to do with attention to, and recognition of, objects during the passive condition could arise because there is no forward model against which to automatically compare object attributes, it might be more parsimonious to suggest that a more detailed examination of each object was required.

In conclusion, our findings do not support the view that instances of superiority of passive-guided exploration over active exploration can be attributed to limits of the haptic system's ability to cope with inputs. The implausibility of this idea is demonstrated by the fact that we can use all fingers simultaneously when palpating an object and no "sensory overload" results. We maintain that, based on our results, many comparisons of so-called active and passive performance with tactile stimuli fail to take account of the higher cognitive load that some active exploration entails. We plan further studies to test the hypothesis that active and passive comparisons of tactually presented raised line drawings may not be valid tests of touch, but are instead tests with varying degrees of cognitive load.

References

1. Gibson, J.J.: Observations on active touch. Psychological Review 69(6), 477–490 (1962)
2. Blakemore, S.J., Rees, G., Frith, C.D.: How do we predict the consequences of our actions? A functional imaging study. Neuropsychologia 36(6), 521–529 (1998)
3. Symmons, M.A., Richardson, B.L., Wuillemin, D.B.: Components of haptic information: Skin rivals kinaesthesis. Perception 37(10), 1596–1604 (2008)
4. Grezes, J., Decety, J.: Functional anatomy of execution, mental simulation, observation, and verb generation of actions: A meta-analysis. Human Brain Mapping 12, 1–19 (2001)
5. Blakemore, S.J., Oakley, D.A., Frith, C.D.: Delusions of alien control in the normal brain. Neuropsychologia 41, 1058–1067 (2002)
6. Christensen, M.S., Lundbye-Jensen, J., Geertsen, S.S., Petersen, T.H., Paulson, O.B., Nielsen, J.B.: Premotor cortex modulates somatosensory cortex during voluntary movements without proprioceptive feedback. Nature Neuroscience 10(4), 417–419 (2007)
7. Büchel, C., Friston, K.J.: Modulation of connectivity in visual pathways by attention: cortical interactions evaluated with structural equation modelling and fMRI. Cerebral Cortex 7, 768–778 (1997)
8. Blankenburg, F., Ruff, C.C., Deichmann, R., Rees, G., Driver, J.: The cutaneous rabbit illusion affects human primary sensory cortex somatotopically. PLoS Biology 4, e69 (2006)
9. Magee, L.E., Kennedy, J.M.: Exploring pictures tactually. Nature 283, 287–288 (1980)
10. Richardson, B.L., Wuillemin, D.B.: Can passive tactile perception be better than active? Nature 292, 90 (1981)
11. Richardson, B.L., Wuillemin, D.B., MacKintosh, G.J.: Can passive touch be better than active touch? A comparison of active and passive tactile maze learning. British Journal of Psychology 72(3), 353–362 (1981)
12. Symmons, M.A., Richardson, B.L., Wuillemin, D.B.: Active versus passive touch: Superiority depends more on the task than the mode. In: Ballesteros, S., Heller, M.A. (eds.) Touch, Blindness, and Neuroscience, pp. 179–185. Universidad Nacional de Educacion a Distancia, Madrid (2004)
13. Simões-Franklin, C., Whitaker, T.A., Newell, F.N.: Active touch vs. passive touch in roughness discrimination: An fMRI study. Poster presented at the 9th International Multisensory Research Forum, Hamburg, Germany (2008)
14. Mima, T., Sadato, N., Yazawa, S., Hanakawa, T., Fukuyama, H., Yonekura, Y., Shibasaki, H.: Brain structures related to active and passive finger movements in man. Brain 122(10), 1989–1997 (1999)
15. Weiller, C., Jüptner, M., Fellows, S., Rijntjes, M., Leonhardt, G., Kiebel, S., Muller, S., Diener, H.C., Thilmann, A.F.: Brain representation of active and passive movements. NeuroImage 4, 105–110 (1996)
16. Colebatch, J.G., Deiber, M.P., Passingham, R.E., Friston, K.J., Frackowiak, R.S.: Regional cerebral blood flow during voluntary arm and hand movements. Journal of Neurophysiology 65, 1392–1401 (1991)
17. Kawashima, R., Roland, P.E., O'Sullivan, B.T.: Activity in the human primary motor cortex related to ipsilateral hand movements. Brain Research 663, 2511–2516 (1994)
18. Brooks, D.J.: The role of the basal ganglia in motor control: Contributions from PET. Journal of Neurological Science 128, 1–13 (1995)

19. Stephan, K.M., Fink, G.R., Passingham, R.E., Silbersweig, D., Ceballos-Baumann, A.O., Frith, C.D., Frackowiak, R.S.: Functional anatomy of the mental representation of upper extremity movements in healthy subjects. Journal of Neurophysiology 73(1), 373–386 (1995)
20. Marchand, W.R., Lee, J.N., Thatcher, J.W., Hsu, E.W., Rashkin, E., Suchy, Y., Chelune, G., Starr, J., Barbera, S.S.: Putamen coactivation during motor task execution. Neuroreport 19(9), 957–960 (2008)
21. Mattingley, J.B., Hussain, M., Rorden, C., Kennard, C., Driver, J.: Motor role of human inferior parietal role revealed in unilateral neglect patients. Nature 392, 179–182 (1998)
22. Bush, G., Luu, P., Posner, M.I.: Cognitive and emotional influences in anterior cingulate cortex. Trends in Cognitive Science 4(6), 215–222 (2000)
23. Shima, K., Tanji, J.: Both supplementary and presupplementary motor areas are crucial for the temporal organization of multiple movements. Journal of Neurophysiology 80, 3247–3260 (1998)
24. Oldfield, R.C.: The assessment and analysis of handedness: The Edinburgh inventory. Neuropsychologia 9(1), 97–113 (1971)
25. Brown, J.W., Braver, T.S.: Learned predictions of error likelihood in the Anterior Cingulate Cortex. Science 307, 1118–1121 (2005)
26. Brown, J.W.: Conflict effects without conflict in anterior cingulate cortex: Multiple response effects and context specific representations. NeuroImage 47, 334–341 (2009)
27. Bechara, A., Damasio, H., Damasio, A.: Emotion, decision making and the orbitofrontal cortex. Cerebral Cortex 10, 295–307 (2000)
28. Garavan, H., Ross, T.J., Stein, E.A.: Right hemispheric dominance of inhibitory control: An event-related functional MRI study. Proceedings of the National Academy of Science USA 96, 8301–8306 (1999)
29. Pardo, J.V., Fox, P.T., Raichle, M.E.: Localization of a human system for sustained attention by positron emission tomography. Nature 349, 61–64 (1991)
30. Rosen, A.C., Rao, S.M., Caffarra, P., Scaglioni, A., Bobholz, J.A., Woodley, S.J., Hammeke, T.A., Cunningham, J.M., Prieto, T.E., Binder, J.R.: Neural basis of endogenous and exogenous spatial orienting: A functional MRI study. Journal of Cognitive Neuroscience 11(2), 135–152 (1999)
31. Reddy, H., Floyer, A., Donaghy, M., Matthews, P.M.: Altered cortical activation with finger movement after peripheral denervation: Comparison of active and passive tasks. Experimental Brain Research 138(4), 484–491 (2001)
32. Gao, J.H., Parsons, L.M., Bower, J.M., Xiong, J., Li, J., Fox, P.T.: Cerebellum implicated in sensory acquisition and discrimination rather than motor control. Science 272, 482–483 (1996)
33. Slotnick, S.D., Schacter, D.L.: A sensory signature that distinguishes true from false memories. Nature Neuroscience 7, 664–672 (2004)
34. Rizzolatti, G., Sinigaglia, C.: Mirrors in the brain: How we share our actions and emotions. Oxford University Press, Oxford (2006)
35. Grill-Spector, K., Kourtzi, Z., Kanwisher, N.: The lateral occipital complex and its role in object recognition. Vision Research 41(10-11), 1409–1422 (2001)
36. James, T.W., Servos, P., Kilgour, A.R., Huh, E.J., Lederman, S.: The influence of familiarity on brain activation during haptic exploration of 3-D facemasks. Neuroscience Letters 397, 269–273 (2006)

37. Kitada, R., Johnsrude, I., Kochiyama, T., Lederman, S.J.: Functional specialization and convergence in the occipito-temporal cortex supporting haptic and visual identification of human faces and body parts: An fMRI study. Journal of Cognitive Neuroscience 21, 1–19 (2009)
38. Thangavel, R., Sahu, S.K., Van Hoesen, G.W., Zaheer, A.: Modular and laminar pathology of Brodmann's area 37 in Alzheimer's disease. Neuroscience 152(1), 50–55 (2008)
39. Klatzky, R.L., Lederman, S.J.: Do intention and exploration modulate the pathways to haptic object identification? Behavioural and Brain Sciences 30(2), 213–214 (2007)
40. Zelano, C., Bensafi, M., Porter, J., Mainland, J., Johnson, B., Bremner, E., Telles, C., Khan, R., Sobel, N.: Attentional modulation in human primary olfactory cortex. Nature Neuroscience 8(1), 114–120 (2005)
41. Philips, J.R., Johnson, K.O., Browne, H.M.: A comparison of visual and two modes of tactual letter resolution. Perception & Psychophysics 34, 243–249 (1983)
42. Voss, M., Ingram, J.N., Haggard, P., Wolpert, D.M.: Sensorimotor attenuation by central motor command signals in the absence of movement. Nature Neuroscience 9(1), 26–27 (2006)

Haptics in Between-Person Object Transfer

Satoshi Endo[1,*], Geoff Pegman[2], Mark Burgin[2], Tarek Toumi[1,3], and Alan M. Wing[1]

[1] Behavioural Brain Sciences Centre, University of Birmingham, UK
{s.endo,a.m.wing,t.toumi}@bham.ac.uk
[2] R.U. Robots Limited, Manchester, United Kingdom
{geoff.pegman,mark.burgin}@rurobots.co.uk
[3] LaSTIC Laboratory, Computer Science, University of Batna, Algeria

Abstract. Although object handover between people is a commonly performed task, little about underlying control mechanisms is known. The present study examined haptic contributions in object handover. On each trial one participant held an object and passed it to the other participant at self-selected, fixed or randomly varied positions. In some trials, the receiver wore a glove to attenuate tactile information. The results showed that the passer's time of grip release relative to contact was later when the transfer location randomly varied or when the receiver wore the glove. On the other hand, forces at contact dropped across trials with negligible effects of glove or transfer location. In conclusion, the present study demonstrated that the dyad reduced redundant forces at contact by forming a stereotypical handover movement in a feedforward manner, while the sensory feedback modulates timing of object handover to avoid premature release of grip by the passer.

Keywords: Joint action, Cooperation, Motor control, Object Manipulation.

1 Introduction

People hand over objects to one another smoothly and effortlessly as part of daily life. Coordinating action with another person is not a trivial task, however. In one-person lifting of an object using precision grip (opposed index finger and thumb), grip force normal to each contact surface allows the development of frictional resistance against the vertical load force tangential to the contact surface. In order to prevent the object from slipping, the product of the grip force and the coefficient of friction between the digits and the object must exceed the load force with a small safety margin [1]. In two-person object transfer, in addition, position and time of handover is not certain, and there is a risk of the receiver dropping the object if the passer releases the object too soon. On the other hand, if the passer hesitates in releasing the grip at the appropriate time, the receiver's grip may slip due to unexpected drag from the passer. Despite such complexities in coordinating action with another, casual observation suggests that people are easily able to hand over an object. How do humans perform an object handover when there is uncertainty in the partner's action? One possibility is that people learn the spatio-temporal cues of object handover to prevent miscommunication

P. Isokoski and J. Springare (Eds.): EuroHaptics 2012, Part I, LNCS 7282, pp. 103–111, 2012.
© Springer-Verlag Berlin Heidelberg 2012

between the partners. When a person singly manipulates an object, his/her gaze position proceeds the actual action, reflecting unfolding of the motor plan [2]. A study has shown that people also predictively orient their gaze towards the other person's grasping behaviour as though the observer is performing that action by him/herself [3]. Thus, it is plausible that humans learn and predict movement characteristics of their cooperative partner in order to perform object handover.

Thus, the present study investigated the effects of uncertainly about the partner's movement on grip forces during transfer of object possession between a passer and receiver. In this study, participants were asked to pass an object from one to the other while uncertainty in the task was manipulated. In order to examine tactile effects, we reduced tactile sensitivity for the receiver in grasping object, by the receiver wore a glove on the receiving hand. The predictability of the passer's movement was varied by instructing the passer to transfer the object to a fixed location repetitively or to varying locations in a random order. It was hypothesised that partial loss of haptic feedback about the object or increased uncertainty about the partner's movement would result in higher interaction force, higher grip force at contact and increased contact period of the dyad.

2   Method

2.1   Participants

10 right handed participants were recruited at the University of Birmingham. The average age of the participants was 31.4 years (SD = 5.6 years) and five were female. The participants were randomly paired and assigned to the roles of Passer or Receiver for the duration of the task. The Passer was defined as the person who brought the object from a starting location to a transfer location. The Receiver then took the object out of the Passer's hand at this transfer location and placed it on the final location.

2.2   Apparatus

The object was a custom-made 3D printed symmetric plastic structure in which were mounted three 6 DoF force/torque (FT) sensors (see Fig. 1a). The object was 13 cm in length, 6 cm in height and 2.5 cm in width at the ends and its total weight was 150 g. Pairs of participants were asked to use precision grip (thumb pad opposing pads of index and middle fingers) to grasp the sides at each end of the object indicated by a 3 cm x 3 cm square. Two FT sensors (ATI Nano17, USA) mounted under the grip surfaces at each end recorded grip force of each partner. A third FT sensor (ATI Nano43, USA) was placed in the middle of the object to record the interaction force between the two partners. The data from the FT sensors were sampled at 1000 Hz. A 12 camera Oqus motion-tracking system (Qualisys, Sweden) tracked three light-weight spherical markers (3mm in diameter) placed on the object surface at 200 Hz to record the position and orientation of the object. Markers were also placed on the wrist of the Passer and Receiver to track their motions.

Pairs of participants sat facing each other across a table (Figure 1b). The table surface measured 70 cm x 70 cm and on it three lines (midway between the dyad and 10 cm on either side) were drawn each indicating a possible object transfer location. These transfer locations were chosen from a pilot study which revealed that people were likely to pass an object around the midpoint of the workspace with SD of less than 5.0 cm. Thus, the 10 cm shift of a transfer location was expected to be perceptible to both partners. The transfer locations were indexed with the number ascending from the closest to the furthest from the Passer in order to indicate to him/her the transfer location for each trial. There were also two lines used as start and transfer locations for the object 5 cm from either edge of the table.

In each trial, the transfer location was communicated to the Passer via a computer monitor which displayed the location number. The monitor was placed behind the Receiver so as to be only visible to the Passer. The object handover was paced using a metronome which played eight tones at 1 s intervals. The first four tones were for preparation, and on the fifth tone the Passer grasped the object and initiated the transfer. On the sixth tone, the Passer handed the object to the Receiver at a designated location. On the seventh tone, the Receiver placed the object on the final location. The

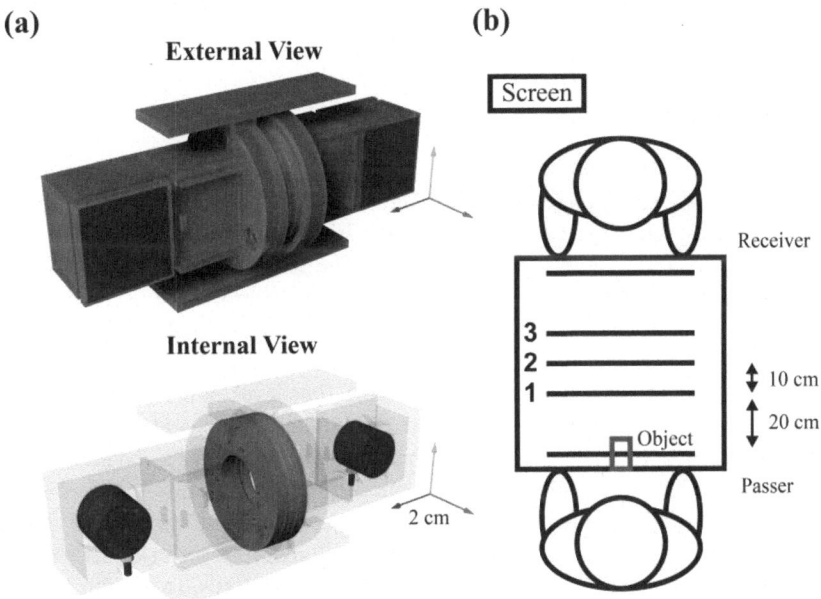

Fig. 1. (a) Drawing of the device. Each participant grasped the object on the square pads at each end. (b) Workspace viewed from the above. On a table, there were three lines indicated possible transfer locations with the designated number written next to these lines. A line at each end of the table indicated a starting/final location.

eighth and final tone indicated the end of trial. The metronome was generated using a custom-made program in MATLAB (MathWorks, USA).

2.3   Procedure

Pairs of participants sat on height-adjustable chairs and faced each other across the table. Instructions were given to both partners at the same time, however conversation or other explicit interaction between the partners was discouraged. The participants were instructed to transfer the object from a starting location to a final location in time with a metronome (see Design for detail). To begin each trial, the experimenter placed the test object on the starting location in front of the Passer. The object was centered at the longitudinal axis on the centre of the starting location. At the beginning of each trial, the Passer placed thumb and index finger of the right hand near the grasping surfaces so as to perform a tripod grip (thumb opposing the index and middle fingers) and waited for the metronome tones. The Receiver rested the right hand on the final location. The participants were allowed to practice with a small hollow cube (5 cm x 5 cm x 5 cm) until they felt comfortable with the keeping their movements in time with the metronome. When the participants were ready, a transfer location was displayed on a monitor and the metronome started. At the end of the trial, the experimenter returned the object to the starting location for another trial. The experiment took approximately 1h.

2.4   Design and Analyses

This study was a 2 x 3 within-subject design. The dependent variables comprised the peak interaction force applied on the longitudinal axis of the object during dyad contact (see Fig. 2). The first contact time of the Receiver was detected from the first moment of grip force increase by this person. The release of the grip by the Passer was defined as the first frame at which the force became zero after the contact.

For the analyses, the first factor was Use of Glove (by the Receiver) which was expected to affect his/her haptic sensitivity to the object. The second factor was Transfer location indicated to the Passer. In the first level of this factor, the Passer was instructed to transfer an object repetitively to the same location (Fixed transfer). A dyad performed 10 trials for each of three transfer locations 3 separate blocks of trials. In the second level, the transfer location pseudo-randomly changed between trials so that the total number of location occurrences was the same as for the Fixed transfer but the order of locationss was random (Random transfer). The total of 30 trials in this level was also separated into 3 blocks of 10 trials. In the third level, the participants performed the task naturally without the transfer location being specified (Natural transfer). This condition was run first to prevent any carryover effect from the conditions which instructs the Passer about the specific transfer location. The order of the bare hand and glove conditions was counterbalanced and presented as two separate blocks. The remaining blocks were randomly administered subsequently. In total, a dyad performed 140 trials administered over 14 blocks.

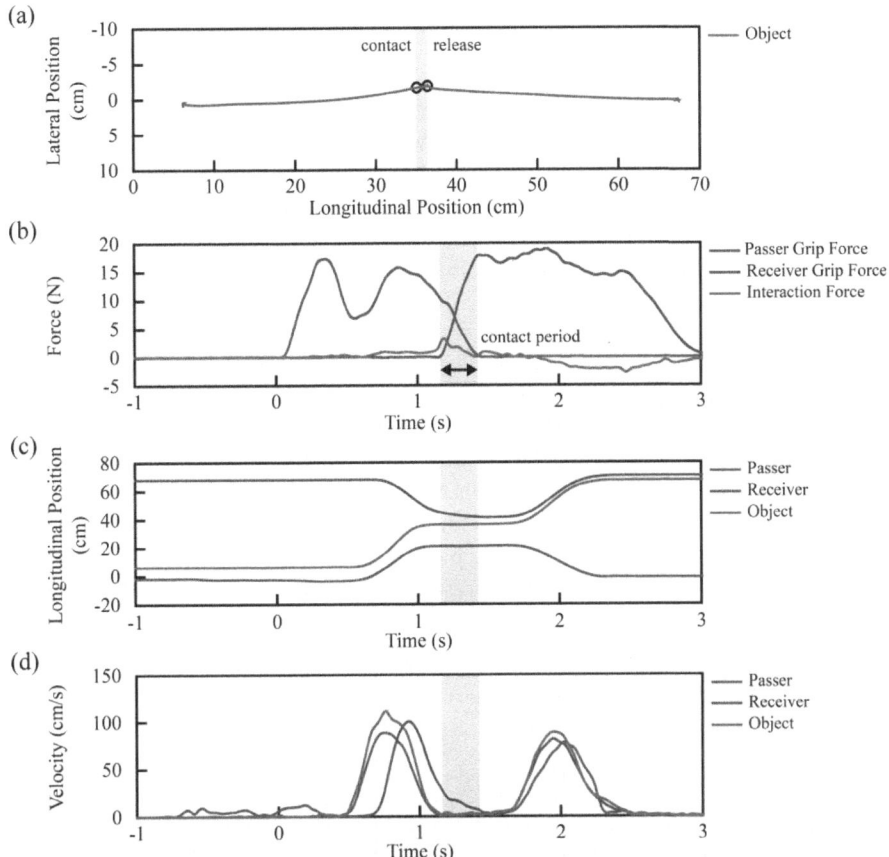

Fig. 2. (a) A single trial example of the workspace as viewed from above. Shaded areas indicate the contact period of the dyad which was defined as the time between the grip force increase by the Receiver and full grip release by the Passer. Values are set to the edge and center of the table for the longitudinal and lateral axis, respectively. (b) Grip forces and interaction force profiles over time. Positive force traces indicates compression. (c-d) Positions and velocities of the Passer, Receiver and object.

3   Results

3.1   Grip Force Modulation and Release Time of the Passer

Fig. 3 depicts average grip force of the Passer and Receiver at around the time of the initial contact. A start of grip force descent was observed 168.1 ms (SD = 39.3 ms) after the initial contact with the Receiver. During the contact period, the partners maintained the net grip force of 16.6 N (within trial SD = 2.8 N) between the partners.

A difference between the experimental manipulations was found in the variance of the grip force profile (shaded area in Fig. 3). A careful inspection shows that in each condition there is an increase in variability shortly before the grip was fully released by the Passer. This is greatest in the Random transfer and likely reflects variability in the release time of the Passer, given that the standard deviation of the release time was larger with the Random transfer (SD = 198.4 ms) compared to the Fixed transfer (SD = 172.8 ms). Passing an object naturally was found to be least variable (SD = 141.5 ms). Friedman's test indicate a trend in the sizes of standard deviations (p = .07). On the other hand, there was no difference in release time variability due to Use of Glove (p = .66).

As Fig. 3 shows, there was a strong reciprocal relation between the grip forces of the partners such that the Passer's grip force reduced as the Receiver's increased to complete the object handover during the contact period (r = - .97). To understand the efficiency of the grip force modulations between the partners, this negative relationship of grip forces between the Passer and Receiver was evaluated using a simple linear regression. The slope coefficients were then analysed using a repeated-measures ANOVA. The statistical test indicated that there was a main effect for Use of Glove in slope size, $F(1, 4) = 7.43$, p = .05. The slope coefficient for with and without Gloves were -1.81 and -1.34, respectively, meaning that the Receiver was more responsive to grip force change of the Passer when he/she was wearing a glove.

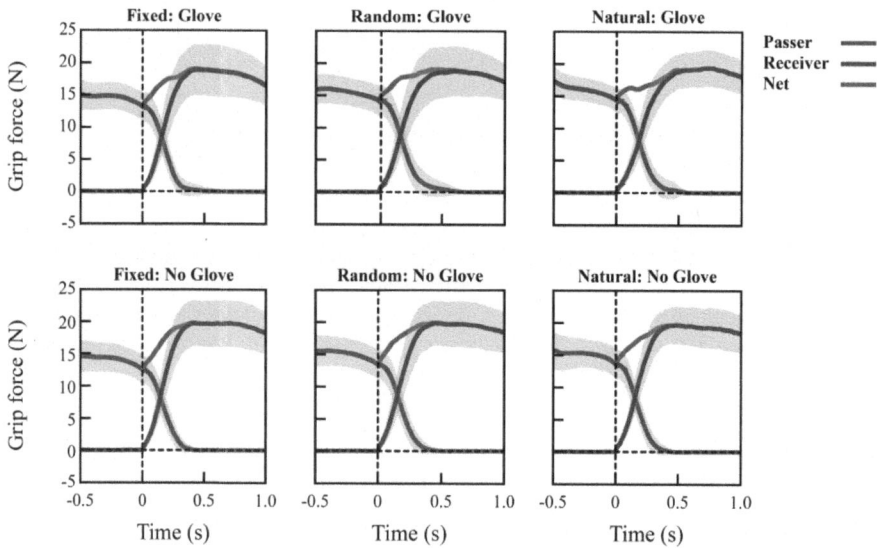

Fig. 3. Average grip force profiles of the Passer and Receiver during contact period. The grip force profiles are time-locked at the moment of a contact with the Receiver (Time = 0 seconds). Shaded areas indicate 1 standard error. The dotted lines are the summed grip force of the Passer and Receiver. Note there is a considerable larger variability before the Passer fully releases the grip when the transfer location was random (middle column) and when the Receiver was wearing a glove (top row).

3.2   Grip Force, Interaction Force and Contact Period

The grip force of passer at the moment of contact by the Receiver was analysed using a 2 x 3 repeated-measures ANOVA with Use of Glove and Transfer location as factors. This indicated that there was a trend towards a main effect (p = .06), with grip force during Natural transfer being highest (Fig. 4a). Use of Glove (p = .17) and Interaction effects (p = .88) were not statistically significant. Given that the Natural transfer was administered first in the experiment, we further investigated the order effect on the strength of the grip force. The total of 140 trials from each dyad was averaged to create 10 epochs of 14 trials each and analysed using one-way ANOVA. The results revealed a reliable practice effect over the course of experiment, F (9, 36) = 2.71, p < .02, such that the grip force gradually reduced across Epochs (Fig. 4b).

The peak interaction force change due to Transfer location and Use of Glove was analysed using a 2 x 3 repeated-measures ANOVA (see Fig. 4c). The statistical test indicated that there was a main effect of Transfer location, F(2, 8) = 9.76, p < .01. The interaction force was highest in Natural transfer (1.08 ± 0.26 N), and less during Random transfer (0.78 ± 0.26 N) and Fixed transfer (0.77 ± 0.9 N). A post-hoc test confirmed a difference between Natural and Fixed transfer (p < .05). No main effect for Use of Glove (p = .24) or interaction effect (p = .48) was observed. Similar to the grip

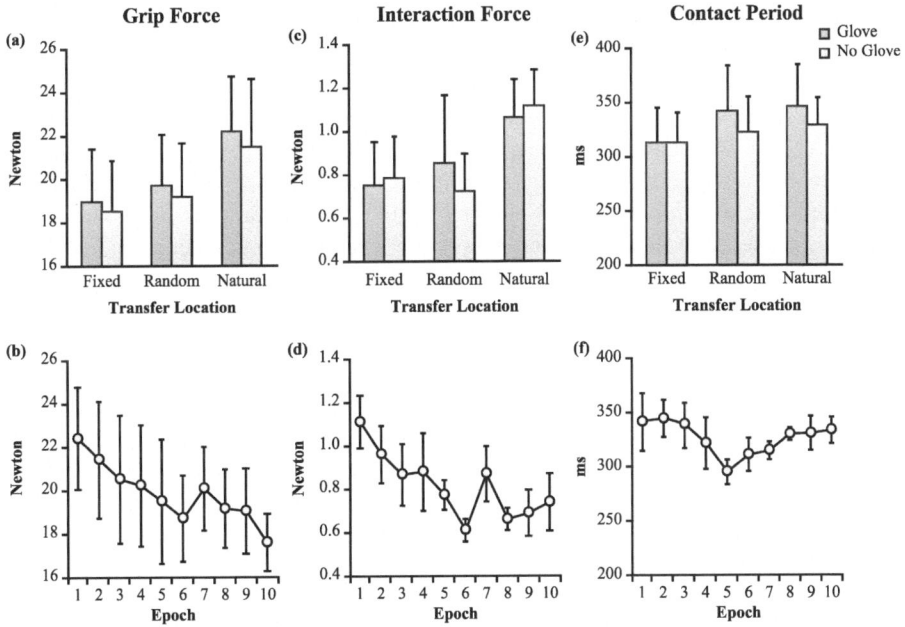

Fig. 4. (a) Grip force of the Passer at the moment of contact with the Receiver. (b) Grip force change over the course of the experiment when divided into 10 epochs. (c) Peak interaction force during contact period. (d) Change of peak interaction over 10 epochs. (e) Contact period of the partners. (f) Change of contact period over 10 epochs. All error bars indicate one standard error.

force measure, a reduction of the interaction force over the course of experiment was observed, $F(9, 36) = 4.597$, $p < .001$ (Fig. 4d).

Concerning the contact period, on average, the Passer fully released his/her grip 323.1 ms (SD = 24.1 ms) after the initial contact with the Receiver (Fig. 4e). A repeated-measures ANOVA showed that there was no main effect for Transfer location ($p = .52$). On the other hand, there was a main effect for Use of Glove ($p < .05$), such that the contact period was longer when performing the task the gloves (Glove: 334.3 ± 26.3 ms, No Glove: 324.0 ± 22.9 ms). In contrast to the force measures, no reliable change over the course of the experiment was observed ($p = .63$).

4   Discussion

The main objective of this study was to investigate role of tactile feedback and uncertainty about the passer's behavior for object handover. Our results showed a random change in a transfer location and the reduced haptic sensitivity influenced temporal aspects of grip force coordination during the direct contact of the dyad. In addition, we found learning effects on force profiles as their sizes gradually reduced over the time-course of the experiment when the object handover was repeated.

Previously, Mason and MacKenzie [4] studied grip force profiles of an object handover when the movement was initiated by passer or receiver. Their study showed that that the grip force of the passer was relatively insensitive to the experimental conditions. Thus, the contact grip force of the passer was similar whether the passer placed the object on a static receiver's hand or the receiver snatched the object from the static passer's hand. Our results may be seen as consistent with this study, in that neither change in quality of haptic feedback with the use of a glove, nor in transfer locations had a significant effect on the grip force of the passer at contact. However, our cross-trial grip force analysis showed a gradual reduction of the passer's grip force at contact as well as the peak interaction force during contact period over the time-course of the experiment regardless of these changes. These findings indicate that the passer used sensory feedback about the task dynamics to modulate the grip force safety margin in anticipation of collision with the receiver [5].

However, random change in transfer location and reduced haptic feedback due to the use of glove affected temporal aspects of grip force modulation at the end of the contact period of the dyad. One plausible interpretation of the present results is that aspects of initial contact are controlled in a feedforward manner by a passer who forms a stereotypic movement defined by temporal cues about task partner [6]. A larger degree of feedback control by the passer during the contact may have prolonged the contact period when the quality of the haptic feedback was reduced or when the location of handover unpredictably changed.

The detailed analyses of the passer's grip force profile highlighted a noticeable variance around the time when he/she released the grip depending on how the transfer location was specified. In particular, the random specification of transfer location impeded the performance in terms of grip release time variability of the passer, while the natural passing of object was associated with the least variability in release time.

Perhaps, a lack of coordination between the dyad affected the decision of timing at which the passer released the object. We believe that this moment of release is decided based on haptic feedback about the stiffness of the object-receiver linkage. Therefore, we aim to further investigate the force/torque profiles in more detail at the object passing moment to understand the nature of sensory feedback with which a passer perceives he/she can safely release an object in a receiver's hand.

Acknowledgement. This work was supported by the CogLaboration project within the 7th Framework Programme of the European Union, - Cognitive Systems and Robotics, contract number FP7-287888, see also http://www.coglaboration.eu/

References

1. Westling, G., Johansson, R.S.: Factors Influencing the Force Control During Precision Grip. Exp. Brain Res. 53, 277–284 (1984)
2. Land, M., Mennie, N., Rusted, J.: The roles of vision and eye movements in the control of activities of daily living. Perception 28, 1311–1328 (1999)
3. Rotman, G., Troje, N.F., Johansson, R.S., Flanagan, R.J.: Eye Movements When Observing Predictable and Unpredictable Actions. J. Neurophysiol. 96(3), 1358–1369 (2006)
4. Mason, A.H., MacKenzie, C.L.: Kinematics and grip forces when passing an object to a partner. Exp. Brain. Res. 163(2), 173–187 (2005)
5. Turrell, Y.N., Li, F., Wing, A.M.: Grip force dynamics in the approach to a collision. Exp. Brain. Res. 128, 86–91 (1999)
6. Knoblich, G., Jordan, J.S.: Action coordination in groups and individuals: Learning anticipatory control. J. Exp. Psychol. Learn. Mem. Cogn. 29(5), 1006–1016 (2003)

New Control Architecture Based on PXI for a 3-Finger Haptic Device Applied to Virtual Manipulation

Ignacio Galiana, Jose Breñosa, Jorge Barrio, and Manuel Ferre

Universidad Politécnica de Madrid, Centre for Automation and Robotics (UPM-CSIC)
Jose Gutiérrez Abascal 2, 28006 Madrid, Spain
{ignacio.galiana,jordi.barrio,m.ferre}@upm.es,
jose.brenosa@gmail.com

Abstract. To perform advanced manipulation of remote environments such as grasping, more than one finger is required implying higher requirements for the control architecture. This paper presents the design and control of a modular 3-finger haptic device that can be used to interact with virtual scenarios or to teleoperate dexterous remote hands. In a modular haptic device, each module allows the interaction with a scenario by using a single finger; hence, multi-finger interaction can be achieved by adding more modules. Control requirements for a multifinger haptic device are analyzed and new hardware/software architecture for these kinds of devices is proposed. The software architecture described in this paper is distributed and the different modules communicate to allow the remote manipulation. Moreover, an application in which this haptic device is used to interact with a virtual scenario is shown.

Keywords: Haptic devices, Remote manipulation, Teleoperation, Multifinger devices.

1 Introduction

Haptic devices are mechatronic systems designed to exert forces to a human user that mimic the sensation of touching or manipulating real objects. Usually, kinesthetic haptic interfaces not only exert forces to the user, but also capture the position and movements of the hand or fingers. From another point of view, haptic devices can be understood as a bidirectional communication channel between a human and a machine, where the information involved in a manipulation task is exchanged as forces or attitudes. Not only they are able to supply big amounts of information to the end-user, but also this information is delivered in a very intuitive, natural and effective way: forces appeared in the virtual or remote environment are reflected as forces to the end-user. Thanks to this, haptic interfaces offer exceptional opportunities in the field of virtual reality. Immersion in virtual environments gets highly enhanced when taking advantage of the user's haptic perception.

In order to exploit these intrinsic features that haptic interfaces have, it is very important to allow the user to interact with the virtual environment by means of several fingers: It seems obvious that exploring or manipulating a scenario using one finger

P. Isokoski and J. Springare (Eds.): EuroHaptics 2012, Part I, LNCS 7282, pp. 112–123, 2012.
© Springer-Verlag Berlin Heidelberg 2012

results in a much poorer experience than doing it with the whole hand. But allowing multi-finger haptic interactions is not a simple task, so the vast majority of the commercially available devices are designed for one finger which can lead to interesting applications related to exploring the virtual or remote environment [1],[2],[3], but may not be enough for precise manipulation such as grasping.

An immediate solution could be the use of off-the-shelf haptic interfaces. However, it is possible to observe that nearly all of the commercially available devices were conceived for a single finger or contact point. It is a well known problem [4] that, when trying to put together two or more conventional haptic devices, many problems come up because of collisions between mechanical structures, greatly reducing the workspace and making it non-viable. This is a well known problem that led researchers to design specific multifinger devices i.e. [5],[6],[7],[8],[9],[10],[11].

In the following sections, new control architecture that satisfies all requirements for a 3-finger haptic device is presented and described. In section II, a quick introduction to the haptic interface will be done, and the basic requirements will be summarized, then, in section III the chosen electronic hardware for implementing the controllers will be described. An example of a 3-finger virtual manipulation is described in section IV and, finally conclusions are summarized in section V.

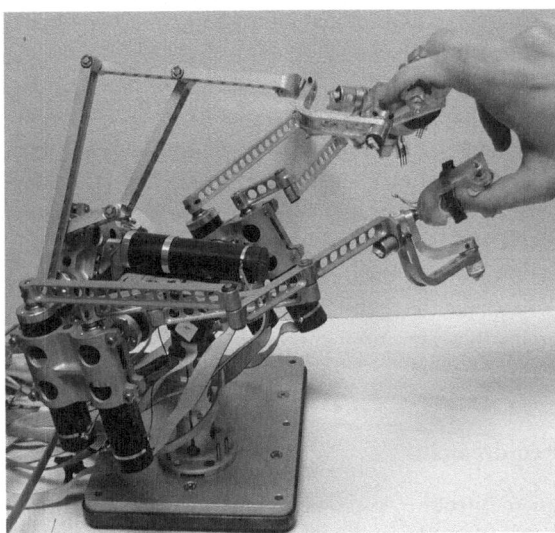

Fig. 1. Multifinger Haptic Interface. Force feedback is generated in all directions for three fingers.

2   System Requirements

In this section, the haptic device used will be briefly described, most relevant requirements for haptic devices in general and some specific requirements for this device will be listed.

2.1   3-Finger Haptic Device

The haptic device that will be used for this study has been designed as a modular system in which each finger is one module. Each module has 6 Degrees of Freedom (DoF) as shown in Fig.1. 3 of them are actuated to exert forces in any direction and the last 3 DoFs, corresponding to the gimbal structure, are just measured and allow the user to orientate the finger in any direction.

The configuration of these three modules to conform the three finger haptic device is a result of an ergonomic study that assures a comfortable workspace for the user minimizing the collisions between each module and guaranteeing a force of 3N in any direction in every point of the total workspace. More details about this study can be found at [11].

The resulting haptic device has a total of 18 DoF, which results in a very high number of signals and in demanding computational requirements to assure a realistic interaction with the virtual environment.

2.2   General Requirements for Haptic Devices

Haptic devices allow users to interact with virtual scenarios and to perceive the changes produced on it by means of their different senses. For the system presented in this paper, users can receive multimodal information as a combination of images (visual), sounds (aural) and force feedback (haptic). In order to interact with the user in a realistic way, different considerations have to be taken into account for each sense or mode (visual, aural and haptic).

It is well known that 50Hz frequency is required to reach smooth transitions in graphics. So, visual information should be refreshed at least at 50Hz to get realistic monocular graphics, if stereoscopic vision is required, this frequency should be increased to at least 100Hz. Force feedback has the higher requirements for the control system not only to provide enough realism but also to assure stability, 1 KHz update rate is necessary to achieve the required haptic fidelity [12].

In order to satisfy these requirements, a real-time control architecture has to be used.

2.3   Specific Requirements

The control design architecture described in next sections is designed for a modular 3-finger haptic device called Masterfinger-3. This device is a three-finger haptic device designed to manipulate objects in virtual environments or to control robotic multi-finger hands [11].

To better understand the chosen control architecture, in this section, all the signals that have to be processed by the controller will be described.

Each finger module consists of a five-bar linkage mechanism with three actuated degrees of freedom (DoF) that allow exerting forces in any direction within the workspace, and three non-actuated DoFs at the end-effector that allow a free orientation of the finger tip with no torque transmission thanks to a gimble configuration [4], [11] shown in Fig.1.

In addition, magnetic encoders are placed in the gimbal to determine the fingers orientation and some force sensors are placed in the thimble to monitor the forces perceived by the user when interacting with the virtual environment [13][14].

To sum up, every module has the following sensors:

- 3 motors each with each its PWM signal, current measurement and encoder.
- 3 magnetic sensors at the gimbal.
- 4 force sensing resistors at the thimble.

Hence, the controller has to manage the following input and output signals to control the system:

PWM.

To actuate the DC motors, a PWM is calculated using the duty cycle provided by the PowerPC and a 10-MHz clock signal, current measurement is also needed to close the current loop.

This PWM is sent to the motors through their power wires (positive and negative in each motor). For this purpose, two output power connections are needed.

Motor Encoder.

The encoder provides a simple square signal that is further processed by the control system. There are three channels (A, B, I) and two power wires (Vcc, GND).

Channels A and B are phase shifted signals that are used to determine the direction of rotation. Channel I (Index) is used as reference point for precise determination of rotation angle.

The line driver produces complementary signals of each channel, so a total of eight wires are needed to manage encoders: two power outputs and six signal inputs.

Gimble encoders.

The orientation of the fingertip is obtained by magnetic encoders located at the gimble's rotational axes.

The encoders used for this system are MA3-P12 from US Digital. The MA3 is a miniature rotary absolute magnetic shaft encoder that reports the shaft position over 360° with no stops or gaps.

MA3-P12 produces a 12-bit PWM output, so it adds two output power connections to the system requirements and one signal input for the PWM angular position.

Force Sensors (output).

The contact sensors located inside the thimbles provide information about the force perceived by the user.

Four Tekscan's FlexiForce A201 are used in each thimble, one of them located under the finger, one at the end of the finger, and one on either side of the finger. The

FlexiForce sensing area is treated as a single contact point to estimate the normal and tangential forces perceived by the user [11].

This resistive sensor needs one voltage input and a voltage output for measuring the force; therefore a power output and one input signal are needed.

As shown in fig.2, a total of 27 signals for sensor information and 3 control signals are required to manage each finger. Three fingers compose this device, so a total of 81 signals have to be managed by the control system. The signals of the 3-finger haptic device will be processed in a single computer.

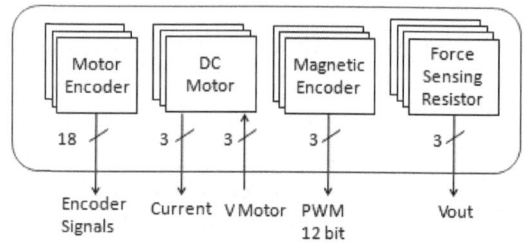

Fig. 2. Control Signals for one module of the 3-finger Haptic Device

As the system interacts with a virtual collaborative environment, a distributed architecture is required. In this architecture, each module is connected using an Ethernet connection with an IP/UDP protocol when in a local network or the IP/BTP [15] protocol when in a nonlocal network. This connection will satisfy the frequency required by MasterFinger3 and should also be easy to install, maintain and update.

The equipment used must be powerful enough to process all the required signals and to close the control loop with a frequency that guarantees a good performance of the system (1kHz) [12].

It has to be also considered that although electronics could be designed specifically for one module and managed with a centralized control, it is preferable to use a single computer where everything can be integrated more easily and establish connections among different degrees of freedom via software.

The following sections will address the hardware chosen for the control equipment and the software architectures designed for that purpose.

3   Electronic Hardware

Due to the high system requirements for Multifinger haptic applications and to the high number of signals (81) that have to be managed and controlled, selecting the correct hardware is essential to guarantee a high performance.

For this three fingered haptic device a PXI chassis system from National Instruments (Fig.3.) [16] was chosen. This system consists on three parts: A Real Time Controller, a FPGA and the power electronics.

Both the FPGA and the Real Time Controller can be programmed in LabView graphical development environment thus reducing the overall programming time for a non-expert in VHDL hardware definition language and C/C++ programming.

While the user is interacting with the virtual scenario by means of the described Multifinger interface, the Scenario Server calculates the resulting interaction force that has to be exerted to the user.

The interaction Force with 'the virtual world' is calculated by a laptop running Windows by using the PhysX module by NVIDIA which has the advantage that most of the physical calculus can be processed very fast by the GPU.

3.1   Real Time Controller

The PXI chassis incorporates a 1.9 GHz Dual-Core RealTime embedded controller NI PXIe-8102 running VxWorks.

This Interaction force calculated by the Scenario Server is sent to the Real-Time Controller by the UDP interface.

The controller is used to program the complex mathematic calculations (Jacobian, Kinematic Equations, Controller Equations, etc.) to transform the interaction force to a current setpoint in the actuators and to calculate the position (x,y,z) from the encoders position of the actuators.

The RT Controller communicates with the FPGA via a MXI high-speed cable connection to receive the encoders' position and to send the current set point.

3.2   FPGA

The PXI Chassis communicates via a high-speed MXI connection with a FPGA Virtex-5 integrated in a compact module that allows connecting up to 14 modules with different functionalities directly to the FPGA (Acquisition of Analog and Digital Signals, power electronics, etc.) (Fig.4.).

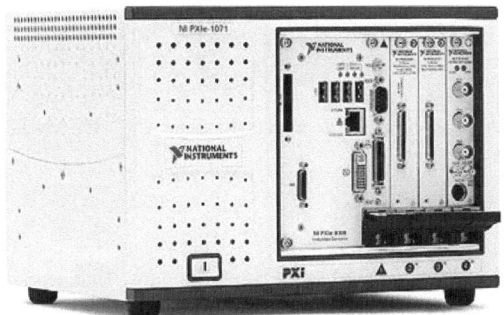

Fig. 3. PXI System with RT Controller

The FPGA Virtex-5 is configured using LabView graphic programming language. For this application, the FPGA is used to:

1. Acquire and process Signals: encoders to calculate position and velocity of the device and current measurement.
2. Actuators' Current Loop: A PI controller is implemented to control the current of the actuators and generate a PWM in each actuator to exert to the user the required force.
3. Communication with the Real Time Controller: The encoder position of the actuators is sent to the RealTime Controller and the Current Command required to exert the interaction force to the user is received from the RT Controller.

Fig. 4. FPGA Virtex-5 with Chassis to connect different modules

3.3   Power Electronics

For each actuator a Full H-Bridge Brushed DC Servo Drive Module (Fig.5.) from NI directly connected to the FPGA was used.

Each of these modules provides a measurement of the current that circulates over the DC actuator, has a data acquisition for the Encoder and provides power to the DC actuators used.

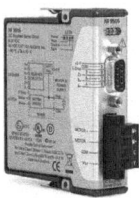

Fig. 5. Full H-Bridge Brushed DC Servo Drive Module

4   Software Architecture

Software architecture design is an important decision that must be taken carefully when a new platform is developed. Decisions as the union between haptic and visual interfaces, or the software election or how the virtual scenarios are recreated, should be analyzed. Also, software architecture should be as independent as possible from the hardware architecture.

There are two important parts in the architecture proposed:

1. Low level software in charge of controlling the haptic device.
2. High level software, responsible for updating the scenario, calculating force feedback and processing visual and aural information.

The suggested solution shown in Fig.6., splits the software into different blocks that run in the different electronic devices as described in the previous section and communicate via an Ethernet connection.

The software solutions proposed for low level blocks have to be developed ad hoc, as they are very dependent on the device's hardware and it was a development by the UPM. Meanwhile, a wide range of commercially available software products can be used for high-level blocks, as they are performed in an abstract level.

In the next paragraphs, explanations on which, why and how this blocks are implemented are shown.

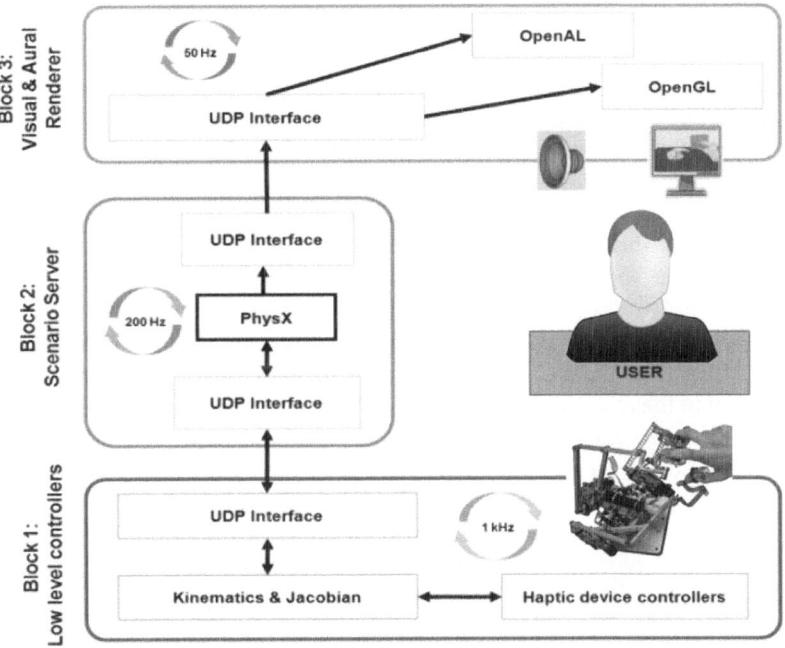

Fig. 6. Software architecture for controlling a 3-finger haptic device

Block 1: Low level controllers.

This block is in charge of processing the measurements from the sensors, calculating the attitudes of the end-effector of each independent module, and controlling the motor currents and torques, and communicating via Ethernet with the immediately upper block.

In order to avoid unwished vibrations in the thimbles, this software must work at 1kHz [12].

Furthermore, this block synchronizes all the other processes and performs kinematic and jacobian calculations. Once the attitudes of the different thimbles are calculated, they are sent to the upper block, the physics engine.

In return, the forces to be reflected are received from the physics engine and the torques to be produced in each motor are calculated by means of the Jacobian calculations.

Delays in this block can lead to inconsistent situations where real fingers are in very different positions from the virtual ones. This can make the haptic interface become unstable, reflect wrong forces, vibrations, etc. Hence, a hard real time system is required here. For this pourpose, National Instruments' hardware runs VxWorks.

The low level controller runs in the PXI platform that, as described in section 3 consists of a Real-time controller to communicate with the other modules via Ethernet and a FPGA to control the system and acquire the different signals.

The Low-level controller has been programmed using LabView graphical development environment (LabView RealTime module for the RT Controller and LabView FPGA module for the Virtex-5 FPGA).

Block 2: Scenario server.

In this module, very heavy calculations are processed. Here, the virtual scenario is created and updated with the data received from the lower block.

Collision detection, interaction calculations and body dynamics are a few of the tasks assigned to this block. It also has to communicate the scenario status to the graphics interface and the interaction forces to be exerted to the virtual fingers to block 1.

The physics engine is in charge of: identifying collisions, estimating external forces resulting from the collisions (including frictional forces) and to produce a correct response to the residual forces [17].

Due to the high requirements of the physics engine, ximplementing ad hoc solutions for this block is very time-consuming and requires big development efforts. To tackle this problem, the adopted solution was to study the software kits commercially available and try to take advantage of their features. In literature, haptic interfaces using commercial physics engines are not very common. This happens mainly because the existing physical engines are often focused on a small set of physical laws and they are usually implemented aiming very accurate but computationally demanding behaviors that cannot be executed in real time.

Some others physics engines can be found working in medical haptic simulators [18]. Some other engines provide a framework where new physical laws can be implemented and included by using software containers.

Related to videogames, many physics engines have been developed recently. They are not optimized for haptics, but they usually have some very interesting features. It is very common to find fluids in videogames, complex rigid bodies and even soft bodies, which gives an idea about the potential of these tools applied in the field of haptic simulation. Some of these engines were tested: ODE, Bullet (which also have the advantage of being open source), Havok and PhysX. Finally PhysX from Nvidia was used because it allocates most of the calculus in the GPU which leads to faster simulations.

Block 3: Graphic and Aural renderer.

This layer's functionality consists on drawing the virtual objects and fingers and reproducing sounds. As it has been said before, it receives the data from the scenario server.

There are several possibilities regarding software for programming the visual interface, for this system two graphic API's were taken into account: OpenGL and DirectX. DirectX is an API developed by Microsoft and full features (as hardware acceleration) are only obtained when running on Windows based OS. However, OpenGL is an open source API that works under nearly any OS. Its features are enough to meet our demands and it also offers high compatibility, something which is very convenient for this project. For the same reasons, OpenAL was chosen for the aural interface.

5   Applications

Following the control structure described in the previous section, this 3-finger haptic device can be also used to manipulate objects in virtual environments or for teleoperating dexterous robotic hands,

In both situations the device control architecture is the same, by just changing the blocks 2 and 3 of the software architecture for a telemanipulator with a 3-finger robotic gripper and their corresponding controllers.

The haptic mounted and connected to a virtual environment allows to control three contact points, interacting with different objects and feeling the shape and reflected forces.

The following figure shows the manipulation of objects with three fingers in the developed virtual environment.

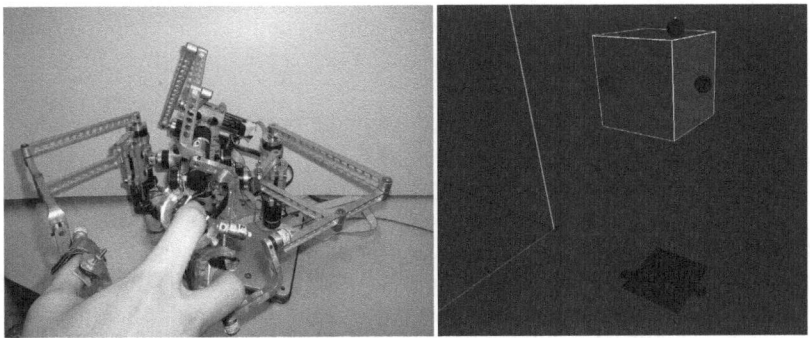

Fig. 7. 3-Finger haptic device to perform advanced virtual manipulation

Future application will focus on using the described device to control a 3-finger robotic hand connected to an industrial manipulator in order to perform precise manipulation tasks¡, an example of this is shown in Fig.8

Fig. 8. 3-Finger hand Robotiq connected to a Kraft Grips manipulator. Future application will focus on using the developed device to control this system.

6   Conclusion

Controlling Multifinger haptic devices adds very high requirements to the system in terms of control frequency, number of signals that are sent over the network and high computation that has to be done in Real Time to assure the required high fidelity of the haptic interaction.

To control this kind of devices, modular software architecture is proposed in this article that divides the problem in different systems: i) the low level controller for each module and the redundant axis considering them as a complete haptic system and ii) the virtual environment supported by the scenario server for calculating the physics interaction and the Visual and Aural rendering module.

The Hardware architecture has also different modules and consists of an FPGA for the low-level controller and a RT controller for managing all the complex calculations (jacobian, kinematics, etc.); this provides a compact and scalable solution for the required high computation capabilities assuring a correct frequency rate for the control loop (1 kHz).

A virtual scenario has been developed using the proposed architecture to manipulate a virtual box by using 3-fingers. Future works will focus on performing real tasks in remote environments by teleoperating dexterous robotic hands.

Acknowledgments. This work has been partially supported by Madrid Community in the framework of The IV PRICIT through the project TECHNOFUSION(P2009/ENE/1679), the TEMAR project under grant DPI2009-12283 from the Spanish Ministry of Science and Innovation (MICINN) and UPM under 'Formación de Personal Investigador'.

References

1. Haption, Inc., Haption–Products (April 2012),
   http://www.haption.com/site/index.php/en/products-menu-en
2. Sensable, Sensable-Products (April 2012),
   http://www.sensable.com/products-haptic-devices.htm
3. Massie, T.H., Salisbury, J.K.: The phantom interface: a device for probing virtual objects. In: Proc. ASME Winter Annual Meeting, Symposium on Haptic Interfaces for a Virtual Environment and Teleoperator Systems (1994)
4. Cerrada, P., Breñosa, J., Galiana, I., López, J., Ferre, M., Giménez, A., Aracil, R.: Optimal Mechanical Design of Modular Haptic Devices. In: IEEE/ASME International Conference on Advanced Intelligent Mechatronics, AIM 2011, Budapest, Hungary (2011)
5. Garcia-Robledo, P., Ortego, J., Barrio, J., Galiana, I., Ferre, M., Aracil, R.: Multifinger Haptic Interface for Bimanual Manipulation of Virtual Objects. In: IEEE International Workshop on Haptic Audio Visual Environments and Games, HAVE (2009)
6. García-Robledo, P., Ortego, J., Ferre, M., Barrio, J., Sánchez-Urán, M.A.: Segmentation of Bimanual Virtual Object Manipulation Tasks Using Multifinger Haptic Interfaces. IEEE Transactions on Instrumentation and Measurement 60(1), 69–80 (2011)
7. Bouzit, M., Popescu, G., Burdea, G., Boian, R.: The Rutgers Master II-ND Force Feedback Glove. In: Proc. IEEE Haptics Symposium, Orlando (March 2002)
8. Frisoli, A., Simoncini, F., Bergamasco, M.: Mechanical Design of a Haptic Interface for the Hand. In: Proc. ASME Design Engineering Technical Conf. and Computer and Information in Engineering Conf., Montreal, Canada, September 29-October 2 (2002)
9. Gosselin, F., Jouan, T., Brisset, J., Andriot, C.: Design of a wearable haptic interface for precise finger interactions in large virtual environments. In: Proc. Int. Conf. World Haptics 2005, Pisa, Italy, March 18-20, pp. 202–207 (2005)
10. Endo, T., Yoshikawa, T., Kawasaki, H.: Collision Avoidance Control for a Multi-fingered Bimanual Haptic Interface. In: Kappers, A.M.L., van Erp, J.B.F., Bergmann Tiest, W.M., van der Helm, F.C.T. (eds.) EuroHaptics 2010. LNCS, vol. 6192, pp. 251–256. Springer, Heidelberg (2010)
11. Breñosa, J., Cerrada, P., Ferre, M., Aracil, R.: Design of an Ergonomic Three-Finger Haptic Device for Advanced Robotic Hands Control. In: IEEE-World Haptics Conference (WHC), Istanbul, Turkey (2011)
12. Hannaford, B., Okamura, M.: Haptics. In: Siciliano, B., Khatib, O. (eds.) Handbook of Robotics, ch. 20, p. 720. Springer (2008) ISBN: 9789-3-540-23957-4
13. Galiana, I., Bielza, M., Ferre, M.: Estimation of Normal and Tangential Manipulation Forces by Using Contact Force Sensors. In: Kappers, A.M.L., van Erp, J.B.F., Bergmann Tiest, W.M., van der Helm, F.C.T. (eds.) EuroHaptics 2010, Part I. LNCS, vol. 6191, pp. 65–72. Springer, Heidelberg (2010)
14. Monroy, M., Ferre, M., Barrio, J., Eslava, V., Galiana, I.: Sensorized Thimble for Haptic Applications. In: IEEE International Conference on Mechatronics, Málaga, Spain (2009)
15. Wirz, R., Marin, R., Ferre, M., Barrio, J., Claver, J.M., Ortego, J.: Bidirectional Transport Protocol for Teleoperated Robots. IEEE Transactions on Industrial Electronics 56(9) (2009)
16. National Instruments (April 2012) (Online), http://www.ni.com/pxi/
17. Melder, N., Harwin, W., Sharkey, P.: Translation and rotation of multi-point contacted virtual objects. In: Proceedings of the World Haptics Conference, pp. 218–277 (2003)
18. Nourian, S., Shen, X., Georganas, N.D.: XPHEVE: An Extensible Physics Engine for Virtual Environments. In: Canadian Conference on Electrical and Computer Engineering, CCECE 2006, pp. 1546–1549 (May 2006)

Combining Brain-Computer Interfaces and Haptics: Detecting Mental Workload to Adapt Haptic Assistance

Laurent George[1,2,4], Maud Marchal[1,2,4],
Loeiz Glondu[3,4], and Anatole Lécuyer[1,4]

[1] INRIA, Rennes, France
[2] INSA, Rennes, France
[3] ENS Cachan, Bruz, France
[4] IRISA, Rennes, France

Abstract. In this paper we introduce the combined use of Brain-Computer Interfaces (BCI) and Haptic interfaces. We propose to adapt haptic guides based on the mental activity measured by a BCI system. This novel approach is illustrated within a proof-of-concept system: haptic guides are toggled during a path-following task thanks to a mental workload index provided by a BCI. The aim of this system is to provide haptic assistance only when the user's brain activity reflects a high mental workload. A user study conducted with 8 participants shows that our proof-of-concept is operational and exploitable. Results show that activation of haptic guides occurs in the most difficult part of the path-following task. Moreover it allows to increase task performance by 53% by activating assistance only 59% of the time. Taken together, these results suggest that BCI could be used to determine when the user needs assistance during haptic interaction and to enable haptic guides accordingly.

Keywords: Brain-Computer Interface, EEG, Force-Feedback, Adaptation, Mental Workload, Guidance.

1 Introduction

A Brain-Computer Interface (BCI) is a communication system that transfers information directly from brain activity to a computerized system [1]. The original goal of BCI is to provide control and communication capabilities for people with severe disabilities. For example, BCI-based spellers enable to spell letters by only using brain activity [2]. Another example is the use of BCI to send commands to prostheses, for example to open and close prostheses by imagination of hand movements [3]. BCI can also be used by healthy users, for instance for enhancing interaction with video games [4]. An approach called "passive BCI" aims at using brain activity information to adapt and enhance the current application without the need for the user to voluntarily control his/her brain activity [5,6]. For example, this approach has been used to adapt virtual environments content [7].

P. Isokoski and J. Springare (Eds.): EuroHaptics 2012, Part I, LNCS 7282, pp. 124–135, 2012.

Adapting interaction modalities according to the user mental state could also be another interesting option offered by passive BCIs.

Haptic feedback has already been used in a BCI system [8–10]. However, to our best knowledge, BCI have never been used for adapting haptic feedback yet. In this paper, we introduce the use of the passive BCI approach in the haptic realm.

We propose to use BCI technologies to adapt force-feedback in real-time. We introduce assistive tools, i.e. haptic guides, which are automatically and continuously adapted to the user's mental workload measured through a passive BCI.

The remainder of this paper is organized as follows. Related work is presented in Section 2. In Section 3 we detail our concept of haptic assistance based on BCI. An implementation of this concept and a user study are proposed in Section 4. We discuss results in Section 5. Finally the main conclusions are summarized in Section 6.

2   Related Work

Our description of related work is subdivided into three parts. First we provide a brief summary of previous work on haptic guidance. Second we present related work in the field of BCI, notably the use of BCI in virtual environments and the passive BCI approach. In the last part we give an overview of how haptic feedback and BCI have already been combined together.

2.1   Haptic Guidance

Haptic guidance can be defined as an interaction paradigm in which the user is physically guided through an ideal motion by a haptic interface [11]. Bluteau et al. [12] have compared different types of guidance and showed that the addition of haptic information plays an important role on the visuo-manual tracking of new trajectories, especially when forces are used for the guidance. Recent work has focused on the adaptation of haptic guidance to maximize learning effect [13].

2.2   BCI

BCI and Virtual Environments: BCI have been early demonstrated as usable for interacting within 3D virtual environments [14]. They have been notably used for navigating in the virtual environment or moving a virtual object [14]. For a comprehensive survey on the combintation of BCI and Virtual Reality and videogames the interested reader can refer to recent surveys [7,14]. BCI have also been used in novel interaction paradigms, where the BCI would not substitute for classical interaction peripherals but complement them. For instance it has been proposed to use the brain activity information for changing the interaction protocol or the content of the virtual world [14], which refers to the "passive BCI" approach.

Passive BCI: The aim of a passive BCI is not to allow the user to send explicit commands to an application but to provide information concerning his/her mental state with the purpose of adapting or enhancing the interaction accordingly [5,6]. For instance, passive BCI have been used in the context of videogames to adapt the way the system responds to commands [15] or to adapt the content of the game itself [7]. In "Alpha WoW" [7] the user avatar form is updated (from elf to wolf) according to the measured level of alpha activity. The aim is to provide the most adequate avatar according to the context (e.g. a detected stress would activate an avatar meant for close-combat). Passive BCI approach has also been used in real driving environment [16]. In their system, Kohlmorgen et al. proposed to use passive BCI to interrupt secondary tasks of a driver when detecting a high mental workload [16].

2.3 Haptic for BCI

Haptic interfaces have already been combined with BCI for providing stimuli and feedback for BCI systems. Vibro-tactile stimuli have notably been demonstrated to be usable for BCI systems. Müller-Putz et al. [8] showed that using vibro-tactile stimuli on fingertips results in brain activity that can be modulated by the user will. Focusing on left-hand or right-hand finger stimulation allows the user to send commands using what is called a "Steady-State Somatosensory Evoked Potential" [8]. Force-feedback has also been used to provide valuable feedback to the user during BCI interaction [9, 10].

However, to the authors' best knowledge, there has been no previous work on the adaptation of haptic feedback thanks to a BCI, using a passive BCI approach.

3 Using Mental Workload to Adapt Haptic Assistance

The concept proposed in this paper consists in using a passive BCI to assess in real-time an index of the user's mental workload and to adapt haptic assistance accordingly.

Mental workload is a generic term which can cover or apply to different and varied cognitive processes and mental states. It could apply for example to a memorization task (e.g. image memorization), driving task, lecture task and cognitive task [17]. In this paper we use the term mental workload to qualify the modification of the user brain activity in relation to the difficulty of a manipulation task. Mental workload index is expected to increase with the manipulation difficulty level. Several EEG markers have been identified as correlated with mental workload, task engagement or attention [16–18]. In [18] the authors proposed to use ratios of activity in specific band-power such as alpha (8 Hz–12 Hz) or theta (3 Hz–7 Hz) bands to compute an index of the user task engagement. More recently, more complex approaches based on machine learning have been used to compute index of user's mental workload based on EEG activity [16,17]. In this paper, we propose to use a BCI system based on these approaches for assessing an index of the user's mental workload during a haptic interaction task.

Haptic assistance systems could benefit from the user mental workload information. Indeed, it could be used to know when the user needs assistance or not. For example, a high mental workload could present a risk in the context of safety-critical haptic tasks. A smart assistance system that interrupts the haptic interaction or toggles specific haptic guides might improve comfort or safety of operations in such conditions. Haptic guides that would be only active when the user presents a high mental workload (e.g. the user is focused on a difficult precision gesture) might improve task performances and learning process.

4   Evaluation

Thereafter, we design and evaluate a proof-of-concept system that toggles a haptic guidance during a path-following task based on the user mental workload. The task consists in following a path in a virtual 2D maze avoiding collisions with borders. An EEG-based BCI is used to compute an online index related to the user mental workload. If the index indicates a high (resp. low) mental workload, the haptic assistance is activated (resp. deactivated).

4.1   Objectives and Hypotheses

Our experiment has two goals:

1. To test the operability of a system that adapts force-feedback based on mental workload measurement;
2. To evaluate the influence of such a system on task performance.

We could make the hypothesis that the adaptation of haptic assistance based on mental workload index would help the user and would result in an increase in task performance.

4.2   Population

Eight participants (7 men, 1 woman) aged between 25-30 *(M = 26.4, SD = 1.9)* took part in the study. None of them had any known perception disorder. All participants were naïve with respect to the proposed techniques, as well as to the experimental setup and purpose.

4.3   Apparatus

The experimental apparatus is shown in Figure 1a. Participants manipulated a 2D cursor through a Virtuose 6D haptic device (Haption, Soulgé sur Ouette, France). A Gtec UsbAmp was used to acquire EEG signals at a sampling rate of 512 Hz. EEG data were measured at positions: Fp1, Fp2, F7, F8, T7, T8, F3, F4, C3, C4, P3, P4, O1, O2, Pz and Cz according to the 10-20 international system. A reference electrode (located at FCz) and a ground electrode (located on the left ear) were also used. This electrode montage allows to cover a large surface of the scalp. Similar electrodes choices were previously used with success for recording mental workload [17].

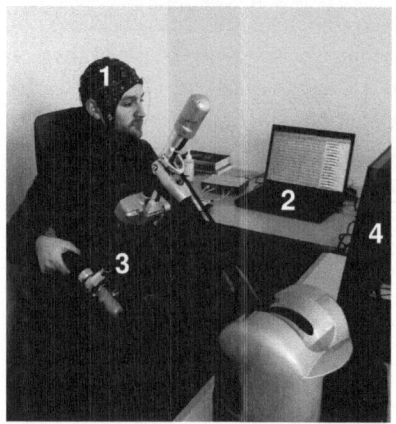

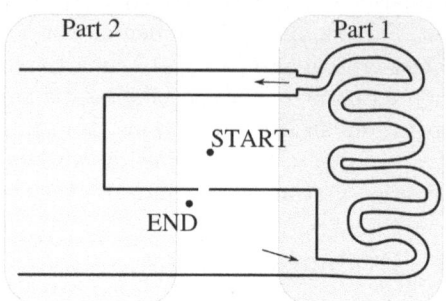

(a) **Apparatus** *1: EEG headset, 2: EEG acquisition, 3: haptic device, 4: LCD screen where the virtual scene is displayed.*

(b) **Virtual scene.** The Path to follow is divided into 2 parts. Part 1 (in red) is more difficult with numerous turns. Part 2 (in blue) is less difficult with less turns and less borders (i.e. less collisions are possible).

Fig. 1. Experimental Apparatus

The Experimental Task consists in following a path by moving a sphere-cursor in a maze avoiding collisions (Figure 1b). The scene is divided in two parts. These two parts aim at exhibiting two different levels of difficulty. The first part is composed of numerous turns and should lead to high difficulty, whereas the second part presents less collision possibilities to show less difficulty.

The Virtual Environment, Haptic Force and Collision Detection. are computed and simulated with the open-source physical engine Bullet. The maze walls are composed of spherical not movable objects.

The Haptic Guide. consists in a repulsive force inversely proportional to the distance of the 2D cursor to the nearest wall (see Figure 2). This haptic guide aims at helping the user to slide between walls avoiding collision. The cursor was colored in blue when the haptic guide was active.

A Mental Workload Index. is computed using OpenViBE software [19]. A technique based on [20] is used. EEG signals are passed through a bank of 4 Hz bandwidth filters centered on all the frequencies between 5 Hz and 12 Hz. A Common Spatial Pattern (CSP) method [21] is then used to compute spatial filters for each of them. Minimum redundancy maximum relevance feature selection is used to select the six most relevant couples of frequency band and spatial filter [22]. A Linear Discriminant Analysis (LDA) classifier is trained on the learning data-set using a moving window of 1 s (overlap=0.9 s).

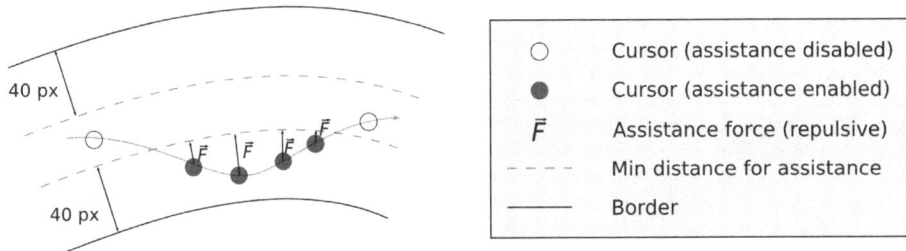

Fig. 2. Haptic assistance. Repulsive force inversely proportional to the distance to the nearest wall: force is null if the distance to the wall is superior to 40 pixels.

The learning data set consists in 2 minutes of EEG activity. The first minute is recorded when the user is performing a simple control task which should lead to a low mental workload: the user has to move the cursor around a rectangle without trying to avoid collisions. The second minute corresponds to a more difficult task: the user has to move the cursor inside a spiral pattern while avoiding collision.

Online values provided by the classifier (1 for high mental workload, -1 for low mental workload) are smoothed on a 1 s window. The median of the last 10 values is provided to the application at a 1 Hz rate. This index is used to activate or inhibit the haptic assistance.

4.4   Procedure

Three conditions were evaluated: **No Haptic Assistance (NO_A)**, **Haptic Assistance based on BCI (BCI_A)** (i.e. assistance activated if the mental workload is above 0), and **Haptic Assistance activated all the time (ALL_A)**. Each participant performed the task in the three conditions and repeated it 3 times. To reduce order effect, the order of presentation was permuted across subjects.

4.5   Collected Data

For each trial and each participant, the mental workload index, the cursor positions and the number of collisions were recorded. The mental workload index and the cursor position were recorded at 10 Hz.

At the end of the experiment, participants were asked to fill a subjective questionnaire in which they had to grade the 3 conditions according to different criteria. We used a Likert-scale where participants could rate the criteria from 1 (very bad) to 7 (very good). Criteria were: efficiency, easiness of execution and physical tiredness. Participants were also asked to grade the correlation between their perceived mental workload and the computed mental workload index.

4.6   Results

Performances (i.e. number of collisions per trial) for each condition are presented in Figure 3a. Friedman test shows a significant effect on assistance condition

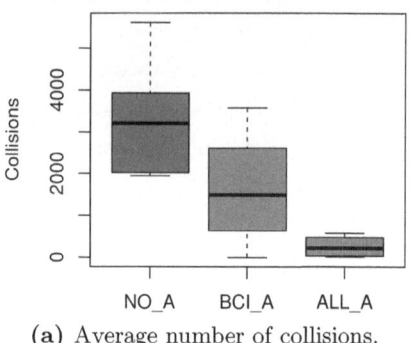
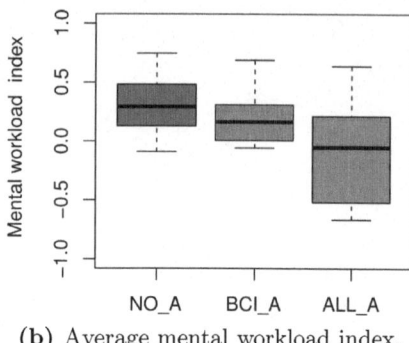

(a) Average number of collisions.   (b) Average mental workload index.

Fig. 3. Performance and mental workload index in each condition (*NO_A: no haptic assistance, BCI_A: haptic assistance activated based on BCI, ALL_A: haptic assistance activated all the time*). (a) Boxplots of collisions. (b) Boxplots of mental workload index. They are delimited by the quartile (25% quantile and 75% quantile) of the distribution of the condition over all the individuals. For each trial the median is shown.

($\chi^2 = 38.7, p < 0.001$). Post-hoc comparisons were performed using Wilcoxon signed-rank test with a threshold of 0.05 for significance. The post-hoc analysis shows a significant difference between condition NO_A and ALL_A ($p < 0.001$), and between condition NO_A and BCI_A ($p < 0.001$). BCI_A and ALL_A did not differ significantly from each other ($p = 0.08$). Activation of assistance enables to reduce the number of collisions. The average decrease over trials is 53% for condition BCI_A and 88% for condition ALL_A.

Percentage of activation of haptic guides during trials for condition BCI_A and each subject is shown in Table 4b. The assistance was in average activated 59% of the time for condition BCI_A (100% for condition ALL_A). During condition BCI_A, assistance was activated more frequently during part 1 (64% of the time) compared to part 2 (46% of the time). Figure 4a displays the positions where the assistance was activated along the path.

The computed mental workload index in each condition is presented in Figure 3b. A Friedman test revealed a significant effect of assistance mode on the index value ($\chi^2 = 16.0, p < 0.001$). A post-hoc analysis revealed a significant difference between condition ALL_A and NO_A ($p < 0.001$), no significant difference between condition NO_A and BCI_A ($p = 0.053$) and no significant difference between condition BCI_A and ALL_A ($p = 0.21$). Mental workload index was lower when the assistance was activated.

Figure 5 shows the evolution of the mental workload through the path followed by participants. During turns (part 1) we can observe a higher mental workload index than during part 2 (Mean over subjects: $M = 0.27$ for part 1, $M = -0.14$ for part 2). It is particularly clear on Figure 5 for condition NO_A.

Average marks of questionnaire answers are provided in Figure 6. Friedman test shows a significant effect of conditions on easiness ($\chi^2 = 16, p < 0.001$) and

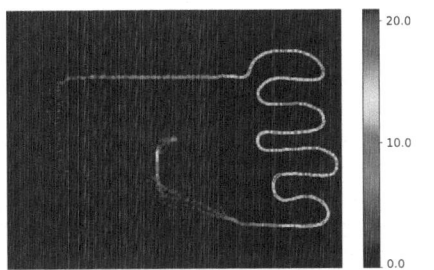

Subjects	1	2	3	4	5	6	7	8	μ	SD
% whole	51	60	55	63	24	71	86	62	**59**	16.6
% part 1	54	62	62	57	26	90	98	65	**64**	20.7
% part 2	38	23	41	63	00	34	99	73	**46**	28.9

(a) Number of activations per position.   (b) Percentage of activation per subject.

Fig. 4. Activation of guides based on BCI (condition BCI_A). (a) Number of activation for each position for all the 8 subjects and 3 trials. (b) Percentage of activation of haptic guides during trials.

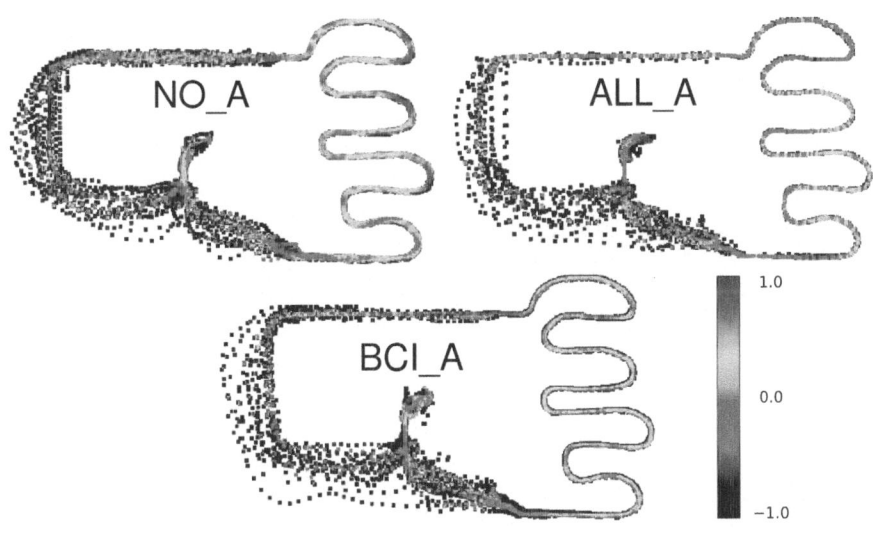

Fig. 5. Mental workload averaged over trials and subjects for each condition (*NO_A: no assistance, BCI_A: assistance activated based on BCI, ALL_A: assistance activated all the time*). Colored squares (9x9 pixels) are used to display mental workload index at each position on a 1024x768 image which represents the virtual scene. For each pixel, mental workload indexes were averaged over all the subjects and trials. A red color reflects a high mental workload index whereas a blue color corresponds to a low workload index.

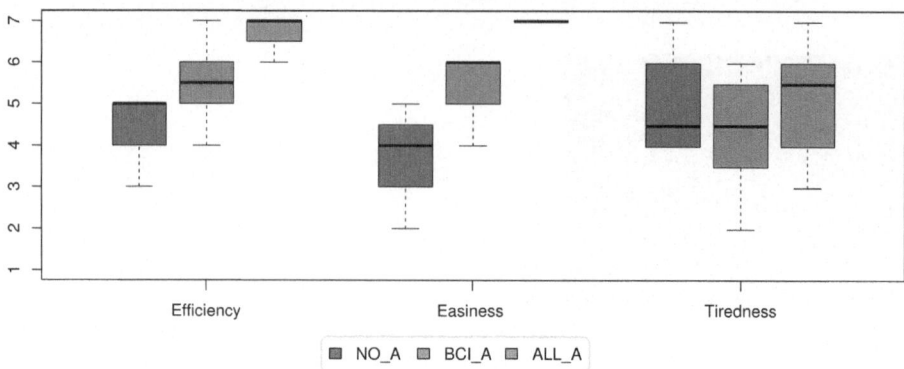

Fig. 6. Questionnaire results. Boxplots of questionnaire reported marks. They are delimited by the quartile (25% quantile and 75% quantile) of the distribution of the condition over all the individuals. The median is shown for each condition. A Likert-scale (1: very bad, 7:very good) was used for criteria efficiency, easiness and physical tiredness in each condition (*NO_A: no assistance, BCI_A: assistance activated based on BCI, ALL_A: assistance activated all the time*).

efficiency ($\chi^2 = 14.5, p < 0.001$). A higher efficiency and a higher easiness was reported when the haptic assistance was activated all the time (ALL_A) compared to no haptic assistance condition ($p < 0.001$). No significant difference was observed between BCI_A and other conditions. This suggests that this condition is located between the two others in terms of subjective easiness and efficiency. Questionnaire results about tiredness seem to indicate that participants felt the experiment was rather tiring.

Participants reported a high correlation between computed and felt mental workload ($M = 71\%$, $sd = 4.7$). This suggests that the BCI system is able to provide a convincing measurement of the mental workload.

5   Discussion

We could test the operability of our proof-of-concept system in a path-following task. Results indicate that the proposed system works and helps the users to achieve the task. Activation of guides based on measured mental workload index allows to increase performance by significantly reducing the number of collisions. No significant difference was observed between condition ALL_A and BCI_A in terms of performance ($p = 0.08$). This suggests that assistance activated based on BCI is almost as helpful as permanent assistance.

Results also suggest that this proof-of-concept system was able to measure a mental workload index that seems well correlated with the difficulty of the task. Indeed the measured mental workload was higher when the task was more difficult (near walls) as shown in Figure 5. Moreover the users reported a high correlation between the computed index and their perceived mental workload (above 70%).

In this study we used a binary adaptation (i.e. activating or deactivating assistance). The system could benefit from a progressive adaptation, e.g. more assistance if the workload is higher. We should note that 25% of participants asked for this feature.

A future system could also use other kinds of haptic assistance. Indeed, adapting the damping level proportionally to the user mental workload or toggling inverted damping [23] only if the user presents a high mental workload are options that should be studied. Concerning the measurements of the mental workload, a combination of EEG with other modalities such as Galvanic Skin Response in a multi-modal measurement system could increase the reliability of the mental workload index. Finally, it would also be interesting to evaluate the role of BCI-based adaptation in real applications such as medical training systems notably in terms of learning performance.

6 Conclusion

In this paper we studied the combination of haptic interfaces and Brain-Computer Interfaces (BCI). We proposed to use BCI technology to adapt a haptic guidance system according to a mental workload index. We designed a proof-of-concept system and conducted an evaluation that revealed the feasibility and operability of such a system. High levels of mental workload could be well identified by the BCI system in the most difficult parts of a path-following scenario. Toggling haptic guides only when the user presented a high mental workload could improve task performance indicators. Taken together our results pave the way to novel combinations of BCI and haptics. Haptic feedback could be fine-tuned according to various mental states of the user, and for various purposes. Future work will focus on the extraction of other mental states, and on real applications such as haptic-based medical simulators.

Acknowledgements. This work was supported by the French National Research Agency within the OpenViBE2 project and grant ANR-09-CORD-017. The authors would also like to thank all the participants who took part in our study for their patience and kindness.

References

1. Wolpaw, J.R., Birbaumer, N., McFarland, D.J., Pfurtscheller, G., Vaughan, T.M.: Brain–computer interfaces for communication and control. Clinical Neurophysiology 113(6), 767–791 (2002)
2. Farwell, L.A., Donchin, E.: Talking off the top of your head: toward a mental prosthesis utilizing event-related brain potentials. Electroencephalography and Clinical Neurophysiology 70(6), 510–523 (1988)
3. Guger, C., Harkam, W., Hertnaes, C., Pfurtscheller, G.: Prosthetic control by an EEG-based brain-computer interface (BCI). In: Proc. European Conference for the Advancement of Assistive Technology (1999)

4. Allison, B., Graimann, B., Gräser, A.: Why use a BCI if you are healthy? In: ACE Workshop - Brain-Computer Interfaces and Games, pp. 7–11 (2007)
5. George, L., Lécuyer, A.: An overview of research on "passive" brain-computer interfaces for implicit human-computer interaction. In: International Conference on Applied Bionics and Biomechanics, Venise, Italy (2010)
6. Zander, T.O., Kothe, C.: Towards passive brain–computer interfaces: applying brain–computer interface technology to human–machine systems in general. Journal of Neural Engineering 8(2) (2011)
7. Nijholt, A., Bos, D.P.O., Reuderink, B.: Turning shortcomings into challenges: Brain–computer interfaces for games. Entertainment Computing 1(2), 85–94 (2009)
8. Müller-Putz, G.R., Scherer, R., Neuper, C., Pfurtscheller, G.: Steady-state somatosensory evoked potentials: suitable brain signals for brain-computer interfaces? IEEE Transactions on Neural Systems and Rehabilitation Engineering 14(1), 30–37 (2006)
9. Cincotti, F., Kauhanen, L., Aloise, F., Palomäki, T., Caporusso, N., Jylänki, P., Mattia, D., Babiloni, F., Vanacker, G., Nuttin, M., Marciani, M.G., Millán, J.d.R.: Vibrotactile feedback for brain-computer interface operation. In: Computational Intelligence and Neuroscience (2007)
10. Chatterjee, A., Aggarwal, V., Ramos, A., Acharya, S., Thakor, N.: A brain-computer interface with vibrotactile biofeedback for haptic information. Journal of NeuroEngineering and Rehabilitation 4(1) (2007)
11. Feygin, D., Keehner, M., Tendick, R.: Haptic guidance: experimental evaluation of a haptic training method for a perceptual motor skill. In: Haptic Interfaces for Virtual Environment and Teleoperator Systems, pp. 40–47 (2002)
12. Bluteau, J., Sabine, C., Yohan, P., Edouard, G.: Haptic Guidance Improves the Visuo-Manual Tracking of Trajectories. PLoS ONE 3(3) (2008)
13. Asseldonk, E.H.F., Wessels, M., Stienen, A.H., van der Helm, F.C., van der Kooij, H.: Influence of haptic guidance in learning a novel visuomotor task. Journal of Physiology 103(3–5), 276–285 (2009)
14. Lécuyer, A., Lotte, F., Reilly, R., Leeb, R., Hirose, M., Slater, M.: Brain-computer interfaces, virtual reality, and videogames. Computer 41(10), 66–72 (2008)
15. Mühl, C., Gürkök, H., Plass-Oude Bos, D., Thurlings, M., Scherffig, L., Duvinage, M., Elbakyan, A., Kang, S., Poel, M., Heylen, D.: Bacteria Hunt: A multimodal, multiparadigm BCI game. In: International Summer Workshop on Multimodal Interfaces, University of Genua (2010)
16. Kohlmorgen, J., Dornhege, G., Braun, M., Blankertz, B., Müller, K.-R., Curio, G., Hagemann, K., Bruns, A., Schrauf, M., Kincses, W.: Improving human performance in a real operating environment through real-time mental workload detection. In: Dornhege, G., Millán, J.d.R., Hinterberger, T., McFarland, D., Müller, K.-R. (eds.) Toward Brain-Computer Interfacing, pp. 409–422. MIT press, Cambridge (2007)
17. Heger, D., Putze, F., Schultz, T.: Online workload recognition from EEG data during cognitive tests and human-machine interaction. In: Advances in Artificial Intelligence, pp. 410–417 (2010)
18. Pope, A.T., Bogart, E.H., Bartolome, D.S.: Biocybernetic system evaluates indices of operator engagement in automated task. Biological Psychology 40(1), 187–195 (1995)
19. Renard, Y., Lotte, F., Gibert, G., Congedo, M., Maby, E., Delannoy, V., Bertrand, O., Lécuyer, A.: OpenViBE: An Open-Source Software Platform to Design, Test and Use Brain-Computer Interfaces in Real and Virtual Environments. Presence 19, 35–53 (2010)

20. Hamadicharef, B., Zhang, H., Guan, C., Wang, C., Phua, K.S., Tee, K.P., Ang, K.K.: Learning EEG-based Spectral-Spatial Patterns for Attention Level Measurement. In: IEEE International Symposium on Circuits and Systems, Taipei, Taiwan, Province of China, pp. 1465–1468 (2009)
21. Blankertz, B., Tomioka, R., Lemm, S., Kawanabe, M., Müller, K.R.: Optimizing spatial filters for robust EEG single-trial analysis. IEEE Signal Processing Magazine 25(1), 41–56 (2008)
22. Peng, H., Long, F., Ding, C.: Feature selection based on mutual information criteria of max-dependency, max-relevance, and min-redundancy. IEEE Transactions on Pattern Analysis and Machine Intelligence 27(8), 1226–1238 (2005)
23. Williams, J., Michelitsch, G.: Designing effective haptic interaction: inverted damping. In: Extended Abstracts on Human Factors in Computing Systems, pp. 856–857. ACM (2003)

Development of Intuitive Tactile Navigational Patterns

Christos Giachritsis[1,*], Gary Randall[1], and Samuel Roselier[2]

[1] BMT Group Ltd, Goodrich House, 1 Waldegrave Road, Teddington,
Middlesex, TW11 8LZ, UK
{cgiachritsis,grandall}@bmtmail.com
[2] CEA, LIST, Sensory & Ambient Interfaces Lab.
18 Route du Panorama, 92265 Fontenay-aux-Roses CEDEX, France
samuel.roselier@gmail.com

Abstract. Developing vibrotactile patterns for mobile navigation is challenging since they have to be able to map effectively navigational behaviour learned through visuo-audio-motor interaction with the environment to specific vibrotactile patterns. Here, we present a method for developing intuitive navigation patterns representing basic *directions*, *landmarks* and *actions*. A group of users familiar with the device delivering the vibrotactile signals are asked to create navigational patterns. Then the patterns are edited and presented to another group of users who are also familiar with the device. They were asked to identify the patterns with/out information about their meaning. It was found that simple *directions* were easier to identify than *landmarks* or *actions*. The identification of *landmarks* or *actions* improved when information about their meaning was available. We discuss implications for the design and development of vibrotactile navigational patterns.

Keywords: mobile navigation, intuitive patterns, tactile interfaces, vibrotactile feedback.

1   Introduction

Mobile devices such as cell phones and *portable navigation devices* (PND) are using vibrations to alert users to oncoming calls and enhance users experience when interacting with the device (e.g., by simulating the tactile properties of a clicked button). Recent research in tactile navigation has investigated the use of vibrotactile signals on smart phones (directly or through another device), wearable devices and steering wheels [1-5]. Most of the designs use navigational patterns defined by the developer and require the user to learn these patterns before being able to use them. Here, we report a method of developing intuitive navigational patterns representing basic *directions*, *landmarks* and *actions*. We asked a group of users familiar with the device to define the patterns using the device. The patterns were edited for clarity and arranged in series representing navigation from a starting point to a destination point. Then they were presented for identification to another group of users also familiar with the

---

* Corresponding author.

P. Isokoski and J. Springare (Eds.): EuroHaptics 2012, Part I, LNCS 7282, pp. 136–147, 2012.
© Springer-Verlag Berlin Heidelberg 2012

device. These users were asked to identify the patterns with and without any information about their meaning.

2   The Study

2.1   The Viflex Device

The ViFlex device uses an electromagnetic actuator to rotate a mobile platform with an octagon shape and area of 45x45 mm^2. The platform can rotate around its two cardinal XY axes and their intermediate axes by magnitude of ±10° (Figure 1, left). Thus, through rotation the platform could signal eight different directions/positions. The device is capable of fast rotations up to 200°s^{-1} and can achieve a torque of 20 Nmm. More detailed description of the technical characteristics of ViFlex can be found in [6].

Fig. 1. The Viflex device with its octagonal platform (left). The device with its battery extension handle can be used with one (middle) or two (right) fingers.

The ViFlex device can be used with one (index) or two (index and middle) fingers (Figure 1, middle and right). By employing only one finger the device will primarily engage the cutaneous system to deliver vibrotactile signals at different locations of the finger pad. When the user employs two fingers positioned at the edges of the X-axis, the device will primarily engage the proprioceptive system at the metacarpal-phalangeal joints when rotating about the Y-axis and the cutaneous system when rotating about the X-axis.

2.2   Establishing Spatiotemporal Thresholds and Optimal Exploration Mode

The purpose of this study was twofold: first, establish spatiotemporal thresholds for detecting the direction of rotations generated by the ViFlex device and, second, find whether lower thresholds could be achieved with one (index) or two fingers (index and middle). In this way we would use patterns that would be easier for the users to perceive.

2.2.1   Methods

Ten (5 females and 5 males) users participated in the study with an average age of 31. All users were right-handed and used their right-hand to explore the vibrotactile signals. Three degree of rotations (1°, 3° and 9°) in both X- and Y- axes were tested. In addition, three transition times (50ms, 250ms and 1250ms) between the initial and the final (after the completion of the rotation) positions of the platform were tested. The platform maintained its final position for 500ms resulting in total stimulus duration of 600ms, 1000ms and 300ms (Figure 2). Moreover, two exploration modes were tested: one-finger passive exploration and two-finger passive exploration. There were 12 trials per pattern resulting in 216 patterns per exploration mode.

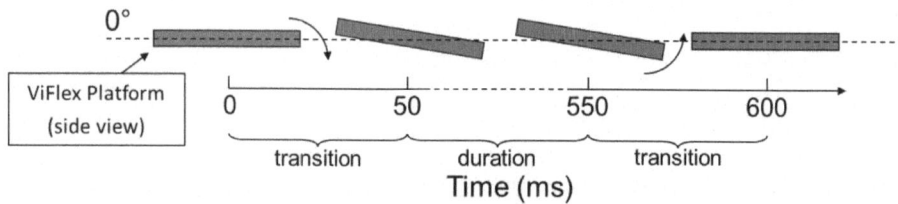

Fig. 2. Example stimulus timeline

2.2.2   Results

Overall data were fitted with the cumulative exponential function $\psi(x;\alpha,b)=\alpha(1-e^{-\beta x})$, where $x \geq 0; \alpha = 1; \beta > 0$ (Figure 3).

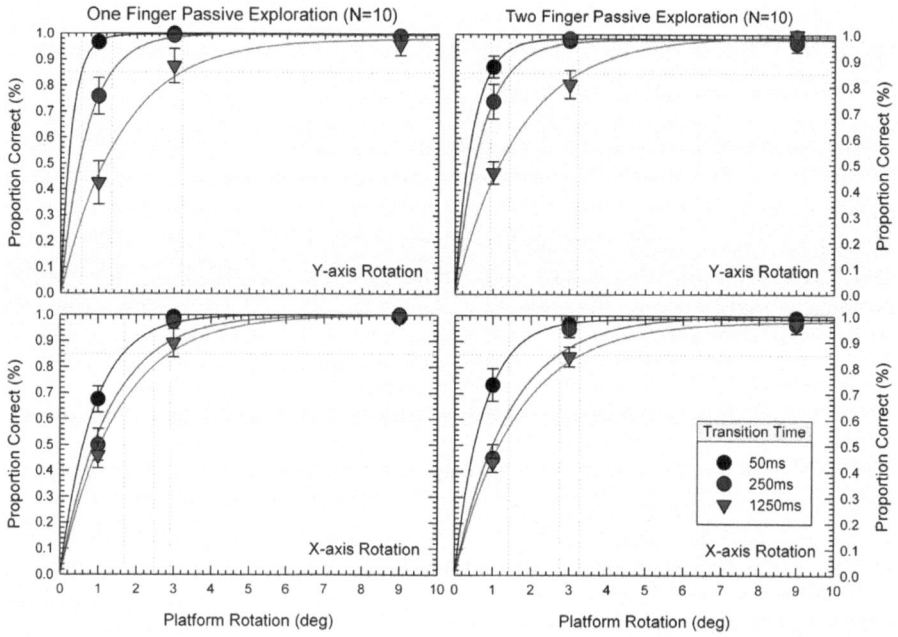

Fig. 3. Overall cumulative exponential curves. Each data point is the average performance of ten users.

Results showed that it was easier for the users to achieve 85% accuracy when the platform rotated about the y-axis, with greater degrees and shorter transition times. In addition, there was no difference between the two exploration modes. However, 70% of the users stated that preferred the use of one rather than two fingers.

2.3   Developing Intuitive Navigational Patterns

The purpose of the experiment was to find out if it is possible to generate intuitive navigational patterns based on patterns defined by users experienced with the functionalities of a device. Seven female and four male (average age of 30) with no sensory impairment took part in the study. Eight of them had taken part in the previous study to establish the spatiotemporal thresholds. The rest of the participants familiarized themselves with the functionality of ViFlex before asked to carry out the required navigational tasks.

2.3.1   Types of Navigational Patterns

Three types of navigational patterns were used: directions, landmarks, and actions. Table 1 shows the navigational patterns used with their abbreviations.

Table 1. The three types of navigational patterns used in the study

Directions	Landmarks	Actions
Forward (F)	Roundabout (RA)	GO
Backward (B)	Crossroad (CR)	STOP
Right (R)	T-Junction (TJ)	Destination Reached (DR)
Left (L)	Y-Junction (YJ)	
Forward-Right (FR)	Uphill (UH)	
Forward-Left (FL)	Downhill (DH)	
Backward-Right (BR)	Stairs-Up (SU)	
Backward-Left (BL)	Stairs-Down (SD)	
Clockwise 180° (CW180)	Obstacle (O)	
Anti-Clockwise 180° (ACW180)		

Similar patterns are commonly used in navigational tasks [7-9] and some of them were previously used in a preliminary ViFlex study [10]. The study included two stages. In the first stage, five users were asked to *create* patterns and in the second stage, the remaining six users were asked to *identify* the patterns created in the first stage.

2.3.2   Creating Navigational Patterns

Directional and landmarks patterns were created by the users as follows. The users were presented with the ViFlex device and a pattern, which was verbally communicated by

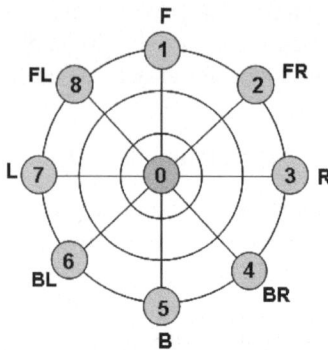

Fig. 4. The circular grid used to record users responses when they were asked to indicate how the platform should move to represent a particular pattern. Position 0 represents the origin of the platform (i.e., horizontal positioning). Position 1 indicates a 'forward' rotation about the X-axis (L-R), position 3 a 'rightward' rotation about the Y-axis (F-B), and so on.

the experimenter. Then, they were asked to move the platform in any direction that they thought it could be used to represent the specific pattern. Their responses (i.e., movement of the platform) were recorded on a circular grid representing the eight different directions/positions of the platform (Figure 4).

2.2.2.1   Results

Table 2 shows the position sequences that users indicated to represent *directional* patterns. Simple directions in the cardinal and intermediate axes were unanimously indicated. *Directional* patterns with multiple positions such as CW180 and ACW180 were indicated as semi-circular patterns by four out of five users.

Table 2. Directional patterns indicated by the users. A final pattern was created to be identified in the next stage.

	Directions									
User	**F**	**B**	**L**	**R**	**FR**	**FL**	**BR**	**BL**	**CW180**	**ACW180**
1	01	05	07	03	02	08	04	06	056781	054321
2	01	05	07	03	02	08	04	06	056781	054321
3	01	05	07	03	02	08	04	06	056781	054321
4	01	05	07	03	02	08	04	06	012345	018765
5	01	05	07	03	02	08	04	06	0812	0218
Final	010	050	070	030	020	080	040	060	0567810	0543210

Table 3 shows the position sequences that users indicated to represent *landmarks*. In general, users agreed as to the structure of the pattern but the position sequences they used were more variable. For example, RA was represented by a complete rotational movement of the platform even though the starting point and the direction of movement varied. Most of the users represented CR by moving along the cardinal axes. Only one participant moved the platform along the intermediate axes forming an '×'. Similarly, most users indicated the same positions, though in different order, to represent the TJ and YJ patterns. The UH and DH were represented by the majority of the users by maintaining (in the table this is shown by repeating a single position; the more repetitions the longer users maintained that position – not actual timing was measured) an 'upward' position (i.e., positive rotation about the X-axis – rotation about the X-axis towards position 5) and a 'downward' position (i.e., positive rotation about the X-axis – rotation about the X-axis towards position 5), respectively. The SU and SD patterns were represented by using the positions 01 or 05, but without consistency. Finally, the O pattern was indicated by the positions 0, 1 and 5 but also without consistency.

Table 3. Landmarks patterns indicated by the users and the final pattern that was created to be identified in the next stage

| User | Landmarks | | | | |
	RA	CR	TJ	YJ	UH
1	0567812345	015037	07305	080205	0555
2	0187654321	015037	05073	050802	0111
3	0543218765	015037	03705	050208	0555
4	0543218765	051073	03705	02070	0555
5	0567812345	026084	07305	050802	0555
Final	0567812340	0150370	037050	0508020	05550

User	DH	SU	SD	O
1	0111	055005500550	011001100110	0151515
2	0555	011001100110	055005500550	05555
3	0111	011001100110	055005500550	05555
4	0111	05050	01010	0515151
5	0111	055005500550	011001100110	05050
Final	01110	055005500550	011001100110	03737370

The final patterns that would be used for identification in the next stage were created on the basis of the most dominant positions indicated by the users. The exception is the final pattern for O. Since positions 0, 1 and 5 were used to represent UH, DH, SU and SD, we decided to use positions 3 and 7 to avoid mapping too many navigational patterns to the same platform positions. *Directions* and *landmarks* were implemented by rotating the platform by 10° towards the corresponding position. This magnitude was reliably perceived by all users during the signal detection study.

The *action* patterns were pre-defined by the experimenter as follows: GO=01010 [normal speed rotations of 9°], STOP=050505050 [fast rotations of 9°], DR=0370150 [fast rotations of 3°].

Following the results of the signal detection study, the navigational patterns were developed using >3° rotations and shorter than 1250ms transition times. Users were also instructed to use their index finger to perceive the navigational patterns.

2.3.3   Identifying Navigational Patterns

Six users were asked to navigate from a starting point to a destination using two series (A and B) of 45 patterns. Both series started with GO and ended with DR and included exactly the same navigational patterns with the same presentation frequency but, different presentation order. For example, Series A was GO-F-L-R...CW-L-DR and Series B was GO-L-F-STOP...L-R-DR. The patterns in both sequences were presented to the users one at a time and the users were asked to identify the current pattern before the next pattern was presented. There was no visual representation of the patterns.

When users were asked to identify the patterns in Series A, they were given no information about the type of patterns presented. They were simply instructed to use the patterns resented by the ViFlex device to navigate from point A to point B. However, with Series B, users were provided with information about the types of patterns and the patterns presented in the list. This information was available to them for referencing during the test. When users were uncertain about the meaning of a pattern the pattern was repeated as many times as it was necessary to obtain an interpretation. If an interpretation or identification was not possible then a 'Do not know' response was recorded.

All users were right handed and used the index of their right hand to perceive the navigational patterns. Also, they navigated through series A and B only once while the order of presentation was always A-B so that all users experienced exactly the same sequence of patterns in the same order.

2.3.2.1   Results

In the 'no information' condition, users found it easier to identify and interpret *directional* patterns rather than *landmarks* and *actions*. Single point *directional patterns* (e.g., F, R, L, FL, FR) were easier to identify than multipoint *directional* patterns. *Landmarks* and *actions* resulted in more variable interpretations. This may be because they involved more complex spatiotemporal signals and, sometimes, similar positions/directions. Performance with both series was measured in terms of overall proportion correct, proportion correct per pattern, mean number of repetitions per pattern, cost effective patterns and least identifiable patterns.

Overall performance was improved from 59% without any information about the meaning of the patterns (Series A), to 72% with information available (Series B) (Figure 5). However, a T-test showed that this improvement was not statistically significant. Single-point *directional* patterns (i.e., F, L, FR, FL) resulted in highly accurate interpretation in Series A and identification with Series B. Multipoint *directional* patterns (i.e., CW180 and ACW180) resulted in less accurate interpretations with Sequence A but, with Sequence B, resulted in over 80% identification accuracy. *Landmarks* resulted in less accurate interpretation than *directions* in Series A. In

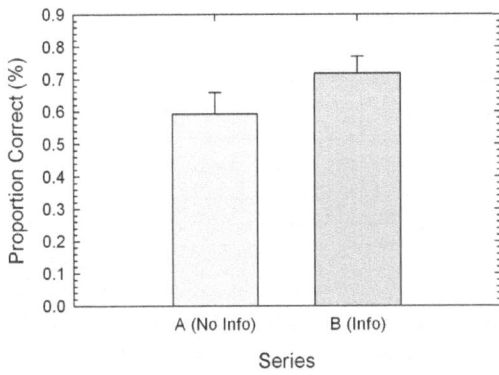

Fig. 5. Overall performance with Series A and B. Error bars represent constant errors.

principle, it seemed easier to interpret *landmarks* involving different points, such as CR and TJ, rather than *landmarks* involving similar points and movements, such as UH, DH, SU, SD and O. Identification of *landmarks* improved in Series B. In particular, all junction-*landmarks* (i.e., CR, TJ and YJ) were identified with an accuracy of > 80%. The identification of the rest of the *landmarks* also improved. *Actions* were the patterns that resulted in the least noticeable changes of interpretation and identification accuracy between Series A and B. While there was some improvement with the patterns GO and STOP, the accuracy with the DR pattern did not change (Figure 6). However, it should be noted that the GO and STOP patterns were the most frequently occurred patterns in both sequences with 10 and 9 appearances, respectively, while the DR pattern appeared only once, in the end.

Mean number of repetition per pattern was also reduced when the meanings of the patterns were available to the participants (Figure 7). In principle, interpretation of patterns with Series A resulted in more repetitions per pattern than identification of patterns with Series B. The greatest reduction in repetitions was observed with the multipoint *directions* CW180 and ACW180 as well as with most of the *landmarks*. The only exception to this trend was the DR pattern which resulted in more repetitions during identification (Series B) rather than interpretation (Series A).

Effectiveness of a pattern was defined as the ratio between the proportion correct achieved and the number of repetitions required for that level of accuracy. An effective pattern should be easily recognisable so that the user should not need to repeat it. The most cost-effective patterns were the single-point *directional* patterns, particularly the ones in the cardinal directions (F, L and R). The single-point in the intermediate directions (FL and FR) and the multipoint *directional* patterns (CW180 and ACW180) were less effective even though their effectiveness improved with the availability of information about their meaning (Series B). The effectiveness of *landmarks* was also improved with Sequence B. The least improvement of effectiveness was observed with the *action* patterns (Figure 8).

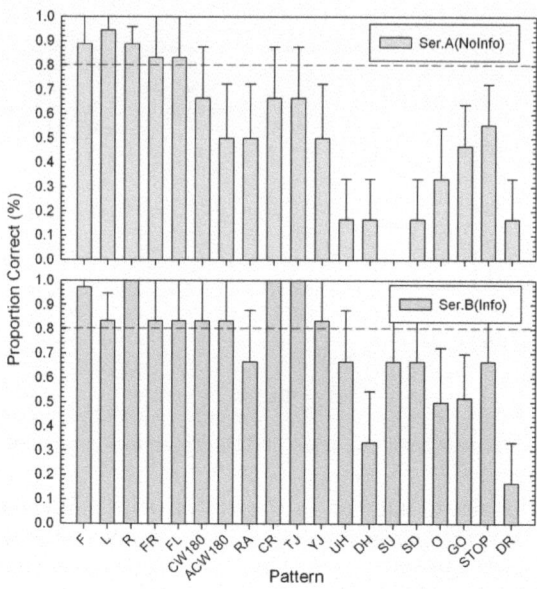

Fig. 6. Percentage correct per pattern in Series A and B. Error bars represent constant errors.

Finally, overall, it seems that *directional* patterns elicited the least number of 'Do not know' responses followed by *landmarks*. *Actions* resulted in more 'Do not know' responses (Figure 9).

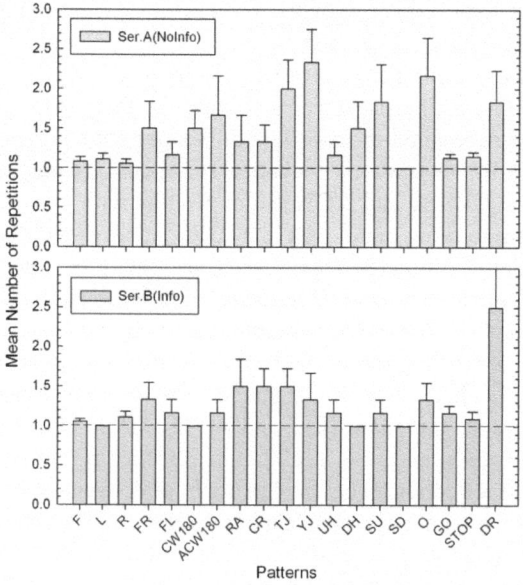

Fig. 7. Mean number of repetitions per pattern (Repetitions/Frequency) with constant errors

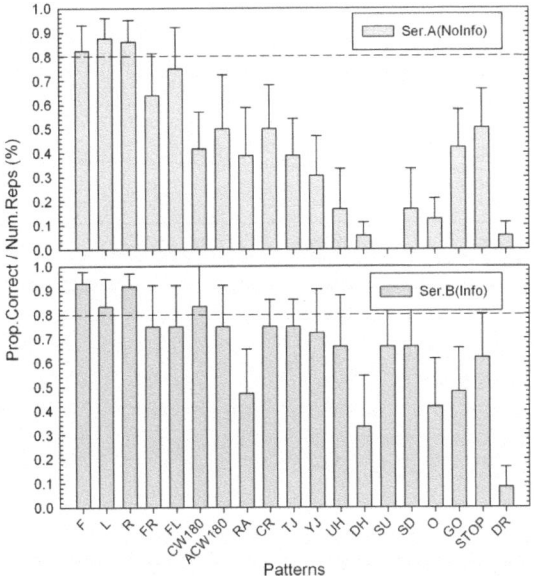

Fig. 8. Effective patterns. Error bars represent constant errors.

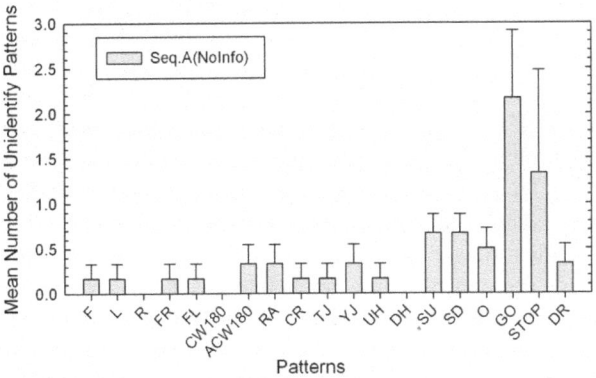

Fig. 9. 'Do not know' responses. Error bars represent constant errors.

3   Discussion

The study showed that users easily interpreted and identified simple, single-point *directional* patterns. Multipoint directional patterns such as CW180 and ACW180 were more difficult to interpret in the absence of any information about the patterns. However, when information was available participants identify them with high accuracy (>80%). One of reason for this result may be that the CW180 and ACW180 could be confused with the *landmark* RA which involved similar pattern of movement of the ViFlex platform. While semicircular (CW180 and ACW180) and circular (RA) movements of the platform could be resolved at a sensory level and therefore perceived accurately by the participants (since they were developed using signals well above threshold), the attributed meaning varied.

Table 4. Navigational patterns involved same platform (and finger pad) positions but had different temporal characteristics

Patterns	Amplitude (deg)	Frequency (Hz)	Cycles
F	10	1.333	1
UH	10	0.286	1
DH	10	0.286	1
SU	10	1.067	4
SD	10	1.067	4
GO	9	1.667	2
STOP	9	5.333	4

Moreover, it seems that this variation of interpretation and identification increased with temporal variation of patterns that involved similar spatial positions. Table 4 shows the amplitudes, frequencies and cycles for patterns stimulating the same finger pad location. For example, the patterns F, DH, SD and GO involved the same platform/fingertip positions (i.e., 0 and 1), similar amplitudes (F=DH=SD=10° and GO=9°) but different number of cycles (F=DH=1, SD=4 and GO=2) and frequencies (F=1.33Hz, DH=0.29Hz, SD=1.33Hz and GO=2Hz). Often, participants seemed to interpret these patterns in a similar way (e.g., F and GO often were similarly interpreted) or identified them incorrectly (e.g., identifying GO as SD). The patterns UH, SD and STOP also resulted in similar responses; for example, STOP was sometimes interpreted as B or identified as O.

In conclusion, the present study has shown that it is possible to develop vibrotactile navigation signals based on users' experience of the functionality of a device. In addition, simple spatiotemporal signals can be interpreted and identified easier than more complex ones. Further research is needed which includes a diverse user group (e.g., pedestrians, drivers, visually-impaired) that can provide insight about how visuo-motor-audio experience during real life navigation can be mapped onto spatiotemporal tactile signals at specific body locations.

References

1. Magnusson, C., Rassmus-Gröhn, K., Szymczak, D.: Scanning angles for directional pointing. In: Proc. MobileHCI 2010, Lisbon, Portugal, September 7-10 (2010)
2. Poppinga, B., Pielot, M., Boll, S.: Tacticycle – A tactile display for supporting tourists on a bicycle trip. In: Proc. MobileHCI 2009, Bonn, Germany, September 15-18 (2009)

3. Kern, D., Marshall, P., Hornecker, E., Rogers, Y., Schmidt, A.: Enhancing Navigation Information with Tactile Output Embedded into the Steering Wheel. In: Tokuda, H., Beigl, M., Friday, A., Brush, A.J.B., Tobe, Y. (eds.) Pervasive 2009. LNCS, vol. 5538, pp. 42–58. Springer, Heidelberg (2009)
4. Pielot, M., Poppinga, B., Boll, S.: PocketNavigator: Vibrotactile waypoint navigation for everyday mobile devices. In: Proc. MobileHCI 2010, Lisbon, Portugal, September 7-10 (2010)
5. Van Erp, J.B.F., Van Veen, H.A.H.C., Jansen, C.: Waypoint navigation with a vibrotactile waist belt. Transactions in Applied Perception 2(2), 106–117 (2005)
6. Roselier, S., Hafez, M.V.: A compact haptic 2D interface with force-feedback for mobile devices. In: Proc. EuroHaptics 2006, Paris, France, July 3-6 (2006)
7. Hile, H., Vedanthm, R., Cuellar, G., Liu, A., Gelfand, N., Grzeszczuk, R., Borriello, G.: Landmark-based pedestrian navigation from collections of geotagged photos. In: Proc. 7th International Conference on Mobile and Ubiquitous Multimedia 2008, Umea, Sweden, December 3-5 (2008)
8. Milloning, A., Schechtner, K.: Developing landmark-based pedestrian navigation systems. In: Proc. 8th International IEEE Conference on Intelligent Transportation Systems, Vienna, Austria, September 13-16 (2005)
9. Su, J., Rosenzweig, A., Goel, A., De Lara, E., Truong, K.N.: Timbremap: Enabling the visually-impaired to use maps on touch-enabled devices. In: Proc. MobileHCI 2010, Lisbon, Portugal, September 10 (2010)
10. Anastassova, M., Roselier, S.: A preliminary user evaluation of vibrational patterns transmitting spatial information. In: Proc. MobileHCI 2010, Lisbon, Portugal, September 10 (2010)

FootGlove: A Haptic Device Supporting the Customer in the Choice of the Best Fitting Shoes

Luca Greci, Marco Sacco, Nicola Cau, and Flavia Buonanno

ITIA-CNR Istituto di Tecnologie Industriali e Automazione – Consiglio Nazionale delle Ricerche, Milan, Italy
luca.greci@itia.cnr.it

Abstract. The core issues of mass customization (MC) are to provide more and more conveniences to meet customer's customized requirements and to ensure near mass production efficiency and quality in manufacturing processes. Since the MC provides to the customer, in the footwear sector, the possibility to choose shoes with different materials, colours, sizes and performances, the shops need big stocks in order to give the possibility to the customer to try the different solutions available. The FootGlove is a haptic device simulating the internal volume of the shoe, through a set of mechatronics mechanism, allowing the customer to try the shoes that are not yet manufactured. One of the advantages is permitting to the shop to have small stock. This paper will discuss how the FootGlove simulates the fitting of the shoes with different sizes and typologies to support the customer in the choice of his best fitting shoe.

Keywords: Mass Customization, Best Fitting, Shoes, Haptic Device, Foot.

1   Introduction

The shoes customization is a fundamental added value in the shoes market because it improves the capability of satisfying customers' new requirements and because it is competitive not only on a cost-base [1]. The websites of the bigger shoes manufacturers (i.e. NikeID) are already providing this kind of service but strictly related to the aesthetic aspect. Therefore, the satisfaction of the customer about the fitting and the performance of his purchase is not ensured.

In the context of the FIT4U Project [2], a different approach in the shoes buying process has been developed; this process provides to the customer the possibility of modelling the shoe on his foot and of satisfying the required performances.

The FIT4U approach concerns a set of tools for the definition of the shoes the customer is looking for in terms of fitting, insole performance and aesthetic aspects. Once the desired shoes have been individuated, the related data are passed to the FootGlove (FG), a mechanical haptic device, developed by ITIA-CNR, able to reproduce the internal volume of the shoes. While wearing the FG, the customer can check if the selected size fits on the feet and if the insole performance answers the expectation. In the case one of the aspects is not satisfied, the customer can tune the dimension of the

P. Isokoski and J. Springare (Eds.): EuroHaptics 2012, Part I, LNCS 7282, pp. 148–159, 2012.

internal volume of the FG respect to her/his feet to find a better fitting or change the insole. Thanks to the FG, the manufacturer can easily find the best existing last (size available) or provide information regarding the shoes that will be manufactured (namely about the last that has to be realised).

The last is the support, typically made in polymer, that provides the dimensions and the shape of the shoes to be manufactured. The dimensions characterizing the last [2] are shown in Fig. 1. Not all the dimensions shown in figure 1 are used for each type of shoe. For the shoes family (safety and trekking) that we have assumed as a case study for the FG, the dimensions involved are: the longitudinal dimension, the high instep girth, the low instep girth, the ball girth and the toe girth.

The longitudinal dimension (1) defines the length fitting and the heel height of the shoe. The high instep girth (4) is the measurement of the girth at the arch of the foot; the low instep girth (6) is the measurement at the waist of the foot and is usually around the midpoint between the ball and the instep. The ball girth (7) is the measurement of the foot circumference at the ball of the foot at the widest part of the foot. This is also the point of the foot where the flexion of the foot occurs; the toe girth (8) is the measurement of the girth at the toe of the foot. This last measurement is relevant for diabetic foot.

The FIT4U project guidelines imposed the choice of safety and trekking shoes, this typology of shoes has negligible variations both for the heel height and the insole shape. These shoes characteristics reduce the complexity of the FG device in the simulation of the last. Different insole shapes are provided by the use of a physical internal insole (footbed).

In further works, the simulation of different heel heights and of different insole shape will be also taken into account.

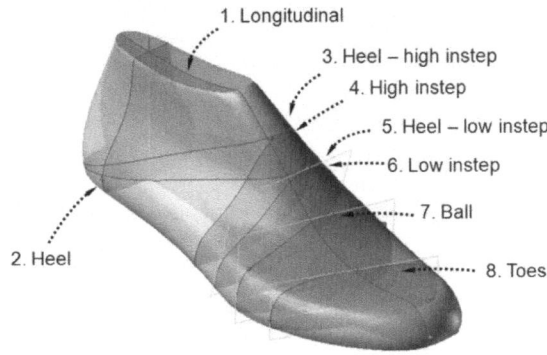

Fig. 1. The fundamental dimensions to build a last

To give the customer a complete experience during the shoes try-on, ITIA-CNR has developed the MagicMirror (MM), a system providing the aesthetic feedback on just customized shoes that are not yet manufactured [4].

2   The FootGlove system

The customer goes to a shop looking for a new pair of shoes. The shop is a FIT4U shop where a set of computer applications provide the capability to customize the aesthetic, the performance and the fitting of the shoes (Fig.2). Once the shoes have been defined in each detail, the data are sent to a CAD, linked with a database of 3D model of lasts, in order to extract the size and the section's length of each dimension. The communication between the different tools is managed through a chain of tools where the CAD PLM is the server and the client makes call on demand. The data type format used for exchanging the information is XML. The XML-files contain all the data about the customer choices, the lengths of the sections and the size that the FG has to reproduce to simulate the fitting of the selected shoe size.

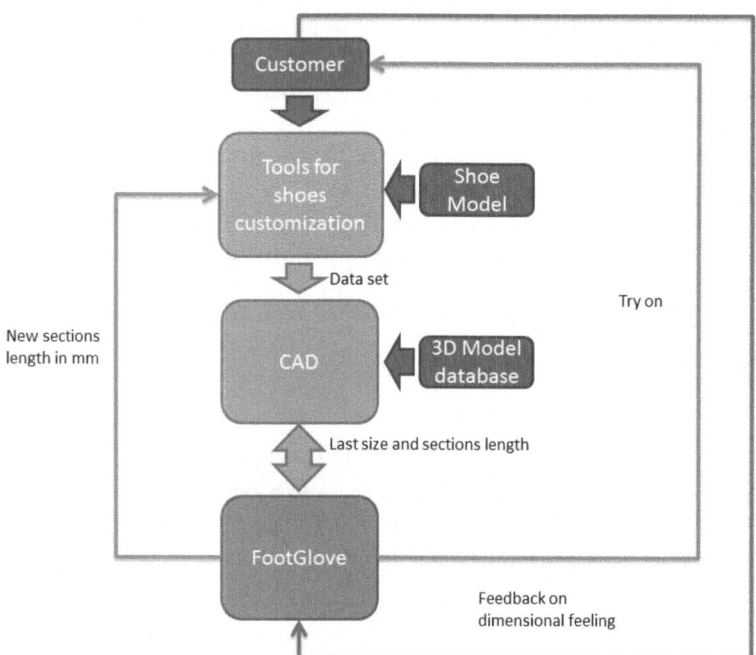

Fig. 2. The interaction among various actors involved in the FootGlove system

The length of the sections is subdivided in bottom length and upper length (Fig.3): the bottom length defines the width of the insole of each section while the upper length defines the volume available for the feet between the insole and the upper.

The configurator tool reads the xml data and converts the lengths in turns for the stepper motor. The choice of a stepper motor, as the configurator for the FG, guarantees a faster and more precise setup than a manual one. Since there are many parts to be set, the configurator tool has been built to minimize the human error through a step-by-step procedure. When the FG is properly configured the customer wears the devices, one for foot, and makes the fitting and walking test he usually does with the real shoes. The difference between the normal and the FG try-on is the re-configurability of the FG device that allows finding the best fitting. By using the con-figurator, the customer can indeed decide to modify the sections length until he finds "his/her personal" best fitting. At this point, the system sends the new data from the FG configurator to the shoes customisation tools to find a new last on which to build the shoes with the dimensions similar to those chosen by the customer. If no lasts are available with the features required, the data are passed to the CAD that generates a custom made last for the customer. An appropriate innovative machine can rapidly and cheaply provide customised lasts [2].

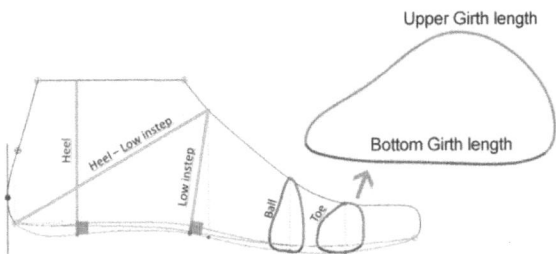

Fig. 3. An example of the data represented in the XML regarding the section length: it has been split in upper girth length and Bottom girth length

3   The FootGlove

The FG (Fig. 4) is a mechanical haptic device able to simulate the four fundamental dimensions of a last (the high instep girth, the low instep girth, the ball girth and the toe girth), the insole shape and the shoes length.

The device requirements are a dimension similar to those of the correspondent real shoes, a weight lower than 500gr for device, an easy and quick setup (less than 3 minutes for device), the capability to simulate sizes from 36 to 45 and the possibility of mounting a real footbed on the device.

Fig. 4. Pictures of the FootGlove prototype standalone and of a foot wearing the FootGlove

The rule to generate the data related to the position of the section is coming from other tools developed in the FIT4U project: it is the distances between the heel and each section for all the sizes respect to the average of the European feet dimension.

The device is composed by three distinct and cooperating mechanisms: one for the simulation of the growing in length, one for the simulation of the growing in width of the insole (bottom girth length) and one for the simulation of the upper girth length; this last mechanism has been called "Net".

The simulation of the insole shape is very complex. To provide a better simulation of its shape and a feeling similar to the real one, it has been decided to directly mount on the device the physical footbed. This solution has made the set-up of the device easier and has indeed enhanced the feeling/perception of wearing a real shoe (thus the customer experience). The footbeds are coupled with a Velcro in the lower part both in the toe and the heel area. This solution allows to interchange the footbed in short time and to test different footbed solutions increasing the capability of the customer to find the desired footbed. Different footbeds have different performances but also different thicknesses. Changing the footbed means also to change the volume available for the foot inside the device improving the customer possibilities to find the best fitting.

3.1   "Growing in Length" Mechanism Description

The FG is built with a sole mechanism able to reproduce different foot sizes, representing as a real continuous curve the sole shape and supporting the foot in the zones (heel and ball) where the weight of the user is more concentrated. The mechanism consists of several rigid metal parts linked with deformable rubber elements and it reproduces every foot size by applying just one axial force to it, as shown in Figure 5(a). The dimensions, position and quantity of the elastomeric components determine the displacements of each metal part of the device itself. The mechanism providing the force is achieved by tie-rod strand of steel running longitudinally through the sole and drives acting on a screw to regulate it.

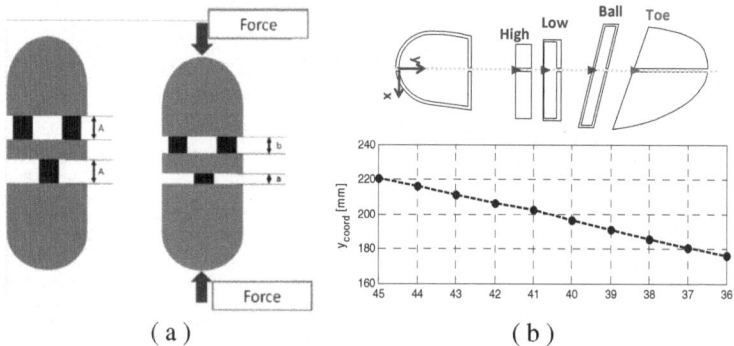

(a)   (b)

Fig. 5. Sole length mechanism: (a) Schematic description, (b) Design specification for the toe

For each foot size, the various sections (toe, ball, low, high) have to be set in specific positions. Figure 5(b) shows as an example the specification for the toe section. In order to cope with these design specifics, the rubber parts of the mechanism had to be dimensioned.

3.2   Mechanism Dimensioning

Since one of the materials involved in the mechanism is rubber, the first step was represented by a material characterization, in order to adequately model its constitutive behaviour. For this scope, uniaxial testing of the rubber was performed according to the specifics given by ASTM D412. Figure 6(a) shows the test, while Figure 6(b) shows one example of the experimental curves obtained. With the experimental data, different hyper-elastic material models [6-8] were fitted and evaluated. In Figure 6(c) is shown the selected third order Yeoh for modelling the rubber behaviour [9].

For modelling the mechanism, finite elements analysis (FEM) was required due to the complex characteristics of the design problem. The problem presents: complex shape, large deformation effects and deformation-based elastic module of the rubber. A finite elements model of the sole mechanism was constructed in ANSYS® including both the material model as all the necessary aspects (i.e constraints, contacts) for an adequate simulation of reality (see Figure 7).

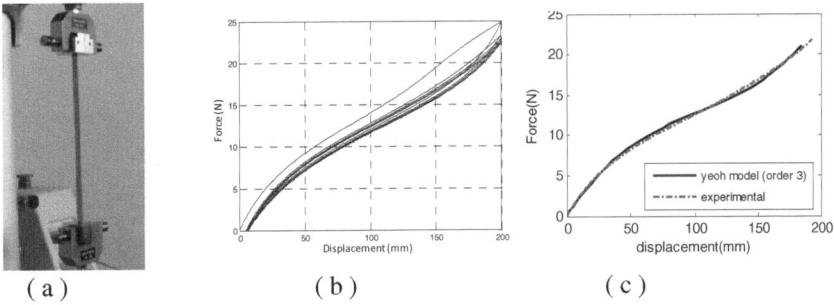

(a)   (b)   (c)

Fig. 6. Rubber model identification: (a) uniaxial test, (b) example of obtained experimental data, (c) model fitting (3rd order Yeoh)

Fig. 7. FootGlove's Finite Elements Model (FEM)

Finally, the dimensions of the rubber were obtained through the formulation and the solution of a design optimization problem, whose goal was to obtain the values for the rubber geometry that best cope with the design specifications. A design optimization problem can be formulated as [9] (1)

$$\min_{\underline{x} \subset R_{ndv}} \underline{f}(\underline{x})$$
$$\text{subject to } g_i(\underline{x}) \leq 0; \ i = 1, \dots, n_c \tag{1}$$
$$x_{ilow} \leq x_i \leq x_{iup} \ ; \ i = 1, \dots, n_{dv}$$

The main elements constituting the design optimization problem are summarized in Figure 8. The variables of the design are the parameters defining the rubber geometry, while the constraints are related to the geometrical feasibility of the design. The first three objective functions are the errors in position of the three displaced sections of the FG with respect to the design specifics, while the last two objective functions are the mean rotations and the displacements of the sections.

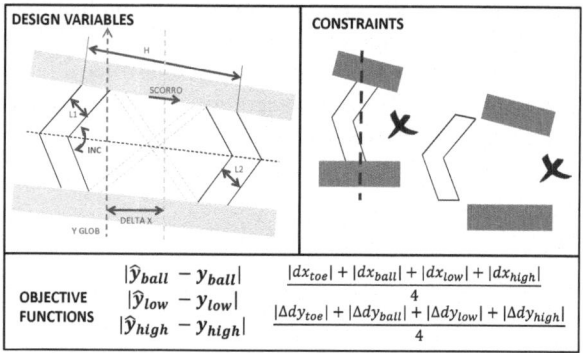

Fig. 8. The optimization problem formulated for the design of the sole length mechanism

The design optimization problem was solved through the implementation of an optimization routine where ANSYS® was used for the calculation of the model outputs, and MatLab® for the optimization part. The problem was addressed as multiobjective and in Figure 9 is shown the solution that was.

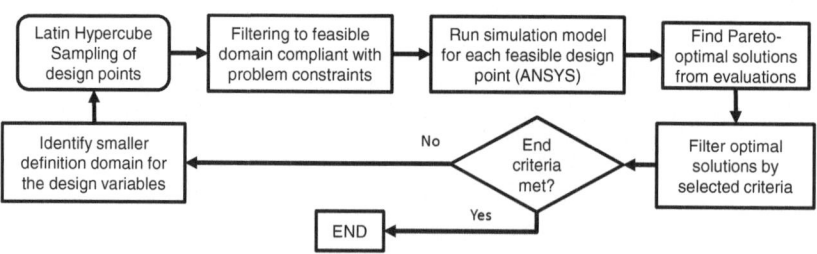

Fig. 9. Procedure employed for optimization

The results obtained for the optimization problem are shown below: , Figure 11 shows the direct comparison with respect to the position in the design specifications, while Figure 10 shows the results in terms of absolute errors. The results obtained are within the acceptable tolerances.

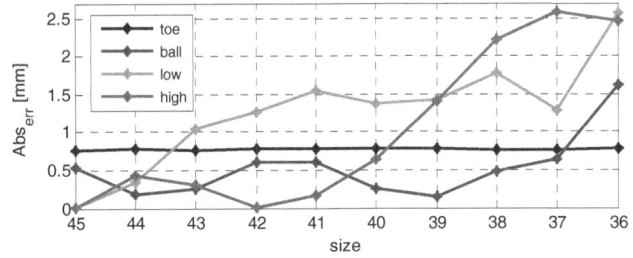

Fig. 10. Absolute errors (mm) between design specific and optimized results (for each section)

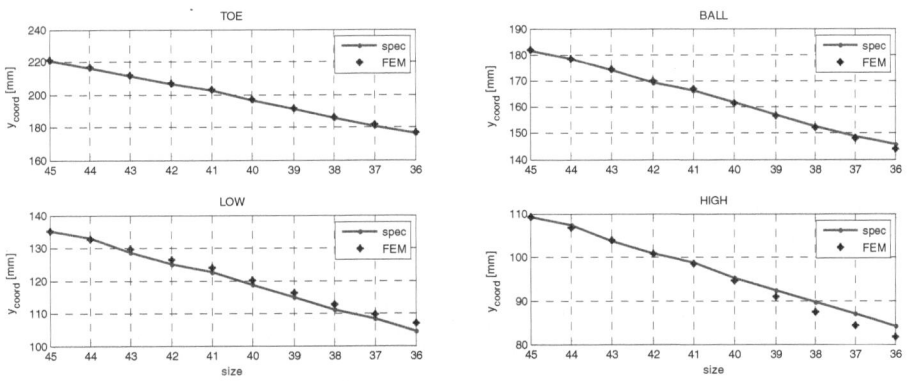

Fig. 11. Comparison between optimized results and design specifics

In order to assess the validity of the FEM model for the FG, specific measurements were done on the constructed prototype. For each size, the position of each section was measured and compared to the value obtained from the model. The absolute errors computed are shown in Figure 12. The results show a very good alignment between the experimental data and the model, with errors bounded within a 1mm tolerance, indicating a good level of adequacy of the model to the problem in consideration. Preliminary wearability tests conducted with the users confirm that those differences are not perceived in terms of comfort. Thus, it can be concluded that the prototype obtained performs well in terms both of geometrical specifications and wearability.

3.3   "Net" and "width Mechanisms" Description

As mentioned in the §3.1 the device is composed by several rigid metal parts (body) linked with deformable rubber elements. The metal parts have the function to provide

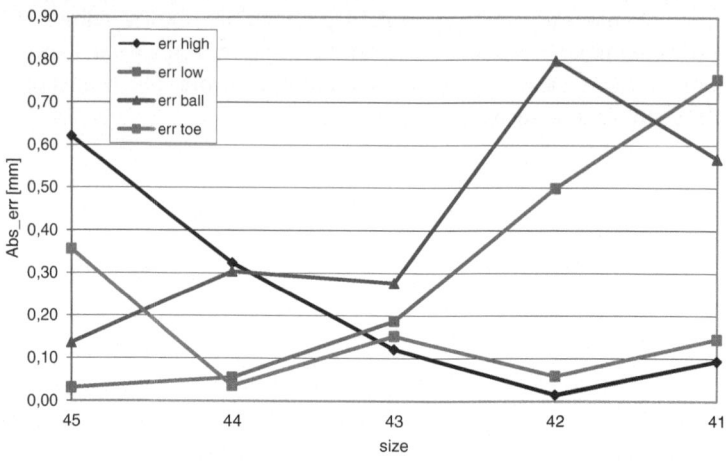

Fig. 12. Errors between FEM and measurements on the prototype

a solid support to the foot in the area where it applies the maximum pressure during the standing or the walking. These parts also contain the mechanisms to set up the "Net" and to regulate the width of the device in each section. For each section, as shown in Figure 13, the metal parts have been machined in the upper side in order to simulate the shape of the insole of the shoes family used as case study.

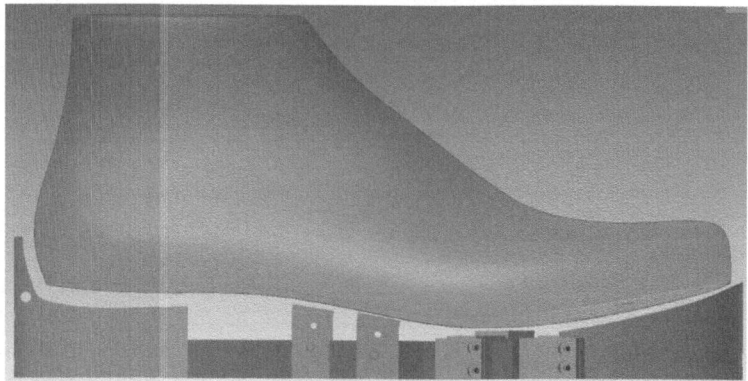

Fig. 13. The FootGlove metal parts (in orange) are machined following the insole shape (red line) of the reference last

The mechanism for the "growing in width" is composed by two elements moved by a screw with both the left and the right thread (Figure 14 a). By screwing or unscrewing the two elements become closer or farther until the right displacement is reached. The left and right thread have different pitches because during the grading the internal part of the insole grows less than the external one, this choice has permitted to maintain the outer parts at the proper distance with respect to the longitudinal plane (Figure 1). Besides setting the width of the sections, the two elements guided by the screw (figure 14 b) are also involved in the managing of the net: on the element in yellow two threaded holes have been created, having the task to fix one of the ends of the net while the parts in green present a hole guiding the net. The dimension of the hole allows the net to slide without friction.

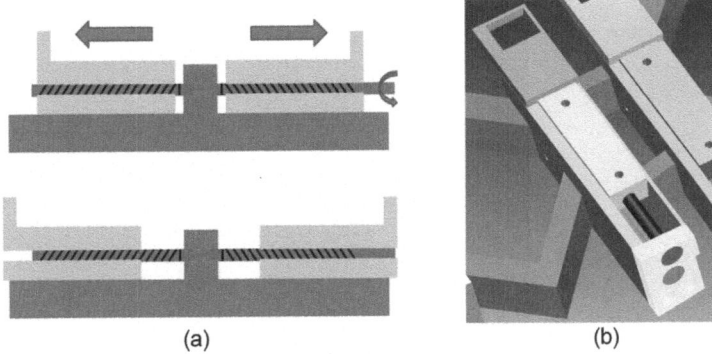

(a)   (b)

Fig. 14. "Growing in width" mechanism: (a) schematic description, (b) zoom on the components

A semi-rigid rubber constitutes the "Net" (figure 15). It is linked at one end to the slider for the "growing in width" mechanism and, at the other end, to another slider, driven by a screw, mounted in the lower part of the body. The choice of a semi-rigid rubber is dictated by the requirement that the net cannot collapse on itself when the device is not worn and to prevent tilt modifications due to the foot insertion in the device. This rubber has elastic properties similar to the material used for the upper of the shoes of the case study thus enhancing the feeling of wearing real shoes.

Drive acting on a screw regulates the mechanism; by screwing or unscrewing, the length is modified.

Fig. 15. Pictures of the FootGlove highlighting the *"Net"* during the no wear phase

The last aspect to be described is the insertion of a lace in the heel area, which is shown Figure 16 by the red circle. The lace has the double function to improve the perception of wearing a real shoe and to maintain the FG tied to the foot during the walk.

Fig. 16. The particular of the lace seat in the heel area

4   Conclusion

The choices made during the design phase of the FG have allowed to reach all the requirements described in the §3. Tests have been performed both on the wearability and on the walkability of the device, obtaining the results that were expected. Through the compilation of a questionnaire, the testers have also validated that the FG provides the proper dimensional feeling about the shoe dimensions and that the FG is feasible to be used for finding the best fitting shoes.

5   Further Work

In order to improve the feeling perception, a solution has to be studied to link the net with a cover providing continuity to the upper surface that now is more similar to a sandal than to a closed shoe. The possibility of coupling the FG and the MagicMirror socks has already been tested [4].

In order to provide a continuous surface for the foot-rest, the sole will be re-designed as a single object made by a mixture of aluminium and rubber which will mount on board the stepper motor.

Furthermore, the prototype will be equipped with a new mechanism for the heel height simulation to increase the FG capability to simulate real heels shoes.

References

1. Boer, C.R., Dulio, S.: Mass customization and footwear: myth, salvation or reality? A comprehensive analysis of the adoption of the mass customization paradigm in the footwear, from the perspective of the EUROShoE (Extend User Oriented Shoe Enterprise) Research Project. Springer- Verlag London limited (2007) ISBN-13-9781846288647
2. FIT4U, Framework of Integrated Technologies for User Centred Products. Collaborative Research Project, project number 229336, http://www.fit4u.eu
3. Blattner, M.: Il mondo della Scarpa. Schweizerischer Schuhhändler-Verband (2001) ISBN 3-9522096-4-3
4. Redaelli, C., Pellegrini, R., Mottura, S., Sacco, M.: Shoe customers behaviour with new technologies: the Magic Mirror case. 15th International Conference on Concurrent Enterprising, ICE 2009, Leiden, Holland, June 22-24 (2009) ISBN 978-0-85358-259-5, http://www.ice-proceedings.org
5. Ali, A., Hosseini, M., Sahari, B.B.: A review of constitutive models for rubber-like materials. American Journal of Engineering and Applied Sciences 3(1), 232–239 (2010)
6. Mase, G.T., Mase, G.E.: Continuum Mechanics for Engineers, 2nd edn. CRC Press, Boca Raton (1999)
7. Arruda, E.M., Boyce, M.C.: Constitutive Models of Rubber Elasticity: A review. Rubber Chem. Technol. 81, 837–848 (2000)
8. Miller, K.: Testing Elastomers for Hyper-elastic Material Models in FE Analysis (2000), http://www.axelproducts.com/downloads/TestingForHyperelastic.pdf
9. Mastinu, G., Gobbi, M., Miano, C.: Optimal design of complex mechanical systems. Springer, New York (2006)

Discrimination of Springs with Vision, Proprioception, and Artificial Skin Stretch Cues

Netta Gurari[1], Jason Wheeler[2,3], Amy Shelton[4], and Allison M. Okamura[1,2]

[1] Department of Mechanical Engineering,
Johns Hopkins University, Baltimore, MD, 21218, USA
[2] Department of Mechanical Engineering, Stanford University,
Stanford, California, 94305, USA
[3] Intelligent Systems Controls, Sandia National Laboratories,
Albuquerque, NM, 87185, USA
[4] Department of Pyschological and Brain Sciences,
Johns Hopkins University, Baltimore, MD, 21218, USA

Abstract. During upper-limb prosthesis use, proprioception is not available so visual cues are used to identify the location of the artificial limb. We investigate the efficacy of a skin stretch device for artificially relaying proprioception during a spring discrimination task, with the goal of enabling the task to be achieved in the absence of vision. In this study, intact users perceive the location of a virtual prosthetic limb using each of four sensory conditions: Vision, Proprioception, Skin Stretch, and Skin Stretch with Vision. For the conditions with skin stretch, a haptic device stretches the forearm skin by an amount proportional to the angular rotation of a virtual prosthetic limb. Sensory condition was not found to significantly influence task performance, exploration methods, or perceived usefulness. We conclude that, in the absence of vision, artificial skin stretch could be used by prosthesis wearers to obtain position/motion information and identify the behavior of a spring.

Keywords: Proprioception, Compliance, Sensory Substitution, Prosthetics.

1 Introduction

Proprioception, the perception of the location of one's limbs in space and how they are moving, is necessary for intuitive and fluid motor control [5,24]. Trained movements can be performed with minimal or no proprioception, possibly using internal models, low-level controllers (an open-loop control strategy), and/or visual cues. However, when users interact with new environments, particularly in the absence of sight, proprioception is necessary to close the feedback loop and provide real-time haptic information about the limb.

Currently when an upper-extremity amputee controls an artificial limb, proprioceptive information is primarily relayed through sight (Figure 1). However, for scenarios such as when lights are turned off in a room or at night, vision is not available. Also, upper-limb amputees desire that the visual demand of a prosthesis be reduced [1]. State-of-the-art in commercially available upper-extremity prostheses enables only minimal non-visual position feedback, via auditory cues and socket force/torque cues. Several groups are investigating methods for relaying proprioceptive information neurally

P. Isokoski and J. Springare (Eds.): EuroHaptics 2012, Part I, LNCS 7282, pp. 160–172, 2012.

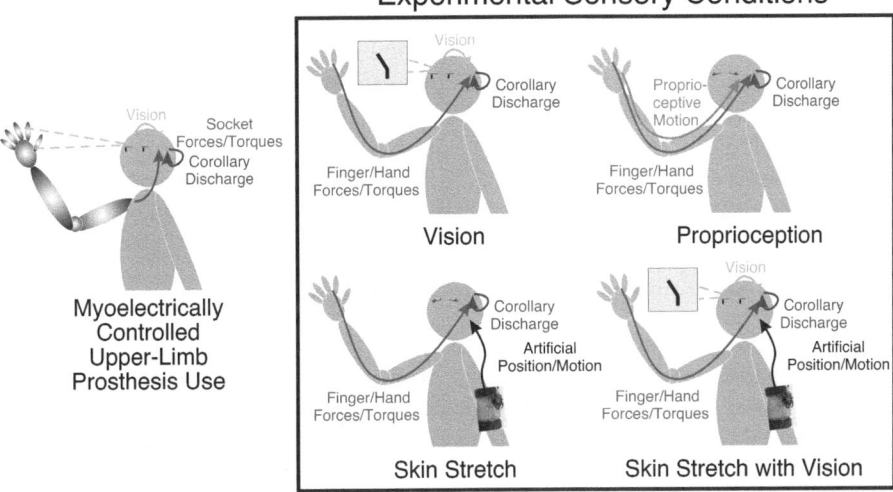

Fig. 1. Description of how users of myoelectric upper-limb prostheses and our system may perceive the location of the artificial limb

[8,14], but decades may pass before such technology is available to safely convey information to humans. Researchers are also investigating methods for relaying proprioceptive information noninvasively using sensory substitution [2,4,20,21,22]. For example, Bark, et al. developed a novel skin stretching device for this purpose [2,29].

Here, we test the feasibility of the skin stretch device [2,29] to artificially convey proprioception during an object interaction task: pressing on a linear spring. This simple task characterizes the effect of proprioception during activities such as identifying the ripeness of a piece of fruit or the air pressure in a tire. The stiffness of a spring is defined by the relationship between the amount the object compresses and the applied force, or from the human's perspective, the relationship between the amount a body part travels and the force the human applies. Prior work showed that when position/motion cues were perceived proprioceptively, spring discrimination performance did not significantly differ from when position/motion cues were perceived visually [12]. This study is the first to compare task performance in a scenario relaying artificial proprioception (via skin stretch feedback), natural proprioception, and visual feedback. Our goal is to determine whether the skin stretch device is suitable for relaying positional cues to an upper-extremity amputee using an artificial limb, so that a prosthetic arm can be used in the absence of sight.

2 Background

2.1 Spring Perception

Proprioceptive, force, cutaneous, and/or visual cues combine to define an object's stiffness [15,19,25,27,28] by integrating information from sources such as muscle spindle fibers, Golgi tendon organs, joint angle receptors, cutaneous mechanoreceptors, and corollary discharges (a copy of the efferent command that is sent to the sensory perception area of the brain) [6,7,9,10,17]. To discriminate springs, humans use both force and position cues when the distance the limb travels is variable, whereas humans use only force cues when the distance the limb travels is fixed [18,23,26]. Position cues are always sensed to identify the distance traveled. Spring discrimination capabilities may improve if visual cues are relayed in addition to proprioception [28]; however, the visual cues may dominate over kinesthetic cues when a visual-haptic mismatch exists [25].

2.2   Sensory Substitution for Upper-Limb Prosthesis Users

To relay proprioceptive information artificially, it is necessary to consider which sensory modality to stimulate (e.g., audio, visual, haptic) and the location to stimulate (e.g., foot, arm). Additionally, the sensory display should be unobtrusive and provide background information that does not interfere with one's daily activities [16]. We aim to provide haptic feedback on a part of the body that is not actively used, and provide noninvasive stimulation for reasons of safety and user comfort.

Electrocutaneous and vibrotactile stimulation have been explored for haptically and noninvasively relaying information during upper-limb prosthesis use [2,4,13,20,21,22]. A major concern for electrocutaneous feedback is that the cues may be uncomfortable, or even painful, if the stimulus is not properly designed. For vibratory feedback, a major concern is that the stimulation may be uncomfortable and/or ignored if the signal is continuous. More recently, skin stretching devices were designed for artificially relaying information [2,11]. These devices can relay positional cues by giving a direct mapping between skin stretch amount and the artificial limb's location. Skin stretch feedback was shown to give superior targeting capabilities to vibratory feedback and no feedback [2].

3   Experimental Setup

3.1   Hardware

A custom user interface and skin stretch device, as shown in Figure 2, are used for testing. Haptic and visual rendering updates occur at 1 kHz and 33 Hz, respectively.

Custom User Interface. The apparatus was first described in [12] and was slightly modified for this study. It is a one-degree-of-freedom, impedance-controlled haptic device that allows control of a virtual spring by rotating one's right index finger. The mechanical workspace of the device is approximately $-10°$ to $60°$ of rotation. The user's right index finger attaches to the finger plate by a velcro strap, and the user grasps the cylindrical tube, with outside diameter of 3.2 cm, so that motion is solely about the metacarphophalangeal (MCP) joint. Visual feedback is provided on a Dell 1908WFPt flat panel monitor (Texas, United States) with resolution of 0.2 mm/pixel. The horizontal placement of the monitor was chosen by visual inspection so that the real finger is aligned with the virtual finger, and the vertical placement was chosen so that the real finger is offset from the virtual finger by approximately 0.09 m.

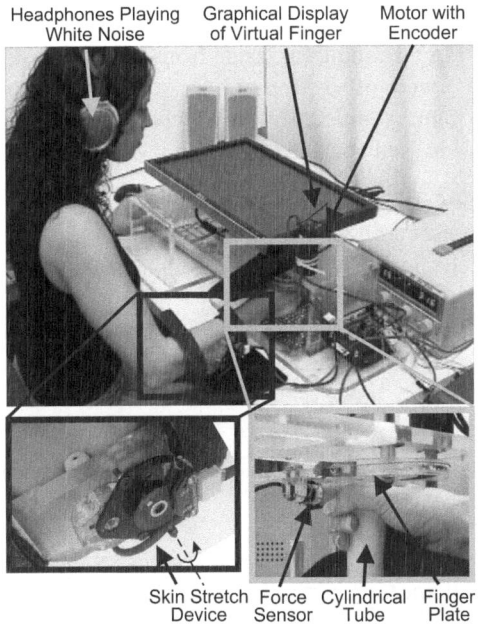

Headphones Playing   Graphical Display   Motor with
White Noise   of Virtual Finger   Encoder

Skin Stretch Force   Cylindrical Finger
Device   Sensor Tube   Plate

Fig. 2. Setup for spring discrimination experiment on a custom haptic interface using skin stretch feedback to the ipsilateral arm

An ATI Nano-17 6-axis force/torque sensor (North Carolina, United States), with resolution of 0.0017 N along the testing axis, measures applied finger force. It is affixed to the finger plate at an adjustable distance of $l_f - 0.015$ m, where l_f is the distance between the MCP joint and finger tip. The finger plate is connected to a capstan drive with a ratio of 10:1, which is driven by a non-geared, backdrivable Maxon RE 40 DC motor (Sachseln, Switzerland). The capstan drive minimizes effects of friction and backlash in the system, and increases the maximum torque output to 1.8 N·m. A HEDS 5540 encoder, with resolution of 0.018°, is attached to the motor.

Skin Stretch Device. A characterization of the skin stretch device and results on its performance during a targeting task are given in [29]. Here, the device conveys proprioception by stretching the skin clockwise and counterclockwise. The maximum comfortable range-of-motion in each direction is 40°. A Shinsei USR30-B3 non-backdriveable ultrasonic motor (Himeji, Japan), with a capstan/cable transmission that has a speed ratio of 6:1, applies a torque to the end-effector, and an optical encoder measures the angular rotation. The end-effector is comprised of two cylindrical disks with diameter of 1.4 cm spaced 2.6 cm apart. The body of the device was created using Shape Deposition Manufacturing [3] and is fastened to the arm with velcro straps.

3.2 Sensory Conditions

The angular rotation of the virtual prosthetic limb is relayed by Vision, Proprioception, Skin Stretch, or Skin Stretch with Vision. For each condition, 11 virtual springs are rendered, displaying stiffnesses of approximately $k_{des_{1,...,11}} = 250$ N/m to 330 N/m in increments of 8 N/m. For all conditions, aside from Vision, the stiffnesses commanded to the motor, $k_{cmd_{1,...,11}}$, were offset from $k_{des_{1,...,11}}$ to compensate for friction and inertial forces in the system. Additionally, for all conditions the low-pass-filtered force measurement, $\hat{F}_{fp,meas}$, is used in the controllers.

Proprioception. The user's finger is permitted to rotate (Proprioception On), or it is held stationary by mechanically locking the finger plate at $0°$ (Proprioception Off). The former motion is elastic and the latter is isometric. The user presses on the finger plate and feels a rotation of the finger about the MCP joint by an amount, θ_f, as measured by the encoder, and a resistive torque from the finger plate. The desired torque output by the motor to the finger plate, $\tau_{fp,output}$, is:

$$\tau_{fp,output} = \frac{\pi(l_f - 0.015)}{180°} \cdot k_{cmd_i} \theta_f (l_f - 0.015), \tag{1}$$

where the commanded stiffnesses are $k_{cmd_{1,...,11}} = (k_{des_{1,...,11}} + 5)$ N/m. A low-level proportional controller with error gain of $k_p = 5.0$ enforces the desired haptic response; thus, the commanded motor torque, τ_{fp}, is:

$$\tau_{fp} = \tau_{fp,output} + k_p[\tau_{fp,output} - \hat{F}_{fp,meas}(l_f - 0.015)]. \tag{2}$$

Skin Stretch. The skin stretch device rotates on the user's forearm to indicate the location of the virtual finger (Skin Stretch On), or it is held stationary (Skin Stretch Off). When Skin Stretch is On, the user's real finger is fixed in place by mechanically locking the finger plate at $0°$. The desired rotation amount for the skin stretch device, $\theta_{ss_{des}}$, is defined by the user-selected maximum force, F_{max}, and the commanded stiffness, $k_{cmd_{1,...,11}} = (k_{des_{1,...,11}} - 7)$ N/m, giving:

$$\theta_{ss_{des}} = \begin{cases} \frac{40° \hat{F}_{fp,meas}}{F_{max}} \frac{k_{cmd_1}}{k_{cmd_i}}, & \text{if } \hat{F}_{fp,meas} \leq F_{max} \\ 40°, & \text{if } \hat{F}_{fp,meas} > F_{max} \end{cases} \tag{3}$$

The rotation amount of the device, $\theta_{ss_{meas}}$, is obtained from encoder measurements, and a low-level proportional controller with error gain of $k_{p_{ss}} = 2.0 \ s^{-1}$ commands the rotational speed, $\omega_{ss_{com}}$, as:

$$\omega_{ss_{com}} = \begin{cases} 0, & \text{if } |\theta_{ss_{des}} - \theta_{ss_{meas}}| < 0.08° \\ \omega_{ss_{com}} = k_{p_{ss}}(\theta_{ss_{des}} - \theta_{ss_{meas}}), & \text{otherwise} \end{cases}, \tag{4}$$

where a deadband of $0.08°$ was implemented to eliminate chatter.

For the Skin Stretch with Vision condition, angular rotation of the virtual finger, θ_{vf}, on the monitor is displayed as:

$$\theta_{vf} = \frac{\theta_{ss_{meas}} F_{max}}{40° k_{cmd_1}} \frac{180°}{\pi l_f}. \tag{5}$$

Vision. A graphical representation of the virtual finger is either displayed on the monitor (Vision On), or it is not displayed (Vision Off). The virtual finger is rendered as a vertical line the length of the user's real finger, l_f. The lower endpoint of the virtual finger is connected to a stationary vertical line that represents the subject's hand, and both lines are solid black with a thickness of 2.0 mm. The commanded stiffnesses are $k_{cmd_{1,...,11}} = k_{des_{1,...,11}}$ N/m, and the amount the virtual finger rotates about its lower endpoint in the counter-clockwise direction from the $0°$ vertical positions, θ_{vf}, is:

$$\theta_{vf} = \frac{\hat{F}_{fp,meas}}{k_{cmd_i}} \frac{180°}{\pi(l_f - 0.015)}. \tag{6}$$

3.3 Relationship to Myoelectrically Controlled Upper-Limb Prosthesis Use

The goal of this study was to assess the feasibility of using skin stretch for relaying proprioceptive information under idealized control conditions. The primary application we focus on is myoelectrically controlled upper-limb prosthesis use; however, this study can extend to the role of artificially relaying proprioception for other scenarios including teleoperating a system. Figure 1 displays how a myoelectrically controlled upper-limb prosthesis user and a user of our experimental setup perceive the location of the artificial limb. The efferent commands are similar for all scenarios – for our study, the force sensor measures the user's intentions and for upper-limb prosthesis use, EMG sensors measure the user's intentions. Measurements of the user's intended motion are cleaner and clearer with our system than with myoelectric sensing, and the ability to control the virtual limb with our system is less variable to real-world effects of friction and inertia that are witnessed when using a real artificial limb. For this reason, we conduct testing with intact individuals and not upper-extremity amputees.

Differences in sensing occur in the afferent pathways, specifically how motion is relayed. For all scenarios, corollary discharges convey position/motion information. Additionally, forces convey location; for our study, forces are perceived at the right index finger and palm of the hand, which is grasping the cylindrical tube, and for artificial limbs, forces are perceived throughout the socket, or the location where the prosthesis attaches to the user. Motion is also relayed visually, proprioceptively, and/or through skin stretch. During upper-limb prosthesis use and our Vision condition, the motion of the artificial limb and of the virtual line are respectively perceived through sight. During Proprioception, motion is perceived through the movement of the real finger. During Skin Stretch, motion is perceived through the stretching of the user's skin. And last, during Skin Stretch with Vision, motion is perceived through both skin stretching and sight. For this study, the real finger is fixed in space, and for upper-limb prosthesis use, the phantom limb may be at a fixed location. Thus, for all scenarios, aside from Proprioception, stationary positional cues are relayed and sensory integration likely occurs.

4 Methods

4.1 Task

The method of constant stimuli was used to quantify perceptual performance. Participants completed a two-alternative, forced-choice task by pressing on two springs – a comparison ($k_{cmd_{1,\cdots,5,7,\cdots,11}}$) and standard ($k_{cmd_6}$) – and indicating which is more compliant. Pilot testing indicated that discriminating compliance was more intuitive than stiffness (the more compliant the spring, the more the skin stretch device rotates).

To begin a trial, the participant presses a specified key on the keyboard. All keyboard presses were made using the left hand. The participant uses the right index finger to press on the finger plate, rotating about the MCP joint only. To switch between springs, he or she applies less than a threshold force and then presses the space bar. The threshold force ensures that when the springs change, the participant does not perceive a visual and/or haptic change. The speed of rotation of the natural and artificial fingers is limited to less than $100°$/sec to ensure that linear virtual springs with comparable stiffnesses are rendered across all conditions. Also, the maximum applied force allowed is set to F_{max} to ensure that the whole workspace of the skin stretch device is used. If the participant's speed or applied force exceeds a limit, a visual indicator is posted on the monitor and is only removed by pressing the space bar. Once finished exploring the springs, the participant presses the '1' or '2' key on the keyboard to identify the spring that is more compliant; this marks the end of the trial. The participant is limited to 30 s of exploration time for each trial, excluding time during which error messages are displayed. This was shown to be a reasonable length of time to comfortably explore the springs and respond [12].

4.2 Procedures

The experiment was conducted in two sessions to minimize fatigue and boredom. Sessions were held on consecutive days and each lasted approximately 90 min. During each session, the participant was presented with two of the four conditions. Presentation order of conditions, presentation order of comparison springs, and placement location of the comparison spring (i.e., Spring 1 versus 2) was randomized for all participants across all conditions.

The first session began by adjusting the haptic interface for the participant's comfort and use. The participant was outfitted with the skin stretch device, which was to be worn throughout the entire experiment, on the right upper forearm. Red-e Tape^{TM}, a strong skin-safe adhesive, maintained contact between the participant's arm and the device's end-effector. Next, the participant calibrated the device by pressing the force sensor, which was locked in place, with a moderately high force, F_{max}, that could be sustained throughout the session. Then the skin-stretch controller was turned On, and the participant pressed on the force sensor to learn the mapping between the input force and the output device rotation. Four trials were performed for each condition, for a total of 16 trials, with the softest spring presented twice followed by the stiffest spring. Presentation order of the conditions during this learning phase was always Proprioception, Vision, Skin Stretch, and Skin Stretch with Vision. During Skin Stretch with Vision, the participant was instructed to attend to both sensory modalities equally so that the weighting of each would be equal.

At this time, the participant began the practice trials. His/her hand was concealed so that visual feedback from the real finger's motion was not possible. The participant's measured finger speed was indicated by a visual display. The discrimination task was performed once for each comparison spring for a total of ten trials, and feedback was

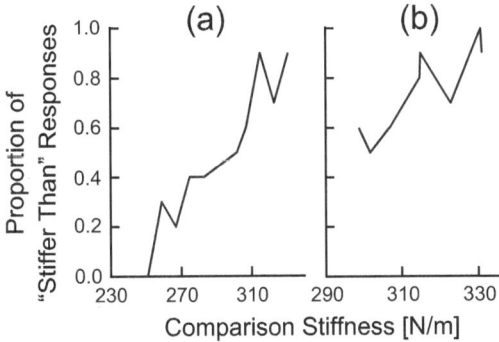

Fig. 3. Psychometric curve for an example condition of an example participant. Proportion of "stiffer than" responses for each comparison spring (a) as data was originally collected, and (b) in the flipped- and folded-over condition.

given on whether the response was correct so that the range of stiffnesses was learned. The participant then began the testing trials. Headphones playing white noise were worn to mask auditory cues from the apparatus and distractions in the room. The discrimination task was performed 10 times for each of the 10 comparison springs, for a total of 100 trials, with a 90 s rest break after every block of 25 trials. Feedback was not provided as to whether the participant responded correctly so that learning no longer occurred. Last, the participant completed a questionnaire, commenting on methods used to discriminate between springs (e.g., pressing the spring and holding at a constant force, pressing continuously) and rating the difficulty of the task. After completing the questionnaire for the first condition, the participant then completed the practice, testing, and questionnaire phases for the second condition. This marked the completion of the first session.

The second session was nearly identical to the first, with the participant completing the practice, testing, and questionnaire phases for the two remaining conditions. Then the participant filled out an additional questionnaire, giving his/her age, gender, experience with haptic virtual environments, health concerns, experience using the right finger in dexterous motions, ranking of the usefulness of each condition for discriminating, and additional comments.

4.3   Subjects

Approval was obtained from the Johns Hopkins University Homewood Institutional Review Board to collect data from human subjects. Six female and two male volunteers, who were all healthy with no neurological illnesses or right hand impairments, participated with informed consent. They ranged in age from 19 to 27 years, in right index finger length from 7.8 cm to 9.4 cm, and in self-reported virtual environment experience from 'None' to 'Some'. Participants were monetarily compensated for their time.

4.4   Data Analysis

Participant responses were first converted from compliance to stiffness to be consistent with prior work in which the study is discussed using stiffness. The proportion of responses in which the user states that the ith comparison spring is stiffer than the standard spring, PS_i, was then calculated. The average stiffness of all springs, $k_{1,...,11}$, was estimated using the method of total least squares, in which a best-fit line quantified the relationship between the force and position data acquired when pushed the spring [19]. Figure 3(a) plots the psychometric curve of PS_i versus k_i for an example participant. Next, we quantified perceptual performance by calculating Area Under the Normalized Curve (AUNC), since it gives a cumulative description of discrimination capabilities and allows perceptual performance to be compared across conditions. The more traditional estimation of Weber fraction to compare perceptual performance was not used since we were not able to a get a goodness-of-fit for a number of the conditions and subjects with this analysis method.

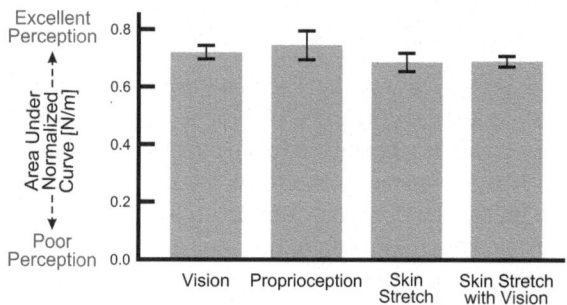

Fig. 4. Mean and standard error of AUNC across all sensory conditions for all subjects

To obtain AUNC, the psychometric curve, e.g., Figure 3(a), was flipped- and folded-over (FFO), e.g., Figure 3(b), to account for perceptual biases about the standard spring. Thus, $k_{FFO_{1,...,5,7,...,11}}$ and $PS_{FFO_{1,...,5,7,...,11}}$ were obtained respectively by:

$$k_{FFO_i} = \begin{cases} k_6 + (k_6 - k_i), & \text{for } i = 1,...,5 \\ k_{i+1}, & \text{for } i = 6,...,10 \end{cases}, \text{ and} \quad (7)$$

$$PS_{FFO_i} = \begin{cases} \frac{1}{2} + (\frac{1}{2} - PS_i), & \text{for } i = 1,...,5 \\ PS_{i+1}, & \text{for } i = 6,...,10 \end{cases}. \quad (8)$$

Then $k_{FFO_{1,...,10}}$ and the corresponding $PS_{FFO_{1,...,10}}$ were sorted from lowest to highest stiffness, and AUNC, A_{UNC}, was estimated by:

$$A_{Rectangle_j} = PS_{FFO_j}(k_{FFO_{j+1}} - k_{FFO_j}) \quad (9)$$

$$A_{Triangle_j} = \frac{1}{2}(PS_{FFO_{j+1}} - PS_{FFO_j})(k_{FFO_{j+1}} - k_{FFO_j}) \quad (10)$$

$$A_{UNC} = \sum_{j=1}^{9}(A_{Rectangle_j} + A_{Triangle_j}). \quad (11)$$

5   Results

5.1   Task Performance

A within-subjects one-way analysis of variance (ANOVA) with a Geisser-Greenhouse epsilon-hat adjustment to correct for violations of sphericity evaluated the effect of condition on task performance and exploration methods. AUNC results for all subjects across all conditions are summarized in Figure 4; sensory condition was not found to significantly affect perceptual performance [$F(3,21) = 0.84$, $\hat{\varepsilon} = 0.5134$, $p = 0.43$]. This finding indicates that the spring discrimination task may be achieved using position/motion feedback from the skin stretch device. Sensory type was also not found to affect number of spring presses [$F(3,21) = 0.42$, $\hat{\varepsilon} = 0.5800$, $p = 0.64$], total time spent pressing springs [$F(3,21) = 1.54$, $\hat{\varepsilon} = 0.6964$, $p = 0.25$], or maximum penetration depth [$F(3,21) = 1.18$, $\hat{\varepsilon} = 0.5191$, $p = 0.33$]. This suggests that the manner in which the discrimination task is performed may not significantly change if the skin stretch device is used to convey positional cues.

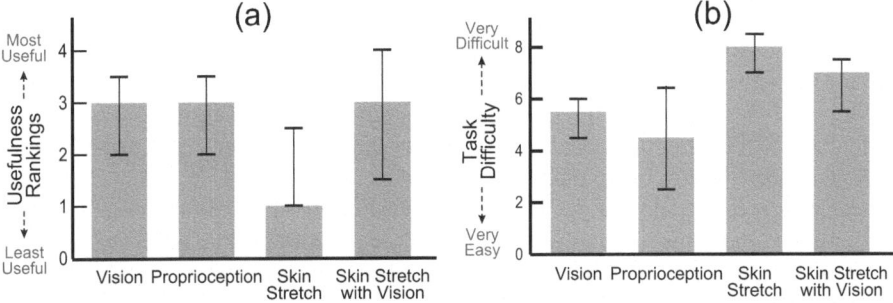

Fig. 5. Median, lower, and upper quartiles across all sensory conditions and subjects for (a) usefulness rankings and (b) task difficulty ratings

5.2   Questionnaire

The Friedman test, a nonparametric version of the within-subjects one-way ANOVA, identified the effect of condition on perceived task difficulty and usefulness rankings. User ratings for task difficulty across all subjects and conditions are given in Figure 5(a). Sensory condition was found to have a significant effect [$\chi^2(3) = 9.64$, $p = 0.022$], however a post-hoc analysis using the Wilcoxon Signed-Rank Test and a Bonferroni correction (significance level of $p < 0.008$), did not find any pairs to significantly differ. Usefulness rankings across all conditions and subjects are shown in Figure 5(b); sensory condition was not found to have an effect [$\chi^2(3) = 4.95$, $p = 0.176$].

6   Discussion

The feasibility of using a skin stretch device to replace visual cues during upper-limb prosthesis use was investigated. The skin stretch device was not found to alter task performance or exploration method when compared to natural vision or proprioception. A possible reason for why statistical significance was not observed is because of the limited number of subjects tested. However, the means in AUNC differ by such a small amount across the conditions, that practically, in the absence of sight, the force cues felt at the right index finger and hand along with the positional cues relayed by the skin stretch device provide sufficient information for discriminating springs. Thus, the results indicate that the device is suitable for relaying position information during the tested object manipulation task.

Subjective responses gave similar results to the quantitative findings. Usefulness rankings for each condition were not found to significantly differ. The lack of significant differences may be because of the limited information conveyed from the data collected; the more informative ratings (e.g., 1 through 10), rather than rankings, may have been more descriptive. A statistically significant difference was found for task difficulty, but the post-hoc test used for nonparametric data is very conservative and did not find any significantly different pairs. Based on the median values in Figure 5, it seems that the task is perceived as more difficult when using the skin stretch device than when using sight or proprioception; additional subject testing will illuminate whether this is true.

The addition of vision to skin stretch was not found to enhance perceptual performance or the subjective experience over skin stretch alone. Even though a goal is to reduce the visual demand while using a prosthesis [1], based on our results we do not recommend the skin stretch device for use in addition to sight during similar manipulation tasks. Obtaining positional cues from the skin stretch device may make the task feel more difficult than using sight alone.

Future research could test the effect of longer training time with the device on task performance and subjective ratings. Also, additional studies could investigate the feasibility of using several skin stretch devices simultaneously to convey a richer and higher dimensionality of hand configuration.

Acknowledgments. This work was supported by the JHU Applied Physics Laboratory under the DARPA Revolutionizing Prosthetics program, contract N66001-06-C-8005, a NSF Graduate Research Fellowship, Johns Hopkins University, Brain Science Institute, a travel award from the IEEE Technical Committee on Haptics, and Stanford University. We thank Mark Cutkosky and Karlin Bark for their help.

References

1. Atkins, D.J., Heard, D.C.Y., Donovan, W.H.: Epidemiologic overview of individuals with upper-limb loss and their reported research priorities. Journal of Prosthetics and Orthotics 8(1), 2–11 (1996)
2. Bark, K., Wheeler, J.W., Premakumar, S., Cutkosky, M.R.: Comparison of skin stretch and vibrotactile stimulation for feedback of proprioceptive information. In: Proceedings of the 16th International Symposium on Haptic Interfaces for Virtual Environments and Teleoperator Systems, pp. 71–78 (2008)
3. Binnard, M., Cutkosky, M.R.: Design by composition for layered manufacturing. Journal of Mechanical Design 122(1), 91–101 (2000)

4. Cipriani, C., Zaccone, F., Micera, S., Carrozza, M.C.: On the shared control of an EMG-controlled prosthetic hand: Analysis of user-prosthesis interaction. IEEE Transactions on Robotics 24(1), 170–184 (2008)
5. Cole, J.: Pride and a Daily Marathon. The MIT Press (1995)
6. Collins, D.F., Prochazka, A.: Movement illusions evoked by ensemble cutaneous input from the dorsum of the human hand. Journal of Physiology 496(3), 857–871 (1996)
7. Collins, D.F., Refshauge, K.M., Todd, G., Gandevia, S.C.: Cutaneous receptors contribute to kinesthesia at the index finger, elbow, and knee. Journal of Neurophysiology 94(3), 1699–1706 (2005)
8. Dhillon, G.S., Horch, K.W.: Direct neural sensory feedback and control of a prosthetic arm. IEEE Transactions on Neural Systems and Rehabilitation Engineering 13(4), 468–472 (2005)
9. Edin, B.B.: Quantitative analyses of dynamic strain sensitivity in human skin mechanoreceptors. Journal of Neurophysiology 92(6), 3233–3243 (2004)
10. Gandevia, S.C., Smith, J.L., Crawford, M., Proske, U., Taylor, J.L.: Motor commands contribute to human position sense. Journal of Physiology 571(3), 703–710 (2006)
11. Gleeson, B.T., Horschel, S.K., Provancher, W.R.: Communication of direction through lateral skin stretch at the fingertip. In: Third Joint EuroHaptics Conference and Symposium on Haptic Interfaces for Virtual Environment and Teleoperator Systems, pp. 172–179 (2009)
12. Gurari, N., Kuchenbecker, K.J., Okamura, A.M.: Stiffness discrimination with visual and proprioceptive cues. In: Proceedings of the Third Joint Eurohaptics Conference and Symposium on Haptic Interfaces for Virtual Environment and Teleoperator Systems, pp. 121–126 (2009)
13. Kaczmarek, K.A., Webster, J.G., Rita, P.B., Tompkins, W.J.: Electrotactile and vibrotactile displays for sensory substitution systems. IEEE Transactions on Biomedical Engineering 38(1), 1–16 (1991)
14. Kuiken, T.A., Marasco, P.D., Lock, B.A., Harden, R.N., Dewald, J.P.A.: Redirection of cutaneous sensation from the hand to the chest skin of human amputees with targeted reinnervation. Proceedings of the National Academy of Sciences of the United States of America 104(50), 20061–20066 (2007)
15. Kuschel, M., Luca, M.D., Buss, M., Klatzky, R.L.: Combination and integration in the perception of visual-haptic compliance information. IEEE Transactions on Haptics 99(4), 234–244 (2010)
16. MacLean, K.E.: Putting haptics into the ambience. Transactions on Haptics 2(3), 123–135 (2009)
17. McCloskey, D.I.: Kinesthetic sensibility. Physiological Reviews 58(4), 763–820 (1978)
18. Paljic, A., Burkhardt, J.M., Coquillart, S.: Evaluation of pseudo-haptic feedback for simulating torque: a comparison between isometric and elastic input devices. In: Proceedings of the 12th International Symposium on Haptic Interfaces for Virtual Environment and Teleoperator Systems. Stiffness dscrimination isometric elastic haptic device, pp. 216–223 (2004)
19. Pressman, A., Welty, L.J., Karniel, A., Mussa-Ivaldi, F.A.: Perception of delayed stiffness. International Journal of Robotics Research 26(11-12), 1191–1203 (2007)
20. Pylatiuk, C., Kargov, A., Schulz, S.: Design and evaluation of a low-cost force feedback system for myoelectric prosthetic hands. Journal of Prosthetics and Orthotics 18(2), 57–61 (2006)
21. Riso, R.R., Ignagni, A.R.: Electrocutaneous sensory augmentation affords more precise shoulder position command generation for control of FNS orthoses. In: Proceedings of the Annual Conference on Rehabilitation Technology, pp. 228–230 (1985)
22. Rohland, T.A.: Sensory feedback for powered limb prostheses. Medical and Biological Engineering and Computing 13(2), 300–301 (1975)
23. Roland, P.E., Ladegaard-Pedersen, H.: A quantitative analysis of sensations of tensions and of kinaesthesia in man. Brain: a Journal of Neurology 100(4), 671–692 (1977)

24. Sainburg, R.L., Ghilardi, M.F., Poizner, H., Ghez, C.: Control of limb dynamics in normal subjects and patients without proprioception. Journal of Neurophysiology 73(2), 820–835 (1995)
25. Srinivasan, M.A., Beauregard, G.L., Brock, D.L.: The impact of visual information on the haptic perception of stiffness in virtual environments. In: Proceedings of the 5th International Symposium on Haptic Interfaces for Virtual Environment and Teleoperator Systems. American Society of Mechanical Engineers Dynamic Systems and Control Division, vol. 58, pp. 555–559 (1996)
26. Tan, H.Z., Durlach, N.I., Beauregard, G.L., Srinivasan, M.A.: Manual discrimination of compliance using active pinch grasp: the roles of force and work cues. Perception & Psychophysics 4(57), 495–510 (1995)
27. Tiest, W.M.B., Kappers, A.M.L.: Cues for haptic perception of compliance. IEEE Transactions on Haptics 2(4), 189–199 (2009)
28. Varadharajan, V., Klatzky, R., Unger, B., Swendsen, R., Hollis, R.: Haptic rendering and psychophysical evaluation of a virtual three-dimensional helical spring. In: Proceedings of the 16th International Symposium on Haptic Interfaces for Virtual Environments and Teleoperator Systems, pp. 57–64 (2008)
29. Wheeler, J., Bark, K., Savall, J., Cutkosky, M.: Investigation of rotational skin stretch for proprioceptive feedback with application to myoelectric prostheses. IEEE Transactions on Neural Systems and Rehabilitation Engineering 18(1), 58–66 (2009)

Augmentation of Material Property by Modulating Vibration Resulting from Tapping

Taku Hachisu[1,2], Michi Sato[1,2], Shogo Fukushima[1,2], and Hiroyuki Kajimoto[1,3]

[1] The University of Electro-Communications
1-5-1 Chofugaoka, Chofu, Tokyo 182-8585, Japan
[2] JSPS Research Fellow
[3] Japan Science and Technology Agency
{hachisu,michi,shogo,kajimoto}@kaji-lab.jp

Abstract. We present a new haptic augmented reality system that modulates the perceived stiffness of a real object by changing the perceived material with vibratory subtraction and addition. Our system consists of a stick with a vibrotactile actuator and a pad with an elastic sheet. When a user taps the pad, the innate vibration resulting from the tapping is absorbed by the elastic surface. Simultaneously, the vibrotactile actuator provides the intended vibration, which represents a modulated perceived material property such as rubber, wood, or aluminum. The experimental results showed that the participants were able to discern the three materials by tapping.

Keywords: haptic augmented reality, material, stick-type interface, stiffness modulation, vibrotactile sensation.

1 Introduction

Haptic augmented reality (haptic AR) is an emerging haptic research area where a user can touch an augmented or untouchable environment [1, 2]. The system generally consists of a haptic display and a sensor to measure the environment.

Stiffness, which is one of the most fundamental haptic properties, has been successfully modulated by a haptic AR system. Nojima et al. proposed SmartTool, which is composed of a stylus with an active force feedback device and a sensor attached at the tip of the tool [3]. The active force feedback provides a reactive force according to the information detected by the sensor, which lets the user touch and know an untouchable boundary, such as the interface between oil and water, as if the interface got a stiff wall. Jeon and Choi proposed a haptic AR system that modulates the stiffness of a real object [2]. The system is composed of an active force feedback device and force sensor to measure the reaction force from the surface of the real object and control the device.

However, both systems require an active force feedback device, which is generally expensive and complicated. Therefore, in contrast to visual and audio AR systems, it is difficult to apply their systems to daily life, such as for entertainment.

P. Isokoski and J. Springare (Eds.): EuroHaptics 2012, Part I, LNCS 7282, pp. 173–180, 2012.

This paper thus proposes a new haptic AR system to modulate the perceived stiffness of a real object, but with a simple implementation that focuses on a "tapping by a stick" situation. The paper first begins with a review of previous work on stiffness representation by simulation of material vibrations resulting from tapping and stylus-type haptic AR devices. Next, we describe our proposal, which modulates the perceived stiffness of a real object by changing the perceived material via vibratory subtraction and addition. We then present a material identification experiment to demonstrate the efficacy of our proposal. Finally, the paper ends with the conclusion and a description of potential applications.

2   Previous Work

2.1   Haptic Simulation of Tapping Object

When tapping the surface of a hard object, we can discern the material by using haptic cues without requiring visual or acoustic ones. The haptic sensation consists of a kinesthetic sensation (i.e., reactive force from the surface of the object) and vibrotactile sensation (i.e., cutaneous mechanical deformations and vibrations).

Various methods have been proposed to present kinesthetic and vibrotactile sensations. Wellman and Howe proposed mounting of a vibrator on an active force feedback device [4]. Okamura et al. presented both kinesthetic and vibrotactile stimuli solely through an active-force feedback device [5]. Both groups employed the following decaying sinusoidal waveform to simulate the vibration resulting from tapping:

$$Q(t) = A(v)\exp(-Bt)\sin(2\pi ft) \tag{1}$$

The vibratory acceleration Q is determined by the amplitude A as a function of the impact velocity v, decay rate of sinusoid B, and sinusoid frequency f, where A, B, and f are dependent on the type of material. Okamura et al. used their vibration model to simulate three materials (rubber, wood, and aluminum) and demonstrated that users could successfully discern the materials. However, both proposals require an expensive haptic display.

Hachisu et al. proposed a technique using pseudo-haptic feedback to provide kinesthetic sensation instead of an active-force feedback device [6, 7]. Pseudo-haptic feedback is a haptic illusion where visual cues create a haptic sensation without a physical haptic stimulus [8]. However, full substitution of kinesthetic sensation by this illusion is still not possible.

2.2   Stylus-Type Haptic AR Devices

The Haptic Pen [9] and Ubi-Pen [10] are both stylus-type devices with embedded vibrators and tactile actuators. In these devices, the kinesthetic sensation is naturally presented by real contact, whereas the vibrotactile sensation is added to present the geometrical properties on the touch panel.

3   Proposal

3.1   Concept

To deal with the high cost issue of kinesthetic feedback, we apply the real reactive force generated by contact with a real object like previous stylus-type haptic AR devices. We also modulate the perceived stiffness by adding a decaying sinusoid vibration at the moment of contact. However, the addition of the vibration is not enough for modulation because the innate vibration still exists.

We previously proposed the idea of subtracting the vibrotactile sensation by simply using an elastic sheet and then presenting the intended vibration through a vibrotactile actuator embedded in the stick (details are described in the following section) [11].

Furthermore, to achieve real-time superposition of the vibrotactile actuation on the real contact, we use conduction to detect contact between the stick and the surface of a real object; this is simply implemented by using an I/O port on a microcontroller.

3.2   Implementation

Stick. The stick and its internal configuration are shown in Fig. 1. The head is made of acrylonitrile butadiene styrene (ABS) resin, and its surface is covered with conductive coating material; it is connected to a power-supply voltage from the microcontroller. The handle is made of aluminum and has an embedded voice-coil type vibrotactile actuator (Tactile Labs Inc., Haptuator [12]). The length and weight of the stick are 200 mm and 90 g, respectively.

Fig. 2 shows the frequency response of the stick. The data was collected by providing a sinusoidal input (1 Vrms; from 10 Hz to 500 Hz in increments of 10 Hz) to the actuator and measuring the acceleration at the handle using an acceleration sensor (Kinonix Inc., KXM52-1050).

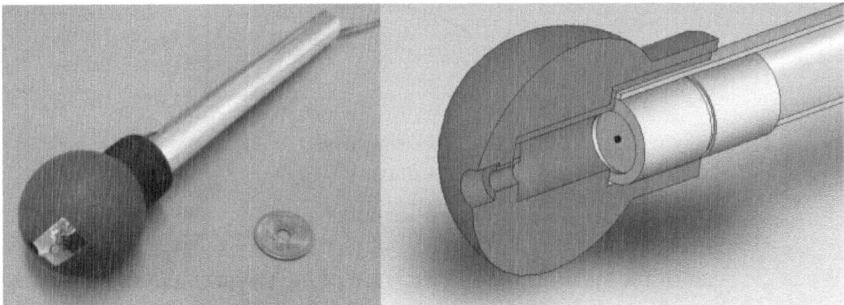

Fig. 1. Stick: The left image shows the exterior, and the right shows the internal configuration. The voice-coil type actuator is embedded.

Pad. The pad is composed of an elastic sheet and a thin conductive sheet on top. The conductive sheet is made of aluminum, and its thickness is less than 0.05 mm. The conductive sheet is connected to an I/O port of a microcontroller (NXP Semiconductors,

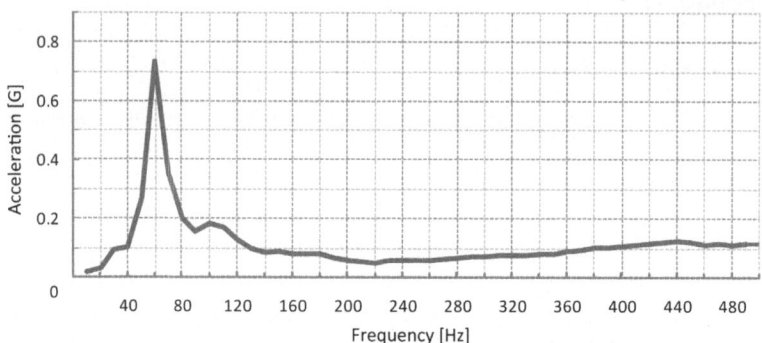

Fig. 2. Frequency response of the stick to various input frequencies

mbed NCP LPC1768) and to a signal ground through a pull-down resistor. The elastic sheet is made of styrene elastomer, which is generally used for impact and vibration absorption. The thickness of the elastic sheet is 3 mm.

System and Principle. Our proposed devices are implemented in a system consisting of the microcontroller, an audio amplifier (Rasteme Systems Co., Ltd., RSDA202), the stick, and the pad (Fig. 3).

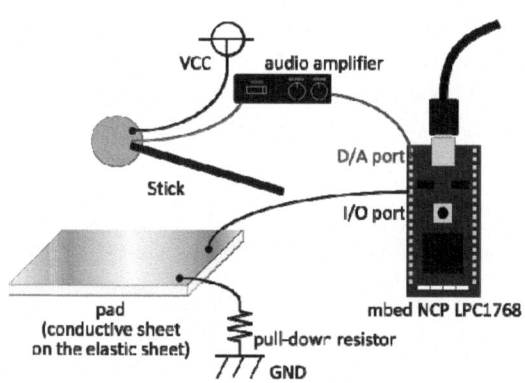

Fig. 3. System configuration

When the head of the stick contacts the surface of the pad, the pad first absorbs the vibration resulting from the collision. At the same time, the voltage of the I/O port changes from low to high. This allows the microcontroller to detect the collision instantly. Then, the microcontroller outputs the decaying sinusoid waveform from its D/A port to the actuator in the stick through the audio amplifier. Finally, the user feels the stiffness modulated by subtracting and adding vibrations. The refresh rate of the D/A port is 10 kHz.

Note that our system employs the decaying sinusoid model as described in equation 1. However, at present, the strength of the impact is not considered, i.e., the initial amplitude is set to be constant.

4   Experiment

We verified the efficacy of our proposal by testing whether participants could discern the materials.

4.1   Experimental Setup and Procedure

In this experiment, we used three pads on an acrylic board, as shown in Fig. 4. Three vibration models (rubber, wood, and aluminum) were applied to each pad in random order. We also prepared real samples of each material for comparison. The size of the pads and real samples was 50 mm × 50 mm × 3 mm. We employed Okamura's parameters [5] as the vibration models, as shown in Table 1. As mentioned in the previous section, our current system fixes the initial amplitude.

The participants were asked to select the perceived materials for each pad from three candidates (rubber, wood, and aluminum). They were informed that there was no overlap. They were allowed to tap the pads and the real samples freely during trials but were asked to do so lightly; this was because our current system cannot handle intense vibrations, and we wanted them to discern the materials via vibrotactile rather than kinesthetic cues. The participants were unaware of the correct answer. There was no limit on the time taken to respond. Each participant performed this identification experiment three times.

Twelve participants—eleven men and one woman—aged between 20 and 35 (mean = 24.4; SD = 3.8) took part in the experiment. All participants were right-handed. None of them was familiar with the research.

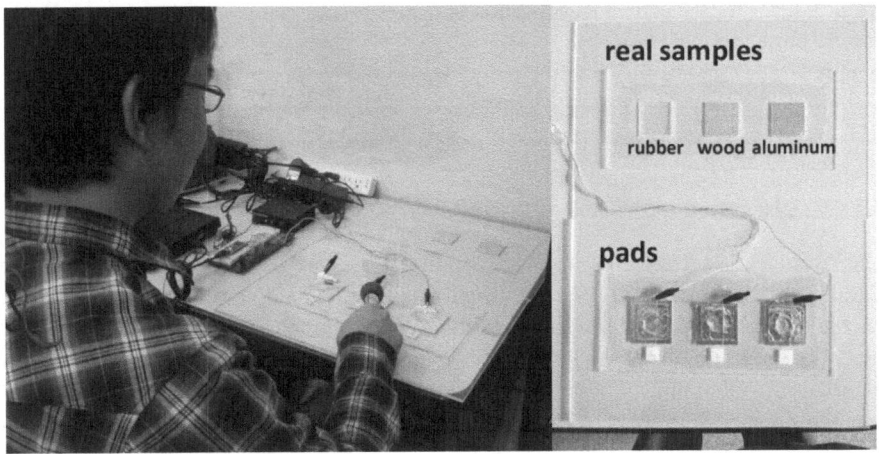

Fig. 4. Setup for the experiment

Table 1. Vibration parameters

	A [m/s^2]	B [s^{-1}]	f [Hz]
Rubber	1.0	60	30
Wood	0.7	80	100
Aluminum	1.3	90	300

4.2   Results

One participant's data was eliminated due to his comment that he tried to identify the material by the reactive force (kinesthetic sensation) and ignored the vibrotactile sensation.

The answer rates for three vibration models on the pads are listed in Table 2. The correct answer rates (cells highlighted in yellow) were higher than 70%. In particular, aluminum was successfully identified in most of the trials (90.9%). Most errors occurred by confusion between rubber and wood.

As shown in Table 3, the correct identification rate (i.e., the correct identification of all three materials by a participant) was 72.7%. The rates increased as the trials progressed, and most of the participants correctly identified all three materials by the third trial (90.9%).

Table 2. Results of the experiment: Answer rates for three vibration models. Yellow cells represent the correct answer rates.

	Vibration models		
Answer	Rubber	Wood	Aluminum
Rubber	81.8%	18.2%	0.0%
Wood	18.2%	72.7%	9.1%
Aluminum	0.0%	9.1%	90.9%

Table 3. Rates for the correct identification of materials (i.e., correct identification of all three materials by a participant)

Fist trial only	Second trial only	Third trial only	Overall
54.5%	72.7%	90.9%	72.7%

4.3   Discussion

The overall correct rate for material identification was lower than the results from Okamura's experiment (83.3%) [5], which employed an active-force feedback device. On the other hand, the rate of the third trial was similar to Okamura's result (85.7%). The tendency for errors owing to confusion between rubber and wood was also reported by Okamura. These observations show that it is possible to represent perceived material properties by employing the decaying sinusoid model in a haptic AR environment as well as a virtual one.

After the experiment, three of the participants reported that they were able to identify the materials through a vibrotactile cue, but the absence of a repulsive sensation induced an uncomfortable sensation. This was because the elastic sheet of the pad absorbed the impact as well as vibration. Employing an elastic sheet with a high coefficient of restitution might be one solution to providing a repulsive force. However, this solution may induce a similar uncomfortable sensation when the user taps the pad for the rubber model. To deal with this difficulty, as future work, we would like to vibrate the pad as well as the stick in order to control the repulsive force.

In addition, we must consider the materials of the conductive sheet and head of the stick because they are also related to the default haptic cue. Thus, it is necessary to adopt the optimal materials for the stick and pad.

Interestingly, two of the participants reported that they felt a magnet-like force that attracted or repelled the stick when tapping the pad. This phenomenon was only observed when the rubber model was adopted. Currently, we cannot explain why this force sensation was generated. Future work will involve investigation of the behavior between the stick and pad using a high-speed camera.

5   Conclusion

This paper describes a haptic AR system that focuses on tapping, which modulates the perceived stiffness of a material by subtracting and adding vibrations. The system consists of a microcontroller, audio amplifier, stick, and pad. We carried out an experiment to demonstrate the efficacy of our proposal. The experimental results showed that the participants were able to discern the three materials that we simulated. Notably, almost all of the participants correctly identified all three materials in the third (final) trial (90.9%).

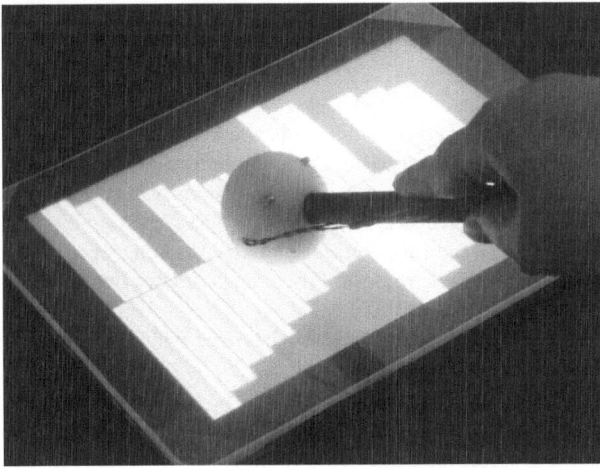

Fig. 5. Example application. Enhancing the touch interface enables haptic interaction with musical instruments (xylophone and glockenspiel) on a tablet PC [11].

The current system employs Okamura's parameters for the decaying sinusoidal waveform [5], which was obtained with a different haptic device. Therefore, the parameters should be obtained with our system for optimization. Furthermore, as described in the previous section, optimization of the materials (elastic sheet, head of the stick, etc.) is another part of future work.

We have considered several applications for the stick-type haptic AR system. Touching and feeling the image of a visual AR is one possibility, which can be achieved by superimposing the visual image on the conductive sheet with a projector (in this case, the aluminum sheet should be replaced by another sheet that will work both as a screen and as a conductive sheet). Enhancing the touch interface with tactile feedback is another possibility. This can be achieved by employing a transparent electrode sheet as the conductive sheet on top of the multitouch interface. This can be applied to musical instruments, especially chromatic percussions such as the xylophone and glockenspiel, as proposed in [11] (Fig. 5).

References

1. Bayart, B., Kheddar, A.: Haptic Augmented Reality Taxonomy: Haptic Enhancing and Enhanced Haptics. In: Proceedings of EuroHaptics, pp. 641–644 (2006)
2. Jeon, S., Choi, S.: Haptic Augmented Reality: Taxonomy and an Example of Stiffness Modulation. Presence 18, 87–408 (2009)
3. Nojima, T., Sekiguchi, D., Inami, M., Tachi, S.: The SmartTool: A System for Augmented Reality of Haptics. In: Proceedings of IEEE Virtual Reality Conference, pp. 67–72 (2002)
4. Wellman, P., Howe, R.D.: Towards Realistic Display in Virtual Environments. Proceedings of the ASME Dynamic Systems and Control Division 57, 713–718 (1995)
5. Okamura, A.M., Cutkosky, M., Dennerlein, J.: Reality Based Models for Vibration Feedback in Virtual Environments. IEEE/ASME Transactions on Mechatronics 6, 245–252 (2001)
6. Hachisu, T., Cirio, G., Marchal, M., Lécuyer, A., Kajimoto, H.: Pseudo-haptic Feedback Augmented with Visual and Tactile Vibration. In: Proceedings of IEEE International Symposium on VR Innovation, pp. 327–328 (2011)
7. Hachisu, T., Cirio, G., Marchal, M., Lécuyer, A., Kajimoto, H.: Virtual Chromatic Percussions Simulated by Pseudo-haptic and Vibrotactile Feedback. In: Proceedings of ACM International Conference on Advances on Computer Entertainment Technology, 20 (2011)
8. Lécuyer, A.: Simulating Haptic Feedback using Vision: A Survey of Research and Applications of Pseudo-haptic Feedback. Presence 18, 39–53 (2009)
9. Lee, J.C., Dietz, P.H., Leigh, D., Yerazunis, W.S., Hudson, S.E.: Haptic Pen: A Tactile Feedback Stylus for Touch Screens. In: Proceedings of User Interface Software and Technology, pp. 291–294 (2004)
10. Kyung, K.U., Lee, J.Y.: Ubi-Pen: A Haptic Interface with Texture and Vibrotactile Display. IEEE Computer Graphics and Applications 29, 56–64 (2009)
11. Hachisu, T., Sato, M., Fukushima, S., Kajimoto, H.: HaCHIStick: Simulating Haptic Sensation on Tablet PC for Musical Instruments Application. In: Proceedings of User Interface Software and Technology, pp. 73–74 (2011)
12. Yao, H.–Y., Hayward, V.: Design and Analysis of a Recoil-Type Vibrotactile Transducer. Journal of the Acoustical Society of America 128, 619–627 (2010)

Feel the Static and Kinetic Friction

Felix G. Hamza-Lup[1] and William H. Baird[2]

[1] Computer Science and Information Technology,
[2] Chemistry and Physics
Armstrong Atlantic State University, Savannah, USA
{Felix.Hamza-Lup,William.Baird}@armstrong.edu

Abstract. Multimodal simulations augment the presentation of abstract concepts facilitating theoretical models understanding and learning. Most simulations only engage two of our five senses: sight and hearing. If we employ additional sensory communication channels in simulations, we may gain a deeper understanding of illustrated concepts by increasing the communication bandwidth and providing alternative perspectives.

We implemented the sense of touch in 3D simulations to teach important concepts in introductory physics. Specifically, we developed a visuo-haptic simulation for friction. We prove that interactive 3D haptic simulations – if carefully developed and deployed – are useful in engaging students and allowing them to understand concepts faster. We hypothesize that large scale deployment of such haptic-based simulators in science laboratories is now possible due to the advancements in haptic software and hardware technology.

Keywords: Haptics, Friction, Physics, e-Learning.

1   Introduction

Simulators are often used to illustrate abstract concepts that are generally difficult to grasp. Students may gain a deeper understanding of these concepts when using simulators that provide one or many accurate contexts for them [1]. Due to the flexibility of simulators in terms of configuration and range of options, they are sometimes superior to traditional laboratory experiments. For instance, simulators can be used to illustrate concepts that would otherwise require expensive equipment. Even in cases where the equipment itself is inexpensive, such as the wooden blocks and inclined planes commonly used in laboratory exercises studying friction, there is a limit to the number of different physical realizations of the block and board that can be either purchased or stored. Students can also manipulate components of a simulated environment in ways that are impossible in some traditional experiments. In the aforementioned case of the wooden block and inclined plane, the values of the frictional coefficients can be varied smoothly and over an arbitrary range at will within a simulator. Furthermore, one of the problems with the physical study of friction is the relative lack of reproducibility; a student who places the block in a slightly different location on the plane (or who happens to put the block down on a different side) may

P. Isokoski and J. Springare (Eds.): EuroHaptics 2012, Part I, LNCS 7282, pp. 181–192, 2012.

get significantly different results between trials. In this case, the laboratory exercise may cause confusion rather than enhance concept understanding.

Haptics is the science of applying the tactile sense to computer applications, enabling users to receive tangible feedback, in addition to receiving other cues (e.g., auditory and/or visual). The tactile sense is frequently employed to understand the world around us [2]. With haptic devices, students are able to experience tactile sensations in the simulated environment, enabling a potentially deeper understanding of concepts and phenomena.

The paper is structured as follows. In Section 2 we present research and development work related to our visuo-haptic simulator. In Section 3 we focus on the user interaction from the visual and haptic perspective. In Section 4 we present the experimental setup deployed in a classroom environment and provide an analysis of the results. We conclude with a few remarks regarding the development of the visuo-haptic simulator and assessment in Section 5.

2   Related Work

Interest in the field of haptics has increased in recent years, mainly due to the potential applications in entertainment (e.g., games) and medical training. Our current focus is to develop and assess the efficiency of haptic applications in education.

There are several research programs focusing on applications of haptics into higher education. Stanford University has developed a low-cost haptic device, the haptic paddle, to augment teaching undergraduate dynamic systems courses [3]. The system was adopted and modified by Rice University researchers to fit their undergraduate course needs [4]. A group from Ohio University has developed several haptics-based activities to demonstrate concepts from physics to undergraduate engineering students [5]. At the University of Michigan, two haptics interfaces, the iTouch Motor and the Box, were designed for use in a system dynamics course and an embedded control systems course [6].

Haptics use has also expanded into K-12 education. For example, an atomic force microscope allows middle and high school students to physically manipulate live viruses over the Internet, enhancing their understanding of virus morphology [7] and significantly increasing their interest in science. Haptic Virtual Manipulatives [8] have been developed to help teach mathematics to students with learning disabilities. The group at the Ohio University pushed haptics even further by developing a set of downloadable tutorials for high school physics students [9].

An interesting haptics-based system for modeling complex molecular structures has been developed, which allows students to study molecules that are too difficult to represent in a textbook using the traditional ball-and-stick method [10]. Users are able to feel forces at the molecular level using the Interactive Molecular Dynamics system by manipulating molecules in a haptic simulation [11]. Johns Hopkins University has promoted the incorporation of haptics into all levels of education. For instance, they suggested the installation of haptic interfaces in museums to help demonstrate scientific and mathematical phenomena [12]. The University of Patras in Greece developed simulators to provide instruction to children in various areas of science, including space exploration and Newton's laws [13]. What all these simulations have in common is a framework of forces that can be simulated using haptics to emphasize and enhance abstract concept understanding. In the following section we draw the spotlight on the static and kinetic friction model.

3   Simulating Friction

When developing visuo-haptic simulators, we look for concepts that involve forces, so we can present these concepts from a novel perspective. We chose friction since we observed that students have difficulty applying the theoretical concepts to problems. To provide a different perspective on the forces that act on a block on an inclined plane, we developed a 3D visuo-haptic simulator.

The theoretical framework defines three types of friction forces: *static*, which prevents the initial movement of an object along a surface; *kinetic*, which replaces static friction once the object is in motion; and *rolling*, which acts on a rolling object.

Static friction is defined by the inequality $F_s \leq \mu_s N$, where F_s is the force of static friction, μ_s is the coefficient of static friction, and N is the normal force. The maximum value of the static friction F_s^{max} is equal to $\mu_s N$. Fig. 1 illustrates the forces that act on an object being pushed up an inclined plane. We can visualize the normal force N (vector pointing up perpendicular to the plane), the user-applied force F (pointing right), static friction F_s (pointing in the opposite direction of F in this case), and the force of gravity G.

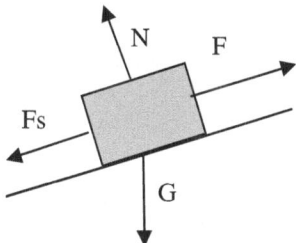

Fig. 1. Forces acting on a block pushed up on an inclined plane

The fact that the static frictional force is described by an *in*equality is the ultimate source of difficulty for many students. Since all forces they have seen before are described by ordinary equalities, they tend to set $F_s = F_s^{max}$. Depending on the problem, students may not be able to realize their mistake (e.g., when an object is pushed with a force greater than F_s^{max}). If the force applied to an object is less than F_s^{max}, the use of the incorrect equality $F_s = F_s^{max}$ yields the nonphysical result that an object will move in the opposite direction of the force being applied. This dynamic component necessary to understand this phenomenon, however, cannot be illustrated in a textbook. Assume a student is given a problem like the one illustrated in Fig. 2, where the she must determine if and which way the blocks will move, and with what acceleration.

Because – for the right values of each mass, angle of inclination, and coefficient of static friction – the system can move in either direction or be in equilibrium, looking only in a textbook figure, there is nothing to help the student realize that she is making an error by setting $Fs = Fs_{max}$. An interactive haptic simulation where she could feel and see the forces, however, could complement the in-classroom teaching material. This would be especially beneficial while the concepts are fresh in memory, before experimenting in the laboratory.

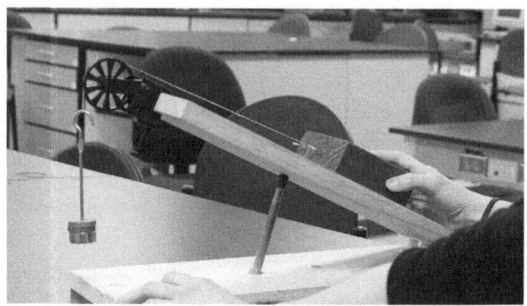

Fig. 2. A typical friction problem – *traditional* laboratory experiments

In the following sections provide a description of the visual component (the graphical user interface) and the haptic component (the haptic user interface – HUI). The complexity of the system comes from the requirement to obtain an ideal perceptual integration of the visual and haptic cues while maintaining high levels of interactivity.

3.1   The Visual Component

We employed the H3D API [14], Extensible 3D (X3D) [15], and the Python scripting language [16] to develop the simulator.

As illustrated in Fig. 3, the visual component of the simulator consists of an inclined plane, a set of floating menus for the configuration of the experiment, and the visual pointer of the haptic device (shown as a small dot).

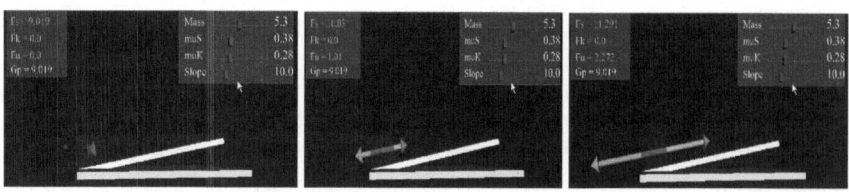

Fig. 3. Force magnitudes represented as arrows, as the user pushes the block up. The dot on the leftmost image near the block represents the position of the haptic pointer.

By using the menus in the heads-up display, the students can control parameters such as the block's mass, the coefficients of friction, and the slope of the plane. Such configuration changes allow students to *see* and *feel* the effects each parameter has on the forces. The magnitude of the force vectors are displayed in the other menu, enabling students to observe how these forces vary in response to configuration changes. Furthermore, the force vectors are displayed dynamically as small arrows of varying length during the interaction with the block. To obtain a different perspective of the scene, the student may change the viewpoint by rotating a disk at the bottom of the screen. The interaction can be recorded in a sequence of screenshots or small movies and used later to complement course material or laboratory sessions.

3.2   The Haptic Component

The HUI relies on the Novint hardware. We employed the Novint Falcon haptic device [17] because of its compatibility with the H3D API, its haptic resolution characteristics, and its affordability. The cost becomes an important aspect, as we intend to equip classrooms of thirty to forty students with one haptic interface per computer. Most physics laboratories can also be enhanced by connecting these devices to available computers (using a plug-n-play USB connection). Students can now use the Novint Falcon to interact with the virtual block and plane, and feel the resulting forces, as illustrated in Fig. 4.

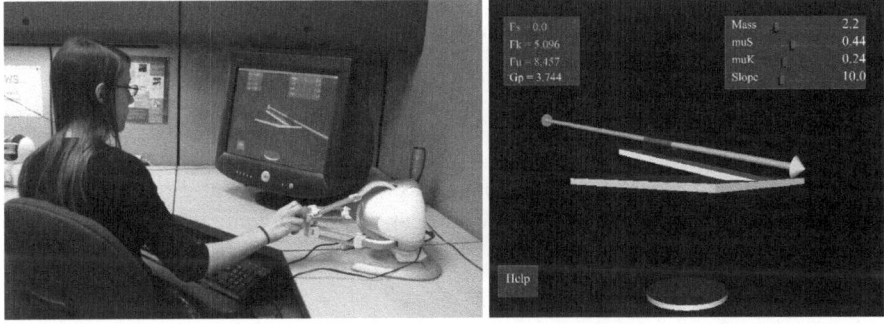

Fig. 4. Student using the haptic friction simulator: room view (left) and screen snapshot (right)

There are several challenges to implementing these haptic components. Because the Falcon device has a limited range of movement (i.e., physical working volume), it is possible to push the virtual block to an unreachable physical area if enough force is applied. Since the H3D API does not provide any tools for boundary implementation, we impose special boundary conditions on the virtual objects in the scene. We accomplished this by monitoring the block's momentum and position, and defining a range for the block movement in either direction. If the student attempts to push the block beyond these limits, a net force of zero is sent to the block to keep it stationary. If, however, enough force is applied, the block will continue moving according to its momentum and may go beyond the set boundaries. To deal with this case, we invert the block's momentum. The user will interpret this as the cube running into an invisible wall when the boundaries are reached. To simplify user interaction, we also constrained the movement of the block to one axis, movement up and down on the inclined plane.

4   Experimental Design and Results

The main goals of the simulator are to enhance student learning, capture student attention, and involve undergraduates in interdisciplinary research. We also want to promote the widespread educational use of this simulator, so we carefully considered the cost factor. A detailed discussion regarding cost will be presented in the conclusion section. In what follows, we describe the experimental framework used to objectively and subjectively measure the simulator's efficiency.

4.1   Simulator Efficiency Assessment

In the spring and summer of 2011, we performed several sets of experiments to determine the impact of the simulator in an introductory college level Physics course. We had a total of 86 participants in the experiments.

Before participating in the learning activity, the students took a pre-test, which aimed to evaluate their prior knowledge for learning the subject unit. The pre-test showed that most students had only a rudimentary knowledge of static and kinetic friction, with the average score being 36.7% (random chance would yield a score of 19.7%).

After the pre-test, the students received a 50-minute conventional lecture about static and kinetic friction. The lecture was followed by a post-test about static and kinetic friction.

Post-test results were used to divide the students into two groups (A and B, illustrated in Fig.5) such that each group had equivalent post-test performance. A t-test on the post-test scores of the two groups showed no significant difference (t=1.49, p>.05), implying that the groups had equivalent theoretical knowledge before participating in the laboratory activity.

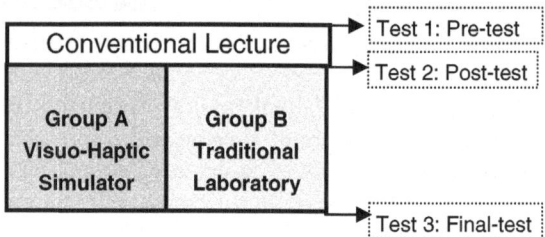

Fig. 5. Assessment – groups and tests

After the division into groups, group A performed lab experiments using the visuo-haptic simulator while students in group B performed similar experiments in a traditional laboratory setup (see Fig. 2).

Both the traditional physics laboratory and the visuo-haptic lab had a paper laboratory handout which provided the students with explanations on how to set up and interact with the blocks on an inclined plane.

Fig. 6. Students in group A, experimenting with the simulator

Each student had 15 minutes of hands-on work with the simulation (as illustrated in Fig. 6) and 15 minutes of observation. A final test was administered to all students at the same time. The final test results are provided in Fig. 7.

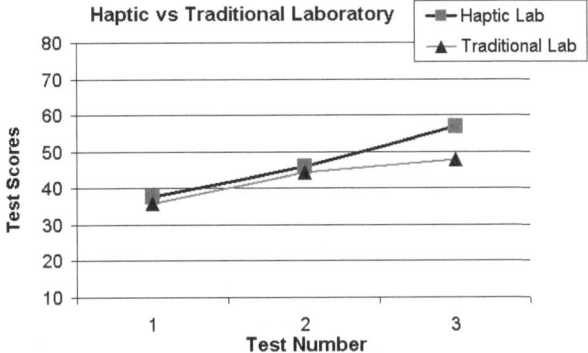

Fig. 7. The average test scores for group A and B

For each group, the normalized gain between the second and third tests was calculated as (Test 3 − Test 2)/(100-Test 2). This metric, therefore, provides the gain as a fraction of the maximum possible gain that could have been achieved between the two tests.

When averaged across all participants, the normalized gain for students using the haptic simulation was 0.182. For students in the traditional physics group, the normalized gain was actually slightly negative at -0.011. Through the use of a t-test, we determined that the chance of an outcome like this occurring if the null hypothesis were correct is 1.2%.

4.2 Attitude Surveys and Student Attention Stimulation

An important side-effect of the simulator is student attention. The unfamiliarity with the haptic user interface stirs the students' curiosity and stimulates their attention.

In spring and summer of 2011, we performed an attitude survey with group A, the one involved in the haptic simulation. To better understand the students' perception of the use of the haptic learning system, this study also collected the students' feedback in terms of perceived *usefulness* and perceived *usability* (i.e. ease of use and learnability) of the simulator.

The physics students interviewed enjoyed the simulator's capability to provide novel perspectives on friction. Most students agreed that the simulator effectively demonstrated both static and kinetic friction from a novel perspective.

The attitude surveys were very helpful in improving the simulator's user interface. From the survey, we concluded that navigating the 3D environment was the main problem students had. At first, many students had trouble aligning the Falcon's virtual pointer with the side of the block in order to push it up or down the plane. Some of the students suggested that the color of the pointer should change when it comes in contact with the block, providing additional visual cues in parallel with the haptic ones.

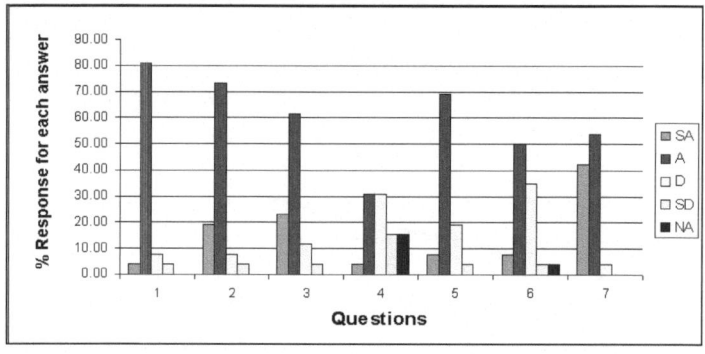

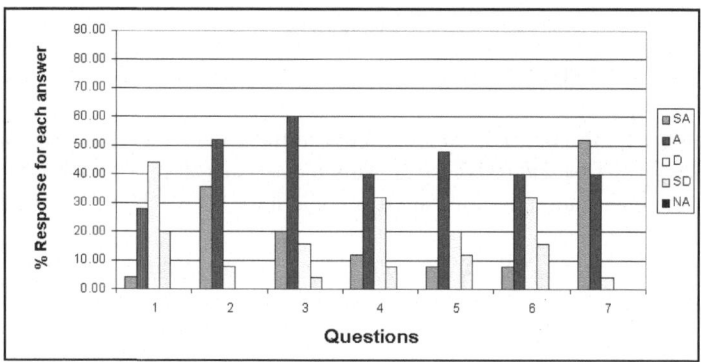

Fig. 8.   Attitude surveys from Spring (top) and Summer (bottom)

A copy of the attitude surveys is available in the Appendix. While these surveys are heavily subjective, we observed an increase in the students' interest in haptics.

4.3   Interdisciplinary Research and Development

Students in Computer Science and Physics were involved in the project from the design to the implementation and testing stages. Since the specific focus of our physics program is Applied Physics and the department has only four full-time physics professors, opportunities for interdisciplinary research efforts with allied departments (e.g., computer science) are especially valuable for the students. Involvement in this project provided the students with a deeper understanding of the illustrated concepts. One of the primary areas of instructional focus in our program is the growing importance of interaction between the physical and virtual worlds. Collection, processing, and analysis of data are the key components of this project as well as modern Applied Physics in general.

5   Conclusions

The current cohort of students, known as Millennials or Generation Y, has grown up around technology that is far more sophisticated and abundant than that of their predecessors. Expectations of multimedia entertainment coupled with low attention spans increase the challenge of engaging a student in learning activities by conventional pedagogical methods. Cognitive studies have shown that students are more apt to learn when engaged by the method of exposure. If students could apply their familiarity with modern technology to their learning objectives, they could more easily understand abstract and/or difficult concepts to better relate new information to what they already understand.

Research in psychology demonstrates that learning styles vary from student to student, and that students have diverse learning needs depending on their cognitive styles and abilities. Various brain regions involved in spatial tasks are activated by the synthesis of multiple sensory inputs. Kinesthetic learners make up about 15% of the population and struggle to learn by just reading or listening [18]. We strongly believe that the application of haptic technology to enhance learning of difficult or abstract concepts in science will improve not only the student's laboratory experience, but also a student's attention and retention in the field. Visuo-haptic applications can improve student learning if simulations are carefully chosen by interdisciplinary teams. Moreover, haptics may provide a medium to learn by doing, through first-person experience.

5.1   The Cost Factor

The cost of haptic devices is now significantly lower than a few years ago, which makes them affordable augmentations to existing science laboratories. Since the majority of these laboratories are already equipped with computers, the addition of a haptic hardware interface is often as trivial as installing a mouse would be. We chose the Novint Falcon due to its low cost and device characteristics (small working volume and maximum force values) sufficient for simulating friction. In terms of hardware for visualization, one solution is inexpensive 3D red-and-blue or polarized glasses. The software components are also attainable due to their low cost. The X3D standard, supported by several plug-ins, allows for rapid development of 3D virtual scenes in a Web browser and keeps the graphical user interface and 3D environment navigation intuitive since most students are already experts at Web browsing. The H3D API developed by SenseGraphics is a freely available library for developing haptic applications and is closely related to the X3D standard, thus providing interoperability.

5.2   Goals

It is important to remember that our goal is not the replacement of traditional learning tools that work well. We explore concepts and paradigms for which a visuo-haptic simulation will enable a better understanding. We envision such environments augmenting rather than replacing existing teaching methods. We are strongly convinced that there are many abstract ideas in science which cannot be cheaply or easily realized physically in a pedagogically useful manner, but that would be well-illustrated through the visuo-haptic approach.

An efficient learning environment must provide excellent perceptual integration, which is not only task-dependent, but might be even more difficult to attain than the technical integration. Discovering and defining simulators and training tools that would benefit from the haptic feedback are also challenging tasks. One must identify the concepts that lend themselves best to such simulation, and then design a learning experience rather than merely a simulation.

While the technical integration of the haptic sensation is important, so is the measurement of its impact on learning. After two years of experimentation we have identified several problems in the practical and objective assessment of the simulator. A balanced group composition in terms of test scores is required, and sufficient warm-up trials with the haptic devices are also necessary.

We experienced several setup issues (e.g., time constraints were important as the haptic devices had to be attached to laptops and the applications preconfigured to be ready during class). Scheduling was complicated by the demands on student and instructor time, as well as the large number of other classes using the physics classrooms. Regardless of the technical integration issues, we have proven that carefully designed and deployed haptic simulators can have a positive role in physics education.

References

1. Dede, C., Salzman, M., Loftin, B., Sprague, D.: Multisensory Immersion as a Modeling Environment for Learning Complex Scientific Concepts. In: Feurzeig, W., Roberts, N. (eds.) Computer Modeling and Simulation in Science Education, pp. 282–319. Springer, New York (1999)
2. Klemmer, S.R., Hartmann, B., Takayama, L.: How bodies matter: five themes for interaction design. In: Proceedings of the 6th Conference on Designing Interactive Systems (DIS 2006), pp. 140–149. ACM, New York (2006)
3. Richard, C., Okamura, A.M., Cutkosky, M.R.: Feeling is Believing: Using Force-Feedback Joystick to Teach Dynamic Systems. ASEE Journal of Engineering Education 92(3), 345–349 (2002)

4. Bowen, K., O'Malley, M.K.: Adaptation of Haptic Interfaces for a LabVIEW-based System Dynamics Course. In: Proceedings of the Symposium on Haptic Interfaces for Virtual Environment and Teleoperator Systems (HAPTICS 2006). IEEE Computer Society, Washington, DC (2006)

5. Williams, R.L., He, X., Franklin, T., Wang, S.: Haptics-Augmented Engineering Mechanics Educational Tools. World Transactions on Engineering and Technology Education 6(1) (2007)

6. Gillespie, R.B., Hoffman, M.B., Freudenberg, J.: Haptic Interface for Hands-On Instruction in System Dynamics and Embedded Control. In: Proceedings of the 11th Symposium on Haptic Interfaces for Virtual Environment and Teleoperator Systems (HAPTICS 2003), p. 410. IEEE Computer Society, Washington, DC (2003)

7. Jones, M.G., Bokinsky, A., Tretter, T., Negishi, A., Kubasko, D., Taylor, R., Superfine, R.: Atomic Force Microscopy with Touch: Educational Applications. In: Mendez-Vilas, A. (ed.) Science, Technology and Education of Microscopy: An Overview, Madrid, Spain, vol. 2, pp. 686–776 (2003)

8. Singapogu, R.B., Burg, T.C.: Haptic virtual manipulatives for enhancing K-12 special education. In: Proceedings of the 47th Annual Southeast Regional Conference (ACM-SE 47). ACM, New York (2009)

9. Williams II, R.L., Chen, M.-Y., Seaton, J.M.: Haptics-Augmented High School Physics Tutorials. International Journal of Virtual Reality 5(1) (2005)

10. Sankaranarayanan, G., Weghorst, S., Sanner, M., Gillet, A., Olson, A.: Role of Haptics in Teaching Structural Molecular Biology. In: Proceedings of the 11th Symposium on Haptic Interfaces for Virtual Environment and Teleoperator Systems (HAPTICS 2003), Los Angeles, CA (2003)

11. Stone, J.E., Gullingsrud, J., Schulten, K.: A system for interactive molecular dynamics simulation. In: Proceedings of the 2001 Symposium on Interactive 3D graphics (I3D 2001), pp. 191–194. ACM, New York (2001)

12. Grow, D., Verner, L.N., Okamura, A.M.: Educational Haptics. In: AAAI 2007 Spring Symposia- Robots and Robot Venues: Resources for AI Education (2007)

13. Pantelios, M., Tsiknas, L., Christodoulou, S.P., Papatheodorou, T.S.: Haptics technology in Educational Applications, a Case Study. Presented at JDIM, 2004, pp. 171–178 (2004)

14. SenseGraphics AB, website (August 2011), http://www.h3dapi.org/

15. Web3D Consortium, website (July 2011), http://www.web3d.org/

16. Python Software, website (August 2011), http://www.python.org/

17. Novint Technologies: Novint Falcon haptic device, website (July 2011), http://www.novint.com/

18. Dunn, R., DeBello, T.C. (eds.): Improved test scores, attitudes, and behaviors in America's schools: Supervisors' success stories. Bergin and Garvey, Westport (1999)

Appendix: Attitude Survey

Check one:

- Freshman
- Sophomore
- Junior
- Senior

For each question, please select one of the following:

- SA   - Strongly Agree,
- A   - Agree,
- D   - Disagree,
- SD   - Strongly Disagree,
- NA   - Not Applicable.

1. The Novint Falcon haptic device was easy to use.

                    SA   A   D   SD   NA

2. The simulator was effective in demonstrating the behaviour of static friction.

                    SA   A   D   SD   NA

3. The simulator was effective in demonstrating the behaviour of kinetic friction.

                    SA   A   D   SD   NA

4. The simulator was more effective in illustrating friction than a conventional laboratory experiment.

                    SA   A   D   SD   NA

5. The environment was intuitive and easy to understand.

                    SA   A   D   SD   NA

6. It was easy to navigate in the environment.

                    SA   A   D   SD   NA

7. It was easy to adjust the parameters that affect the force of friction.

                    SA   A   D   SD   NA

Mechanical Impedance as Coupling Parameter of Force and Deflection Perception: Experimental Evaluation

Christian Hatzfeld and Roland Werthschützky

Technische Universität Darmstadt, Institute for Electromechanical Design
Merckstr. 25, DE-64283 Darmstadt
c.hatzfeld@emk.tu-darmstadt.de

Abstract. This paper investigates the mechanical impedance of a human subject as a potential coupling parameter between force and deflection perception. Measurements of the force perception threshold of 27 subjects at the fingertip and the mechanical impedance of 29 subject at the same location in the frequency range of 5 ... 1000 Hz were conducted. From the results, a model for the impedance was fitted and thresholds for the perception of deflections were calculated. These were compared to already published thresholds from other research groups. The results show a good fit of both data sets, therefore confirming the mechanical impedance as coupling parameter between these two dimensions of perception.

Keywords: force perception, deflection perception, mechanical impedance.

1 Motivation

While many haptic applications rely on commercial off-the-shelf haptic displays, more and more systems with task-specific haptic interfaces emerge in research and industry. Example applications range from surgical procedures [20,21] to communication means [23] and many more. In the design of these task-specific haptic interfaces, human haptic perception is one of the most important sources for system requirements [24]. To obtain requirements for system components like actuators, sensors and kinematics, a variety of studies which are investigating psychophysical parameters is available.

Looking closer at these studies and the apparatuses used, two aspects attract attention: Firstly, the majority investigates the perception of stimuli that are defined by the deflection of the stimulus-presenting part of the setup [6,11,12]. Secondly, most studies dealing with forces investigate kinesthetic [13,22] and/or forces with frequencies less than 10 Hz [1]. This is probably due to the fact, that the measurement setups for dynamic deflections are easier to design than setups that use stimuli defined by the dynamic force that is coupled into the subject's skin. In this case, additional safety measures have to be provided to

P. Isokoski and J. Springare (Eds.): EuroHaptics 2012, Part I, LNCS 7282, pp. 193–204, 2012.

prevent physical injuries for the subjects. Furthermore, force displaying systems are prone to load changes at near-threshold levels and dynamic properties like inertia of the measurement setup have to be considered in data processing [8], all increasing cost and complexity.

Nevertheless, force perception data can be very useful in designing and evaluating systems, it is essential for compression algorithms [10] and admittance-controlled (displacement-feedback) systems. To fill this gap, more force perception measurements would be needed, depending on body locus, reference forces and several other parameters. On the other hand, a large number of measurements of deflection perception parameters is available. The use of this data could supply many force perception parameters without any further subject tests, when a defined coupling parameter between the two perceptual dimensions of force and deflection can be asserted. It would reduce the complexity of obtaining new parameters and allow for new perception-inspired approaches.

From a pure mechanical point of view, the relation between a force defined and a deflection defined stimulus coupled into the same structure is pretty simple and commonly known as the mechanical impedance $\underline{z}_m$ [1] of this structure. This impedance is defined as the ratio between forces ($\underline{F}$) and motion parameters like deflections ($\underline{x}$) or velocity ($\underline{v}$) as shown in eq. (1)

$$\underline{z}_m = \frac{\underline{F}}{\underline{v}} = \frac{\underline{F}}{j\omega \cdot \underline{x}} \tag{1}$$

From an engineering point of view, it seems obvious, that this ratio between physical dimensions should also be valid for psychophysical dimensions. This would mean that mechanical properties of skin and the tissue surrounding the tactile receptors could describe the relation between the perception of forces and the perception of deflections alone, despite of sensory and decision processes.

This relation is already used in several publications [11,12], but not yet investigated thoroughly enough. All data presented until now is based on deflection-defined stimuli, calculating force perception thresholds based on impedance measurements and deflection thresholds.

To fully verify this relation, the opposite direction, calculating deflection perception thresholds from impedance measurements and force perception thresholds, should to be verified. Under the assumption, that all psychophysical and mechanical parameters are measured at the same body location, this formulates the main hypothesis of this work. With $\underline{x}_{th}$ and $\underline{F}_{th}$ as absolute thresholds for deflection and force perception respectively and $\underline{z}_m$ as mechanical impedance, the hypothesis can be formulated as in eq. 2.

$$\underline{x}_{th} \overset{?}{=} \frac{\underline{F}_{th}}{j\omega \underline{z}_m} \tag{2}$$

The evaluation of this hypothesis is the scope of the study presented here. In the following section, measurement results of the mechanical impedance and

---

[1] Underlined characters denote complex parameter according to DIN 5483-3 / ISO 31-2.

absolute force perception thresholds are reported and the absolute threshold of deflection perception is predicted according to eq. 2. Additionally, a network theory model is calculated for the measured mechanical impedance.

The values of the predicted thresholds are compared to deflection perception thresholds determined in well-known publications (see sec. 2.3) of another research group. Results are analyzed using an error propagation approach and discussed.

2   Measurements

The tip of the index finger was chosen as investigation body site, because of its importance in haptic interaction. Measurements of the mechanical impedance and of force perception thresholds were made at the frequencies 5, 10, 20, 40, 80, 160, 320, 500 and 1000 Hz. As a contact situation, the grip shown in fig. 1 was used. A circular, concave contactor (Ø 19 mm) embedded in a rigid surrounding with a 1 mm gap was chosen to ensure the excitation of Pacinian corpuscles, that exhibit a large receptive field and are sensitive for high frequency vibrations [5]. The contactor size ensures stimulation of most of the tactile receptors in the subjects' skin regardless of actual size of the finger. Contactor and rigid surrounding were made from aluminum because of small material density (leading to lightweight structures and higher dynamics) and good thermal conductivity. The contactor and the surrounding were incorporated in the two different

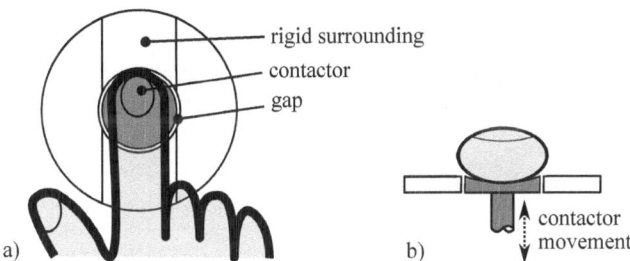

Fig. 1. Contact situation used in the experiments, test persons used the index finger of the dominant hand, a) top view, b) sectional side view

measurement setups used in this study - one for the measurement for the mechanical impedance and one for the measurement of force perception thresholds. The same contactor is also used in the measurement setup detection threshold (see sec. 2.3).

2.1   Mechanical Impedance

To verify the hypothesis given in eq. (2), mechanical impedances of the fingertip of 29 subjects (21 male, 8 female, aged 24.0 ± 2.8 years) were measured. Because of the non-negligible influence of contact situation and contact force on the impedance measurements [15], an explicit measurement was chosen over the use of a model or previous recorded data, since existing publications [11,12,16] do not incorporate the contact situation as shown in fig. 1.

Apparatus. Impedance measurements were conducted using an impedance head *(model 8001, Brüel&Kjær, Nærum, Denmark)* mounted on a shaker *(model 4810, power amplifier model 2706, both Brüel&Kjær)* as shown in fig. 2. The impedance head measures acceleration and force simultaneously using two piezo-electric sensors. A signal conditioner *(model nexus 2692, Brüel&Kjær)* was used for integration of the acceleration signal, delivering force and velocity measures to a signal analyzer *(model 35670A, Santa Clara, CA, USA)*, which calculates the impedance according to eq. (1). The movement of the shaker was controlled by the built-in source of the signal analyzer in such a way, that a constant force amplitude of 0.316 N was coupled into the test persons finger.

Procedure. Measurements were made at 51 logarithmically distributed points in the frequency range between 3 and 5000 Hz. Frequencies were swept upwards, three sweeps were measured for each subject. An initial calibration measurement of the impedance of the measurement setup was used to eliminate the impedance parts of the setup in the measured data and the results of the three data sets of each subject were averaged.

During the test, subjects were required to maintain a constant contact force of 1 N, measured with a force sensor *(model 85075, Burster, Gernsbach, Germany)* mounted under the shaker. The force output was shown on an oscilloscope display, and subjects were asked to keep it inside a given tolerance band of ± 10 %.

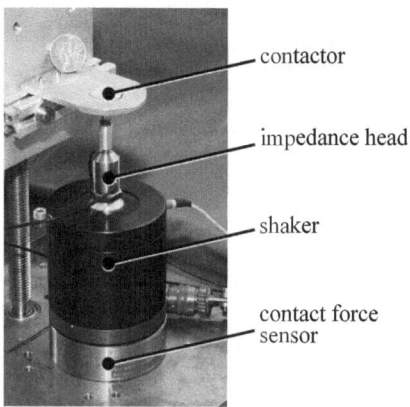

Fig. 2. Impedance measurement setup. Signal analyzer, signal conditioning and power amplifier are not shown. For size reference, a 1 euro coin is placed next to the contactor.

Results. The results are shown in fig. 3. For each frequency, the median and the 0.95 and 0.05 percentiles of the measured impedances are given, since results exhibit a large variance between test persons. Plotted graphs therefore do not show the results from one individual subject. The measured impedances exhibit a compliance characteristic in the lower frequency range up to about 50 ... 80 Hz. A viscous damping characteristic is dominant from these frequencies up to about 1000 Hz, where a mass characteristic becomes prevalent for higher frequencies. The viscous damping part also exhibits several resonances, most distinct a parallel resonance around 250 Hz and a serial resonance in the bandwidth of 600 ... 700 Hz.

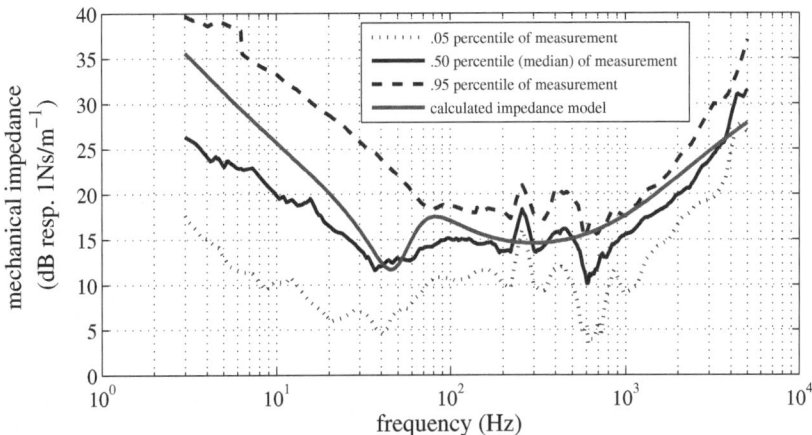

Fig. 3. Mechanical impedance of the index finger of 29 test persons. Shown are the 0.05, the 0.5 and the 0.95 percentile and the values of the calculated impedance model.

Modelling. From the given data, the parameters for a network model with concentrated elements [18] were fitted. A general network model with eight independent elements for human fingers was introduced in [15]. This model is capable to incorporate different contact forces and grasp situations over a larger frequency range [16] than simpler models based on second-order models [7]. Furthermore it can be directly used for frequency-domain-based modeling of haptic systems with concentrated network parameters [14].

Since no parameter set for the above given contact situation (see fig. 1) is reported, a new set was fitted to the model based on the data measured in this study. Each subject's data set was fitted to the model using a least-square-algorithm. This results in 29 different parameter sets for the network model shown in fig. 4, one for each subject. To obtain a general model, the average for each parameter over the 29 subjects was calculated. The such obtained model parameters are given in fig. 4 and the resulting mechanical impedance is depicted in fig. 3.

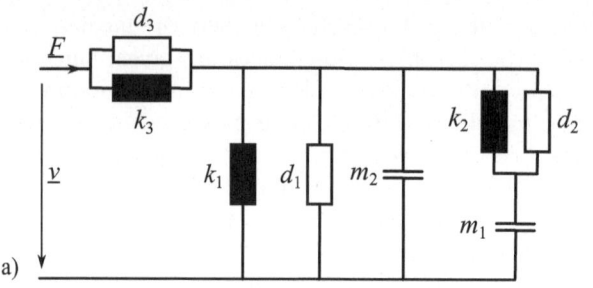

parameter		value
k_1	(N/m)	1648
m_1	(kg)	0.0092
d_1	(Ns/m)	2.857
d_2	(Ns/m)	2.995
k_2	(N/m)	1736
k_3	(N/m)	3533
d_3	(Ns/m)	41.38
m_2	(kg)	0.0011

Fig. 4. Network theory model of the mechanical impedance of a human finger with concentrated elements, a) network representation taken from [16], b) model parameters for the model shown in fig. 3

2.2   Perception of Forces

The absolute perception thresholds for forces were determined for 27 subjects (20 male, 7 female, aged 24.8 ± 2.61 years). These test persons were not identical with the test persons taking part in the measurement of the mechanical impedance as described above. This approach was chosen to obtain a higher statistical independence between the data sets for force perception and mechanical impedance. Instructions how to do the test were given in written form.

Apparatus. The apparatus used consists of a electrodynamic force source based on the magnetic system of a commercial loudspeaker (*model TIW-300, Visaton, Haan, Germany*). Forces were measured by a 6 DoF force sensor (*model nano17, ATI Industrial Automation, Apex, NC, USA*) that was equipped with custom-made secondary electronics to reduce latency and to provide an external offset correction of the force signal. To ensure high accuracy and to adapt the system to different loads, an analog PID-controller was used to establish a closed-loop force control.

Displacement and velocity were recorded using a laser triangulation system (*model LK-G32, Keyence, Osaka, Japan*) and a custom-made voice-coil assembly. The force source was driven by a high-bandwidth linear amplifier (*model BAA-1000, BEAK electronic, Mengersgereuth-Hämmern, Germany*). The measurement system was controlled by a data acquisition board (*model PCI-7833R, National Instruments, Austin, TX, USA*). Force, velocity and displacement data were recorded with 16-bit resolution at a sampling rate of 10 kHz. Additional information about the measurement system can be found in [8]. The setup was already used in a preliminary study of absolute force perception thresholds [9] and is depicted in fig. 5.

Procedure. An adaptive staircase method [17] with a 2 down-1 up progression rule combined with a 3 interval forced choice answer paradigm (3IFC) was used [2]. This procedure converges at a detection probability of 70.7 % [19]. Step sizes

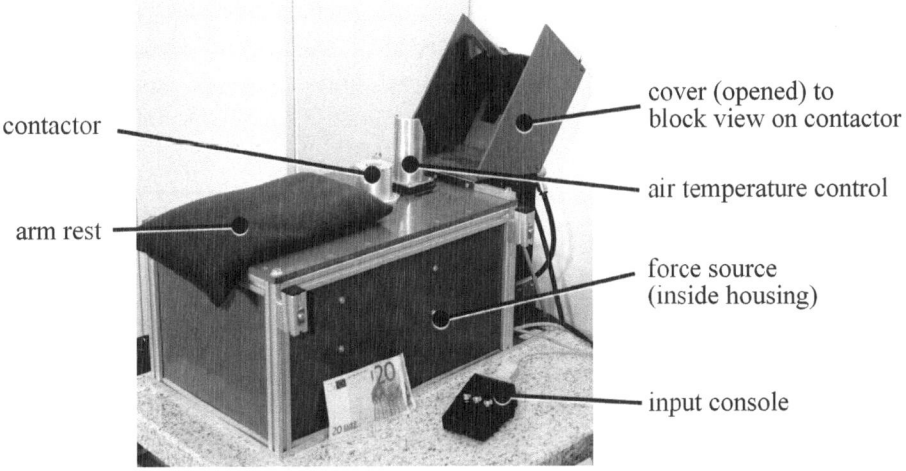

Fig. 5. Force perception measurement setup. Control PC, secondary electronics and power amplifier are not shown. For size reference, a 20 euro note is placed in front of the housing.

were set to 3 dB for the first four reversals and to 1 dB for all the following. To determine the starting values of the staircases, a prior measurement run with a Method of Limits [17] was conducted for all nine frequencies. Runs were terminated after 12 reversals.

With a 3IFC paradigm, subjects are required to identify in which one out of three intervals a stimulus was present. Intervals lasted 2 seconds each. Force stimuli were presented after a random intermission (500 ms at longest) for a duration of 1 second. The first and the last 100 ms of the stimulus were superimposed with a linear rise resp. fall to avoid transient effects. Subjects wore earplugs (*model Howard Leight Max, Sperian Protection, Lübeck, Germany*) and headphones, latter delivering white noise during the three presentation intervals. After the three intervals, a 1 second brake preceded the answering period of 2 seconds. Answers were given by choosing the corresponding key to an interval on a small console. No feedback was given, whether the answer was correct. The whole procedure lasted about 2 h per subject with required intermissions of at least 1 minute between the runs.

Results. Staircases were inspected visually by the experimenter after termination and repeated if no convergence was shown. Thresholds were calculated individually for each subject and for each frequency as the arithmetic mean of the stimulus sizes at the last eight reversals. The results are shown in fig. 6 as a boxplot, since this kind of display provides more data about the thresholds. The large data variances are due to general different perception capabilities of the test persons. Since a χ^2-test confirmed normal distribution of the thresholds for each frequency, descriptions using mean and standard deviation could also be used. This is especially important for further statistical analysis, since common methods like analysis of variances (ANOVA) or confidence interval tests (i.e. F-test) assume normal distribution of analyzed data.

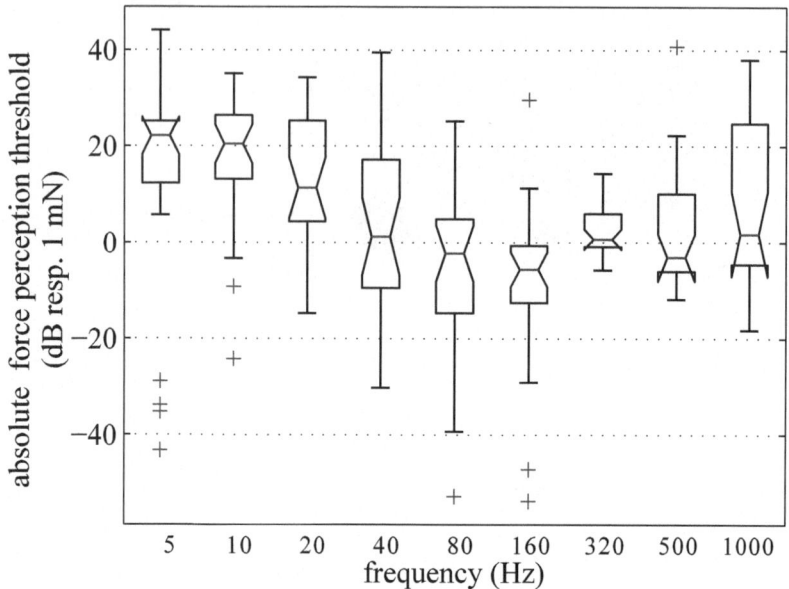

Fig. 6. Results of the force perception measurement shown as a boxplot indicating median (horizontal lines), marker for median differences (medians are significantly different at a 0.95 confidence level, if notches do not overlap), interquartile range (IQR, distance between 0.25 and 0.75 percentile of data, denoted by the outline of the box), data range (lines) and outliers (defined by a distance of more than 1.5 IQR lower or higher than the 0.25 or 0.75 percentile respectively, denoted by crosses)

2.3   Reference Deflection Perception Thresholds

Data about the perception of vibrations defined by deflections were taken from the studies of GESCHEIDER ET AL.. The studies are considered very reliable, since they are based on a large number of test persons and major scientific experience of the research group. Two studies were considered as relevant for the above described contact situation: The first study investigates displacement thresholds at the thenar eminence [3] with the same contactor as described above, the second study investigates displacement thresholds at the fingertip with a smaller contactor, but obtaining similar values for the detection thresholds [4]. Both studies use procedures that converge at 75 % detection probability.

Since the studies originally address topics like frequency-dependence of channels and spatial and temporal summation, only the thresholds are given in the publications with no further information about variances. Since no other data with similar contact situation could be found, data from [4] was used to investigate the main hypothesis of this study, although no information about the standard deviation of the measurements is given in the publications.

3   Results

To assess the hypothesis, predicted perception thresholds for deflections were calculated according to eq. (2) for the frequencies 5, 10, 20, 40, 80, 160, 320 and 500 Hz. Since no deflection thresholds for 1000 Hz could be found, this data was omitted. For force thresholds, mean values $\underline{F}_{\text{th,mean}}$ were used and the standard deviation σ_F was considered as a description of the stochastic errors of the measurement. Values for the mechanical impedance were taken from the model obtained in sec. 2.1, since this will probably be the common approach in the application of the hypothesis.

Based on this data, the predicted deflection detection threshold $\underline{x}'_{\text{th}}$ was calculated based on eq. (2). According to the error propagation of measurement results with stochastic description of the data, the mean $x'_{\text{th,mean}}$ and the standard deviation $\sigma_{x'}$ of the assumed deflection thresholds can be calculated according to eqs. (3) and (4). The standard deviation of the impedance model was neglected in eq. (4), assuming $\sigma_z^2 = 0$. This was done for practical reasons, since it is probably the most common use of models like this. Additionally, this reduces

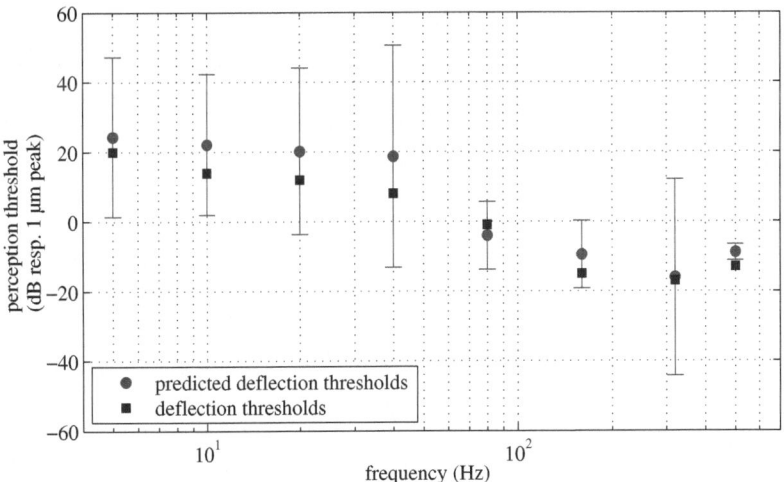

Fig. 7. Comparison of the predicted deflection thresholds according to the hypothesis (circles with error bars) and directly measured deflection thresholds (squares). Most measured deflection thresholds are inside the error bar region of the predicted thresholds, confirming the hypothesis of mechanical impedance as coupling parameter between force and deflection perception.

the standard deviation $\sigma_{x'}$ and is therefore a stronger criterion on the main hypothesis.

$$x'_{\text{th,mean}} = \left| \frac{F_{\text{th,mean}}}{j\omega \underline{z}_{\text{m}}} \right| \tag{3}$$

$$\sigma_{x'} = \sqrt{ \left| \frac{\partial x'_{\text{th,mean}}}{\partial F_{\text{th,mean}}} \right| \cdot \sigma_F^2 + \left| \frac{\partial x'_{\text{th,mean}}}{\partial \underline{z}_{\text{m}}} \right| \cdot \sigma_z^2 }$$

$$= \sqrt{ \frac{1}{|\underline{z}_{\text{m}}^2|} \cdot \sigma_F^2 } = \frac{1}{|\underline{z}_{\text{m}}|} \cdot \sigma_F \tag{4}$$

The such calculated predicted deflection $x'_{\text{th}} = x'_{\text{th,mean}} \pm \sigma_{x'}$ is shown in fig. 7 together with the values for the directly measured deflection perception thresholds according to sec. 2.3. The predicted range for deflection perception thresholds shows large variances inherited from the force perception measurement, but includes all directly measured threshold values except an outlier at 500 Hz. However, predicted thresholds seem to be slightly higher than the directly measured thresholds.

4   Conclusion

The above presented data suggests the validity of the initial hypothesis, that mechanical impedance can be considered as coupling parameter between force and deflection perception in the range of the standard deviation of measurements. Two reasons for the slight mis-estimation of higher calculated deflection perception thresholds are likely: First the mechanical impedance could be underestimated by the model used, secondly force perception thresholds could be over-estimated. While the impedance model could be improved by including more measurement data quite easily, the acquisition of more force perception data is considerably more time-consuming. Another approach could be the inclusion of statistical variance parameters of the measured deflection thresholds, which would allow the declaration of a confidence interval for the hypothesis.

Further Work. The above mentioned measurements of force perception thresholds and mechanical impedances were made with different subjects, aiming at a statistical independence of both data sets. Therefore the calculations of the predicted deflection perception is an inter-dataset calculation, using statistical descriptions of both data sets. This leads to large standard deviations of the predicted deflection perception according to eq. (4).

If both parameters (force perception and mechanical impedance) of the same persons were known, inter-person deflection thresholds could be calculated, which could result in clearer evidence of the hypothesis with less errors. In this case, only the measurement errors for force thresholds and impedances have to be considered in the calculation of the predicted deflection thresholds. This results in smaller errors for the individually predicted threshold, probably leading to a narrower distribution of the predicted deflection thresholds as in fig. 7.

Another approach would be the investigation of other psychophysical parameters like differential thresholds (JND) or masking parameters with similar measurements. These parameters can be investigated based on deflection or force defined stimuli as well, hence mechanical impedance could be investigated as a coupling parameter, too.

Practical Aspects. For the practical application of the hypothesis, i.e. the calculation of force perception parameters from deflection parameters and models for the mechanical impedance, the above discussed mis-estimation leads to a conservative estimation. Assuming the same contact situation (contactor size and shape, contact force, temperature, etc.) force perception thresholds can be calculated using the relations in eq. (1). The resulting force perception parameters based on this calculation will be lower than the actual force perception threshold, such giving a conservative estimation of these thresholds for the development of haptic devices.

The presented study therefore confirms the validity and the feasibility of the usage of the mechanical impedance as a coupling parameter for the perception dimensions of force and deflection. This result will reduce the amount of psychophysical studies about force perception and open up new applications of deflection perception data.

Acknowledgments. This work was funded by Deutsche Forschungsgemeinschaft (DFG) under grant WE2308/7-1. The authors greatly appreciate the work of Ms. Siran Cao in conducting the experiments and the helpful discussions with Ms. Tina Felber of the statistics department of Technische Universität Darmstadt.

References

1. Abbink, D.A., van der Helm, F.C.: Force Perception Measurements at the Foot. In: IEEE Conf. on Systems, Man and Cybernetics (2004)
2. Buus, S.: Psychophysical methods and other factors that affect the outcome of psychoacoustic measurements Genetics and the Function of the Auditory System. In: Proceedings of the 19th Danavox Symposium (2002)
3. Gescheider, G.A., Bolanowski, S.J., Hardick, K.R.: The frequency selectivity of information-processing channels in the tactile sensory system. Somatosensory and Motor Research 18 (2001)
4. Gescheider, G.A., Bolanowski, S.J., Pope, J.V., Verillo, R.T.: A four-channel analysis of the tactile sensitivity of the fingertip: frequency selectivity, spatial summation and temporal summation. Somatosensory and Motor Research (2002)
5. Gescheider, G.A.: Psychophysics. Lawrence Erlbaum Associates (1997)
6. Gescheider, G.A., Wright, J.H., Verillo, R.T.: Information-Processing Channels in the Tactile Sensory System. Psychology Press (2009)

7. Hajian, A.Z., Howe, R.D.: Indentification of the Mechanical Impedance at the Human Finger Tip. Journal of Biomechanical Engineering 119 (1997)
8. Hatzfeld, C., Kern, T.A., Werthschützky, R.: Design and Evaluation of a Measuring System for Human Force Perception Parameters. Sensors and Actuators A: Physical 162 (2010)
9. Hatzfeld, C., Werthschützky, R.: Vibrotactile Force Perception Thresholds at the Fingertip. In: EHC (2010)
10. Hinterseer, P., Hirche, S., Chaudhuri, S., Steinbach, E., Buss, M.: Perception-Based Data Reduction and Transmission of Haptic Data in Telepresence and Teleaction Systems. IEEE Transactions on Signal Processing 56 (2008)
11. Israr, A., Choi, S., Tan, H.Z.: Detection Threshold and Mechanical Impedance of the Hand in a Pen-Hold Posture. In: Proceedings of the 2006 IEE/RSJ International Conference on Intelligent Robots and Systems (2006)
12. Israr, A., Choi, S., Tan, H.Z.: Mechanical Impedance of the Hand Holding a Spherical Tool at Threshold and Superthreshold Stimulation Levels. In: WHC (2007)
13. Jones, L.A.: Matching forces: constant errors and differential thresholds. Perception 18 (1989)
14. Kassner, S.: Transparenz von Teleoperationssystemen - Anwendung der Netzwerktheorie auf mehrdimensionale, parallel-kinematische haptische Bedienelemente. Jahrestagung der Deutschen Gesellschaft fr Akustik, DAGA (2012)
15. Kern, T.A., Werthschtzky, R.: Studies of the Mechanical Impedance of the Index Finger in Multiple Dimensions. In: EHC (2008)
16. Kern, T.A. (ed.): Engineering Haptic Devices. Springer (2009)
17. Leek, M.R.: Adaptive procedures in psychophysical research. Perception & Psychophysics 63 (2001)
18. Lenk, A., Ballas, R.G., Werthschtzky, R., Pfeifer, G.: Electromechanical Systems in Microtechnology and Mechatronics. In: Electrical, Mechanical and Acoustic Networks, their Interactions and Applications. Springer (2011)
19. Levitt, H.: Transformed Up-Down Methods in Psychoacoustics. JASA 49 (1971)
20. McMahan, W., Gewirtz, J., Standish, D., Martin, P., Kunkel, J., Lilavois, M., Wedmid, A., Lee, D., Kuchenbecker, K.: Tool Contact Acceleration Feedback for Telerobotic Surgery. IEEE Transactions on Haptics (2011)
21. Meiss, T., Budelmann, C., Kern, T.A., Sindlinger, S., Minamisava, C., Werthschützky, R.: Intravascular Palpation and Haptic Feedback during Angioplasty. In: World Haptics Conference (2009)
22. Pang, X., Tan, H., Durlach, N.: Manual Resolution of Length, Force and Compliance. In: ASME DSC Advances in Robotics, vol. 42 (1992)
23. Prescher, D., Weber, G., Spindler, M.: A tactile windowing system for blind users. In: ACM SIGACCESS (2010)
24. Tan, H.Z., Srinivasan, M.A., Eberman, B., Cheng, B.: Human Factors for the design of force-reflecting haptic interfaces. In: ASME DSC (1994)

Acquisition of Elastically Deformable Object Model Based on Measurement

Koichi Hirota[1] and Kazuyoshi Tagawa[2]

[1] Graduate School of Frontier Sciences, The University of Tokyo
k-hirota@k.u-tokyo.ac.jp
[2] Ritsumeikan Global Innovation Research Organization, Ritsumeikan University
tagawa@cv.ci.ritsumei.ac.jp

Abstract. This paper describes an approach to acquiring impulse response deformation model (IRDM) through measurement on a real deformable object. A step-wise input force is applied onto a node by an air jet; the responding deformation is recorded by measuring the node's motion using a stereo camera. In addition, the impulse response is computed from the step response. Measurement of actual object in the shape of dome and rectangular prism was performed. Also, an experiment that evaluates stiffness and temporal deformation of the obtained model was carried out, and similarity of the model to the real object was confirmed.

Keywords: deformation model, deformable object, impulse response.

1 Introduction

This paper discusses an approach to construct a deformable object model using the measurements of a real object based on the impulse response deformation model (IRDM) which has been investigated in our previous studies [1]. IRDM is a model that represents dynamic behavior of an object using deformation in response to impulse force. Hence, the process of making this model is measuring the impulse response of the object. An advantage of this approach is that it can deal with dynamic characteristic of deformation without the process of seeking for parameters such as damping.

2 Related Research

Haptic feedback has been regarded as an essential factor of VR systems from an early stage. A lot of research has been carried out from both the perspective of hardware and software [2]. Methods of computing force in interactions or haptic rendering algorithms have been recognized as an important part of haptic research [3]. Also, the presentation of deformable objects is an important area of versatile interactions in haptic rendering.

P. Isokoski and J. Springare (Eds.): EuroHaptics 2012, Part I, LNCS 7282, pp. 205–217, 2012.

2.1   Model-Based Deformation

Most research on the model-based approach employs the FEM and the spring network model to represent the force-deformation relationship and realize interaction based on simulation. Although the computation time and precision of the model is in a trade-off relationship, the computation cost of this approach is relatively high. Also, it is still not easy on currently prevalent personal computers to perform real-time simulation. Many approaches to accelerate the computation have been investigated from both algorithms and computation hardware. There is some research that investigates the reduction of computation cost of FEM by modeling St.Venant-Kirchhoff Material using second- or fourth- order springs. Thus the redundancy in FEM computation is eliminated [4,5]. The authors proposed an algorithm that reduces the cost of computing the interaction force in linear elastic model [6]. Also there is some research that utilizes FPGA [7] and GPU [8] for computation acceleration.

Another approach to acceleration is changing the topology of the mesh dynamically in the progress of simulation; cells are divided or unified depending on the distribution of stress in the object. An interesting method is investigated by Tanaka et al. where the process of mesh generation is accelerated by taking advantage of the similarity in recursive division/unification of tetrahedral cells [9].

2.2   Measurement-Based Modeling

MacLean et al. proposed the idea of measuring and recording the force-deformation relationship which they termed 'haptic camera' [10]. Pai et al. presented the concept of creating a virtual object model based on measurement [11]; models of stiffness and texture were included in the subject of measurement, however, the dynamic aspect of deformation was not dealt with. Ueda et al. have investigated the identification of the viscoelastic model which is composed of a spring and a damper based on experimental distortion on actual material [12]. The aim of the research was acquiring parameters for a model-based presentation. Hoever et al. have investigated the identification and presentation of elasticity based on the Maxwell model [13]. Although this approach is expected to provide precise modeling of force at an action point, it is not immediately applicable to the implementation of a deformable object. Weir et al. proposed a method of visualizing the haptic feature in interaction by plotting the relationship among displacement, velocity and force [14]. This research was focusing on understanding the feature of deformation rather than presenting it. Bickel et al. proposed a method of modeling non-linear heterogeneous object based on measurement [15]; material parameters for FEM simulation were computed by interpolation among parameters from deformation examples. However, the method was not dealing with the dynamic aspect of deformation.

2.3   Recording and Reproducing Vibration

Vibration on the surface of an object is difficult to simulate in real time because of the high frequency of the phenomenon; hence a recording-and-reproducing approach has been investigated to improve reality. Pioneering research has been carried out by Wellman et al. who have revealed that reality of the sensation is improved by adding vibration in the virtual tapping operation [16]. Okamura et al. applied this approach to the presentation of texture and rupture of membranes [17]. Later, they proposed the concept of Reality-Based Modeling [18]. Niemeyer et al. have brought progress in the tapping operation [19]; they developed a library of transient profiles and computed the force and vibration by interpolating profiles from the library; they also evaluated the efficacy of the approach.

2.4   Recording / Pre-computing Deformation

Presentation of a deformable object is different from the presentation of vibration in that it usually deals with the shape in addition to reacting force. Also, the degree of freedom of interaction tends to be large in the sense that the user is allowed to interact at any point on the surface. Therefore, different approaches have been investigated. James et al. proposed the representation of dynamic deformation using state space; the sequence of deformation which was termed 'impulse palette' is computed beforehand and activated in the interaction [20]. Although the state space model is generic, the degree of freedom of actual interaction is restricted by the variety of the impulse palette. Fong presented the method of modeling deformation by actively probing on the surface of object [21]; a force-filed model was employed for the haptic interaction.

The authors have also studied the recording-and-reproducing approach and proposed a model termed impulse response deformation model (IRDM) [1]. The model defines the relationship between force and deformation using impulse response assuming linearity. This assumption is significantly effective in expanding degrees of freedom and reducing the order of computational complexity.

The idea of the recording-and-reproducing approach has been applied to interactions other than deformation which is in some sense specific to application. Chial et al. investigated the profile of cutting force while using scissors [22]. Dobashi et al. presented paddling force in a canoe simulator using a pre-computed result [23]; table of force depending on water velocity and attack angle are created by pre-computation and the interaction force in the simulator is determined by looking up the table.

2.5   Model-Based Approach vs. Measurement-Based Approach

In the presentation of virtual deformable objects the degree of freedom of interaction is a critical issue. In the model-based approach interaction with the object is dealt with by changes of boundary conditions and there is no need to change the model itself. In the measurement-based approach change of interaction means the change of referenced data, hence higher freedom of interaction requires a larger set of data that covers the interaction. Difficulty of modeling is another point of view. In the model-based approach, if the material and structure of the object is known, dominant equations and parameters in the model are determined relatively easily. A problem occurs when the material is unknown or difficult to model. The measurement-based approach is advantageous in that it does not require knowledge on the material and the structure, however, there were few investigations on interaction with large number of degree of freedom using the measurement-based approach. Considering this background and related research, our research focused on the construction of an IRDM using measurement of a real object.

3   Method of Measurement

IRDM is a model that defines the relationship between force input and displacement output as impulse response. It comprises impulse response data of all combinations of force input and displacement output nodes where nodes are the representative points of computation on the surface of an object. Hence creation of IRDM based on measurement means acquisition of entire impulse response data. Precisely this means each node has 3 degrees of freedom for force input and also 3 degrees of freedom for displacement output, and 9 impulse responses are defined for each combination.

This feature of the model suggests that it is inefficient to measure the impulse response of one pair of input/output nodes at a time; it requires actions of $O(n^2)$ where n is the number of nodes. Our study investigated measuring the response of multiple nodes at the action of force on one node by using cameras for the measurement of displacement. This approach is expected to reduce the order of actions to $O(n)$ [1]. In the implementation that is stated below, markers are put on the nodes locations and their motion is obtained by stereo measurement.

An air jet was used to apply force to the object; air is invisible and can keep the marker visible for the cameras. A straight forward method of acquiring impulse response is measuring displacement in response to impulse force. Actually it is difficult to measure the intensity of impulse force, because the object is in dynamic state, hence the inertial force is superimposed onto the measured value. Our alternative is measuring step response and computing impulse response by deconvolution; step force is measured while the air is blowing stably and displacement transition is measured after shutting the air off. A disadvantage of using air jet is that information on contact including friction is not obtained.

3.1   Measurement System

The structure of the measurement system is shown in Figure 1. Two high-speed cameras (MotionScope M5, IDT and Ai Nikkor 50mm f/1.2S, Nikon) were used as shooting markers on the surface of the object. Each camera was connected to a PC (Intel Core i7, 8GB of memory, Windows XP 64bit) that is equipped with a frame grabber card (Xcelera-CL PX4, Dalsa) by camera-link, and controlled by capturing software (Motion Studio, IDT). The two cameras are operated synchronously by connecting sync and trigger signals. LED lights (LEL-RB9N-F, Toshiba, equivalent to 100W incandescent lamp) were used for illumination. Approximate configuration of the cameras and the object is illustrated in Figure 2. Precise geometry among them is (elaborated below) determined based on camera parameters that are obtained by calibration.

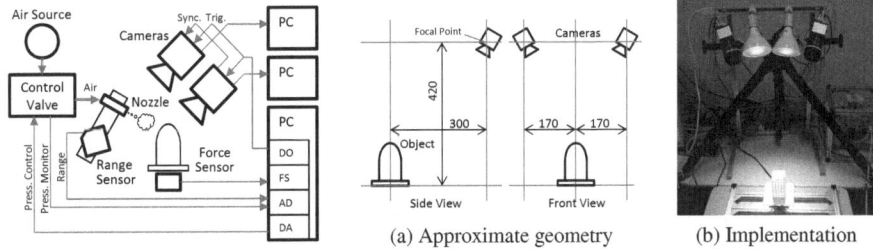

Fig. 1. Measurement System

(a) Approximate geometry   (b) Implementation

Fig. 2. Camera Configuration

Since the configuration is fixed, measurement of the marker position is possible only on markers that are visible from both of these two cameras, hence it is impossible to acquire impulse response data on the entire combination of all nodes. In the experiment that is stated below, the entire set of impulse response is obtained from measurable data by taking advantage of the symmetry of the target object. In a future study, relative motion of the cameras to the object should be enabled in order for an asymmetric object to be dealt with.

The control of force by the air jet and sending the trigger signal to cameras were performed by a controller PC (AMD Athlon 64X2, Windows XP). Compressed air from the air source is sent to the nozzle (KN-Q06-100, SMC) through an air valve (VY1B, SMC). The valve controls output pressure proportionally to its input signal voltage and actual pressure before and after the valve is measured by pressure sensors (PSE540, SMC). As will be explained below, the nozzle is combined with a laser range finder (ZX-SD104, Omron). Analog inputs from those sensors and output to the valve are connected to an analog interface card (PCI-360116, Interface). The trigger signal for the cameras is also outputted from the card. Force that is caused by the air jet is measured by a six-axis force sensor (IFS-67M25A15-I40, Nitta) installed at the base of object and read by the PC through a dedicated interface card (PCI-2184T, Nitta).

3.2   Method of Acting Force

In our current implementation, the air nozzle is manipulated by a human operator. In the operation, it is necessary to precisely locate the center of the air jet on the marker. Since the air is invisible some guiding interface for the operator must be provided. This system employed a laser range finder. The laser spot indicates the intersection point of the beam line with the object and it gives the distance from the spot. The operator moves the nozzle set and the range finder (denoted nozzle set below) to take the distance while keeping the spot on a target point. Configuration of the nozzle set is shown in Figure 3, where the laser spot is located at the center of the air flow when the distance is kept at 100 mm.

The force vector has 3 degrees of freedom therefore the measurement operation must be performed three times per node by applying force of different orientations. Ideally these three forces should be orthogonal or at least they must not be dependent. As a rough guideline, operator was instructed to keep the nozzle set so that the laser beam was roughly perpendicular to the surface of object, and to rotate the nozzle set around the beam by an angle of approximately 120 degree. This guideline helps to keep the forces independent each other. In the following experiment, however, there were some cases where the guideline could not be complied because of the problem of occluding markers by the nozzle set.

The distribution of pressure on a plane surface caused by the air is plotted in Figure 4. The laser beam was kept perpendicular to the surface and the air jet was applied obliquely upward at 45 degrees of the x axis. The pressure was measured in a small hole on the plane surface using a pressure sensor (ASDXL10D44R, Honeywell). The measurement was performed at an air pressure of 0.1 MPa because of the limitation of the range of pressure sensor even though the air pressure was set to 0.4 MPa in the actual experiment, which can cause some difference in the area of distribution.

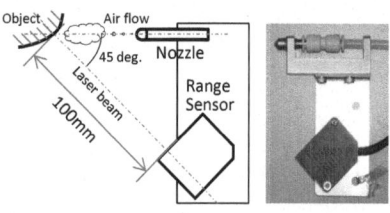   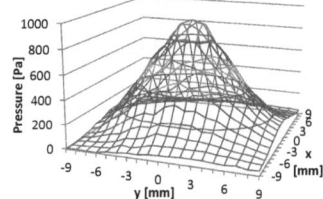

Fig. 3. Configuration of the Nozzle Set   **Fig. 4.** Pressure Distribution on Plane Surface

It should be noted that the impulse response comprehends the characteristic of contact; for example, in the case where the model is expected to present interaction using a fingertip then impulse response must be measured using a contactor that has similar characteristic to the finger. The distribution of pressure by air differs from that of a finger and it can cause difference in the interaction. The evaluation of the effect is one of our future works and will not be discussed in this paper.

3.3   Measurement of Displacement

Intrinsic and extrinsic camera parameters were obtained by camera calibration. The process was supported by camera calibration functions of OpenCV where a checker board pattern is used as for reference [24]. The calibration was performed once before the experiments because the parameters are valid provided the geometry of the cameras and the object is not changed. During the entire experiment the shutter speed and the aperture were kept to 1900 μs and F5.6 respectively considering the depth of field.

The spatial position of each marker is obtained by stereo matching. The region of each marker in the image is extracted by adaptive threshold processing and its position on the image is computed at the barycenter of the region. The process of matching markers on left and right images was automated by tracking each marker over the sequence of images. Since the image of the final frame of all measurements should be identical to each non-deformed image by manually identifying markers on non-deformed image, it becomes possible to track them back from the final frame. Using the position of markers on the left and the right images and the camera parameters spatial position is computed. Finally, the sequence of displacement is obtained by removing the undeformed position of each node which is obtained from the final frame.

3.4   Deconvolution

Deconvolution was performed assuming that the transition of force by the air is step-wise. Actually some delay and ramp is expected, however it was not possible to measure the precise transition because the responsiveness of the pressure sensor was not sufficient. Also the experimental computation of deconvolution assuming ramp transition of force tends to cause the resulting response to be unstable or divergent. Hence, in this paper, step transition was assumed. Finally, the response to orthogonal unit impulse force was obtained by decomposing the impulse response assuming linear relationship among force and displacement.

4   Measurement and Modeling

Measurement on an actual object was carried out, and IRDM was created based on the process discussed in the previous section.

4.1   Measurement

The duration of impulse response was assumed to be 1000 ms based on preliminary experiment on deformation therefore the operator was instructed to keep the nozzle set as stationary for 1000 ms before shutoff. Since the force sensor causes drift over time, force was measured for 500 ms both before shutoff and after attenuation of object motion; step height of force input was obtained as the remainder of the average of those forces. Images from the cameras were recorded at every 2 ms for 1000 ms, or 500 frames.

Two models were created for experiments; both of them were made of silicon rubber (KE-1308, Shinetsu silicone) and markers (3mm, black) were arranged at the position of the nodes. The arrangement of the markers and appearance of the objects are presented in Figure 5. In 'dome' shape, 121 nodes after excluding nodes in the lower part were subject to action of force and 31 nodes that are visible from two cameras were used for measurement of displacement. In the 'rectangle prism' shape similarly 145 and 55 nodes were used. Air pressure of the valve output was set to 0.4MPa, room temperature was approximately 30 degrees Celsius.

Figure 6(a) shows an example of measured displacement of a node that is located on the side of the dome (marked red in Figure 5) in the case when force was applied on the same node; three plots indicate the response to three forces respectively.

As stated above, measurement was performed manually. In the first round, measurements for 363 (121×3) times for the dome model and 435 (145×3) times for the rectangle prism model were performed. Re-measurement rounds were carried out on some nodes; because of the lack of independency in forces mainly by occlusion of a marker by the nozzle set. In both models, the first round measurement took approximately 5 hours, and re-measurement rounds also took similar time in total.

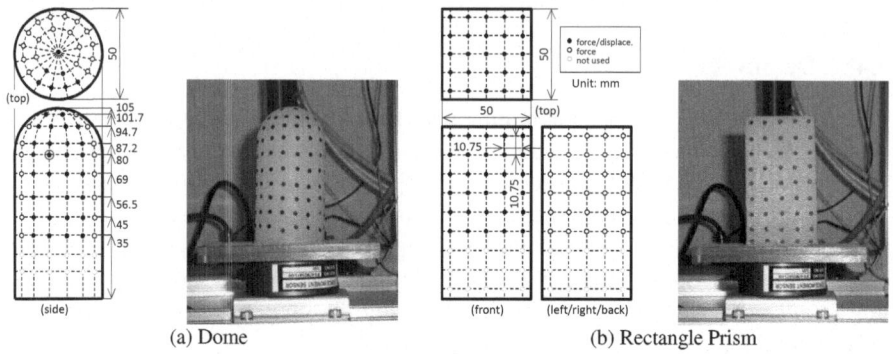

(a) Dome   (b) Rectangle Prism

Fig. 5. Models used in the experimentsModeling

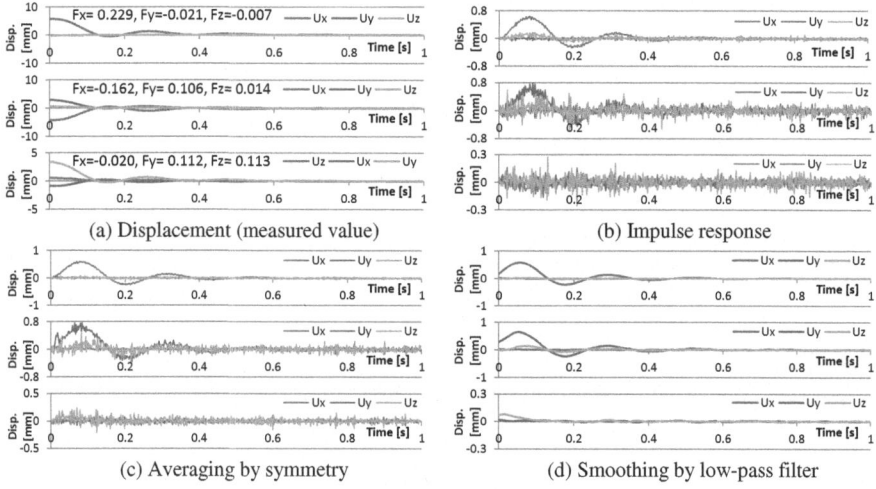

(a) Displacement (measured value)   (b) Impulse response

(c) Averaging by symmetry   (d) Smoothing by low-pass filter

Fig. 6. Process of computing IRDM

Firstly, with the deconvolution process as stated in section 3.4, 3 × 3 impulse response data is obtained for all combinations of force and displacement nodes. Figure 6(b) shows the impulse responses obtained after the process. Higher noise level on y and z axes compared to the x axis is considered to be due to the configuration of the cameras.

Next, impulse responses for all node combinations were computed from the limited combination of force and displacement nodes assuming symmetry. Let us consider of an impulse response from force node i and displacement node j. It is possible to find a set of nodes (i', j') by rotating (i, j) around the symmetrical axis, and (i'', j'') by reflecting (i', j') by a symmetry plane. If the impulse response for (i', j') or (i'', j'')

are given as measured data, then it can be transformed into the impulse response for (i, j) by applying transformations of rotation and reflection. This approach also has the effect of averaging the equivalent impulse responses in the measured data. Therefore, the effect of stabilizing the data by averaging multiple data is attained.

In the case of the dome, equivalent combinations were sought by rotation at 22.5 degree steps and reflection and data of 4 to 16 were averaged depending on the combination of nodes. In the case of a rectangle prism, rotation was performed at 90 degree steps, and 2 to 8 equivalent combinations were found. Figure 6(c) shows the impulse response after the symmetry is taken into account. Regarding the combination, four equivalent impulse response data were averaged and the noise level of the wave was reduced.

Finally, smoothing in temporal variation was performed. A low-pass filter based on FTT and inverse FFT was used. Cut-off frequency was determined empirically considering the stability of the resulting IRDM; 19.6Hz for the dome model and 4.9 Hz for the rectangle prism model. Figure 6(d) shows the smoothed impulse response.

The process of tracking markers on the images required approximately 30 hours for the dome model, and also similar time was required for the rectangle prism model, using a PC (AMD Athlon 64X2, Windows XP). The process of deconvolution, construction of entire model by symmetry, and smoothing took relatively negligible time.

5   Interaction and Evaluation

The acquired IRDM was evaluated through interaction in a virtual environment. IRDM gives displacement in response to force, and haptic device (PHANToM Omni, Sensable Technologies) output force depending on stylus position, hence a virtual coupling [25] was used to integrate the model and the device; constant for the coupling was set to 0.7 N/mm, which is relatively much stiffer compared with usual contact with the object.

5.1   Interaction Algorithm

Firstly, the position of interaction point p_t is obtained from the device. The contact point of the interaction point with the surface of the non-deformed shape of the object, denoted by p_c, is obtained by collision detection computation. It is supposed here that the contact point is on a patch that has vertices on node i_1, i_2, i_3, and area coordinate of the contact point is (k_1, k_2, k_3).

Next, the interaction force is computed. Current displacement on node i_1, i_2, i_3 are computed by convolution of past force with the impulse response which are denoted by $\widetilde{U}_{i_1}, \widetilde{U}_{i_2}, \widetilde{U}_{i_3}$, respectively. Current displacement at the contact point is assumed to be a weighted average of these displacements by the area coordinate: $\widetilde{U}_c = k_1 \widetilde{U}_{i_1} + k_2 \widetilde{U}_{i_2} + k_3 \widetilde{U}_{i_3}$. In the virtual coupling, it was assumed that the interaction force is proportional to the intrusion depth, hence, the force on the contact point is computed $F_c = K(p_t - p_c - \widetilde{U}_c)$ where K is the stiffness constant of the coupling.

Finally, forces on nodes are computed. Since IRDM has impulse response data from a node to another node, forces for convolution must be defined on nodes. In our current implementation, the forces are determined by distributing F_c to node i_1, i_2, i_3 proportionally to the value of the area coordinate as follows: $F_{i_1} = k_1 F_c$, $F_{i_2} = k_2 F_c$, and $F_{i_3} = k_3 F_c$.

5.2   Deforming Operation

Difference in deformation between a real object and the IRDM model was visually observed by overlaying graphic image of IRDM on the video image of the real object. The real object was deformed by pushing it using a rod. The motion of the rod was recorded by stereo cameras and the identical operation was performed on the model by simulation. Since the friction model has not been introduced into the simulation, the operation was performed carefully not to cause tangential force. Figure 7 shows some frames from the result; the simulation result is superimposed on the camera image using camera parameters. The result suggests similarity in deformed shapes in general. The simulation of the dome is showing shrinking behavior which is not observed in the real object. The difference is considered to be derived from the linearity of the IRDM model.

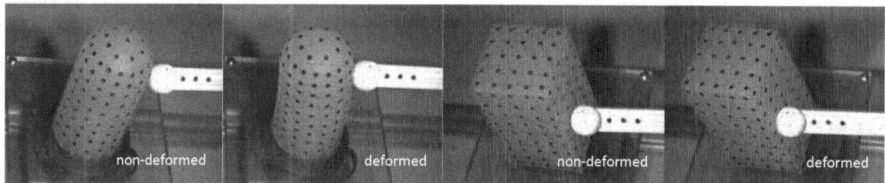

Fig. 7. Comparison of Deformation

5.3   Subjective Evaluation

The similarity of the model to the real object was evaluated by an experiment. The characteristic of force and deformation is changed by scaling the impulse response in both time and intensity, and haptic interaction with the scaled model is compared with the real object (see Figure 8(a)). The temporal and intensity scales were changed in 0.5, 0.57, 0.65, 0.75, 0.87, 1.0, 1.14, 1.31, 1.51, 1.74, 2.0, respectively; a larger value means quicker object deformation in temporal scale, and more rigid object in intensity scale. In the experiment system, scales were changed by pressing keys that step up and down the scales. Both graphic and haptic feedback is provided to the user. The real object was deformed using another stylus; tip of the stylus was sphere whose radius was 20 mm. Subjects were asked to tune the scales so that the interaction with the object is most similar to the real object. Each subject performed the task 30 times for both the dome and rectangle prism models respectively. Initial scales of each task were randomly determined. Participants of the experiment were 8 persons (3 female and 5 male) from 22 to 47 years old.

The result of the experiment is shown in Figure 8(b)(c) where distribution of the frequency of answer is plotted over the scales. The distribution was rather spread especially regarding in the scale of intensity. One reason is considered to be that friction of the force feedback device was confused with the resistance of deformation. Averages of time- and intensity-scales were 1.15 and 1.11 for dome and 1.24 and 1.17 for rectangle prism. The result suggests that the non-scaled IRDM is slightly slow in response and soft in deformation. Although the reason is not clear, it is provable that difference of temperature has affected the stiffness of real object; the experiment was carried out under room temperature of 18-22 degree Celsius, which is lower than the temperature of measurement.

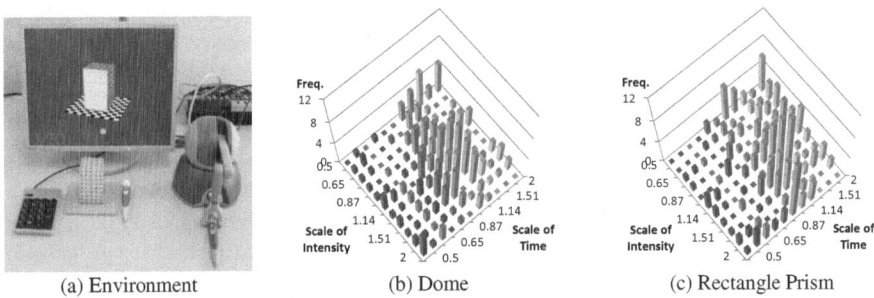

(a) Environment   (b) Dome   (c) Rectangle Prism

Fig. 8. Experiment on Similarity to Real Object

6   Conclusion

This paper discussed an approach to generating IRDM through measurement of force-displacement relationship. The measurement system that consists of a force acting mechanism using an air jet and also displacement sensing using stereo cameras were proposed, and feasibility of the approach was proved through evaluation of the resulting model. The result of the experiment proved that both static and dynamic (or, intensity and temporal) aspects of the object were reflected onto the model.

There are some works to be done regarding evaluation of our approach; accuracy of deformed shape must be qualitatively evaluated, and reason for the difference must be made clear. Also, there are some problems to be investigated. Transition of force by air jet must be examined for more precise measurement of impulse response. Since workload and time of measurement is relatively high, automation of the process needs be studied before our approach can be practically useful.

References

1. Tagawa, K., Hirota, K., Hirose, M.: Impulse Response Deformation Model: an Approach to Haptic Interaction with Dynamically Deformable Object. In: Proc. IEEE Haptics 2006, pp. 209–215 (2006)

2. Burdea, G.C.: Force and Touch Feedback for Virtual Reality. Wiley-Interscience (1996)
3. Salisbury, K., Brock, D., Massie, T., Swarup, N., Zilles, C.: Haptic rendering: Programming touch interaction with virtual objects. In: Proc. Symp. Interactive 3D Graphics, pp. 123–130 (1995)
4. Delingette, H.: Biquadratic and Quadratic Springs for Modeling St Venant Kirchhoff Materials. In: Bello, F., Edwards, E. (eds.) ISBMS 2008. LNCS, vol. 5104, pp. 40–48. Springer, Heidelberg (2008)
5. Kikuuwe, R., Tabuchi, H., Yamamoto, M.: An Edge-Based Computationally Efficient Formulation of Saint Venant-Kirchhoff Tetrahedral Finite Elements. ACM Trans. Graphics 28(1), 8:1-8:13 (2009)
6. Hirota, K., Kaneko, T.: Haptic representation of elastic objects. Presence 10(5), 525–536 (2001)
7. Tomokuni, S., Hirai, S.: Real-time Simulation of Rheological Deformation on FPGA. Journal of the Virtual Reality Society of Japan 10(3), 443–452 (2005)
8. Rasmusson, A., Mosegaard, J., Sørensen, T.S.: Exploring Parallel Algorithms for Volumetric Mass-Spring-Damper Models in CUDA. In: Bello, F., Edwards, E. (eds.) ISBMS 2008. LNCS, vol. 5104, pp. 49–58. Springer, Heidelberg (2008)
9. Takama, Y., Tsujino, Y., Hori, T., Tanaka, H.: Online Remeshing of Tetrahedral Adaptive Mesh for Deformation of Soft Objects. Trnas. VRSJ 13(1), 69–78 (2008)
10. MacLean, K.E.: The "Haptic Camera": A Technique For Characterizing And Playing Back Haptic Properties of Real Environments. In: Proc. ASME DSC, vol. 58, pp. 459–467 (1996)
11. Pai, D.K., van den Doel, K., James, D.L., Lang, J., Lloyd, J.E., Richmond, J.L., Yau, S.H.: Scanning Physical Interaction Behavior of 3D Objects. In: Proc. ACM SIGGRAPH 2001, pp. 87–96 (2001)
12. Ueda, N., Hirai, S., Tanaka, H.T.: Extracting Rheological Properties of Deformable Objects with Haptic Vision. In: Proc. ICRA 2004, pp. 3902–3907 (2004)
13. Hoever, R., Kosa, G., Szekely, G., Harders, M.: Data-Driven Haptic Rendering - From Viscous Fluids to Viscoelastic Solids. Trans. Haptics 2(1), 15–27 (2009)
14. Weir, D.W., Peshkin, M., Colgate, J.E., Buttolo, P., Rankin, J., Johnston, M.: The Haptic Profile: Capturing the Feel of Switches. In: Proc. Haptics 2004, pp. 186–193 (2004)
15. Bickel, B., Bacher, M., Otaduy, M.A., Matusik, W., Pfister, H., Gross, M.: Capture and modeling of non-linear heterogeneous soft tissue. ACM Trans. Graph. 28(3), Article 89 (2009)
16. Wellman, P., Howe, R.D.: Towards realistic vibrotactile display in virtual environments. In: Proc. ASME DSC, vol. 57(2), pp. 713–718 (1995)
17. Okamura, A.M., Dennerlein, J.T., Howe, R.D.: Vibration Feedback Models for Virtual Environments. In: Proc. IEEE ICRA, vol. 3, pp. 2485–2490 (1998)
18. Okamura, A.M., Cutkosky, M.R., Dennerlein, J.T.: Reality-Based Models for Vibration Feedback in Virtual Environments. IEEE/ASME Trans. Mechatronics 6(3), 245–252 (2001)
19. Niemeyer, G., Kuchenbecker, K.J., Fiene, J.: Improving Contact Realism through Event-Based Haptic Feedback. Trans. Visualization and Computer Graphics 12(2), 219–230 (2006)
20. James, D.L., Fatahalian, K.: Precomputing Interactive Dynamic Deformable Scenes. In: Proc. SIGGRAPH 2003, pp. 879–887 (2003)
21. Fong, P.: Sensing, Acquisition, and Interactive Playback of Data-based Models for Elastic Deformable Objects. Intl. J. Robotics Research 28, 630–655 (2009)

22. Chial, V.B., Greenish, S., Okamura, A.M.: On the Display of Haptic Recordings for Cutting Biological Tissues. In: Proc. Haptics 2002, pp. 80–87 (2000)
23. Dobashi, Y., Hasegawa, S., Kato, M., Sato, M., Yamamoto, T., Nishita, T.: A Fluid Resistance Map Method for Realtime Haptic Interaction with Fluids. In: Proc. VRST 2006, pp. 91–99 (2006)
24. OpenCV Reference Manual, v2.2 (December 2010)
25. Colgate, J.E.: Issues in the haptic display of tool use. In: Proc. Intelligent Robots and Systems 1995, vol. 3, pp. 140–145 (1995)

Tradeoffs in the Application of Time-Reversed Acoustics to Tactile Stimulation

Charles Hudin[1], José Lozada[1], Michael Wiertlewski[1,2], and Vincent Hayward[2]

[1] CEA, LIST, Sensorial and Ambient Interfaces Laboratory, 91191, Gif-sur-Yvette Cedex, France
[2] UPMC Univ. Paris 6, ISIR, Institut des Systèmes Intelligents et de Robotique, 75005, Paris, France
{charles.hudin,jose.lozada}@cea.fr,
{michael.wiertlewski,vincent.hayward}@isir.upmc.fr

Abstract. The creation of active tactile surfaces through electromechanical actuation is an important problem. We describe here the application of time-reversed acoustics to the creation of deformations localized in time and in space in a stretched membrane that can be touched. We discuss the basic physical and engineering tradeoffs of this approach and describe the results obtained from an experimental mock-up device.

Keywords: Surface Haptics, Time-Reversed Acoustics.

1 Introduction

The electromechanical stimulation of the fingertip has attracted the interest of researchers since the early works of Gault [1,2]. Recently there has been a lot of interest in providing active surfaces that can stimulate the fingertips mechanically, while permitting free exploration.

This question has been approached by vibrating the entire surface being touched, in the normal or tangential directions, as is commonly done is consumer devices. Another approach is to modulate the interaction force between a finger and a surface during sliding. Surface acoustic waves can be employed to this end, but require the use of an thin intermediary sheet between the finger and the surface [3,4]. Nonlinear acoustic pumping applied to tactile stimulation was pioneered by Watanabe and Fukui [5]. This approach, which modifies the finger-surface interaction in a controlled manner, and without intermediary, has been recently further developed [6,7,8,9]. Electro-vibration, where the interaction force modulation is achieved by electrostatic attraction between the surface and the skin, originally demonstrated by Strong and Troxel [10], remains popular today owing to its implementation simplicity [11], despite its inherent weakness and sensitivity to physiological and environmental factors [12]. Another direction of investigation is the transport of mechanical energy through waves. Acoustic phase arrays can create significant non-contact stimulation from remotized transducers, as shown by Iwamoto et al. [13]. They were also further developed to create specific tactile patterns [14].

P. Isokoski and J. Springare (Eds.): EuroHaptics 2012, Part I, LNCS 7282, pp. 218–226, 2012.

Here, we describe a new surface actuation mode, which combines the remotization of the actuators with an active tactile surface delivering the stimulation. It is based on the concept of 'computational time reversal' and is able to stimulate to one or several regions, and hence several fingers, independently with a high temporal and spatial resolution, using a membrane as an acoustic propagation medium. We discuss in this article the basic physical and realization tradeoffs and report on a proof-of-concept device that can displace a surface out-of-plane by 200 micro-meters in a 1 cm² region.

2 Time Reversal

2.1 Principle

Time-reversal is a computational technique that takes advantage of a basic property of waves which, at first sight, seems to run counter to the principle of irreversibility [15]. For instance, owing to the irreversibility of certain processes, it is not possible to reconstruct the exact configuration of a broken glass from scattered pieces, even if each single atom on each side of the cracks can be brought back its original place. Macroscopic waves, however, obey a fundamental symmetry arising from the wave equation $\partial^2 u/\partial t^2 = c^2 \nabla^2 u$. The equation is invariant under substitution of t by $-t$, which means that the initial and final conditions can be interchanged. This property holds even with complex, anisotropic, inhomogenous, nonlinear media, where the right-hand-side, involving the Laplacian operator, can be arbitrarily complicated, provided that it has no memory and does not depend on time. If the medium is dissipative, with terms involving $\partial u/\partial t$, the symmetry is broken, but the losses can be corrected for [16].

Computational time-reversal was originally developed to focus ultrasound waves in inhomogeneous, scattering media such as tissues [17]. Since then, it has been applied to surface waves in water, electromagnetic waves, sound waves within the audible range, and other cases. It has applications in medical imaging, lithotripsy, non-destructive testing, communications as well as many other areas.

2.2 Basic Theory

In a bounded domain, Ω, let $h_{AB}(t)$ represent the measured velocity of a material point, B, resulting from the application of an impulse of force applied at time $t = 0$ at point A. The reciprocity principle lets us interchange the sensor and the actuator, so that the signal recorded at A due to an impulse of force applied at B is the same, that is, $h_{AB}(t) = h_{BA}(t)$. Suppose now that the force applied at B is the response at A to an impulse applied at B, but inverted and shifted in time by T, that is, $f_B(t) = h_{AB}(T - t)$, the velocity at a point C is given by

$$v_C(t) = f_B(t) \otimes h_{BC}(t) = h_{AB}(T - t) \otimes h_{BC}(t)$$

$$= \int_0^t h_{AB}(T - \xi)\, h_{BC}(t - \xi)\, \mathrm{d}\xi, \tag{1}$$

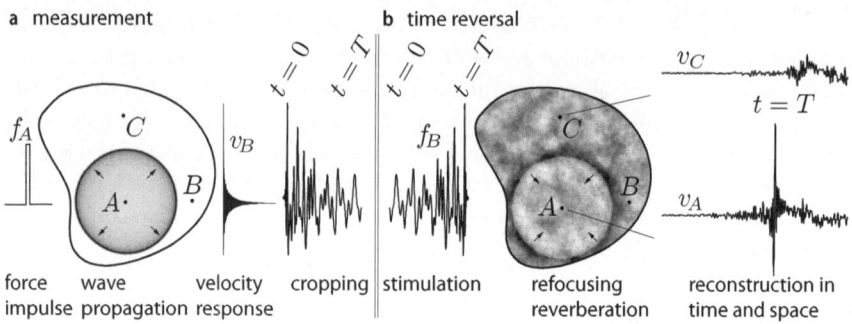

Fig. 1. Time-reversal applied to a reverberant cavity. **a**) The velocity response at B of a force impulse applied at A is recorded and the initial portion of the signal length T is cropped. **b**) The signal is time-reversed and used an actuator signal in B. Waves propagate and reverberate to eventually refocus in A. Perfect reconstruction would entail an infinitely long window and the absence of transducer noise. In practice, the response separates into a signal, as in A, and a background noise, as in A and C.

where $\otimes$ denotes the convolution operation. If h_{AB} and h_{BC} are not correlated, waves interfere non-constructively, giving a background noise that can be modeled by a random signal with zero-mean velocity and standard deviation σ [18]. Applying the reciprocity principle, setting $C = A$, and time $t = T$ in (1) gives,

$$v_A(T) = \int_0^T h_{AB}^2 (T - \xi) \, \mathrm{d}\xi. \tag{2}$$

The interference is now constructive, yielding a peak of signal localized in space *and* in time. This process is graphically represented in Fig. 1 for a two dimensional domain.

3   Physical and Engineering Tradeoffs

A time-reversal set-up typically involves an array of transducers located at the periphery of the domain of interest. The number of transducers can even be reduced to one when the domain has reverberant properties and if the transducer has sufficient bandwidth. The reduction of the number of transducers, which comes with possible gains in implementation complexity, also comes with a cost in performance as illustrated in Fig. 2a. Sets, connected or not, and of any measure, can theoretically be reconstructed. Multi-digit, whole hand stimulation, etc, thus does not differ in principle from the single region example illustrated in Fig. 2b. Repetitions in time are easily achieved by convolution of the signals by a Dirac comb. All these possibilities are crucially dependent on achieving sufficient resolution in time, space, and signal magnitude. It is therefore important to develop, from first principles, the basic tradeoffs involved.

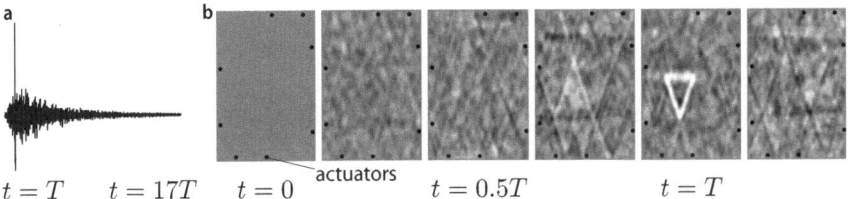

Fig. 2. a) A highly reverberant cavity with a single actuator can yield good reconstruction quality in short time, but with a long tail of noise. **b)** Example of a triangle reconstructed from eight impulse responses.

3.1   Contrast Ratio

The ratio $C = v_A(T)/\sigma$ is called contrast, or signal-to-noise ratio, at point A. Intuitively, this ratio increases with the introduction of higher modes that have smaller wavelengths. According to the time-frequency uncertainty principle, the time needed to resolve two frequencies Δt, is inversely proportional to their difference, Δf, that is, $\Delta t \Delta f \sim 1$. A longer time-reversed window, T, makes it possible to resolve mode separated by smaller Δf's, thus increasing contrast. With one actuator in a chaotic cavity [19], or in a scattering medium [18], contrast increases with $\sqrt{T}$. Higher modes become impossible to resolve with the consequence that, when T approaches their decay characteristic time, τ, contrast no longer follows a square root law and reaches a limit. With N actuators, however, simultaneously reproducing inverted impulse responses sum their contributions at any point. Contrast, then, follows the trend $C \propto \sqrt{NT}$, as long as $T < \tau$, which is equivalent to increasing the length of the time-reversed window.

3.2   Cavity Reverberation

Cavity having the longest characteristic time possible would seem to be advantageous at first sight. During the reconstruction process, waves converge toward the reconstructed set and then diverge once $t > T$. The time during which diverging waves reverberate before dying out is determined by τ. Repetition of the process at rate faster than $1/\tau$, will cause noise to accumulate, lowering contrast. The reverberation characteristic time, τ, should therefore be of the same order than T/N to achieve good contrast ratio while raising the reconstruction frequency. A larger number of actuators makes it possible to design the cavity with a shorter reverberation characteristic time.

3.3   Spatial Resolution

A factor which must be considered when applying the time-reversal approach to tactile stimulation is that of spatial resolution. It is driven by the smallest reconstructed spot of diameter, ϵ, which is bounded by the smallest wavelength, $\lambda^- = c/f^+$, where c is the medium's wave propagation celerity, and f^+ the frequency used in the re-focusing process. For a same displacement amplitude, smaller wavelengths can be achieved either by increasing the range of frequencies or by lowering propagation celerity. The focused impulse duration, ζ, corresponds to the inverse of the largest frequency used in the re-focusing process giving, $\zeta < \epsilon/c$. A low celerity has an additional benefit. A given amplitude can be achieved with lower frequencies and hence with less power.

3.4 Medium Deformation

The focused point size is given by the smallest wavelength used in the reconstruction process and amplitude must be sufficient to fall into the range tactile sensibility. The propagation medium must therefore be able sufficient displacement gradients, that is, deformation. As exemplified in Section 4, the Young's modulus of a material and the smallest achievable bending radius of a membrane are not independent quantities, suggesting that more elastic materials employed in manufacturing of the propagation medium are favorable, lowering propagation velocity, shortening wavelengths, and achieving higher deformations.

3.5 Power Transmission to a Finger

Stimulation techniques must transfer power into the load. Here, the stimulated finger perturbs wave propagation when it is coupled to the medium, causing phase shifts and wave front attenuation. To gauge these effects, we developed a simple model of interaction between a finger and flexural waves, see Fig. 3. The finger is lumped into a mass-spring-damper system and waves propagate in an infinite strip of width, l, representing the width of the contact. An incident wave of wavelength, $\lambda = 2\pi/k$, amplitude, A_I, propagates at celerity, $c = \omega/k$, along the x-axis, $u_I(x,t) = A_I \, e^{j(\omega t - kx)}$. There are two other waves, $u_R(x,t) = A_R \, e^{j(\omega t - kx)}$ and $u_T(x,t) = A_T \, e^{j(\omega t - kx)}$, the reflected wave and the transmitted wave, respectively. The power transported by u_I is $P_I = c^2 k^2 Z_S |A_I|^2$, where $Z_S = f_{\text{act}}/v_I$ is impedance of the strip at the point of actuation and $v_I = \partial u_I / \partial t$ is the out-of-plane velocity [20].

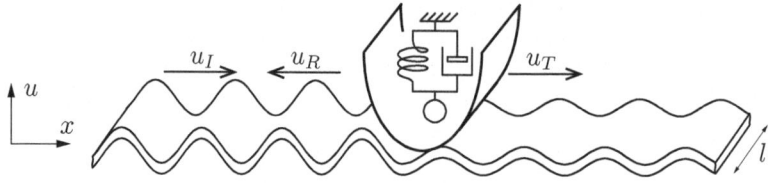

Fig. 3. Model of wave-finger interaction

At $x = 0$, finger displacement has the amplitude of the transmitted wave, $u_F(t) = u_T(0,t)$. Enforcing continuity of displacement and conservation of momentum at $x = 0$ gives,

$$A_F = A_T = A_I \frac{2Z_S}{Z_F + 2Z_S}, \qquad A_R = -A_I \frac{Z_F}{Z_F + 2Z_S}, \tag{3}$$

Table 1. Strip material properties and power needed to displace a fingertip by 10 μm

Properties	E	h	ρ_l	R_l	P
	GPa	mm	kg/m^3	N	W
Glass	69.0	1.0	2300	0.0	500
BOPET	4.0	0.125	1400	2.0	1.5

where Z_S is the lumped impedance of the strip at the contact and $Z_F = \beta + j\left[\mu\left(ck\right) - \kappa/(ck)\right]$ is the impedance of the finger where, μ, β, and κ are mass, damping, and stiffness respectively. Expressing the power of the incident wave as a function of the finger displacement gives,

$$P_I = \frac{c^2 k^2}{4} \frac{|Z_F + 2Z_S|^2}{|Z_S|} |A_F|^2, \qquad (4)$$

showing that slow propagation celerity achieves the same displacement for less power. It shows also that Z_S should be larger than Z_F to prevent muffling. On the other hand, as is well known, optimal transmission is when $Z_F = Z_S$. For an isotropic thin strip under uniform traction,

$$Z_s = \sqrt{\rho_l\left(R_l + D_l k^2\right)}, \quad \text{and} \quad c = \sqrt{1/\rho_l\left(R_l + D_l k^2\right)}, \qquad (5)$$

where the lineic bending stiffness is, $D_l = \frac{1}{12}Eh^3 l$, and where, E, is the strip Young's Modulus, l, the width, h, the thickness, ρ_l, the lineic mass density, and R_l, the lineic tension. This simple model, while ignoring scattering and finite contact size effects, indicates that if a membrane is used as transmission medium, its material should have high density and low stiffness. Assuming reasonable values for a fingertip impedance, $\mu = 0.2$ g, $\beta = 1.0$ N·s/m and $\kappa = 1.0$ N/mm, see [21], $\lambda = 15$ mm, $l = 10$ mm, we can evaluate the appropriateness of some materials for the application of time-reversed acoustics to tactile stimulation in terms of the power needed to displace the skin of a finger, see Table 1.

4   Preliminary Validation

A 150 × 150 × 0.125 mm sheet of BOPET (biaxially-oriented polyethylene terephthalate) was stretched and glued to a rigid frame with a surfacic tension of $R_s = 30$ N/m. For a membrane, D_l must be replaced in (5) by $D_s = \frac{1}{12}Eh^3/(1 - \nu^2)$, where ν is the Poisson's ratio of the material. Setting $E = 4$ GPa and $\nu = 0.3$, we obtain a wave celerity varying from 34 to 43 m/s for wavelengths varying from ∞ to 15 mm, that is frequencies from 0 to 3 kHz. The first mode was at 150 Hz and the characteristic time was about 70 ms. Eight custom-made, miniature, moving-magnet electromagnetic devices (6.2 Ω, 1.9 mH, $bl = 6$ N/A, 1.5 × 0.5 mm) impinged on the membrane. They were placed in a configuration that avoided symmetries as indicated in Fig 4. The devices were used as force transducers in actuation mode and as velocity transducers in sensing mode.

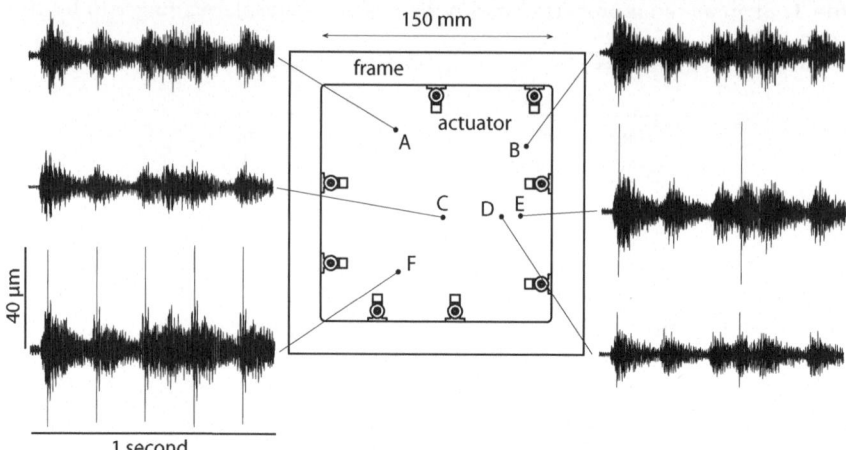

Fig. 4. Experimental set-up and focusing results, at 5 Hz in E and at 2 Hz in D

A two-dimensional cavity of surface area S, perimeter P and wave celerity $c(f)$, has M mode below a frequency f according to $M(f) = \pi S (f/c)^2 + (Pf)/(2c)$, see [22]. It follows that the separation, Δf, between two modes is given by $2c^2/(4\pi Sf + Pc) = 2c^2/(4\pi Sck + Pc) = 2c/(2Sk + P)$. Following the discussion of Section 3.1, the time reversal window duration was set at $T \sim (2Sk_{max} + P)/(2Nc_{min}) \simeq 30$ ms.

Waves were re-focused onto two points, E and F at 2 and 5 Hz, respectively to re-create time-space impulses. Displacements were recorded by a laser vibrometer and synchronized with the emissions. The results are plotted in Fig 4 for various locations on the membrane. The pulses in E and F achieved 40 μm in amplitude. At any other point, background noise was present with same amplitude as in E and F.

For the same contrast ratio, the peak amplitude could reach 200 μm by scaling up the signal. When a finger touched the surface the amplitude of the signal and of the noise was attenuated by a factor 5. It was also possible to scale down the signals so that the background noise fell below the tactile sensitivity while keeping the sensation experienced in E and F above threshold. The signal bandwidth fell within the audible range and could be heard, but since the sound was a reproduction of an actual impact, it did sound exactly like an impact. The tactile and auditory experiences were therefore mutually coherent.

5   Conclusion

Computational time reversal was successfully applied to focusing flexural waves to a spot of similar size to a finger contact, and with amplitudes compatible with the tactile sensitivity range. Further research will aim at increasing the contrast ratio to obtain more vivid stimuli, keeping in mind that the stretched membrane approach is only one among many other options for a propagation medium.

The ability of reconstructing arbitrary waveforms within arbitrary sets can be approached either by pre-computing response libraries or in a discrete manner or applying the appropriate interpolation laws, or by the development of a model which is sufficiently accurate and computationally acceptable to calculate impulse responses offline, or even online, using advanced computational methods.

References

1. Gault, R.H.: Tactual interpretation of speech. The Scientific Monthly 22(2), 126–131 (1926)
2. Gault, R.H.: Recent developments in vibro-tactile research. Journal of the Franklin Institute 221, 703–719 (1936)
3. Takasaki, M., Nara, T., Tachi, S., Higuchi, T.: A tactile display using surface acoustic wave with friction control. In: International Workshop on Micro Electro Mechanical Systems, pp. 240–243 (2001)
4. Takasaki, M., Kotani, H., Nara, T., Mizuno, T.: Transparent surface acoustic wave tactile display. In: Proceedings of the IEEE/RSJ International Conference on Intelligent Robots and Systems, pp. 1115–1120 (2005)
5. Watanabe, T., Fukui, S.: A method for controlling tactile sensation of surface roughness using ultrasonic vibration. In: Proceedings of the IEEE International Conference on Robotics and Automation, pp. 1134–1139 (1995)
6. Winfield, L., Glassmire, J., Colgate, J.E., Peshkin, M.: T-PaD: Tactile pattern display through variable friction reduction. In: Proceedings of the Second Joint EuroHaptics Conference and Symposium on Haptic Interfaces for Virtual Environment and Teleoperator Systems, World Haptics 2007, pp. 421–426 (2007)
7. Biet, M., Giraud, F., Lemaire-Semail, B.: Squeeze film effect for the design of an ultrasonic tactile plate. IEEE Transactions on Ultrasonics, Ferroelectrics and Frequency Control 54(12), 2678–2688 (2007)
8. Chubb, E.C., Colgate, J.E., Peshkin, M.A.: Shiverpad: A glass haptic surface that produces shear force on a bare finger. IEEE Transactions on Haptics 3(3), 189–198 (2010)
9. Amberg, M., Giraud, F., Semail, B.: Interface tactile vibrante transparente. French patent 1153963, Laboratoire d'électrotechnique et d'électronique de puissance (L2EP) (L2EP) (May 2011)
10. Strong, M.S., Troxel, D.E.: An electrotactile display. IEEE Transactions on Man-Machine Systems 11(1), 72–79 (1970)
11. Bau, O., Poupyrev, I., Israr, A., Harrison, C.: TeslaTouch: electrovibration for touch surfaces. In: Proceedings of the 23nd Annual ACM Symposium on User Interface Software and Technology, pp. 283–292 (2010)
12. Tang, H., Beebe, D.J.: A microfabricated electrostatic haptic display for persons with visual impairments. IEEE Transactions on Rehabilitation Engineering 6(3), 241–248 (1998)
13. Iwamoto, T., Akaho, D., Shinoda, H.: High resolution tactile display using acoustic radiation pressure. In: Proceedings of SICE Annual Conference, pp. 1239–1244 (August 2004)
14. Hoshi, T., Iwamoto, T., Shinoda, H.: Non-contact tactile sensation synthesized by ultrasound transducers. In: Proceedings of the World Haptics Conference, pp. 256–260 (2009)

15. Fink, M.: Time reversed acoustics. Physics Today 50, 34 (1997)
16. Montaldo, G., Tanter, M., Fink, M.: Real time inverse filter focusing through iterative time reversal. Journal of the Acoustical Society of America 115(2), 768–775 (2004)
17. Fink, M.: Time reversal of ultrasonic fields. i. basic principles. IEEE Transactions on Ultrasonics, Ferroelectrics and Frequency Control 39(5), 555–566 (1992)
18. Derode, A., Tourin, A., Fink, M.: Limits of time-reversal focusing through multiple scattering: Long-range correlation. The Journal of the Acoustical Society of America 107, 2987 (2000)
19. Quieffin, N., Catheline, S., Ing, R.K., Fink, M.: Real-time focusing using an ultrasonic one channel time-reversal mirror coupled to a solid cavity. The Journal of the Acoustical Society of America 115, 1955 (2004)
20. Morse, P.M.C., Ingard, K.U.: Theoretical acoustics. Princeton University Press (1986)
21. Wiertlewski, M., Hayward, V.: Transducer for mechanical impedance testing over a wide frequency range through active feedback. Review of Scientific Instruments (in press, 2012)
22. Blevins, R.D.: Modal density of rectangular volumes, areas, and lines. The Journal of the Acoustical Society of America 119(2), 788 (2006)

Error-Resilient Perceptual
Haptic Data Communication Based
on Probabilistic Receiver State Estimation

Julius Kammerl, Fernanda Brandi, Florian Schweiger, and Eckehard Steinbach

Technische Universität München, Institute for Media Technology,
Munich, Germany
{kammerl,fernanda.brandi,florian.schweiger,eckehard.steinbach}@tum.de

Abstract. We present an error-resilient perceptual haptic data compression scheme based on a probabilistic receiver model. While the previously proposed perceptual deadband approach successfully addresses the challenges of high packet and data rates in haptic real-time communication, packet loss in the network leads to perceivable distortion. To address this issue, a sender-driven transmission scheme for low-latency packet loss compensation is proposed. In this scheme, packet transmissions are adaptively triggered only if the reveicer state is likely to deviate from the error-free signal by more than the applied perception thresholds. Conducted experiments validate that the proposed haptic communication scheme successful compensates for packet loss with low computational complexity and without the need of acknowledgments.

Keywords: Haptics, Communication, Compression, Error-resiliency.

1 Introduction

Haptic data communication is an emerging research area in the context of telepresence and teleaction (TPTA) systems. These systems enable the execution of remote explorative and manipulative tasks by immersing a user into an environment that is distant, inaccessible, scaled to macro- or nano-dimensions or hazardous for a human being. A TPTA system can be decomposed into three main subcomponents: the human operator (OP) connected to a human system interface device (HSI), a teleoperator (TOP) which receives control commands for the execution of remote operations, and the communication link which bidirectionally transmits the multimodal information over a communication channel (see Fig. 1). In order to provide physical access to the remote environment, particularly the exchange of haptic velocity and force feedback signals plays a key role. In contrast to the transmission of audio and video signals, the haptic signals are bidirectionally transmitted and thereby close a global control loop between the human OP and the TOP. To maintain control loop stability and a high degree of system transparency, haptic real-time communication in TPTA systems is characterized by strict delay constraints. As the communication and processing latency must be kept at an absolute minimum, haptic samples are typically

P. Isokoski and J. Springare (Eds.): EuroHaptics 2012, Part I, LNCS 7282, pp. 227–238, 2012.

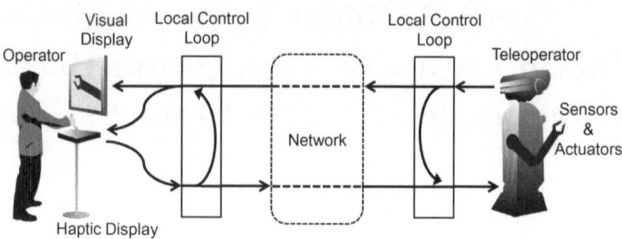

Fig. 1. Illustration of a telepresence and teleaction system and its main subcomponents: the human-system interface (HSI), the communication system and the teleoperator (TOP). Figure adopted from [1].

transmitted immediately upon their generation. In packet-switched networks, e.g., the Internet, this transmission behavior leads to high packet rates of up to the applied haptic sampling rates at the HSI and the TOP (typically 1000 Hz). Furthermore, modern TPTA systems deploy an increasing number of degrees of freedom (DoF) to provide intuitive and flexible remote manipulation. As every DoF is individually sampled and controlled, the bidirectional exchange of haptic signals leads to high data rates which are particularly challenging in large-distance or wireless TPTA scenarios.

Several approaches addressing the challenges of haptic real-time communication have been presented in literature. Early work on haptic data compression can be found in [2,3], where adaptive sampling and quantization techniques for haptic signals are discussed and compared with each other. In [4], predictive coding for haptic signals based on Differential Pulse Code Modulation (DPCM) and Adaptive Differential Pulse Code Modulation (ADPCM) combined with Huffman coding are used to reduce the load on the haptic channel. First research explicitly targeting the high packet rates in networked control loops can be found in [5]. Here, the concept of a deadband-based packet rate reduction is proposed where sensor readings are only transmitted if their magnitude changes by a fixed threshold. First work specifically exploiting limitations in human haptic perception can be found in [6,7]. Here, the so-called "perceptual deadband" (PD) compression approach (or in short "deadband approach") has been presented which is further investigated in terms of stability criteria in [8,9]. It explicitly extends the deadband-based packet rate reduction scheme in [5] with a model of human haptic perception. It reduces the packet rate by detecting and transmitting only perceptually relevant haptic samples over the haptic communication channel.

Consequently, as transmitted haptic samples are considered to describe perceivable information, packet loss in the network immediately leads to perceivable distortion in the haptic feedback. Research from the field of real-time video streaming [10,11] proposes a technique for robust low-latency video streaming, which has been shown to also apply to the challenges of lossy haptic communication [12,13]. It is based on a Markov decision tree which considers every triggered network transmission to either successfully arrive at the receiver or to be lost.

In this way, it allows the consideration of all possible receiver states at any time. If possible signal disturbances become likely, additional redundancies can be added to the haptic channel for compensating the network errors. However, the exponential growth of the tree structure that occurs with every triggered network transmission limits the applicability due to its expensive memory and computation requirements. Brandi *et. al* address this issue [13] and show that the exponentially growing Markov tree can be significantly simplified to an adaptive Markov chain model which reduces the complexity to linear time. However, the constantly growing data structures require techniques to limit their size such as the transmission of acknowledgment feedback or the detection and removal of unlikely receiver states.

In this work, a novel probabilistic receiver state model using a multivariate Gaussian distribution is presented. In contrast to previous approaches, it does not require any feedback information (ACKs or NACKs) and operates with low computational complexity. Conducted experiments validate that the proposed approach can successfully restore the perceived quality of haptic feedback when operating a TPTA system on a highly erroneous network channel.

This paper is organized as follows. In Section 2, the perceptual deadband scheme for haptic real-time compression is explained. The novel error-resilient haptic communication scheme which is based on probabilistic receiver state estimation is introduced in Section 3 and experimentally evaluated in Section 4. Section 5 concludes the paper.

2   Perceptual Haptic Data Compression

The challenges of high packet and high data rates in haptic real-time communication are successfully addressed by the perceptual deadband (PD) compression scheme proposed by Hinterseer et al. [6]. It allows a tremendous reduction in network transmissions, imperceptible to the human. To this end, two key components are used: a psychophysical model at the sender, and a signal predictor deployed both at the sender and the receiver (see Fig. 2). Hinterseer et al. further show that a simple yet effective psychophysical model of human haptic perception suitable for haptic real-time communication can be built based on Weber's Law of Just Noticeable Differences (JND) [14]. Weber's research revealed that over a large dynamic range, the perception of relative changes in stimulus is linearly proportional to the stimulus intensity itself. This implies a constant ratio between the intensity of a pending stimulus and the maximum change in intensity that is just not noticeable.

Weber's Law of the JND can be mathematically described in the following way:

$$\frac{\Delta I}{I} = k = const. \quad \text{or} \quad \Delta I = kI \tag{1}$$

where I is the stimulus intensity, ΔI is the so called Difference Threshold or the JND and k is the constant *Weber fraction*. Integrated into the proposed perceptual coding scheme, Weber's Law of JND can be used to estimate haptic

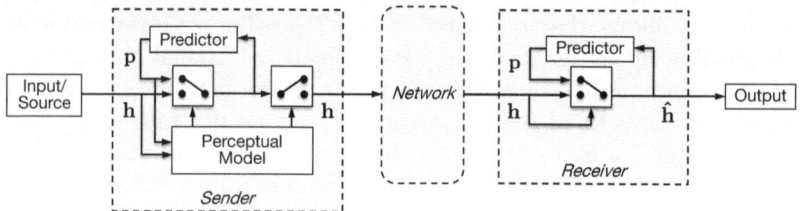

Fig. 2. Illustration of the perceptual predictive coding scheme for haptic real-time transmission introduced in [6]

discrimination bounds. As long as the difference between the incoming signal **h** and the predicted signal **p** stays within imperceptible bounds, no network transmissions are required and the predicted samples at the receiver can be output. If the difference between the incoming and the predicted haptic sample exceeds the applied perception thresholds at the sender, additional signal information is sent over the network which also updates the predictors. To this end, the haptic signals are perceptually downsampled while keeping the introduced compression distortion below human haptic perception thresholds.

For the prediction of haptic signals, typically linear predictors of very low order (e.g. order one) are deployed to comply to the strict delay constraints which are required to maintain control loop stability. They successfully model the low frequency characteristics of haptic signals while still being able to quickly adapt to transients during contact events due to their low group delay.

Fig. 3 visualizes the process of PD compression using a zeroth-order predictor (hold-last-sample algorithm) and with perceptual thresholds defined by Weber's Law of JND. Samples illustrated with solid vertical lines are considered to be relevant for transmission. They also define the discrimination thresholds based on Weber's Law of JND represented by PDs, illustrated as gray zones. Note that the size of the applied PD is a function of a deadband parameter k and the magnitude of the most recently predicted haptic sample value $|\mathbf{p}_{i-m}|$, where m samples back in time the last violation of the then applicable perception thresholds occurred. This can be described by

$$\Delta_i = \Delta_{i-1} = \ldots = \Delta_{i-m} = k \cdot |\mathbf{p}_{i-m}|. \tag{2}$$

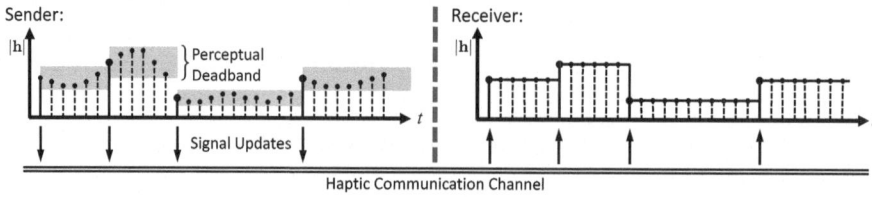

Fig. 3. Perceptual deadband compression using a zeroth-order prediction (hold-last-sample) algorithm

Once the PD is violated by a new input sample, this input sample is transmitted and redefines the applied perception thresholds. Samples with dotted lines fall within the currently defined deadband and can be dropped as their change in signal is too small to be perceptible.

Similarly, the first-order haptic prediction scheme proposed in [6] takes the two most recently transmitted haptic signal updates $\mathbf{h}_i$ and $\mathbf{h}_j$ at time i and j into account. This allows for the calculation of a gradient $\mathbf{g}_j$

$$\mathbf{g}_j = \frac{\mathbf{h}_i - \mathbf{h}_j}{i - j} \qquad \text{with } i > j > 0 \tag{3}$$

Accordingly, the haptic signal prediction $\mathbf{p}_i$ can be calculated by

$$\mathbf{p}_i = \mathbf{h}_j + \mathbf{g}_j \cdot (i - j) \tag{4}$$

3   Error-resilient Perceptual Haptic Compression

As the perceptual haptic data compression approach transmits only perceptually relevant haptic samples, lost network packets consequently lead to noticeable disturbances. Furthermore, failed haptic data transmission leads to inconsistent predictor states which also results in disturbing artefacts in the decoded haptic signal.

A common strategy for detecting and compensating for packet loss in a network is to acknowledge successfully transmitted packets to the sender. If the sender does not receive an acknowledgment message from the receiver within a fixed period of time, it assumes the packet was lost and triggers retransmissions until a successful transmission is achieved. In haptic real-time communication, however, the strict latency constraints prevent any acknowledgment-based waiting strategy. Furthermore, since the transmission of compressed haptic packets is temporally irregular, the receiver does not know when it is supposed to receive a network packet. This poses a serious challenge for the detection and corresponding concealment of network errors on the haptic channel.

Brandi et al. [12,13] address this challenge (inspired by [11]) and show that robust low-latency streaming of compressed haptic signals can be achieved with the help of a binary Markov decision tree, which reflects all possible receiver states after n packet transmissions. Whenever a network transmission is triggered, the corresponding data packet either successfully updates the receiver or it is lost due to packet loss. If the actual sender state and possible receiver states deviate by more than the applied perception thresholds, additional data transmissions can be adaptively triggered in order to compensate for the packet loss at the cost of increased packet rates on the haptic channel. As the Markov tree structure grows exponentially with every transmitted haptic packet, the corresponding computational complexity and memory requirements are limiting its application. To address this issue, acknowledgments with receiver state information are typically signaled from the receiver back to the sender (see [12,13,10,11]).

In the following, a novel receiver state estimation technique is presented which does not require the transmission of acknowledgment feedback, and is characterized by low computational complexity.

3.1   Probabilistic Receiver State Estimation

Experiments reveal that the distribution of the signal error in the PD compression scheme originating from packet loss exhibits a Gaussian-like distribution (see Fig. 4). This motivates us to probabilistically model the uncertainty of the receiver state with a multivariate normal distribution (see Fig. 5).

Initially, the sender and the receiver are initialized with state $\mathbf{h}_0$. From then on, a successful transmission of packet $\mathbf{h}_{n+1}$ updates the receiver state ($\hat{\mathbf{h}}_{n+1} = \mathbf{h}_{n+1}$) with probability p. Otherwise, the receiver holds its most recently successfully received update at time $i \leq n$, hence $\hat{\mathbf{h}}_{n+1} = \hat{\mathbf{h}}_i$ with probability $(1 - p)$. Hence,

$$\Pr\{\hat{\mathbf{h}}_{n+1} = \mathbf{h}_i\} = \begin{cases} \Pr\{\hat{\mathbf{h}}_n = \mathbf{h}_i\} \cdot (1 - p) & : 0 \leq i \leq n \\ p & : i = n + 1 \end{cases} \tag{5}$$

By using (5), we can iteratively calculate the sample mean $\overline{\mathbf{h}_{n+1}}$ of all possible receiver states after $n + 1$ transmissions which then serves as an estimate of the expected mean of the receiver state.

$$\overline{\mathbf{h}_{n+1}} = \sum_{i=0}^{n+1} \Pr\{\hat{\mathbf{h}}_{n+1} = \mathbf{h}_i\} \cdot \mathbf{h}_i \tag{6}$$

$$= \sum_{i=0}^{n} \underbrace{\Pr\{\hat{\mathbf{h}}_{n+1} = \mathbf{h}_i\}}_{\Pr\{\hat{\mathbf{h}}_n = \mathbf{h}_i\} \cdot (1-p)} \cdot \mathbf{h}_i + \underbrace{\Pr\{\hat{\mathbf{h}}_{n+1} = \mathbf{h}_{n+1}\}}_{p} \cdot \mathbf{h}_{n+1} \tag{7}$$

$$= (1 - p) \cdot \overline{\mathbf{h}_n} + p \cdot \mathbf{h}_{n+1} \tag{8}$$

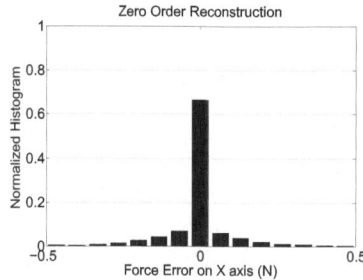

(a) Force error when using a zeroth-order PD compression.

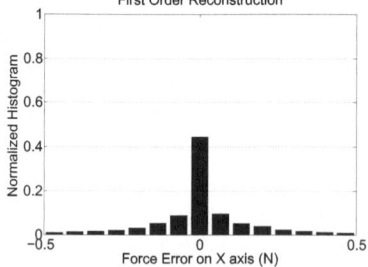

(b) Force error when using a first-order PD compression.

Fig. 4. Distribution of haptic signal errors originating from packet loss. In this experiment, a uniform packet loss with probability 0.5 and PD compression with $k = 10\%$ is used.

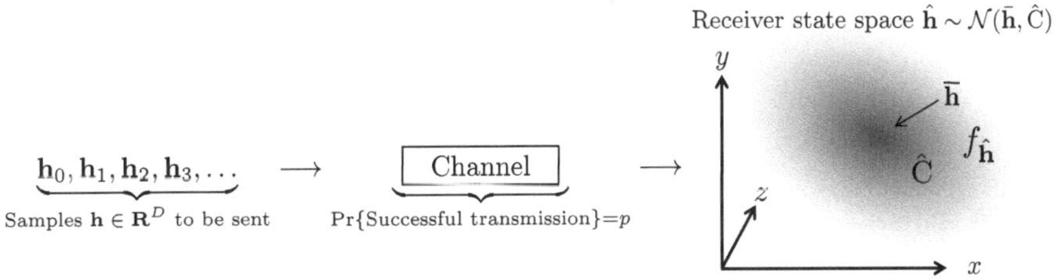

Receiver state space $\hat{\mathbf{h}} \sim \mathcal{N}(\bar{\mathbf{h}}, \hat{\mathbf{C}})$

$\mathbf{h}_0, \mathbf{h}_1, \mathbf{h}_2, \mathbf{h}_3, \ldots$ $\longrightarrow$ | Channel | $\longrightarrow$

Samples $\mathbf{h} \in \mathbf{R}^D$ to be sent   Pr{Successful transmission}=p

Fig. 5. Illustration of a multivariate Gaussian distribution describing the receiver state

Hence, the sample mean $\overline{\mathbf{h}_{n+1}}$ can be successively updated by only taking the previous sample mean $\overline{\mathbf{h}_n}$ and the current haptic input sample $\mathbf{h}_{n+1}$ into account. Similarly, the sample mean of all vector products $\mathbf{h}_i \mathbf{h}_i^{\mathsf{T}}$ can be iteratively updated with every triggered network transmission:

$$\overline{\mathbf{h}_{n+1} \mathbf{h}_{n+1}^{\mathsf{T}}} = \sum_{i=0}^{n+1} \Pr\{\hat{\mathbf{h}}_{n+1} = \mathbf{h}_i\} \cdot \mathbf{h}_i \mathbf{h}_i^{\mathsf{T}} \tag{9}$$

$$= \sum_{i=0}^{n} \Pr\{\hat{\mathbf{h}}_{n+1} = \mathbf{h}_i\} \cdot \mathbf{h}_i \mathbf{h}_i^{\mathsf{T}} + \Pr\{\hat{\mathbf{h}}_{n+1} = \mathbf{h}_{n+1}\} \cdot \mathbf{h}_{n+1} \mathbf{h}_{n+1}^{\mathsf{T}} \tag{10}$$

$$= (1-p) \cdot \overline{\mathbf{h}_n \mathbf{h}_n^{\mathsf{T}}} + p \cdot \mathbf{h}_{n+1} \mathbf{h}_{n+1}^{\mathsf{T}} \tag{11}$$

With the help of the sample mean values $\overline{\mathbf{h}_{n+1}}$ and $\overline{\mathbf{h}_{n+1} \mathbf{h}_{n+1}^{\mathsf{T}}}$, an estimate of the covariance matrix $\hat{\mathbf{C}}_{n+1}$ can be determined which expresses the uncertainty of the current receiver state estimation. It is defined by:

$$\hat{\mathbf{C}}(\mathbf{h}_{n+1}) = \overline{\mathbf{h}_{n+1} \mathbf{h}_{n+1}^{\mathsf{T}}} - \overline{\mathbf{h}_{n+1}} \cdot \overline{\mathbf{h}_{n+1}}^{\mathsf{T}} \tag{12}$$

$$= \left[(1-p) \cdot \overline{\mathbf{h}_n \mathbf{h}_n^{\mathsf{T}}} + p \cdot \mathbf{h}_{n+1} \mathbf{h}_{n+1}^{\mathsf{T}} \right] - \tag{13}$$

$$\left[(1-p) \cdot \overline{\mathbf{h}_n} + p \cdot \mathbf{h}_{n+1} \right] \cdot \left[(1-p) \cdot \overline{\mathbf{h}_n}^{\mathsf{T}} + p \cdot \mathbf{h}_{n+1}^{\mathsf{T}} \right]$$

Hence, the receiver state can be statistically modeled with every network transmission by successively updating the mean and covariance estimates of $\hat{\mathbf{h}}_{n+1}$. This iterative estimation procedure can be performed with low computational effort and at high update/transmission rates at the sender.

3.2 Error-resilient Transmission Scheme

The estimated covariance matrix $\hat{\mathbf{C}}$ incorporates correlations within the transmitted signal $\mathbf{h}$. The principal axes of the recently transmitted haptic samples $\mathbf{h}_i$ are reflected by the eigenvectors of the covariance matrix $\hat{\mathbf{C}}(\mathbf{h})$. Hence, the uncertainty described by the covariance matrix increases towards the direction of current signal change and, consequently, stays within smaller bounds along an orthogonal direction to the current signal trajectory. If $\hat{C}$ is nonsingular and positive definite (which is usually the case after a few iterations), its determinant $\det(\hat{C})$ can be used to obtain a measure of uncertainty of the current receiver state estimation process. If $\det(\hat{C})$ falls below a small threshold ϵ, the sample mean $\overline{\mathbf{h}}_{n+1}$ can be considered to be a precise estimate of the current receiver state $\hat{\mathbf{h}}_{n+1}$. Consequently, a perceptual evaluation of the sample mean $\overline{\mathbf{h}}_{n+1}$ at the sender allows a reliable decision over the transmission of additional packets (see Section 2). However, if $\det(\hat{C})$ exceeds the uncertainty threshold ϵ, additional steps are required to decide over possibly required redundancy on the haptic channel which are discussed in the following (see Fig. 6).

As the exact receiver output is unknown at the sender at time i, the region in the receiver state space is of interest, within which haptic output samples $\hat{\mathbf{h}}$ are in PD compliance with the current input sample $\mathbf{h}_i$ at the sender. If $\mathbf{h}_i$ falls into this region, the difference in signal between sender input $\mathbf{h}_i$ and receiver output $\hat{\mathbf{h}}_i$ can be considered to be imperceptible and no additional network transmissions are required. Interestingly, this PD compliance assuring region is of circular/spherical shape with center $\mathbf{s}$ and radius r, as shown in the following equations:

$$|\hat{\mathbf{h}} - \mathbf{h}| \leq k|\hat{\mathbf{h}}|$$

$$(\hat{h}_x - h_x)^2 + (\hat{h}_y - h_y)^2 + (\hat{h}_z - h_z)^2 \leq k^2\hat{h}_x^2 + k^2\hat{h}_y^2 + k^2\hat{h}_z^2$$

$$\vdots$$

$$\left(\hat{h}_x - \underbrace{\frac{h_x}{1-k^2}}_{=:\,s_x}\right)^2 + \left(\hat{h}_y - \underbrace{\frac{h_y}{1-k^2}}_{=:\,s_y}\right)^2 + \left(\hat{h}_z - \underbrace{\frac{h_z}{1-k^2}}_{=:\,s_z}\right)^2 \leq \underbrace{\frac{k^2}{(k^2-1)^2}(h_x^2 + h_y^2 + h_z^2)}_{=:\,r^2}$$

$$|\hat{\mathbf{h}} - \mathbf{s}| \leq r \tag{14}$$

Hence, if $|\hat{\mathbf{h}} - \mathbf{s}| \leq r$, the current haptic input $\mathbf{h}$ at the sender is PD compliant with the receiver output $\hat{\mathbf{h}}$. Consequently, the probability of being in PD compliance can be calculated by integrating over the PD assuring spherical region within the probability density function $f_{\hat{\mathbf{h}}}$ of the modeled receiver state distribution

$$\Pr[\text{PD Compliance}] = \int_{\{\mathbf{x}|r \geq |\mathbf{x}-\mathbf{s}|\}} f_{\hat{\mathbf{h}}}(\mathbf{x})\,d\mathbf{x} \tag{15}$$

using the probability density function $f_{\hat{\mathbf{h}}}$ of a D-dimensional multivariate Gaussian distribution:

$$f_{\hat{\mathbf{h}}}(\mathbf{x}) = (2\pi)^{-\frac{D}{2}}\det(\hat{C})^{-\frac{1}{2}}\exp\left(-\tfrac{1}{2}(\mathbf{x}-\overline{\mathbf{x}})^{\mathsf{T}}\hat{C}_{\mathbf{x}}^{-1}(\mathbf{x}-\overline{\mathbf{x}})\right) \tag{16}$$

As long as the probability of the signal error being in PD compliance with the input signal $\mathbf{h}$ stays below a target probability ψ, additional packet transmissions need to be triggered until the desired PD compliance probability ψ is retained.

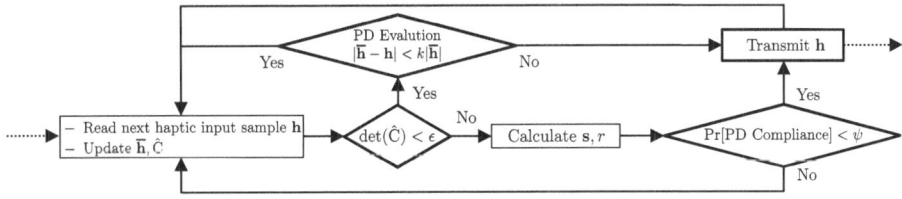

Fig. 6. Flow diagram at the sender of the error-resilient transmission scheme

3.3 Extension to First-Order PD Compression

In the first-order PD compression scheme (see Section 2), the prediction coefficient/gradient $\mathbf{g}$ is typically jointly transmitted together with the current absolute haptic sample value $\mathbf{h}$ [13]. This allows the separate modeling of not only the statistical distribution of the absolute signal $\mathbf{h}$, but also of the prediction gradient $\mathbf{g}$. However, transmission errors in predictive coding schemes propagate over time which requires updating of the sample mean and covariance estimates of $\hat{\mathbf{h}}$ and $\hat{\mathbf{g}}$ at every sampling instance (and not only when network transmissions are triggered). According to linear regression theory in statistics, the absolute sample amplitude and the corresponding gradient can be understood as independent random variables. As the sum of normally distributed independent random variables is again normally distributed, the signal prediction step can be easily applied to the probabilistic models during periods of network inactivity:

$$\overline{\mathbf{h}_{n+1}} = \overline{\mathbf{h}_n} + \overline{\mathbf{g}_n} \tag{17}$$

$$\hat{C}(\hat{\mathbf{h}}_{n+1}) = \hat{C}(\hat{\mathbf{h}}_n) + \hat{C}(\hat{\mathbf{g}}_n) \tag{18}$$

Hence, the modeled receiver state and its covariance are linearly modified over time. However, each triggered update reduces again the uncertainty in the haptic communication system. Consequently, the presented approach does not require acknowledgments from the receiver. However, feedback can be optionally used to support the receiver state estimation process.

4 Experimental Evaluation

A psychophysical study has been conducted to investigate the performance of the proposed robust haptic data communication scheme. 16 test subjects participated in the experiment. The experimental setup consists of a Phantom Omni HSI device from SensAble and a simulated virtual TPTA environment. It contains a touchable virtual sphere with a smooth surface which can be approached and haptically touched from any direction. During contact with the remote environment, collision detection and contact force rendering are performed by the *CHAI3D* library at 1000 Hz. In the experiment, a simulation of a uniform packet loss is applied to the transmission of the force feedback signal. The position/velocity signals stay unmodified during the experimental study.

The uncertainty threshold ϵ applied to the determinant of $\hat{C}$ is set to $10^{-15} N^2$. The PD parameter k is kept at 10% during the experiment. The integral in (15) is numerically evaluated with a resolution of 0.01 N in every dimension. The experiment consists of two runs which investigate the performance of the proposed packet loss concealment scheme applied to zeroth-order and first-order PD compression, respectively. In each run, ten different system configurations with varying packet loss and deadband compliance probability ψ are evaluated which are randomly presented to the subjects. Specifically, the impact of network packet loss of 0%, 20%, 40%, 60% on PD compressed force feedback signals is investigated. In addition, for each of the test configurations with a packet loss probability $> 0\%$, the influence of the proposed packet loss compensation scheme using the multivariate Gaussian receiver model on subjective quality and packet rates are observed. Here, three PD compliance probabilities $\psi = 0\%, 70\%, 95\%$ are used to configure the sender-driven transmission scheme. The subject can freely choose between the ten test configurations by pressing the numbers 0–9 on a keyboard. As soon as a different test configuration is selected, the corresponding feedback is immediately displayed to the subject which allows for pairwise comparison of the test items. In addition, a reference configuration is provided to the subject which deactivates the simulated packet loss on the haptic channel. This enables the direct comparison and subjective rating of the different test configurations against the presented reference. The subjects are asked to perceptually evaluate any perceived difference in the haptic feedback to the reference setting in accordance to a predefined subjective rating scale from 0 (strong difference) to 100 (no difference perceivable).

The experimental results are shown in Fig. 7. They demonstrate the strong performance of the PD compression scheme. Here, the test configurations without applied packet loss achieve a packet rate reduction of $> 90\%$ while keeping introduced coding artefacts below haptic discrimination thresholds. The upper figures show the mean packet rates and mean subjective ratings for the zeroth-order PD compression scheme. It can be observed that with increasing packet loss probability, the subjective quality of the haptic feedback becomes clearly impaired. The light and the dark gray bars illustrate the results with applied packet loss compensation using the multivariate Gaussian receiver model. In order to conceal the packet loss on the haptic channel, additional packet transmissions are triggered as soon as disturbing signal distortion at the receiver becomes likely, which can be seen in Fig. 7(b). The subjective ratings in Fig. 7(a) show that by adding redundancy to the haptic channel, the disturbing packet loss effects in the TPTA system can be successfully compensated. Even at a PD compliance probability of 70%, the transparency of the TPTA system is almost completely restored. The lower figures illustrate the mean packet rates and the mean subjective rating of the first-order PD compression scheme. The results clearly show that first-order PD compression is significantly more sensitive to data loss on the communication channel. Here, already a low packet loss probability significantly impairs the subjective quality of haptic feedback. This is mainly due to incorrect signal predictions which occur as soon as haptic predictor updates are lost

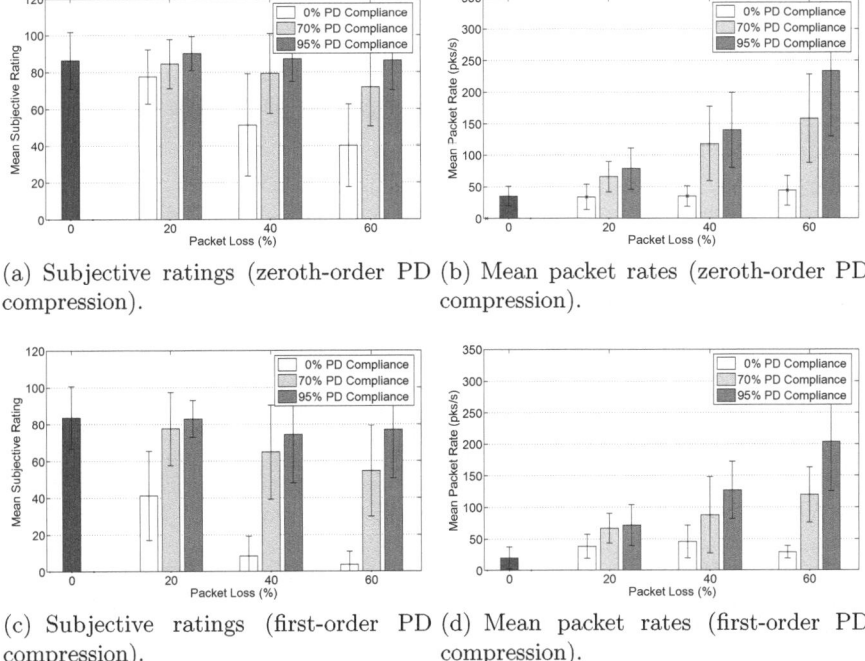

(a) Subjective ratings (zeroth-order PD compression).

(b) Mean packet rates (zeroth-order PD compression).

(c) Subjective ratings (first-order PD compression).

(d) Mean packet rates (first-order PD compression).

Fig. 7. Experimental results of perceptually robust haptic communication based on the multivariate Gaussian receiver state model. In all test configurations, PD compression with a PD parameter of $k = 10\%$ is applied to the force feedback.

during transmission. However, the results in Fig. 7(c) demonstrate that also for the first-order PD compression scheme, the sender-driven transmission scheme successfully compensates for the network data loss. Interestingly, the additional packet rates required for error-resilient haptic communication are similar for the zeroth-order and first-order PD compression schemes.

5   Conclusion

In telepresence and teleaction systems, the communication of compressed haptic signals is sensitive to data loss on the communication channel. In this paper, a novel error-resilient perceptual communication scheme has been proposed which employs a probabilistic model of the perceptual deadband receiver. Whenever the actual sender state and the estimated receiver state are expected to deviate by more than the applied perception thresholds, additional data transmissions are adaptively triggered in order to compensate for the packet loss. Experimental results reveal that the novel multivariate Gaussian receiver model is able to compensate even high packet loss up to a loss probability of 60%. This can be achieved with low computational complexity and in real time, without the need of acknowledgment feedback.

Acknowledgment. This work has been supported, in part, by the European Research Council under the European Union's Seventh Framework Programme (FP7/2007-2013) / ERC Grant agreement no. 258941 and, in part, by the German Research Foundation (DFG) under the project STE 1093/4-1.

References

1. Ferrell, W.R., Sheridan, T.B.: Supervisory control of remote manipulation. IEEE Spectrum 4(10), 81–88 (1967)
2. Ortega, A., Liu, Y.: Lossy compression of haptic data. In: Touch in Virtual Environments, ch. 6, pp. 119–136. Prentice Hall (2002)
3. Shahabi, C., Ortega, A., Kolahdouzan, M.R.: A comparison of different haptic compression techniques. In: Proc. of the Int. Conf. on Multimedia & Expo., Lausanne, Switzerland, pp. 657–660 (August 2002)
4. Borst, C.W.: Predictive coding for efficient host-device communication in a pneumatic force-feedback display. In: Proc. of the First Joint Eurohaptics Conf. and Symposium on Haptic Interfaces for Virtual Environment and Teleoperator Systems, World Haptics, Pisa, Italy, pp. 596–599 (March 2005)
5. Otanez, P.G., Moyne, J.R., Tilbury, D.M.: Using deadbands to reduce communication in networked control systems. In: Proc. of the American Control Conf., Anchorage, Alaska (2002)
6. Hinterseer, P., Hirche, S., Chaudhuri, S., Steinbach, E., Buss, M.: Perception-based data reduction and transmission of haptic data in telepresence and teleaction systems. IEEE Trans. on Signal Processing 56(2), 588–597 (2008)
7. Hinterseer, P., Steinbach, E., Hirche, S., Buss, M.: A novel, psychophysically motivated transmission approach for haptic data streams in telepresence and teleaction systems. In: Proc. of the Int. Conf. on Acoustics, Speech, and Signal Processing, Philadelphia, PA, USA, vol. 2 (March 2005)
8. Hirche, S., Hinterseer, P., Steinbach, E., Buss, M.: Transparent data reduction in networked telepresence and teleaction systems. part i: Communication without time delay. Presence: Teleoperators & Virtual Environments 16(5), 523–531 (2007)
9. Hirche, S., Hinterseer, P., Steinbach, E., Buss, M.: Network traffic reduction in haptic telepresence systems by deadband control. In: Proc. of the IFAC World Congress, Int. Federation of Automatic Control. Prague, Czech Republic (2005)
10. Chakareski, J., Girod, B.: Computing rate-distortion optimized policies for streaming media with rich acknowledgments. In: Proc. of the Data Compression Conf., Snowbird, Utah, USA, pp. 202–211 (2004)
11. Chou, P., Miao, Z.: Rate-distortion optimized streaming of packetized media. IEEE Trans. on Multimedia 8(2), 390–404 (2006)
12. Brandi, F., Kammerl, J., Steinbach, E.: Error-resilient perceptual coding for networked haptic interaction. In: Proc. of the ACM Multimedia, Firenze, Italy (October 2010)
13. Brandi, F., Steinbach, E.: Low-complexity error-resilient data reduction approach for networked haptic sessions. In: Proc. of the IEEE Int. Symposium on Haptic Audio-Visual Environments and Games, Nanchang, China (October 2011)
14. Weber, E.: Die Lehre vom Tastsinn und Gemeingefühl, auf Versuche gegründet. Vieweg, Braunschweig (1851)

What Feels Parallel Strongly Depends on Hand Orientation

Astrid M.L. Kappers and Bart J. Liefers

Helmholtz Institute, Physics of Man
Padualaan 8, 3584 CH Utrecht, The Netherlands

Abstract. Parallel in the outside world is not necessarily perceived as parallel. Previous studies have shown that what is felt as parallel can deviate significantly from what is physically parallel. In a new set-up, the influence of hand/arm orientation is investigated in detail by systematically varying the angle between the two hands, while the participants have to make a test bar parallel to a reference bar. Large positive deviations were found of about 32 % of the angle between the hands. The deviations were always in the direction of the rotation of the right hand with respect to the left hand. These findings are consistent with the hypothesis that the haptic perception of spatial relations is biased in the direction of the egocentric reference frame connected to the hand.

Keywords: Reference frames, haptic spatial perception, egocentric, allocentric, hand orientation, parallel.

1   Introduction

It has already long been known that our haptic spatial perception of the world around us is not veridical (e.g. [1], [16]). Extensive recent investigations confirmed this finding by using a parallel setting task (e.g. [4], [5], [6], [7], [10], [13]). In a typical version of this task, blindfolded participants are required to match the orientation of a reference bar by rotating a test bar at a different location. Large (up to and above 90°) but systematic deviations are the result (e.g. [7]). The right bar always deviates clockwise with respect to the left bar.

Currently, the deviations are understood in terms of biases due to the influence of egocentric reference frames, such as the hand or body reference frames [8], [9]. An allocentric reference frame is a reference frame external to the body, such as the reference frame of the experimental room or the table on which the set-up is placed. For the current discussion, the allocentric reference frame will be equated to a physical reference frame fixed to the set-up with cardinal directions 0° and 90° (see Fig. 1). An egocentric reference frame is not fixed to the external world, but to a body part such as the hand. If the hand rotates, the reference frame rotates with it (see Fig. 1).

In all versions of the parallel setting task, the participant is required to match the orientations of two bars in an allocentric reference frame, either unimanually or bimanually. So if the reference bar has an orientation of 60°, the test bar also

P. Isokoski and J. Springare (Eds.): EuroHaptics 2012, Part I, LNCS 7282, pp. 239–246, 2012.
© Springer-Verlag Berlin Heidelberg 2012

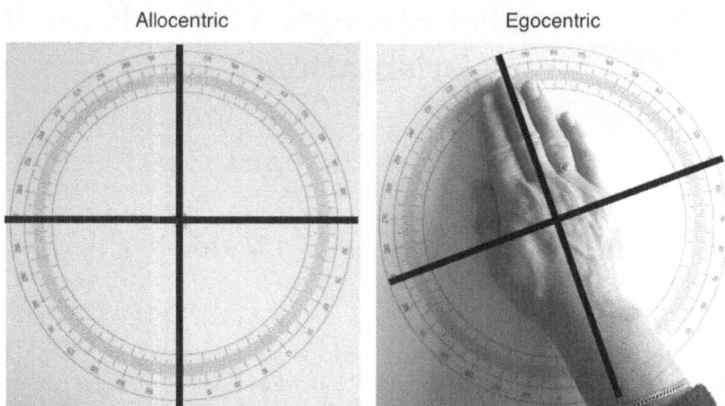

Fig. 1. Illustration of different reference frames. Left: allocentric reference frame. The cardinal orientations are aligned with 0° and 90°. Right: egocentric reference frame fixed to the hand. The cardinal orientations are aligned with the hand and perpendicular to it. The egocentric reference frame in this example is rotated 20° anticlockwise with respect to the allocentric reference frame.

needs to be orientated in 60°, irrespective of its location. However, if the locations of the reference and test bars are far apart, the orientation with respect to the egocentric reference frame of the hand touching the bars will be quite different. "Parallel" in the egocentric reference frame could lead to completely different allocentric orientations. In practice, participants do not completely rely on an egocentric reference frame, but their deviations are significantly biased in that direction. Different participants are influenced to a different degree by their egocentric reference frames, leading to large individual differences in average deviation [7], [9]. This explanation of a biasing influence of an egocentric reference frame has also been proposed by other researchers for other tasks like pointing and grasping (e.g. [2], [14]). There is also some neurophysiological evidence that there are two separate pathways for the somaesthetic modalities, one of which supposedly employs a more allocentric frame of reference, whereas the other is more egocentric [15].

There is not just one single egocentric reference frame, but for different tasks, different egocentric reference frames might apply. Think of vision, where both a retina-centred and a head-centred reference frame might play a role. In the haptic parallel setting task, a hand-centred reference frame is quite an obvious candidate, but also a body-centred reference frame could play a role. In a body-centred reference frame, "parallel" would be defined as having a fixed orientation with respect to concentric circles around the body midline [5].

In a previous study [12], the influence of the egocentric hand reference frame was systematically investigated by forcing participants to use a more or less fixed hand orientation during the parallel setting task. The hands were placed symmetrically around the midsagittal plane, 60 cm apart on a horizontal setup. The hands could either be parallel, converging or diverging with respect to each other. The arm angle between the hands had a significant influence on performance: when the hands diverged, so that the influence of the egocentric hand frame would be increased, the deviations indeed increased, whereas in the converging condition, the deviations decreased. However, also in the parallel condition there were significant deviations, that could not be attributed to the egocentric hand influence. These deviations were consistent with an influence of a body reference frame, so it was concluded that both hand and body reference frames play a role in the haptic perception of parallelity.

The present study is a sequel to [12], where the aim is to further investigate the influence of a hand-centred frame, while excluding the possible influence of a body-centred reference frame. In order to do so, the hands need to be positioned right in front of the body at the same location. As it has been shown that a delay between touching the reference and test bars improves performance by presumably strengthening the influence of the allocentric reference frame [18], any delay had to be avoided. Therefore, a set-up was designed, where the two hands would be at the same position in the horizontal plane, but with only a slight difference in height. The angle between the arms will be controlled and varied in a systematic way.

2 Methods

2.1 Participants

Twelve participants (6 males and 6 females) took part in the experiment on a voluntary basis. One male was left-handed, all other participants were right-handed as assessed by means of a standard questionnaire [3]. Their age was on average 21 years (age range 19-23). All participants were naïve about the purpose of the experiment.

2.2 Set-Up

The set-up consisted of two circular iron disks fixed on wooden plates, one 12 cm above the other (see Fig. 2). The iron plates were covered with a plastic layer on which protractors with a radius of 10 cm were printed. Aluminium bars (length 20 cm, diameter 1.1 cm) could be placed on the protractors by means of a small pin that fitted in a hole in the centre. Small magnets attached at the bottom of the bars caused some resistance against movement, so that the bars could still be rotated but unintended rotations were avoided. One end of the bars was arrow-shaped, so that the orientation of the bar could be read off with an accuracy of 0.5°. By means of a hinge at the back side, the upper wooden plate could be unfolded, so that the experimenter could adjust the reference bar on the lower protractor (see Fig. 2 upper right).

To each wooden plate an arm rest was attached, that could be fixed in steps of 10° ranging from -100° to 100° (0° corresponds to the position closest to the participant's body). The following arm angles were used: 40°, 60°, 100°, 120°, 160° and 180°. In the current experiment, these arm angles were always symmetrical around the midsagittal plane (see Fig. 3).

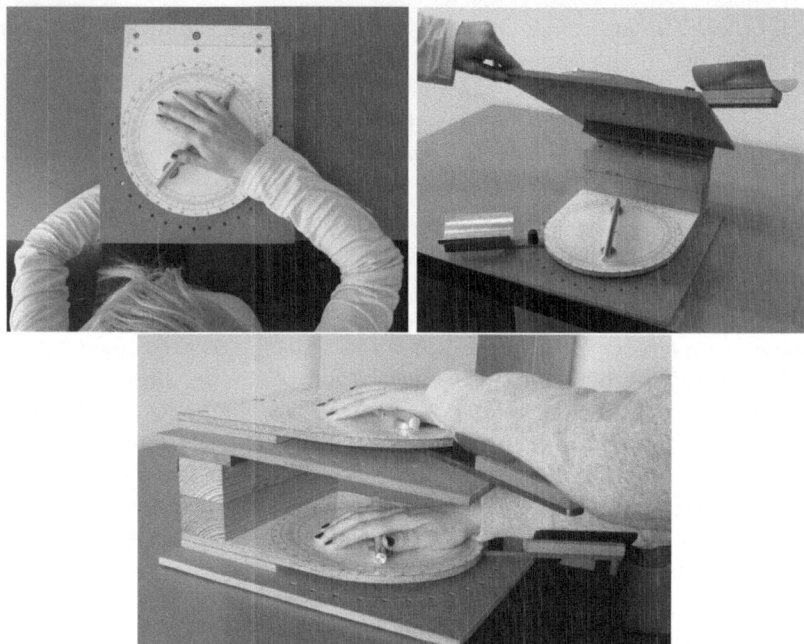

Fig. 2. Set-up used in the experiment. Top left: top view. The arm angle of the participant is 120°. Top right: the upper plate could be unfolded to get easy access to the lower plate. Bottom: side view. The arm angle of the participant is 60°. Notice in the top right and bottom pictures the arm rests, that could be adjusted in order to fix the orientation of the arms of the participant.

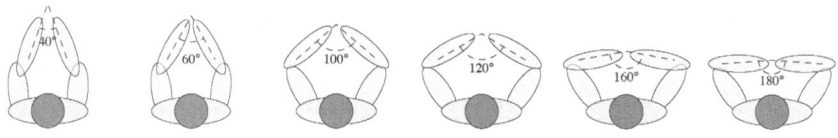

Fig. 3. The six different arm angles used in the experiment. The arm orientations are determined by the arm rests of the set-up, that can be fixed in various orientations (see Fig. 2).

2.3 Stimuli

The use of different reference orientations was only meant to provide variation in stimulus presentation. As cardinal (0° and 90°) and "oblique" (45° and 135°) orientations might cause so-called "oblique effects" [7], [8], these orientations were avoided as much as possible. Four orientations (22.5°, 67.5°, 112.5° and 157.5°) were used as reference, both in allocentric and in egocentric reference frames, resulting for most arm angle conditions in 8 different reference orientations in allocentric coordinates. However, for the arm angle of 180°, the egocentric and allocentric reference frames are aligned (see Fig. 3) resulting in just four different reference orientations. The reference orientation was always presented on the lower protractor. The test bar (always the bar on the upper protractor) was set by the experimenter in a random orientation between 0° and 180° (i.e., the arrow-shaped end of the bar was always pointing to the right side).

The 6 arm angle conditions were presented in random order and within a condition, the order of the 8 reference orientations was also randomized. The 6 conditions were performed once with the right hand on top of the reference bar and once with the left hand on top.

2.4 Procedure

Before the experiment started, it was checked whether the participants had the correct notion of the term "parallel". In order to do so, they were asked a few times to position two pens in such a way that they were parallel. Next they were blindfolded and seated on a stool in front of the set-up. The experimenter fixed the arm rests in the correct orientation, positioned the first reference bar and the participant would place his/her arms in the arm rests. The participants were asked to keep their hands more or less aligned with their arm, although small movements were allowed in order to rotate the test bar to the desired orientation. Their task was to orientate the test bar in such a way that it felt, as being parallel to the reference bar. In other words, the test bar had to match the direction of the reference bar. There was no time constraint, but participants were urged not to take more than a few seconds per trial. Participants did not receive any feedback.

2.5 Data Analysis

Deviations are defined as the orientation difference between the bar touched by the right hand and that touched by the left hand. A positive deviation corresponds to a situation where the bar touched by the right hand is rotated anticlockwise with respect to the left hand bar.

3 Results

In Fig. 4 (left) the deviations averaged over all reference orientations and arm angles are shown for each participant. As in all previous studies (e.g.[7]), the deviations are systematic (all positive) and the size of the deviations is clearly participant dependent, in the current case ranging from 16° to 37°. A two-sided t-test did not reveal a significant difference between male and female participants.

Another representation of the same data is given in the right panel of Fig. 4. Here the deviations are averaged over all reference orientations and all participants and given for the 6 different arm angles. Clearly, the deviations increase as a function of arm angle. A linear fit through these data points results in the following highly significant relation between deviation d and arm angle a: $d = -5.23 + 0.32a$ ($R^2 = 0.998, p < 0.00001$). The 95 % confidence interval for the offset is (-7.61, -2.85) and that for the slope is (0.30, 0.34).

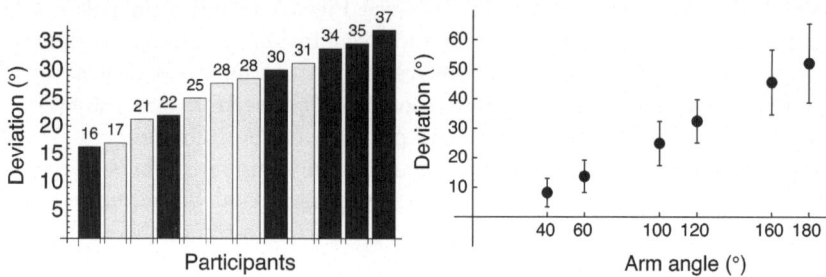

Fig. 4. Deviations. Left: deviations for each participant averaged over the 6 different arm angles. Black bars indicate male participants, grey bars females. Right: deviations for each arm angle averaged over participants. Error bars indicate standard deviations.

In Fig. 5 deviations averaged over participants are given as a function of reference orientation for the 6 arm angle conditions. Egocentric and allocentric reference orientations (both expressed in allocentric orientations) are indicated with open and closed circles, respectively. Not surprisingly, the same as in the right panel of Fig. 4 can be observed: deviations increase with arm angle. However, these graphs do not show any indication that the deviations depend in a systematic way on the reference orientations, nor on whether the reference was egocentric or allocentric, so there was no need for further statistical analyses.

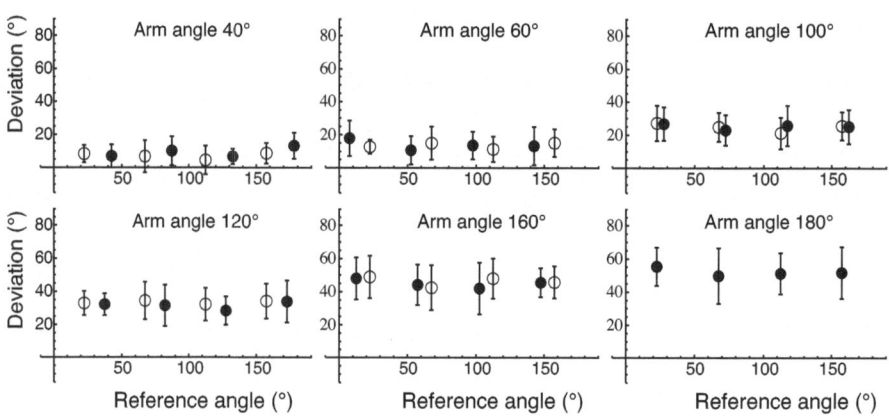

Fig. 5. Deviations as a function of reference orientation for all arm angles. Open circles indicate the egocentric reference orientations (expressed in allocentric orientations) and black disks the allocentric reference orientations. Note that the egocentric reference orientations vary with arm angle. For an arm angle of 180° the egocentric and allocentric orientations coincide, as the hand is aligned with one of the allocentric cardinal orientations (see Fig. 3). Error bars indicate standard deviations.

4   Discussion and Conclusions

Also in this new set-up large and systematic deviations were found. All deviations were positive indicating that the bar touched by the right hand had to be rotated anticlockwise with respect to the bar touched by the left hand in order to be haptically perceived as parallel. As in this set-up the right hand itself is always rotated anticlockwise with respect to the left hand, this finding is consistent with the hypothesis that the deviations are caused by a biasing influence of the egocentric reference frame, in this case the hand. Moreover, by presenting the stimuli right in front of the body midline at exactly the same position in horizontal space, the influence of a body-centred reference frame can be ruled out in this case.

Like in all previous studies, the settings are not completely determined by the egocentric reference frame. If that had been the case, the settings would result in bars parallel in the hand reference frame and that is not what was found. The linear relationship between deviation and arm angle had a slope of 0.32, which means that the deviation is 32 % of the arm angle and not 100 %. This is within the range of percentages reported in a previous study using a whole field of bars made parallel on a horizontal plane [9]. In an experimental study where haptic mental rotation was investigated, the influence of the egocentric reference frame was even larger, ranging from 40 to 80 % [17].

As in all previous studies (e.g. [8]) the deviations are strongly dependent on the participant. Apparently, some participants rely more on their egocentric reference frame than others. Or, alternatively, some participants are less able to ignore the influence of their biasing egocentric reference frame. However, it should be stressed that these influences are not necessarily and even not very likely caused by conscious processing. A study in which participants were given various kinds of feedback in order to teach them how to correctly set two bars parallel, showed that this was very hard to learn [11]. Haptic feedback only was not sufficient as most participants simply did not believe that what the experimenter told them was parallel was indeed parallel.

In the current study, there was no significant difference between the deviations of the male and the female participants. In most of the previous studies, female participants had larger deviations than males (e.g. [7]). As deviations are strongly dependent on the participant, this is most likely due to the limited number of participants in this study.

Extrapolation of the data points in the right panel of Fig. 4, predicts a 0-degree deviation for an arm angle of 16°. One reviewer remarked that this is the angle between the arms when the hands are touching as far away from the body as possible. This is an interesting and intriguing suggestion, as "parallel" is mathematically also defined as lines that touch in infinity. However, as we did not measure the deviations with this arm angle in the current set-up, it remains to be seen whether this extrapolation is indeed allowed.

In conclusion, this study provides once again strong evidence for the hypothesis that the large and systematic deviations found in a haptic parallel setting experiment are caused by the combined influence of egocentric and allocentric reference frames. Because of the current experimental design, the egocentric reference frame here is that aligned with the hand and/or arm.

Acknowledgement. This research was supported by the Netherlands Organisation for Scientific Research (NWO).

References

1. Blumenfeld, W.: The relationship between the optical and haptic construction of space. Acta Psychol. 2, 125–174 (1937)
2. Carrozzo, M., Lacquaniti, F.: A hybrid frame of reference for visuo-manual coordination. NeuroReport 5, 453–456 (1994)
3. Coren, S.: The left-hander syndrome. Vintage Books, New York (1993)
4. Fernández, M., Travieso, D.: Performance in haptic geometrical matching tasks depends on movement and position of the arms. Acta Psychol. 136, 382–389 (2011)
5. Kaas, A.L., Van Mier, H.I.: Haptic spatial matching in near peripersonal space. Exp. Brain Res. 170, 403–413 (2006)
6. Kappers, A.M.L.: Large systematic deviations in the haptic perception of parallelity. Perception 28, 1001–1012 (1999)
7. Kappers, A.M.L.: Large systematic deviations in a bimanual parallelity task: further analysis of contributing factors. Acta Psychol. 114, 131–145 (2003)
8. Kappers, A.M.L.: The contributions of egocentric and allocentric reference frames in haptic spatial tasks. Acta Psychol. 117, 333–340 (2004)
9. Kappers, A.M.L.: Haptic space processing – Allocentric and egocentric reference frames. Can. J. Exp. Psychol. 61, 208–218 (2007)
10. Kappers, A.M.L., Koenderink, J.J.: Haptic perception of spatial relations. Perception 28, 781–795 (1999)
11. Kappers, A.M.L., Postma, A., Viergever, R.F.: How robust are the deviations in haptic parallelity? Acta Psychol. 128, 15–24 (2008)
12. Kappers, A.M.L., Viergever, R.F.: Hand orientation is insufficiently compensated for in haptic spatial perception. Exp. Brain Res. 173, 407–414 (2006)
13. Newport, R., Rabb, R., Jackson, S.R.: Noninformative vision improves haptic spatial perception. Curr. Biol. 12, 1661–1664 (2002)
14. Paillard, J.: Motor and representational framing of space. In: Paillard, J. (ed.) Brain and Space, pp. 163–182. Oxford University Press, Oxford (1991)
15. Rossetti, Y., Gaunet, F., Thinus-Blanc, C.: Early visual experience affects memorization and spatial representation of proprioceptive targets. NeuroReport 7, 1219–1223 (1996)
16. Von Skramlik, E.: Psychophysiologie der Tastsinne. Akademische Verlagsgesellschaft, Leipzig (1937)
17. Volcic, R., Wijntjes, M.W.A., Kappers, A.M.L.: Haptic mental rotation revisited: multiple reference frame dependence. Acta Psychol. 130, 251–259 (2009)
18. Zuidhoek, S., Kappers, A.M.L., Van der Lubbe, R.H.J., Postma, A.: Delay improves performance on a haptic spatial matching task. Exp. Brain Res. 149, 320–330 (2003)

A Masking Study of Key-Click Feedback Signals on a Virtual Keyboard

Jin Ryong Kim[1,2], Xiaowei Dai[1,3], Xiang Cao[1], Carl Picciotto[4], Desney Tan[1], and Hong Z. Tan[1,2]

[1] Microsoft Research Asia, Beijing, P.R. China
[2] Purdue University, West Lafayette, IN, USA
[3] Beihang University, Beijing, P.R. China
[4] Microsoft Corporation, Redmond, WA, USA
`jessekim,hongtan@purdue.edu,`
`{v-xidai,xiangc,Carl.Picciotto,desney}@microsoft.com`

Abstract. We study masking of key-click feedback signals on a flat surface for ten-finger touch typing with localized tactile feedback. We hypothesize that people will attribute tactile feedback to the key being pressed, even with global tactile feedback, provided that the tactile signal on other parts of the surface is sufficiently attenuated. To this end, we measure the thresholds at which a tactile signal is barely perceptible to a finger that is resting passively on a surface while another finger actively presses on the surface and receives a key-click feedback signal. Combinations of the index and middle fingers of both hands are tested. The results indicate that the thresholds are independent of the signal amplitude on the active finger. Larger signal attenuation is needed when the index fingers of both hands are involved than when two fingers of the same hand are involved. Future research will extend the current experimental design to ten fingers and typing-based tasks.

Keywords: touch surface, virtual keyboard, zero-travel keyboard, key-click feedback, tactile masking, attenuation threshold.

1 Introduction

As touch screens become increasingly pervasive on computing devices, finger typing on soft (or virtual) keyboards has become a part of our daily activities. Despite its prevalence, typing on a non-reactive glass surface can be challenging due to the need to 1) visually place the fingers in the correct "home" locations above the keyboard and 2) confirm key entries without the familiar tactile feedback of a mechanical keyboard. The visual search can cause frequent gaze shifts between the keyboard and the text display areas on a touch screen, degrading the typing experience and performance [1-3]. However, we believe we can alleviate this problem in the case of a large surface that supports ten-finger typing by leveraging people's existing experience of eyes-free touch typing on traditional keyboards. Findlater et al. studied unconstrained text entry patterns of twenty touch-typists on a flat surface with or without a visual keyboard

P. Isokoski and J. Springare (Eds.): EuroHaptics 2012, Part I, LNCS 7282, pp. 247–257, 2012.
© Springer-Verlag Berlin Heidelberg 2012

[4]. They found that even without the reference of a visual keyboard, touch typists were still able to make key presses at relatively consistent locations within an individual typist, even though the "natural" distribution of key-press locations deviated from those of a rectangular *qwerty* keyboard (e.g., the "natural" rows tend to be curved, there is a larger gap between hands, etc.). Their finding is promising for developing personalized key-press classification systems that support eyes-free touch typing without tactile cues for key locations [5, 6], especially for larger form factor displays.

The other challenge of typing on a flat glass surface is the need for confirmation of key entries. Without the tactile feedback that usually accompanies the depression and release of keys with moving parts, other forms of feedback (e.g., visual enlargement of a key, an auditory ring, or a vibration) are helpful for faster and more accurate key entries [7-10]. Many mobile phones and devices have built-in vibrotactile feedback for confirming touch-screen virtual-key presses. There is typically one tactor that delivers the same tactile signal to the entire device, providing *global* feedback. However, such global feedback may not be appropriate for multi-touch interactions on a larger screen such as on a tablet or slate, because all contacted fingers might receive feedback simultaneously. At any time during ten-finger typing, several fingers rest on a touch screen while one finger does the typing. In order for the tactile feedback to emulate what happens on a mechanical keyboard, the key-click feedback signal should be clearly perceivable to the typing finger but not to any of the fingers that are simply resting on the virtual keys. The present study attempts to address this second challenge of providing *localized* key-click feedback to enhance the experience of typing on a zero-travel virtual keyboard on large form factor display surfaces.

There have been numerous inventions for providing localized tactile feedback on a touch screen. Instead of vibrating the entire device, actuators can be mounted on the display glass plate or to the touch-sensitive glass only [11, 12]. By mechanically isolating the vibrating glasses from the casing, vibrotactile feedback can be directed towards the touching finger instead of the hand holding the device. Other mechanisms such as ultrasound-based air squeeze film effect [13] and electrovibration [14] can also effectively confine haptic effects on a touch surface without vibrating the whole unit. These solutions work well as long as only one finger touches the surface at a time. However, for multi-touch interactions such as full-size ten-finger touch-typing keyboards, the aforementioned solutions cannot offer localized tactile feedback for each of the fingers in contact with the screen. One way to achieve localized feedback is to restrict tactile feedback to a small area [15] or to construct multiple actuators with each actuator affecting only a small part of a screen [16]. The latter approach could be applicable to a full-size virtual keyboard (i.e., by mounting one actuator on each key of the keyboard), albeit at the cost of increased hardware complexity and decreased reconfigurability of the keyboard layout. This motivates us to investigate the optimal (and most economical) placement of actuators for delivering localized tactile feedback that can support touch typing on a virtual keyboard.

The present study takes a perception-based approach. Instead of aiming to achieve a surface with *physically* localized tactile feedback, we ask the question of *when a global tactile feedback is perceived to be local*. We hypothesize that with sufficient attenuation, a tactile feedback signal "leaked" from the touched location on a surface can become imperceptible to the fingers holding or resting on the device. This way, the system can effectively provide localized tactile feedback without instrumenting one actuator for every single key of the keyboard. Our hypothesis is based on observations made during a previous study of key-click feedback using a piezoelectric actuator mounted on the bottom half of a cellphone mockup [17]. When the user pressed on a virtual key at the center of the piezo, the key-click feedback signal appeared to originate from the virtual key underneath the thumb. However, when the user held the phone and let another user press the surface, the user holding the phone could perceive key-click feedback on the hand. In both cases, the mechanism of tactile feedback was the same, but the perception of the location of tactile feedback was different. This is a classic example of sensory *masking* which means "the reduced ability to detect the target signal in the presence of a background, or masking, stimulus" [18]. In the present study, we assume that the presence of a stronger tactile feedback on the finger actively pressing on the surface can make it harder for the passively resting fingers to feel a weaker version of the same feedback signal. We believe that we can take advantage of this phenomenon in constructing a typing surface on which a global tactile feedback signal *feels* local.

The goal of the present study is to measure the thresholds at which key-click feedback signals on passive fingers (the fingers resting on the screen) are masked by the signal on an active finger (the finger interacting with the screen). This initial investigation starts with two fingers, one actively pressing on a surface and the other passively resting on the surface. As a first step, we limit our study to the index and middle fingers of both hands using simple clicks. The results will inform the design of follow-up studies in which we plan to extend the experimental design to realistic ten-finger typing tasks.

2   Methods

2.1   Participants

Twelve volunteers (P1-P12; 6 males and 6 females; average age 27 years old, std. dev. 3.4 years old) participated in the experiments. Eight of the participants were right-handed and four were left-handed by self-report. The participants were not remunerated for their participation.

2.2   Apparatus and Stimuli

We constructed two identical stimulators with sensing and actuation capabilities. The stimulators resemble keys on a zero-travel virtual keyboard that are common on most mobile phones and touch screens. Unlike keys on most virtual keyboards, however, these stimulators could emulate the tactile feedback that the user would experience with mechanical keys. Each key in our experimental setup consisted of a two-layered piezoelectric actuator (a 22-mm ceramic disk mounted concentrically on a 35-mm metal disk; Figure 1a) sandwiched between two clear plastic layers with two force-sensing resistors (Standard 400 FSR with 4 mm diameter active area, Interlink Electronics, Inc., USA) attached to the bottom of the structure. Figures 1(b) and 1(c) show the top and bottom views of one key, respectively. In order to help participants place and maintain the positions of their fingers on the two keys, a red dot was glued on top of each flat surface (Figure 1b). Each key was then placed on a thick foam pad that served to isolate the vibrations of the key from the tabletop underneath it. The foam pads supporting the two stimulators were mechanically isolated with a 3 mm gap between them. The distance between the two red dots measured 32 mm. The keys and the foam pads were housed inside a clear box measuring 360 mm (length) × 270 mm (height) × 270 mm (depth) with an open front for the participant's hand and forearm. During the experiment, the top and the front of the box were covered to prevent the participant from viewing the fingers. In addition, the participant listened to pink noise and wore noise-reduction circumaural headphones (Peltor H10A Optime105 with 29 dB attenuation, 3M Corporation, USA) to block any auditory cues from the experimental apparatus. The clear side panels of the box allowed the experimenter to observe and reinforce the placement of the participant's fingers on the two red dots. Figure 1(d) shows the experimental setup without the visual covers.

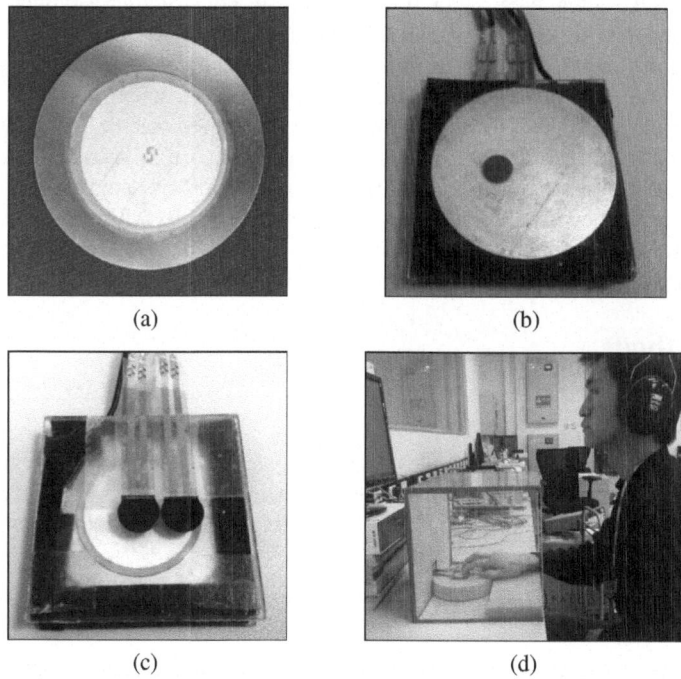

(a)   (b)

(c)   (d)

Fig. 1. (a) One side of the piezoelectric actuator showing both the ceramic and metal disks. (b) Top view of one stimulator key. (c) Bottom view of one stimulator key. (d) Experimental setup without the visual covers.

When a finger pressed a key, the FSRs were triggered and a waveform was sent to the piezoelectric actuator to deliver a key-click feedback signal to the finger. We used the two output channels of a sound card (SoundBlaster SB0100, Creative Technology, Ltd., Singapore) to independently control the waveforms sent to the two keys. The outputs of the sound card went through a voltage amplifier with a gain of 100 (Dual

Channel High Voltage Precision Power Amplifier, Model 2350, TEGAM Inc., USA) before driving the piezoelectric actuators on the two keys. The experimental application was coded in Visual C++ and OpenGL.

The waveform sent to each piezo consisted of one cycle of a raised sinusoidal pulse at 500 Hz. In a previous study, it was found that the piezo response to a single-cycle 500-Hz input signal felt like a "crisp" key click [17]. Acceleration at the red dot (Figure 1b) was calibrated using a triaxial accelerometer (8688A50, Kistler Group, Switzerland) under unloaded condition (i.e., without the finger pressing or resting on the instrumented key). Note that finger loading was expected to have a negligible effect on the piezo response at the relatively high frequency of 500 Hz (e.g., see [19]). The peak acceleration $(m \cdot s^{-2})$ changed linearly with the peak-to-peak (ptp) input voltage (V) with a gain of 0.0683 $(m \cdot s^{-2} \cdot V^{-1})$. At an input voltage of 100 V ptp, this corresponded to a peak acceleration of 6.83 $m \cdot s^{-2}$. The waveform for the finger that passively rested on the key was similar except that the amplitude changed according to an adaptive procedure (see the next section). In the rest of this article, we specify signals in terms of the input voltage (ptp) to the piezoelectric actuator.

2.3   Procedure

We used a well-established psychophysical method called the "three-interval, one-up and one-down method" that adapts to the participant's performance level [20, 21]. On each trial, the participant was asked to "press down on the pad as if typing on a keyboard" three times (the three "intervals"). Each time, a tactile feedback signal with amplitude A_{active} was sent to the active finger as soon as the FSRs detected a key strike. During one randomly-selected interval, the passive finger received a tactile feedback signal with amplitude $A_{passive}$. The passive finger received no feedback signal during the other two intervals. The participant was asked to indicate during which interval (first, second, or third) the passive finger felt a signal by saying "one," "two" or "three". The experimenter then entered this response on a computer keyboard. It was necessary for the experimenter to enter the responses for the participant because some of the experimental conditions required the participant to place both hands inside the box containing the two keys. This was a forced choice paradigm and the participants had to make a guess if they were not sure.

According to the "one-up, one-down" adaptive rule, $A_{passive}$ was increased after each incorrect response (to make the task easier) and decreased after each correct response (to make the task harder). A_{active} was kept the same at 228 V ptp after each key press. For each series of trials, the initial value of $A_{passive}$ was always 200 V ptp. It changed by 2 dB during the first 4 reversals (a reversal is defined as $A_{passive}$ changing from increasing to decreasing, or vice versa) and by 1 dB during the remaining 12 reversals. The larger step size (2 dB) allowed the $A_{passive}$ level to converge to the expected threshold quickly, and the smaller step size (1 dB) ensured the resolution of the estimated threshold.

There were six experimental conditions that differed in the fingers used and the assignment of active and passive fingers (see Table 1). Conditions C1 and C2 involved the index and middle fingers of the right hand, C3 and C4 involved the index fingers of both hands, and C5 and C6 involved the index and middle fingers of the left hand. At the beginning of each condition, the participant was told which finger was the active finger and which was the passive finger. Training was provided so the participant could become familiar with the task. During the training, the amplitude of $A_{passive}$ was kept constant at 200 V ptp. Correct-answer feedback was provided after each trial. The order of conditions was randomized for each participant. The total number of trials for each condition was between 25 to 48 trials. It took each participant up to 50 minutes to complete all six conditions.

Table 1. Experimental conditions

Condition	C1	C2	C3	C4	C5	C6
Hand	Right		Both		Left	
Active finger	Index	Middle	Index (L)	Index (R)	Middle	Index
Passive finger	Middle	Index	Index (R)	Index (L)	Index	Middle

Note: Filled and open circles on the fingertips indicate the active and passive fingers, respectively.

Prior to the main experiment, we ran a pilot test to investigate the possible effect of feedback signal strength on masking thresholds. Three of the twelve participants took part in the pilot study. In addition to the six conditions listed in Table 1 where A_{active} was kept at 228 V ptp, the three participants were also tested with another set of the same six conditions with an A_{active} of 100 V ptp. The order of the twelve conditions was randomized for each participant. The results indicated that the amplitude A_{active} did not have a significant effect on the thresholds. Therefore, the remaining nine participants were tested using one amplitude value of 228 V ptp.

2.4   Data Analysis

Figure 2 shows a series of trials for one participant under condition C2 that is reasonably representative of all participants across all conditions. The local maximum values (peaks) and minimum values (valleys) of the last 12 reversals at the 1-dB step size were extracted from the recorded values of $A_{passive}$ in dB re 1V. The 12 peaks and valleys were then averaged to obtain an estimate of the threshold at which the passive finger could barely detect the feedback signal. The threshold was then converted to the corresponding attenuation from A_{active} to $A_{passive}$. For example, in the plot shown in Figure 2, the threshold for $A_{passive}$ was 36.3 dB re 1V. Since A_{active} was 228 V ptp (i.e., 47.2 dB re 1V), this corresponded to a minimum attenuation of 10.9 dB in order for the feedback signal on the passive finger to be unnoticeable. We report experimental results in terms of this ***attenuation threshold***, calculated as $20 \times \log_{10}(A_{active}/A_{passive})$. We ran an analysis of variance (ANOVA) and *post hoc* Tukey tests, all with $\alpha = .05$.

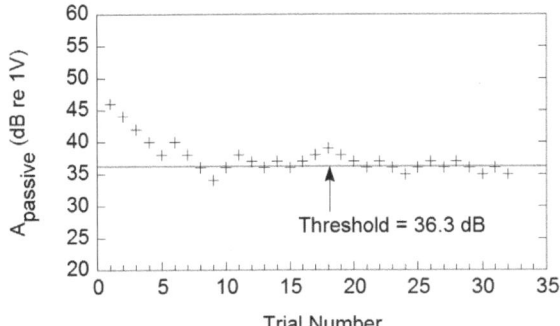

Fig. 2. A typical series of trials for condition C2 for one participant. Shown are the levels of $A_{passive}$ (dB re 1V) that changed adaptively to the participant's responses, while A_{active} was kept at 228 V ptp (47.2 dB re 1V).

3   Results

Figure 3 shows the attenuation thresholds obtained from all twelve participants under the six experimental conditions at a feedback signal strength of A_{active} = 228 V ptp on the active finger. We observe a trend that thresholds for conditions C3 and C4 were higher than those from the other four conditions. A one-way ANOVA confirmed that condition was indeed a significant factor ($F_{5,426}$ = 79.34; p < .0001). A post hoc Tukey test showed three groups of thresholds: C4 (μ = 19.7) and C3 (μ = 19.2); C1 (μ = 12.5), C6 (μ = 12.1) and C2 (μ = 11.0); C6, C2 and C5 (μ = 10.2). The fact that C2 and C6 belong to both of the latter two groups indicates that the main difference among the six experimental conditions has to do with whether one hand or both hands were tested. This was further confirmed with a one-way ANOVA with the factor hand combination (right hand alone, both hands, left hand alone). It was found that hand combination was a significant factor ($F_{2,429}$ = 187.61; p < .0001), and a post hoc Tukey test showed two groups of thresholds: both hands (μ = 19.4), right hand alone (μ = 11.7) and left hand alone (μ = 11.1). Therefore, when both hands are involved, the passive finger is more sensitive and larger signal attenuation is needed from the active to the passive finger.

   Among the four conditions C1, C2, C5 and C6, there is an apparent trend for attenuation thresholds to be higher when the middle finger is passive (C1 and C6) than when the index finger is passive (C2 and C5). This suggests a higher sensitivity of the middle finger than the index finger when the active finger is on the same hand. Could this trend reflect a bias in the way the participants allocated their attention to the two fingers? Note that we did not explicitly instruct the participants to focus more on an active vs. a passive finger or an index vs. a middle finger. To gain insight into attention allocation, the participants were debriefed after the experiment. They were asked to describe the strategies they used to perform the experimental task. The participants reported that they tried to maximize the focus on the passive finger while ignoring the sensations on the active finger. They also commented that the task appeared more difficult when one hand instead of both hands were involved.   These anecdotal notes support the general trend of the data, but do not explain the apparent higher sensitivity of the middle finger as suggested by the data in Figure 3.

There were significant inter-participant differences, as confirmed by a one-way ANOVA ($F_{11,420}$ = 11.59; p < .0001). A post hoc Tukey test showed mainly two groups of thresholds: P2 and P3 (μ = 8.3 and 9.8, respectively); P1 and P4-12 (μ = 12.8–17.8). Note that ten of the twelve participants required a higher attenuation threshold in order for the key-click feedback signal not to be noticed by the passive finger. If we can develop a virtual keyboard that can satisfy the more demanding requirements of this majority, then it will be guaranteed that the key-click feedback signals will feel localized to the remaining two participants (P2 and P3).

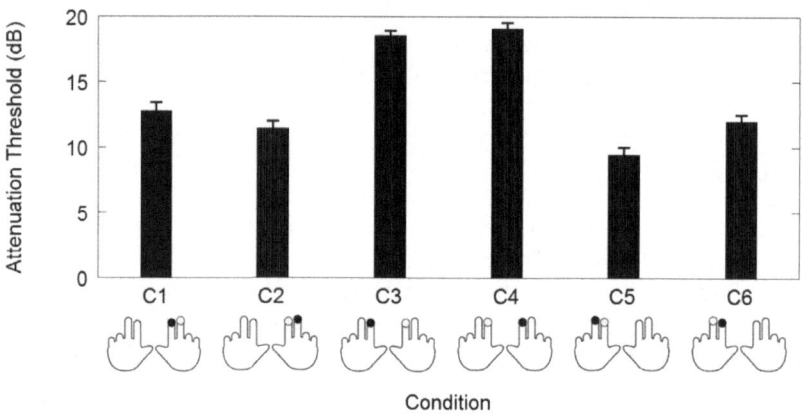

Fig. 3. Attenuation thresholds for all twelve participants from the main experiment. Shown are the average attenuation thresholds and the standard errors.

4   Discussion and Conclusions

In this initial study on global vs. local tactile feedback on a large form factor soft keyboard, we asked the question of whether it is necessary to instrument the individual areas occupied by each key in order to achieve localized tactile feedback. We hypothesize that people will attribute tactile feedback to the key being pressed, even with global tactile feedback, provided that the tactile signal on other parts of the surface is sufficiently attenuated. To this end, we measure the thresholds at which a tactile signal is barely perceptible to a finger that is resting passively on a surface while another finger actively presses on the surface and receives a key-click feedback signal. We report this threshold in terms of $20{\times}\log_{10}(A_{active}/A_{passive})$, called the attenuation threshold, that specifies the difference in signal strength between the active and passive fingers, for two reasons. We chose to use a log scale instead of a linear scale to report signal amplitude because it is well established that perceived magnitude of vibrations grows linearly with the log of vibration amplitude [18]. We also chose to report attenuation threshold, instead of $20{\times}\log_{10}(A_{passive})$, because the former appears to be independent of the overall signal strength, based on the results of a pilot study (see below).

Before the main experiment, we conducted the pilot study with three participants to investigate the effect of feedback signal strength using two A_{active} levels: 100 V ptp and 228 V ptp. For a 7.2 dB [i.e., $20 \times \log_{10}(228/100)$] change of reference signal strength, the average difference in attenuation threshold across all six experimental conditions changed by only 0.1 dB. The same average difference specified in $20 \times \log_{10}(A_{passive})$ would have been 7.1 dB. In other words, when A_{active} increased by 7.2 dB, the measured threshold $20 \times \log_{10}(A_{passive})$ increased by almost the same amount (7.1 dB), whereas the corresponding attenuation threshold $20 \times \log_{10}(A_{active}/A_{passive})$ remained almost constant. This finding supports the use of the relative measure, *attenuation threshold*, as a more parsimonious way to specify the conditions under which a tactile feedback signal on the passive finger can be effectively masked by the signal on the active finger. Based on the results of the pilot study, we were also able to decide that subsequent data collection could proceed at one feedback signal level on the active finger.

From the results of the main experiment, the attenuation threshold averaged over all experimental conditions was about 14.1 dB, corresponding to an A_{active} over $A_{passive}$ ratio of roughly 5. That such a threshold exists, as opposed to an $A_{passive}$ value of 0 before the key-click feedback on the passive finger could no longer be detected, confirms that people can indeed attribute a global tactile feedback to the key being pressed by the active finger, provided that the intensity of the signal "leaked" to the passive fingers are below the measured thresholds.

Another major finding of the present study is that larger signal attenuation is needed when the index fingers of both hands are involved (average of C3 and C4 = 19.5 dB) than when the index and middle fingers of the same hand are involved (average of C1, C2, C5 and C6 = 11.5 dB). This finding is consistent with the optical imaging data reported in [22]. Li *et al.* studied the neuromechanism for the tactile funneling illusion, where stimulation to two locations on the skin (e.g., index and middle fingertips) can be felt at a point between the two locations (e.g., a point in space between the two fingers) where no physical stimulus exists. They showed that stimulation to two adjacent fingers on squirrel monkeys resulted in one fused activation area between the known topographic locations for the two fingers on the primary somatosensory cortex, but stimulation to two non-adjacent fingers resulted in two distinct activation areas at the expected topographic locations. The stimulation method used in our present study was not conducive to eliciting the tactile funneling illusion because of the key-pressing action required of the active finger. However, it is conceivable that, for conditions C1, C2, C5 and C6 where two adjacent fingers received tactile feedback, the neural representation might have been less distinct in their locations on the somatosensory cortex than those for conditions C3 and C4 where two non-adjacent fingers were stimulated. This would help to explain why masking of tactile feedback signal is less effective for fingers of different hands.

In light of Li *et al.*'s (2003) findings, it may be predicted that the attenuation thresholds for two non-adjacent fingers on the same hand (e.g., index and ring fingers) should be similar to those of conditions C3 and C4 (index fingers of both hands) rather than those of the remaining four conditions (adjacent fingers of the same hand). This however turned out not to be the case. A follow-up experiment was conducted with three of the twelve participants under four conditions that were almost the same as C1, C2, C5 and C6 in the main experiment except that the middle finger was replaced with the ring finger. The attenuation thresholds involving the index and ring fingers were very similar among the four conditions tested and averaged 13.1 dB, which was similar to the average of 11.5 dB obtained with the index and middle fingers under four similar conditions in the main experiment. Therefore, whereas Li *et al.*'s study predicts a marked difference in perception based on whether the stimulated fingers are adjacent, our results show a difference in masking thresholds based on whether the stimulated fingers are on the same hand. Without further investigation, it is not exactly clear why the results of our follow-up experiment do not follow the prediction of Li *et al.*'s study. The discrepancy may be attributed to the fact that Li *et al.* used a passive perception task whereas our present study required the participants to actively press on a key.

The present study examined only one passive finger, but touch typing on a full-size soft keyboard may involve up to nine passive fingers. In the future, we will investigate how attenuation thresholds may change when, for example, the index finger is active and any combination of the rest of the fingers of the two hands are passive. An increase in the number of passive fingers to be monitored may enhance the masking effect (i.e., lower attenuation thresholds). Furthermore, the simple click task used in the present study may not reflect the differential attention allocation to different fingers during a real typing task. It is conceivable that when people are asked to type a pre-specified passage of text, they will focus their attention mostly on the active finger for the typing task, making the passive fingers less sensitive to "leaked" tactile feedback signals, further lowering attenuation thresholds. Any decrease in attenuation thresholds is welcomed news because it will decrease the amount of signal attenuation needed for tactile feedback on a virtual keyboard.

Ultimately, the results of the present and future studies will provide quantitative engineering specifications for the development of a tactile feedback system that supports faster and more enjoyable touch typing experience on flat surfaces with strategically placed actuators. For example, based on the result that people are more sensitive to leaked tactile feedback when the active and passive fingers are on different hands, we can propose the use of two actuators with maximum mechanical isolation, one for each hand, to be placed under the left and right halves of a soft keyboard. The attenuation thresholds obtained from the present and future studies can provide the quantitative specifications for how much signals can "leak" on each half of the proposed virtual keyboard before they become noticeable. Additional actuators and mechanical isolation among them can be deployed if sufficient signal attention cannot be realized with one actuator per hand. We believe that in the near future, we can take advantage of sensory masking and use human perception threshold data to construct a typing surface on which a global tactile feedback signal *feels* local, thereby supporting a more natural and pleasant typing experience on slate surfaces.

Acknowledgment. The authors thank Jaeyoung Park of Purdue University for his assistance with statistical data analysis. We thank Koji Yatani, Darren Edge, and Gunhyuk Park for comments and discussions on an earlier version of this article.

References

1. Zhai, S., Kristensson, P.-O., Smith, B.A.: In search of effective text input interfaces for off the desktop computing. Interacting with Computers 17, 229–250 (2005)
2. Ryall, K., Morris, M.R., Everitt, K., Forlines, C., Shen, C.: Experience with and observations of direct-touch tabletops. In: Proceedings of the First IEEE International Workshop on Horizontal Interactive Human-Computer Systems (TABLETOP 2006), pp. 89–96 (2006)

3. Bonner, M.N., Brudvik, J.T., Abowd, G.D., Keith Edwards, W.: No-Look Notes: Accessible Eyes-Free Multi-Touch Text Entry. In: Floréen, P., Krüger, A., Spasojevic, M. (eds.) Pervasive 2010. LNCS, vol. 6030, pp. 409–426. Springer, Heidelberg (2010)
4. Findlater, L., Wobbrock, J.O., Wigdor, D.: Typing on flat glass: Examining ten-finger expert typing patterns on touch surfaces. In: Proceedings of CHI 2011, pp. 2453–2462 (2011)
5. Murase, T., Moteki, A., Ozawa, N., Hara, N., Nakai, T., Fujimoto, K.: Gesture keyboard requiring only one camera. In: Proceedings of UIST 2011 (demonstration), pp. 9–10 (2011)
6. Sax, C., Lau, H., Lawrence, E.: LiquidKeyboard: An ergonomic, adaptive QWERTY keyboard for touchscreens and surfaces. In: Proceedings of The Fifth International Conference on Digital Society (ICDS 2011), pp. 117–122 (2011)
7. Hoggan, E., Brewster, S.A., Johnston, J.: Investigating the effectiveness of tactile feedback for mobile touchscreens. In: Proceedings of CHI 2008, pp. 1573–1582 (2008)
8. Bender, G.T.: Touch screen performance as a function of the duration of auditory feedback and target size. Doctoral Dissertation, Wichita State University (1999)
9. Brewster, S., Chohan, F., Brown, L.: Tactile feedback for mobile interactions. In: Proceedings of CHI 2007, pp. 159–162 (2007)
10. Fukumoto, M., Sugimura, T.: Active click: tactile feedback for touch panels. In: Proceedings of CHI 2001 (extended abstracts), pp. 121–122 (2001)
11. TouchSense 1000 Haptic System (January 30, 2012),
    http://www.immersion.com/products/touchsense-tactile-feedback/1000-series/index.html
12. Poupyrev, I., Maruyama, S.: Tactile interfaces for small touch screens. In: Proceedings of UIST 2003, vol. 5, pp. 217–220 (2003)
13. Winfield, L., Glassmire, J., Colgate, J.E., Peshkin, M.: T-PaD: Tactile pattern display through variable friction reduction. In: Proceedings of World Haptics Conference 2007, pp. 421–426 (2007)
14. Bau, O., Poupyrev, I., Israr, A., Harrison, C.: TeslaTouch: Electrovibration for touch surfaces. In: Proceedings of UIST 2010, pp. 283–292 (2010)
15. Luk, J., Pasquero, J., Little, S., MacLean, K., Levesque, V., Hayward, V.: A role for haptics in mobile interaction: Initial design using a handheld tactile display prototype. In: Proceedings of CHI 2006, pp. 171–180 (2006)
16. Jansen, Y., Karrer, T., Borchers, J.: MudPad: localized tactile feedback on touch surfaces. In: Adjunct Proceedings of UIST 2010, pp. 385–386 (2010)
17. Chen, H.-Y., Park, J., Dai, S., Tan, H.Z.: Design and evaluation of identifiable key-click signals for mobile devices. IEEE Transactions on Haptics 4, 229–241 (2011)
18. Verrillo, R.T., Gescheider, G.A.: Perception via the sense of touch. In: Summers, I.R. (ed.) Tactile Aids for the Hearing Impaired, ch. 1, pp. 1–36. Whurr Publishers, London (1992)
19. Tan, H.Z., Rabinowitz, W.M.: A new multi-finger tactual display. In: Proceedings of the International Symposium on Haptic Interfaces for Virtual Environment and Teleoperator Systems, vol. 58, pp. 515–522 (1996)
20. Levitt, H.: Transformed up-down methods in psychoacoustics. Journal of the Acoustical Society of America 49, 467–477 (1971)
21. Leek, M.R.: Adaptive procedures in psychophysical research. Perception & Psychophysics 63, 1279–1292 (2001)
22. Chen, L.M., Friedman, R.M., Roe, A.W.: Optical imaging of a tactile illusion in area 3b of the primary somatosensory cortex. Science 302, 881–885 (2003)

Saliency-Driven Tactile Effect Authoring for Real-Time Visuotactile Feedback

Myongchan Kim[1], Sungkil Lee[2], and Seungmoon Choi[1]

[1] Pohang University of Science and Technology
[2] Sungkyunkwan University
{billkim,choism}@postech.ac.kr, sungkil@skku.edu

Abstract. New-generation media such as the 4D film have appeared lately to deliver immersive physical experiences, yet the authoring has relied on content artists, impeding the popularization of such media. An automated approach for the authoring becomes increasingly crucial in lowering production costs and saving user interruption. This paper presents a fully automated framework of authoring tactile effects from existing video images to render synchronized visuotactile stimuli in real time. The spatiotemporal features of video images are analyzed in terms of visual saliency and translated into tactile cues that are rendered on tactors installed on a chair. A user study was conducted to evaluate the usability of visuotactile rendering against visual-only presentation. The result indicated that the visuotactile rendering can improve the movie to be more interesting, immersive, appealing, and understandable.

Keywords: Tactile Effect, Visual Saliency, 4D Film, Multimedia.

1 Introduction and Background

Recent advances in haptics technologies have proven its worth as an effective communicative source in a wide variety of applications. While the majority of multimedia contents are being mediated through visual and auditory channels, recent research and industrial applications, such as 4D films, are extending beyond the bimodal interaction to encompass diverse physical experiences including vibration, breeze, smell, mist, or tickler [6,13]. Such new-generation films provide better experiences in terms not only of immersion and entertainment, but also of better content delivery through unallocated haptic sensory channels. As the commercialization of stereoscopic TV was perceived as a particularly successful occasion, it is not far away to *feel physical movies* in everyday life. In creating such immersive experiences, the haptic sensation is one of the crucial components to bring about pervasive changes in multimedia.

Generating a haptic film requires the haptic contents that are coordinated with the existing semantics presented audiovisually; otherwise, they cause confusions in understanding the director's intention. If there exists an explicit computational model that can be used to reproduce the audiovisual signals, the authoring of haptic contents becomes easier to a greater extent. However, most

P. Isokoski and J. Springare (Eds.): EuroHaptics 2012, Part I, LNCS 7282, pp. 258–269, 2012.
© Springer-Verlag Berlin Heidelberg 2012

of real-world contents are directly captured without recognizing their sources. This poses a challenge in new media creation. A straightforward approach is to rely on the insights of content designers, but non-trivial efforts are necessary to be put for the coordination with other modalities. For instance, in the early 1970s, a static black-and-white picture was used to generate tactile cues for 400-points tactile stimulator mounted on the chair [3]. Recently, 3D videos with depth videos were used to produce force feedback synchronized with visual signals [1]. Kim et al. used a manual line-drawing interface to author tactile motion segments in a video for their tactile glove system [9]. Intuitive GUIs have often been helpful for pre-encoding lengthy haptic media with a number of actuators [12,9]. However, such manual authoring are laborious and time-consuming tasks, or do not account for what is important in the content. Also, it is obviously not the case in the situations that require complex scenarios and a tremendous amount in real time. This challenge inspired us to explore an automated approach for haptic film authoring, while being in accordance with the visual media.

While some of the 4D media are already popular in a limited extent, it is not clear how to derive and involve haptic cues into the existing media. It is particularly challenging in an automated approach, without being aware of semantics and spatiotemporal structure of a scene—object recognition is still ongoing research in the computer vision community. In the absence of particular context information, one of the mainstreams in vision study attempting to find visual importance suggests to use *visual saliency* as a key aspect to generate tactile stimuli. Salient inhomogeneous structures of visual features are prioritized and perceived first to humans. Various features including color, brightness, and edge were reported as significant in such perception [7,14,4]. Such visual saliency is the basis of our approach for extracting tactile contents from visual information. This computational autonomy is enabled by the *feature integration theory*, one of the most influential theories on bottom-up (feature-driven) visual perception [15,10,5]. Using the theory, the streaming media are processed spatiotemporally, and its corresponding saliency map can be generated. In this work, we note the possibility of automatic translation of the saliency map to a tactile map that can drive an array of tactors.

This paper reports our research on algorithms to transform streaming visual signals to tactile cues using the visual saliency, a real-time tactile display built upon an array of tactors installed on a chair, and a user study that evaluated the usability of our system. Furthermore, our system is aimed at a real-time interaction system unlike the previous researches, having the great benefit of distributing haptic contents without manual pre-encoding. To our knowledge, this is the first attempt for an automatic, real-time tactile effect authoring system making use of movies. Our system can play synchronized visuotactile effects in real time directly from a movie source. In addition, it has a potential advantage for human-aided design. An initial seed can be provided rapidly by our system, and then designers can take it over and enhance the tactile scenes, thereby, reducing production costs significantly. This can be a viable alternative considering no semantics is taken into account in our system. In addition, our study uses low-cost vibration motors for tactile rendering. This choice greatly elevates the practicability of our system, but it comes with a large actuation latency to take care of. For synchronous visuotactile stimulation, our system uses asynchronous commanding, that is, issues tactile commands earlier than visual commands by pre-calibrated differences between the display latencies.

2   Overview of Framework

In this section, we provide a brief perspective on our system. Fig. 1 illustrates the pipeline of our system. In the system, visual and tactile renderings are asynchronously executed using two different threads. For every frame, the thread for visual display runs as usual, but meanwhile, the thread builds the saliency map that spatiotemporally abstracts perceptual importance in a visual scene. The resulting saliency map is translated and stored in tactile buffers, the resolution of which is identical to that of a physical tactor array. In the other thread for tactile rendering, the tactile buffers are read into a tactile map at a lower frame rate, e.g., 5 Hz. This tactile map is mapped to the actuation commands to be sent to the tactors. In particular, the tactile commands are issued in advance to the visual commands with compensation of vibration latency. Each step is detailed in the following sections.

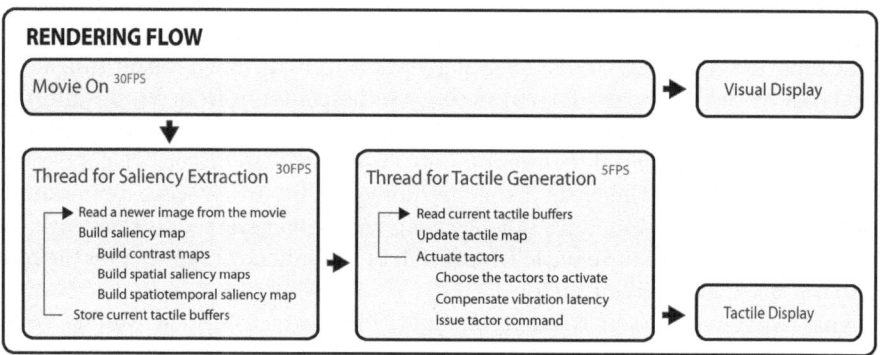

Fig. 1. Rendering flow of our system. Two threads for visual and tactile display run simultaneously but in different frame rates.

3   Tactile Movie Generation Based on Visual Saliency

First, we briefly review the neuroscientific background on visual saliency and its implementation. It is well known that attentional allocation involves the reflexive (bottom-up) capture of visual stimuli, in the absence of user's volitional shifts [14,4]. Albeit humans are generally efficient in searching visual information from complex scenes, this does not necessarily mean that everything is perceived simultaneously. Preattentive primitives such as color and lightness are first detected in parallel and then separately encoded into feature maps. A slow serial conjunctive search is followed to integrate the feature maps into a single topographical *saliency map* [15,5]. Neuronal mechanisms of early vision underlying bottom-up attention give us an important insight for detecting salient areas. The periphery of retinal zone (surround) suppresses neuronal activation in narrow receptive fields of the highest spatial acuity (center). Therefore, visual structures are particularly well-visible when they occupy a region popping out of its local neighborhood.

The extraction of visual saliency from an input image basically relies on the typical computational method proposed by Itti et al. [10,8], which has been distinguished for its effectiveness and plausible outcome in analyzing gaze behaviors. The key idea of their algorithm is finding salient regions by subtracting a pair of images each other, spatially convolved over different kernel sizes; they further accelerate this procedure using the image pyramid. The image averaged with a smaller blur kernel preserves finer structures than that with a larger blur kernel. Therefore, the image difference between them effectively captures spatially salient areas differing from local neighborhood, which simulates the biological process in the human visual system. Such image difference is called the *center-surround* difference. We first describe their basic algorithm in more details, and then describe our extensions on the use of CIE L*a*b* color space, temporal saliency, and binary thresholding. See also Fig. 2 for the whole pipeline of our algorithm.

The basic algorithm of Itti et al. is as follows. Given an input RGB image, visual feature maps (e.g., color, luminance, and orientation) are first extracted. The image pyramid of each feature map is built by successively downsampling an image to the $1/4$ size of its predecessor until reaching the coarsest image of 1×1. For example, for an image with a resolution of $2^N \times 2^N$, the levels of its image pyramid ranges from 0 (the finest image) to N (the coarsest image). For each image pyramid, six pairs of center (c) and surround (s) levels are defined; we used the common configuration from the previous studies, $c \in \{2, 3, 4\}$ and $s = c + \delta$, $\delta \in \{3, 4\}$ [8]. For all the center-surround pairs, cross-scale image differences (i.e., *center-surround* differences) are computed (we call the result as *contrast maps*). Computationally, the center-surround is realized by upsampling a coarser surround image to the finer center image and subtracting each other. Finally, the contrast maps are linearly combined to yield single topographical saliency map. Further details can be found in [8,11]. We note that the parameters used here are commonly accepted when using a visual saliency map, and the choice of them is decoupled from a particular tactile rendering algorithm and hardware. Since optimal parameters to further enhance tactile sensation are unknown, finding such parameters would be an interesting direction for future research.

We improve the definition of visual features (e.g, color, luminance, and orientation) using CIE L*a*b* color space (in short, Lab space), a widely known perceptual color space [2]. One common challenge in using the saliency map is to find appropriate weights for linear combination of different contrast maps. One

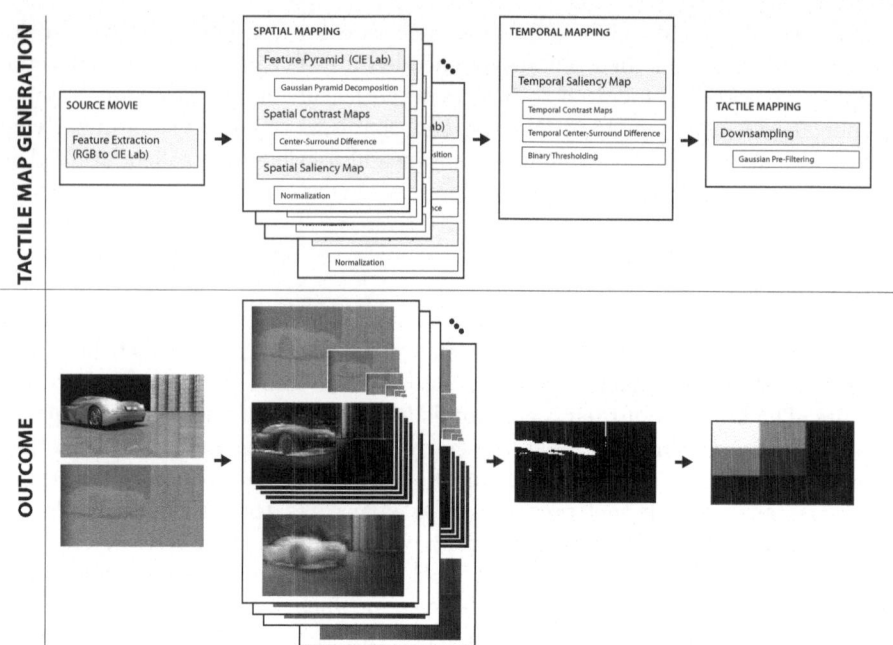

Fig. 2. Overall pipeline of the visuotactile mapping algorithm

could use unit weights, but there is no guarantee that this is an optimal selection. To cope with this, we use the Lab color space wherein a Euclidean distance between points roughly scales with their perceived color difference. Instead of multiple feature maps, we only define a single Lab image as a feature map. This allows us to efficiently evaluate the perceptual color difference at a single step without the linear combination issue.

Also, we augment the previous definition of the visual saliency along the *temporal* dimension. Since the saliency map was initially designed for static image analysis, it only deals with spatial dimension. Therefore, directly applying it to streaming images with dynamic scenes may not be appropriate. For instance, salient yet static objects may be less significant than dynamic objects in a scene. Hence, in order to preclude such static spots and track dynamic motions, we also use temporal saliency. The principle is the same as that of the spatial saliency, but the *temporal* image pyramids are built along the time. In other words, a level-n image averages the spatial saliency maps of 2^n previous frames including the current frame. We note that the temporal image pyramids have the same image size as the spatial saliency map has, and are built on the fly. The final temporal saliency map is computed using the *temporal* center-surround differences, whereas the temporal center levels $\{0, 1, 2\}$ were used instead of $\{2, 3, 4\}$.

The resulting spatiotemporal saliency map was globally scaled using the nonlinear normalization operator as was done in [8], which uses the ratio between the mean of local and global maxima. While fine details of a saliency map promote the gaze analysis of a static image, they often inhibit the maximum saliency response in an image. To draw more focus on the most salient spots, it is more effective to discard excessive details. Thus, we applied binary thresholding under a certain cutoff value (in our case, 50 percentage of the maximum level) [16].

The last step is relating the final saliency map to tactile cues to actuate tactors. To abstract tactile display hardware, a tactile map such as a 3×3 array is used. The resolution of the saliency map is usually much higher than tactors, and hence, we need to define the mapping between the saliency map and tactile map. In our implementation, we used a simple linear downsampling with Gaussian prefiltering, leaving room for better mapping that considers scene semantics and the expectation of user's volitional factors.

At the tactile rendering stage, the intensities of the tactile map are interpreted as the levels of vibration, and actuate the tactors whose dimension is the same as the tactile map. The actuation is performed in a continuous form, since the tactile map is also streaming along with the source video. The mapping between the tactile map and commands is straightforward, and thus, software commands for issuing haptic signals are virtually negligible. However, the mechanical latency takes longer than time for a single video frame, and thus, it requires to be compensated and this will be discussed in the next section.

We here report the computational performance of our haptic content generation algorithm. Our system was implemented on an Intel i5 2.66 GHz with Intel OpenCV library. For most movie clips, up to the resolution of 1000×1000, our system performed more than 30 FPS, the common requirement for real-time rendering. For HD resolutions such as 1080p, parallel GPU processing can be exploited on demand, as was done in [11].

4   Tactile Rendering

In order to test our saliency-based algorithm for visuotactile mapping, although it is independent of particular hardware platforms, we have built a test platform for tactile display. The display is designed to provide vibrotactile stimuli onto the lower back of a user sitting on a chair. Coin-type eccentric-mass vibration motors, one of the most popular and inexpensive actuators, are installed on a chair in a 3×3 array. Each tactor is 10 cm apart from neighboring tactors and independently connected to a customized control circuit. The maximal voltage to actuate tactors is 3 V, which can provide the vibration intensity up to 49 G at a 77-Hz vibration frequency. When tactors are fastened on a solid chair, the tactor's vibration may be propagated onto neighboring tactors and be dislocated when they vibrate. To alleviate this problem, the tactors are installed in the tactor housings of a chair cushion cover made of thin nylon, and used the cover to wrap the sponge fabric of the chair cushion (see Fig. 3).

The procedure of tactile rendering is as follows. The tactile map for rendering read tactile buffers which store the translated results of the video thread. Since visual and tactile renderings run at different threads with different refresh rates, latency from vibrotactile actuators requires suitable compensation to avoid confusions in visuotactile presentation. The lag is mainly caused by

Fig. 3. Our test platform for tactile display

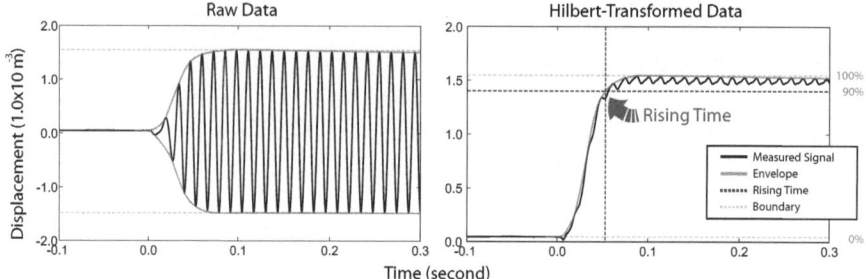

Fig. 4. Example of a vibration intensity response for the definition of rising time

the mechanical actuation delay of triggering vibration on the tactors. When a delay falls short of the duration required to play a single visual frame (e.g., 33 ms), the latency is in an acceptable range in practice and does not need to be compensated.

A simple remedy for the latency problem is pre-issuing tactile commands a few frames earlier than the corresponding visual signals. To do so, we measured the latency values for a number of starting and target vibration voltages. When a target voltage was smaller than a starting voltage, i.e., when the motor was decelerated, we observed the maximum latencies were in the acceptable range and thus, decided not to consider those falling times further. However, the acceleration process required for the target voltage larger than the starting voltage showed a noticeably longer delay. Hence, we defined the rising time of actuation as the time period required to reach the 90 percent of steady-state vibration level, as shown in Fig. 4, and compensated it in tactile rendering.

The vibration intensity was noninvasively measured by looking at the tactor's vibration displacement using a laser vibrometer (SICK, model: AOD5-N1). During the measurements, the vibration motor was fixed on a flat sponge 30 mm next to the vibrometer. Staring voltages and offsets to the target voltages were sampled in the range from 0 to 3.0 V by a step size 0.1 V. The recorded data were fed to Hilbert transform to reconstruct their signal envelope for accurate amplitude estimation (Fig. 4).

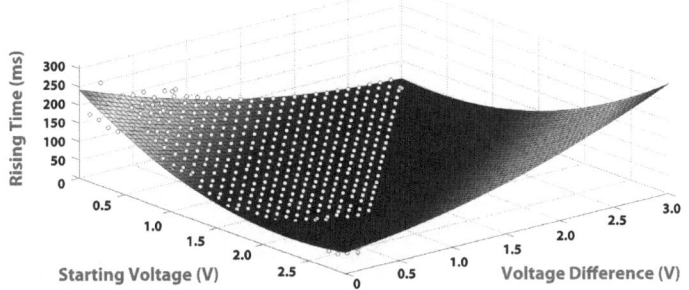

Fig. 5. Example of vibration displacement measured using a laser vibrometer

For the rendering purpose, a function of parametric form is more convenient than interpolating the latency data in real time. Thus, we regressed the rising time data to a quadratic form ($R^2 = 0.98$) as, such that:

$$f_r(V_s, V_d) = 241.9 - 175.7V_d - 117.7V_s + 38.86V_d^2 + 49.59V_dV_s + 18.07V_s^2, \quad (1)$$

where $f_r(V_s, V_d)$ is the estimated rising time, and V_s and V_d are the starting voltage and the voltage offset, respectively. In practice, we do not have to consider all the target voltages. Weak voltage commands less than a certain threshold (e.g., 0.5 V) can be discarded. By excluding such inputs, we have a set of rising times with 150 ms or less, in which the maximum resolution of tactile cues can be up to five video frames. Accordingly, a single data block for tactile signals is filled out of five video frames. For instance, when 60 ms is found as the rising time for the next tactile data block, the first three frames of the data block are written with the previous input voltage and the remaining two frames are filled with the target voltage at the next data block. A concern about force discontinuity might arise here in issuing discrete force commands within the tactile data block. However, it does not manifest itself, since the low-bandwidth dynamics of the actuator is likely to interpolate the abrupt changes of the vibration intensity. Hence, this simple strategy can effectively compensate for the motor delay of vibrotactile rendering to synchronize the visual and tactile stimulations, while avoiding torque discontinuity.

5   Evaluation

We conducted a user experiment to assess the usability of our system. Six usability items comparing visual-only and visuotactile presentations for different types of movies were collected via questionnaire. This section reports the methods used in our evaluation and experimental results.

Twelve paid undergraduate students (6 males and 6 females; 19–30 years old with average 22.3) participated in the experiment. The participants were asked to wear a thin T-shirt, leaning back on a chair to directly feel the vibration. Also, they wore earplugs and a headset to isolate them from tactor noises.

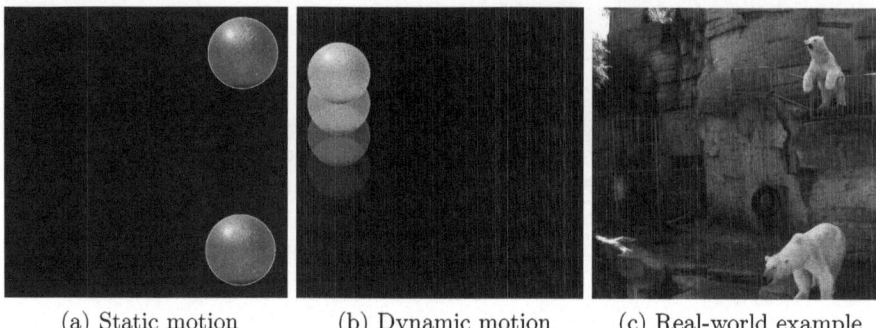

(a) Static motion   (b) Dynamic motion   (c) Real-world example

Fig. 6. Examples of 6a static-motion movie (two balls appear in various locations), 6b dynamic-motion movie (a single ball moves in various speed), and 6c real-world movie (a documentary film of two bears in a zoo)

The experiment used a two-factor within-subject design. The first factor was the presence of tactile cues while playing a movie. The other factor was the type of movies to be presented. Three movies, two synthetic and one natural, were used in the experiment (see Fig. 6 for each). The video resolution of 1000×1000 at 30 Hz was used commonly. One synthetic movie showed the static motions in that one or two balls repeatedly appeared at various locations and remained at the same position more than 1 second. The other synthetic movie contained dynamic motions with a ball moving around at various speeds with sudden stalling motions. The last one was a real-world movie that shows bears in a zoo. By combination, each participant went through the total of six successive experimental sessions. Their presentation order was balanced using Latin squares.

After each session, a break longer than two minutes was provided to the participant to fill out the questionnaire. The questionnaire consisted of six questions. Four questions were common to all the sessions, and the other two were tactile-specific questions given only in the conditions where tactile cues were presented. One additional survey asking a free evaluation of the overall system was followed. The common questions included: Q1. How interesting did you find the system? Q2. How much did you like the whole system? Q3. How much were you immersed in the movie with the given system? Q4. How well did you understand the contents of the movie? The tactile-specific questions were: QA. How well were the vibrations matched with the movie? QB. How much did the vibrations improve the immersion into the movie? Each question except the survey used a 100-point continuous scale, where 100 represents the strong agreement to the question, 50 a neutral opinion, and 1 the strongest disagreement.

The subjective ratings of the participants obtained in the experiment are plotted with standard errors in Fig. 7. Overall, the presence of tactile stimulation elicited much positive responses. The participant preferred the tactile-enabled movies to the original movies. Further, the tactile cue supports significantly enhanced the immersion and content delivery as well. We applied ANOVA to see the statistical significances of these differences between the visuotactile and

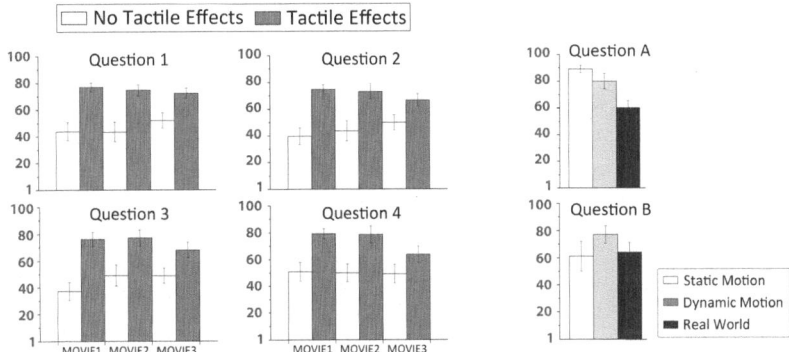

Fig. 7. Average subjective ratings measured in the user experiment

visual-only presentations. Statistical significances were found for all the four common questions; all the p values were less than 0.001 and their F-statistics were $F_{1,11} = 39.95$, $F_{1,11} = 33.05$, $F_{1,11} = 30.87$, and $F_{1,11} = 21.79$, respectively. In contrast, none of the responses resulted in statistical significance for the factor of movie type. The tactile-specific question, QA, examined the system performance of the motion extraction; as expected, static-motion movie was evaluated as the most well matched (mean score 89, SE=2.68), while the real-world movie had the weakest performance (mean score 60, SE=5.48). The question on the absolute measurement of immersion, QB, showed that the real-world movie was the least favored over the other two movies.

6 Discussion

The user study proved that the synchronized tactile display system improved the multimedia presentation; the participants were more engaged and immersed in the movie experience; tactile cues were effective in improving content delivery. The positive responses elucidated that the tactile cues were translated relatively well and it would emphasize the motions of salient spots in images.

More qualitative evaluation, regarding the overall system, was also collected via additional survey. The participants saw the system innovative and new, while the tactile cues are fairly well generated according to the salient spots of the movies. On the other hand, some of them saw the system abnormal or difficult to follow when they first tried it out. One suggestion for revising the rendering algorithm was limiting the duration of tactile cues, which makes them more memorable and avoids excessive tactile events as well.

As seen in the experimental results, the quality of translation in creating tactile cues is the key in conveying better experiences. The simple configurations of the synthetic movies were translated almost perfectly and preferred in terms of immersion with higher ratings, while the real-world movie was evaluated rather unorganized.

The visuotactile translation can be enhanced by tweaking our system. For instance, simple linear mapping between the visual and tactile viewports can be dynamically adapted. Cinematography commonly tries to locate salient ones in the middle of the scene to draw attention. However, this may cause tactile effects occurring in the same location for a long period, and generate an unpleasant concentrated feeling to the user. In addition, the periphery of the screen is often of less importance, and small movements in the center often require to be exaggerated. To consider this, region of interests can be dynamically widen and narrowed down according to the present context, and the provision of tactile cues can be limited in a short period, solely for important shots in scene semantics.

On the other hand, excessive and wrong translation would be a natural consequence of our system because we do not hold the high-level semantics of objects present in the scene. Without such scene semantics, the translation is hardly perfect. One good way to improve this problem is combining automated processing and manual authoring in the postprocessing. In the automated step, we aim at generating rather excessive tactile cues to some extent without filtering. Given tactile cues automatically generated from the visual information, the tactile content designer tries to prune out redundant cues to provide more focused feedback or to add some missing semantics during the translation. Also, diverse tastes from different cultures and favors can be incorporated in this postprocessing stage. We envision the tactile authoring will be automated in this fashion to provide sophisticated tactile cues.

7   Conclusion

We presented an automated framework of tactile effect authoring to provide synchronized visuotactile stimuli in real time. The visual saliency served as a basis for extracting spatiotemporal importance from existing visual media and translating the visual importance to tactile cues. Vibrotactors installed on a chair were used to render tactile effects synchronously, along with the compensation of vibrotactile latency. The user study found that visuotactile rendering were preferred to visual-only presentation, eliciting more immersion and involvement, and better understandings of content.

Since our algorithm is independent from particular rendering methods, our approach has special importance for haptic content creators and interaction designers, who strive to create online or offline physical contents inducing spatially-present experience. Haptic cues extracted automatically from the existing media can facilitate the rapid manipulation of a tactile movie in the postproduction stage. In the future, we are planning to include more sophisticated supports for the semi-automatic postproduction.

Acknowledgements. This work was supported in parts by a BRL program 2011-0027953 and a Pioneer program 2011-0027995 from NRF and by an ITRC program NIPA-2012-H0301-12-3005 from NIPA, all funded by the Korean government.

References

1. Cha, J., Kim, S.Y., Ho, Y.S., Ryu, J.: 3D Video Player System with Haptic Interaction based on Depth Image-Based Representation. IEEE Trans. Consumer Electronics 52(2), 477–484 (2006)
2. C.I.E: Recommendations on uniform colour spaces, colour difference equations, psychometric colour terms, supplement No.2 to CIE publication No.15 (E.-1.3.1) 1971/(TC-1.3.) (1978)
3. Collins, C.: Tactile Television-Mechanical and Electrical Image Projection. IEEE Trans. Man-Machine Systems 11(1), 65–71 (1970)
4. Connor, C.E., Egeth, H.E., Yantis, S.: Visual attention: Bottom-up versus top-down. Current Biology 14(19), R850–R852 (2004)
5. Desimone, R., Duncan, J.: Neural mechanisms of selective visual attention. Annual Review of Neuroscience 18(1), 193–222 (1995)
6. Hamed, H.M., Hema, R.A.: The Application of the 4-D Movie Theater System in Egyptian Museums for Enhancing Cultural Tourism. Journal of Tourism 10(1), 37–53 (2009)
7. Hoffman, D., Singh, M.: Salience of visual parts. Cognition 63(1), 29–78 (1997)
8. Itti, L., Koch, C., Niebur, E.: A model of saliency-based visual attention for rapid scene analysis. IEEE Trans. Pattern Analysis and Machine Intelligence 20(11), 1254–1259 (1998)
9. Kim, Y., Cha, J., Ryu, J., Oakley, I.: A Tactile Glove Design and Authoring System for Immersive Multimedia. IEEE Multimedia 17(3), 34–45 (2010)
10. Koch, C., Ullman, S.: Shifts in selective visual attention: towards the underlying neural circuitry. Human Neurobiology 4(4), 219–227 (1985)
11. Lee, S., Kim, G.J., Choi, S.: Real-Time Tracking of Visually Attended Objects in Virtual Environments and Its Application to LOD. IEEE Trans. Visualization and Computer Graphics 15(1), 6–19 (2009)
12. Lemmens, P., Crompvoets, F., Brokken, D., van den Eerenbeemd, J., De Vries, G.: A body-conforming tactile jacket to enrich movie viewing. In: Third Joint Eurohaptics Conference and Symposium on Haptic Interfaces for Virtual Environment and Teleoperator Systems, pp. 7–12 (2009)
13. Oh, E., Lee, M., Lee, S.: How 4D Effects cause different types of Presence experience? In: Proc. the 10th International Conference on Virtual Reality Continuum and Its Applications in Industry, pp. 375–378 (2011)
14. Parkhurst, D., Law, K., Niebur, E.: Modeling the role of salience in the allocation of overt visual attention. Vision Research 42(1), 107–123 (2002)
15. Treisman, A.M., Gelade, G.: A Feature-Integration Theory of Attention. Cognitive Psychology 12, 97–136 (1980)
16. Walther, D., Itti, L., Riesenhuber, M., Poggio, T., Koch, C.: Attentional Selection for Object Recognition - A Gentle Way. In: Bülthoff, H.H., Lee, S.-W., Poggio, T.A., Wallraven, C. (eds.) BMCV 2002. LNCS, vol. 2525, pp. 472–479. Springer, Heidelberg (2002)

Stable and Transparent Bimanual Six-Degree-of-Freedom Haptic Rendering Using Trust Region Optimization

Thomas Knott, Yuen Law, and Torsten Kuhlen

Virtual Reality Group, RWTH Aachen University,
Seffenterweg 23, 52074 Aachen, Germany
http://www.vr.rwth-aachen.de

Abstract. In this paper we present a haptic rendering algorithm for simulating the interaction of two independently controlled rigid objects with each other and a rigid environment. Our penalty based approach is based on a linearization model of occurring forces, and employs, for the computation of object positions and orientations, an iterative trust-region-based-optimization method. At this, the combination of a per step passivity condition and an adaptively controlled maximal object displacement achieves a stable and transparent rendering in free space and in contact situations.

Keywords: bimanual, six degree-of-freedom haptic rendering, numerical integration, stability, virtual coupling, passivity.

1 Introduction

The haptic sense is an important feedback channel in humans' everyday life interaction with objects. To transfer the benefits and possibilities of tactile and force cues to virtual environments, force and torque feedback has to be synthesized computationally. This process, commonly referred to as haptic rendering, has many applications; the most common are virtual prototyping, games, and surgical training simulations.

Two important goals in the development of haptic rendering algorithms are the creation of a (1) stable and (2) transparent behavior: a haptic interface should feel free in free space and contacts with virtual rigid objects should feel stiff. Achieving both at the same time is challenging.

The two main contributions of this paper are:

- The extension of an implicit integration approach for 6-DoF haptic rendering based on the concept of virtual coupling and penalty forces [24] to a *bimanual* interaction scenario. Thereby, our approach allows the user a stable simultaneous control of two virtual tools, which can interact with the environment and each other.
- The adaptation of an iterative optimization method based on trust regions to the haptic rendering problem. At this, the utilized adaptive control of

P. Isokoski and J. Springare (Eds.): EuroHaptics 2012, Part I, LNCS 7282, pp. 270–281, 2012.

maximal tool movement per time step allows transparent rendering in contact and in free space situations, while the explicitly calculated energy in the system is used to enforce passivity of each simulation step and thereby enables a stable simulation.

The remainder of this paper is organized as follows. First we will discuss the related work and afterwards, in section 2, describe our two main contributions. Here we will review the idea of implicit integration for haptic rendering, define the needed formulations of energies, forces and their derivatives, and then extend them to the bimanual case. In section 2.4 we show how we adapt the trust region method to our usage scenario and end section 2 by describing the complete haptic rendering algorithm. Afterwards, an analys of the approach is described in section 3, and section 4 summarizes our work and gives future research directions.

1.1 Related Work

One of the most important goals in haptic rendering is stability. Stability analysis on the problem of rendering a stiff virtual wall has shown that high force update rates are necessary to maintain a stable system [8]. At this, theoretically sufficient conditions for stable rendering of such unilateral constraints have been derived. The assumption for a guaranteed stability is the passivity of each part of the system. More recently, these criterion has been relaxed to cyclo-passivity by [15], and [10] proposed the usage of passivity-observers and controllers for an adaptive damping to achieve net passivity seen over multiple time steps.

Another popular approach to enable a stable rendering is the so-called virtual coupling, here the state of the controlled virtual tool is separated from the state of the haptic input point (HIP)[1]. While the latter is controlled directly by the user, the first is computed and may be constrained in its movements by other virtual objects, whereas the coupling tries to realign the tool and HIP. For the computation of a collision response, preventing or minimizing interpenetrations in the virtual environment, several strategies have been developed and analyzed. The most common approaches can be categorized into penalty based, impulse based, or constraint-based methods [12, 19].

In penalty based approaches, a violation of constraints, e.g. object interpenetration, induces an according force which tries to restore a state of non penetration [16]. Often a dynamic simulation based on ordinary differential equations considering object inertia is used to simulate the behavior of the tool under the influence of external forces [14, 18, 24]. [7] points out that implicit integration with penalty based approaches lead to a passive rigid body simulation. A semi-implicit backward Euler integration was then proposed in [18], allowing stiff contacts and low tool masses. Another slightly different approach was introduced in [24] and extended in [3], it omits dynamic effects and directly computes an approximate state of static-equilibrium between coupling and penalty forces, under consideration of a linearization of the forces.

Research on the special case of bimanual haptic interaction is rather rare, much work is either about perceptional or design aspects [5, 22] or haptic device design [5, 21]. Some work on the topic of bimanual haptic rendering exists but either only on 3-DoF haptic rendering [9, 17] or cases in which inter-tool collisions are neglected [9, 23] or both. A reason for the lack of research might be that some approaches to haptic rendering would not differ significantly when going from an unimanual to a bimanual scenario, e.g., [14?]. Other approaches, e.g. using collision response based on constraints or on penalty methods combined with implicit integration, would need to be adapted. In this work we will extend the 6-DoF static-equilibrium approach mentioned above to support bimanual interaction respecting also inter-tool collisions.

Notation. Throughout the paper we will use lowercase bold-face letters to represent vectors and quaternions, and upper case bold-face letter for matrices. Unless explicitly mentioned, all values are defined with respect to the global coordinate system. The expression $\mathbf{u}^*$ corresponds to the matrix enabling the representation of a cross product as a matrix-vector product: $\mathbf{u}^*\mathbf{x} = \mathbf{u} \times \mathbf{x}$.

2   Methods

Otaduy and Lin give in [19] a possible general definition of the haptic rendering problem:

> *Given a configuration of the haptic device $\mathcal{H}$, find a configuration of the tool that minimizes an objective function $f(\mathcal{H}, \mathcal{T})$, subject to environment constraints. Display to the user a force $\mathbf{F}(\mathcal{H}, \mathcal{T})$ dependent on the configurations of the device and the tool.*

In case of the non-dynamic virtual coupling and penalty based approach we are using, the objective function could be defined as the sum of the potential-energy stored in the springs of the virtual coupling $E_{vc}(\mathcal{H}, \mathcal{T})$ and the potential energy of the penetration penalties $E_p(\mathcal{T})$:

$$f(\mathcal{H}, \mathcal{T}) = E_{vc}(\mathcal{H}, \mathcal{T}) + E_p(\mathcal{T}) \tag{1}$$

In the static virtual coupling approach introduced by [24] and extended by [3], in each step penalty and coupling forces, as well as their gradients regarding $\mathcal{T}$, are computed and used to find a state in which all forces are approximately in equilibrium. Embedding this method into the rendering problem definition from above, forces and their gradients correspond to first and second derivative of f. Solving for a force equilibrium then corresponds to the usage of a single Newton solver iteration per step to find an approximate local minimum of the objective function (or to be more precise, a stationary point). At this, in each step, f is implicitly approximated by $L(\mathbf{h})$, the second order Taylor expansion around $\mathcal{H}, \mathcal{T}$ (2), whose minimum is then found with (3):

$$f(\mathcal{H}, \mathcal{T} + \mathbf{h}) \approx \mathbf{L}(\mathbf{h}) = f(\mathcal{H}, \mathcal{T}) + \mathbf{h} f'(\mathcal{H}, \mathcal{T}) + \frac{1}{2}\mathbf{h}^2 f''(\mathcal{H}, \mathcal{T}) \qquad (2)$$

$$\mathbf{h} = -f'(\mathcal{H}, \mathcal{T})/f''(\mathcal{H}, \mathcal{T}) \qquad (3)$$

In our work we want to extend this approach to the bimanual case, by means of simulating two tools, controlled by two haptic devices, and their interaction with the virtual environment and each other. Additionally, we exchange the Newton solver with a different optimization method with the goal to increase the transparency of the haptic rendering as well as its passivity and stability. We will start, in the next section, by describing how the energy functions E_{vc} and E_p and their first and second derivative can be calculated, first for the unimanual and then for the bimanual case. Finally, in section 2.4 we bring everything together in a description of our proposed complete haptic rendering algorithm.

We formulate the states $\mathcal{H}, \mathcal{T}$ of the haptic device and the tool by their positions $\mathbf{x}_\mathcal{H}$, $\mathbf{x}_\mathcal{T}$ and their orientations by quaternions $\mathbf{q}_\mathcal{H}$, $\mathbf{q}_\mathcal{T}$. Similar to [2, 18, 24] we use Euler vectors to describe rotations in case of the second derivative.

2.1   Single Static-Virtual Coupling

The Tool and haptic device are connected via a generalized spring inducing forces and torques which try to them. In this context it has to be considered that haptic devices are only able to display limited forces and torques to the user. In stiff environments these are easily saturated, as a result, the user can move the haptic device deeply into a surface. If linear coupling forces were to be used, this could lead, in the virtual environment, to unwanted deep penetrations. To circumvent this problem, non-linear coupling springs could be employed [2, 24]. We use an approach similar to [2]; a linear force function for deviations below a specified threshold r_{lin} and above an exponential saturation towards a maximal force $F_{vc,max}$. Virtual coupling force $\mathbf{F}_{vc}$ and torque $\mathbf{T}_{vc}$ are then defined via the following equations [2]:

$$\mathbf{F}_{vc} = \begin{cases} k_{VC}(\mathbf{x}_\mathcal{H} - \mathbf{x}_\mathcal{T}) & : r \le r_{lin} \\ \mathbf{e}\left(F_{vc,max} - \frac{1}{2}\exp\left(\dfrac{k_{vc}(r_{lin} - r)}{F_{vc,max}}\right)\right) & : r > r_{lin} \end{cases} \qquad (4)$$

$$\mathbf{T}_{vc} = k_{VC}\,\mathrm{vec}(\mathbf{q}_\mathcal{H} \cdot \mathbf{q}_\mathcal{T}^{-1}) \qquad (5)$$

with the coupling deviation $r = \|\mathbf{x}_\mathcal{H} - \mathbf{x}_\mathcal{T}\|$, the coupling force direction $\mathbf{e} = (\mathbf{x}_\mathcal{H} - \mathbf{x}_\mathcal{T})/r$, and the stiffness k_{vc} of the linear part. Here, $\mathrm{vec}(\mathbf{q})$ is the vector part of the quaternion q, therefore, also the torque magnitude is a non-linear function of the rotational deviation: $\|\mathbf{T}_{vc}\| = sin(\alpha/2)$. For the according gradients of force and torque we refer the interested reader to [2].

For the trust region method (see sec. 2.4) we need the potential energy stored in the coupling $E_{VC} = E_{vc,Tr} + E_{vc,Rot}$ as well. Where $E_{vc,Tr}$ and $E_{vc,Rot}$ respectively correspond to the rotational part and translation parts.

$$E_{vc,Tr} = \begin{cases} \frac{1}{2}k_{VC}(\mathbf{x}_{\mathcal{H}} - \mathbf{x}_{\mathcal{T}})^2 & : r \le r_{lin} \\ \frac{1}{2}k_{VC}r_{lin}^2 - F_{vc,max}\left(2r + \frac{F_{vc,max}}{2k_p}\left(e^{\left(\frac{k_p(r_{lin}-r)}{F_{vc,max}}\right)} + 1\right)\right) & : r > r_{lin} \end{cases}$$

$$E_{vc,Rot} = 2k_{VC}(1 - cos(\alpha/2))$$

2.2   Penalty-Based Collision Response for Environment

From a collision detection we get a set of collision triples, each consists of a contact normal $\mathbf{n}_i$ pointing outside the environment, a contact point $\mathbf{p}_{c,i}$, and the penetration distance d_i. Additionally, we calculate the according moment arm $\mathbf{r}_{c,i} = (\mathbf{p}, \mathbf{i} - \mathbf{x}_{\mathcal{T}})$ with respect to the tool. We interpret each collision triple i as planar constraint and calculate its penalty force $\mathbf{F_{p,i}}$ and torque $\mathbf{T_{p,i}}$ as:

$$\mathbf{F_{p,i}} = -k_p\mathbf{n_i}d_i \qquad\qquad \mathbf{T_{p,i}} = \mathbf{r}_{c,i} \times \mathbf{F_{p,i}} \qquad (6)$$

where k_p is the contact stiffness. As mentioned, also the gradients of the penalty forces w.r.t the state of the virtual tool have to be derived in each frame. Similar to [24] we consider that the constraint normal and the moment arm are constant during one simulation step. Therefore, in (6), the penetration distance d_i is the only variable which changes with respect to the movement of the tool. As each contact is modeled as a planar constraint, we get the penetration depth change by projecting the movement of the contact point onto the normal n_i of the plane. With $\frac{\partial \mathbf{d}_i}{\partial \mathbf{x}_{\mathcal{T}}} = \mathbf{n}_i$ and $\frac{\partial \mathbf{d}_i}{\partial \omega_{\mathcal{T}}} = \mathbf{n}_i\mathbf{r}_{c,i}^*$, the derivatives then look like:

$$\frac{\partial \mathbf{F}_p}{\partial \mathbf{x}_{\mathcal{T}}} = -k_p\mathbf{n}_i^T\mathbf{n}_i, \quad \frac{\partial \mathbf{F}_p}{\partial \omega_{\mathcal{T}}} = \frac{\partial \mathbf{F_p}}{\partial \mathbf{x}_{\mathcal{T}}}\mathbf{r}_c^*, \quad \frac{\partial \mathbf{T}_p}{\partial \mathbf{x}_{\mathcal{T}}} = \mathbf{r}_c^*\frac{\partial \mathbf{F_p}}{\partial \mathbf{x}_{\mathcal{T}}}, \quad \frac{\partial \mathbf{T}_p}{\partial \omega_{\mathcal{T}}} = \mathbf{r}_c^*\frac{\partial \mathbf{F_p}}{\partial \omega_{\mathcal{T}}}. \quad (7)$$

The potential penalty energy of each contact i is defined as $\frac{1}{2}k_pd_i^2$. To get the total values of energy and first and second derivatives, the according values for all collision triples have to be summed up respectively.

2.3   Extension to Bimanual Interaction

In this section we describe the extension of the static-virtual coupling approach to the bimanual case We will refer to each haptic device and tool states with $\mathcal{H}1$, $\mathcal{H}2$, $\mathcal{T}1$, and $\mathcal{T}2$. The fundamental idea is to define the objective function for the bimanual case by taking the unimanual function f twice, once for each device/tool, and extend it by an additional term reflecting the potential penalty energy in inter-object collisions E_{obj}:

$$f_{bim}(\mathcal{T}1, \mathcal{T}2, \mathcal{H}1, \mathcal{H}2) = f(\mathcal{H}1, \mathcal{T}1) + f(\mathcal{H}2, \mathcal{T}2) + E_{obj}(\mathcal{T}1, \mathcal{T}2) \qquad (8)$$

Now we only have to define the function E_{obj} and provide its first and second derivatives with respect to $\mathcal{T}1$ and $\mathcal{T}2$. Like in the case of collisions with the environment in the unimanual case, we get from the collision detection a set of contact triples containing penetration depth, contact normal, and contact point. Like in the unimanual case, the energy is then defined as $E_{obj} = \sum \frac{1}{2} k_p d_i^2$. For derivation of first and second derivative we interpreted collisions again as planar constraints and will assume that the contact normals point into the inside of tool one. The according moment arms of a contact with respect to each tool are defined as: $\mathbf{r}_{c,i,\mathcal{T}1} = (\mathbf{p} - \mathbf{x}_{\mathcal{T}1})$ and $\mathbf{r}_{c,i,\mathcal{T}2} = (\mathbf{p} - \mathbf{x}_{\mathcal{T}2})$. Forces and torques are calculated straight forward by:

$$\mathbf{F}_{\mathbf{p},i,\mathcal{T}1} = -k_p \mathbf{n}_i d_i, \qquad \mathbf{T}_{\mathbf{p},i,\mathcal{T}1} = \mathbf{r}_{c,i,\mathcal{T}1} \times \mathbf{F}_{\mathbf{p},i,\mathcal{T}1}. \qquad (9)$$

with negative sign for the second tool. For the derivation of the gradients with respect to the movement of both tools, like before we consider the normals and moment arms to be constant during one frame. Therefore, again only the penetration depth d_i changes with respect to movements of tool one and two: $\frac{\partial d_i}{\partial \mathbf{x}_{\mathcal{T}1}} = \mathbf{n}_i$, $\frac{\partial d_i}{\partial \omega_{\mathcal{T}1}} = \mathbf{n}_i \mathbf{r}_{c,\mathcal{T}1}^*$, $\frac{\partial d_i}{\partial \mathbf{x}_{\mathcal{T}2}} = -\mathbf{n}_i$, and $\frac{\partial d_i}{\partial \omega_{\mathcal{T}2}} = -\mathbf{n}_i \mathbf{r}_{c,\mathcal{T}2}^*$. This results in the same equations for the partial derivatives of forces and torques of each tool with respect to its own movement as in the unimanual case (7). In case of the partial derivatives describing the inter tool dependencies, e.g. how the force on tool one changes with respect to movements of tool two, we get:

$$\frac{\partial \mathbf{F}_{p,i,\mathcal{T}1}}{\partial \mathbf{x}_{\mathcal{T}2}} = k_p \mathbf{n}_i^T \mathbf{n}_i \qquad \frac{\partial \mathbf{F}_{p,i,\mathcal{T}1}}{\partial \omega_{\mathcal{T}2}} = \frac{\partial \mathbf{F}_{p,i,\mathcal{T}1}}{\partial \mathbf{x}_{\mathcal{T}2}} \mathbf{r}_{c,i,\mathcal{T}2}^* \qquad (10)$$

$$\frac{\partial \mathbf{T}_{p,i,\mathcal{T}1}}{\partial \mathbf{x}_{\mathcal{T}2}} = \mathbf{r}_{c,i,\mathcal{T}1}^* \frac{\partial \mathbf{F}_{p,i,\mathcal{T}1}}{\partial \mathbf{x}_{\mathcal{T}2}} \qquad \frac{\partial \mathbf{T}_{p,i,\mathcal{T}1}}{\partial \omega_{\mathcal{T}2}} = \mathbf{r}_{c,i,\mathcal{T}1}^* \frac{\partial \mathbf{F}_{p,i,\mathcal{T}1}}{\partial \omega_{\mathcal{T}2}} \qquad (11)$$

and vice versa for the analogous case. Again the first and second derivatives of all collisions have to be summed up.

2.4 Adaptive Optimization

As described in the beginning of the section, the haptic rendering problem could be mapped to a minimization problem regarding the objective function f. The searched minimum corresponds to a state of static-equilibrium of all forces in our penalty-based approach. The static-virtual coupling, or Newton-method-based approach, then uses the local approximation $L(\mathbf{h})$ of f, based on the second Taylor expansion, to iteratively search the root of f' (a point with zero net forces). At this, convergence depends on the quality of the approximation $L(\mathbf{h})$ with respect to a step size h. Now, $L(\mathbf{h})$ is based on (i) our non-linear coupling forces and torques, and on (ii) the result of a discrete collision detection. (i) leads, especially in case of large tool device deviations, to a bad approximation for large steps due to the exponential relation of deviation and force. (ii) leads as well to a decrease of the approximation quality, e.g., when a tool hits on a wall and an already violated constraint has to be minded. If $L(\mathbf{h})$ is a bad approximation of f it is possible that a chosen step leads to an increase in f. In our haptic rendering context, this means that the simulation is not passive anymore but increases the energy of the system which induces instabilities [8]. To reduce the influence of such cases, [2] introduced a limitation of the step size and an additional static damping factor to dissipate eventually created energy. Both values are fixed during simulation, and therefore damping and movement limitations have to be chosen to fit all possible situations.

We propose a haptic rendering approach to circumvent the problem of having fixed damping parameters for all situations. Instead, it uses in each step a metric to value the quality of the current approximation $L(\mathbf{h})$, and adapt the taken steps accordingly. Our approach is based on the trust region (TR) optimization method [13]. In this, also a local approximation $L(\mathbf{h})$ of the objective function is used, but additionally, as $L(\mathbf{h})$ is accurate in only a small but variable area, the TR method permits only steps within an adaptively specified trusted region.

In the following we will first describe the basic idea of TR methods and its adaptation to the haptic rendering scenario. Afterwards, we briefly describe the so-called dogleg method which we applied to efficiently perform a trust-region-method step.

Trust Region Method. As already mentioned, trust region methods use in each step a local approximation $L(\mathbf{h})$ of the objective function to calculate the next step. Now the fundamental idea behind TR methods is the assumption that there exist a step size Δ for which the approximation is accurate enough, i.e. results in a decrease in f [13]. Therefore, instead of computing a step $\mathbf{h}$ freely like in the Newton method, in TR methods a constrained problem is solved in each step:

$$\mathbf{h}_{tr} = argmin_{\|\mathbf{h}\| \leq \Delta} \{L(\mathbf{h})\}. \tag{12}$$

Herein, the constraint enforces that the step stays inside a sphere with radius Δ, the trust region. The TR method starts with an initial value for Δ and updates it after each iteration. If, for example, after a step has been calculated, it is noticed that $L(\mathbf{h})$ is not sufficiently accurate inside the current trust region, it is shrinked for the next step. Furthermore, in case the step would even lead to an increase in the objective function, it is rejected. In the context of haptic rendering this means that the passivity of each simulation step can be enforced.

A usual way to define the accuracy of the approximation L regarding the step $\mathbf{h}$ is the so-called gain ratio:

$$\varrho = \frac{f(T+h) - f(T)}{L(\mathbf{h}) - L(\mathbf{0})} \tag{13}$$

describing the relation between actual change in the objective function and the one predicted by L. Based on this value the following popular update rule could be used to change the trust region radius in each step [13]:

```
if (ϱ  <  0.25) Δ  :=  Δ/2
else if (ϱ  >  0.75) Δ  :=  max(Δ,  3  *  ‖h‖)
```

The predicted energy increase is always negative, therefore the ratio value would be smaller than zero if the step increases the objective function. For an early prevention of such bad steps the trust region radius is already decreased in case the gain ratio sinks below the threshold of 0.25.

Application to Haptic Rendering. The method described above creates a situation where each dimension of $\mathcal{T}$ is handled equally, i.e. we have the same trust region bounds for translational and rotational degrees of freedom. This creates two problems, (first) we are not able to define distinct upper bounds, i.e. different maximal rotational and translational speeds; (second), as the dimensions have different units and physical meanings, a specific trust region radius may result e.g. in a responsive behaviour of the tool regarding translations while at the same time rotational movements are highly damped. To resolve these problems we integrated a so-called affine scaling formulation, enabling us to define elliptical trust regions [6]. These allow to define different bounds for each dimension and respect their different meanings. The basis of the approach is to exchange the spherical constraint by an elliptical one in (12):

$$\mathbf{h}_t r = argmin_{\|\mathbf{S}h\| \leq \delta_k} \{L(\mathbf{h})\} \tag{14}$$

where S is a diagonal matrix specifying the extent of the TR in each dimension.

As we want to be able to define a maximum rotational and translational velocity[1], (first) we set the entries of the diagonal of $\mathbf{S}$ to the according max values, and (second) we extend the update rule of the TR radius (in sec. 2.4) by a limitation of Δ to the current simulation time step length ($0.001s$). We thereby achieve an upper bound for the rotational and translational velocities of our tools. Furthermore, we limit Δ to a lower threshold (10^{-6}), as we noticed that, otherwise, the transparency in a complex contact situation could be decreased. Additionally, in case we are in an equilibrium state, the step size and the predicted energy gain become zero. In such cases the result of (13) is undefined and we do not change the trust region radius. The rationale behind not changing Δ to, e.g., the max or min value, is that equilibrium states could be reached in free space motion, which is normally approximated well by L, but also in complex contact situations, for which this is usually not the case.

Calculating a trust region step (12) is a constrained optimization problem. To circumvent the cost-intensive calculation of the exact solution, the so-called *dogleg* method can be employed to find an approximate solution $\tilde{\mathbf{h}}$ [20]. At this, the final step is calculated taking into account the classic Newton step and the steepest descent direction (the force direction), resulting in only a small increase in computational costs. To respect elliptical trust regions, [4] adapted the approach, by just using the original dogleg method on the already scaled problem:

$$\tilde{\mathbf{h}} = argmin_{\|\tilde{\mathbf{h}}\| \leq \delta_k} \left\{ L(\mathbf{S}^{-1}\tilde{\mathbf{h}}) \right\} \text{ where } \tilde{\mathbf{h}} = \mathbf{S}h \tag{15}$$

---

[1] As described in section 3 we are using a collision detection which utilizes temporal coherence and needs information about fixed maximal object velocities.

2.5 Algorithm Description

Now we describe how everything can be combined into a haptic rendering algorithm. Step 0 is an initialization step and performed once before the simulation starts; during the simulation, steps 1 to 9 are performed sequentially in every haptic frame.

0. Initialization
   - Set initial tool states $\mathcal{T}1_0$, $\mathcal{T}2_0$, trust region radius Δ_0 to 0.001, k to 0, and $\mathbf{S}$ diagonal to each dimension maximum velocities (see 2.4).
   - Perform collision detection (CD) to get contact list CL_1 for $\mathcal{T}1_0$, $\mathcal{T}2_0$.
1. Calculate pre-step objective function value f_k and first and second derivative f'_{bin}, f''_{bin} regarding current contact list CL_k, and current tool and device states $\mathcal{T}1_k$, $\mathcal{T}2_k$, $\mathcal{H}1_k$, $\mathcal{H}2_k$.
2. Caculate potential step $\mathbf{h}_{k'}$ with elliptical trust region dogleg method (see sec. 2.4) and set potential tool states $\mathcal{T}1_{k'}$ and $\mathcal{T}2_{k'}$ accordingly.
3. Perform CD to get the contact list $CL_{k'}$ for potential tool states $\mathcal{T}1_{k'}$, $\mathcal{T}2_{k'}$.
4. Calculate potential objective function value $f_{k'}$ with respect to potential tool states $\mathcal{T}1_{k'}$, $\mathcal{T}2_{k'}$ and current device state $\mathcal{H}1_k$, $\mathcal{H}2_k$
5. Caculate the potential gain ratio $\varrho_k = (f_{k'} - f_{k-1})/(-\mathbf{h}_{k'}F'_k - \mathbf{h}^{tr}_{k'}F''_k\mathbf{h}_{k'})$.
6. Check if step $\mathbf{h}_{k'}$ is accepted, i.e. if $\varrho_k > 0$ (no artificial energy gain).
   - if accepted: $\mathcal{T}1_{k+1} := \mathcal{T}1_k$, $\mathcal{T}2_{k+1} := \mathcal{T}2_{k'}$, and $CL_{k+1} = CL_{k'}$.
   - else: $\mathcal{T}1_{k+1} := \mathcal{T}1_k$, $\mathcal{T}2_{k+1} := \mathcal{T}2_k$, and $CL_{k+1} = CL_{k-1}$.
7. Compute trust region radius for next frame Δ_{k+1} (see sec. 2.4).
8. Compute coupling force based on $\mathcal{T}1_{k+1}$, $\mathcal{T}2_{k+1}$, $\mathcal{H}1_k$, and $\mathcal{H}2_k$ and send to haptic devices (see sec. 2.1).

3 Experiments and Results

In this section we will describe the experiment we carried out to validate the proposed algorithm. The computer we used was a quadro Intel Xeon 2.53 GHz with 12.0GB of memory and Windows 7 OS. We used a 6-DoF Phantom Premium 1.5 and a 3-DoF Phantom Omni haptic device which were controlled via the Open-Haptics API with a constant frequency of 1kHz. The trajectories were recorded with haptic feedback and the proposed trust-region-based method, and reproduced with both methods: static damping (SD) and trust region (TR). In case of SD we used the values suggested by [3]: max translational velocity $0.6m/s$, max rotational velocity $10rad/s$, static damping factor 0.5. In case of TR we used: max translational velocity $3m/s$, max rotational velocity $70rad/s$. For both methods we used: $k_p = 50kN/m$, $k_{vc,tr} = 200N/m$, and $k_{vc,rot} = 1N/rad$, and as saturation force $F_{vc,max} = 2.5N$. The employed collision detection algorithm uses point shells and distance fields as object representations, and utilizes temporal coherence and parallelization. The used point shells for the environment and the tools have $40,000$ and $10,200$ points respectively; the distance fields are defined analytically.

Transparency Analysis. To validate that the proposed method increases the transparency, we recorded a trajectory with a collision between both tools, bracketed by two fast free space movements. The top left graph of Fig. 1 shows the penetration distance of the two tools; we see that (1) with SD the objects penetrate more than twice as deep and (2) the impact is later due to the static damping. This is also reflected in the top right graph, here we see that using SD, already before even a contact exists, the feedback force is large due to the highly damped rapid movement. Furthermore, after the tools leave the colliding state, feedback forces are non zero. Using TR, (see Fig. 1 top center graph) the force has a smooth curve and is only non zero during contact. From this we conclude that with the TR method we achieved our goal of transparent and stable bimanual haptic rendering for this scenario.

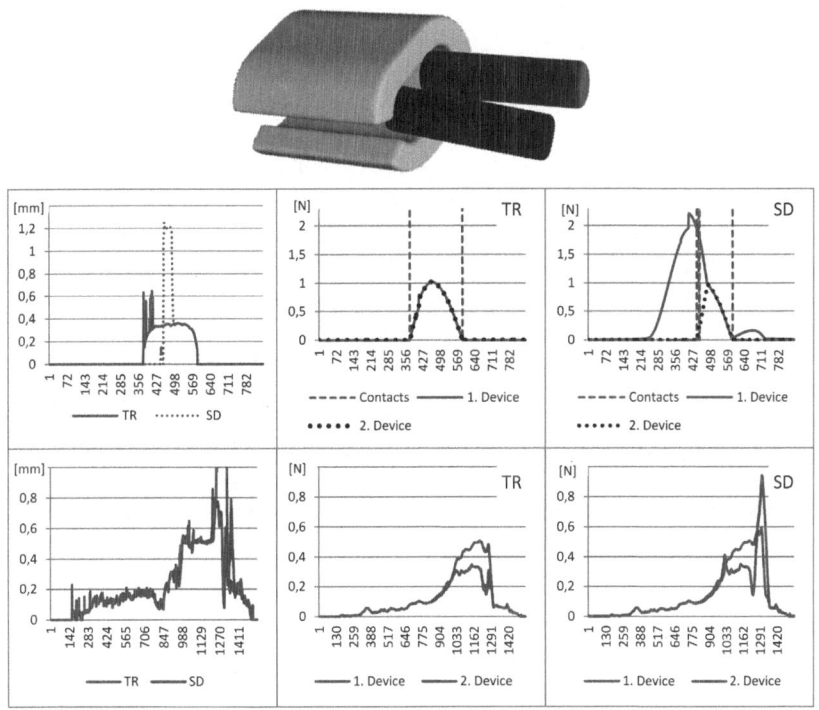

Fig. 1. Analysis Two-pegs-in-one-hole scenario (top). The upper graphs are the results of transparency analysis and the lower of the stability analysis for a complex contact scenario. (left graphs) Comparison of maximum local penetration depth in mm, (center graphs) feedback forces in N using trust region approach, (right graphs) feedback forces in N using static damping approach.

Stability Analysis for a Complex Contact Scenario. To show the stability of our approach in a complex contact scenario, we used a simple two-pegs-in-one-hole scenario matching our bimanual rendering(see Fig. 1 top). The bottom left graph of Fig. 1, shows the maximal penetration depth of each approach.

In the beginning both methods behave similar: are stable and stay below one millimeter; but towards the end, while TD values keep low, the penetration depth of SD makes a jump up to over five millimeters. This indication of instability is also reflected in the SD feedback forces of device two (see Fig. 1 bottom right graph), which makes a sharp oscillation at that point. The results for TR (see Fig. 1 lower center graph), on the other side, show nearly smooth behavior for the whole trajectory, which indicates that the TR method is also stable in case of complex contact situations with high contact stiffness.

4   Conclusion and Future Work

We have presented a novel approach for bimanual 6-DoF haptic rendering based on the static-virtual coupling and penalty method, and adapting a sophisticated numerical method. In this, we regard the problem of haptic rendering as an interactive optimization problem and apply a so-called trust region method on it. Herein, an approximate problem is defined based on a linearization of coupling and contact forces, which also consider inter-object collisions. In each step a metric on quality of the approximation is used to regulate the maximal displacement of the grasped objects in the step. To further increase stability, we ensure passivity of each simulation step by calculating the potential energy in the system before and after the step and rejecting steps which would increase this energy. Finally, we performed two experiments, which indicate that the proposed haptic rendering algorithm achieves our goals of transparency and stability.

As the tested scenarios are rather artificial, we plan for the future to validate our approach more application-related, i.e. in a medical training simulation, and perform studies with human subjects. Moreover, we want to integrate further physical aspects like friction or compliant environments.

Acknowledgments. This work is funded by the German Research Foundation (DFG) under grant KU 1132/6-1.

References

[1] Adams, R., Hannaford, B.: Stable haptic interaction with virtual environments. Trans. on Robotics and Automation, 465–474 (1999)

[2] Barbic, J.: Real-time reduced large-deformation models and distributed contact for computer graphics and haptics. Ph.d. dissertation, Carnegie Mellon University (2007)

[3] Barbic, J., James, D.: Six-DoF Haptic Rendering of Contact Between Geometrically Complex Reduced Deformable Models. IEEE Trans. on Haptics 1(1), 39–52 (2008)

[4] Bellavia, S., Macconi, M.: Diagonally Scaled Dogleg Methods for nonlinear systems with simple bounds. Tech. Rep. 6

[5] Bernstein, N., Lawrence, D., Pao, L.: Design of a uniactuated bimanual haptic interface. In: Haptic Interfaces for Virtual Environment and Teleoperator Systems, pp. 310–317 (2003)

[6] Coleman, T.: An interior trust region approach for nonlinear minimization subject to bounds. Tech. rep. Cornell University, Ithaca (1993)

[7] Colgate, J., Stanley, M., Brown, J.: Issues in the haptic display of tool use. In: Int. Conf. on Intelligent Robots and Systems., vol. 3 (1995)

[8] Colgate, J.: Factors affecting the z-width of a haptic display. In: Robotics and Automation 1994, pp. 3205–3210 (1994)

[9] Devarajan, V., Scott, D., Jones, D., Rege, R., Eberhart, R., Lindahl, C., Tanguy, P., Fernandez, R.: Bimanual haptic workstation for laparoscopic surgery simulation. Studies In Health Technology and Informatics (2001)

[10] Hannaford, B., Ryu, J.: Time-domain passivity control of haptic interfaces. IEEE Trans. on Robotics and Automation (2002)

[11] Johnson, D., Willemsen, P.: Six degree-of-freedom haptic rendering of complex polygonal models. In: Haptic Interfaces for Virtual Environment and Teleoperator Systems, pp. 229–235 (2003)

[12] Laycock, S., Day, A.: A Survey of Haptic Rendering Techniques. Computer Graphics Forum 26(1), 50–65 (2007)

[13] Madsen, K., Bruun, H., Tingleff, O.: Methods for non-linear least squares problems, 2nd edn. (2004)

[14] McNeely, W., Puterbaugh, K., Troy, J.: Six degree-of-freedom haptic rendering using voxel sampling. In: Computer Graphics and Interactive Techniques, New York, pp. 401–408 (1999)

[15] Miller, B., Colgate, J., Freeman, R.: Guaranteed stability of haptic systems with nonlinear virtual environments. IEEE Trans. on Robotics and Automation, 712–719 (2000)

[16] Moore, M.: Collision detection and response for computer animation. ACM Siggr. Computer Graphics 22(4), 289–298 (1988)

[17] Murayama, J., Bougrila, L., Luo, Y., Akahane, K., Hasegawa, S., Hirsbrunner, B., Sato, M.: SPIDAR G&G: A two-handed haptic interface for bimanual VR interaction. In: Eurohaptics, Munich, pp. 138–146 (2004)

[18] Otaduy, M., Lin, M.: A modular haptic rendering algorithm for stable and transparent 6-DOF manipulation. Trans. on Robotics 22 (2006)

[19] Otaduy, M., Lin, M.: Haptic Rendering: Foundations, Algorithms, and Applications. AK Peters (2008)

[20] Powell, M.: A hybrid method for nonlinear equations. In: Numerical Methods for Nonlinear Algebraic Equations, London, pp. 87–114 (1970)

[21] van Rhijn, A., Mulder, J.: Spatial input device structure and bimanual object manipulation in virtual environments. In: ACM Symp. on Virtual Reality Software and Technology, p. 51 (2006)

[22] Ullrich, S., Knott, T., Law, Y.: Influence of the bimanual frame of reference with haptics for unimanual interaction tasks in virtual environments. 3D User Interfaces, 39–46 (2011)

[23] Ullrich, S., Rausch, D., Kuhlen, T.: Bimanual Haptic Simulator for Medical Training: System Architecture and Performance Measurements. JVRC, 39–46 (2011)

[24] Wan, M., McNeely, W.: Quasi-static approximation for 6 degrees-of-freedom haptic rendering. In: IEEE Visualization, p. 34 (October 2003)

The Effect of the Stiffness Gradient on the Just Noticeable Difference between Surface Regions

Umut Koçak, Karljohan Lundin Palmerius, Camilla Forsell,
and Matthew Cooper

C-Research, Linköping University, Sweden
{umut.kocak,karljohan.lundin.palmerius,
camilla.forsell,matthew.cooper}@liu.se

Abstract. Numerous studies have considered the ability of humans to perceive differences in forces and how this affects our ability to interpret the properties of materials. Previous research has not considered the effect of the rate of change of the material stiffness in our ability to perceive differences, however, an important factor in exploration processes such as a doctor's palpation of the skin to examine tissues beneath. These effects are the topic of this research which attempts to quantify the effects of stiffness gradient magnitude and form on the discernment of changes in stiffness.

Keywords: Perception, stiffness, gradient, exploratory procedures, JND.

1 Introduction

Many disciplines have contributed to research in haptics in the last decade. One of them is psychophysics which has focussed on understanding our perception mechanism in order to facilitate the design of more effective haptics hardware and software solutions in a wide range of application areas. An ever increasingly important application for haptic technologies is surgery simulation where the feedback increases the chance of transferring knowledge and skills between the simulated and real environment.

One of the major aspects of exploring tissues and understanding the properties of the material at hand is its hardness/softness. Yet surgery simulation is but one example of haptic interaction with compliant surfaces, so it is to be expected that stiffness/compliance will be one of the most frequently surveyed material properties with several studies (for example [12]) being conducted on just noticeable difference (JND) within different scenarios and ranges. In our opinion, however, the multifaceted nature of stiffness exploration has only been explored for one type of interaction. The continuity of the contact and stiffness changes during continuous contact have not been surveyed comprehensively, despite their frequent importance in real life scenarios. During tissue palpation, for example searching for malign growth beneath the skin, the perception of stiffness change during contact is crucial since the changes reflect the material properties of the different tissues present. Several studies in the medical field [4,15,16] have

P. Isokoski and J. Springare (Eds.): EuroHaptics 2012, Part I, LNCS 7282, pp. 282–292, 2012.

been conducted showing the direct relationship between tissue characteristics and the stiffness gradient, yet its implications for human perception of stiffness are largely untested.

In this paper we present a study of the effect of stiffness gradient on the perception of stiffness in materials. Such a study can capture the effects of the rate of change of the stiffness during continuous contact as experienced in such exploratory procedures as drilling, cutting or palpation. Previous work [5] has demonstrated the effects of continuous contact, and axis of contact, on the JND in the stiffness but the effects of the rate of change have not been previously considered in perception studies, to the authors' knowledge. Our experiments, carried out within a virtual environment, have been conducted in order to demonstrate and measure the effect of stiffness gradient on discrimination. The results show not only an effect upon the observable difference between stiffness levels but also that the stiffness gradient affects the discrimination during continuous contact.

2   Related Work

Stiffness (or compliance) is one of the most studied properties, representing the hardness or softness of an object. The most common means to explore stiffness perception is to present a measure, showing how well humans can perceive the varying levels of hardness or softness, in the form of a JND. The results found vary depending on the differences between the methods employed. Effects of other factors such as multi-modality, and cutaneous and kinaesthetic cues are also being surveyed.

There are a number of studies (for example [12]) which have examined the JND in the stiffness. The effects of force and work cues on compliance discrimination were surveyed in [13] and the significant effect of force cues on discrimination was evident. In [14], the effect of surface deformation cues was examined and it was shown that the subjects' ability to discriminate the difference in stiffness was reduced by a factor of more than three without deformation cues.

The effect of visual information on stiffness perception was explored in the studies [10,17]. The dominance of visual feedback over kinaesthetic sense of hand position was demonstrated in [10]. Compliant objects that are further away were perceived to be softer in the case of haptic feedback alone [17], while the addition of the visual information reduced this bias.

Further studies [3,6,11] have examined the effects of some of the exploratory procedures [7] on stiffness perception. These exploratory procedures directly affect which properties of the object can be observed and how we perceive them. In [11], the contribution of tactile and kinaesthetic cues were explored for deformable and non-deformable objects. It was shown that the tactile information alone is sufficient for discrimination capacity of deformable objects while additional kinaesthetic feedback is necessary for non-deformable compliant objects. When a tool was used for exploration, additional kinaesthetic cues were found to be necessary for all types of objects [6]. Squeezing a deformable object between thumb and index finger was explored in [3,9] and tactile information was found to be negligible for this scenario [9].

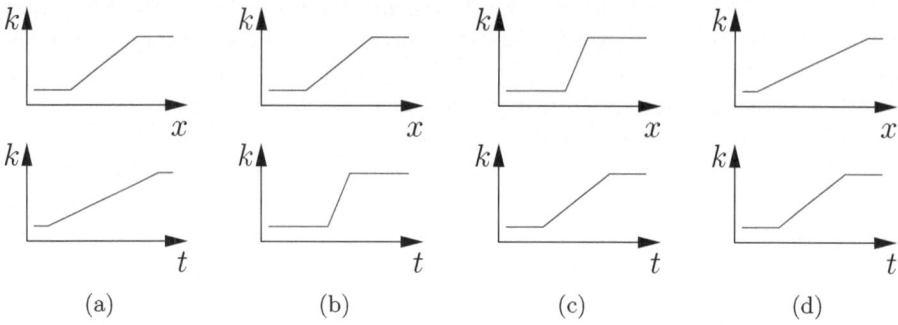

Fig. 1. The stiffness gradient over space (top) and over time (bottom). In subfigures (a) and (b) the position dependent stiffness over the transition region is the same, with low hand velocity (a) and high hand velocity (b) producing different perceived temporal changes in stiffness. In subfigures (c) and (d), the time dependent stiffness over the transition region is the same, with low hand velocity (c) and high hand velocity (d) producing different perceived spatial changes.

The studies described above surveyed the perception of stiffness and some affecting factors such as the visual information or exploratory procedure. The discrimination during continuous contact has been surveyed in [5] but, to the knowledge of the authors, the stiffness gradient has never been considered in discrimination studies.

3   Stiffness Change during Contact

The exploration of stiffness change during retained contact considered in this paper requires at least a C^0 continuous stiffness function. There are, however, two ways to describe the stiffness change during contact: stiffness as a function of position and stiffness as a function of time. The former is naturally what a real interaction would express. Nevertheless, for the proper understanding of the actual effect time may have a significant impact, as is discussed below.

3.1   Time-Dependent and Position-Dependent Change

The velocity of the hand during contact is one of the factors affecting the stiffness characteristics such that different hand velocities result in different stiffness-time or stiffness-position functions, as illustrated in figure 1. One can think of the scenario in figures 1(a) and 1(b) as stiffness perceived during a lateral movement across a flat surface composed of two regions exhibiting different stiffnesses with a transition region between them. The graphs at the top refer to the stiffness change across the flat surface. If one makes a lateral movement on the surface from one side to the other, lower hand velocity would create an apparently smoother stiffness change with time, as shown at the bottom graph in figure 1(a), while a higher hand velocity would correspond with the bottom graph in figure 1(b) which shows a sharper stiffness change in time. The position-stiffness characteristics are not affected by the hand velocity in this case.

Alternatively, if the stiffness is controlled by time, and independent of position, another scenario arises (see figures 1(c) and 1(d)). For a continuous lateral motion across the flat surface with a reference hand velocity, the stiffness-time and stiffness-position graphs would be the same. A lower velocity would create the stiffness-position graph at the top in figure 1(c) with an apparently sharper stiffness change over space while a higher velocity would create the stiffness-position graph at the top in figure 1(d) with an apparently more gradual stiffness change over space.

The effect of these parameters need to be understood for the design of the full study presented in this paper so a pilot study was conducted, exploring the effects of the two scenarios in figure 1 with respect to the JND. During the pilot studies the subject was asked to make a lateral movement across a virtual surface while matching the speed indicated by an animation of a set of balls moving between sides with a constant velocity, figure 3(b). The subject was not expected to exactly match the velocity of the moving balls, the aim was instead to provide an approximate guidance of the movement of the subject's hand. A JND of stiffness was found for the position-dependent and time-dependent stiffness conditions for two different hand velocities, 2 cm/sec and 8 cm/sec, by one subject and with 3 repetitions. In the case of the position-dependent stiffness the JND was smaller for higher hand velocity cases, creating a sharper stiffness change in the time domain, in 2 out of 3 repetitions. For the time dependent stiffness the JND was smaller in lower velocity cases, creating a sharper stiffness change in the position domain, in all repetitions. As a result, the pilot studies hinted at possibly better discrimination performance with a higher stiffness gradient and showed that we need to consider the hand velocity during contact since both the stiffness-time and stiffness-position functions have the potential to affect the perception.

3.2   Transition Region

To examine the effect of the stiffness gradient on stiffness discrimination, three conditions with different position dependent stiffness functions were considered. All of the stiffness functions included two regions exhibiting different, constant stiffness values with a transition region between them. The two stiffness values and the width of the transition region were chosen to be the same for all three conditions. The difference between the stiffness levels, Δ_k, the transition width, Δ_x and the stiffness gradients are illustrated in figure 2. The three conditions differ in the interpolation function used in the transition region. The three interpolation functions for the conditions are *linear*, *cosine*, and *tanh* each of which have different derivatives. This results in a different maximum magnitude of stiffness gradient with transition region size and stiffness difference kept the same.

In the case of linear interpolation the gradient is a constant function with a value of Δ_k/Δ_x in the transition region. In the second condition a cosine function is mapped by

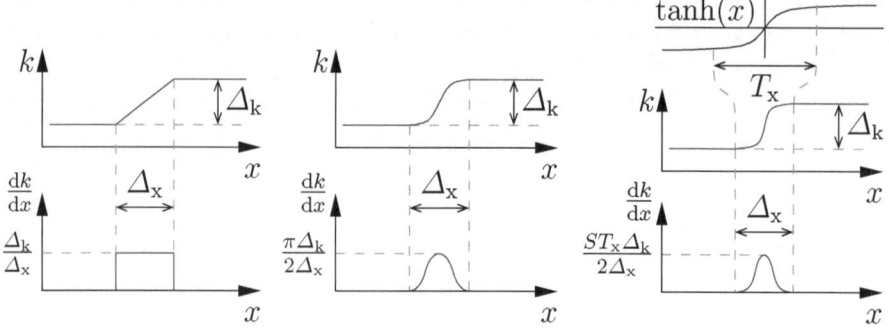

(a) Linear interpolation.   (b) Cosine interpolation.   (c) Tanh interpolation.

Fig. 2. The interpolation functions used in the study, with their respective spatial derivatives

$$k(x) = \frac{\Delta_k}{2} \cos\left(\pi \left(\frac{x - x_0}{\Delta_x}\right) - \pi\right) \tag{1}$$

with a derivative

$$\frac{dk}{dx}(x) = \frac{-\Delta_k}{\Delta_x} \frac{\pi}{2} \sin\left(\pi \left(\frac{x - x_0}{\Delta_x}\right) - \pi\right) \tag{2}$$

where x_0 is the beginning point of the transition region. It can be seen from equation 2 that the maximum stiffness gradient in the transition region is $\pi/2$ times the linear case. The tanh function, which converges to 1 and -1 at infinity and negative infinity respectively, was chosen as a third condition. By mapping a limited range of the tanh function around the origin to the transition region one can adjust the maximum stiffness gradient in the transition region. The function must be stretched by a factor, S, to make it C^0 continuous in the edge between the transition region and the respective adjacent constant regions. The tanh function is thus mapped to the transition region by

$$k(x) = \frac{S\Delta_k}{2} \tanh\left(T_x \left(\frac{x - x_0}{\Delta_x}\right) - \frac{T_x}{2}\right) \tag{3}$$

where

$$S = \frac{1}{\tanh\left(\frac{T_x}{2}\right)} \tag{4}$$

with a derivative

$$\frac{dk}{dx}(x) = \frac{S\Delta_k}{\Delta_x} \frac{T_x}{2} \frac{1}{\cos^2\left(T_x \left(\frac{x - x_0}{\Delta_x}\right) - \frac{T_x}{2}\right)} \tag{5}$$

Deploying three different interpolation functions allows comparison of three different stiffness gradient magnitudes while keeping the transition width and the

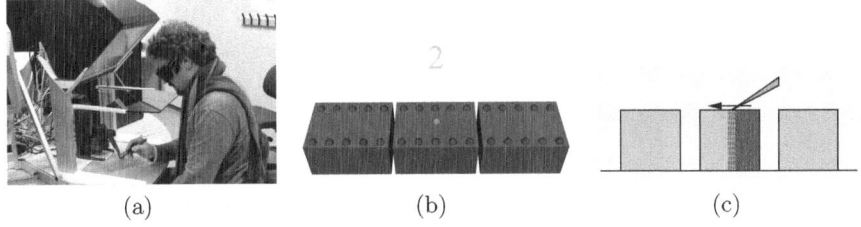

Fig. 3. (a) A 3D virtual image registered with the real hand position is obtained with the help of a semi-transparent mirror and stereo glasses. (b) An animation of a series of red balls moving to the left and to the right were rendered at the front and back of the boxes, respectively. (c) Trials included three boxes, only one of which included a stiffness change along lateral motion across the surface.

stiffness difference identical for all three conditions. The maximum stiffness gradient is $\pi/2$ times and $ST_x/2$ times the gradient of the linear condition in the cosine and tanh conditions, respectively.

4   Evaluation

To explore the effect of stiffness gradient we performed a user study in two stages: gradient magnitude and transition width. In the gradient magnitude case, we aimed to test the effect of gradient magnitude by comparing the three different interpolation functions in the transition region. In the second stage, different transition widths were presented and the psychometric function of the gradient was measured from the subjects' responses.

4.1   Method

The experiments were performed in a virtual environment and a Desktop Phantom and a semi-transparent framework was used as the equipment, as illustrated in figure 3(a). Each stimulus was composed of three virtual boxes that were visually rendered as in figure 3(b). The orientation of the boxes was adjusted such that the palpation occurs on the axis perpendicular to the desk. To prevent visual cues relating to the strain applied, the appearance of the boxes did not change in response to the compression and the haptic probe was rendered as a sphere which remained on the surface of the boxes during contact in all situations. The subject was also prevented from seeing the real hand position under the semi-transparent mirror by installing a sheet of white paper under the mirror and setting the background colour to bright white. In order to have the subjects adopt approximately the same hand velocity during lateral movement, an animation of a set of red balls moving to right and left were rendered at the front and back of the boxes, respectively. Two of the boxes were identical while one of them, selected at random, had a non-uniform stiffness as described below. The force feedback from the boxes was evaluated by multiplying the stiffness and the depth of probe from the surface of the box, based on Hooke's Law. The subjects were told to make 'sweeping' movements sideways (left-to-right-to-left) across the surface, following the speed of the ball animation, and select the box which exhibited non-uniform stiffness in each trial.

In all trials in both experiments, the subjects were presented with three different boxes two of which presented only the reference stiffness while the third, selected at random, presented the reference stiffness on the left half side and a harder stiffness on the right with a transition region in between them, see figure 3(c). The height of the boxes was set to 3 cm throughout the experiments. In the literature a wide range of reference stiffnesses, varying from 100 N/m to 16900 N/m [3], have been surveyed. During several pilot studies various stiffness values had been tested. It was observed that continuous use of the haptic device with higher stiffness values can result in overheating of the motors, requiring a break for the system to cool down. Finally, 100 N/m was determined as a suitable reference stiffness.

Twelve subjects took part in the experiments, 9 male and 3 female. They were all undergraduate or graduate students aged between 23 and 40 years (mean age was 28). 10 of the subjects had tried a haptic device on a few occasions previous to the experiments and 2 had used them quite often. All subjects had normal or corrected to normal vision. They received no compensation for taking part in the experiments.

Before the experiments began background information was obtained from each subject. They then reviewed written instruction material and were instructed about the equipment and the tasks to be performed. Before the real experiments they also completed a set of practice trials. For each individual trial the task was to identify which box, out of the three, exhibited the different stiffness levels and give a response by pressing a button placed on the haptic device while pointing to that box. Total participation time lasted 1 hour, on average, including the introductory part.

4.2   Gradient Magnitude Evaluation

The JNDs of three conditions with different interpolation functions in the transition region of the non-uniform box, figure 3(c), were compared. Having a better discrimination performance in one of these conditions, being able to detect smaller Δ_k, would indicate the contribution from another component. We suspected that the presence of a higher stiffness gradient would contribute to the sense of discrimination and so result in a smaller JND. Therefore we compared the JNDs of the three conditions depending on the interpolation function used in the transition region: *Linear, cosine* and *tanh* functions (as described in section 3.2).

For all the three conditions, the magnitude of the harder stiffness in the non-uniform box was changed depending on the subjects' previous responses, while the reference stiffness was kept constant. A one-up two-down adaptive staircase procedure was used [8] while changing the harder stiffness. An adaptive staircase starts with an initial difference magnitude and, depending on an individual subject's responses, changes the magnitude of the difference such that it converges to the perception limit of discrimination for that subject. In the case of a one-up two-down staircase, the magnitude is decreased following two consecutive correct responses and increased after each single incorrect response. This procedure converges to a stimulus level at which subjects can make accurate responses with a certainty of 70.7%. In our case each session started with a stiffness difference of 66.7 N/m (2/3 of the reference stiffness value). Initially, the stiffness difference was changed by 9 N/m per response and then by 4.5 N/m after the third reversal and by 2.25 N/m after the sixth reversal. The session was terminated after nine reversals and the average of the peaks and valleys of the last six reversals were calculated to be the JND. The width of the transition region was set to 10 mm for all conditions. For the *tanh* condition, the mapped range of the tanh function, T_x, was set to 6.

The evaluation was performed as a within-subjects design with one independent variable (stiffness) having three interpolation functions (*Linear* vs. *Cosine* vs. *Tanh*) or conditions. The experiment was performed over three separate sessions where each condition was carried out once. The presentation order of the conditions for each subject was balanced by using a Latin-square procedure. The placement of the box with variable stiffness was randomized for each trial.

4.3   Transition Width Evaluation

In the second stage, an experiment was performed by each subject in order to calculate the probabilities of the correct guesses for different transition region widths. In this stage, the stiffness difference was kept constant at the JND value for each subject, as found using the *linear* condition in the first stage at a transition width of 10 mm. The width of the transition region was set to four different values: 1.25, 5, 20 and 80 mm. Each transition width was repeated 10 times in a random order, with the linear transition function used throughout. These repetitions were used to calculate the probability of a correct response for each width.

5   Results

The values for each of the 12 subjects for all three conditions in the first stage were analyzed. According to the Kolmogorov-Smirnov and Shapiro-Wilks tests, the data did not fit a normal curve therefore we employed a logarithmic transformation which corrected the fit. Hence we used a parametric test for further analysis. A repeated measures ANOVA with a decision criterion of 0.05 showed that there was a significant difference between the three conditions, $F(2,22) = 9.059$, $p=0.001$.

To determine which conditions significantly differed from each other, Bonferroni corrected pairwise comparisons were performed at a 0.0167 (0.05/3) level of significance as a post-hoc test. A significant difference was observed between condition *tanh* and the other two conditions, $p<0.05$, while no significant difference was observed between the conditions *cosine* and *linear*, $p>0.05$. The

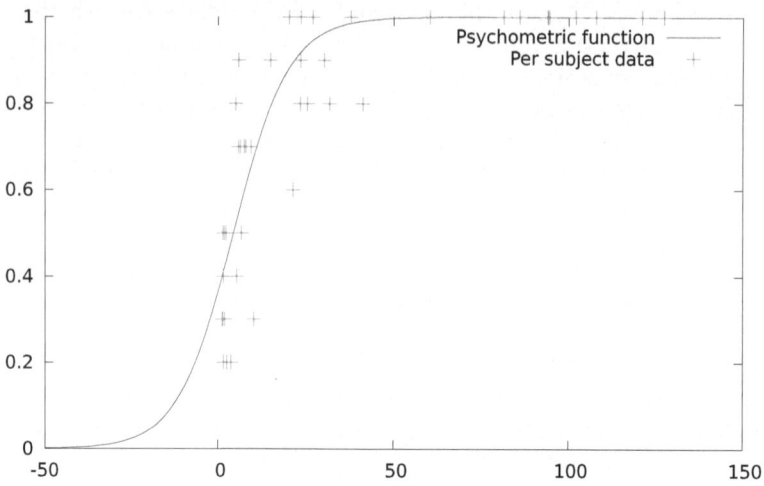

Fig. 4. The probability of a correct response as a function of the stiffness gradient (N/m per cm)

significant difference for the *tanh* condition was more likely since the maximum stiffness gradient is $ST_x/2$, approximately three times the gradient in the *linear* condition, where S is close to 1 and T_x was chosen to be 6. Not having a significant difference between the *cosine* and *linear* conditions might be explained by the closer maximum stiffness gradient values. Maximum gradient of the *cosine* condition was $\pi/2$ (approximately 1.6) times that of the *linear* condition.

The analysis of the results shows that the discrimination performance for the condition with the highest stiffness gradient magnitude in the transition region, *tanh*, is significantly better than the other two conditions. The mean value of the JND for the *cosine* condition is 14.7±8.9%, while the *linear* and *tanh* conditions have a mean JND of 17.95±9.62% and 10.86±6.24% respectively.

The probabilities of correct responses found in the second stage experiment are illustrated in figure 4. One can observe the characteristic psychometric function [8] for a range of the stiffness gradient showing strong correlation between gradient and discrimination. The graph extends into negative (imaginary) widths because of the risk of missing the small stiffness difference even at a small transition width. A least-squares fitted sigmoid curve shows that the average 50 % probability lies at approximately 4,5 N/m per cm, but also that the individual data differs substantially compared to this average.

6   Discussion

We performed experiments in order to observe the effect of stiffness gradient on discrimination of the stiffness change during continuous contact. We compared the JNDs of three conditions with different interpolation functions deployed in the transition between two stiffness levels. The interpolation functions differed in their stiffness gradient magnitude and a significantly better discrimination was obtained for the one with highest gradient magnitude. Among the other two conditions the discrimination appeared to be better for the higher gradient magnitude condition but the results were not statistically significant.

The results being consistent with the relation that the higher the stiffness gradient the better the discrimination, support the idea that it is not only stiffness difference but also the gradient affecting the perception. This was also supported by the second stage of the experiment, performed to achieve the psychometric function showing the probability of correct responses in the discrimination task as a function of stiffness gradient.

In [2], it was discussed that observing a Weber fraction for a continuous signal, the percentage change in the signal that can be barely noticed, depends on the type of the haptic signal. Stiffness was among the signals for which the Weber fraction has often been observed, making the stiffness perception context-dependent. This context-dependency is also well-known for the perception of colour [1], due to perceiving the same colour differently depending on the surrounding colours within which it is presented. Our results, by showing the relationship between the stiffness gradient and the discrimination, also indicate the importance of the context – the type of the change in addition to the magnitude of the change.

Discrimination of stiffness change might be crucial in real life scenarios such as exploring tissues for malign growth, needle insertion, and tissue cutting and bone drilling in surgeries. Understanding our perception mechanism and its limits has the potential of improving the procedures in real life, in addition to improving virtual medical applications such as surgery simulators. The knowledge of the perception of continuous stiffness change can also be useful for piecewise linear modelling of nonlinear stiffness for providing smoother representations of the nonlinearity.

We think of this result as a starting point for further examination of stiffness perception for scenarios closer to real life situations. Studies [3,6,11] of the effect of exploratory procedures on perception have shown the need to design more realistic scenarios in perception studies to achieve useful results. In addition to surveying different aspects of perception during continuous touch, we intend to explore ways to combine these results in our haptic software solutions as well.

Acknowledgments. This work has been funded by the Swedish Science Council through grant number 621-2005-3609, the Foundation for Strategic Research (SSF) under the Strategic Research Center MOVIII, and the Swedish Research Council Linnaeus Center CADICS.

References

1. Adelson, E.H.: Lightness Perception and Lightness Illusions. In: The New Cognitive Neurosciences, 2nd edn., pp. 339–351. MIT Press, Cambridge (2000)
2. Cholewiak, S.A., Tan, H.Z., Ebert, D.S.: Haptic Identification of Stiffness and Force Magnitude. In: Haptic Interfaces for Virtual Environment and Teleoperator Systems, pp. 87–91. IEEE Press (2008)

3. Freyberger, F.K., Frber, B.: Compliance discrimination of deformable objects by squeezing with one and two fingers. In: Eurohaptics, pp. 271–276 (2006)

4. Isenberg, B.C., Dimilla, P.A., Walker, M., Kim, S., Wong, J.Y.: Vascular Smooth Muscle Cell Durotaxis Depends on Substrate Stiffness Gradient Strength. Biophysical Journal 97, 1313–1322 (2009)

5. Koçak, U., Palmerius, K.L., Forsell, C., Ynnerman, A., Cooper, M.: Analysis of the JND of Stiffness in Three Modes of Comparison. In: Cooper, E.W., Kryssanov, V.V., Ogawa, H., Brewster, S. (eds.) HAID 2011. LNCS, vol. 6851, pp. 22–31. Springer, Heidelberg (2011)

6. Lamotte, R.H.: Softness discrimination with a tool. J. Neurophysiology 83(4), 1777–1786 (2000)

7. Lederman, S., Klatzky, R.: Haptic perception: A tutorial. Attention, Perception, and Psychophysics 71(7), 1439–1459 (2009)

8. Levitt, H.: Transformed up-down methods in psychoacoustics. J. Acoustical Society of America 49, 467–477 (1971)

9. Roland, P., Ladegaard-Pedersen, H.: A quantitative analysis of sensations of tension and of kinasthesia in man: Evidence for a peripherally originating muscular sense and for a sense of effort. Brain 100(4), 671–692 (1977)

10. Srinivasan, M.A., Beauregard, G., Brock, D.: The impact of visual information on the haptic perception of stiffness in virtual environments. In: ASME Dynamic Systems and Control Division, pp. 555–559 (1996)

11. Srinivasan, M.A., Lamotte, R.H.: Tactual discrimination of softness. J. Neurophysiology 73(1), 88–101 (1995)

12. Tan, H.Z., Pang, X.D., Durlach, N.I.: Manual resolution of length, force and compliance. In: ASME Dynamic Systems and Control Division, pp. 13–18 (1992)

13. Tan, H.Z., Durlach, N.I., Beauregard, G., Srinivasan, M.A.: Manual discrimination of compliance using active pinch grasp: The roles of force and work cues. Perception and Psychophysics 57(4), 495–510 (1995)

14. Tiest, W.M.B., Kappers, A.M.: Cues for haptic perception of compliance. IEEE Transactions on Haptics 2(4), 189–199 (2009)

15. Tse, J.R., Engler, A.J.: Stiffness Gradients Mimicking In Vivo Tissue Variation Regulate Mesenchymal Stem Cell Fate. PLoS ONE 6(1), e15978 (2011)

16. Wong, J.Y., Velasco, A., Rajagopalan, P., Pham, Q.: Directed Movement of Vascular Smooth Muscle Cells on Gradient-Compliant Hydrogels. Langmuir 19(5), 1908–1913 (2003)

17. Wu, W.C., Basdogan, C., Srinivasan, M.A.: Visual, haptic, and biomodal perception of size and stiffness in virtual environments. In: ASME Dynamic Systems and Control Division, pp. 19–26 (1999)

Tactile Apparent Motion between Both Hands Based on Frequency Modulation

Soo-Chul Lim[1], Dong-Soo Kwon[2], and Joonah Park[1]

[1] Future IT research center, Samsung Advanced Institute of Technology, San14, Nongseo-dong, Giheung-gu, Yongin-si, Gyeonggi-do, Republic of Korea
{soochul.lim,joonah}@samsung.com
[2] Telerobotics & Control Laboratory, KAIST, Daejeon, Republic of Korea
kwonds@kaist.ac.kr

Abstract. In this paper, the effects of the frequency modulation of vibration elements on the representation of dynamic tactile apparent motion between both hands will be proposed. The sensation level difference due to the different frequencies that result when using vibrating motors on the right and left fingers causes a phantom sensation that is perceived as if the stimuli were between the fingers. The change of sensation level difference between both hands due to the frequency modulation creates a somatosensory illusion using this phantom sensation, which occurs in such a way as to feel like a vibration flow from one hand to the other hand. We conducted experiments to evaluate whether the tactile flow and the phantom sensation can be perceived. Participants reported sensing the vibrotactile flow and the phantom sensation.

Keywords: Tactile apparent motion, vibrotactile feedback, sensation level, frequency modulation.

1   Introduction

Vibrotactile feedback has recently become widely used to allow for haptic interaction with mobile devices. Touch feedback can be regarded as one of the dominant factors that can increase the degree of realism or immersion because it is not easy to increase the size of a visual display unit to a level at which users are truly immersed by the size of the re-created objects. In game consoles, vibrotactile feedback is used due to the increase in realism or game interest that such feedback can provide. These previous sorts of tactile feedback have been involved in simply transferring temporal information. However, in recent research, some researchers have started to study vibrotactile feedback that can provide to the user not only temporal information but also spatial information. Much research has focused in various ways on the idea of the tactile flow. Kim investigated a vibrotactile flow on a rigid body (Kim et al. 2009). Using multiple motors and actuating them at different times, it was found to be possible to synthesize a new vibration by combining multiple vibrations at a specific location. Seo and Choi investigated the idea of a linear vibrotactile flow on a user's palm with two voice-coil actuators attached to a device mockup similar in size to a mobile

P. Isokoski and J. Springare (Eds.): EuroHaptics 2012, Part I, LNCS 7282, pp. 293–300, 2012.
© Springer-Verlag Berlin Heidelberg 2012

device (Seo and Choi 2010). They showed the magnitude difference of the motors at a specific frequency between two motors, which are attached to each side of the mock-up, causing the vibrotactile flow. These two studies generated ideas about the movement of vibrotactile feeling between motors in a rigid body. Some researchers have studied the movement of vibrotactile feeling psychophysically. Barghout investigated the spatial resolution of vibrotactile perception on a human forearm when applying multiple stimuli that describe a phantom sensation midway between the multiple stimuli with amplitude modulation when the multiple stimuli are present simultaneously at adjacent locations on the human skin. Their psychophysical experiments were designed to look at the human spatial perception ability on the human forearm for stationary and moving vibrotactile stimuli (Barghout et al. 2009). Miyazaki showed that a cutaneous rabbit can "hop out of the body" onto an external object held by the subject. They delivered rapid sequential taps to the subjects' left and right index fingers. When the subjects held a stick in such a way that it was laid across the tips of their index fingers and when they received the taps via the stick, they reported sensing illusory taps in the space between the actual stimulus locations (Miyazaki et al. 2010). The localization of vibrotactile feeling, such as a tactile apparent motion or a cutaneous rabbit based on two tactile stimuli on the skin, is similar to a sound being heard between the two ears. The interaural time difference (ITD) is based on the fact that there is a difference in time at which a sound reaches the left and right ears (Wightman and Kistler 1998). This is important component in the localization of sounds, as it provides a cue to the direction or angle of the sound source from the head. Another cue, the interaural level difference (ILD), is based on the difference in the sound pressure level between the ears.

In this paper, we investigate the effects of the frequency modulation of vibration elements on the representation of dynamic tactile feedback between both hands. We prove that the sensation level difference in vibrating motors with different frequencies on the right and left finger cause a phantom sensation that is perceived as if the stimuli were between the fingers. To utilize phantom sensation, a tactile flow from one hand to the other hand with frequency modulation was generated.

2   Tactile Apparent Motion between Both Hands

Tactile flow can be generated with a sensation level difference according to the magnitude difference. In previous research, we showed that frequency modulation causes sensation level change(Lim et al. 2011). A frequency change can cause changes in the sensation level because the threshold and the sensation level of the index finger are a function of the frequency and amplitude, showing a U-shaped curve. Based on the sensation level difference according to the frequency difference, the change of sensation level in both hands may cause vibrotactile flow between the two vibration regions.

2.1   Participants

Seven participants, aged 23 to 32 year olds, participated in the experiments. All participants were right-handed and reported no known cutaneous or kinesthetic sensing problems. In this study, participants were selected so as to be younger than 32 were selected due to the known decreasing sensitivity of mode elderly humans(Manning et al. 2006). All of the participants were paid for their participation and were unaware of the purposes of the experiments.

2.2   Apparatus

Two force reactors (AF series L-type, ALPS Co.Ltd) which are impact type linear vibrators provided tactile stimuli directly to the index finger. The vibrator can be controlled with sinusoidal vibratory output at desired frequency as controlling electrical sinusoidal input at the frequency. The vibratory sinusoidal output amplitude can be controlled with modifying electrical sinusoidal input amplitude. The stimuli were generated so that one stimulus was directed to the left index finger and another was directed to the right index finger. The distance between the force reactors was 20 cm. Each force reactor was mounted on a vibration absorbing material to reduce the effect of vibration transfer through the desk (Fig. 1(a)).

2.3   Procedure

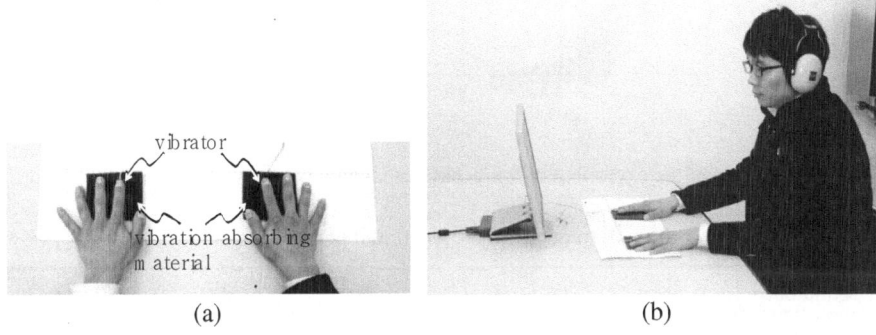

(a)   (b)

Fig. 1. (a) Stimulation of vibrotactile feedback on both hands (b) Presentation of vibrotactile feedback through the vibrator

The participants were seated in front of the tactile stimulus device and were instructed to maintain their finger, hand, and arm position while each index finger was on a different vibration motor; participants were asked to sit naturally and hold their hands in a natural position in order to minimize the modulation of sensation due to body position(Fig. 1(b)) (Azañón and Soto-Faraco 2008; Medina and Rapp 2008). Many studies have proved that the threshold and sensation level curve of the index finger is of a U-shape with a minimum level around the frequency of 250 Hz. A transition of the sensation level according to the frequency modulation was noted (Verrillo 1985; Lim et al. 2011). Based on the findings, we decided on an 11 frequency set that can show the tactile flow between two hands. The preliminary experiment attempted to find out whether a tactile flow with a feeling of hopping out of the finger could be sensed when a stimulus was presented for 200 msec at each set while increasing the set number from 1 to 11 or decreasing the set number from 11 to 1 with intervals at each set as in Table 1. Fig. 2 provides an illustration how the frequency modulation was exploited to create a tactile apparent motion. Participants answered "rightward", "leftward" or "indiscernible" with alternative forced choices. Each subject performed a total of 20 trials (2 directions × 10 trials). For randomly repeated measurements, all stimuli were presented ten times with a uniform random distribution.

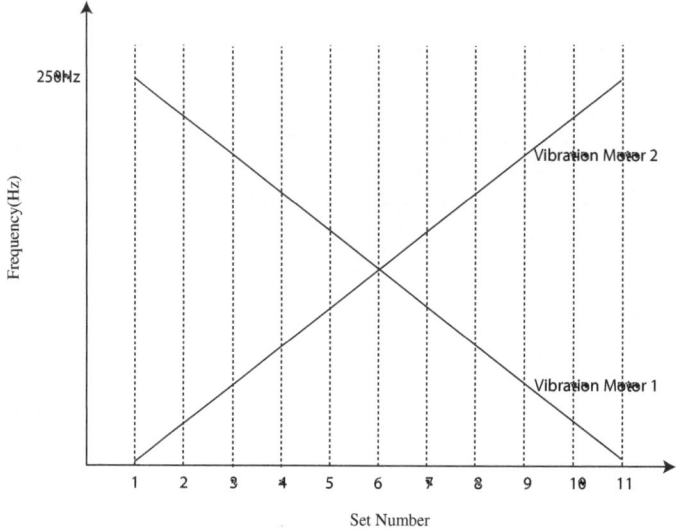

Fig. 2. Illustration of the exploiting of frequency modulation to create a tactile apparent motion

Table 1. Stimuli frequencies set for stimulation of each index finger

Set number	1	2	3	4	5	6	7	8	9	10	11
Frequency of Motor 1 (Hz)	1	26	51	76	101	126	150	175	200	225	250
Frequency of Motor 2 (Hz)	250	225	200	175	150	126	101	76	51	26	1

The main experiment was designed to see if participants were able to sense mechanical vibration at the finger pad of the right and left index finger. Fig. 3 shows the stimuli sequence. The standard set and the comparison set were delivered to the two index fingers. A comparison vibration of one second was followed by a reference vibration of one second. The stimuli frequencies that were presented to each index finger with the different standard and comparison vibration set were selected randomly from the eleven sets in Table 1. After the stimulus delivery, participants were asked to answer a question about the relative perceived position of the comparison set in order to compare it with the standard set. Participants answered "left", "right" or "indiscernible" with alternative forced choices. Each subject performed a total of 1100 trials (11 standard sets × 10 comparison sets × 10 trials). For randomly repeated

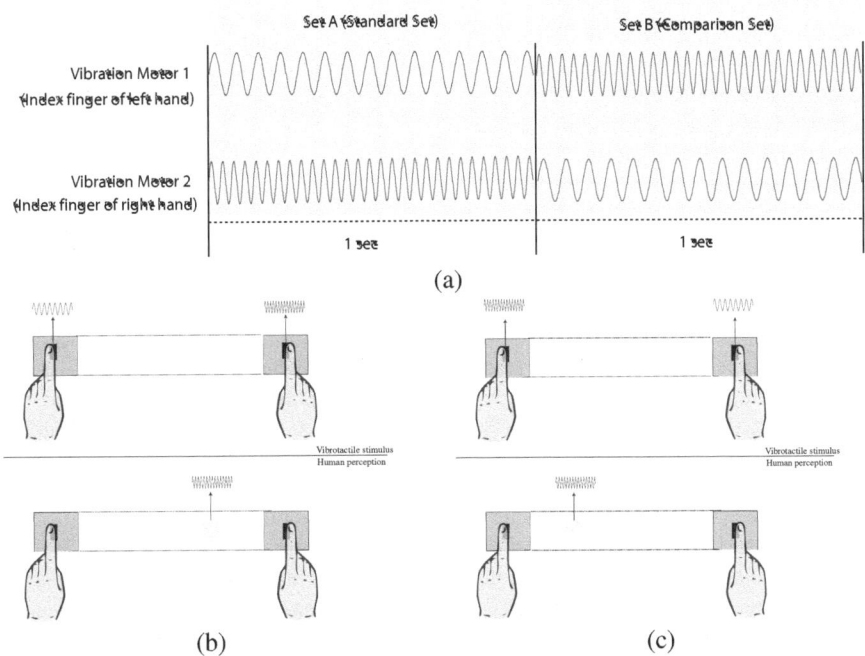

Fig. 3. Example of main experiment (a) Time chart of stimulus pairs presented to each index finger (b) tactile stimulus and expected stimulus position that was perceived in standard set (c) tactile stimulus and expected stimulus position that was perceived in comparison set

measurements, all stimuli were presented ten times with uniform random distribution. An initial 30s adaptation period was given at the beginning of experiment. Participants took a one minute break every five minutes.

3   Result and Discussion

This experiment examined the tactile apparent motion with frequency modulation between the right and left index fingers. The preliminary experiment was designed to see whether a tactile flow with a feeling like a hopping out of a finger could be sensed when the stimulus was presented for 200 msec at each set while increasing the set number from 1 to 11 or decreasing the set number from 11 to 1. In the experiment, all participants in all trials were able to discern the direction of the flow. This means that the participants sensed the tactile apparent motion between the right and left index finger at a 20 cm distance. With an increase of the frequency, the sensation level also increased up to 250 Hz. The sensation level difference between the left and right finger pad caused the illusion of tactile stimuli. When the frequency of one stimulus increased and the frequency of other stimulus decreased on each vibrating finger pad, the sensation level difference gave the feeling of tactile apparent motion with a feeling of hopping out of the fingers.

The main experiment investigated whether participants could discern the relative positions of stimuli perceived between the standard set and the comparison set when there was a frequency difference in the stimuli given to both index fingers. Fig. 4 provides a graphical representation of the correct answer rate for each reference and comparison set. The horizontal axis is the standard stimulus set and the vertical axis is the comparison stimulus set. The gray scale represents the correct answer rates. A correct answer means that the stimuli in the conditions of the comparison set was perceived as vibrating on the left side compared to the stimuli in the conditions of the standard set when the comparison set number is larger than the standard set number, or the stimuli in the condition of the comparison set was perceived as vibrating on the right side compared to the stimuli in the condition of the standard set when the comparison set number was smaller than the standard set number. In cases in which the stimuli was vibrated to both hands with a large frequency difference, participants perceived the stimuli as if it has been made in a location near the one finger that had been stimulated with the higher frequency vibration when two vibrotactile motors generated tactile stimuli simultaneously with a large frequency difference at each index finger; this was the case in set such as Set 1, 2, 10 and 11, as shown in Table 1. These results mean that inter-tactile level difference is an important cue for tactile localization when two different tactile stimuli are presented to the left and right index fingers. However, some participants were confused about the relative position between the standard and comparison set when the frequency change of the stimuli between the sets was not large. Furthermore, the correct answer rate was low near the center of the space between the two fingers. Participants were confused about the position of virtual stimuli between hands when stimuli with a small frequency difference on both hands were delivered. This means that participants were confused when attempting to discern stimuli positions when the stimuli positions were located near the center or when the differences of stimuli frequency between the standard and comparison set were not large.

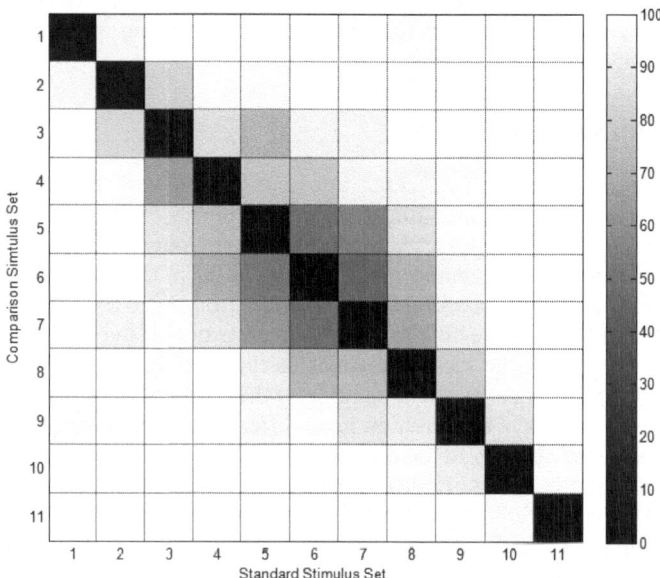

Fig. 4. Correct answer rate for each sample in reference and comparison sets

Previous researchers observed the cutaneous rabbit illusion only on the body in (Blankenburg et al. 2006; EIMER et al. 2005; Flach and Haggard 2006). Such tactile apparent motion has also been observed only on the body (Kirman 1974, 1983; Seo and Choi 2010; Barghout et al. 2009). Our results, however, show that tactile apparent motion with tactile stimuli "hopped out of the finger" when stimulation was given to participants' index fingers while changing stimuli frequencies. Our findings suggest that the effect involves not only the currently accepted idea of somatotopic representation but also the idea of representation of an external object that interacts with the body.

4   Conclusion and Further Work

The effects of the frequency modulation of vibration elements on the representation of dynamic tactile feedback in both hands were investigated. We proved that the sensation level difference in vibrating motors with different frequencies on the right and left finger caused a phantom sensation that is perceived as if the stimuli had been between the fingers. To utilize this phantom sensation, tactile flow from one hand to the other hand with frequency modulation is generated.

For future research, we will implement this idea of tactile apparent motion based on frequency modulation in both hand to handheld game devices when the person is holding the consoles with both hands. We believe that we can expect that the flow feeling will be able to provide a more interesting game experience, especially in such devices as driving simulators and archery games.

References

Azañón, E., Soto-Faraco, S.: Changing Reference Frames during the Encoding of Tactile Events. Current Biology 18(14), 1044–1049 (2008)

Barghout, A., Jongeun, C., El Saddik, A., Kammerl, J., Steinbach, E.: Spatial resolution of vibrotactile perception on the human forearm when exploiting funneling illusion. In: IEEE International Workshop on Haptic Audio visual Environments and Games, HAVE 2009, pp. 19–23 (2009)

Blankenburg, F., Ruff, C.C., Deichmann, R., Rees, G., Driver, J.: The cutaneous rabbit illusion affects human primary sensory cortex somatotopically. PLoS Biol. 4(3), e69 (2006), doi:05-PLBI-RA-0938R2 [pii] 10.1371/journal.pbio.0040069

Eimer, M., Forster, B., Vibell, J.: Cutaneous saltation within and across arms: A new measure of the saltation illusion in somatosensation. Perception & Psychophysics 67(3), 458–468 (2005)

Flach, R., Haggard, P.: The Cutaneous Rabbit Revisited. Journal of Experimental Psychology: Human Perception and Performance 32(3), 717–732 (2006)

Kim, S.Y., Kim, J.-O., Kim, K.Y.: Traveling vibrotactile wave - a new vibrotactile rendering method for mobile devices. IEEE Transactions on Consumer Electronics 55(3), 1032–1038 (2009)

Kirman, J.H.: Tactile apparent movement: the effects of number of stimulators. Journal of Experimental Psychology 103(6), 1175–1180 (1974)

Kirman, J.H.: Tactile apparent movement: the effects of shape and type of motion. Perception & Psychophysics 34(1), 96–102 (1983)

Lim, S.-C., Kyung, K.-U., Kwon, D.-S.: Presentation of Surface Height Profiles Based on Frequency Modulation at Constant Amplitude Using Vibrotactile Elements. Advanced Robotics 25(16), 2065–2081 (2011)

Manning, H., Tremblay, F.: Age differences in tactile pattern recognition at the fingertip. Somatosensory & Motor Research 23(3), 147–155 (2006)

Medina, J., Rapp, B.: Phantom Tactile Sensations Modulated by Body Position. Current Biology 18(24), 1937–1942 (2008)

Miyazaki, M., Hirashima, M., Nozaki, D.: The "Cutaneous Rabbit" Hopping out of the Body. J. Neurosci. 30(5), 1856–1860 (2010), doi:10.1523/jneurosci.3887-09.2010

Seo, J., Choi, S.: Initial study for creating linearly moving vibrotactile sensation on mobile device. In: 2010 IEEE Haptics Symposium, pp. 67–70 (2010)

Verrillo, R.T.: Psychophysics of vibrotactile stimulation. The Journal of the Acoustical Society of America 77(1), 225–232 (1985)

Wightman, F., Kistler, D.: Of vulcan ears, human ears and 'earprints'. Nat. Neurosci. 1(5), 337–339 (1998), doi:10.1038/1541

Comparing Direct and Remote Tactile Feedback
on Interactive Surfaces

Hendrik Richter, Sebastian Loehmann, Florian Weinhart, and Andreas Butz

University of Munich, Amalienstrasse 17, 80333 Munich, Germany
{hendrik.richter,sebastian.loehmann,andreas.butz}@ifi.lmu.de

Abstract. Tactile feedback on touch surfaces has shown to greatly improve the interaction in quantitative and qualitative metrics. Recently, researchers have assessed the notion of remote tactile feedback, i.e., the spatial separation of touch input and resulting tactile output on the user's body. This approach has the potential to simplify the use of tactile feedback with arbitrary touch devices and allows the design of novel tactile stimuli by stimulus combination. However, a formal comparison of direct and remote tactile feedback during touch input is still missing. Therefore, we conducted three consecutive laboratory studies. First, we compared the effects of both direct and remote tactile stimuli on the user's performance during touchscreen interactions. No difference was found in the positive effects of both types of feedback. Second, we evaluated the impact of the remote tactile actuator's position on the user's body. For remote tactile stimuli, we found improved accuracy and interaction speed, regardless of the body location. Third we analyzed remote tactile feedback under additional cognitive load. The results support the positive effects of tactile feedback on user performance and subjective evaluation. These findings encourage us to further exploit the potential of remote tactile feedback to simplify and expand the multimodal interaction with arbitrary touch interfaces.

Keywords: remote tactile feedback, interactive surfaces, touch input.

1 Introduction

Using the direct touch of our fingertips or hands to activate or manipulate digital information is currently becoming a widespread interaction paradigm. Touchscreen interfaces are ubiquitously used from mobile devices or medical systems to vending machines or in-vehicle interfaces. They are cheap, flexible and easy to use, and will continue to form the de facto standard for interaction with multi-functional systems [5]. Touch interfaces heavily rely on visual feedback to indicate the results of a user's input, but nevertheless are also used in dynamic multi-tasking and heavy-visual-load scenarios such as driving a vehicle. Their interactive screens only present a flat surface to the interacting fingertips. Non-visual feedback up to now is mostly restricted to an audible beep or an ambiguous tactile buzz, but researchers and designers have started to think about more meaningful tactile feedback for direct touch interactions. Active tactile stimuli can inform the user about the position, form and function of

P. Isokoski and J. Springare (Eds.): EuroHaptics 2012, Part I, LNCS 7282, pp. 301–313, 2012.
© Springer-Verlag Berlin Heidelberg 2012

visual elements and confirm input actions. This additional feedback channel has shown to reduce the errors made, to improve interaction speed and to boost subjective appraisal of direct touch interactions as a whole [3,15,23]. Consequently, the use of the full subtlety and potential of our sense of touch has become a main interest of researchers and engineers of touch interfaces. A recent and promising approach to providing rich tactile feedback is the spatial separation of tactile and visual displays, i.e. remote tactile feedback. To communicate supplemental stimuli, actuators are integrated into the user's direct environment or wearable interfaces. In comparison to direct tactile stimuli, the use of this remote tactile feedback has some innate advantages: It can be used to provide haptic stimuli for touch interactions, which do not depend on the size, form or material of the interactive surface or to create novel haptic stimuli by applying different actuators on several locations on the user's skin. However, to this day, no formal comparison of the effects of directly and remotely applied tactile feedback for touch interactions has been provided. In this paper, we present the results of three consecutive user studies we conducted for the following reasons:

1. To compare the effects of solely visual, visual + direct tactile and visual + remote tactile feedback on objective measures (accuracy, total task time) and subjective measures (naturalness, ranking of modalities) of touch-based text input.
2. To analyze the effects of remote tactile feedback and the body location in which it is applied on task performance during drag-and-drop interaction on a tabletop.
3. To assess the effects of remote tactile feedback on task completion time during multi-touch input under increased cognitive load.

2   Tactile Feedback on Interactive Surfaces

Adding tactile feedback to interactive surfaces such as the touchscreen of a mobile phone or an interactive tabletop display has been a topic for researchers and engineers for over a decade [8, 19]. Tactile information can be conveyed on the form, surface structure, malleability, state, meaning, function or distance of a depicted interactive virtual element. Studies in multimodal interaction already show the importance of non-visual feedback in dynamic scenarios entailing noise, movement, distraction, attention-shifts or cognitive load such as entering text on a mobile device or handling in-vehicle infotainment systems [4,14]. Primarily tactile feedback is investigated as a feedback mechanism on touch surfaces and it results in significantly increased speed and accuracy of input [3,8]. It can even bring the performance of touchscreen keyboards close to the level of real, physical keyboards [11]. In addition, haptic stimuli during touch input reduce visual and cognitive load [20,15] and prevent occlusion [7]. In addition to such objective measures, palpable stimuli on otherwise flat touch surfaces also increase the user's subjective appraisal of the interaction. Studies show an enhanced feeling of realism and naturalness [3,9,23]. Technically, the actuators for the presentation of direct tactile stimuli in touch surfaces can be electrical, electromechanical, pin-based, piezo-driven and even pneumatic [6]. Accordingly, prototypical implementations mostly fall into one of three categories: In the first, the screen or the mobile device is actuated as a whole [25,2]. However, such touch devices can only provide a single touch stimulus, which is the same for every finger. The second category uses tangible user interfaces (TUI) to give tactile feedback. These devices offer great flexibility in the design of stimuli and interactions [13,17]. However, the interaction itself loses the beneficial characteristics of direct manipulation, suffers from visual occlusion and lacks scalability. The third approach is the segmentation of the display into multiple individually movable elements, i.e., 'tactile pixels' [10, 23]. With individual electromechanical actuators for every tactile pixel, this approach is still hardly scalable und lacks visual and tactile resolution.

3   Remote Tactile Feedback

Remote tactile feedback (RTF) can potentially eliminate some of the drawbacks mentioned above. The more general approach of tactile sensory relocation has been extensively analyzed and reproduced in the fields of accessibility and sensory substitution [1, 12]. Few researchers have incorporated relocated tactile stimuli on touch surfaces before, but could show promising effects on usability and performance: McAdam et al. [18] used the vibration motor in users' mobile phones to provide haptic feedback during interactions with a touchscreen. Results showed significantly increased text entry rates when remote tactile feedback was given. Richter et al. [24] proposed the use of remote tactile stimuli during touch interactions as a way to simplify the design and implementation of actuator technology. Tactile stimuli were applied to the non-dominant forearm. The authors characterize their work as a way to easily create synchronized remote tactile stimuli during touch interactions. Other recent papers show the potential of remote tactile feedback to create novel forms of haptic stimuli by combining different types of actuators (e.g., vibrotactile and linear moving) on the body. A haptic armrest was used to communicate tactile surface characteristics and forms of buttons on touch surfaces [26]. Users indicated the high hedonic and pragmatic quality of the created stimuli. We can think of the following additional benefits that result from the spatial separation of manual input and tactile output:

 **Multi Haptics**: Individual tactile feedback for each point of contact with the interactive surface can be given. In contrast to conventional approaches (see section 2), this enables simultaneous, but different haptic feedback for each hand touching the surface.

 **Arbitrary Surfaces:** The interactive surface is not restricted to a specific form and size. For example, organic interactive surfaces (i.e., made from clay [21]) could be enriched with additional tactile stimuli.

 **Stimuli before and after interaction:** With tactile actuators being in permanent contact with the user's skin, tactile cues describing proximity or acknowledgement of a touch interaction can be given before or after the finger actually touches the screen.

Using remote tactile stimuli as a form of multimodal feedback raises the following question: *Can remote tactile stimuli on touch surfaces be compared to direct tactile feedback in terms of benefits in objective and subjective measures?* If remote tactile feedback has the potential to reduce error rates, increase interaction speed or user satisfaction with touchscreen input to a degree comparable to direct tactile stimuli, this would further support our approach of exploiting the inherent potentials of RTF.

4   Evaluations

We conducted three user studies comparing the effects of both direct and remote tactile stimuli on the user's performance during touchscreen interactions, evaluating the impact of the remote tactile actuator's position on the user's body and analyzing remote tactile feedback under additional cognitive load.

4.1   Effects of Direct vs. Remote Tactile Feedback

So far, we assumed that the communication of direct and remote feedback has comparably positive effects on the touch interaction as a whole in terms of improving accuracy, increasing interaction speed and the positive effect on subjective ratings. Therefore, we had the following hypotheses:

H1.1: Accuracy is higher with tactile feedback during touch input tasks than without tactile feedback.
H1.2: Total task time is lower with direct and remote feedback than without tactile feedback.
H1.3: Less keys are missed when typing with direct and remote feedback than without tactile feedback.
H1.4: Users will prefer interactions with tactile feedback to interactions without tactile feedback.

Study Design. The study had a within subject/repeated measures design. A text input task with a given phrase set had to be performed. There were three feedback conditions: no feedback, direct tactile feedback and remote tactile feedback. They were presented in counterbalanced order to avoid unwanted training effects. The dependent variables were accuracy, total task time and key misses.

Fig. 1. (a) screenshot of text input task; (b) setup for *no feedback* and *direct tactile feedback*; (c) setup for *remote tactile feedback*

Apparatus. To analytically compare directly and remotely applied tactile stimuli we used a rather artificial, purpose-built device. Our goal was to make psychophysical conditions such as the distribution of mechanoreceptors and perception of tactile stimuli as comparable as possible for both feedback types. Thus, we chose the fingertips of the index fingers on the dominant and non-dominant hand as the stimulus area.

The prototype is depicted in figures 1 and 2. The horizontal tactile touchscreen consists of two transparent capacitive touch-sensing panels[1], which are mounted to modified voice coil actuators[2]. Speakers are a common method to communicate tactile stimuli [6]. Both touchscreens are freely movable in vertical direction. A standard 15 inch screen is installed under one of the panels. When the user touches the panel

---

[1] 3M MicroTouch SCT3250EX 15.68" Surface Capacitive USB Touch System.
[2] Dynavox DY-166-9A, 4 Ohm, 80W, resonance frequency 50 Hz.

above the display, direct tactile feedback is given by shaking or moving the panel vertically. In contrast, for remote tactile feedback, the user's index finger of the non-dominant hand is rested in the center of the additional touch panel. When the user touches the other touch panel over the display with his dominant hand, the remote touch panel is shaken or moved vertically. This setup is exclusively used for the comparison of both feedback types.

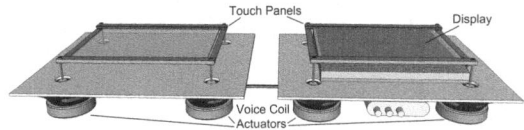

Fig. 2. The purpose-built tactile touch screen device used in the first evaluation

Experimental Setup and Task. A screenshot of the display during the evaluation is shown in figure 1. The participants were sitting and wore headphones with music to block external noise. The task was to enter the depicted text phrase using the on-screen interface. In order to obtain generalizable results, we used 10 phrases out of the established 500 phrases published by MacKenzie [16]. The input method was designed as a targeting task. We chose a text-input based on drag-and-drop, incorporating the take-off metaphor by Potter et al. [22]. Dragging-and-dropping is a common task on interactive surfaces; an extensive amount of information can be communicated during the long contact with the surface. For entering a character, the user had to put the finger on a start area on the bottom of the graphical user interface. The start area turned red to indicate that it was activated. Then the user could drag his or her finger over the screen on to a key of the QWERTY keyboard. Each key gave a short feedback impulse, when it was entered. The sine wave for regular keys has a frequency of 170 Hz and is enabled for 40 milliseconds. According to physical keyboards, the keys F and J are marked by sine waves with a frequency of 70 Hz enabled for 80 milliseconds. A letter was typed by lifting off the finger from the respective key. The resulting letter was shown on the screen. This very artificial text-entry method is comparable to methods such as take-off on standard touchscreens [22]. To ensure that mistakes had no effect on the task time, there was no error correction. Participants were told that the correct letter should be entered at the correct position. When they made a mistake they should go on with the next letter. Furthermore, they were told to enter the phrases as accurate and as fast as possible.

Participants. Twelve participants (five female) with an average age of 22 years took part in the study. All participants stated that they had experiences with touchscreens before. All were right-handed.

Procedure. Each participant was introduced to the prototype and typed three phrases in a demo application to get used to the typing method and the three different stimuli. Subsequently participants received the task and the measurement started. The order of the three feedback types was counterbalanced. For each modality, the order of the 10 phrases was randomized. After each feedback modality, subjects were asked to fill out a questionnaire containing a semantic differential with five-point Likert scales to evaluate the resemblance to reality, signal communication and usability for direct and remote tactile stimuli, as well as personal preference.

Results. From the study described above, we obtained the following results:

Accuracy: We define the accuracy rate as the number of characters correctly entered divided by the overall length of the phrases entered. The median of the accuracy rate is 0.963 in the *no feedback* condition, 0.930 in the *local feedback* condition and 0.955 in the *remote feedback* condition. An analysis of variance (ANOVA) showed no significant differences between the three modalities ($F(2, 33) = 0.87$, $p = 0.43$). Accordingly, H1.1 cannot be affirmed.

Total Task Time: Similarly to the accuracy an ANOVA shows that the measured times for each feedback condition have no significant differences ($F(2, 33)= 0.44$, $p = 0.65$). The median for the *no feedback* condition is 310.5 seconds, for the *direct feedback* condition 354.0 seconds and for the *remote feedback* condition 316.1 seconds. Accordingly, H1.2 cannot be affirmed either.

Missed Keys: An ANOVA shows no significant differences between all three modalities ($F(2, 33)= 0.29$, $p = 0.76$). The median is 14.5 key misses without feedback, 12 key misses with local feedback and 8.5 key misses with remote feedback. Although H1.3 cannot be affirmed, the median values are clearly in favor of remote feedback.

Subjective user ratings: The ratings for realism, signal design and usability on the five-point Likert scale were highly positive, especially for understandability and simplicity (see figure 3). Again, we found no larger difference between the values for direct and remote tactile stimuli. Seven out of twelve subjects voted for local tactile feedback as the most pleasant type of interaction in the experiment, four voted for remote tactile feedback and one for the *no feedback* condition. Subjective ratings are in favor of tactile feedback. In summary, H1.4 can be affirmed.

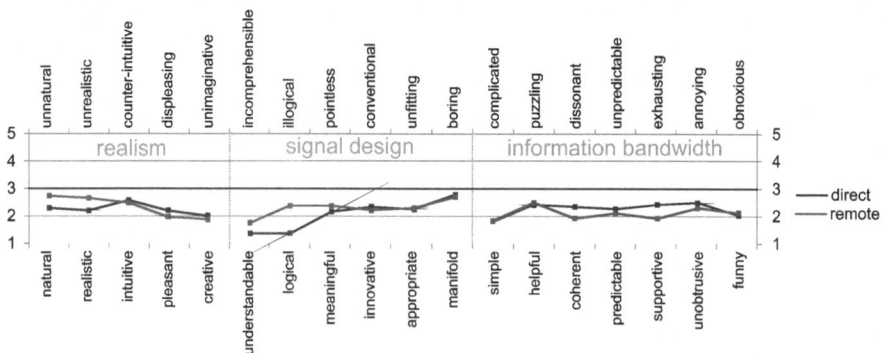

Fig. 3. Subjective evaluation of direct and remote tactile stimuli. Discrete values connected for readability.

4.2   Effects of RTF on Performance during Standard Touch Interaction

In the second study, we evaluated the influence of remote tactile feedback on interactions performed on a tabletop. One goal was to verify the assumption that RTF has the potential to improve interaction on touch surfaces in terms of speed and errors [18]. We also had the goal of comparing the effects of dominant and non-dominant RTF:

H2.1: Remote tactile feedback in touch input tasks will increase interaction speed and reduce error rate.

H2.2: When moving remote tactile feedback to the non-dominant hand, the benefits will be reduced.

Study Design. The independent variable of the study was the feedback given to participants during touch interactions. In addition to the omnipresent visual feedback, RTF was applied to the dominant or the non-dominant wrist. Using a within-subject design, each participant performed the interaction tasks with all three types of feedback (no RTF, non-dominant RTF and dominant RTF). The monitored dependent variables were interaction speed as well as the number of errors made.

Apparatus. Tactile feedback was applied using vibrating pancake motors, which were attached to the participants' arms by wristbands (see figure 4). This body position was chosen according to McAdam and Brewster's results on testing distal tactile feedback on different body locations, identifying the wrist as most promising for improving interaction speed [18].

Fig. 4. Remote tactile feedback provided by wrist-worn actuators during tabletop interactions

Interaction Task. The drag-and-drop gesture was chosen due to its constant use in touch interfaces. A blue square with an edge length of 50 pixels had to be dragged into an only slightly larger red target area (54 pixels wide). Time measurement started as soon as the blue square was touched and stopped with the fulfillment of the task. Dropping the square outside of the target area caused the error counter to be increased by one. Participants received a tactile cue whenever the square was dragged completely into the target area and thus ready to be dropped. The same information was given visually, the square turned green when it is ready to be dropped. In consequence, the tactile information was redundant. We chose this setting in order to identify the benefits of additional tactile feedback.

Procedure. After receiving information about the functionality of the tabletop and the vibrotactile wristbands, participants had the chance to practice the drag-and-drop task without, with non-dominant and with dominant remote tactile feedback. The actual study consisted of 90 drag-and-drop tasks, 30 for every type of feedback. To avoid order effects, the feedback types were counterbalanced. To conclude the study, participants were interviewed for impressions and opinions about remote tactile feedback.

Participants. 18 Participants took part in the study, six of them female, non of them color-blind. The average age was 28 ranging from 22 to 38. All of them had used touch interfaces before. Three participants were left-handed.

Results. After removing outliers, at least 460 data items remained for each type of feedback. The arithmetic mean of the time was calculated for the drag-and-drop tasks, resulting in 2.590 milliseconds for trials without tactile feedback, 2.333 milliseconds for non-dominant tactile feedback and 2.345 milliseconds for dominant feedback. A one-way repeated measures ANOVA showed that the independent feedback variable had a significant influence on the time participants needed for the drag-and-drop tasks, $F(1.25, 21.19) = 5.49$, $p < .05^3$. Bonferroni post-hoc tests ($p < .05$) could not confirm a significant difference between any mean times ($p < .09$ for none/non-dominant tactile feedback and $p < .08$ for none/dominant tactile feedback). However, these low p-values show a tendency towards faster interactions with remote tactile feedback. Participants were approximately 10% faster with this kind of feedback (see figure 5). Concerning the number of errors made, an ANOVA showed that remote tactile feedback had no significant influence. In summary, H2.1 must be rejected for error rates but can be affirmed concerning interactions speed. No difference in task completion time or error rates was discovered when non-dominant or dominant feedback was applied. This result can be supported by qualitative results. 12 participants stated that the body position in which tactile feedback was received had no influence on task performance. Thus, H2.2 cannot be confirmed. The advantages of remote tactile feedback are relevant even if it is moved further away from the actual touch input. This could allow a more flexible positioning of tactile actuators on the body.

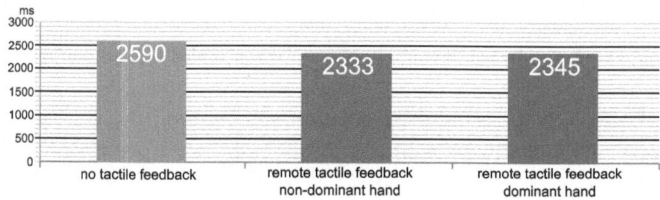

Fig. 5. Mean times in *ms* for the single-touch drag-and-drop task. Drag-and-drop is approx. 10% faster when remote tactile feedback is given, regardless of the body side of application.

4.3   Effects of RTF during Multi Touch Interaction under Cognitive Load

For the third study using the same tabletop, multi-touch gestures for scaling and rotating virtual objects were added to the interaction task. Consistently to the second study, this was tested for both non-dominant and dominant feedback on users' wrists.

H3.1:   Remote tactile feedback during multi-touch input increases interaction speed.

H3.2:   When moving remote tactile feedback to the non-dominant hand, the benefits will be kept.

---

[3] Mauchly's test revealed a violation of sphericity, $\chi2(2) = 14.86$, $p < .05$. Degrees of freedom were corrected using Greenhouse-Geisser ($\varepsilon = 0.62$).

Furthermore, an additional auditory task was assigned to participants, which had to be completed simultaneously to the touch interactions. This was due to the research of Leung et al. [15] who stated that the advantages of tactile feedback are enforced under high cognitive load. The second goal was to find out if this result can be also verified for remote tactile feedback.

H3.3:   Additional cognitive load will increase the benefits of remote tactile feedback.

Study Design. The first independent variable was the feedback provided during the touch interactions. Conditions were visual feedback only, additional remote tactile feedback on the non-dominant wrist and feedback on the dominant wrist. The other independent variable was the presence or absence of the additional task creating cognitive load. Using a within-subject design, the interaction tasks were divided into six blocks: two for each type of feedback, once without and once with additional cognitive load. The sequence of the blocks was counterbalanced. The dependent variable was the task completion time for the touch interactions.

Interaction Task. The task was a combination of three single-touch and multi-touch gestures for manipulating a square on the tabletop display. First, the size of a square with an edge length of 50 pixels had to be doubled using a scaling gesture. Second, the resulting square had to be rotated by 180 degrees clockwise using the corresponding rotating gesture. For both gestures, participants were instructed to use the index fingers of both hands. Finally, the square had to be moved into a slightly larger target area using drag-and-drop. Task completion time was measured from the first contact to the square until the square was dropped into the target area. Additionally, we calculated the times needed to perform every single gesture. To signal if the correct size, orientation or position was reached, the square's color changed from red to green. Remote tactile feedback was applied according to the visual cues, so that no additional information was provided. The auditory task added to create additional cognitive load was designed according to Leung et al. [15]. A random sequence of letters from O to V was read out loud with a speed of 100 letters per minute. Approximately five times per minute, the same letter appeared three times in a row. Participants were instructed to signal this event by speaking out the word "now". This auditory task had to be completed simultaneously to the interaction task. To ensure an acceptable task fulfillment, it was tracked how many occurrences of repeated letters were noticed.

Procedure. After completing a demographic questionnaire, participants were familiarized with the functionality of the tabletop and the vibrotactile wristbands. Next, they had time to practice both interaction and auditory task. During the main part of the study, the six blocks with different types of feedback and the additional cognitive load were completed. In each block, the interaction task was repeated ten times to obtain an adequate number of task completion times. A qualitative interview about experiences made with the remote tactile feedback concluded the study.

Participants. The study was completed by 18 participants (five female) with an average age of 28 ranging from 26 to 32. None were left-handed or color-blind.

Results. After removing outliers, we calculated arithmetic means to gain an average task completion time for each of the six conditions (see figure 6). The first result shows the additional auditory task causing participants to need significantly more time for the touch interactions. This was the case for all gestures, independent from the type of feedback. Thus, the auditory task caused the intended additional cognitive load. This was also supported by the qualitative feedback. Considering the time needed for completing the interaction task without additional cognitive load, an ANOVA was performed for every gesture. The results show a significant impact of the type of feedback on task completion time of the scaling gesture, $F(2, 34) = 5.28$, $p < .05$. Bonferroni post-hoc tests identified remote tactile feedback on the dominant wrist to decrease the duration of the task significantly, $p < .05$. Compared to the condition without tactile feedback, a time advantage of 18% was reached. Although non-dominant feedback reduced task completion time by 10%, no significance was found. Thus, hypothesis 3.1 can only partially be confirmed. For drag-and-drop and rotation, no significant influences of the feedback were found. This can be explained by the additional visual feedback: when the square was ready to be dropped, it changed color from red to green, which was not the case in the earlier study. This indicates that the provided visual feedback was sufficient in this case. This was conformed by 16 participants, who stated that they did not need additional tactile feedback here.

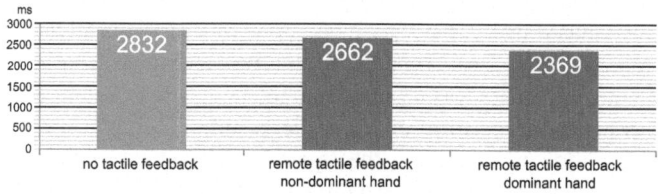

Fig. 6. Mean times in *ms* for multi-touch scaling with three feedback modes. Scaling is significantly faster with dominant remote tactile feedback.

For none of the gestures a significant difference between task completions times of the two conditions with remote tactile feedback was discovered. This underlines the results from the second study. In other words: remote tactile feedback can be applied on both the dominant and the non-dominant arm without changing the influence on task completion times when performing single- and multi-touch gestures on a tabletop. This is also true for conditions with increased cognitive load. Accordingly, participants agreed that the body side did not matter for the application of the tactile cues. Thus, hypothesis 3.2 can be confirmed. After adding the auditory task and thus increasing the cognitive load, ANOVAs did not reveal any significant influence of the type of feedback on the task completion times. This result is true for all three gestures. The observation could be explained by the domination of the visual and auditory communication channels, which causes the tactile channel to lose importance. This assumption was supported by two participants after the study, stating that they were not sure if there even was any tactile feedback during the conditions with the auditory task. Thus, hypothesis 3.3 has to be rejected.

5   Limitations and Discussion

The limited sophistication of our tactile devices and the resulting loss of tactile bandwidth might be the main reason for the statistical non-significance of some of our results. Both actuators used in the evaluations were purpose-built for these occasions. We used inexpensive off-the shelf materials such as voice-coil actuators and vibration motors. Thus, unwanted noise, latency and unnatural posture of subjects might have created unwanted effects. In the evaluations, we provided simple tactile stimuli as location information (first study) and semantic affirmation of action (second and third study). The stimuli were given redundantly to visual feedback; no extra information was encodeded haptically.

6   Conclusion and Future Work

We compared the effects of direct and remote tactile feedback, analyzed the implications of the location of actuators on the body, and assessed multi touch feedback under increased cognitive load. The results are in favor of remote tactile feedback: users could decrease the number of missed keys during text input on touchscreens. When given in addition to redundant visual cues, remote tactile feedback on the dominant arm significantly decreased the task completion time for scaling gestures on a tabletop. When applied to the non-dominant arm, the total task time could be decreased by 10%. When asked about their subjective opinion, users are in favor of tactile feedback in general. Over 90% of the participants preferred tactile feedback to no tactile feedback. For remote tactile stimuli, over 30% even preferred it to direct tactile stimuli. We also found that the location of the feedback on the user's body had no effect on subjective appraisal. For future implementations of remote tactile feedback, we are working on incorporating other locations of the human body such as the back (e.g., by the seat of a car) or the forearm (e.g., by wearable devices or the frame of the touch surface). Thus, the diverse density of mechanoreceptors in the skin has to be taken into account. With remote tactile feedback, a dedicated tactile display for every single pixel of the interactive surface is not necessary. Thus, we will incorporate tactile feedback into non-flat or large touch interfaces. Finally, we are working on the creation of novel tactile stimuli by combining thermal and vibrotactile remote actuators on the skin. In summary, the results of our evaluations back our assumption of the positive effects of remote tactile feedback on the performance of the user. Advantages of additional non-visual stimuli are observable regardless of the location of application on the body. The unique potential of RTF will extend and enrich the design and use of direct touch interfaces as a whole.

References

1. Bach-y-Rita, P., Kercel, S.: Sensory substitution and the human-machine interface. Trends in Cognitive Sciences 7(12), 541–546 (2003)
2. Bau, O., Poupyrev, I., Israr, A., Harrison, C.: TeslaTouch. In: Proc. UIST 2010, pp. 283–292 (2010)

3. Brewster, S., Chohan, F., Brown, L.: Tactile feedback for mobile interactions. In: Proc. CHI 2007, pp. 159–162 (2007)

4. Burke, J., et al.: Comparing the effects of visual-auditory and visual-tactile feedback on user performance: a meta-analysis. In: Proc. ICMI 2006 (2006)

5. Buxton, W.: Multi-Touch Systems that I Have Known and Loved. Microsoft Research (2007)

6. Chouvardas, V.G., Miliou, A.N., Hatalis, M.K.: Tactile displays: Overview and recent advances. Displays 29, 185–194 (2008)

7. Forlines, C., Balakrishnan, R.: Evaluating tactile feedback and direct vs. indirect stylus input in pointing and crossing selection tasks. In: Proc. CHI 2008, pp. 1563–1572 (2008)

8. Fukumoto, M., Sugimura, T.: Active click: tactile feedback for touch panels. In: CHI 2001 EA, pp. 121–122 (2001)

9. Hale, K.S., Stanney, K.M.: Deriving Haptic Design Guidelines from Human Physiological, Psychophysical, and Neurological Foundations. IEEE Computer Graphics and Applications 24, 33–39 (2004)

10. Harrison, C., Hudson, S.E.: Providing dynamically changeable physical buttons on a visual display. In: Proc. CHI 2009, pp. 299–308 (2009)

11. Hoggan, E., Brewster, S.A., Johnston, J.: Investigating the Effectiveness of Tactile Feedback for Mobile Touchscreens. In: Proc. CHI 2008, pp. 1573–1582 (2008)

12. Kaczmarek, K.A., Webster, J.G., Bach-y-Rita, P., Tompkins, W.J.: Electrotactile and vibrotactile displays for sensory substitution systems. IEEE Transactions on Bio-Medical Engineering 38(1), 1–16 (1991)

13. Lee, J., Dietz, P., Leigh, D., Yerazunis, W.: Haptic pen: a tactile feedback stylus for touch screens. In: Proc. UIST 2004, pp. 291–294 (2004)

14. Lee, J.-H., Poliakoff, E., Spence, C.: The Effect of Multimodal Feedback Presented via a Touch Screen on the Performance of Older Adults. In: Altinsoy, M.E., Jekosch, U., Brewster, S. (eds.) HAID 2009. LNCS, vol. 5763, pp. 128–135. Springer, Heidelberg (2009)

15. Leung, R., MacLean, K., Bertelsen, M., Saubhasik, M.: Evaluation of Haptically Augmented Touchscreen GUI Elements under Cognitive Load. In: Proc. ICMI 2007, pp. 374–381 (2007)

16. MacKenzie, I.S., Soukoreff, R.W.: Phrase sets for evaluating text entry techniques. In: CHI 2003 EA, p. 754 (2003)

17. Marquardt, N., Nacenta, M., Young, J., Carpendale, S., Greenberg, S., Sharlin, E.: The Haptic Tabletop Puck: Tactile Feedback for Interactive Tabletops. In: Proc. ITS 2009, pp. 93–100 (2009)

18. McAdam, C., Brewster, S.: Distal tactile feedback for text entry on tabletop computers. In: Proc. BCS-HCI 2009, pp. 504–511 (2009)

19. Nashel, A., Razzaque, S.: Tactile virtual buttons for mobile devices. In: CHI 2003 EA, pp. 854–855 (2003)

20. Oakley, I., McGee, M.R., Brewster, S., Gray, S.: Putting the Feel in Look and Feel. In: Proc. CHI 2000, pp. 415–422 (2000)

21. Piper, B., Ratti, C., Ishii, H.: Illuminating clay: a 3-D tangible interface for landscape analysis. In: Porc. CHI 2002, pp. 355–362 (2002)

22. Potter, R.L., Weldon, L.J., Shneiderman, B.: Improving the accuracy of touchscreens: an experimental evaluation of three strategies. Sparks of Innovation in Human-Computer Interaction, 161 (1993)

23. Poupyrev, I., Maruyama, S., Rekimoto, J.: Ambient touch: designing tactile interfaces for handheld devices. In: Proc. UIST 2002, pp. 51–60 (2002)

24. Richter, H., Blaha, B., Wiethoff, A., Baur, D., Butz, A.: Tactile Feedback without a Big Fuss. In: Proc. UbiComp 2011, pp. 85–89 (2011)
25. Richter, H., Ecker, R., Deisler, C., Butz, A.: HapTouch and the 2 + 1 State Model. In: Proc. Automotive UI 2010, pp. 72–79 (2010)
26. Richter, H., Löhmann, S., Wiethoff, A.: HapticArmrest: Remote Tactile Feedback on Touch Surfaces Using Combined Actuators. In: Keyson, D.V., Maher, M.L., Streitz, N., Cheok, A., Augusto, J.C., Wichert, R., Englebienne, G., Aghajan, H., Kröse, B.J.A. (eds.) AmI 2011. LNCS, vol. 7040, pp. 1–10. Springer, Heidelberg (2011)

Sensorimotor Feedback for Interactive Realism: Evaluation of a Haptic Driving Paradigm for a Forklift Simulator

Pierre Martin, Nicolas Férey, Céline Clavel, Françoise Darses, and Patrick Bourdot

VENISE & CPU teams, CNRS/LIMSI, B.P. 133, 91403 Orsay cedex, France
firstname.name@limsi.fr

Abstract. The Virtual Environments are more and more applied in industry to learn the manipulation of some complex and/or dangerous equipments. In this context, we are developing a virtual simulator dedicated for learning the manipulation of a specific forklift. In real use, this kind of forklift requires a physical involvement of people to manipulate it. This paper proposes thus an innovative haptic paradigm for the virtual driving of such a forklift. Within the virtual simulation, it aims to provide a sensorimotor stimulation which is close to the one that users have on the real forklift. Taking in account the physical behaviours of the real forklift (inertia, damping, turning radius), we haptically simulate the mechanical hinge constraints of the forklift handle to control the forklift direction. In addition, a specific haptic push/pull technique was designed to control the forklift velocity. Our hypothesis is that our paradigm is more realistic compared to classical interactive techniques, such as joystick without force feedback. To evaluate this realism, we conducted an ergonomic study of a driving task. The same driving task has been performed in three conditions: in large room with the real forklift (R), and in a Virtual Environment with a joystick (J), and with our haptic paradigm (H). We conclude that our hypothesis is verified when revisiting different acceptations of the realism concept in Virtual Environment, such as performance transfers, but also the behaviour and psychological processes transfers.

Keywords: Virtual Reality, Haptic, Driving simulation, Realism.

1 Introduction

Virtual Reality (VR) and immersive technologies provide solutions for simulating realistic situations in a controlled and secure context. That is especially required for learning applied to industry field, for instance when novice users are starting to learn the use of complex and dangerous equipments. In this context, we designed an interactive paradigm for the driving of a forklift in a virtual simulator. Our hypothesis is that more the interactive paradigm is realistic, more the transfer skills between virtual and real conditions is efficient [1]. However, the dimensions of the realism are many. The most studied ones are relating to the visual immersion, to the affordance of the virtual objects, and to the involvement of the users in the immersive task [2].

P. Isokoski and J. Springare (Eds.): EuroHaptics 2012, Part I, LNCS 7282, pp. 314–325, 2012.

In this paper we mainly question the role of sensorimotor stimulations for the realism of the interaction. This channel is crucial in tasks that require physical involvement of the subject. That is exactly the case for driving our targeted forklift, because it is a non-motorised vehicle. Indeed, the forklift is controlled by manipulation of its handle to push/pull it, and control its direction (Figure 1). Actually, haptic technologies are supposed to provide solutions to stimulate the sensorimotor channels of the users. However, haptic interactions have some limitations in terms of workspace, precision, range of the movements. We explain how we overcome these drawbacks to design haptic paradigm taking in account the mechanical specificities of the real forklift, to provide a realistic driving control on the forklift in virtual context, in terms of physical engagement of the user and perception of the handle.

Fig. 1. The real and virtual non-motorised forklift

To evaluate the realism of our paradigm, we compare it to classical navigation interactive technique using joystick. This evaluation focused on the current acceptations of the realism concept [2]. A same driving task was proposed to subjects in three situations: in the real world with the targeted lift truck (R); in the virtual simulation of the scene in an immersive context using head tracking stereoscopy with a joystick interface to control the forklift (J); and in the same visual stereoscopic immersive environment but using the haptic driving paradigm we propose (H). Significant results are presented from analyses of subjects' performances, but also data on their behaviours, and their answers to subjective questionnaires. We discuss these results to put them in perspective, and then conclude.

2   Related Work

Haptic devices have limited workspace, relative precision, and no complete range for rotational movements. In order to overcome these drawbacks, a basic interactive paradigm is the *clutching/declutching* technique which is however time consuming and which distracts the users' attention from the focus of their task. To avoid these drawbacks, several approaches use haptic devices to control navigation during interaction, using the relative position of an object of interest, the position of the haptic device, and the user position [4], to compute a more suitable point of view for the user to interact with the object. Other solutions deal with the hardware limitations of the interaction device, such as the *Bubble technique* [5] or the *Haptic Hybrid Rotation* method [6] to navigate with haptic device. These techniques provide a rate-control for object manipulation, based on an isomorphic mapping when the device is far from its workspace boundaries, and on a non-isomorphic mapping near the boundaries, also used by [7] or [8]. On the contrary to these methods our haptic paradigm is a non-isomorphic rate-control based on the comparisons of a tracked referential with a neutral one, such as the navigation control based on 6DoF tracking proposed by [3]. This concept is augmented by an elastic force to retrieve this neutral referential, since it has been proved by [9] that such function is more suitable for rate control paradigms.

Some works also focus specifically on VR applications dedicated to forklifts. Safety is one of the application fields. [10] studied the impacts on the forklift occurring on drive-in racking structures. They proposed a general method for calculating forces generated under forklift truck impact. In [11] the prevention of forklift capsize is considered. Based on the fact that drivers' training and proper safety procedures are not sufficient enough to reduce accidents, they developed an intelligent control system, embedded on the forklift, which analyzes onboard sensors data and proposes some corrections. There are also some studies on real-time simulator with a forklift truck model, and on evaluation of numerical integrators to propose a simulation as realistic as possible [12]. Finally, [13] and [14] described a prototype of a full-immersive simulation of forklift truck operations for safety training.

Within this context, we propose below a haptic paradigm to provide sensorimotor stimulations which increase the realism of the interaction for tasks requiring physical involvement of the subjects.

3   System Description

This section describes first the physical behaviours of the virtual forklift. Then we detail the main features of the haptic driving paradigm itself.

3.1   Forklift Physical Behaviours

The inertia and damping of the vehicle in the real condition was obtained empirically, by velocities and deceleration measurements between two points. Actually, three forces are involved to move this kind of forklift: the weight of the vehicle (400 kg), its friction on the floor, and the force applied by the user to move it. We measured an empirically friction of wheels (stopping distance 4.25 m, initial velocity 1 m/s).

The realism of the interaction also supposes that the user has to feel the force on the forklift through the handle to make the virtual forklift move. In the one hand, this force is captured by the haptic arm. In the other hand, this is linked to the amplitude of the pressure exerted on the joystick. In both cases we need to express the new velocity of the vehicle.

Equation 1 is the expression of the new velocity according to F, while Δt is the time variation, t the current time, $V(t)$ the velocity at t, β the damping factor, m the object weight, and α a control constant.

$$V(t + \Delta t) = \left(V(t) e^{\frac{-\beta}{m} \Delta t} + \alpha \frac{F \Delta t}{m} \right) \tag{1}$$

All parameters of Equation 1 are known, except β: this variable takes into account all the forces applied to the object, including the friction forces. In our physical simulation of the forklift we assume that β is constant, its variation being negligible. To determine its value, we ran the simulation with the forklift moving at a speed of 1 m/s, and adapted the value of that constant to make this virtual vehicle stopping at the right distance (the one measured on the real forklift) from its starting point. The only parameter that one must define is α, a constant factor that we could define arbitrarily. That is actually useful, because we cannot apply on the haptic device the same force as the one used to move the real forklift.

In addition, we determined the curvature radius of the forklift, which only depends on the handle angle with the vehicle motion axis. We empirically find the linear mapping between the handle's angle and the inverse of the curvature radius. By this way we were able to compute the next position and orientation of the virtual forklift.

3.2   Haptic Driving Paradigm

Apart the rendering of the inertia and the damping of the forklift as external force feedbacks, our haptic driving paradigm simulates the mechanical behaviours of the real forklift's handle. Moreover, it provides an elegant solution for having an intuitive perception of the push and pull actions, while the base of the haptic device cannot move like the real forklift.

To simulate the real forklift's handle, we designed a virtual mechanism based on a rigid body with hinge constraints (Figure 2 - right). A first hinge (blue sphere, red arrow) controls the orientation of the forklift's wheels. A second hinge (green arrow), allows user actions to raise or lower the forklift's handle.

The haptic driving paradigm includes a push/pull haptic technique to control the velocity of the virtual forklift. This solution aims to provide a sensorimotor stimulation to the users during the driving task, by introducing a non-isomorphic rate-control of a virtual vehicle. However, on the contrary of the real truck, the base of the haptic device is fixed. Accordingly, we integrated a system of relative motion.

The blue sphere (Figure 2 - right) is constraint along the green bar located under the blue sphere. This green bar follows the orientation's wheels that are control by the virtual handle. More the blue sphere goes forward, more the force applied by the user on the vehicle will be large and vice versa. Thus, the velocity of the virtual forklift is computed from the force applied by the user on the haptic devices, except it is constrained by the virtual mechanism. However, it is not possible to provide the same intensity of force feedback through the haptic device (device limitation) than in the real context. That's why we applied a linear scaling on the forces computed by the physical simulation (cf. Equation 1).

A principle for designing a relative motion system is to provide a non-isomorphic rate-control manipulation based on the position and angular distances with a neutral referential [3]. In our case, it was sufficient to consider a neutral zone at the middle of the green bar, where we apply an elastic force which tends the blue sphere to return into this zone. Thus, more the user moves the sphere away from the neutral zone, more the strength to bring it back will be strong. Apart it provides the feeling of pushing or pulling harder, the user is navigating forward or backward with the virtual forklift in the full scene, in spite of moving only close to the device. A dead zone without elastic force around this neutral referential is also used, to avoid force oscillation and vibration, and to increase user comfort.

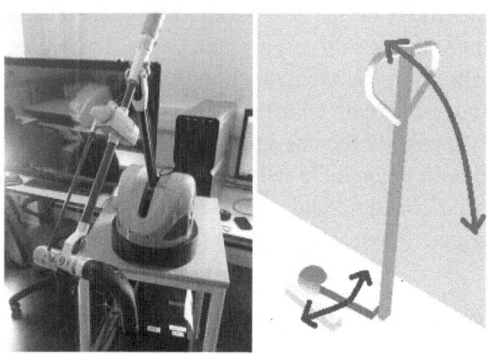

Fig. 2. The 6 DoF haptic arm that we used, a Virtuose 6D 35-45 of Haption (left). The virtual mechanism of the forklift handle (right) implemented with the Interactive Physical Pack (IPP) developed by Haption for the Virtools platform that we used for the visual immersion.

4   Evaluation

We evaluated a task of forklift driving in three experimental conditions: one with the real forklift (R), two others in a virtual context: with our haptic paradigm (H), and with a joystick interaction (J). Actually, a joystick can be considered as the most commonly used interface to perform navigation tasks. However it doesn't provide sensorimotor feedbacks such as haptic devices. Thus, our fundamental questioning is the contribution of the sensorimotor feedbacks to the realism of the interaction. More specially, we hypothesis that our paradigm is better than joystick use for learning the driving task we wanted to address, because it makes possible a physical involvement of the user which is close to real situation. Accordingly, the physical rendering of the forklift collisions with the objects in the virtual scene is not our focus here. We mainly use the haptic device for simulating the motion constraints of the forklift handle, for rendering the inertia of this vehicle (non-motorized forklift), and for providing to the user a relative motion system. These three haptic feedbacks provide a sensorimotor stimulation, which is supposed to allow a physical involvement of the user in the driving task.

4.1   Method

For the real condition (R), the non-motorized forklift was manipulated with its articulated handle in a 13x6 meters room (Figure 1, left). For the virtual context of the H and J conditions, we used a CAVE-like system with 4 screens around the user (left, front, right and floor), with adaptive stereoscopy based on the tracking of the subjects' head to provide the best visual immersion. Realistic 3D models of the real experimental room and the forklift have been designed (Figure 1, right). Moreover, the inertia behaviour on the virtual forklift was equally applied in H and J conditions. In this way, we can only focus on sensorimotor realism of our haptic paradigm, because the same level of visual realism was provided, whatever is the interactive device in the virtual context.

In the virtual conditions (H and J), we easily get the trajectory (Figure 3) of the forklift by logging the tracking events, the position and orientation of the virtual forklift. To get the same kind data in the real condition (R), all the trails were video captured, on which we applied classical image processing techniques based on the colour segmentation of the yellow and green spheres (top of the forklift - Figure 1), and a homographic projection to get a 2D trajectory on the floor. Using this data, we extracted a number of features on the driving tasks trajectories.

Concerning the subjective measures, we submitted to the users a questionnaire to analyze their perception of the task and their feedbacks. After each experimental condition, participants completed a set of rating-scale questions. Users had to report their level of agreement according to an eight-point Likert scale. We designed our questionnaire from different studies of presence [15, 17, 18]. Presence is a multifaceted concept [16], defined as the subjective experience of being in one place or environment, even when one is physically situated in another. We have set up a questionnaire based on four presence factors identified by [15], namely: control, sensory, realism, and distraction factors. We used the same questionnaire in the different conditions, with adaptations and deletions of some claims for the real world version.

18 subjects participated in the experiment: 14 males, 4 females, ranging between 24 and 57 years old (average = 35). All participants filled out a background questionnaire. A large number of participants were people with a computer science background. Most of them have previously experienced interacting with 3D user interfaces (gamepad or joystick) in virtual environments (video games). Among these participants, 16 are right-handed and 2 are left-handed. Everyone used his dominant hand to operate with the proposed devices (real handle, joystick, haptic arm).

We asked orally to subjects to perform trajectories 8-shaped according to landmarks (traffic cones), 4 times starting from the right (Figure 3 - left), 4 times starting from the left (Figure 3 - centre). This driving task was performed after a training session based on a simpler driving task (strength line, left turn, right turn) to learn the manipulation of the real forklift, as well as the virtual one with the joystick or with the haptic paradigm. After the end of each experimental condition, participants completed a set of rating-scale questions. Then they moved on to the next condition, following the same protocol, until all three conditions were executed.

We performed a within-subject study (repeated measures) with the three forklift driving conditions. The execution order of the three conditions was randomized between subjects. Each subject performed the driving tasks successively without waiting time, except the time needed to complete the questionnaire after each condition. This setup was necessary for study the learning effects coming from the order of condition.

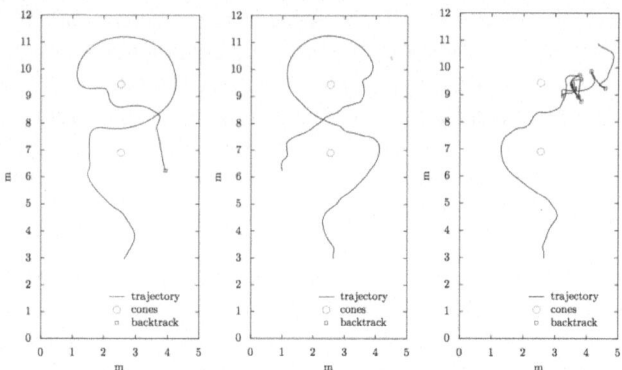

Fig. 3. Driving task through trajectory log analysis: success left path (left), success right path (centre), and a failed left path with a completion ratio of 50% and backtracks (right). Orange circles are the positions of landmarks (traffic cones), red rectangles represent the backtracks.

4.2   Performance Analysis

We compared the number of failures and success of the driving task according to experimental conditions (R, J, H). The user fails when there is a collision between the forklift and a landmark or a wall. This outcome of the task was analyzed with chi-square (χ^2) test. We observed that the driving task depended on the experimental conditions (χ^2 (2) = 22.44; p < 0.00001). Subjects failed the task more often in the virtual condition (J and H) than in the real one (R).

In order to compare experiment conditions, we had to separately analyze the trajectories features (time, distance, velocity, and backtrack) according to the success levels of the subjects. The results presented are statistically significant (p<0.05). Results are explicitly referred as a trend if p is between 0.05 and 0.1. We applied the Kolmogorov-Smirnov test to verify that the variables succeed in satisfying normality assumptions. We performed an ANOVA with experimental conditions as within-subject variables. Fisher's LSD was used for post-hoc pair-wise comparisons. All the analyses were performed with Statistica 9.

For subjects who succeed (S cluster), a number of the trajectories features depend of the experimental conditions. That is the case for: distances (F(2, 162) = 8.03; p = 0.00047), total duration of the task (F (2, 162) = 23.551; p = 0.00001), and number of backtracks (F (2, 162) = 3.60; p = 0.029). A post hoc analysis revealed that distances are shorter in R than in H and J. In addition, the task duration is faster in R and J than in H (mean durations: 48 seconds for H, 30 seconds for R, and 28 seconds for J). And finally, a higher number of backtracks appeared for H, but no significant differences between J and R. However, all these results concern the S cluster, while for learning we are more interested by the subjects who failed.

For subjects who failed (F cluster), we observed that the distance varies according to the experimental conditions (F(2, 114) = 7.65; p = 0.00076). Before failing, the users covered a greater distance in R condition (20.55 m) than in virtual conditions (16.51m for J, 16.19m for H). In addition, the completion time varies according to the experimental conditions (F (2, 114) = 13.41; p = 0.00001). The mean duration of a driving task was 27 seconds for J condition whereas it was 43 seconds for R condition and 46 seconds for H condition. The users failed faster in J condition than in R and H conditions.

Table 1. Set of correlations of the trajectories features between the different experimental conditions, for S and F clusters, and for all subjects

	Time	Distance	Velocity	Backtrack
S cluster				
HR	0.63	0.23	0.72	0.30
JR	0.39	ns	0.63	ns
HJ	0.35	ns	0.56	ns
F cluster				
HR	0.45	ns	0.76	ns
JR	ns	ns	ns	ns
HJ	ns	-0.33	ns	ns
All subjects				
HR	0.42	ns	0.78	0.22
JR	ns	ns	0.50	ns
HJ	0.27	ns	0.48	ns

Finally, we analyzed if the measures of the trajectories features were correlated in different conditions. The results show that for subjects who have succeed or failed (S and F clusters), and globally for all the subjects of the experiments, there is a strong or moderate correlation between H and R for completion time (blue - Table 1). We also notice that, except for subjects who have failed (F cluster), there is a weak correlation between H and R for backtracks (red - Table 1). Finally for velocity, the strongest correlations are observed between H and R (green - Table 1).

4.3   Subjects Behaviour Analysis

One acceptation of realism is the fact that behaviours or events observed in the real condition are reproduced in the virtual one [2]. To study the global behaviour of all subjects against condition task, we first focused on global standard deviation on curvilinear distance for each condition (H, J, R) when subject success 100% of the task, we obtain this relation: $SD_{Distance}(H)=5.05>SD_{Distance}(R)=3.66> SD_{Distance}(J)=2.19$.

Moreover, we looked at the curvilinear distance of the trajectories related to the 8 trails performed by each subject for each condition. For each condition, Figure 4 – left presents the standard deviation of the mean curvilinear distance of the 8 trajectories of the 18 subjects, while Figure 4 – right shows the mean of the 18 standard deviations of the curvilinear distances of the 8 trajectories. Then we studied the ranking of subject for each condition in terms of velocity and completion time. Figure 5 shows that the average difference of the variations of position compared to the ranking R, is twice larger for the ranking J than for the ranking H.

All together these results suggest that the subjects' behaviour in H condition is closer to the R condition than to the J one.

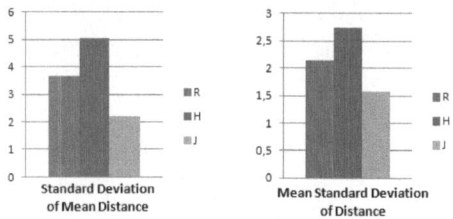

Fig. 4. The H condition increases inter-individual performance differences in term of curvilinear distance in metter (left), as well as inter-trial performance of subjects (right), both also observed in the R condition

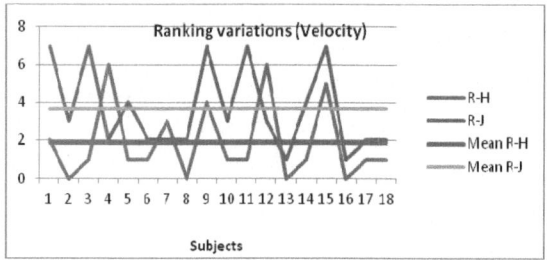

Fig. 5. Variations of subjects' positions in rankings of their velocities: 'Haptic' ranking is closer to 'Real' ranking than to the 'Joystick' one. The same occurs for the completion time.

We performed an ANOVA with experimental conditions as within-subject variables. Fisher's LSD was used for post-hoc pair-wise comparisons.

On one hand, whatever is condition (R, H or J) no significant difference is observed concerning subjects' motivation for the task, while they judge that the interactive devices are all relatively intuitive. No difference is observed concerning a number of realism aspects or interface awareness: the subjects reported having lost consciousness of the screens, not being affected by the visual display quality, not having the feeling of seeing a movie, and have seen many 3D objects. Concerning the sensory factor, the perception of the degree of movement was not affected by the experimental condition: the subjects said they were able to move in the scene, to change of perspective, and explore easy the scene in the real as well as in virtual conditions. All these results indicate that the virtual scene was realistic and well perceived by the subjects.

On the other hand, significant differences are observed for the control factor, and particularly for the mode of control. That is also interesting, because these features positively highlight natural aspect of the interaction with the environment. Firstly, the reported perception of external force feedbacks, of the inertia of the forklift, and of the physical implication varies according to experimental conditions (external force feedbacks: $F(2.34)=38.92$ $p<0.00001$; inertia $F(2.34)=21.56$ $p<0.0001$; physical implication: $F(2.34)=20.48$ $p<0.001$). Post hoc comparison showed that users perceive less external force feedbacks and inertia of the forklift, and being less physical implicated for the J condition than for the H and R conditions. Secondly, the perception of the behaviours of the forklift's handle varies according to the experimental condition (F(2.34)=9.53 P<0.00005). Post hoc comparison showed that subjects consider more that the forklift is manipulated by a handle in the R and H conditions than in the J condition. Thirdly, the J condition was rated as globally providing less natural interaction than the H condition. The J interaction was perceived as being less close of the R interaction than H interaction is closed to the R interaction. These results confirm our hypothesis. Lastly, the subjective measures show significant differences concerning the involvement in the forklift driving. The results suggest that the users are more focused on the manipulation paradigm in the R and H conditions than in the J condition. Indeed the subjects reported being more aware of the situation events (F(2.34)=5.52 p<0.008) and being more focused on the manipulation device (F(2.34)=6.16 p<0.005) in the R and H than in the J condition.

5   Discussion

One of the definitions of realism - through the action fidelity or functional fidelity concepts [2], is that performance similarities between the virtual condition and real condition should correlate. It is especially the case for trajectories features (Table 1), which globally shows a better correlation of the performance between H and R than between J and R. On this field, the most important correlation between R and H is concerning velocity. Actually, except orientation control, velocity is the main parameter that user may control during the task. Velocity is a measure that characterizes a choice and strategy to succeed the task, and thus an individual specific behaviour. That means that in R and H conditions, the users adopt a same velocity strategy. This result corroborate with another observation: the user performance ranking on mean velocity in H condition is twice closer to the R condition ranking, than the J condition ranking is (Figure 5), which means that H seems to keep ranking on velocities.

The fact that there is the strongest correlation on velocities is an argument in favour of a better realism of H condition, according to [1] that supposes realism transfers behaviours of the users between virtual and real condition. The way that users perform the task in the J condition is completely different than in the R condition. That also corroborates the subjects' answers to the questionnaires regarding the fidelity of the H condition with respect to the R one, when they declare that they have the feeling to manipulate the forklift handle in H condition, not in J.

Moreover the global standard deviation on curvilinear distance (Figure 4, left) highlights the fact that haptic condition reproduces and increases the high interindividual differences observed in the R condition whereas these differences are blurred in the J condition. That is also a behaviours property, which is globally observed within the full group of subjects in real condition, and seems to be reproduced in haptic condition. In addition, at the individual scale, one must notices the weak homogeneity of performance between trajectories (Figure 4, right) performed by each user observed in R, was more observed in H than in J.

Another aspect of the realism is the learning ability that virtual environment may offer [1]. In that field we observed that when the users don't succeed the task, they have failed earlier in J condition than in H and R conditions. That is important because this forklift virtual simulator is basically dedicated for non-expert users that are supposed not really succeed the driving task when they are novices.

6   Conclusion

We designed an innovative driving paradigm for the virtual learning of forklifts, when a physical involvement of drivers is necessary in real context. On one hand, a rigid body model allows the users to perceive haptically the physical and mechanical behaviour of the forklift and its handle. On the other hand, the velocity control of the virtual forklift is simulated with a push/pull interactive technique. To evaluate this haptic paradigm, 18 subjects performed driving tasks in three conditions: with a non-motorized forklift within a real room, with virtual simulator of the forklift and the room, and for this second context, with a joystick interaction, and with our haptic paradigm. We covered all the conditions orders, and realised statistical analysis on trajectories performances, subjects' behaviours, and on subjective data.

One dimension of realism supposes that a performance transfer must be observed between real and virtual contexts. On this acceptation, we basically found some interesting correlations between real driving condition and our haptic driving paradigm, mainly for completion time and velocity. But our haptic technique clearly appeared more realistic than the joystick one, when considering another dimension of realism, namely the behaviour and psychological processes transfers. Actually, real and haptic conditions are correlated on several behaviours: velocity chosen by the subject, conservation of subjects' rankings for velocity, completion time, and behaviours at individual and group scales (inter-individual and inter-trial performance differences). This dimension of the realism in favour of our haptic paradigm has been also highlighted within the subjective data analysis: good perception of external forces and inertia of the forklift, feeling to be physically involved, feeling to manipulate the forklift by a handle, better focus on the driving task. Accordingly, we may globally conclude that our haptic paradigm provides more realism than the joystick interaction, for the virtual driving of such kind of forklifts.

References

1. Patrick, J.: Training: Research and practice. Academic Press, San Diego (1992)
2. Stoffregen, T.A., Bardy, B.G., Smart, L.J., Pagulayan, R.J.: On the nature and evaluation of fidelity in virtual environments. Virtual and Adaptative Environments: Applications, Implications, and Human Performance Issues, 111–128 (2003)
3. Bourdot, P., Touraine, D.: Polyvalent Display Framework to Control Virtual Navigations by 6DOF Tracking. In: Proc. of IEEE Virtual Reality, pp. 277–278 (2002)
4. Otaduy, M.A., Lin, M.C.: User-Centric Viewpoint Computation for Haptic Exploration and Manipulation. In: Proc. of IEEE Visualization, pp. 311–318 (2001)

5. Dominjon, L., Lécuyer, A., Burkhardt, J.M., Andrade-Barroso, G., Richir, S.: The "Bubble" Technique: Interacting with Large Virtual Environments Using Haptic Devices with Limited Workspace. In: Proc. of the World Haptics Conference (2005)
6. Dominjon, L., Richir, S., Lécuyer, A., Burkhardt, J.M.: Haptic Hybrid Rotations: Overcoming Hardware Rotational Limitations of Force-Feedback Devices. In: Proc. of the IEEE Virtual Reality (2006)
7. Poupyrev, I., Weghorst, S., Fels, S.: Non-isomorphic 3D rotational techniques. In: Proc. of the SIGCHI Conference on Human Factors in Computing Systems, pp. 540–547 (2000)
8. LaViola, J.J., Katzourin, M.: An Exploration of Non-Isomorphic 3D Rotation in Surround Screen Virtual Environments. In: Proc. of 3D User Interfaces, pp. 49–54 (2007)
9. Zhai, S.: Human Performance in Six Degree of Freedom Input Control. Ph.D thesis. University of Toronto (1995)
10. Gilbert, B.P., Rasmussen, K.J.R.: Determination of accidental forklift truck impact forces on drive-in steel rack structures. Engineering Structures 33(5), 1403–1409 (2011)
11. Rinchi, M., Pugi, L., Bartolini, F., Gozzi, L.: Design of control system to prevent forklift capsize. International Journal of Vehicle Systems Modelling and Testing 5(1), 35–58 (2010)
12. Iriarte, X., Gil, J., Pintor, J.M.: Evaluation of different numerical integrators applied to a forklift truck real-time simulator. In: Multibody Dynamics, ECOOMAS Thematic Conference (2005)
13. Yuen, K.K., Choi, S.H., Yang, X.B.: A full-immersive CAVE-based VR simulation system of forklift truck operations for safety training. Computer-Aided Design and Applications 7(2), 235–245 (2010)
14. Bergamasco, M., Perotti, S., Avizzano, C.A., Angerilli, M., Carrozzino, M., Facenza, G., Frisoli, A.: Fork Lift truck simulator for training in industrial environments. In: Research in Interactive Design: Proceedings of Virtual Concept (2005)
15. Witmer, B.J., Singer, M.J.: Measuring Presence in Virtual Environments: A Presence Questionnaire. Presence 7(3), 225–240 (1998)
16. Sanchez-Vives, M.V., Slater, M.: From presence to consciousness through virtual reality. Nature Neuroscience 6, 8–16 (2005)
17. Slater, M., Usoh, M.: Presence in Immersive Virtual Environments. In: Proc. of the IEEE International Conference on Virtual Reality, pp. 90–96 (1993)
18. Schuemie, M.J., van der Straaten, P., Krijn, M., van der Mast, C.A.P.G.: Research on presence in Virtual Reality: a survey. Cyber Psychology & Behavior 4(2), 183–201 (2001)

Spectral Subtraction of Robot Motion Noise for Improved Event Detection in Tactile Acceleration Signals

William McMahan and Katherine J. Kuchenbecker

Haptics Group, GRASP Laboratory
Department of Mechanical Engineering and Applied Mechanics
University of Pennsylvania, Philadelphia, USA
{wmcmahan,kuchenbe}@seas.upenn.edu
http://haptics.seas.upenn.edu

Abstract. New robots for teleoperation and autonomous manipulation are increasingly being equipped with high-bandwidth accelerometers for measuring the transient vibrational cues that occur during contact with objects. Unfortunately, the robot's own internal mechanisms often generate significant high-frequency accelerations, which we term ego-vibrations. This paper presents an approach to characterizing and removing these signals from acceleration measurements. We adapt the audio processing technique of spectral subtraction over short time windows to remove the noise that is estimated to occur at the robot's present joint velocities. Implementation for the wrist roll and gripper joints on a Willow Garage PR2 robot demonstrates that spectral subtraction significantly increases signal-to-noise ratio, which should improve vibrotactile event detection in both teleoperation and autonomous robotics.

Keywords: haptic feedback for teleoperation, vibrations, tactile accelerations, noise suppression.

1 Introduction

Dynamic acceleration signals provide humans with discernible tactile cues up to 1000 Hz [3], and these vibrations are known to play a significant role in a wide range of manual tasks, including tool-mediated perception of surface roughness [13] and dexterous object manipulation [12]. Given the recent proliferation of tiny MEMS-based accelerometers and the importance of these signals for human perception and motor control, one practical approach to improving operator awareness in telemanipulation systems is to provide the operator with haptic feedback of the dynamic acceleration signals experienced by the remote robot, typically by recreating the signals via a voice coil actuator.

This feedback approach has been demonstrated via bench-top research devices [14,20], and has shown promise for both industrial [7] and medical applications [25,15,18]. This approach has also been investigated for improving the tactile sensitivity of hand-held (non-robotic) tools [26]. Additionally, researchers

P. Isokoski and J. Springare (Eds.): EuroHaptics 2012, Part I, LNCS 7282, pp. 326–337, 2012.

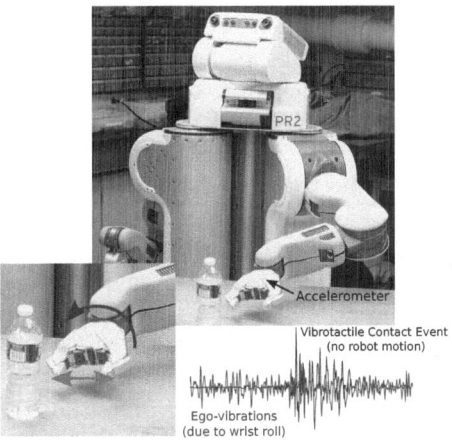

Fig. 1. The Willow Garage PR2 is an example of a modern robotic platform that has high-bandwidth acceleration sensors embedded in its grippers. Gripper joint and wrist roll joint motion are indicated with green and red arrows, respectively. Time series acceleration plots show examples of a clean vibrotactile contact event signal and robot ego- vibrations; the two signals have similar magnitude and spectral content, reducing signal to noise ratio during robot movement.

have developed methods for autonomous robotic manipulation systems to use the perceptual cues provided by vibrotactile acceleration signals (e.g., [9,22]).

Unfortunately, measurement of tactile acceleration signals can easily be masked or degraded by vibrations that are generated by a robot's own motion, as illustrated in Fig. 1. These "ego-vibrations" often lie within the same frequency range as the external contact signals that one wants to detect for presentation to the human operator or for use in an autonomous robot's controller. Thus the noise-reduction performance of traditional filters (e.g., high-pass, low-pass, notch) is severely limited.

Some previous research in this area has recognized that ego-vibrations mask desired signals and degrade system performance. For example, surgeons using our VerroTouch feedback system occasionally comment that feeling the motion of the da Vinci surgical robot distracts from the vibration cues caused by contact [18]. The primary means of addressing ego-vibrations has been through electromechanical system design. Some researchers have designed custom hardware to mechanically isolate the acceleration sensors from robot motion vibrations [14,9,7]. Others have used robotic hardware that is specifically designed for smooth motion, such as Sensable's Phantom Omni haptic device [20,19]. Others have used their accelerometer only in limited contexts, when they knew that robot motion noise would be small [22].

To improve the performance of vibration feedback systems for teleoperation and to enable measurement of useful high-frequency tactile accelerations on robotic platforms that are not mechanically optimized, we propose a signal processing approach that mirrors human neuropsychology. The human central nervous system is believed to use the motor commands sent to the muscles to predict the sensory consequences of movement. These predictions allow one to distinguish self-produced sensations from those arising from external events [5].

Thus, we propose to use knowledge of robot motion to predict the contribution of ego-vibration noise to the measured acceleration signal, and to remove this contribution through spectral subtraction, a method that was originally developed for noise suppression in speech processing [6]. The basic idea of spectral subtraction is that noise in signals can be removed by transforming to the frequency domain and subtracting out an estimate of the noise spectrum. Additionally, Ince et al. have successfully applied this technique to the similar problem of audible robot motion noise in robot audition [10,11].

The mathematics and signal processing pipeline of spectral subtraction are detailed in Section 2. Section 3 describes our implementation of this approach on a Willow Garage PR2 humanoid robot, which experiences significant ego-vibrations from its wrist roll and gripper joints. We evaluate the performance of this approach in Section 4, and we conclude with Section 5.

2   Spectral Subtraction

It is natural to compare vibrotactile acceleration signals to audio signals; the primary way in which these signals differ is in bandwidth. While human skin can perceive vibrotactile cues up to approximately 1000 Hz [3], audio signals are detectable up to 20,000 Hz. Compared to the study of audio signals, the study of vibrotactile acceleration signals is quite immature. Thus we are inspired to look to the audio processing literature, as many of their methods can be directly adapted to handling vibrotactile signals.

The problem of robot ego-vibrations seems most similar to the problem of suppressing additive background noise from a single audio channel of speech. Research into this problem is generally classified as speech enhancement or noise reduction; many of the best methods in this area are reviewed in [23,4]. Among these approaches, spectral subtraction seems particularly promising for dealing with ego-vibrations because of its straightforward implementation and inexpensive computational requirements that allow for implementation in real-time applications with minimal processing latency.

This section describes our proposed adaptation of spectral subtraction to the problem of ego-vibration suppression in tactile acceleration measurements. Fig. 2 provides a block-diagram illustration of the algorithm. The mathematical notation used here is primarily adapted from [23].

Block Processing. Mathematically, noisy observations from an accelerometer can be modeled as

$$y[k] = x[k] + n[k], \tag{1}$$

where $x[k]$ is the vibrotactile event signal, $n[k]$ is additive noise from ego-vibrations, and k is the discrete time index. Following the methods of [16],

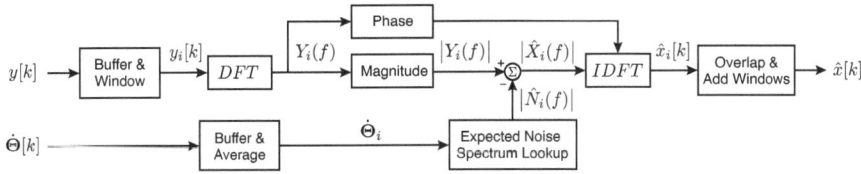

Fig. 2. Block diagram of the spectral subtraction method adapted for ego-vibration suppression in vibrotactile signals

the multiple orthogonal axes of accelerometer output are combined into a single channel by addition, a computationally simple approach that yields a good spectral match and temporal match with the original signals without introducing any time delay. A band-pass filter then removes both low-frequency cues pertaining to robot motion and high-frequency signals that are not detectable to humans.

The one-dimensional filtered signal is then subjected to block processing [6] as follows:

1. The incoming signal $y[k]$ is buffered into small time buffers $y_i[k]$ of length L that overlap by M samples.
2. Each buffered block $y_i[k]$ is multiplied by a window function $w[k]$ to reduce discontinuities at the end points during discrete Fourier transform (DFT).
3. Each windowed buffer is subjected to spectral subtraction as described in the next subsection.
4. The resultant output signals $\hat{x}_i[k]$ are recombined into the full $\hat{x}[k]$ using the overlap-add method [2].

Typical parameter choices for block processing are 50% buffer overlap ($M = L/2$), and a normalized Hamming window function for $w[k]$. Buffer length choice is a trade-off between frequency resolution and time delay; block processing implementation introduces an algorithm delay of one buffer length.

Magnitude Subtraction. After the noisy input signal has been buffered and windowed, each short segment $y_i[k]$ is transformed into the frequency domain using the discrete Fourier transform (DFT),

$$y_i[k] \xrightarrow{DFT} Y_i(f) = X_i(f) + N_i(f). \tag{2}$$

The operation of spectral subtraction can be described by the equation

$$|\hat{X}_i(f)| = |Y_i(f)| - |\hat{N}_i(f)|, \tag{3}$$

where $|\hat{X}_i(f)|$ and $|\hat{N}_i(f)|$ are the estimated magnitude spectra of the restored vibrotactile event signal and the noise spectrum respectively. Derivation of the ego-vibration's spectral magnitude estimate $|\hat{N}_i(f)|$ will be discussed in the next subsection. Given the random nature of $N_i(f)$, this operation will sometimes result in negative values for a discrete frequency subband f; negative values are consequently set to zero.

Using the phase of the original noisy signal $\Phi_{Y_i}(f)$, which is optimal with regard to minimum mean square error [8], the estimated vibrotactile signal $\hat{X}_i(f)$ can brought back into the time domain as $\hat{x}_i[k]$ via the inverse discrete Fourier transform ($IDFT$),

$$|\hat{X}_i(f)|e^{j\Phi_{Y_i}(f)} \xrightarrow{IDFT} \hat{x}_i[k]. \tag{4}$$

Noise Spectrum Magnitude Estimate. While traditional spectral subtraction research has involved developing methods to adaptively estimate the spectrum of an unknown background noise in real-time, when dealing with ego-vibrations we have the luxury of knowing the state of our noise source. A priori observation of the vibrations that occur during unimpeded robot motion can be used to model the expected ego-vibrations during subsequent actions. Depending on the hardware, ego-vibrations might be expected to vary with the robot's joint positions Θ, joint velocities $\dot{\Theta}$, and/or joint accelerations $\ddot{\Theta}$. Thus the ego-vibration magnitude spectrum estimate $|\hat{N}_i(f)|$ in Equation (3) could be a function of all of the robot's present joint states. However, the process of gathering data that fully explores and models the noise spectra that occur in this high-dimensional configuration space would require considerable time and storage capacity. Thus, we seek to reduce the dimensionality of the problem.

Joint acceleration was reported to be unimportant for spectrum estimation of audible joint noise with a Honda ASIMO [10]. Assuming a smooth robotic motion controller, it is reasonable for us to assume that joint acceleration also has little effect on ego-vibrations. For this initial investigation, we have also chosen to neglect the possible influence of joint position. Given our short-time-window implementation, we mark each recorded acceleration sample $y_i[k]$ using the element-wise means of the buffered joint velocity vector $\dot{\Theta}_i$.

Because we decided to focus on the effects of joint velocity, we designed our estimate of the ego-vibration magnitude spectrum to be

$$|\hat{N}_i(f)| = |\hat{N}(f, \dot{\Theta}_i)| = \mu_N(f, \dot{\Theta}_i) + \sigma_N(f, \dot{\Theta}_i). \tag{5}$$

Here, $\mu_N(f, \dot{\Theta}_i)$ and $\sigma_N(f, \dot{\Theta}_i)$ are the mean and standard deviation of the magnitude spectra recorded during a long test at joint velocity vector $\dot{\Theta}_i$. Other approaches to spectral subtraction sometimes use only the mean value $\mu_N(f, \dot{\Theta}_i)$. However, empirical results show that some amount of over-subtraction tends to yield better noise suppression in low signal-to-noise ratio (SNR) speech signals [23]. Adding in the standard deviation $\sigma_N(f, \dot{\Theta}_i)$ allows the degree of over-subtraction to be determined by the variance in a particular frequency subband [23].

3   PR2 Implementation Details

We tested our full approach to ego-vibration suppression (Fig. 2) using a Willow Garage PR2 humanoid robot.

Hardware and Algorithm Parameters. The PR2 was chosen for use in this project because it includes high-bandwidth accelerometers in its sensor suite [1]; as marked in Fig. 1, a three-axis 10-bit digital accelerometer (Bosch BMA150) is embedded within each gripper assembly. Unlike some custom research devices, this robot does not have a wrist-mounted force/torque sensor or all-over pressure-sensitive skin. Instead, the accelerometers are the system's best sensors for dynamic tactile measurements, as demonstrated in [22].

For these initial algorithm-evaluation experiments, we focused on signals from the accelerometer in the right hand. It is subjected to significant ego-vibrations from the nearby wrist rotational joint (roll clockwise/counter-clockwise) and gripper translational joint (close/open). Because the wrist and gripper joints are heavily geared and located so close to the PR2's accelerometer, we believe they are the most challenging and beneficial joints to deal with for ego-vibration suppression. Thus, we define our joint velocity vector as $\dot{\Theta} = [\dot{\theta}\ \dot{d}]$, where $\dot{\theta}$ and $\dot{d}$ are the wrist roll rotation and gripper translation velocities, respectively.

Experimental data was recorded from the three channels of the accelerometer at a sampling rate of 3000 Hz and a measurement range of ± 80 m/s^2 per axis, the maximum specifications of the accelerometer. The corresponding joint velocities, $\dot{\theta}$ and $\dot{d}$, are derived from optical encoder readings and recorded at 1000 Hz.

As described in Section 2, the three acceleration channels were first summed together and band-pass filtered. For the PR2, we employ a fourth-order Butterworth band-pass filter from 150 Hz to 750 Hz; though narrower than the range of human vibrotactile perception, this filtering was chosen to be sure to remove strong signal content observed at 1000 Hz, which likely results from a structural resonance. The just-noticeable difference of human vibrotactile frequency discrimination scales with frequency, resulting in the poorest discrimination at the highest perceivable frequencies; pilot testing of our algorithms with various filter bandwidths indicated that the best perceptual performance occurred when filtering out the 1000 Hz resonance rather than trying to cancel it through spectral subtraction.

Although we evaluated our spectral subtraction algorithm off line, the processed accelerometer readings $y[k]$ and the joint velocity vectors $\dot{\Theta}[k]$ are fed sequentially into the block process implementation in a manner that is consistent with real-time processing. Block length was chosen to be 64 samples ($L = 64$) with 50% overlap ($M = 32$). At a sampling rate of 3000 Hz, this block length corresponds to an algorithm processing delay of about 21 ms. Studies on human perception of vibrotactile textures [21] and force feedback [24] indicate that delays less than 40 ms are imperceptible, so our processing should not adversely affect the quality of vibrotactile feedback in teleoperated systems.

PR2 Ego-Vibration Estimation. First, we gathered the data set necessary to compute the velocity-dependent ego-vibration magnitude spectrum estimates $|\hat{N}(f, \dot{\Theta}_i)|$ defined in Equation (5). Joint velocities were obtained from the ROS Diamondback *pr2_mechanism_model* class, which reports velocities using first-order differentiation of encoder readings. The maximum speed for the wrist roll and gripper joints are 3 rad/s and 0.04 m/s, respectively. All combinations of

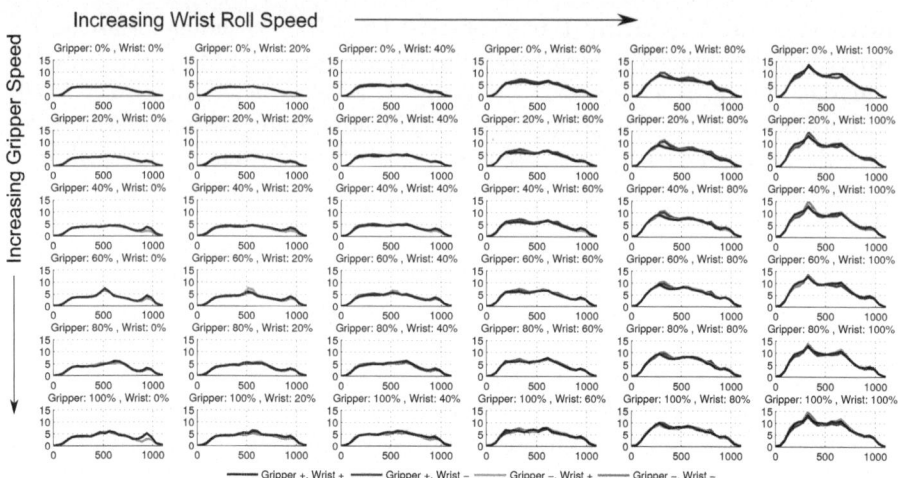

Fig. 3. Plots of the estimated ego-vibration magnitude spectra caused by combinations of gripper and wrist roll speeds. Speeds are defined as a percentage of the joint's maximum speed. The x-axes are frequency (Hz) and the y-axes are amplitude (m/s²). The colored lines show small variations in noise at the four joint velocity direction combinations noted in the legend. Note that the top-left plot shows the background accelerometer noise that is present when the robot is stationary.

wrist roll velocity and gripper velocity were sampled at intervals of 10% of the joint's maximum speed, ranging from -100% (wrist rolling clockwise and gripper closing at maximum speed) to 100% (wrist rolling counter-clockwise and gripper opening at maximum speed). Note that these tests included 0% velocities, when the wrist and/or gripper are stationary, to ensure measurement of background accelerometer noise. The data was gathered in semi-continuous 15-second chunks; while the PR2's wrist joint is capable of continuous rotation, the gripper velocity command had to be reversed when the gripper reached the limits of its 86 mm translation workspace, so we concatenated multiple acceleration recordings together for most wrist-gripper velocity combinations.

We calculate the magnitude spectrum for the many overlapping short time windows of each recording using the same input buffering and windowing approach described in Section 2, These data points are then used to calculate the mean $\mu_N(f, \dot{\Theta}_i)$ and standard deviation $\sigma_N(f, \dot{\Theta}_i)$ of the magnitude spectrum, which are combined to find $|\hat{N}(f, \dot{\Theta}_i)|$. This procedure provides equal spacing of ego-vibration magnitude spectrum estimates throughout the entire wrist roll and gripper joint velocity space.

Fig. 3 shows the estimated ego-vibration magnitude spectra for a subset of the sampled joint velocities, calculated from equation (5). The mean and the standard deviation of the noise both increase as joint speeds increase. Visual inspection of the recorded time series data and the noise residuals seems to indicate that it is reasonable to assume that vibrations depend only on joint

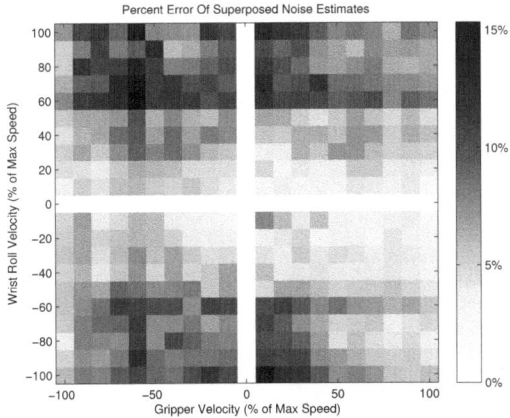

Fig. 4. Percent error between the superposed and observed estimates at different joint velocity combinations

velocities at lower speeds. However as speeds increase, we observed changes in the vibration signal that appear to depend on joint position; we plan to investigate this extension in future work.

Reducing Dimensionality. It is important to understand that the size of the velocity configuration space increases exponentially with every new joint that is included in the modeling process. To reduce the testing, calculation, and storage burden of generating spectral magnitude ego-vibration models, we propose that vibration spectra may be superposed (added) across joints. This assumption mirrors the superposition of audio noise assumption made in [11].

Thus for the two-joint case presented in this paper, the noise spectrum is

$$
\begin{aligned}
|\hat{N}(f, \dot{\mathbf{\Theta}}_i)| &= |\hat{N}(f, [\dot{\theta}\ \dot{d}])| \\
&= \mu_N(f, [\dot{\theta}\ 0]) + \mu_N(f, [0\ \dot{d}]) - \mu_N(f, [0\ 0]) \\
&\quad + \sqrt{\sigma_N(f, [\dot{\theta}\ 0])^2 + \sigma_N(f, [0\ \dot{d}])^2 + \sigma_N(f, [0\ 0])^2}.
\end{aligned}
\tag{6}
$$

This expression represents the noise spectrum at joint velocities $\dot{\theta}$ and $\dot{d}$ as the sum of the noise spectra measured when each joint was moving alone at these velocities, minus the noise spectrum measured when both joints were stationary (since this background noise is presumably present in both of the added signals).

We examined the validity of this assumption by calculating the mean error between the magnitude estimate calculated from equation (6) and the observed magnitude estimate summed over the frequency subbands. Fig. 4 graphically shows the percent error for all 441 joint velocity combinations. The additive models have low error for wrist roll velocities up to about 50% of maximum, but the assumption starts to break down at high wrist roll velocities, reaching errors of 15%. The remainder of this paper employs the superposed noise assumption to test its reasonableness.

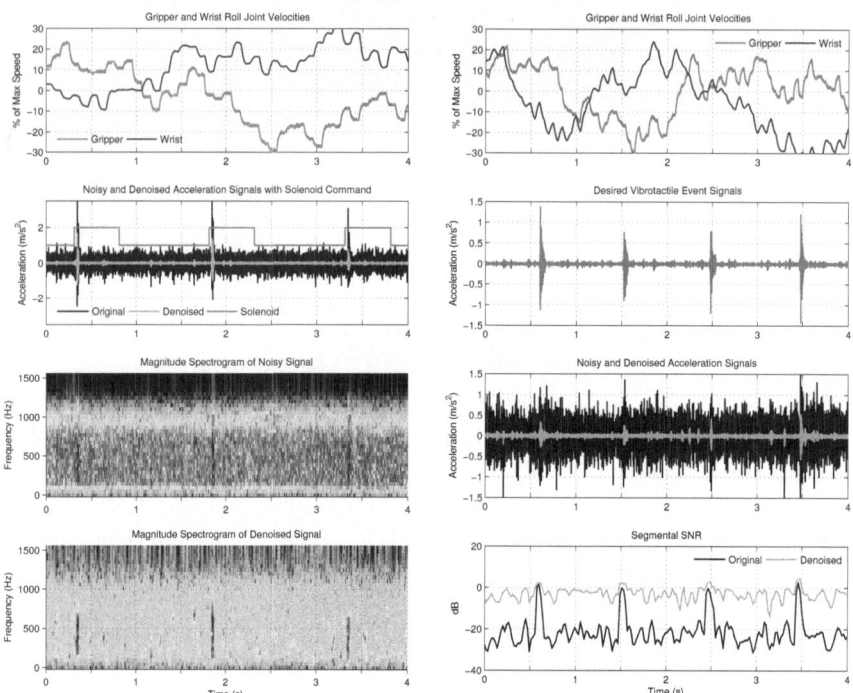

Fig. 5. Results from the solenoid tapper experiment (left) and from adding recorded tap signals to recorded ego-vibration signals (right)

4   Performance Experiments

This section presents the two experiments we conducted to evaluate the performance of our spectral subtraction noise suppression algorithm during dynamic joint motions. The wrist roll and gripper joints were given random velocity commands that had been low-pass filtered to meet our smooth controller assumption.

First, a small solenoid (Ledex 174534-035) was used to consistently tap on the PR2's lower arm, near the wrist joint. This setup provides the robot with a consistent vibrotactile event stimulus that simulates the signal it would experience when making contact with its environment.

Results from a trial of this experiment are shown in Fig. 5 (Left). The solenoid taps are visually apparent in the time domain signal and magnitude spectrogram of both the noisy and denoised acceleration signals. However, the denoised signal shows a significant reduction in the noise ceiling of the signals, from a mean signal magnitude of 0.27 m/s^2 to 0.02 m/s^2. Albeit, this improvement comes at some expense to the magnitude of the tap signal, which features a smaller percentage fall from an average peak magnitude of 3.41 m/s^2 to 1.27 m/s^2. Using these values, we estimate an improvement in peak signal-to-noise ratio (SNR) from 22 dB to 36 dB.

However, accurately quantifying signal-to-noise ratio improvements requires ground truth knowledge of the desired signal. Unfortunately, the nature of the ego-vibration problem makes it impossible to accurately determine ground truth. For this reason, we generated a data set that replicates the previous experiment using a denoised solenoid event signal that was pre-recorded while the robot was stationary. We then recorded ego-vibrations from random robot joint motions and added this noise signal to the pre-recorded solenoid tap signals.

The results of processing one of these summed signals are shown in Fig. 5 (Right). In this case, the magnitude of the original taps is similar to the maximum noise magnitude. This test simulates the contact signals that the robot experiences when making very light contact with its environment. Note that using a simple noise threshold here would also destroy the tap signal.

Segmental SNRs were calculated for the noisy and denoised signal for 64 sample segments with 50% overlap. During contact events, the two signals show similar SNRs. However, the denoised signal shows a ~20 dB improvement in SNR when no contact event signal is present. Trials with larger joint velocities showed similar results, but fast joint motion was prone to generating noise spikes that may be falsely perceived by human operators and autonomous magnitude thresholding as tactile event signals.

5   Conclusion

This paper developed a spectral subtraction approach to suppressing ego-vibration noise in robotic high-frequency acceleration signals; we used pre-recorded data to estimate the magnitude spectrum of the noise during a range of robot joint velocities and a block processing procedure to remove the estimated noise spectrum from the measured signal. To our knowledge, this is the first work to address the robot motion noise problem in tactile acceleration signals via advanced signal processing rather than mechanical optimization.

This approach was implemented and tested using the gripper translational and wrist roll rotation joints of a Willow Garage PR2. Our results demonstrate that spectral subtraction can significantly improve signal-to-noise ratio (SNR) in high-frequency vibrotactile acceleration signals. This increase in SNR should lead to improved detection of vibrotactile events for both human operators receiving vibrotactile feedback and autonomous robots using a quantitative event-detection criterion, such as signal magnitude or power. Anecdotally, we feel that this algorithm improves the quality of the vibrotactile feedback.

From our experience with telerobotic systems that provide haptic feedback of measured vibrotactile signals [20,18], we believe that a low noise ceiling can be especially important for usability. We have anecdotally observed that users tend to choose vibrotactile amplification levels that limit the perceptibility of vibration signals during free motion. We attribute this trend to the human sensitivity to haptic noise. As often stated in haptic device design, "free space should feel free" [17]. With a reduced noise floor, users will be more willing to use higher feedback gains, further improving the perceptibility of the signals they feel.

Secondarily, we found that an estimate of the ego-vibration magnitude spectrum that relied only on joint velocities could provide good noise suppression performance, at least at lower joint velocities. A further simplifying step of superposing (adding) noise estimate models across joints also yielded good performance. At larger joint speeds, a higher dimensional model is needed to more fully capture the behavior of the ego-vibration noise. In future work, we will further seek to improve the noise suppression capabilities of this method through improved modeling of the ego-vibration noise. We will also seek to formally quantify the effects of this approach through task performance experiments by human subjects with and without spectral subtraction of robotic ego-vibrations.

Acknowledgments. Research was sponsored by the Army Research Laboratory and was accomplished under Cooperative Agreement Number W911NF-10-2-0016. The authors thank Joseph M. Romano and Benjamin J. Cohen for their assistance with the PR2 and Willow Garage for their loan of a PR2 to the University of Pennsylvania GRASP Laboratory via the PR2 Beta Program.

References

1. Willow Garage, Overview of the PR2 (2011)
   http://www.willowgarage.com/pages/pr2/overview (accessed on February 14, 2011)
2. Allen, J.: Short term spectral analysis, synthesis, and modification by discrete fourier transform. IEEE Transactions on Acoustics, Speech and Signal Processing 25(3), 235–238 (1977)
3. Bell, J., Bolanowski, S., Holmes, M.H.: The structure and function of Pacinian corpuscles: A review. Progress in Neurobiology 42(1), 79–128 (1994)
4. Benesty, J., Chen, J., Huand, Y., Cohen, I.: Noise Reduction in Speech Processing, Springer Topics in Signal Processing, vol. 2. Springer (2009)
5. Blackemore, S.J., Wolpert, D., Frith, C.: Why can't you tickle yourself? Neuroreport 11, R11–R16 (2000)
6. Boll, S.: Suppression of acoustic noise in speech using spectral subtraction. IEEE Transactions on Acoustics, Speech and Signal Processing 27(2), 113–120 (1979)
7. Dennerlein, J.T., Millman, P.A., Howe, R.D.: Vibrotactile feedback for industrial telemanipulators. In: Proc. of the Sixth Annual Symposium on Haptic Interfaces for Virtual Environment and Teleoperator Systems, ASME International Mechanical Engineering Congress and Exposition, pp. 107–113 (November 1997)
8. Ephraim, Y., Malah, D.: Speech enhancement using a minimum-mean square error short-time spectral amplitude estimator. IEEE Transactions on Acoustics, Speech and Signal Processing 32(6), 1109–1121 (1984)
9. Howe, R.D., Popp, N., Akella, P., Kao, I., Cutkosky, M.R.: Grasping, manipulation, and control with tactile sensing. In: Proc. of the IEEE International Conference on Robotics and Automation, vol. 2, pp. 1258–1263 (1990)
10. Ince, G., Nakadai, K., Rodemann, T., Hasegawa, Y., Tsujino, H., Imura, J.I.: Ego noise suppression of a robot using template subtraction. In: Proc. of the IEEE/RSJ International Conference on Intelligent Robots and Systems, pp. 199–204 (October 2009)

11. Ince, G., Nakadai, K., Rodemann, T., Hasegawa, Y., Tsujino, H., Imura, J.i.: A hybrid framework for ego noise cancellation of a robot. In: Proc. of the IEEE International Conference on Robotics and Automation, pp. 3623–3628 (May 2010)
12. Johansson, R.S., Flanagan, J.R.: Coding and use of tactile signals from the fingertips in object manipulation tasks. Nature Reviews Neuroscience 10, 345–359 (2009)
13. Klatzky, R.L., Lederman, S.J.: Perceiving object properties through a rigid link. In: Lin, M., Otaduy, M. (eds.) Haptic Rendering: Algorithms and Applications, ch. 1, pp. 7–19. A. K. Peters (2008)
14. Kontarinis, D.A., Howe, R.D.: Tactile display of vibratory information in teleoperation and virtual environments. Presence: Teleoperators and Virtual Environments 4(4), 387–402 (1995)
15. Kuchenbecker, K.J., Gewirtz, J., McMahan, W., Standish, D., Martin, P., Bohren, J., Mendoza, P.J., Lee, D.I.: VerroTouch: High-Frequency Acceleration Feedback for Telerobotic Surgery. In: Kappers, A.M.L., van Erp, J.B.F., Bergmann Tiest, W.M., van der Helm, F.C.T. (eds.) EuroHaptics 2010, Part I. LNCS, vol. 6191, pp. 189–196. Springer, Heidelberg (2010)
16. Landin, N., Romano, J.M., McMahan, W., Kuchenbecker, K.J.: Dimensional Reduction of High-Frequency Accelerations for Haptic Rendering. In: Kappers, A.M.L., van Erp, J.B.F., Bergmann Tiest, W.M., van der Helm, F.C.T. (eds.) EuroHaptics 2010, Part II. LNCS, vol. 6192, pp. 79–86. Springer, Heidelberg (2010)
17. Massie, T.H., Salisbury, J.K.: The PHANToM haptic interface: A device for probing virtual objects. In: Proceedings of the ASME Winter Annual Meeting, Symposium on Haptic Interfaces for Virtual Environment and Teleoperator Systems, Chicago, Illinois (1994)
18. McMahan, W., Gewirtz, J., Standish, D., Martin, P., Kunkel, J.A., Lilavois, M., Wedmid, A., Lee, D.I., Kuchenbecker, K.J.: Tool contact acceleration feedback for telerobotic surgery. IEEE Transactions on Haptics 4(3), 210–220 (2011)
19. McMahan, W., Kuchenbecker, K.J.: Haptic display of realistic tool contact via dynamically compensated control of a dedicated actuator. In: Proc. of the IEEE/RSJ International Conference on Intelligent Robots and Systems, pp. 3171–3177 (October 2009)
20. McMahan, W., Romano, J.M., Rahuman, A.M.A., Kuchenbecker, K.J.: High frequency acceleration feedback significantly increases the realism of haptically rendered textured surfaces. In: Proc. of the IEEE Haptics Symposium, pp. 141–148 (March 2010)
21. Okamoto, S., Konyo, M., Saga, S., Tadokoro, S.: Detectability and perceptual consequences of delayed feedback in a vibrotactile texture display. IEEE Transactions on Haptics 2(2), 73–84 (2009)
22. Romano, J.M., Hsiao, K., Niemeyer, G., Chitta, S., Kuchenbecker, K.J.: Human-inspired robotic grasp control with tactile sensing. IEEE Transactions on Robotics (99), 1–13 (2011)
23. Vaseghi, S.V.: Advanced Digital Signal Processing and Noise Reduction. John Wiley & Sons Ltd. (2000)
24. Vogels, I.M.L.C.: Detection of temporal delays in visual-haptic interfaces. Human Factors: The Journal of the Human Factors and Ergonomics Society 46(1), 118–134 (2004)
25. Waldron, K.J., Enedah, C., Gladstone, H.: Stiffness and texture perception for teledermatology. Studies In Health Technology and Informatics 111, 579–585 (2005)
26. Yao, H.Y., Hayward, V., Ellis, R.E.: A tactile enhancement instrument for minimally invasive surgery. Computer-Aided Surgery 10(4), 233–239 (2005)

Haptic Invitation of Textures: An Estimation of Human Touch Motions

Hikaru Nagano, Shogo Okamoto, and Yoji Yamada

Department of Mechanical Science and Engineering, Graduate School of Engineering,
Nagoya University, Japan

Abstract. Some textures invite human touch motions in daily life, but studies on the methodology of designing such textures have just been initiated [1,2,3]. However, there is still no method of identifying various touch motions invited by these textures. For example, some textures are likely to invite stroking, while others are likely to invite pushing. We developed a Bayesian network model that represents the probabilistic relationships between texture-invited touch motions and properties of textures. We interpreted the constructed model and confirmed that the model is potentially useful for the estimation of human motions.

Keywords: Bayesian Network, Stochastic Reasoning.

1 Introduction

Some textures appeal to human touch. Examples of such textures are smooth and comfortable surfaces such as silk, elastic sponges, and finely woven cloths. Occasionally, such textures invite human touch; that is, people feel compelled to touch them. These textures are potentially useful for designing products that invite human touch motion and interfaces that stimulate human interest. Nevertheless, such textures and human responses have rarely been investigated. Nagano et al. [1,2], who investigated properties of textures that appeal to human touch, revealed that the linear combination of physical factors of textures described the degrees of haptic invitation with accuracies of 70–80%. Although glossiness and surface shape types strongly affected the haptic invitation, surface colors barely affected them. Klatzky and Peck [3] also investigated relationships between objects' properties and their appeal to human touch and reported that simple objects invited human touch rather than complex objects, and that people wanted to touch moderate objects more than rough objects. Based on the continuity of these studies, the methodologies for determining the best combinations of physical factors in terms of appeal to human touch will be established.

The results of past studies mentioned above will enable us to design textures and objects that appeal to human touch. However, exactly how people touch these textures is still not known. For example, hard pushing and soft stroking are just two of many touch motions. The touch motion for elastic rubber may be different from that of smooth fur. Some researchers have analyzed hand motions

P. Isokoski and J. Springare (Eds.): EuroHaptics 2012, Part I, LNCS 7282, pp. 338–348, 2012.

of touch. For example, Lederman and Klatzky [4] investigated relationships between hand movements during haptic object exploration and desired properties of objects, such as heaviness and hardness, and Peine et al. [5] analyzed finger motions during surgical palpation. However, touch motions invited by textures have not yet been investigated. If the relationships between touch motions and properties of textures are revealed, we can design not only textures that appeal to human touch but also texture-invited touch motions. For example, we could design textures that are more likely to invite specific touch motions such as stroking or pushing.

The objective of this study was to develop a stochastic model that represents the relationships between texture-invited touch motions and properties of textures. We observed such human touch motions through experiments; thereafter, we developed a Bayesian network, which enabled us to estimate the properties of textures from observed touch motions and vice versa.

2   Experiments

We conducted two types of experiments. In Experiment 1, participants evaluated textures consecutively using a semantic differential (SD) method. Sensory properties of textures acquired here were used in constructing the stochastic model. In Experiment 2, participants touched textures that had a strong appeal to human touch. Human touch motions were measured using a camera and a six-axis dynamic force sensor. Features of touch motions were also applied to the stochastic model. The details of each experiment are described below:

2.1   Experiment 1: Sensory Properties of Textures

Participant Five laboratory students, excluding the authors, approximately twenty years of age participated in Experiment 1.

Task: Sensory Evaluation. The participants evaluated the textures using five-point scales in terms of five adjective pairs: "rough-smooth," "uneven-flat," "hard-soft," "warm-cold," and "sticky-slippery," without touching them. The evaluation sheets provided both English and Japanese terms.

As shown in Fig. 1a, a large white plate with an 80 mm × 80 mm square window was placed on a texture so that participants could see only the surfaces and not the sides of the samples. The participants were instructed to keep their head positions fixed in order to retain the relative position between the head and textures. The textures and adjective pairs were presented to each participant in random order.

Stimuli. Preliminary experiments were conducted in order to measure the degrees of affinity for various textures. For details of measurement method of degrees, refer to our articles [1,2]. The textures whose degrees of haptic invitation varied significantly depending on individuals were eliminated through experiments. This process led to the final thirty textures used in Experiments 1 and 2.

a) Sample displayed in Experiment 1

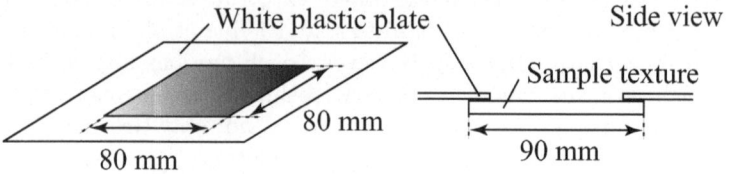

b) Samples displayed in Experiment 2

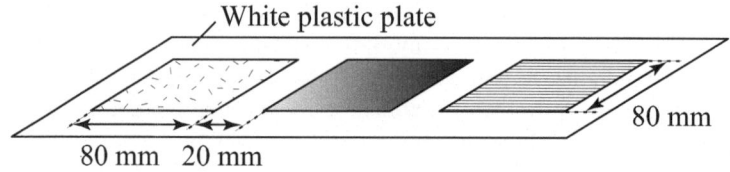

Fig. 1. Sample textures and presentation method

Table 1. List of textures

Group	Texture	Group	Texture	Group	Texture
	Crumpled paper		Fake alligator hide	Cloth	Fine woven cotton
	Fine Japanese paper	Leather	Fake suede 1		Coarse woven cotton
	Coarse Japanese paper		Fake suede 2	Towel	Towel
Paper	Embossed paper 1		Long hair fake fur	Felt	Felt
	Embossed paper 2	Fur	Short hair fake fur	Boa	Fake boa
	Fake paper		Soft fake fur	Artificial grass	Artificial grass
	Wall paper	Straw and rush grass	Fine woven straw	Metal	Woven wire mesh
	Walnut wood		Coarse woven straw	Fiber	Coarse polyester fiber
Wood	Cypress wood		Woven rush grass		
	Chestnut wood	Sponge	Soft sponge		
	Cork board		Hard sponge		

These textures, which are listed in Table 1 and shown in Fig. 2, include a fake fur with long hair and a flat Japanese paper. The textures were 90 mm × 90 mm squares. Flexible textures, such as cloths, were attached to a plastic plate with double-sided tape.

Analysis: Adjective Ratings for Thirty Textures. Ratings of 1 to 5 were assigned to the five-point adjective scales measured in Experiment 1. The rating of each adjective pair was normalized within a single participant to reduce the influence of individual differences in criteria for judgment. Ratings were then averaged across all the participants.

2.2   Experiment 2: Observation of Touch Motions to Textures

Participant. Another five laboratory students, excluding the authors, aged approximately twenty years, participated in Experiment 2.

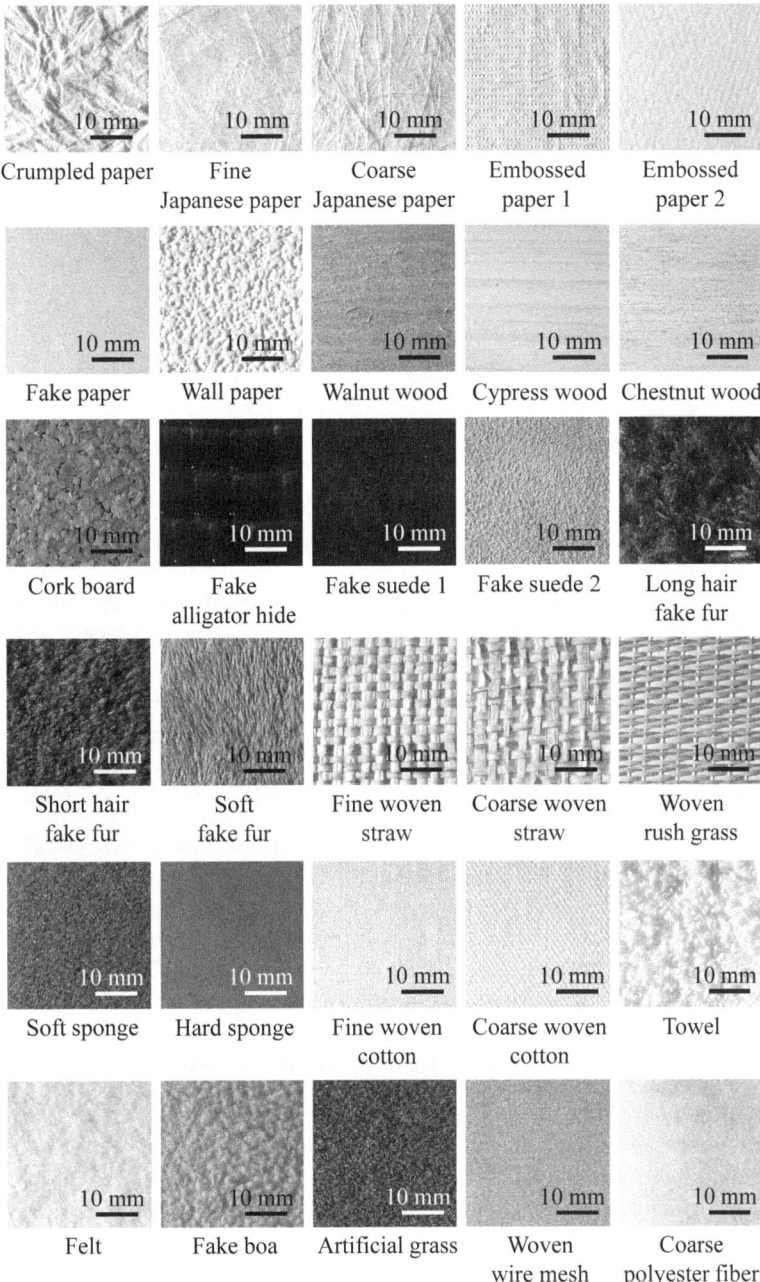

Fig. 2. Thirty textures used in experiments

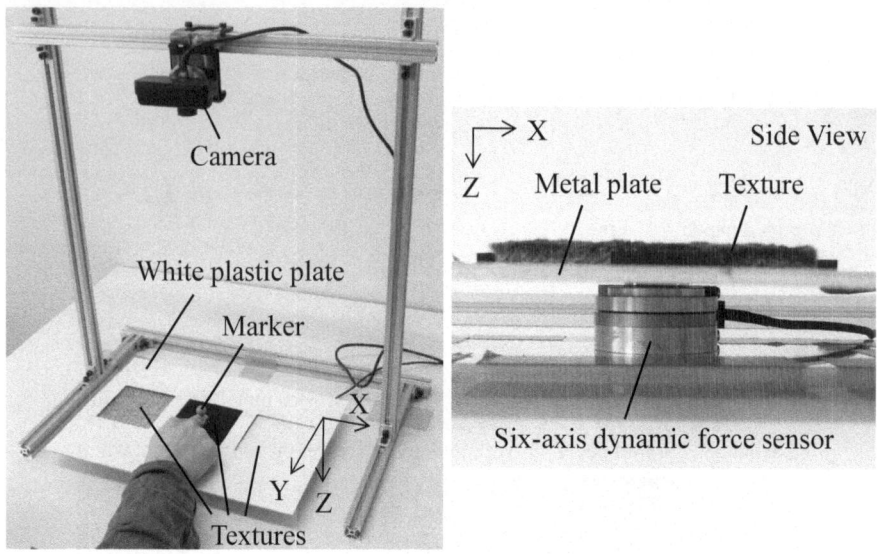

Fig. 3. Measurement of touch motions to textures

Task. In Experiment 2, participants touched one texture among three textures that most strongly appealed to them. These textures were placed under the white plastic plate, as shown in Fig. 1b. If only one texture was shown in the experiments, participants were forced to touch the texture. In order to avoid this unnatural situation, three textures were shown to the participants. Participants were instructed to keep their eyes closed until they heard a beep sound, and then they were free to touch one of three textures.

Stimuli. Three textures were selected through preliminary experiments that were described in Sec. 2.1. The preliminary experiments revealed the ranking of thirty textures in terms of measured degrees of haptic invitation. One of the three textures was selected from ten textures that exhibited the highest degrees of haptic invitation. Another texture was selected from the lowest ten textures, and the last was selected from the remaining ten textures. In total, ten combinations were presented to each participant in random order. Each texture was used only once, and each participant received a unique set of combinations.

Measurement Method for Touch Motions. Human touch motions to textures were measured in Experiment 2. Tip positions of the index finger were measured using a camera (Playstation Eye, Sony co., Tokyo, Japan), as depicted in Fig. 3. The position was detected using a red marker fixed on the nail of the index finger. The camera resolution was 320 × 240 pix, which corresponded to a position resolution of 1.03 mm, and the frame rate was 30 fps. Contact forces were measured by a six-axis dynamic force sensor (MINI 2/10, BL AUTOTEC. LTD., Kobe, Japan), as shown in Fig. 3. The sensor was fixed under a metal plate on which textures were placed. Contact force was recorded at a sampling frequency of 50 Hz.

3   Construction of Bayesian Network Model

In order to estimate touch motions from the properties of textures, we constructed a model that represents relationships among them. Considering that a certain texture frequently invites stroking while another may not, the probabilistic relationships are appropriate for representing connections. In addition, the causal connections are not previously established in relationships. Therefore, we adopted a Bayesian network to represent relationships between human touch motions and sensory properties of textures. Bayesian networks enable us to estimate unobserved states of variables using observed variables. In our case, we estimated properties of textures that are likely to invite specific touch motions.

First, we extracted feature variables that constituted a network from sensory properties of textures and measured touch motions. Second, a Bayesian network model was constructed using these variables as nodes. Finally, the constructed model was interpreted.

3.1   Feature Extraction

The feature variables that became nodes of the Bayesian network are described, and all nodes were discrete variables.

Material Type. Thirty textures were classified into thirteen groups on the basis of their materials, as shown in Table 1. As a result, the node representing material types was quantized into thirteen levels in the following manner:

Material Type: Paper, wood, leather, fur, straw and rush grass, sponge, cloth, towel, felt, boa, artificial grass, metal, or fiber.

Sensory Properties of Textures. We produced five nodes on the basis of adjective ratings "rough-smooth," "uneven-flat," "hard-soft," "warm-cold," and "sticky-slippery," which were acquired in Experiment 1. These ratings were quantized into two levels across their averages. For example, the standardized "rough-smooth" rating of short-hair fake fur was -1.8 (+: rough; -: smooth), which was lower than the average 0. Therefore, the *Micro Roughness* node of short-hair fake fur was labeled "smooth." On the other hand, the standardized "hard-soft" rating of walnut wood was 1.3 (+: hard; -: soft), which was higher than the average 0. Thus, the *Hardness* node of walnut wood was assigned the "hard" label. The five adjective ratings were quantized in the following manner:

Macro Roughness: Flat or bulky

Micro Roughness: Smooth or rough

Hardness: Soft or hard

Warmness: Cold or warm

Friction: Sticky or slippery

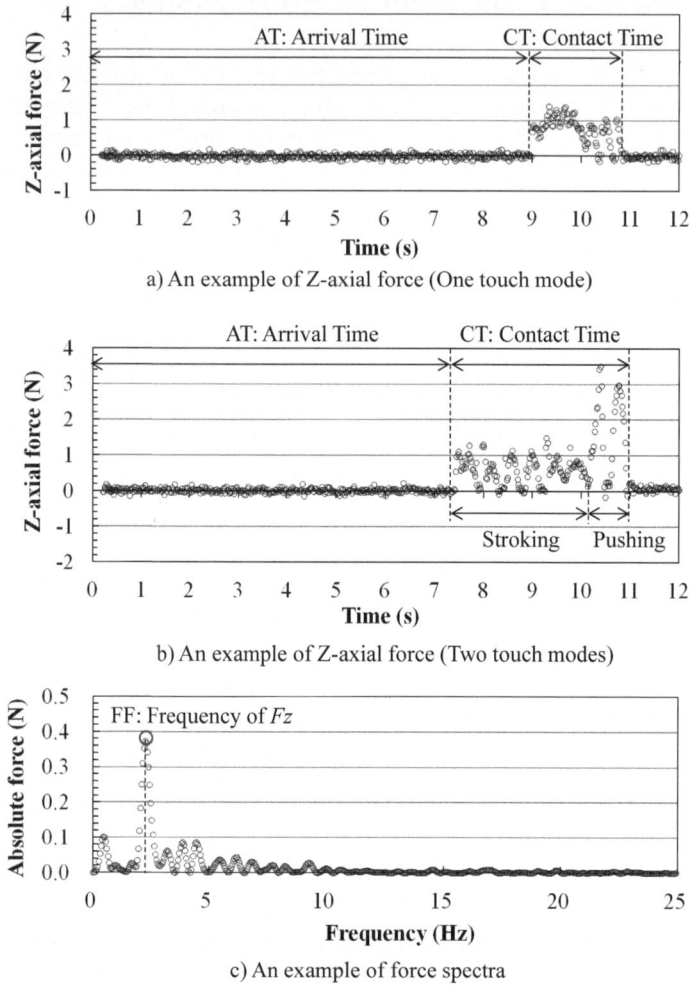

a) An example of Z-axial force (One touch mode)

b) An example of Z-axial force (Two touch modes)

c) An example of force spectra

Fig. 4. Examples of Z-axial force and force spectra

Properties of Touch Motions. Human touch motions were measured using the camera and force sensor. From measured data, we produced eight feature variables. The ratings of each feature variable were normalized within a single participant and were then averaged across all participants. All nodes were discretized into two qualitative states across each average.

Two examples of Z-axial force and an example of force spectra are presented in Figs. 4a, b, and c, respectively. An example of distribution of fingertip position in a two-dimensional space (X-Y space) is depicted in Fig. 5. The following nodes were extracted from these force and position data:

Arrival Time: Short or long

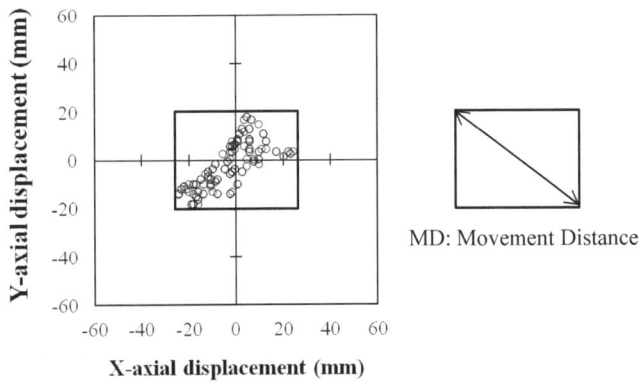

Fig. 5. An example of distribution of finger position

As depicted in Fig. 4a, the *Arrival Time* node was determined from a period between the starting time of the experiment ($t = 0$) at which a sound cue was presented to participants and the time at which the force signal began changing. The *Arrival Time* node was quantized into two levels across its average of normalized values 0. For example, the normalized value of *Arrival Time* for coarse Japanese paper was 2.0 (+: long; -: short), which was higher than zero. Therefore, the *Arrival Time* node of coarse Japanese paper was assigned the "long" label.

Contact Time: Short or long

As depicted in Fig. 4a, the *Contact Time* node was determined as the period during which the texture was touched. The time was quantized into two levels: long and short. For example, for cypress wood, the standardized contact time of -1.2 (+:long; -:short) was assigned the "short" label.

Average Z-axial Force: Weak or strong

The *Average Z-axial Force* node was determined from the average Z-axial force while in contact with the textures.

Maximum Z-axial Force: Weak or strong

The maximum Z-axial force during contact determined the *Maximum Z-axial Force* node.

Frequency of Z-axial Force: Low or high

The *Frequency of Z-axial Force* node was determined from a peak frequency of force spectra, as depicted in Fig. 4c.

Movement Distance: Short or long

As depicted in Fig. 5, we produced a minimum rectangle area surrounding a distribution of the fingertip position. The *Movement Distance* node was determined from a diagonal of the area.

Average Hand Velocity: Slow or fast

We calculated fingertip velocities from time series data of the fingertip position. The *Average Hand Velocity* node was determined from the average fingertip velocity while in contact with the textures.

Maximum Hand Velocity: Slow or fast

The maximum fingertip velocity while in contact with each texture determined the *Maximum Hand Velocity* node.

Mode of Touching Textures

Touch Mode: Soft touch, stroke, push, or scrub

We produced the *Touch Mode* node in order to quantize the modes of touch into four levels. This node was determined from combinations of the *Maximum Z-axial Force* and *Movement Distance* nodes. If the *Maximum Z-axial Force* node was strong and the *Movement Distance* node was long, we determined that the *Touch Mode* node was "scrub." Similarly, we determined three modes of touching: "push" (*Maximum Z-axial Force*: strong, *Movement Distance*: short), "stroke" (*Maximum Z-axial Force*: weak, *Movement Distance*: long), and "soft touch" (*Maximum Z-axial Force*: weak, *Movement Distance*: short).

Change in Touch Mode: Not changed or changed

As depicted in Fig. 4b, the touch mode changed during exploration for some textures and participants. We produced the *Change in Touch Mode* node in order to differentiate examples in which a touch mode changed from those in which the mode did not change. This node was quantized into two levels by the majority mode.

3.2   Structure Learning

We developed a Bayesian network model using the measured data in Experiments 1 and 2. For learning the network model, K2 algorithm and Bayesian information criteria were used. The constructed network is presented in Fig. 6.

3.3   Interpretation of Structure

We briefly interpreted the constructed network on the basis of some examples. The network that obtained the evidence of stroke touch mode estimated the probability of the *Friction* node. The probability of the *Friction* node being slippery was 0.75. The examples of corresponding textures are the short-hair fake fur and felt, which are likely to invite the stroking motion with weak *Maximum Z-axial Force* and long *Movement Distance*.

The network that obtained the *Touch Mode* node of "scrub" estimated that the probability of the *Micro Roughness* being rough was 0.88. The corresponding textures, such as coarse woven straw, are likely to invite the scrubbing motion with strong *Maximum Z-axial Force* and long *Movement Distance*.

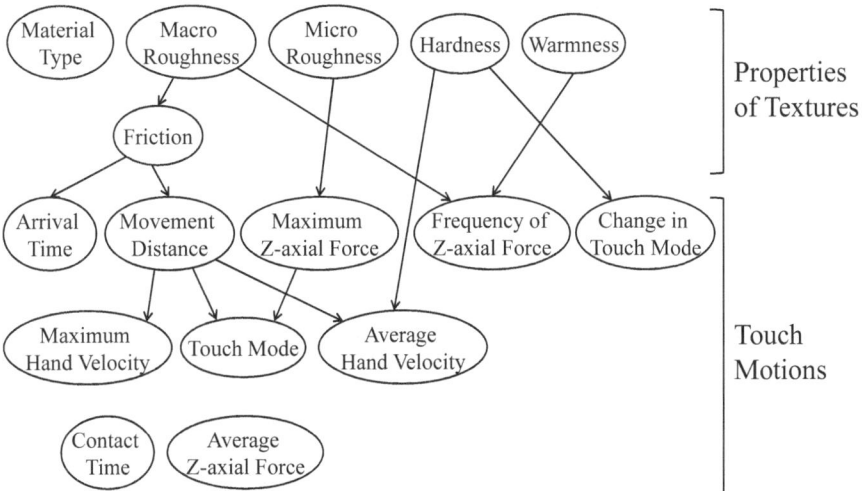

Fig. 6. Bayesian network model

The network that obtained the *Frequency of Z-axial Force* node of "high" estimated that the probability of the *Macro Roughness* being flat was 0.70. The walnut wood and fake suede 1 are examples of corresponding textures, which are likely to invite the high *Frequency of Z-axial Force*.

The network that obtained the *Maximum Hand Velocity* node of "fast" estimated that the probability of the *Macro Roughness* being flat was 0.69. The corresponding textures, such as cypress wood and artificial grass, are likely to invite the fast *Maximum Hand Velocity*.

When the network obtained the *Hardness* node of "soft" and the *Warmness* node of "cold," the probability of the *Touch Mode* being a soft touch mode was estimated to be 0.57. The crumpled paper is an example of a corresponding texture.

The above estimations suggested that the network is potentially useful for designing textures that invite specific human touch motions. However, a detailed validation of the networks is the next challenge.

In this constructed network, the *Material Type* node was isolated perhaps because the even textures belonging to the same material category are very different. For example, although the crumpled paper and embossed paper 2 are both in the paper category, their four sensory property nodes excluding the *Micro Roughness* node were very different. Further, the *Contact Time* and *Average Z-axial Force* nodes were not connected with the other nodes. These nodes have the potential to connect with the nodes we did not use in this study.

4   Conclusion

We investigated the stochastic relationships between touch motions and the properties of thirty textures that invite human touch. The sensory properties of textures were measured through sensory evaluation, and the features of human touch motions were determined from experiments in which participants freely touched the textures they felt compelled to touch. The Bayesian network model was constructed using these features. A brief interpretation suggested that the network is potentially useful for designing textures that invite specific human touch motions, such as a stroke or push. A re-construction of networks with a sufficient number of data and a detailed validation of the networks are the next challenges.

References

1. Nagano, H., Okamoto, S., Yamada, Y.: Physical and sensory factors of textures that appeal to human touch. In: Proceedings of the 4th International Conference on Human System Interactions, pp. 324–329 (2011)
2. Nagano, H., Okamoto, S., Yamada, Y.: What appeals to human touch? effects of tactual factors and predictability of textures on affinity to textures. In: Proceedings of the IEEE 2011 World Haptics Conference, pp. 203–208 (2011)
3. Klatzky, R.L., Peck, J.: Please touch: Object properties that invite touch. IEEE Transactions on Haptics (to appear, 2012)
4. Lederman, S.J., Klatzky, R.L.: Hand movements: A window into haptic object recognition. Cognitive Psychology 19, 342–368 (1987)
5. Peine, W.J., Wellman, P.S., Howe, R.D.: Temporal bandwidth requirements for tactile shape displays. In: Proceedings of the IMECE Haptics Symposium, pp. 107–113 (1997)

Haptically Induced Illusory Self-motion and the Influence of Context of Motion

Niels C. Nilsson, Rolf Nordahl, Erik Sikström, Luca Turchet, and Stefania Serafin

Medialogy Department of Architecture, Design and Media Technology
Aalborg University Copenhagen
{ncn,rn,es,tur,sts}@create.aau.dk

Abstract. The ability of haptic stimuli to augment visually and auditorily induced self-motion illusions has in part been investigated. However, haptically induced illusory self-motion in environments deprived of explicit motion cues remain unexplored. In this paper we present an experiment performed with the intention of investigating how different virtual environments – contexts of motion – influences self-motion illusions induced through haptic stimulation of the feet. A concurrent goal was to determine whether horizontal self-motion illusions can be induced through stimulation of the supporting areas of the feet. The experiment was based on the a within-subjects design and included four conditions, each representing one context of motion: an elevator, a train compartment, a bathroom, and a completely dark environment. The audiohaptic stimuli was identical across all conditions. The participants' sensation of movement was assessed by means of existing measures of illusory self-motion, namely, reported self-motion illusion per stimulus type, illusion compellingness, intensity and onset time. Finally the participants were also asked to estimate the experienced direction of movement. While the data obtained from all measures did not yield significant differences, the experiment did provide interesting indications. If motion is simulated through implicit motion cues, then the perceived context does influence the magnitude of displacement and the direction of movement of self-motion illusions as well as whether the illusion is experienced in the first place. Finally, the experiment confirmed that haptically induced illusory self-motion in the horizontal plane is indeed possible.

1 Introduction

During our everyday interaction with the world the sensation of self-motion remains largely unnoticed. However, we become increasingly concious of this sensation during those rare moments where we experience a sensation of movement despite being stationary. A well-known example is the incorrect motion perception one may experience when being on a motionless train, looking out the window at the adjacent track where another stationary train is located. When this second train departs from the station, one may experience a transient, yet compelling, illusion of being on the train which is moving. This experience is a naturally occurring instance of visually induced illusory self-motion, also referred to as vection [6]. Our susceptibility to such illusions may at least in part be explained by the misleading nature of visual motion stimuli [3]. That is to say, visual motion stimuli are open to not one, but two perceptual interpretations [1].

P. Isokoski and J. Springare (Eds.): EuroHaptics 2012, Part I, LNCS 7282, pp. 349–360, 2012.
© Springer-Verlag Berlin Heidelberg 2012

Either the sight of the moving train leads to to egocentric motion perception if the train passenger correctly perceives himself as being stationary while the train in the adjacent track is moving, or else the visual stimuli lead to exocentric motion perception if the train passenger falsely perceives the surroundings as being stationary while he is moving. Self-motion illusions occurring along some line are refereed to as linear illusory self-motion, while the erroneous sensation of rotation about one or more of the three bodily axes is refereed to as circular illusory self-motion [16].

Self-motion illusions are influenced by the properties of the physical stimuli (bottom-up factors) as well the perceiver's expectations to, and interpretation of, the stimuli (top-down factors) [13]. Riecke and colleagues [11] summarize a number of the bottom-up factors that may influence the onset time, duration, and intensity of the self-motion illusion. These factors include, but are not limited to, the movement speed of the stimulus, the area of the visual field occupied by the display, and the perceived depth structure of the visual stimulus. While the influence influence of bottom-up factors have been studied extensively (e.g.[2,4,18]), evidence suggesting that top-down factors are consequential does exist. To exemplify, it has been shown [13,9,19] that both circular and linear self-motion illusions may be influenced by whether participants are seated in a chair that potentially could move as opposed to one that is immovable. Moreover, it has been been demonstrated that self-motion illusions in some circumstances may be influenced by whether the participants,before being exposed to visual motion stimuli, are asked to attend to the sensation of self-motion or object motion [8]. Auditory motion stimuli is, just as their visual counterparts, open to not one, but two perceptual interpretations, and they may thus lead to either exocentric or egocentric motion perception. Indeed, a sensation of self-motion may be experienced by blinded listeners exposed to sound sources moving relative to their position [16].

In this paper we describe an experiment performed with the intention of investigating how different contexts of motion influence haptically induced illusory self-motion on behalf of individuals exposed to virtual environments devoid of any explicit motion cues. To be more exact, we compared four scenarios involving identical implicit motion cues (auditory and haptic stimuli), but different contexts of motion (visual stimuli depicting an elevator, a train, a bathroom, and a completely dark environment).

2   Related Work

Research on haptically induced illusory self-motion is rather scarce and with a few exceptions [14,7] the experiments have generally focused on whether this form of stimuli positively influences an illusion of movement facilitated by stimulation of another modality [12,17].

Väljamäe and colleagues [17] describe a study performed with the aim of investigating whether sensation of auditorily induced linear illusory self-motion may be intensified by the addition of vibrotactile feedback delivered by means of low frequency sound and mechanical shakers. The authors of that study found that the self-motion illusion was significantly higher during exposure to the mechanically induced vibration. Notably their results also showed that the auditory-tactile simulation of a vehicle engine was as effective as illusions induced via auditory feedback including explicit motion cues, i.e., moving sound fields. Riecke et al. [12] similarly describe an experiment investigating whether physical vibrations of the perceivers' seat and footrest enhance visually induced circular vection. They found that the addition of this form of vibrotactile feedback entailed a slight, yet significant, enhancement of the self-motion illusion.

As it is the case for the influence of haptic feedback on illusory self-motion, also vertical self-motion illusions, that is, perceived movement along the longitudinal axis, remains almost unexplored. One such study, performed by Wright and colleagues [19], aimed at investigating the vestibular and visual contributions to vertical illusory self-motion.

Inspired by the study described by Roll et al. [14], Nordhahl and colleagues [7] performed an experiment intended to determine if it is possible to facilitate vertical illusory self-motion on behalf of unrestrained participants exposed to a immersive VE by haptically stimulating the main supporting areas of their. The dominance of vertical self-motion illusions in the experiment described by Nordahl et al. [7] is arguably a testament to the influence of top-down factors. The participants' past experiences entailed that the context of motion (the virtual elevator) may have coloured their interpretation of the implicit motion cues delivered through auditory and haptic feedback. We hypothesize that when no explicit motion cues are present, then the context of motion may fluencine self-motion illusions induced through implicit motion cues. That is to say, 1) Self-motion illusions are more likely to occur during exposure to virtual environments where the context of motion suggests that movement indeed is possible. 2) The experienced magnitude of displacement is likely to correspond to the magnitudes of displacement associated with the particular context of movement. 3) If the context of motion suggests that movement in a particular direction is possible, then illusory self-motion in that directions is more likely to be experienced.

The experiment described in the current paper should to a large extent be considered as a continuation of work described by Nordahl et al. [7] since it was performed with the intention addressing these three claims. Moreover this implies that it was an implicit goal to determine whether it is possible to induce horizontal self-motion illusions within the context of a virtual environment by haptically stimulating the feet of unrestrained participants.

3   Experiment Design

A within-subjects design was used in order to minimize the effects of the high between-subject variability which often is found in studies of illusory self-motion [10]. The experiment included four conditions, each one representing a different contexts of motion. The virtual environments used to represent the four contexts of motion were the interior of an elevator, a train carriage, a bathroom, and a completely dark environment. The elevator and the train were chosen because they sere as contexts suggesting linear, vertical and horizontal movement, respectively. The particular bathroom was chosen on grounds that it was regarded as unlikely that individuals associate this room with movement. Finally the dark environment was included since it did not impose a context of motion upon the participants. While the visual stimuli differed across the four conditions, the auditory and haptic stimuli were identical. The auditory feedback comprised sounds reminiscent of those produced by an engine. However, these were not identifiable as any particular vehicle or machine. The signal used to control the haptic feedback was a sawtooth waveform. This signal was chosen based on the findings of Nordahl et al. [7]. The intention was for the auditory and haptic stimuli to serve as implicit motion cues. All the stimuli used for the experiment were were devoid of any explicit motion cues. The elevator had opaque walls, the windows of both the train and the bathroom were covered by blinds, and the dark environment did not include visual feedback of any kind. The sound was similarly not spatialized.

3.1   Environment Simulation

The four virtual environments were simulated using the same multimodal architecture used by Nordahl et al. [7]. This architecture was originally developed for the purpose of simulating walking-based interactions through visual, auditory and haptic stimuli [15].

Simulation Hardware. The user interacts with the system by performing natural movements which in turn are registered by the system. The position and orientation of the users head is tracked by means of a 16 cameras Optitrack motion capture system (Naturalpoint) and the forces exerted during foot-floor interactions are registered by a pair of customized sandals augmented with actuators and pressure sensors [15]. Two FSR pressure sensors (I.E.E. SS-U-N-S-00039) are placed inside the sole of each sandal at the points where the toes and heel come into contact with the sole. The analogue values of each of these sensors were digitalized by means of an Arduino Diecimila board. The actuators responsible for delivering the haptic feedback are placed at roughly the same positions. Each sandal is embedded with four of these electromagnetic recoil-type actuators (Haptuator, Tactile Labs Inc., Deux-Montagnes, Qc, Canada), which have an operational, linear bandwidth of 50 to 500 Hz and can provide up to 3 G of acceleration when connected to light loads. Figure 1 illustrates the placement of the pressure sensor and actuators in the heel of one sandal. The visual feedback is delivered through a nVisor SX head-mounted display, with a resolution of 1280x1024 in each eye and a diagonal field of view of 60 degrees. While the multimodal architecture in its original form is capable of delivering auditory feedback using a surround sound system composed by 12 Dynaudio BM5A speakers, a set of headphones (Ultrasone HFI-650) were used during the current experiment. The reason being, that the actuators generate sound while activated, which might make up an undesirable bias. Thus the headphones both served the purpose of providing auditory feedback and masking out the undesired sounds.

Simulation Software. The visual representations of the four environments (see Figure 2) were produced in the multiplatform development environment Unity 3d which facilitates stereoscopic viewing by the placement of two cameras within one environment. Dynamic eye convergence and divergence was simulated by means of a simple raycasting algorithm ensuring that the cameras are always aimed at the closest object immediately in front of the user. The auditory feedback was based on a recording of an industrial fan (freesound.org). The recording was edited into a loop which is 7.3 seconds long and loops seamlessly. The audio loop was played back at 30% reduced speed. A

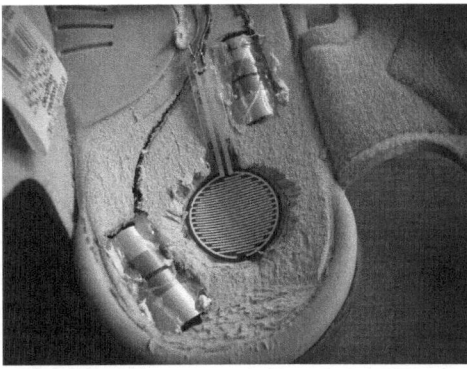

Fig. 1. Placement of a pressure sensor and two actuators in the heel of one sandal

high-pass and a low-pass filter was added to allow for fine tuning of the playback during the final preparations of the experiment. The auditory feedback was delivered using the Max/MSP realtime synthesis engine, which also was used for the synthesis and delivery of the signal used to control the actuators providing haptic feedback. The signal in question was a sawtooth waveform with a frequency of 50 Hz and a symmetric trapezoidal envelope. This signal was chosen since it was the one that Nordahl et al. found the most effective at eliciting self-motion illusions [7]. The data obtained from the pressure sensors was used to ensure that vibration only was activated when the foot is in contact with the ground. A schematic drawing the multimodal architecture used to simulate the virtual elevator can be seen on Figure 3.

3.2 Measures of Illusory Self-motion

The participants' experience of illusory self-motion was assessed by means of existing measures of self-motion illusions, namely, reported self-motion illusion per stimulus type, illusion compellingness, intensity and onset time [16].

The reported self-motion illusion per stimulus type simply corresponds to a binary measure of whether illlusory sef-motion were experienced or not. The compellingness (or convincingness) of the illusion was assessed by asking the participants to rate their sensation on a magnitude estimation scale from '0' to '5' where '0' signified no perceived movement and '5' corresponded to fully convincing movement.

The intensity of the illusion was measured by asking the participants to estimate the magnitude of the displacement on a scale familiar to them (meters or feet). No experienced movement would correspond to a displacement of zero meters. It should be noted that past experiments where intensity has been operationalized as the magnitude of the displacement [19], have included stimuli providing information about the distance to, or size of, objects based on which estimates of distance could be made. The illusion onset time (or latency) was measured as the time elapsed from the onset of the stimuli until the onset of the illusion. The measures of both compellingness, and intensity were adapted from [19]. Finally the participants were asked to estimate the direction in which the believed to have moved.

Fig. 2. Screen shots of three of environments used for the experiment: the train, the elevator, and the bathroom

3.3   Participants and Procedure

A total of 18 participants partook in the experiment (15 men and 3 women) aged between 19 and 40 years (mean = 25.8, standard deviation = 5.4). Before exposure to the VE, the participants were introduced to the scenarios they were about to experience and were asked to attend to the sensation of movement. Moreover it was stressed that we were interested in the participants' honest opinion rather than answers brought about by any assumptions regarding the demand characteristics of the experiment. During the exposure to the four virtual environments the participants were placed on a wooden platform, which they were made to believe might move during one or more of the conditions. The participants were unable to see the experimental setup for the duration of the experiment. This was done since it has been shown that the convincingness of self-motion illusions significantly increases when subjects believe that actual motion may occur [13]. Before the beginning of each trial the participants were placed at the same

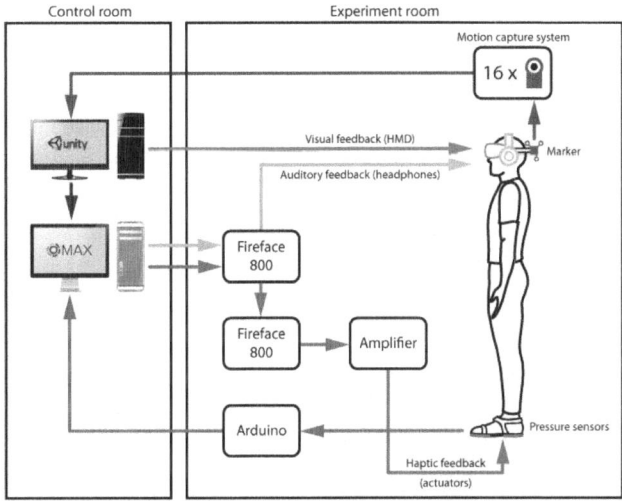

Fig. 3. A schematic drawing of the multimodal architecture used to simulate the virtual environments

position and were asked to face the same direction. The participants were exposed to all four conditions for one minute and after each exposure the participants were asked to answer the provided questions verbally. The order of the conditions was randomized so as to control potential order effects.

4   Results

Table 1 shows the results pertaining to the reported self-motion illusion per stimulus type, that is, the number of participants who experienced a self-motion illusion across the four conditions. However a comparison by means of a Cochran's Q test did not yield any significant difference between the four conditions ($Q(3) = 6.0567, p = 0.11$).

Table 2 summarizes the results obtained from the measures of illusion onset time, compellingness, and intensity. The bar charts presented in figures 6 and 5 provide a graphical overview of these three sets of results. One-ways analyses of variance (ANOVAs) were used to compare the averages obtained from the measures of the compellingness and intensity of the self-motion illusion across the four conditions Significant differences was found in relation to illusion intensity ($F(3,41) = 5.28, p = 0.003$). While the analysis of the results pertaining to illusion compellingness was borderline significant ($F(3,68) = 2.38, p = 0.07$) the same cannot be said of the results related to

Table 1. Reported self-motion illusion per stimulus type

Elevator	Train	Bathroom	Dark
13	15	8	11

Table 2. Mean ± one standard deviation pertaining to three of the measures of illusory self-motion. Values in parenthesis indicate the number of reports based on which the mean and standard deviations were determined.

	Compellingness	Intensity (meters)	Onset time (sec.)
Elevator	2.3 ± 1.8 (18)	27.8 ± 33.8 (14)	19.9 ± 13.57 (10)
Train	2.7 ± 1.7 (18)	443.5 ± 623.1 (10)	22.8 ± 17.9 (14)
Bathroom	1.2 ± 1.5 (18)	25.0 ± 62.2(12)	22.7 ± 9.3 (5)
Dark	1.9 ± 1.9 (18)	12.8 ± 33.1 (9)	26.1 ± 13.4(8)

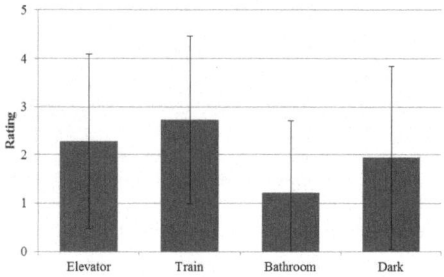

Fig. 4. Mean compellingness ratings. Error bars indicate ± one standard deviation.

the illusion onset time ($F(3,33) = 0.25, p = 0.86$). The latter did not come as a surprise since the number of registered onset times differed greatly from condition to condition, since a large number of participants neglected to report the onset time and no times recorded when no illusion was experienced. Subsequently post-hoc analysis of the results pertaining to illusion intensity was performed by means of Tukey's procedure. This pairwise comparison of the means revealed that conditions the Train condition differed significantly from the remaining three while none of the three differed significantly from one another.

Finally, Table 3 summarizes the results pertaining to the question of what direction the elevator was moving in. It is worth mentioning that three participants experienced movement in directions which differed from the norm. When exposed to the virtual

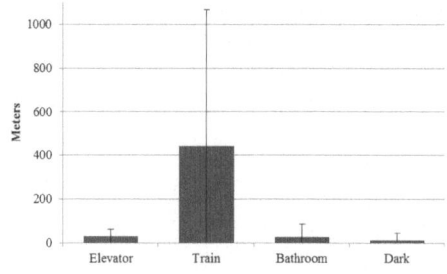

Fig. 5. Mean illusion intensity in meters. Error bars indicate ± one standard deviation.

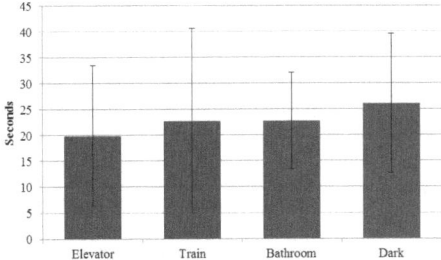

Fig. 6. Mean illusion onset time in seconds. Error bars indicate ± one standard deviation.

Table 3. Frequency of the participants' estimates of the direction of movement across the four conditions. The directions are relative to the participants' orientations at the beginning of each trial, which were identical for all participants.

	Elevator	Train	Bathroom	Dark
Upwards	7	0	0	2
Downwards	6	0	0	3
Forwards	0	9	5	1
Backwards	0	3	0	1
Other	0	0	2	1
Unsure	0	3	1	3

bathroom one participant experienced leftwards movement, while another experienced illusory full-body leaning, alternating from one direction to another. Finally, one participant experienced "roller-coaster like movement" when exposed to the dark condition, that is, the participant had a sensation of moving forward while simultaneously either moving up or downwards.

5   Discussion

Interestingly the reported self-motion illusion per stimulus suggests that all four contexts of motion may elicit self-motion illusions. Indeed more participants experienced illusory self-motion when exposed to the train, compared to the elevator. Thus it would seem that haptically induced illusory self-motion is possible. While no significant differences between the four groups were found, it is notable that the bathroom elicited the lowest number of illusions on behalf of the participants. This indication is to some extent also mirrored in the results obtained from the employed measure of illusion compellingness. That is, exposure to the bathroom gave rise to the least compelling illusions of movement. Previously we suggested that self-motion illusions may be more likely to occur during exposure to virtual environments where the context of motion suggests that movement indeed is possible. At first glance, the obtained results seems to contradict this claim. The context of motion suggesting no movement (the bathroom) did yield self-motion illusions on behalf of some participants. However five of the eight participants who experienced illusions in this environment remarked that they had become convinced that they were on a ship. Considering that the bathroom was purposely selected because it did not appear like a bathroom one would find on a normal ship, plane or other moving vehicle, it is interesting that close to a third of the participants made up exactly this explanation when attempting to make the seemingly conflicting information meaningful. With some caution, one may argue that this indicates just how far our brains are will to go in order to integrate conflicting multimodal stimuli into one meaningful percept. So it would seem that some of the participants after all did rely on the top-down factors since the illusion may have been made possible by their expectations to, and interpretation of, the stimuli.

A significant difference was found between the mean illusion intensity related to the train condition and the remaining three conditions. However, this does not necessarily imply that the self-motion illusions experienced during exposure to the train are superior to the ones elicited by the elevator or the other two conditions for that matter. It suggests that the audiohaptic stimuli in average leads to an experience of a larger displacement when paired with virtual train. This does arguably lend some credence to the claim that the experienced magnitude of displacement is likely to correspond to the magnitudes of displacement associated with the particular context of movement. It is interesting to note the large standard deviation pertaining to this mean. However, it seems possible that the disagreement amongst the participants may be explained by the large range of possible speeds achievable when on one is a train.

The vast majority of the participants who experienced illusory self-motion during exposure to the elevator and the train did so in the vertical and horizontal plane, respectively. The few who did not follow this pattern were did not experience illusions were not meaningful given the supplied context of movement, but were unsure about the direction of movement. Notably the results seemed to correspond with the ones reported by Nordahl et al. [7] since no tendencies seem readily apparent in regards to the perceived direction of movement elicited by the virtual elevator. That is to say, while they all experienced vertical movement, an almost equal number experienced forwards and backwards movement. The same is not true in regards to the train. When exposed to this virtual environment most of the participants experienced forward movement. Anecdotal evidence obtained from one participant may provide a possible answer. This participant explained that the cable connected to the head-mounted display had caused him to experience forward movement. The subtle resistance provided by the cable may in some capacity have been experienced as the gravitational force experienced during acceleration and since this cable was connected behind the participants during the beginning of each trial this may have lead to the interpretation of the train moving in that particular direction. Moreover it is interesting to not that most of the participants who experienced movement inside the virtual bathroom did so in the horizontal plane. The directions of movement experienced during exposure to the dark environment were less consistent. It would seem that this data is in support of the claim that if the context of motion suggests that movement in a particular direction is possible, then illusory self-motion in that directions is more likely to be experienced. This is particularly evident from the results related to the virtual elevator and train. However, it is interesting to note that the bathroom condition, which were interpreted as a the interior of a ship, primarily left to illusions in the horizontal plane while the dark environment, which were open to more interpretations, also lead to less consistent answers.

6 Conclusion

In this paper we have described an experiment performed with the intention of investigating how different contexts of motion influences self-motion illusions induced through haptic stimulation of the feet. The experiment was based on the a within-subjects design and all 18 participants thus experienced the same four conditions. The audiohaptic stimuli was identical across all conditions but the context of motion was varied. The participants experienced the interior of an elevator, a train compartment, a bathroom, and a completely dark environment. The four virtual environments were devoid of any explicit motion cues and the resulting self-motion illusions were thus the consequence of implicit motion cues. The participants' sensation of movement was assessed by means of self-reported measures of illusory self-motion, namely, reported self-motion illusion per stimulus type, illusion compellingness, intensity and onset time. Finally the participants were also asked to estimate the experienced direction of movement. While the data obtained from all measures did not yield significant differences the experiment did provide interesting indications. It would seem that if motion is simulated through implicit motion cues, then the perceived context does influence the magnitude of displacement and the direction of movement of self-motion illusions as well as whether the illusion is experienced in the first place.

References

1. Brandt, T., Dichgans, J., Koenig, E.: Differential effects of central versus peripheral vision on egocentric and exocentric motion perception. Experimental Brain Research 16(5), 476–491 (1973)
2. Dichgans, J., Brandt, T.: Visual-Vestibular Interaction: Effects on Self-Motion Perception and Postural Control, vol. VIII, pp. 756–804. Springer (1978)
3. Goldstein, E.: Sensation and perception. Wadsworth Pub. Co. (2009)
4. Hettinger, L.J.: Illusory Self-motion in Virtual Environments, pp. 471–492. Lawrence Erlbaum (2002)
5. Lappe, M., Bremmer, F., Van den Berg, A.: Perception of self-motion from visual flow. Trends in Cognitive Sciences 3(9), 329–336 (1999)
6. Lowther, K., Ware, C.: Vection with large screen 3d imagery. In: Conference Companion on Human Factors in Computing Systems: Common Ground, pp. 233–234. ACM (1996)
7. Nordahl, R., Nilsson, N.C., Turchet, L., Serafin, S.: Vertical illusory self-motion through haptic stimulation of the feet. In: Proceedings of IEEE VR 2012 Workshop on Perceptual Illusions in Virtual Environments PIVE (2012)
8. Palmisano, S., Chan, A.: Jitter and size effects on vection are immune to experimental instructions and demands. Perception-London 33, 987–1000 (2004)
9. Riecke, B., Feuereissen, D., Rieser, J.: Auditory self-motion illusions (circular vection) can be facilitated by vibrations and the potential for actual motion. In: Proceedings of the 5th Symposium on Applied Perception in Graphics and Visualization, pp. 147–154. ACM (2008)
10. Riecke, B., Feuereissen, D., Rieser, J., McNamara, T.: Self-motion illusions (vection) in vr are they good for anything? In: Proceedings of IEEE VR 2012 (2012)

11. Riecke, B., Schulte-Pelkum, J., Avraamides, M., von der Heyde, M., Bülthoff, H.: Scene consistency and spatial presence increase the sensation of self-motion in virtual reality. In: Proceedings of the 2nd Symposium on Applied Perception in Graphics and Visualization, pp. 111–118. ACM (2005)
12. Riecke, B., Schulte-Pelkum, J., Caniard, F., Bülthoff, H.: Influence of auditory cues on the visually-induced self-motion illusion (circular vection) in virtual reality. In: Proceedings of Eigth Annual Workshop Presence (2005)
13. Riecke, B., Västfjäll, D., Larsson, P., Schulte-Pelkum, J.: Top-down and multi-modal influences on self-motion perception in virtual reality. In: HCI International (2005)
14. Roll, R., Kavounoudias, A., Roll, J.: Cutaneous afferents from human plantar sole contribute to body posture awareness. Neuroreport 13(15), 1957 (2002)
15. Turchet, L., Nordahl, R., Serafin, S., Berrezag, A., Dimitrov, S., Hayward, V.: Audio-haptic physically-based simulation of walking on different grounds. In: 2010 IEEE International Workshop on Multimedia Signal Processing (MMSP), pp. 269–273. IEEE (2010)
16. Valjamae, A.: Auditorily-induced illusory self-motion: A review. Brain Research Reviews 61(2), 240–255 (2009)
17. Väljamäe, A., Larsson, P., Västfjäll, D., Kleiner, M.: Vibrotactile enhancement of auditory induced self-motion and presence. Submitted to Journal of Audio Engineering Society (2005)
18. Warren, R., Wertheim, A.H.: Perception & Control of Self-Motion. Erlbaum, London (1990)
19. Wright, W., DiZio, P., Lackner, J.: Perceived self-motion in two visual contexts: dissociable mechanisms underlie perception. Journal of Vestibular Research 16(1), 23–28 (2006)

Rendering Stiffness with a Prototype Haptic Glove Actuated by an Integrated Piezoelectric Motor

Pontus Olsson[1], Stefan Johansson[2], Fredrik Nysjö[1], and Ingrid Carlbom[1]

[1] Centre for Image Analysis, Uppsala University, Sweden
{pontus.olsson,fredrik.nysjo,ingrid.carlbom}@cb.uu.se
[2] Dept. of Engineering Sciences, Uppsala University, Sweden
stefan.johansson@angstrom.uu.se

Abstract. Bi-directional haptic devices incorporate both sensors and actuators. While small and compact sensors are readily available, actuators in haptic interfaces require a significant volume to produce needed forces. With many actuated degrees of freedom, the mass and size of the actuators become a problem in devices such as haptic gloves. Piezo-technology offers the possibility of compact actuators which can be controlled with high accuracy. We describe a prototype admittance-type haptic device for the hand with a compact integrated piezoelectric motor. The current implementation provides one degree of freedom, but it could be extended with more motors for additional degrees of freedom. We demonstrate both the accuracy with which the device can reproduce force-displacement responses of non-linear elastic material stiffness and the device's fast and stable response to an applied load.

Keywords: Haptics, Haptic Glove, Piezoelectric Motor, Actuator.

1 Introduction

A haptic device must incorporate both sensors and actuators to provide a bi-directional force interface. Force or position sensors are commonly used to sense input from the user. Similarly, force or position actuators are central to haptic output, the former by exerting forces onto the human operator, the latter by controlling the position of the parts of the interface in contact with the operator's body. While small and compact sensors are readily available and suitable for integration into a haptic device, actuators occupy a significantly larger volume in order to deliver the force levels required by many haptic applications. In devices that have few degrees of freedom and that are physically attached to ground, bulky actuators may be acceptable. However, for complex devices with many degrees of freedom, especially body-worn devices such as haptic gloves, the mass and volume of the actuators quickly becomes a problem. Piezo-technology offers the possibility to build very compact actuators which make them suitable for integration into haptic interfaces. Furthermore, the actuator position can be controlled accurately which in turn makes accurate feedback possible.

P. Isokoski and J. Springare (Eds.): EuroHaptics 2012, Part I, LNCS 7282, pp. 361–372, 2012.
© Springer-Verlag Berlin Heidelberg 2012

We describe a haptic device that gives one degree of freedom (DOF) proprioceptive feedback to the index finger, actuated by a compact integrated piezoelectric motor. The device can be attached to existing commercial haptic hardware, such as a Sensable Phantom [1], to provide additional degrees of freedom and the attachment to ground required for weight sensation. We evaluate the device's load-deflection response using target load-deflection curves acquired in a compression testing machine from physical samples with varying degree of non-linear stiffness. We use one synthetic load-deflection curve to measure the haptic device's ability to render rigidity. We also measure the dynamic response of the device when a load is applied.

2   Related Work

There are several examples of complex haptic devices with high degrees of freedom. The commercially available haptic interface CyberGrasp [2] for the hand provides five actuators (one for each finger) working remotely via mechanical tendons. The Rutgers Haptic Master II-ND [3] provides four actuated degrees of freedom with pneumatic pistons powered by pressurized air supplied via pneumatic tubes. Unlike the CyberGrasp, that is built as an exoskeleton around the hand, the Haptic Master works from the palm of the hand. High force magnitudes can be achieved since the actuators are not worn by the user but are placed adjacent to the workspace, making their size and weight less critical. However, remote actuation presents limitations and challenges. Wires, pulleys or pneumatic tubes are bulky and can limit the free movement of the device. And, precise and stable control is hard to achieve since the dynamics of the force transmission has to be taken into account, for instance friction, backlash of the wire transmission, air compressibility and flow resistance in the pneumatic tubes. The Hiro-III [4] uses compact integrated DC-motors for actuation, providing three degrees of actuated feedback to each of the five fingers. But integrated DC-motors require a transmission mechanism in order to provide sufficient torque, which introduces space-demanding moving parts that are sensitive to overload and may cause backlash.

An interesting alternative approach is a passive haptic system that uses controllable brakes. Such systems often utilize magnetorheological fluids [5] where the viscosity is controlled by an electromagnet. And, because these brakes are dissipative, i.e., they remove energy, the system is passive and inherently stable. But due to their passive nature, these devices are unable to render active behavior in a haptic simulation such as a beating heart or the expansion of a compressed spring.

Piezoelectric materials have been employed in haptic devices in the past, but their short actuation stroke has restricted their use primarily to tactile displays where they are used to generate vibrations [6] or to actuate pin-arrays [7]. Vibrations from piezoelectric elements have also been used to control the perceived friction of, and generate lateral forces on, a touched surface [8]. We demonstrate how a compact, yet large-stroke piezoelectric motor can be used as an actuator in a proprioceptive haptic device.

3   Device Design

3.1   Piezoelectric Motor

The core component in our haptic glove is a linear piezoelectric motor (Figure 1) that moves a drive rod with a piezoelectric unit in full friction contact with the rod at all times. There are two motor stators pressed against the drive rod with leaf springs in a twin, symmetric arrangement. The active part in each stator is a unit consisting of four piezoelectric legs. The legs can both be elongated and bent making it possible to "walk" an object, the drive rod, by driving the four legs in two pairs, 180 degrees out of phase. In the simplest case, employed by our haptic glove, each leg tip follows a rhombic trajectory and is in friction contact with the rod during the upper part of the rhomb while releasing and returning for the next step during the lower part of the rhomb. The drive rod is always in friction contact with at least two leg tips of each stator and, if the legs are not activated, the drive rod is held rigidly by the leg tips of the motor. If the load exceeds the holding force, the rod will simply slip without damaging the motor. The driving frequency is kept within a quasi-static (non-resonant) range which sets an upper limit to about 3 kHz; a higher frequency might damage the motor.

The great advantage with non-resonant operation is that the motor can deliver high forces within a small volume. The disadvantage is that the motor operates within the audible range and that the maximum speed is limited. In comparison with existing actuator solutions our motor has no flexibility or play. The non-resonant operation gives an extremely high dynamic speed range, limited only by the driving electronics, which in combination with the stiffness of the legs makes it possible to control the dynamic stiffness of the upper arm in the glove with high resolution. Motor specifications are shown in Table 1.

Fig. 1. Piezoelectric motor Piezo LEGS® LT2010, linear twin, from PiezoMotor AB

Table 1. Motor Specifications

Size	22x11x21 (LxWxH) mm
Max Force	20 N
Speed	15 mm/s
Weight	29 g
Resolution	<1 nm

3.2   The Haptic Glove Prototype

Our haptic glove (Figure 2) consists of an exoskeleton designed for the human hand. To facilitate rapid design iteration cycles, most parts are made with rapid prototyping plastic, using stereolithography (SLA), which cures a liquid photopolymer with a UV-laser, layer by layer. The thumb is placed in the thumb tube (A) and the index finger is placed in the thimble (B). To account for varying hand sizes, the thimble can be moved back and forth by releasing the screw (C) and the thumb tube can be lowered or raised by releasing the screw (D). The glove is fastened to the hand using the Velcro strap (E). The current design provides one actuated degree of freedom, namely the rotation of the metacarpophalangeal (MP) joint (F) of the index finger, allowing precision gripping using the thumb and index finger, which is one of the principal components of dexterous manipulation [9]. The rotation is implemented mechanically by allowing the upper arm (G), which is controlled by the index finger, to move along a circular arc with its center close to the MP joint. To constrain the upper arm motion to the desired arc, we use a curved rail guided relative to the mainframe of the glove by four ball bearings on each side of the arm, with two ball bearings above and two under each side of the rail, positioned so that the center of the rail curvature lies close to the MP joint axis.

The piezoelectric twin motor (H), LT2010 from PiezoMotor AB [10], placed in the back of the exoskeleton structure, is the actuator unit for the upper arm. The native linear motion of the actuator is mechanically coupled into a rotational movement of the upper arm around the MP joint (F). The leverage magnifies the speed five times (resulting in a maximum speed of 75 mm/s at an index finger length of 90 mm), while the maximum force is reduced with the same factor. The design of the exoskeleton makes it possible to include actuators for the remaining fingers in a later stage.

A magnetic position encoder (J) from Nanos Instruments GmbH [11] detects the movements of a magnetic scale (K) which is mechanically connected to the piezoelectric motor's drive rod (L). A thin film force sensor from Tekscan Inc. [12] measures forces applied in the thimble by the index fingertip in the positive and negative directions orthogonal to the upper arm (G). The sensor's force sensitive area is positioned right above the finger thimble, inside the tip of the upper arm.

Signals from the position encoder and force sensor and to the actuator are routed through a PMD90 communication box [10] connected to the computer via USB.

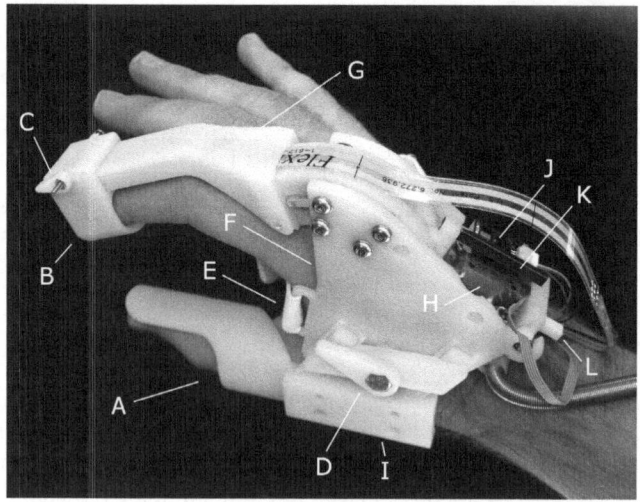

Fig. 2. Haptic glove with an integrated piezoelectric motor

Mechanical adapters can be attached to the connector (I) to allow the glove to be used with other haptic devices. We have built such an adapter for the attachment of the glove to a Phantom Premium [1]. Glove specifications are shown in Table 2. The motion range is dependent on the position of the adjustable thumb tube.

Table 2. Haptic Glove Specifications

Size	155 x 28 x 85 (L x W x H) mm
Force Range	±4 N
Speed Range	±75 mm/s
Motion Range	45 mm (min), 55 mm (max)
Weight	130 g

3.3   Force Sensor

The thin film FlexiForce B201 force sensor from Tekscan Inc. [12] can measure compression forces between 0 and 4 N, but no negative forces, so we preload it with a compressive force of approximately 0.5 N in order to detect forces in both directions with a larger force range in the compressive direction. Sensor specifications are shown in Table 3.

Table 3. Force Sensor Specifications

Size	14 x 14 x 0.2 (L x W x H) mm
Force Range	0 – 4 N
Hysteresis	5%
Weight	0.4 g (including flex cable)

3.4   Magnetic Position Encoder

The compact magnetic position encoder from Nanos Instruments GmbH [11] reads the position of the integrated piezoelectric motor. This position is used to estimate the position of the finger thimble. Encoder specifications are shown in Table 4.

Table 4. Position Encoder Specifications

Size	19 x 3 x 10 (L x W x H) mm
Position Resolution	61 nm
Weight	0.2 g

4   Control

Haptic devices can be classified as admittance or impedance devices [13]. Impedance devices are generally position-controlled, i.e., the user controls the position of the haptic end-effector and the device responds with a force dependent on the haptic context. This type of device is common due to its relatively low cost. Their back-drivability and low friction make them suitable for rendering low impedances, such as compliant virtual objects or free motion. In contrast, admittance devices are force controlled with the force applied by the user as input to a control loop. The actuators control the position of the haptic end-effector which constitutes the feedback. Hybrid designs use both position and force sensors to provide feedback in the control loop, and they may also use velocity sensors and accelerometers for more precise control [14]. The admittance design is often used to render high stiffness, e.g., in virtual assembly training [15]. We classify our proposed device as an admittance-type device because it gives position feedback from an applied force.

The force F applied by the operator's index fingertip onto the finger thimble generates a desired position of the thimble, x_d. The position encoder is part of an inner control-loop that drives the position x of the thimble towards the desired position x_d. The variable x defines the distance between the thimble and the thumb tube. The position encoder does not measure the position of the thimble directly, but it measures the position of the motor drive rod, x_m, from which we estimate the position of the thimble. In the current implementation, we assume a linear relationship with gain K_2 between motor position x_m and actual thimble position x at zero load. However, when a load is applied at the thimble, a non-negligible error, x_{com}, between estimated (x_{est}) and actual thimble position x appears due to some compliance of the prototype plastic and the mechanical couplings. This error is mitigated by modeling the compliance error x_{com} as a linear function with gain K_3 of the applied force F and by adding this deflection to x_{est}. The sensor gains $K_{1,2}$ are determined by calibration. (See Section 4).

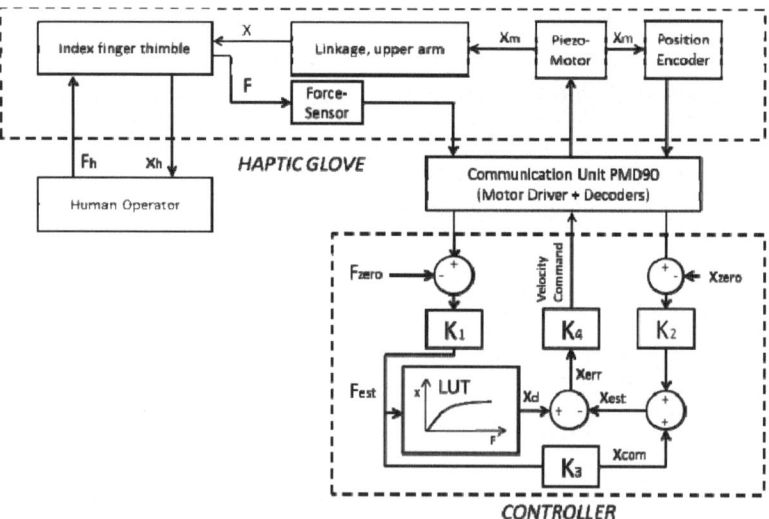

Fig. 3. System Schematics. The operator applies forces to the finger thimble, which gives position feedback actuated by the piezoelectric motor, governed by the controller.

In our experiments we use a direct mapping between the force F applied by the user and the desired opening x_d through a look-up table LUT with linear interpolation between samples. The LUT can store load-deflection curves sampled from real objects or synthetic data. After comparing the desired x_d and estimated x_{est} glove openings, the resulting error signal is fed through the gain K_4 to generate a velocity command for the motor. The gain K_4 is tuned manually for smooth and stable operation. Our controller is designed to render the direct contact and compression of a virtual object with a load-deflection curve that is previously defined and stored in the LUT. The controller could however be replaced by any haptic simulation with force input and position or velocity output. The controller is implemented in C++, and runs at 200 Hz.

5   Calibration

We set the gains and zero levels of the position encoder and force sensor to generate estimates of x_{est} and F_{est} (Figure 3) that are close to the true distance x between the thimble and the thumb tube, and the applied force F, respectively. After applying a series of known loads in the normal direction of the force sensor and recording the corresponding sensor values, we conclude that the force sensor response is linear, and fit a linear function by least squares methods to the data to derive force gain and zero force values (K_1 and F_{zero} in Figure 3). The variable F_{zero} corresponds to the force sensor value when no load is applied to the thimble.

We utilize a similar procedure to calibrate the position encoder. Here, we measure the encoder values with the glove in four known positions; fully closed, fully opened and two intermediate positions. We conclude that the encoder response also is linear, and fit a linear function to the data by least squares methods to obtain the gain and zero position values (K_2 and x_{zero} in Figure 3) which describe the relation between position encoder values and actual distances between the index finger thimble and the thumb tube. The variable x_{zero} corresponds to the encoder value when the glove is fully closed. Note that this value is dependent on the position of the adjustable thumb-tube. Thus, if we adjust the position of the thumb-tube we must re-calibrate.

Finally, we estimate the gain K_3 which relates the applied forces to the deflection due to compliance in the glove, by measuring with a caliper the deflection of the upper arm in its uppermost position with 0, 1, 2, 3 and 4 N loads and by fitting a linear function to the data.

During the calibration procedure, the glove is attached to a fixture to avoid unwanted movements that could contribute to variations in the measurements.

6   Stiffness Measurements

Stiffness, defined as the relationship between load and deflection, is of interest in haptic rendering because it is a mechanical property that can be used to distinguish objects from one another, or guide us how an object should be handled. While an ideal spring obeys a linear relationship between load and deflection, many load-deflection curves of real objects are non-linear. To test the proposed glove's ability to display stiffness, we load the controller with a series of load-deflection curves and measure how closely the glove can follow the curves by measuring the actual position of the index finger thimble while applying a series of known loads. For two of the sampled materials we also measure how fast the thimble reaches the target deflections corresponding to a defined load applied at the thimble.

6.1   Test Set

We use both real and synthetic data to generate load-deflection curves for the look-up table (see Figure 3) in the controller. To generate load-deflection curves from real, physical objects, we used as samples three spherical elastic sponge balls from Luxx Ultra-Tech Inc. [16] with 40 mm diameter and densities 160, 200 and 300 g/dm^3. We made a 10 mm diameter hole through the balls to increase the variation in the load-deflection curve. By compressing the ball along a line perpendicular to the direction of the hole and passing though the ball center, the compression will begin with a compliant (soft) phase corresponding to the collapse of the hole, followed by a stiffer phase when the hole has collapsed and the sponge material of the ball is the main contributor to the slope of the curve. To generate stiffer force-deflection curves, the samples were altered by inserting a metal cylinder through the hole, with the same diameter. We sampled load/deflection values during compression of each ball using a Minimat 2000 compression and tensile testing machine (see Figure 4), at a resolution of 100 µm, and at a rate of 10 mm/min.

In order to test the glove's response to a completely rigid object, we generated a synthetic piecewise linear load-deflection curve with two stiffness intervals, where the second part corresponds to complete rigidity, as illustrated in Figure 6.

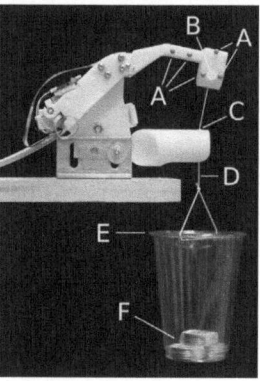

Fig. 4. The load-deflection curve of a physical sample (A), with an inserted metal cylinder (B), is measured in the compression test machine. The sample is compressed between two spherical contact points (C).

Fig. 5. Haptic glove during static and dynamic stiffness experiments. The reference point (B) is tracked with the markers (A). A force is applied at the thimble in the direction of a hole in the thumb-tube (C) via the string (D). The cup (E) holds weights (F).

6.2   Static Measurements of Stiffness Response

To measure the device's response to the forces in the load-deflection curves, we attach the glove to a fixture to avoid movements that could contribute to measurement variations (see Figure 5). We apply loads to the finger thimble via a thread passing through a hole in the thumb-tube. The hole is positioned so that the applied loads are directed approximately along the normal direction of the thimble, independently of the glove opening. At the end of the thread, a free-hanging container holds 35 g weights generating, approximately 0.344 N each.

After calibration, we load a load-deflection curve from the test set into the controller. With no initial load, the thimble moves to the position corresponding to zero load. For each weight added to the container, we measure the closest distance between the index finger thimble and the thumb-tube with a caliper, and record the corresponding load and thimble position. Deflection of the thimble is calculated as the thimble position at zero load minus the position at the current load. We add weights until a force of 5 N is reached. The procedure is repeated three times for all load-displacement curves in the test set to derive mean and standard deviation.

6.3   Dynamic Measurements of Stiffness Response

To determine the dynamic response of the glove when a load is applied, we record the movement of the finger thimble after a load-deflection curve from the test set is loaded into the look-up table (see Figure 3) and a 4 N load is applied at the thimble. Two OptiTrack cameras from NaturalPoint [17] track five optical markers on the thimble (see Figure 5) at 100 fps as we apply the load. The spatial accuracy of the optical tracker is not as high as that of the caliper used in the static response test, but is well suited to capture the motions of the thimble.

7   Results and Discussion

Figure 6 shows the target force-deflection curves (thicker curves) from the test set and the corresponding measured data (with mean and standard deviation). The top plots show samples S1-S3 corresponding to the three sponge balls with empty holes and with densities 160, 200, and 300 g/dm^3, respectively, and the center plots show the measurements for the same samples, but with the metal cylinder. The bottom plots show a synthetic target curve and the corresponding measured curve. The left column shows the measured data without compensation for the glove compliance, and the right column shows the same target load-deflection curves, but with a linear compensation gain in the controller as described in Section 5.

We notice a good correspondence between the target load-deflection curves and the measured actual response, with a mean deflection difference ranging from less than 1mm to about 5 mm. In the right column, with a linear compensation in the controller of $K_3 = 0.5$ mm/N, the error is reduced significantly. The compliance-compensated synthetic curve illustrates the device's ability to render rigidity accurately.

The offset between the target and measured curves appears to follow a consistent pattern, an increasing over-deflection following increasing loads. This can be attributed to compliance in the glove which in turn can be attributed to a rather low Young's modulus, about 2.5 GPa, in the prototyping material, resulting in insufficient stiffness in the exoskeleton.

We observe that the deviations in the measurements are generally larger in regions with lower stiffness, and at a low load. The FlexiForce sensor has a hysteresis of about 5% (0.2 N) and depending on load history the force value will vary. The hysteresis, in combination with for example friction in the load cell, can explain a large part of both the scatter and the absolute errors between measured and target curves at low loads. Force sensors printed on a flexible printed circuit board are very attractive in an application of this type since they occupy little space and can easily be integrated as force arrays in complex mechanical structures. But a stable and reliable force signal is essential to avoid feedback artifacts and more work is necessary to optimize this part of the glove.

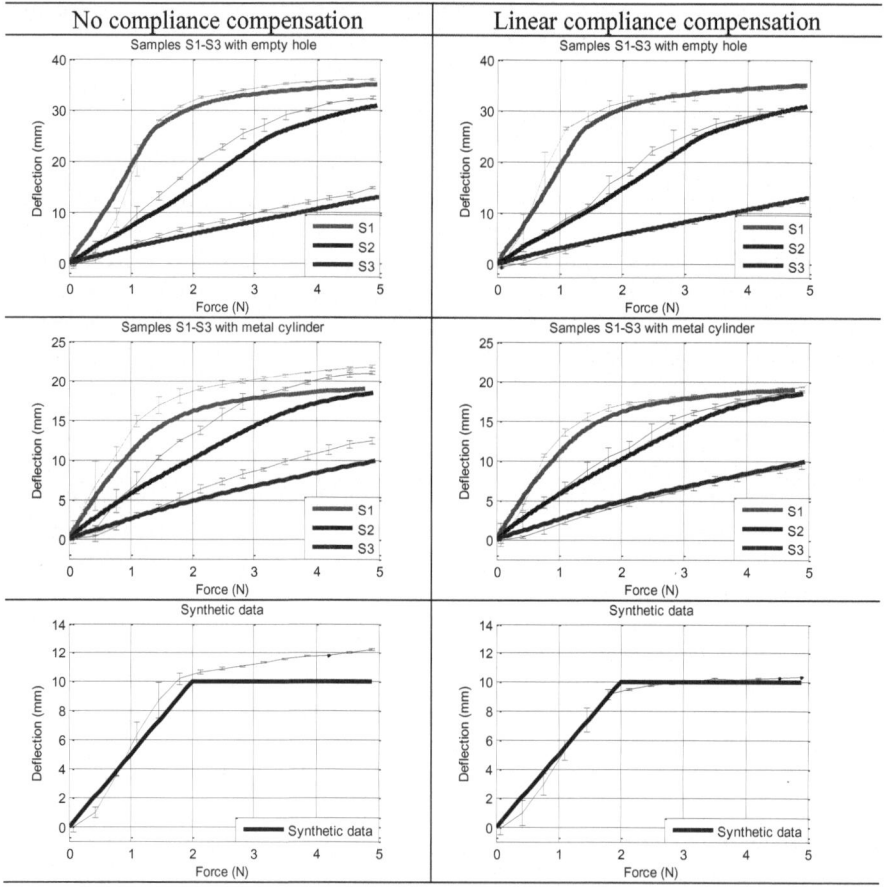

Fig. 6. Target load-deflection curves and actual response from the haptic glove

The position encoder reads the position of the motor and only indirectly the position of the thimble, a design decision related to a desire to make the glove modular and prepared for adding additional degrees of freedom for more fingers and finger joints. But the placement of the encoder introduces errors in the position measurements due to compliance and play in the mechanical structures.

The accuracy of the plastic parts due to the rapid prototyping process is not sufficient, causing play in the joints (motor attachment and curved rail) and unwanted friction, resulting in a non-linear error that is more pronounced at lower loads. Still, we believe that from a user point-of-view, smooth deviations from the target curves are not as critical as sudden force-deflection variations. It is possible to reduce both play and friction in the mechanical parts in a more elaborate version of the glove and hence the mechanics should not become the most crucial point in a future glove version. The piezoelectric motor has no play or backlash making it completely ideal from a static load-displacement point of view.

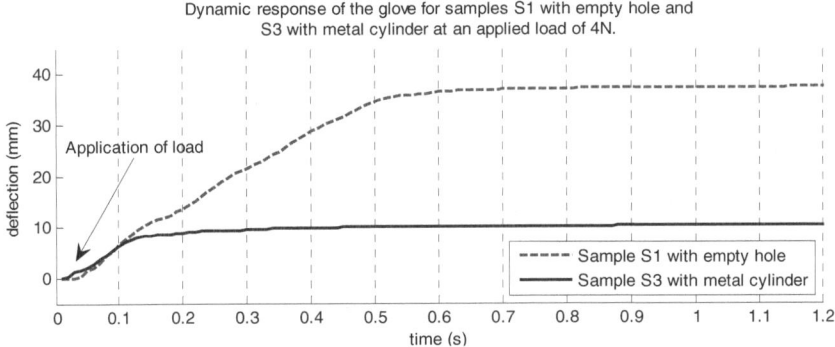

Fig. 7. Movement of the reference point (B in Figure 5) in the up/down direction after a load of 4N was applied at the thimble for the most compliant and the stiffest sampled materials (S1 with empty hole and S3 with metal cylinder)

The dynamic response plot in Figure 7 shows the finger thimble position response after a 4N load is applied to the thimble. The thimble reaches steady state, that is a firm grip, with the stiffest sampled load-deflection curve after a short transition phase of only 0.1s. Even with the lowest stiffness load-deflection curve, the transition phase is only approximately 0.5s, with the increase due to the larger movement of the thimble. In both cases, the thimble reaches its new target position without any overshoot or noticeable oscillation. The slope of the curve is limited by the speed of the motor. If the user tries to exceed the motor speed limit, the response will be a resisting force.

8   Conclusions and Future Work

We have described a prototype haptic glove actuated by a compact integrated piezoelectric motor that gives proprioceptive haptic feedback to the index finger. We have shown that the device can realistically display load-deflection curves obtained from physical samples, and how linear compliance compensation in the control loop can further improve the accuracy. The dynamic response after a load is applied to the thimble is a fast and stable transition to the new target position without overshoot or oscillation.

Since the motor in the current design operates within the audible frequency range (0-3 kHz), we are interested in investigating the use of ultrasonic motors operating within the inaudible frequency range. Ultrasonic operation also provides faster motors which would improve the device's ability to display low impedances such as free motion; however, we expect a tradeoff between maximum motor speed and force/volume ratio. We intend to make a more thorough evaluation of both the static and dynamic behavior of the device, and compare with the results from using a high accuracy force sensor, which we believe could further improve the performance. Other directions for further work include better mechanical structures, and of course additional actuated degrees of freedom.

Acknowledgements. Lars-Erik Lindquist is acknowledged for helping us prepare the physical samples used in the study. We would also like to thank Kristofer Gamstedt and Thomas Joffre for their help in measuring the physical samples.

References

1. Sensable Technologies, Inc., http://www.sensable.com
2. CyberGlove Systems, http://www.cyberglovesystems.com
3. Bouzit, M., Burdea, G., Popescu, G., Boian, R.: The Rutgers Master II-New Design Force-Feedback Glove. IEEE/ASME Trans. Mechatronics 7(2), 256–263 (2002)
4. Endo, T., Kawasaki, H., Mouri, T., Ishigure, Y., Shimomura, H., Matsumura, M., Koketsu, K.: Five-Fingered Haptic Interface Robot: HIRO III. IEEE Trans. Haptics 4(1), 14–27 (2011)
5. Blake, J., Gurocak, H.B.: Haptic Glove with MR Brakes for Virtual Reality. IEEE/ASME Transactions on Mechatronics 14(5) (October 2009)
6. Laitinen, P., Mawnpaa, J.: Enabling Mobile Haptic Design: Piezoelectric Actuator Technology Properties in Hand Held Devices. In: Proc. 2006 IEEE International Workshop on Haptic Audio Visual Environments and their Applications, pp. 40–43 (2006)
7. Kyung, K.U., Kim, S.C., Kwon, D.S.: Texture Display Mouse: Vibrotactile Pattern and Roughness Display. IEEE/ASME Trans. Mechatronics 12(3), 356–360 (2007)
8. Chubb, E.C., Colgate, J.E., Peshkin, M.A.: ShiverPad: A Device Capable of Controlling Shear Force on a Bare Finger. In: Proc. World Haptics Conference 2009, pp. 18–23 (2009)
9. Santello, M., Flanders, M., Soechting, J.F.: Postural Hand Synergies for Tool Use. J. Neurosci. 18, 10105–10115 (1998)
10. PiezoMotor AB, http://www.piezomotor.se
11. NANOS-Instruments GmbH, http://www.nanos-instruments.com
12. Tekscan, Inc., http://www.tekscan.com
13. Kern, T.A.: Engineering Haptic Devices. Springer Publishing (May 2009)
14. Carignan, C., Cleary, K.: Closed-Loop Force Control for Haptic Simulation of Virtual Environments. Haptics-e, The Electronic Journal of Haptics Research 2(2) (February 2000), http://www.haptics-e.org
15. Van der Linde, R.Q., Lammerste, P., Frederiksen, E., Ruiter, B.: The HapticMaster, a New High-Performance Haptic Interface. In: Proc. EuroHaptics, Edinburgh, pp. 1–5 (2002)
16. Luxx Ultra-Tech, Inc., http://www.luxxultratech.com
17. NaturalPoint, Inc., http://www.naturalpoint.com

Two Finger Grasping Simulation
with Cutaneous and Kinesthetic Force Feedback

Claudio Pacchierotti[1,2], Francesco Chinello[2], Monica Malvezzi[1],
Leonardo Meli[1], and Domenico Prattichizzo[1,2]

[1] Department of Information Engineering, University of Siena,
Via Roma 56, 53100 Siena, Italy
[2] Department of Advanced Robotics, Istituto Italiano di Tecnologia,
Via Morego 30, 16163 Genova, Italy
{pacchierotti,chinello,malvezzi,meli,prattichizzo}@dii.unisi.it

Abstract. This paper presents an experiment of two finger grasping. The task
considered is the peg-in-hole and the simulated force feedback is cutaneous or
kinesthetic. The kinesthetic feedback is provided by a commercial haptic device
while the cutaneous one is provided by a new haptic display proposed in this
work, which allows to render at the fingertip a wide range of contact forces. The
device consists of a mobile surface, which interacts with the fingertip, actuated
by three wires directly connected to the motors placed on the grounded structure of the display. This work summarizes the design of the proposed display
and presents the main relationships which describe its kinematics and dynamics.
Results showed that cutaneous feedback exhibits improved performances when
compared to visual feedback only.

1 Introduction

Cutaneous feedback is important to simulate interactions with objects in a virtual environment. Single-contact haptic devices, such as the Omega devices (Force Dimension,
CH), provide haptic feedback, consisting of both cutaneous and kinesthetic forces, to
the user, making him/her aware of the relative position of neighboring parts of the body
by means of sensory organs in muscles and joints [1].

Watanabe *et al.*, in [2], developed a system for controlling cutaneous sensations of
surface roughness by applying ultrasonic vibration to the surface. In [3] the authors
proposed an approach to provide human cutaneous sensation using surface acoustic
wave. A pulse-modulated driving voltage excited temporal distribution of shear force
on the surface acoustic wave substrate. The force-friction distribution was perceived as
cutaneous sensations at receptors in the skin.

A widely-used approach for providing cutaneous sensations is employing dynamic
pin-matrices. Ikei *et al.*, in [4], developed a cutaneous display which has 50 vibrating
pins. The vibratory pin array included 5x10 contact piano-wires 0.5mm in diameter,
aligned in a 2mm pitch with a vibration frequency of 250Hz. In [5] the authors developed a pin-array cutaneous display, composed of a 6x5 pin-array actuated by 30 piezoelectric bimorphs. It was able to display planar distributed and Braille cell patterns.
Pin-arrays were also employed in [6], where the authors used a solenoid, a permanent

P. Isokoski and J. Springare (Eds.): EuroHaptics 2012, Part I, LNCS 7282, pp. 373–382, 2012.

magnet and an elastic spring to develop a miniature pin-array cutaneous module. The elastic springs in the actuators were separated into several layers to minimize the contactor's gap. In [7] the authors used electrostatic force and friction control to render surface roughness sensations. The display consisted of stator electrodes and a thin film slider, on which an aluminium conductive layer was deposited.

Minamizawa *et al.*, in [8,9], presented a wearable and ungrounded haptic display able to simulate weight sensations of virtual objects. The device consisted of two motors and a belt able to deform the fingertip. When motors span in opposite directions the belt applied a perpendicular force to the user's fingertip while, if motors span in the same direction, the belt applied a tangential force on the skin. That device was used in [10] to provide cutaneous feedback in an industrial application involving heavy duty machines, and in [11] for experiences of remote cutaneous interaction. A similar device has been also used in [12], where the authors presented a new approach to sensory substitution in haptics called *sensory subtraction*. They substituted haptic feedback, consisting of both cutaneous and kinesthetic forces, with cutaneous feedback only, in order to achieve the stability of the system and outperform other conventional sensory substitution techinques. More recently, Bau *et al.* developed in [13] a technology to provide cutaneous sensation while moving fingers on touch screens. The touch panel presented has a conductive layer coated with an insulating layer, which the finger rests upon. When voltage difference was applied between the finger and the conductive layer, a normal attractive force was induced. By alternating the voltage, it was possible to modulate the friction force felt by the moving finger. A similar device has been presented in [14], where the authors developed a system, named VerroTouch, for providing cutaneous feedback to surgeons during telerobotic surgery. VerroTouch measured the vibrations caused by tool contact and recreates them on the master handles for the surgeon.

This paper presents a three DoF cutaneous display used to interact with objects in a virtual environment. The device is able to apply contact forces to the fingertip by applying forces to the vertices of a rigid platform by means of three wires. Three servomotors are in charge of moving the platform and applying the requested force to the user's fingertips, ensuring precision, strength, and lightness.

The system provides cutaneous stimuli only and most of the kinesthetic feedback is missing. The proposed device is similar to the wearable haptic device presented in [15] but there are relevant differences which are worth underlining. The cutaneous display here presented can be easily integrated with other systems which provide kinesthetic stimuli (see Sec. 4), it uses three servo motors and can render higher forces at the fingertip. The idea of providing realistic cutaneous sensations while using haptic interfaces has been also discussed in [16]. However, the thimble there presented was only able to provide the cutaneous sensation of making and breaking contact with virtual surfaces.

An important contribution of the paper is to show how this cutaneous device can be used to simulate a pinch grasp and perform a peg-in-hole task.

The paper is organized as follows: the cutaneous device is presented in Sec. 2. In Sec. 3 the statics analysis of the device, represented as a three DoFs parallel mechanism is summarized. An experiment, carried out to evaluate the user experience while using the device in a virtual environment, are presented and discussed in Sec. 4. Finally Sec. 5 addresses concluding remarks and perspectives of the work.

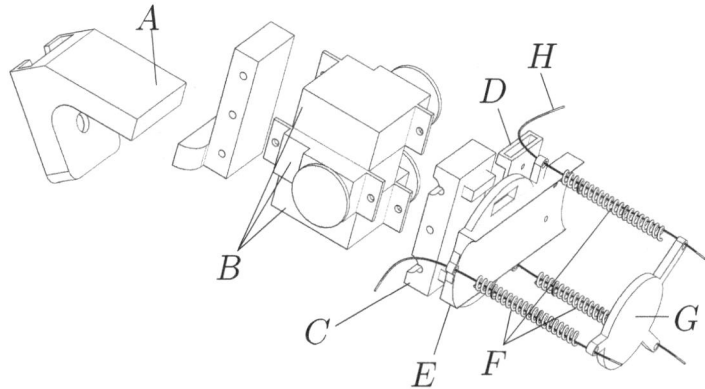

Fig. 1. A sketch of the three DoFs cutaneous device. Three servo-motors control the lengths of three wires in order to tilt the mobile platform according to the virtual surface being touched.

2   Device Description

Fig. 1 sketches the main idea of the proposed three DoFs cutaneous device while a prototype is shown in Fig. 2. It consists of a static part (parts A,C-E in Fig. 1), and a mobile part (part G), able to apply the requested stimuli to the fingertip's volar surface.

Referring to Fig. 1, the user should place the fingertip between part G and part E (see also Fig. 2). Three springs, placed between the mobile platform and the static part, keep the platform horizontally aligned with the rest of the device. Three servo-motors (B) control the length of the three wires (H) connecting the mobile platform vertices to the static platform (E), making the platform able to apply the requested force at the user's fingertip. The mobile platform model is described and discussed in Sec. 3. The actuators used for the device prototype are three HS-55 MicroLite Servo motors [17]. The motors are fixed to part C and D of the device structure. Part A is devoted to connect the cutaneous display to an external support.

In this work the device will be fixed to the end-effector of an Omega 3 haptic device in order to provide kinesthetic feedback, if necessary, and/or track the position of the finger. The mechanical supports for the actuators and the mobile platform are made using a special type of acrylonitrile butadiene styrene, called ABS*Plus*™ (Stratasys Inc., USA). The device uses a velcro strap, fixed to part D, to be fasten tightly to the fingers and make it easier to wear (see Fig. 2). The total weight of the whole device, including actuators, springs, wires, and the mechanical support is about 45g.

The force applied by the device to the user's finger pad is balanced by a force supported by the structure of the device on the back of the finger (part E). This structure has a larger contact surface with respect to the mobile platform (part G) so that the local pressure is much lower and the contact is mainly perceived on the finger pad and not on the back side of the finger. This idea was inspired by the *gravity grabber* presented in [8,9] and previously summarized, where a wearable haptic display was employed to simulate weight sensations of virtual objects. Both devices are able to render cutaneous stimuli and most of the kinesthetic feedback is missed.

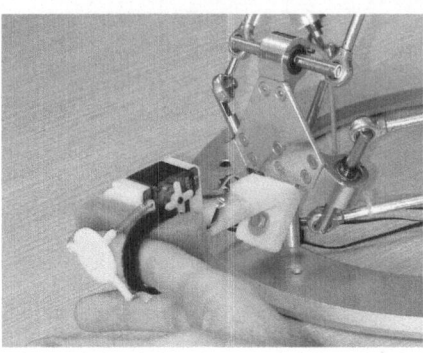

Fig. 2. The three DoFs cutaneous display prototype: the three servo-motors move the platform according to the virtual surface being touched. The device is fixed to the end-effector of the Omega 3 haptic device.

3   Device Model

The kinematic structure of the proposed device is similar to the wearable display described in [15]. The main differences is that the one proposed in this paper is not designed to be portable. The power of the acuators is larger and three passive springs have been included in the design. Similarly to the device proposed in [15], the cutaneous platform can be modeled as a three DoFs parallel mechanism, where the static part is fixed and the mobile platform is in contact with the finger pulp.

The mobile platform is moved acting on three wires connecting its vertices to the actuators. Three springs, which contain the wires, make possibile to fix the platform in a reference configuration. The model of the device presented in this paper differs from the one described in [15] because:

- in this case the wires do not follow the finger shape but a straight line from the static to the mobile platform,
- in the evaluation of actuator forces the compliance of the three springs is taken into account.

Let $w_p = [f_p^\mathrm{T} \ m_p^\mathrm{T}]^\mathrm{T} \in \Re^6$ be the wrench applied to the mobile platform (expressed with respect to S_0), and $Q = [Q_1 \ Q_2 \ Q_3]^\mathrm{T}$ the vector of force (norms) applied to the wires, being their directions defined by the unitary vectors s_1, s_2, and s_3 respectively. We can express the external wrench as a function of the force applied to the wires

$$w_p = J_p^\mathrm{T} Q. \tag{1}$$

where $J_p \in \Re^{3\times 6}$ is the Jacobian matrix and can be evaluated from the analysis of the differential kinematics of the platform. The wire forces Q_i are given by the sum of two components

$$Q_i = Q_{a,i} + Q_{p,i}$$

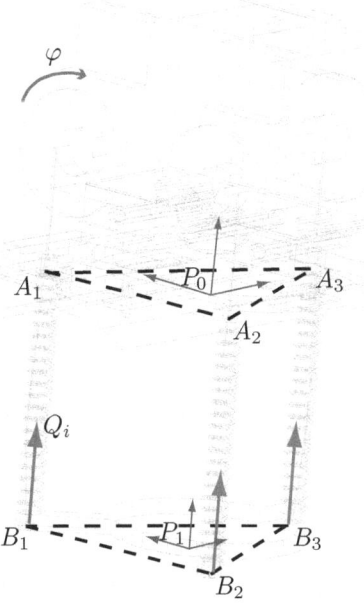

Fig. 3. The three DoFs haptic display kinematic scheme

where $Q_{a,i}$ is the force applied by the i-th actuator, proportional to the motor torque, i.e. $Q_{a,i} = T_i r_i$, T_i is the i-th motor torque and r_i is the i-th motor pulley radius. $Q_{p,i}$ is the contribution generated by the spring deformation

$$Q_{p,i} = k_i \left(\|d_i\| - \|d_{i,o}\| \right)$$

where k_i is the spring stiffness, $\|d_i\|$ is the actual wire length, $\|d_{i,0}\|$ is the nominal spring length.

The described device is underactuated, since it has only three actuators to control the six-dimensional displacement of the mobile platform, so it is not possible to find a one-to-one relationship between the wire lengths and the platform displacement and orientation in the three-dimensional space. If the platform touches the fingertip, the platform displacement $\xi = [p_x \quad p_y \quad p_z \quad \alpha \quad \beta \quad \gamma]^T$ produces a deformation of the fingertip that leads to a contact stress distribution. In quasi static condition the stress distribution on the fingertip is balanced by the wrench applied by the platform w_p [18].

Different mathematical and numerical models of the fingertip have been proposed in the literature. In [19], for example, a 2D continuum fingertip model is described, in which the finger is approximated by an homogeneous, isotropic and incompressible elastic material. Serina *et. al*, in [18], developed a model that incorporates both inhomogeneity and geometry of the fingertip is proposed. In [20] an experimental method for obtaining the 2-dimensional skin tension/extension-ratio characteristics of living human skin is described. In [21] the authors conducted an experiment in order to characterize the response of the *in vivo* fingertip pulp under repeated and compressive loadings,

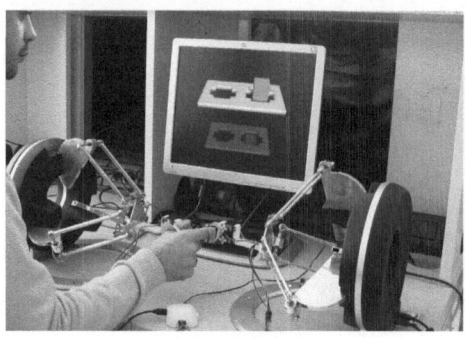
(a) Experimental setup

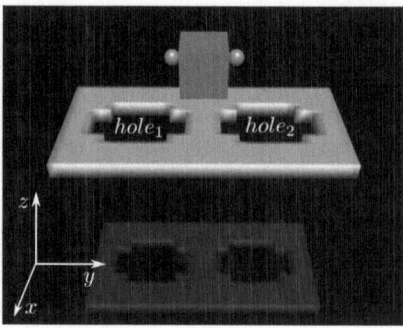
(b) Virtual environment

Fig. 4. Experimental setup and virtual environment. The user had to wear two cutaneous devices, one on the index and one on the thumb finger, and then grasp the virtual cube and complete the peg-in-hole task as fast as possible.

aiming to better understand the force modulation by the pulp. The force/deformation behavior of the fingertips in the lateral, or shearing, direction, is studied in [22]. Actually, the stress/strain behavior of the fingertip under shearing forces is non linear, in fact in [23] the authors experimentally quantified the anisotropic and hysteretic behaviour of the fingertip deformation under the application of tangential forces.

In this paper we consider a linear relationship between the resultant wrench and the platform displacement. In other terms we assume that the platform configuration ξ is proportional to the wrench w_p

$$\xi = K^{-1}w_p \tag{2}$$

where $K \in \Re^{6 \times 6}$ is the fingertip stiffness matrix. In this preliminary study an isotropic elastic behaviour is assumed for all the components of the stiffness matrix: $K = kI$, $k = 2\text{N/mm}$ [24].

From the control point of view, the device can be represented as a non linear, multi-input multi-output (MIMO) coupled system. Different control strategies can be considered, we can control for instance the force applied by the platform to the fingertip or the position and orientation of the mobile platform.

In particular, in the device position control, the motors are regulated so that the mobile platform reaches a reference configuration. The inverse kinematics of the parallel mechanism allows to evaluate the corresponding reference cable lengths. These values are compared to the actual ones and then the error drives the PD controllers of the motors.

4   Experiment

The cutaneous device here presented can tilt the mobile platform according to the reaction force of the virtual object being touched, enhancing users' illusion of telepresence.

This experiment aimed at evaluating user dexterity while using the device here presented in a virtual environment. The cutaneous device was fixed to the end-effector of an Omega 3 haptic device, as shown in Fig. 2. Users were able to interact with virtual objects in a virtual environment built using CHAI 3D [25], an open-source set of C++ libraries for computer haptics and interactive real-time simulation. The experimental setup is shown in Fig. 4.

Nine participants, six males, three females, age range 19–35, took part to the experiment, all of whom were right-handed. Five of them had previous experience with haptic interfaces. None of the participants reported any deficiencies in their perception abilities. The subjects were asked to wear two cutaneous devices, one on the thumb and one on the index finger (see Fig. 4) and complete a peg-in-hole task in a virtual environment [26,27]. The virtual environment was composed by a cube and two holes (named $hole_1$ and $hole_2$, as shown in Fig. 4b). The two holes were 3.5cm deep (x-direction), 3.5cm wide (y-direction), and 0.5cm high (z-direction). The peg was a cube with an edge length of 3cm. Therefore the hole had a tolerance of 0.5cm in the x and y directions.

The task consisted in grasping the cube from the ground, inserting it into the right hole ($hole_2$), then in the left hole ($hole_1$) and then again in $hole_2$ and $hole_1$, therefore the correct sequence was $hole_2$, $hole_1$, $hole_2$, $hole_1$. The task started when the user grasped the object and finished when the user inserted, for the second time, the peg in $hole_1$. At least half of the length of the peg had to be inserted in the hole in order to move to the next hole and the peg had to be inserted from the top to the bottom. When the object was correctly inserted into a hole, the color of the peg changed[1].

Each participant made twelve repetitions of the peg-in-hole task, with three randomized trials for each force feedback modality proposed:

- both kinesthetic and cutaneous feedback provided by the Omega 3 haptic devices and the proposed cutaneous devices (task $K + C$),
- kinesthetic feedback only provided by the Omega 3 haptic devices (task K),
- cutaneous feedback only provided by the cutaneous devices (task C),
- no force feedback (task N).

Visual feedback, as shown in Fig. 4, was always provided to the users. To evaluate the performance of the different force feedback modalities, the time needed to complete the task was recorded, together with the forces generated by the contact between the two proxies, controlled by the user, and the cube. A spring $k_o = 600$N/m is used to model the contact force between the proxies and the object. Data resulting from different trials of the same task, performed by the same subject, were averaged before comparison with other tasks' data.

Fig. 5a shows the average time elapsed between the instant the user grasps the object and the instant it completes the peg-in-hole task. The collected data of each task passed the D'Agostino-Pearson omnibus K2 normality test. Comparison of the means among the feedback modalities was tested using one-way ANOVA (no repeated measures). The means differed significantly among the feedback modalities. Post-hoc analyses (Bonferroni's multiple comparison test) revealed statistically significant difference between all the groups, showing that the time needed to accomplish the task depends on the feedback modality employed in the experiment.

---

[1] A short video of the experiment can be found at http://goo.gl/O3Ax8

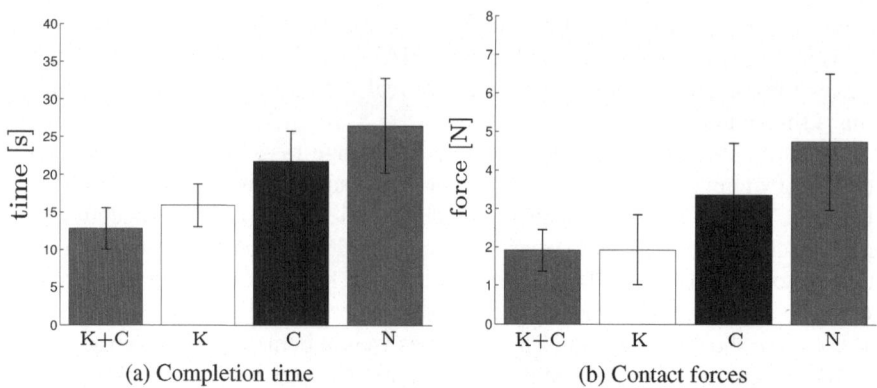

Fig. 5. Time to completion of the peg-in-hole task and force generated by the contact between the two proxies and the object during tests with both kinesthetic and cutaneous feedback (task $K+C$), kinesthetic only (task K), cutaneous only (task C), and no force feedback at all (task N)

The subjects, while receiving both kinesthetic and cutaneous feedback (task $K+C$), completed the task in less time when compared to that obtained while receiving kinesthetic feedback only (task K), and using cutaneous feedback only (task C) yields to significant better results than employing no force feedback at all (task N). This means that employing cutaneous feedback improves subjects' performances in terms of time needed to complete the task proposed. Using kinesthetic feedback (both in task $K+C$ and K) produced better performances, as expected, with respect to employing cutaneous feedback only or no force feedback at all.

Fig. 5b shows the average forces generated by the contact between the two proxies, controlled by the user, and the cube along the y-direction, i.e. the one perpendicular to the object surface (see Fig. 4b). Note that a higher force fed back to the user means a larger penetration into the virtual object and a higher energy expenditure during the grasp. Measuring the average of intensities of the two contact forces is a widely-used approach to evaluate energy expenditure during the grasp [28]. The collected data of each task passed the D'Agostino-Pearson omnibus K2 normality test and a one-way ANOVA test was performed to evaluate the statistical significance of the differences between tasks. The post-hoc analyses (Bonfferroni's multiple comparison test) revealed no statistical significance between the two tasks employing kinesthetic feedback (task $K+C$ and K) while it revealed a difference between the task employing no force feedback (task N) and the one using cutaneous feedback only (task C). It is worth noting that cutaneous feedback yielded to a minor force fed back to the operator and to a minor penetration into the virtual object in comparison to the no-force modality.

5   Conclusion and Future Works

In this work an experiment of pinch grasp with cutaneous feedback only has been presented along with a new device used to exert cutaneous forces at the two finger pads.

A peg-in-hole experiment has been carried out. Nine users had to complete the peg-in-hole task employing four different force feedback modalities: no force feedback at all, kinesthetic feedback, cutaneous feedback, and both kinesthetic and cutaneous feedback. Results showed that employing cutaneous and kinesthetic feedback lead to a higher quality of the grasp (i.e., a smaller energy expenditure) and it improves the performances in terms of time needed to complete the given task with respect to the kinesthetic only feedback.

Future developments will include the analysis of other types of control schemes and the employment of three force sensors placed at the vertices of the mobile platform. The sensors will provide a measurement of the force the platform is applying to the user's fingertip and will allow to modulate correctly the force applied by the cutaneous device.

New experiments of interaction with virtual objects in virtual environments and in augmented reality scenarios will be performed in the next future. Finally, work is in progress to validate the device with more subjects.

Acknowledgment. This work has been partially supported by the European Commission with the Collaborative Project no. FP7-ICT-2009-6270460, "ACTIVE: Active Constraints Technologies for Illdefined or Volatile Environments" and by the Italian Ministry of Education, University and Research (MIUR) with PRIN 2008 Project "Underactuated systems for manipulation in virtual environment".

References

1. Hayward, V., Astley, O., Cruz-Hernandez, M., Grant, D., Robles-De-La-Torre, G.: Haptic interfaces and devices. Sensor Review 24, 16–29 (2004)
2. Watanabe, T., Fukui, S.: A method for controlling tactile sensation of surface roughness using ultrasonic vibration. In: Proc. of IEEE International Conference on Robotics and Automation, vol. 1, pp. 1134–1139 (1995)
3. Takasaki, M., Nara, T., Tachi, S., Higuchi, T.: A tactile display using surface acoustic wave. In: Proc. of 9th IEEE International Workshop on Robot and Human Interactive Communication, pp. 364–367 (2000)
4. Ikei, Y., Wakamatsu, K., Fukuda, S.: Texture presentation by vibratory tactile display-image based presentation of a tactile texture. In: Proc. of Virtual Reality Annual International Symposium, pp. 199–205 (1997)
5. Yang, G., Kyung, K., Srinivasan, M., Kwon, D.: Quantitative tactile display device with pin-array type tactile feedback and thermal feedback. In: Proc. of 2006 IEEE International Conference on Robotics and Automation, pp. 3917–3922 (2006)
6. Yang, T., Kim, S., Kim, C., Kwon, D., Book, W.: Development of a miniature pin-array tactile module using elastic and electromagnetic force for mobile devices. In: Proc. of Eurohaptics and Symposium on Haptic Interfaces for Virtual Environment and Teleoperator Systems, pp. 13–17 (2009)
7. Yamamoto, A., Ishii, T., Higuchi, T.: Electrostatic tactile display for presenting surface roughness sensation. In: Proc. of IEEE International Conference on Industrial Technology, vol. 2, pp. 680–684 (2003)
8. Minamizawa, K., Kajimoto, H., Kawakami, N., Tachi, S.: A wearable haptic display to present the gravity sensation-preliminary observations and device design. In: Proc. of Eurohaptics Conference and Symposium on Haptic Interfaces for Virtual Environment and Teleoperator Systems, pp. 133–138 (2007)

9. Minamizawa, K., Fukamachi, S., Kajimoto, H., Kawakami, N., Tachi, S.: Gravity grabber: wearable haptic display to present virtual mass sensation. In: ACM SIGGRAPH 2007 Emerging Technologies, p. 8–es (2007)
10. Prattichizzo, D., Pacchierotti, C., Cenci, S., Minamizawa, K., Rosati, G.: Using a Fingertip Tactile Device to Substitute Kinesthetic Feedback in Haptic Interaction. In: Kappers, A.M.L., van Erp, J.B.F., Bergmann Tiest, W.M., van der Helm, F.C.T. (eds.) EuroHaptics 2010, Part I. LNCS, vol. 6191, pp. 125–130. Springer, Heidelberg (2010)
11. Prattichizzo, D., Chinello, F., Pacchierotti, C., Minamizawa, K.: Remotouch: A system for remote touch experience. In: Proc. of IEEE International Workshop on Robot and Human Interactive Communication, pp. 676–679 (2010)
12. Prattichizzo, D., Pacchierotti, C., Rosati, G.: Cutaneous force feedback as a sensory subtraction technique in haptics. In: IEEE Transactions on Haptics (2012)
13. Bau, O., Poupyrev, I., Israr, A.; Harrison, C.: Teslatouch: electrovibration for touch surfaces. In: Proc. of the 23nd Annual ACM Symposium on User Interface Software and Technology, pp. 283–292 (2010)
14. Kuchenbecker, K., Gewirtz, J., McMahan, W., Standish, D., Martin, P., Bohren, J., Mendoza, P., Lee, D.: VerroTouch: High-Frequency Acceleration Feedback for Telerobotic Surgery. In: Kappers, A.M.L., van Erp, J.B.F., Bergmann Tiest, W.M., van der Helm, F.C.T. (eds.) EuroHaptics 2010, Part I. LNCS, vol. 6191, pp. 189–196. Springer, Heidelberg (2010)
15. Chinello, F., Malvezzi, M., Pacchierotti, C., Prattichizzo, D.: A three DoFs wearable tactile display for exploration and manipulation of virtual objects. In: Proc. of IEEE Haptic Symposium (2012)
16. Kuchenbecker, K., Ferguson, D., Kutzer, M., Moses, M., Okamura, A.: The touch thimble: Providing fingertip contact feedback during point-force haptic interaction. In: Symposium on Haptic Interfaces for Virtual Environment and Teleoperator Systems, 2008, pp. 239–246 (2008)
17. HiTech Inc.: Hs-55 microlite servo motor data sheet (2012)
18. Serina, E., Mockensturm, E., Mote Jr., C., Rempel, D.: A structural model of the forced compression of the fingertip pulp. Journal of Biomechanics 31, 639–646 (1998)
19. Srinivasan, M., Dankekar, K.: An investigation of the mechanics of tactile sense using two dimensional models of the primate fingertip. Transactions of the ASME, Journal of Biomechanical Engineering 118, 48–55 (1996)
20. Cook, T., Alexander, H., Cohen, M.: Experimental method for determining the 2-dimensional mechanical properties of living human skin. Medical and Biological Engineering and Computing 15, 381–390 (1977)
21. Serina, E., Mote, C., et al.: Force response of the fingertip pulp to repeated compression–effects of loading rate, loading angle and anthropometry. Journal of Biomechanics 30, 1035–1040 (1997)
22. Nakazawa, N., Ikeura, R., Inooka, H.: Characteristics of human fingertips in the shearing direction. Biological Cybernetics 82, 207–214 (2000)
23. Wang, Q., Hayward, V.: In vivo biomechanics of the fingerpad skin under local tangential traction. Journal of Biomechanics 40, 851–860 (2007)
24. Park, K., Kim, B., Hirai, S.: Development of a soft-fingertip and its modeling based on force distribution. In: Proc. of IEEE International Conference on Robotics and Automation, vol. 3, pp. 3169–3174 (2003)
25. Conti, F., Morris, D., Barbagli, F., Sewell, C.: Chai 3d (2006), http://www.chai3d.org
26. Massimino, M., Sheridan, T.: Sensory substitution for force feedback in teleoperation. Presence: Teleoperators and Virtual Environments 2, 344–352 (1993)
27. Hill, J.: Two measures of performance in a peg-in-hole manipulation task with force feedback. In: Proc. of 13 th Annual Conference on Manual Control, pp. 301–309 (1977)
28. Prattichizzo, D., Trinkle, J.: Grasping, ch. 28. Springer (2008)

Comparison of Extensive vs. Confirmation Haptic Interfaces with Two Levels of Disruptive Tasks

Toni Pakkanen, Roope Raisamo, and Veikko Surakka

Tampere Unit for Computer-Human Interaction,
School of Information Sciences, University of Tampere, Tampere, Finland
{Toni.M.Pakkanen,Roope.Raisamo,Veikko.Surakka}@uta.fi

Abstract. In the car environment there are more and more complex infotainment systems, which are used with touchscreens, even by driver while driving the car. While it is known that secondary tasks have a negative impact to the driving safety, there is a lack of information, if haptics can be used to make this interaction safer. In this study we compared two haptically enhanced user interfaces with two levels of user distraction: Commonly used *confirmation haptic* interface, and *extensive haptic* interface, where all possible information was provided with haptics. In the experiment participants entered four-digit numbers, while driving or watching video. Input speed, input error rate, driving errors and subjective experiences were recorded. The results showed that there were no significant performance differences between the user interfaces, but the extensive haptic interface helped to reduce the number of driving errors. Participants did not have significant preference differences between the user interfaces.

Keywords: Haptic feedback, User interaction, Distracted user, Driving user.

1 Introduction

Human being is by nature a multimodal and multitasking being. While we perform a main task, like cooking, we use touch together with short glimpses to perform secondary tasks. It is easy for us, for example, to grab ingredients with the help of touch, while we keep our concentration in the cooking. This unfortunately is not the reality with modern mobile touch devices, where interaction metaphors are similar to real world interaction. A problem arises here, when it is known that interaction with mobile devices in the mobile contexts is usually split in small fragments, while users have to pay attention to the environment [7]. They use devices by direct touch and manipulate virtual elements, but without the haptic feeling of the elements on the screen. However, many of the modern devices have tactile actuators, which could make it possible to create haptic representation for graphical user interface elements. This approach would make it possible to add haptic information cues for the users of the devices to describe the items they are manipulating.

This problem with interaction without help of haptic cues is emphasized in the car environment, where users should have their concentration in the main task: driving. It

P. Isokoski and J. Springare (Eds.): EuroHaptics 2012, Part I, LNCS 7282, pp. 383–394, 2012.

has been shown that using the mobile phone and having additional cognitive load has a negative impact to the driving safety [2][4] and that impact is even larger than impact of natural conversation in the car between the driver and passenger [2]. Thus, under no circumstances it cannot be recommended to use mobile phone or other non-driving related devices while driving a car. However, it is known that people tend to use mobile phones while driving [15] and that even awareness of increased risk of crashing does not predict intention to use the mobile phone while driving [14]. This leads to the question, how use of touchscreen devices in the car could be made safer while the user's attention should be in the driving.

The aim of this study was to find out, if the extensive haptic interface would help the users with the tasks, if the distraction level impacts the usefulness of haptic interface and how the user preference about user interfaces is impacted by conditions. This information would provide a needed knowledge, if the secondary tasks in the car environment can be made safer to perform with the help of haptics.

2   Related Work

While evaluating usefulness of haptics under cognitive load [6], it has been found that when supporting a scrollbar and a progress bar with haptics, performance improvement could be found, but with simple haptics in a button task similar improvement could not be found. Also it has been found that cognitive load does not have significant effect on the performance. This leaves an open question, could supporting extensive haptics help with button-based tasks. Also in the experiment [6] cognitive load was added as a secondary task, and impact of the interaction with device to the other task was not evaluated. Thus, the question remains, if using haptics with touch screen interaction can improve performance in a second task used to distract the interaction and to create cognitive load. Also an open question is, if haptics can positively affect the performance impact caused by more demanding cognitive load.

By using individual haptic textures with number keypad for each number buttons, number input accuracy was improved compared to simple haptic feedback which was the same for all the buttons. However, this was with the cost of input speed [13]. These results support the idea to have informative extensive haptics in the user interface to achieve lower error rate rather than faster interaction speed.

In the car environment, where safety has to be first priority, multimodal feedback could provide possibilities to minimize the need of visual attention to the secondary devices. Especially, haptics play an important role in the car environment by providing eyes-free interaction scenarios [1].

Users also prefer multimodal feedback in touch screen interaction, while driving [9] and have subjective feeling that haptics helps them to drive better, even though this was not seen at actual driving performance [10]. Thus, using haptics together with vision was preferable by the users, but it did not help them actually improve the driving safety. Considering simple confirmation only haptics used in these experiments, the question remains, if with more extensive task specific haptics, the driving task could be supported better.

In the car environment with a simple task, menu selection by using rotating knob, also there couldn't be found beneficial impact with adding haptics to the interaction. Neither task performance nor driving safety was significantly improved, but by using haptics only condition performance was reduced [12]. These results support the approach to experiment with more complex interaction, like number input in our experiment and allowing partial use of gaze to support the task completion. The question remains, if more complex tasks can be supported with haptics so that user performance is impacted positively.

In an experiment, where visual, audio-visual, haptic-visual and audio-haptic-visual feedback was compared [5], it has been found that tri-modal condition reduces the driving errors, while bi-modal conditions did not reach statistical significance. Although, haptic-visual condition reduced the measured workload participants felt (Nasa-TLX workload score). Thus, it could be assumed that while haptic-visual condition reduces the workload of the driver, with more descriptive and helpful haptic interface the gain for driving errors might be larger. This assumption is supported by an experiment [3], where mean standard deviation from given drive line was reduced both with audio-haptic and visual-haptic navigational guidance.

The results seems promising with the more complex haptic input device with number button interface, which provides possibility to find the buttons on the screen with dragging finger on the screen and push them to select. Even though there was no statistical confirmation, a trend in the data indicated that a system mimicking physical buttons could improve user performance and driving safety [11].

These results from earlier experiments support the idea to approach the touch screen interaction in the car environment with task-specific, extensive and informative haptic user interface in demanding primary task and use the touch screen as a secondary task. In our experiment, we compared such an interface with traditional confirmation haptic interface and varied the primary task demand, to find out if more demanding task would benefit more from the more advanced interface.

3   Experiment

In this study, we compared use of two different haptically enhanced user interfaces with two levels of distraction. There was a commonly used *confirmation haptic* number keypad and an *extensive haptic* number keypad, where all possible information was provided also with haptics. Participants had a task to enter four-digit numbers to the phone, while either driving in a simulated environment, or while watching a video.

3.1   Participants

12 voluntary participants (all male) participated in the experiment (mean age 28 years, range 18-42 years). 11 of the participants were right-handed by their own report.

3.2   Experiment Setup

In the experiment the Playstation 3™ game Grand Tourismo 5 Prologue™ was used to simulate driving. A Logitech driving force GT wheel controller was used as the driving wheel. The car from the game used for driving was the Daihatsu Copen '02,

Fig. 1. Experiment setup

which had a top speed approximately of 160 km/h, and the track driven was a simple oval track, Daytona. In the game the guideline for optimal driving track was on and shown on the track. The touch screen device used in the experiment was a Nokia X6 touchscreen phone with a capacitive screen and a rotation motor actuator, which was used for generating the haptic feedbacks. Picture of the experiment setup is shown in Figure 1.

The tasks were to enter four-digit numbers to the phone with two levels of haptics provided with the phone. Other one was currently common confirmation haptics. Thus, a short pulse was provided, when a number was selected. The other was extensive haptic number keypad, where all possible information was provided with rhythmic haptic pulses. Thus, there were feedbacks for the edges of the buttons, for the texture, for the push down event and release event of the button. Also numbers were presented with haptic feedback when participant stopped the finger on top of the number button.

Haptic feedbacks used in this experiment were the same, as has been used earlier in the experiment investigating haptic number representation models [8]. The number model used in this experiment was the best rated model from that experiment, slower speed Arabic number representation. The haptic feedbacks used can be found in Table 1. Numbers were composed by repeating haptic pulses: for number one there was one pulse, for number two there was two pulses, and so on. There was a 100 ms pause between the pulses and to help recognition of larger numbers the pulses were grouped in groups of five by having a 200 ms pause after the fifth pulse. Number representation was repeated with a 500 ms delay.

Table 1. Haptic parameters of haptic feedback used in the experiment. (P) Proportional power, (rd) rotation direction, (t) rotation time, and (Pause) between pulses.

	P (%)	rd	t (ms)	Pause (ms)	P (%)	rd	t (ms)
Number pulse	100	↑	50				
Edge in	100	↑	35	10	100	↓	25
Edge out	100	↑	35				
Texture	5	↑	2				
Pushdown	100	↑	35	50	100	↓	25
Release	100	↑	50	10	100	↓	25

The experiment was videotaped with two cameras to the single video file, so that the mobile phone's interface and participants face were seen in the video. From the video it was controlled that participants interacted with the phone following instructions and did not perform the number entry task by gaze. The driving from the game was saved to the video repeat file and driving errors calculated out of the video. The mobile phone recorded the input to the phone, and input error rates and input speed was collected from the data.

3.3   Procedure

Before the experiment tasks, the participants were introduced to the haptic user interfaces used. They were allowed freely to try out the interfaces, until they told that they understood how the user interfaces work and what kind of haptic feedback they provide.

In the experiment the task for the participants was to enter four-digit numbers to the phone, while both, driving or watching earlier recorded error free driving at the same track with the same car and speed as in driving task from a video repeat file. Before the tasks a baseline driving was done with the same condition without any additional tasks. Tasks were done twice, once with each haptic interface. Participants entered ten four-digit number sequences in each condition and numbers were given for the participants from the laptop screen below the game screen. Numbers were given for input in half way of both straight stretches in the oval track, at the finish line and at the depot entrance. When a new number was given, laptop provided small sound effect to notify participants. Participants were instructed to enter the numbers without any delay. Thus, participants drove five and a half laps in each condition and entered 40 digits to the phone.

Participants were instructed to drive as fast as the car used could and follow the guideline shown in the road as exactly as they could. Driving errors were counted from the deviations from the guide line in the road. They were also asked to hold the phone down on the knee during the experiment, as seen in Figure 1, and not to enter numbers by watching the phone screen. They were however allowed to take a short glimpse down to the phone, if it was necessary for understanding the location of the finger on the number keypad.

In the video task, participants were instructed to keep their eyes in the video and not to perform the number input task by watching the phone screen. They were told that they may glimpse the screen same way as in driving condition, if needed, to check were the finger is on the keypad, but not to input the numbers by using the gaze. Thus, the usage of the gaze was synchronized with the driving task.

All the participants followed the given instructions as asked. Thus, participants did not see the phone screen, otherwise than turning their eyes out of the screen showing the driving interface, i.e. out of the driving path. They also kept the speed asked and in cases of accidents, immediately raised the speed back to the top speed.

Order of the conditions was counterbalanced for the elimination of the learning effect. Also the numbers used were randomized so that all the digits repeated evenly and the same digit sequences were not repeated for the participant. Also same digit was not used within the same four-digit number, so that participants had to move their finger on the screen and find the correct number button every time.

After each task, the participants answered to the questionnaire from that combination of haptic level and distraction level. Nine point bi-polar scales were used for the answers to the questions. Questions asked were: "How well you think, you could enter the numbers", "How pleasant the number keypad was to use", "How easy the use of the number keypad was", "How demanding the task in total was", "How easy the user interface was to understand" and "How much there was haptic feedback to support the task". Also there was asked with three point bi-polar scale a question "Would you use this kind of number keypad in your phone, if it was available".

3.4   Data Analysis

Within-subjects repeated measures analysis of variance (ANOVA) was used for statistical analysis. If the sphericity assumption of the data was violated, Greenhouse-Geisser corrected degrees of freedom were used to validate the F statistic. Bonferroni corrected pairwise t-tests were used for post hoc tests.

The driving errors were categorized to three categories: 1. *Small driving errors*, where the driving was unstable, but the car did not move out of the given track to drive more than a little bit (less than the car's width), *i.e.* "Instability in the driving". 2. *Medium driving errors*, when the driving path was lost and the car moved out of the given track more than the car's width, *i.e.* "A lane switches error". 3. *Large driving errors*, when the control of the car was completely lost, *i.e.* "An accident".

Non-parametric Friedman's rank test was used for statistical analysis for the data from the driving errors. Non-parametric Wilcoxon signed-ranks tests were used for pairwise comparisons of results within each task.

4   Results

Means and standard error of the means (S.E.M.s) for the ratings of the number input errors and number selection speed are presented in Figure 2. For the ratings of the number input errors, a 2×2 two-way (haptic feedback level $\times$ distraction level) ANOVA showed a statistically significant main effect of distraction level $F(1, 11) = 11.5$, $p < 0.01$. The main effect of the haptic feedback level or the interaction of the main effects was not statistically significant. Post hoc pairwise comparisons showed that the participants made less input errors while watching video, than when driving $MD = 9.3\%$, $p < 0.01$. For the ratings of the number selection times, a 2×2 two-way (haptic feedback level $\times$ distraction level) ANOVA showed a statistically significant main effect of haptic feedback level $F(1, 11) = 15.0$, $p < 0.01$. The main effect of the distraction level or the interaction of the main effects was not statistically significant. Post hoc pairwise comparisons showed that the participants entered the numbers faster with confirmation haptic user interface, than with extensive haptic user interface $MD = 1561$ ms, $p < 0.01$.

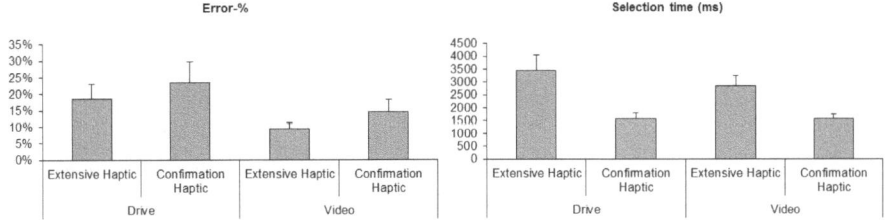

Fig. 2. Number input error-% and selection times

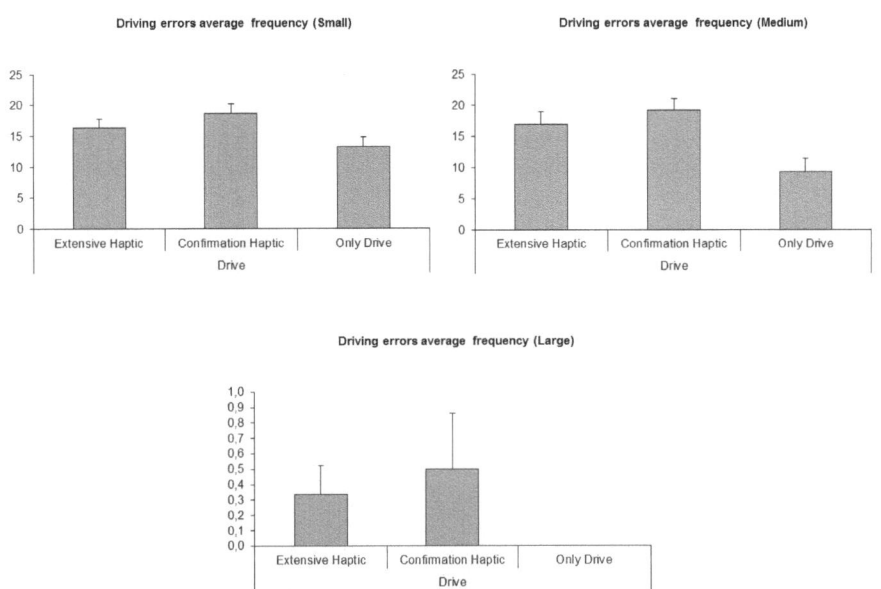

Fig. 3. Driving errors

Means and S.E.M.s for the driving errors are presented in Figure 3. Friedman's rank test showed a statistically significant effect of the haptic level within small driving errors $X^2 = 20.6$, $p < 0.001$ and within medium driving errors $X^2 = 19.7$, $p < 0.001$. The differences within large driving errors were not statistically significant.

For the small driving errors Wilcoxon singed ranks tests showed that there was less driving errors with the extensive haptic interface $Md = 16.5$ than with the confirmation haptic interface $Md = 18.5$, $|Z| = 2.8$, $p < 0.01$, but still more than with the driving only condition, without any distraction $Md = 14.5$, $|Z| = 2.8$, $p < 0.01$. There were more driving errors with the confirmation haptic interface $Md = 18.5$ than with the driving only condition, without any distraction $Md = 14.5$, $|Z| = 3.1$, $p < 0.01$.

For the medium driving errors Wilcoxon singed ranks tests showed that there was less driving errors with the extensive haptic interface $Md = 18.0$ than with the confirmation haptic interface $Md = 21.5$, $|Z| = 2.4$, $p < 0.05$, but still more than with the

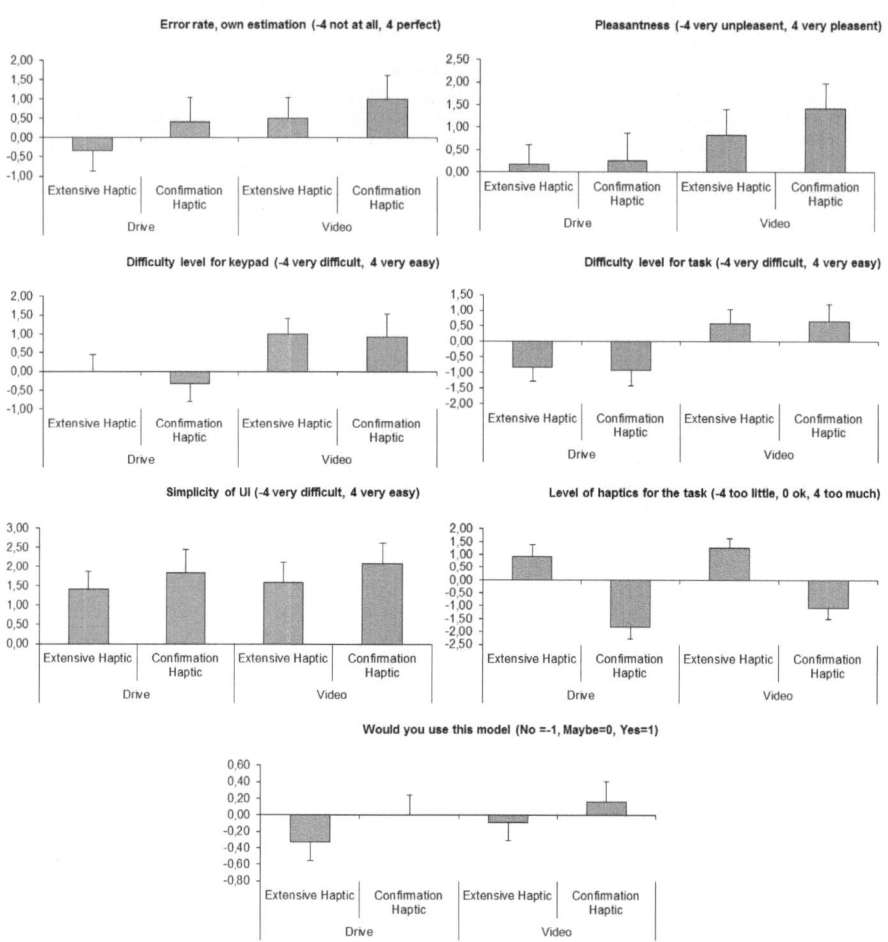

Fig. 4. Subjective evaluations

driving only condition, without any distraction $Md = 11.0$, $|Z| = 3.0$, $p < 0.01$. There was more driving errors with the confirmation haptic interface Md = 21.5 than with the driving only condition, without any distraction $Md = 11.0$, $|Z| = 3.0$, $p < 0.01$.

Means and S.E.M.s for the ratings of the subjective evaluations are presented in Figure 4. For the ratings of the question "How easy the use of the number keypad was", a 2 × 2 two-way (haptic feedback level × distraction level) ANOVA showed a statistically significant main effect of distraction level $F(1, 11) = 13.3$, $p < 0.01$. The main effect of the haptic feedback level and the interaction of the main effects were not statistically significant. Post hoc pairwise comparisons showed that the participants thought that using the keypad was more difficult while driving, than while watching the video $MD = 1.1$, $p < 0.01$.

For the ratings of the question "How demanding the task in total was", a 2 × 2 two-way (haptic feedback level × distraction level) ANOVA showed a statistically significant main effect of distraction level $F(1, 11) = 18.6$, $p \leq 0.001$. The main effect of the haptic feedback level and the interaction of the main effects were not statistically significant. Post hoc pairwise comparisons showed that the participants thought that the driving task was more difficult than the task to watch the video, when using the phone keypad at the same time $MD = 1.5$, $p \leq 0.001$.

For the ratings of the question "How much there was haptic feedback to support the task", a 2×2 two-way (haptic feedback level × distraction level) ANOVA showed a statistically significant main effect of haptics level $F(1, 11) = 35.2$, $p < 0.001$. The main effect of the distraction level and the interaction of the main effects were not statistically significant. Post hoc pairwise comparisons showed that the participants thought that level of the haptic feedback was more sufficient for the task with extensive haptic user interface than with user interface with confirmation haptic interface $MD = 2.5$, $p < 0.001$.

The main effects of the other subjective rating questions were not statistically significant.

5   Discussion and Future Work

In this experiment we evaluated how the level of haptic support and task environment difficulty level impact to user performance and satisfaction. In the present study, the main task was to drive or watch the video. The numeric input with the phone was done as a secondary task.

As our results show, with a descriptive extensive haptic interface, where haptic feedback is supporting the task, the errors in the main task, driving, were reduced. This differs from the result in earlier research [10] [12] [5], where the impact for the driving safety could not be found, while using more simple haptic interfaces than in this experiment. These results support the preliminary view, based on the trend in the data that versatile haptics supporting the task well enough can impact the driving safety [11] and the results that task specific visual-haptic and audio-haptic feedback help drivers to drive more stable [3].

However, it is crucial to notice that in both conditions there were significantly more driving errors, than with the baseline driving without any additional tasks. Thus, these results do not support the view that using the mobile phone should be allowed while driving. But, if the users do it anyway [15][14], using the appropriate haptic interface could make the use of touch screen systems safer, than using them with simple confirmation haptics only. Our results also support adding haptic feedback in in-car touch screens and other systems used while driving.

Regarding the performance results on the numeric input task, the earlier results [6] indicating that cognitive load did not have a significant impact in the performance were not repeated. In the present experiment the input error rate was impacted by the task demand level. However, in this experiment the more demanding task was considerably more demanding than cognitive levels used in Leung's [6] experiment. Thus, in the driving condition, demand is so high that it will reduce secondary task's performance. Based on this, rising error rates should be taken into account in user interface design for the car environment.

The finding that versatile haptic interface will reduce the input rate [13] was repeated in this experiment. The result on better input error rates in Yatani's experiment was not found in this experiment. However, as can be seen in error rate bars in

Figure 2, there is a trend that in extensive haptic interface error rates are getting smaller, even though the difference was not statistically significant. While in this experiment the number of participants was 12, the trend showing smaller error rates might be found significant with a larger sample. Also, the extensive haptic interface was novel and not familiar to the participants, with time and practice the results might get better and users learn interaction strategies to help them to gain more from additional feedback.

From the subjective experience results it can be seen that the participants thought that using the phone while driving was more difficult. Haptic level did not affect to this result, which does not support the initial assumption that haptics might be more useful in a more demanding task environment. The level of haptic support did not affect to the user experience ratings. Interesting in these results was that even though the extensive haptic interface was unfamiliar and provided haptic feedback all the time, it was not considered less pleasant or difficult. Vice versa, the level of haptics was considered to be more sufficient for the task, which supports the result that users prefer multimodal feedback while driving [9].

6   Conclusions

Based on our results, extensive and informative haptic interfaces supporting the task they are designed for would be recommended for the touch screen interfaces, which are used while driving. The results support the assumption that extensive haptic interface helps the driver to drive safer, even though not as safe as by not using the device at all. The cost of that safer drive is the interaction speed with the secondary system. It can be assumed that extensive haptics helps the driver to perform the secondary task more by touch, less by sight, and thus keep the concentration and eyes more on the road, less on the secondary task.

Extensive haptification of user interfaces cannot make the use of the device as safe as not using the device while driving, so under no circumstances based on these results should use of touch screen systems be recommended while driving! However, providing well designed haptifications for the tasks the drivers manage with touch screens, allowed or not, could make it safer.

Acknowledgments. This research was funded by the Finnish Funding Agency for Technology and Innovation (Tekes), decision numbers 40159/09 and 40289/11. For more information, visit http://hapimm.cs.uta.fi and http://hapticauto.sis.uta.fi. We would like to thank all the colleagues who helped with their comments, especially Jani Lylykangas for the mentoring with the statistics.

References

1. Burnett, G.E., Mark Porter, J.: Ubiquitous computing within cars: designing controls for non-visual use. International Journal of Human-Computer Studies 55(4), 521–531 (2001), doi: 10.1006/ijhc.2001.0482, http://dx.doi.org/10.1006/ijhc.2001.0482, ISSN 1071-5819

2. Caird, J.K., Willness, C.R., Steel, P., Scialfa, C.: A meta-analysis of the effects of cell phones on driver performance. Accident Analysis & Amp; Prevention 40(4), 1282–1293 (2008), doi: 10.1016/j.aap.2008.01.009,
http://dx.doi.org/10.1016/j.aap.2008.01.009, ISSN 0001-4575

3. Kern, D., Marshall, P., Hornecker, E., Rogers, Y., Schmidt, A.: Enhancing Navigation Information with Tactile Output Embedded into the Steering Wheel. In: Tokuda, H., Beigl, M., Friday, A., Brush, A.J.B., Tobe, Y. (eds.) Pervasive 2009. LNCS, vol. 5538, pp. 42–58. Springer, Heidelberg (2009), http://dx.doi.org/10.1007/978-3-642-01516-8_5, doi:10.1007/978-3-642-01516-8_5

4. Lamble, D., Kauranen, T., Laakso, M., Summala, H.: Cognitive load and detection thresholds in car following situations: safety implications for using mobile (cellular) telephones while driving. Accident Analysis & Prevention 31(6), 617–623 (1999),
http://dx.doi.org/10.1016/S0001-4575(99)00018-4,
doi:10.1016/S0001-4575(99)00018-4, ISSN 0001-4575

5. Lee, J.-H., Spence, C.: Assessing the benefits of multimodal feedback on dual-task performance under demanding conditions. In: Proceedings of the 22nd British HCI Group Annual Conference on People and Computers: Culture, Creativity, Interaction (BCS-HCI 2008), vol. 1, pp. 185–192. British Computer Society, Swinton (2008)

6. Leung, R., MacLean, K., Bertelsen, M.B., Saubhasik, M.: Evaluation of haptically augmented touchscreen gui elements under cognitive load. In: Proceedings of the 9th International Conference on Multimodal Interfaces (ICMI 2007), pp. 374–381. ACM, New York (2007),
http://doi.acm.org/10.1145/1322192.1322258,
doi:10.1145/1322192.1322258

7. Oulasvirta, A., Tamminen, S., Roto, V., Kuorelahti, J.: Interaction in 4-second bursts: the fragmented nature of attentional resources in mobile HCI. In: Proceedings of the SIGCHI Conference on Human Factors in Computing Systems, CHI 2005, Portland, Oregon, USA, April 02 - 07, pp. 919–928. ACM, New York (2005),
DOI=http://doi.acm.org/10.1145/1054972.1055101

8. Pakkanen, T., Raisamo, R., Salminen, K., Surakka, V.: Haptic numbers: three haptic representation models for numbers on a touch screen phone. In: International Conference on Multimodal Interfaces and the Workshop on Machine Learning for Multimodal Interaction (ICMI-MLMI 2010), article 35, 4 pages. ACM, New York (2010), doi:10.1145/1891903.1891949,
http://doi.acm.org/10.1145/1891903.1891949

9. Pitts, M.J., Williams, M.A., Wellings, T., Attridge, A.: Assessing subjective response to haptic feedback in automotive touchscreens. In: Proceedings of the 1st International Conference on Automotive User Interfaces and Interactive Vehicular Applications (AutomotiveUI 2009), pp. 11–18. ACM, New York (2009), doi:10.1145/1620509.1620512,
http://doi.acm.org/10.1145/1620509.1620512

10. Pitts, M.J., Burnett, G.E., Williams, M.A., Wellings, T.: Does haptic feedback change the way we view touchscreens in cars? In: International Conference on Multimodal Interfaces and the Workshop on Machine Learning for Multimodal Interaction (ICMI-MLMI 2010), article 38, 4 pages. ACM, New York (2010), doi:10.1145/1891903.1891952,
http://doi.acm.org/10.1145/1891903.1891952

11. Richter, H., Ecker, R., Deisler, C., Butz, A.: HapTouch and the 2+1 state model: potentials of haptic feedback on touch based in-vehicle information systems. In: Proceedings of the 2nd International Conference on Automotive User Interfaces and Interactive Vehicular Applications (AutomotiveUI 2010), pp. 72–79. ACM, New York (2010), doi:10.1145/1969773.1969787,
http://doi.acm.org/10.1145/1969773.1969787

12. Rydström, A., Grane, C., Bengtsson, P.: Driver behaviour during haptic and visual secondary tasks. In: Proceedings of the 1st International Conference on Automotive User Interfaces and Interactive Vehicular Applications (AutomotiveUI 2009), pp. 121–127. ACM, New York (2009), doi:10.1145/1620509.1620533,
http://doi.acm.org/10.1145/1620509.1620533

13. Yatani, K., Truong, K.N.: SemFeel: a user interface with semantic tactile feedback for mobile touch-screen devices. In: Proceedings of the 22nd Annual ACM Symposium on User Interface Software and Technology (UIST 2009), pp. 111–120. ACM, New York (2009), doi:10.1145/1622176.1622198,
http://doi.acm.org/10.1145/1622176.1622198

14. Walsh, S.P., White, K.M., Hyde, M.K., Watson, B.: Dialling and driving: Factors influencing intentions to use a mobile phone while driving. Accident Analysis & Prevention 40(6), 1893–1900 (2008) ISSN 0001-4575, 10.1016/j.aap.2008.07.005

15. White, K.M., Hyde, M.K., Walsh, S.P., Watson, B.: Mobile phone use while driving: An investigation of the beliefs influencing drivers' hands-free and hand-held mobile phone use. Transportation Research Part F: Traffic Psychology and Behaviour 13(1), 9–20 (2010), doi:10.1016/j.trf.2009.09.004,
http://dx.doi.org/10.1016/j.trf.2009.09.004, ISSN 1369-8478

A Peer-to-Peer Trilateral Passivity Control for Delayed Collaborative Teleoperation

Michael Panzirsch[1], Jordi Artigas[1], Andreas Tobergte[1], Paul Kotyczka[2],
Carsten Preusche[1], Alin Albu-Schaeffer[1], and Gerd Hirzinger[1]

[1] DLR - German Aerospace Center, Institute of Robotics and Mechatronics,
Oberpfaffenhofen, Germany
michael.panzirsch@dlr.de
http://www.dlr.de/rm/
[2] TUM - Technische Universität München, Institute of Automatic Control Engineering
Garching, Germany
http://www.rt.mw.tum.de/personen/derzeitige/paul-kotyczka/

Abstract. In this paper a trilateral Multi-Master-Single-Slave-System with control authority allocation between two human operators is proposed. The authority coefficient permits to slide the dominant role between the operators. They can simultaneously execute a task in a collaborative way or a trainee might haptically only observe the task, while an expert is in full control. The master devices are connected with each other and the slave robot peer to peer without a central processing unit in a equitable way. The system design is general in that it allows delayed communication and different coupling causalities between masters and slave, which can be located far from each other. The Time Domain Passivity Control Approach guarantees passivity of the network in the presence of communication delays. The methods presented are sustained with simulations and experiments using different authority coefficients.

1 Introduction

Through bilateral teleoperation, where a human operator controls a remote robotic manipulator with a master device, a human gains access to distant evironments or to environments behind a barrier as in minimally invasive surgery. That is also possible over long distances as demonstrated in 2002 with the ZEUS robotic system [1]. Additionally to the remotely operating surgeon there was still a surgeon on site in case of an emergency. This local surgeon could also be integrated into the control system with a local master console, enhancing the bilateral to a trilateral system. Analogous to surgery, potential applications for trilateral systems can be found in deep see or in space.

A trilateral system could either be used in a collaborative way where a local professional gets temporary support by a distanced specialist or as a training system where a trainee learns from a mentor ([2,3,4,5]). At the beginning of such a training the trainee can observe the mentor's action haptically without influence on the slave robot. With increasing experience the trainee should be provided with progressively higher control (see Fig. 1). This procedure is in this paper solved by the variation of an authority factor. In [2] a system with authority allocation and a unilaterally controlled slave is proposed.

P. Isokoski and J. Springare (Eds.): EuroHaptics 2012, Part I, LNCS 7282, pp. 395–406, 2012.

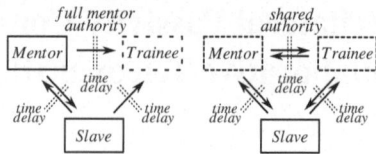

Fig. 1. Trilateral teleoperation including authority allocation and time delay

Visual feedback of the slave's position was provided and time delay in the operators' haptic channel considered. This system was enhanced to a peer-to-peer system in [3] with three equally privileged peers (slave and master devices) in a four channel control architecture (4CH). However the effect of time delay was neglected here. In bilateral systems the time delay as the general challenge in remote control has been tackled with several energy based techniques i.e. the Time Domain Passivity Control Approach (TDPA, [6,7,8]), Raisbeck's passivity criterion [9] and the wave variables technique [10] which is closely related to the scattering formulation [11]. Besides H_∞-control [2], wave variables [4] have been utilised in multi-agent-systems to handle the effects of time delay. Llewellyn's absolute stability criterion which is less conservative than Raisbeck's passivity criterion can not be extended to a trilateral system. Furthermore those two approaches require models of the system's complex mechanical devices. The first trilateral peer-to-peer system respecting time delay is presented in this work whereby the TDPA is applied because of its two major advantages, i.e. the consideration of the ideal case assuming the time delay to be zero ($T_{delay} = 0$) in the design process and the ability to handle non-linearities and unmodeled effects [12].

The focus in this paper is placed (a) on the mechanism to distribute the authority, (b) on how to guarantee passivity despite time delay and (c) how this structure can be generalized. In section 2.1, the signal flow architecture will be discussed with focus on the authority allocation (AA). The network representation is introduced in section 2.2 and the activity analysis of the authority allocation provided in section 3. Based on this the peer-to-peer TDPA is designed and the passivity proof accomplished. Experiments follow in section 4. Conclusions and future research will be discussed in section 5.

2   System Description

2.1   Signal Flow Diagram

Figure 2 shows the signal flow diagram of the proposed peer-to-peer telepresence system. In the depicted position force architecture (PF) velocity (v) and force (F) signals are exchanged between the haptic devices ($Master_1$, $Master_2$) and the robot ($Slave$) through the communication channels represented by time delay elements $e^{-T_i s}$. The PI-controllers (*PI-Ctrl*, virtual damper and spring) are located on the slave's side of the communication channels corresponding to the PF architecture (respectively for the operators' channel on the trainee's side). The factors β_{ME} and β_{TR}, corresponding to mentor and trainee respectively, determine the allocation of authority between the operators through scaling of the delayed forces from the PI-controllers. Those forces correspond to the influence of an agent on the addressed device. The relationship between the two authority variables β_{TR} and β_{ME} is given by:

$$\beta_{TR} = 1 - \beta_{ME} \quad \text{with} \quad \beta_{TR/ME} \in \{0...1\} \tag{1}$$

indicating that a reduction of the mentor's authority β_{ME} leads to a correlated increase of the trainee's authority β_{TR}. Reducing β_{ME} from 1 to 0 progressively assigns consequently higher influence on the system to the trainee. In contrast to [3] the slave's feedback signals (F_{12}, F_{13}) remain unaffected by $\beta_{ME/TR}$ since the slave's position (represented by the feedbacked force) as the main concern should always be presented correctly to the master devices.

2.2 Network Modelling

In this chapter the signal flow of the telepresence system will be transferred into network representation. This electrical modeling provides several useful tools which have been developed for circuit analysis. Since energy is flowing through ports which are described by power conjugated variables the energy observation in the system is heavily simplified. Because of the analogy between the potentials force (F) and voltage and the flows velocity (v) and current the signal flow subsystems can be replaced by so-called network ports. The TDPA utilises passivity observers (PO) which compute the energies at the ports i of a network subsystem in order to analyse the system's activity behavior:

$$E_i(t) = \int_0^t P_i(\tau) \, d\tau,$$

where $P_i(t)$ is the power computed as: $P_i(t) = v_i(t)F_i(t)$. As depicted in Fig. 4 v_i is the velocity flowing through a port i across which the force F_i is applied. The following convention regarding the signs of the port signals is assumed: If the integrated dual

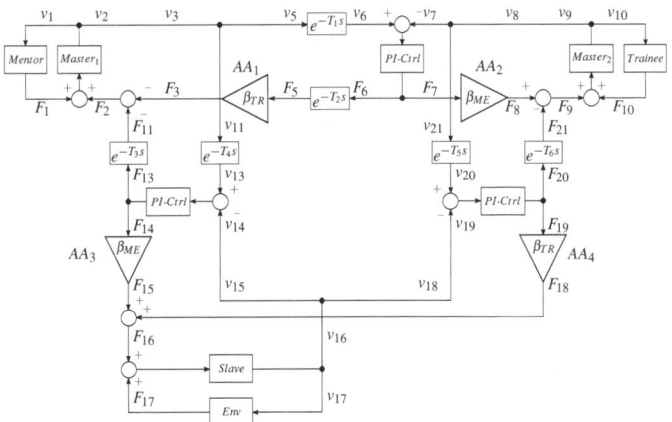

Fig. 2. Signal flow diagram of trilateral PF architecture with authority allocation (β_{TR}, β_{ME}) and time delay ($e^{-T_i s}$)

product $E_i(t)$ of a current entering the network and a positive voltage defined at the first terminal of the port w.r.t. the second one, is positive, the network is passive. Else, if it is negative, it is active. That means that energy flowing into the network results in positive energy. Regarding the sign of the power $P_i(t)$, the direction of flow can be computed (similar to [8]) as:

$$P_{i,in,NP}(t) = \begin{cases} P_i(t), & \text{if } P_i(t) > 0 \\ 0, & \text{if } P_i(t) \leq 0 \end{cases} \tag{2}$$

$$P_{i,out,NP}(t) = \begin{cases} 0, & \text{if } P_i(t) > 0 \\ -P_i(t), & \text{if } P_i(t) \leq 0. \end{cases} \tag{3}$$

The power $P_{i,in,NP}(t)$ flows into a regareded network subsystem at port i on the side of the network subsystem NP. Whereas $P_{i,out,NP}(t)$ stands for the power flowing out of a network subsystem at port i on the side of NP. NP are here the network subsystems terminating the 3-port such that NP can be "M" for Mentor, "T" for Trainee or "S" for Slave. The energy $E_i(t)$ and the power $P_i(t)$ are positive defined and monotone (see eq. (2) and (3)). The passivity controllers (PC) dissipate the amount of energy undesirably generated in an observed network. The subsystem terminated by the devices $Master_1$, $Master_2$ and the $Slave$ robot can be identified as a 3-Port (see Fig. 3), which can be split up in a modular way into three communication channel networks (CC) and three control unit networks (CU). The CUs include the authority allocation (eq. (1)), force distribution and the PI-controllers j, $Z_{PI_j}(s) = K_p/s + K_v$. Depending on the control architecture, different control unit and communication channel blocks can be inserted. In Fig. 5, the CC for a position force (PF) architecture is depicted examplifying the communication between mentor and trainee. The force transmission over the PF communication channel to the mentor can be represented as a voltage source whereas the velocity transmission to the trainee corresponds to a current source [14]. For the study case, i.e. the PF architecture, the network blocks of the mentor and trainee control units are illustrated in Fig. 7. The controllers are represented by an impedance Z_{PI_j}.

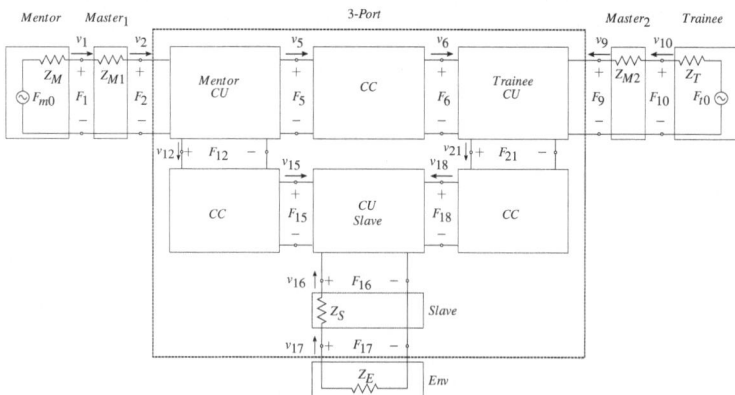

Fig. 3. General electrical analogous network representation of the trilateral system, control unit CU, communication channel CC

2.3 Authority Allocation

The next element which needs to be represented in the electrical scheme is the authority allocation (AA) governed by the coefficients β_{TR} and β_{ME}, as defined in (1). Since the velocities of each device (masters and slave) are not scaled by the authority coefficient ($v_3 = v_5$, $v_7 = v_8$) the AA can be modeled as a dependent force source (see Fig. 6(b)) whose value is given by:

$$F_{\beta 1} = F_5 - F_3 = (1 - \beta_{TR})F_5. \tag{4}$$

The force $F_{\beta 1}$ corresponds to the force which is substracted from F_5 through the AA.

2.4 Force Distribution

The force distribution can be understood by checking the interconnection of the CU's. Taking for instance the trainee side (see Fig. 7(b)) the port 9 attached to the device *master₂* is the result of a series interconnection of port 8 of the authority allocation AA_2 and port 21 next to the CC between trainee and slave.

Thus, the resulting force is given by the sum of each interconnected network. For the case of *master₁*, Fig. 7(a), the sum is given by $F_2 = F_3 + F_{12}$. To verify that the interfaces between the blocks surrounding the force distribution block satisfy the port requirement, it has to be shown that the in- and outflowing velocities at each port are identical. This requirement is fulfilled as can be seen by looking at Fig. 8: $v_2 = v_3 = v_{12}$.

3 Passive Trilateral Control

To examine the influence of the CU on the TDPA design the energy behavior of force distribution and authority allocation has to be studied.

3.1 Energy Analysis of Subsystems

As can be analysed in Fig. 8 the force distribution is a lossless element, since it is designed as a series connection containing no network elements. It follows from the definition of the authority allocation (4) that e.g. AA_1 (see Fig. 6(b)) purely injects

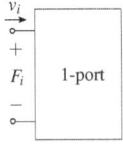

Fig. 4. 1-port network with port i, velocity v_i and force F_i

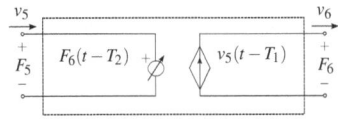

Fig. 5. Network representation of communication channel CC for PF architecture

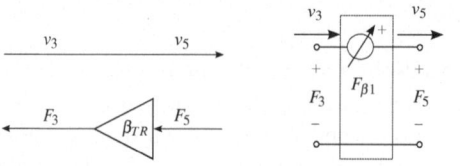

(a) Signal flow diagram   (b) Network representation

Fig. 6. Authority allocation analogy for AA_1

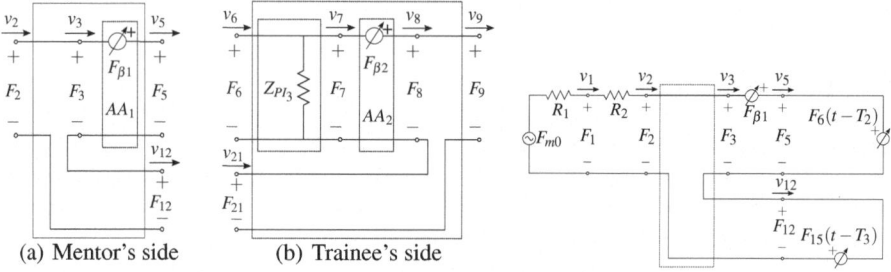

(a) Mentor's side   (b) Trainee's side

Fig. 7. Network representation of control unit CU for PF architecture

Fig. 8. The mentor's electrical circuit in the PF architecture

or dissipates energy depending on the direction of energy flow. AA_1 affects in trainee direction the power $P_{3,in,M}$ and in mentor direction the power $P_{5,in,T}$ ($v_5 = v_3$):

$$P_{5,out,T}(t) = P_{3,in,M}(t) + P_{AA_1,M}$$
$$P_{3,out,M}(t) = P_{5,in,T}(t) + P_{AA_1,T}.$$

Where $P_{AA_1,NP}(t)$ is the positive defined power flowing towards trainee ($NP = M$) and mentor ($NP = T$) respectively. The corresponding energies $A_{S,NP}$ (the energy is injected by network subsystem S from the direction of NP) can be computed as:

$$A_{AA_1,M}(t) = \int_0^t P^{act}_{AA_1,M}(\tau)\,d\tau \quad \text{with} \tag{5}$$

$$P^{act}_{AA_1,M}(t) = \begin{cases} -(P_{3,in,M}(t) - P_{5,out,T}(t)), & \text{if } P_{3,in,M}(t) - P_{5,out,T}(t) \leq 0 \\ 0, & \text{if } P_{3,in,M}(t) - P_{5,out,T}(t) > 0. \end{cases} \tag{6}$$

The power $P^{act}_{AA_1,M}(t)$ accounts in contrast to $P_{AA_1,M}(t)$ only power generated by AA_1 (for the case of the authority allocation $P^{act}_{AA_i,M}(t)$ equals $P_{AA_i,M}(t)$). The positive defined absolut energy dissipation $D_{AA_1,M}(t)$ of a subsystem can be evaluated analogously:

$$D_{AA_1,M}(t) = \int_0^t P^{dis}_{AA_1,M}(\tau)\,d\tau \quad \text{with} \tag{7}$$

$$P^{dis}_{AA_1,M}(t) = \begin{cases} P_{3,in,M}(t) - P_{5,out,T}(t), & \text{if } P_{3,in,M}(t) - P_{5,out,T}(t) > 0 \\ 0, & \text{if } P_{3,in,M}(t) - P_{5,out,T}(t) \leq 0. \end{cases} \tag{8}$$

The power $P_{AA_1,M}^{dis}(t)$ accounts analogously to $P_{AA_1,M}^{act}(t)$ only power dissipated by AA_1. AA_1 shows active behavior in direction of the trainee since the at port 5 outflowing power is always higher than the at port 3 inflowing one. Thus e.g. $D_{AA_1,M}(t)$, $D_{AA_2,T}(t)$ and also $A_{AA_1,T}(t)$ and $A_{AA_2,M}(t)$ are always zero. In contrast energy is e.g. by AA_1 in direction of the mentor and by AA_2 in direction of the trainee $(D_{AA_1,M}(t)$, $D_{AA_2,T}(t))$ partly dissipated. The activated energy must not be dissipated by the TDPA to serve the functionality of the authority allocation.

3.2 Placement of Passivity Observers and Controllers

For the proposed peer-to-peer system three passivity observer (PO) and passivity controller (PC, [6]) placements have been studied. Each of those placements focuses mainly on the passivation of the communication channels. The handling of the 3-port as a black-box surrounded by POs and PCs corresponding to the TDPA controlled 2-Port in the bilateral system is not possible, since the generated and dissipated energy in the system has to be differentiated by its direction of flow [7].

Channel-PO/PC: One possible placement corresponds to the standard bilateral PO/PC system which encloses only the communication channel. Thus in the channel-PO/PC placement one PO/PC system is applied on each of the three CCs in the trilateral system. In contrast to the approach proposed in the following this placement can also be implemented using the wave variables technique.

Track-PO/PC: As already suggested in [13] a bilateral network (track) surrounded by the PO/PC system can include an I-controller (the integral component acts on the position) besides the CC. In the track-PO/PC placement of the trilateral system the authority allocation is added to the TDPA controlled track in addition to the communication channel and the corresponding controllers (track: *AA*, *CC*, *PI*). Thus at each port of the 3-Port one impedance PC is sufficient to dissipate the energy generated in the two tracks in direction to the corresponding device. In Fig. 9 the PO/PC system for the track-PO/PC is depicted. The POs enclose each authority allocation, PI-controller and communication channel. This is the most general approach since it can be applied for all types of control architectures. In this approach the activity of the AAs must be observed and allowed by the PC. Stability is guaranteed e.g. by the Routh-Hurwitz criterion under neglection of the time delay. Furthermore the dissipation of the track subsystems have to be taken into account since they would obscure the activity of other subsystems. Besides the authority allocation each PI-controller j and especially its proportional part as a damper dissipates the energy $D_{PI_j,NP}(t)$ which is calculated analogously to (8). In contrast to the channel-POPC structure this activity is dissipated by the PCs which leads to a more conservative but also more robust system. On the other hand the track-PO/PC enables the conjoint passivation of two tracks leading to one 3-Port termination. Thus not the whole energy generated by an active CC in one track has to be dissipated by the corresponding PC if the CC of the other track is dissipating energy at the same time.

3.3 Passivity Proof

In this section the mentor's track-PO/PC system will be examined representatively (see 9). PO_8 and PO_{15} observe the positive energy flowing into the tracks. PO_7 and PO_8

observe the energy injection of AA_2. PO_7 and PO_6 observe the dissipation of PI_3 in the direction of the mentor. This holds for PO_{13}, PO_{14} and PO_{15}, AA_3 and PI_5 in the same way. The dissipation of AA_1 is observed by PO_5 and PO_3. PO_3 and PO_{12} measure the energy exiting to the mentor. The requirement for passivity of a m-port

$$E_{obs}^{mPort}(t) = \int_0^t F_1(\tau)v_1(\tau) + F_2(\tau)v_2(\tau) + ... + $$
$$+ F_m(\tau)v_m(\tau)\,d\tau + E(0) \geq 0. \tag{9}$$

implies that the amount of energy flowing into the system is higher than the one of the outflowing. The energy $E(0)$ which is stored in the system at $t = 0$ has to be respected. To prove that the mentor's track-PO/PC passivates the communication channels, the energy $E_x^{2tr,M}(t)$ and the energy $E_{obs}^{2tr,M}(t)$ have to be considered. $E_x^{2tr,M}(t)$ is the energy exiting the tracks at port I (see 9(a)) to $master_1$ in a passive system. In an active system the energy $E_{obs}^{2tr,M}(t)$ exits at Port I after dissipation of energy (generated by the tracks) through the PC. The passivity of the tracks is secured if the energy compassed by the

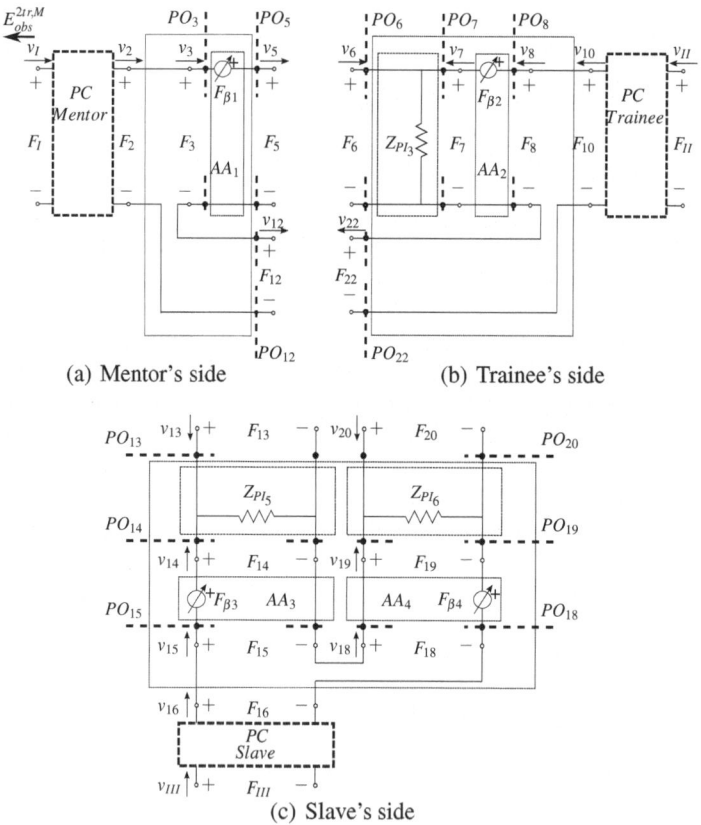

(a) Mentor's side   (b) Trainee's side

(c) Slave's side

Fig. 9. PO/PC system and control units for PF architecture

PC functionality $E_{obs}^{2tr,M}(t)$ is smaller than $E_x^{2tr,M}(t)$:

$$E_x^{2tr,M}(t) - E_{obs}^{2tr,M}(t) > 0. \tag{10}$$

The delay-free energy $E_x^{2tr,M}$ which guarantees the passivity of the tracks is given by:

$$E_x^{2tr,M}(t) = E_{8,in,T}(t) + A_{AA_2,T}(t) - D_{AA_1,T}(t) - D_{PI_3,T}(t) +$$
$$+ E_{17,in,S}(t) + A_{AA_3,T}(t) - D_{PI_5,S}(t) - E_{2,out,M}(t).$$

The separate calculation of energy generation and dissipation presented in section 3.1 serves the observation of the absolute energy generated or dissipated by a subsystem respectively. If instead of the in- and outflowing power flows the in- and outflowing energies are considered (as in the following for the communication channel) the overall energy behavior (since t=0s) is measured. These differing calculations make no difference for the AA_i since these network ports have a constant behavior in each direction of flow. In contrast for the PI-controllers a separate calculation of the generated energy is necessary since those subsystems are at different instants highly dissipating and generating energy. Regarding the overall energy behavior would lead to an energy storage in the PO/PC system. The PC would then react firstly on track activity when this storage is discharged which might result in oscillatory behavior.

The energy $E_{obs}^{2tr,M}(t)$ (observing the activity of CC_2, CC_4, PI_3 and PI_5) is given by:

$$E_{obs}^{2tr,M}(t) = E_{8,in,T}(t - T_2) + A_{AA_2,T}(t - T_2) - D_{AA_1,T}(t) - D_{PI_3,T}(t - T_2) +$$
$$+ E_{17,in,S}(t - T_3) + A_{AA_3,T}(t - T_3) - D_{PI_5,S}(t - T_3) - E_{2,out,M}(t).$$

The PO/PC system designed by this $E_{obs}^{2tr,M}(t)$ leads to the dissipation of the communication channels' and the PI-controllers' activities $A_{PI_j,NP}(t)$. To fulfill (10) in terms of passivity the following inequality must hold:

$$E(t - T_{2/3}) - E(t) < 0.$$

Since the in- and outflowing energies E_{in}, E_{out}, activities A_{AA} and the dissipations D_{AA} and D_{PI} are defined to be purely increasing, never decreasing ($E(t) > E(t - T_{2/3})$) inequality (10) and thus the passivity can be proven. The energy $E_{2diss}^{2tr,M}$ which has to be dissipated by the mentor's PC in the time step T_S results in

$$E_{2diss}^{2tr,M}(t) = E_{8,in,T}(t - T_2) + A_{AA_2,T}(t - T_2) - D_{AA_1,T}(t) - D_{PI_3,T}(t - T_2) +$$
$$+ E_{17,in,S}(t - T_3) + A_{AA_3,T}(t - T_3) - D_{PI_5,S}(t - T_3) - E_{2,out,M}(t)$$
$$- E_{diss}^{2tr,M}(t - T_S)$$

The energy $E_{diss}^{2tr,M}(t - T_S)$ is taken into account which has been dissipated by the mentor PC until the current time step T_S. The passivity proof and PO/PC design for trainee and slave PC is analogous.

4 Experiments

In this section experiments analysing the system's performance in dependence of time delay and authority allocation will be presented. In the following the track-PO/PC has been applied in combination with a position force architecture (PF) on rotatory 1DoF haptic devices (by SensoDrive) which were connected to a QNX system. This hardware was chosen for the masters and the slave likewise. For the experiments every communication channel has been restrained by one unique time delay. The PF control architecture has been implemented on Matlab/Simulink with a sampling time of 1ms. Compiling the model by Real-Time Workshop supported appropriately real-time performance on a QNX machine. The system has been tuned to go unstable with $T_i = 10ms$ (unique PI parameters: damping $B_{PI} = 0.06\frac{Nms}{rad}$, stiffness $K_{PI} = 3.5\frac{Nm}{rad}$).

In the first experiment (see figure 10(a), 10(c)) the mentor has the full authority ($\beta_{ME} = 1$). The mentor guides the slave against a wall (time: 3.5s to 5s) marginally penetrating it. The position plot shows that the slave follows the mentor very well. The trainee though resists the movement. During this resistance the trainee's PC dissipates a high amount of energy (E_{PC}). The effect of the authority allocation can be recognized looking at plot $F_{i,2Sl}$ (see 10(a)). $F_{18,2Sl}$ is the force sent to the slave from the trainee side. This force is completely canceled by the AA ($\beta_{TR} = 0$) whereas the mentor's force $F_{15,2Sl}$ is entirely received by the slave. The passivity proof is accomplished in 10(c) where it can be seen that E_{out} is always smaller than E_{in}. With $\beta_{ME} = 0.75$ the trainee is assigned a little

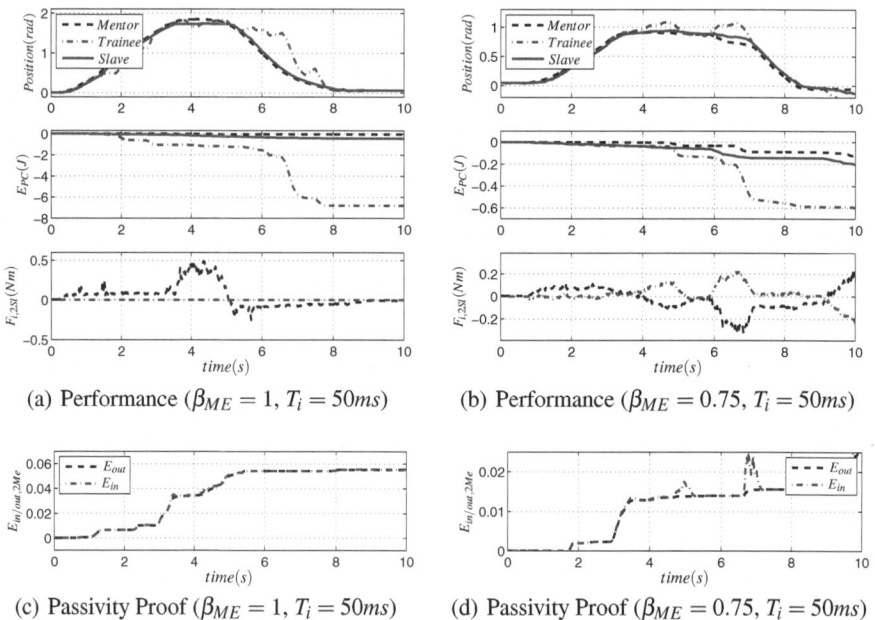

(a) Performance ($\beta_{ME} = 1$, $T_i = 50ms$)   (b) Performance ($\beta_{ME} = 0.75$, $T_i = 50ms$)

(c) Passivity Proof ($\beta_{ME} = 1$, $T_i = 50ms$)   (d) Passivity Proof ($\beta_{ME} = 0.75$, $T_i = 50ms$)

Fig. 10. Performance and Passivity Proof of the track-PO/PC system, with channel delay of 50ms and varying authority allocation in a PF-architecture

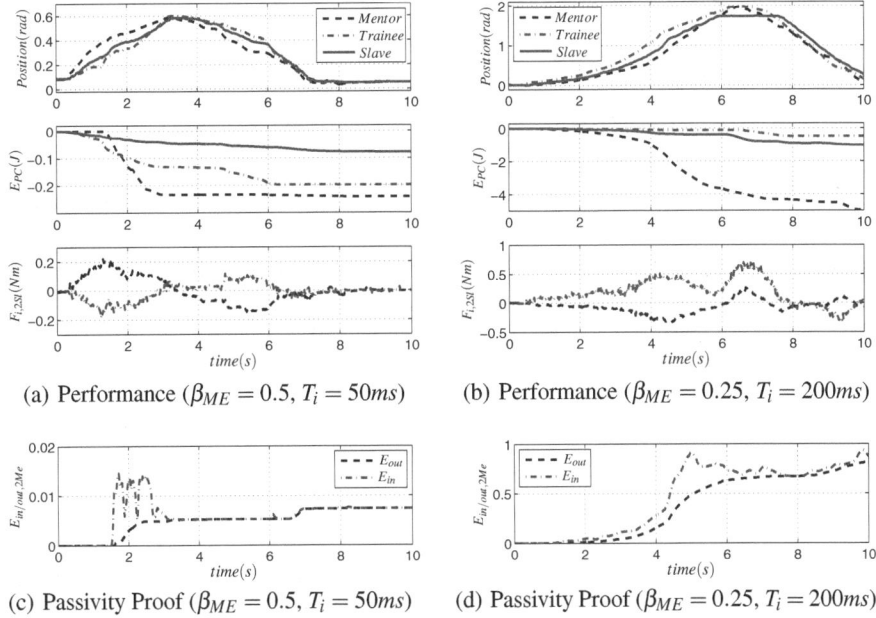

(a) Performance ($\beta_{ME} = 0.5$, $T_i = 50ms$)

(b) Performance ($\beta_{ME} = 0.25$, $T_i = 200ms$)

(c) Passivity Proof ($\beta_{ME} = 0.5$, $T_i = 50ms$)

(d) Passivity Proof ($\beta_{ME} = 0.25$, $T_i = 200ms$)

Fig. 11. Performance and Passivity Proof of the track-PO/PC system, with varying channel delay and authority allocation in a PF-architecture

control in the second experiment as the forces $F_{i,2Sl}$ confirm (10(b)). The position following can be analysed in phases of consistent operator movement and is satisfactory despite the delay of 50ms. The position diagram in figure 10(b) shows that the slave does not stick as much to the mentor as in the first experiment since it is also influenced by the triainee's movement. The passivity proof plot (see 10(d)) shows that the mentor's PC dissipates too much energy in phases of reconvergence of the three devices.

In figure 11(a) a shared authority situation is displayed. The slave is now exactly positioned in the middle of the two operators. The operators' passivity controllers dissipate about the same amount of energy.

In the last experiment the delay was chosen to 200ms per communication channel. The position following of the devices is still satisfactory (see 11(b)). E.g. from time = 8s to 9s the operators have the same intention and thus the same position. The slave's position is delayed by approximately 0.2s as expected. Regarding the experiments conjointly (see 10(a)-11(d)) one can recognize that the amount of dissipated energy (E_{PC}) increases with the delay since the channel's activity rises. Furthermore it can be seen that the PC of the guiding operator (mentor for $\beta_{ME} > \beta_{TR}$ and vice versa) dissipates less energy than the one of the trained operator.

5   Conclusions and Future Research

The TDPA has been applied to a trilateral system in a generic way. It is general in the sense that it allows any control architecture and any communication channel characteristics in the peer-to-peer system. The experiments showed good results for roundtrip delays up to 200ms. The authority allocation system has been optimized resulting in satisfying position following of the three peers. The PF and the PP control architecture (which has not been presented in this paper) are already modeled. Presenting the force sensed by the slave device will in the future improve the perception of the slave's environment. Therefore another control unit and communication channel set for the 4Ch architecture is to be developed. The energy behavior of the introduced authority allocation and the PI controller was analysed and considered in the PO/PC design. The track-PO/PC controlling two tracks conjointly leads to the most robust approach compared to the straight forward usage of the bilateral channel-PO/PC. To avoid high frequency forces generated by passivity controllers with impedance causality a method introducing a virtual mass spring system [8] has already been integrated. This proceeding will in future works be compared to the usage of admittance PCs [6].

References

1. Marescaux, J., Leroy, J., Rubino, F.: Transcontinental Robot-Assisted Remote Telesurgery: Feasibility and Potential Applications. Annals of Surgery, 487–492 (2002)
2. Nudehi, S., Mukherjee, R., Ghodoussi, M.: A Shared-Control Approach to Haptic Interface Design for Minimally Invasive Telesurgical Training. IEEE Trans. on Contr. Sys. Tech. 13, 588–592 (2005)
3. Khademian, B., Hashtrudi-Zaad, K.: Control of Dual-User Teleoperation Systems - Design, Stability Analysis, and Performance Evaluation. Ph.D. thesis, Queen's University Kingston, Canada (2010)
4. Olsson, A.P., Carignan, C.R., Tang, J.: Cooperative control of virtual objects over the Internet using force-reflecting master arms. In: IEEE Proc. on Rob. and Aut., vol. 2, pp. 1221–1226 (2004)
5. Mendez, V., Tavakoli, M.: A Passivity Criterion for N-Port Multilateral Haptic Systems. In: 49th IEEE Conf. on Decision and Control, pp. 274–279 (2010)
6. Hannaford, B., Ryu, J.-H.: Time domain passivity control of haptic interfaces. In: Proc. of the IEEE ICRA 2001, vol. 2, pp. 1863–1869 (2001)
7. Ryu, J.-H., Preusche, C.: Stable Bilateral Control of Teleoperators Under Time-varying Communication Delay: Time Domain Passivity Approach. In: IEEE Int. Conf. on Rob. and Aut., pp. 3508–3513 (2007)
8. Ryu, J.-H., Artigas, J., Preusche, C.: A passive bilateral control scheme for a teleoperator with time-varying communication delay. Mech. 20, 812–823 (2010)
9. Cho, H.C., Park, J.H.: Impedance Controller Design of Internet-Based Teleoperation Using Absolute Stability Concept. In: IEEE Int. Conf. on Int. Rob. and Sys., pp. 2256–2261 (2002)
10. Niemeyer, G.: Using wave variables in time delayed force reflecting teleoperation. Ph.D. thesis, Mass. Inst. of Tech. (1996)
11. Anderson, R.J., Spong, M.W.: Bilateral control of teleoperators with time delay. IEEE Trans. on Aut. Contr. 34, 494–501 (1989)
12. Artigas, J., Ryu, J.-H., Preusche, C.: Time Domain Passivity Control for Position-Position Teleoperation Architectures. Pres. 19, 482–497 (2010)
13. Ryu, J.-H.: Stable Teleoperation with Time Domain Passivity Approach. In: Proc. of the 17th IFAC World Congr., vol. 17, pp. 15654–15659 (2008)
14. Artigas, J., Ryu, J.-H., Preusche, C., Hirzinger, G.: Network Representation and Passivity of Delayed Teleoperation Systems. In: IEEE Int. Conf. on Int. Rob. and Sys., pp. 177–183 (2011)

A Closed-Loop Neurorobotic System for Investigating Braille-Reading Finger Kinematics

Jérémie Pinoteau[1,*,**], Luca Leonardo Bologna[1,*], Jesús Alberto Garrido[2], and Angelo Arleo[1]

[1] Adaptive NeuroComputation Group, Unit of Neurobiology of Adaptive Processes, UMR 7102, CNRS–University Pierre and Marie Curie P6, 75005, Paris, France
{jeremie.pinoteau,luca.bologna,angelo.arleo}@upmc.fr
http://www.anc.upmc.fr
[2] Department of Cellular-Molecular Physiological and Pharmacological Sciences, University of Pavia, 27100, Pavia, Italy
jesus.garrido@unipv.it

Abstract. We present a closed-loop neurorobotic system to investigate haptic discrimination of Braille characters in a reading task. We first encode tactile stimuli into spiking activity of peripheral primary afferents, mimicking human mechanoreceptors. We then simulate a network of second-order neurones receiving the primary signals prior to their transmission to a probabilistic classifier. The latter estimates the likelihood distribution of all characters and uses it to both determine which letter is being read and modulate the reading velocity.

We show that an early discrimination of the entire Braille alphabet is possible at both first and second stages of the somatosensory ascending pathway. Furthermore, 89% of the characters are correctly recognised in a constant-velocity reading task, while a closed-loop modulation of the speed allows for faster scanning and movement kinematics similar to the ones observed in humans –though with a lower classification rate.

Keywords: Dynamic haptic discrimination, Braille reading, Spiking neural networks.

1 Introduction

Braille reading involves haptic texture perception as well as cognitive processing and motor control operations. At the peripheral afferent level, the forces exerted by Braille character dots on the fingertip induce visco-elastic deformations of the skin which stimulate the mechanoreceptors innervating the epidermis. The information conveyed by first afferent neural signals is transmitted to the spinal cord as well as to the cuneate nucleus (CN) of the brainstem. The CN projects to several areas of the central nervous system, including the cerebellum and the thalamus, which in turn transfer the information to the primary somatosensory cortex. Processing along this pathway allows haptic

---

* J.P. and L.L.B. contributed equally to this work.
** Corresponding author.

P. Isokoski and J. Springare (Eds.): EuroHaptics 2012, Part I, LNCS 7282, pp. 407–418, 2012.
© Springer-Verlag Berlin Heidelberg 2012

information to be interpreted and leads to adaptive motor responses affecting the reading finger's movement kinematics. In fact, a Braille readers' fingertip undergoes several accelerations during character scanning. These changes in velocity are common to all subjects and appear to originate mainly from motor control mechanisms. Nonetheless, cognitive (i.e. linguistic) processes are shown to also affect velocity modulation, and the role of lexical, sub-lexical and semantic processing is currently being investigated [1], [2].

Here, we propose a neurorobotic framework to study active sensing during fine haptic discrimination of Braille characters. We simulate skin indentation protocols in which Braille-like tactile stimuli are dynamically scanned by an artificial touch sensor. Deformation signals act as inputs to a network of leaky-integrate-and-fire neurones (LIF), which perform an analogue-to-spike conversion and mimick the role of cutaneous mechanoreceptors. In particular, we model the activity of Slow Adapting type I (SA-I) primary afferents, in terms of both spiking discharge and receptive fields (see [3], for a recent review). LIF neurones project onto a second order network modelling CN responses through a population of Spike Response Model (SRM) [4] units. Downstream from the CN, a naive Bayesian classifier computes the probability distribution of all Braille characters online. The likelihood distributions are ultimately used to discriminate the letter currently scanned and devise an adapted velocity trajectory optimising the scanning/discrimination time.

2   Material and Methods

Figure 1 shows the complete robotic setup. We use a set of 26 different probes, reproducing a scaled version (1:1.67) of all Braille characters, to stimulate an artificial touch sensor in order to simulate the human fingertip deformations exerted by Braille dots. The touch sensor is rubbed over the Braille alphabet and its analogue responses are encoded and decoded by the simulated first- and second-order afferents (mechanoreceptors and cuneate neurones respectively). CN output activity is finally interpreted by a probabilistic classifier to perform character discrimination and adaptive speed control.

2.1   The Artificial Touch Sensor

An artificial skin prototype[1] [6], [7] was initially used to collect and characterise a first dataset of analogue responses to Braille character indentations. This artificial fingertip consists of 24 capacitive square sensors disposed according to a rectangular grid layout. Each sensor has a dimension of approximately 3 mm and the inter-centre distance is 4 mm, for a total sensitive surface of approximately 18×23 mm (Fig. 1B). The array is covered by a 2.5 mm thick neoprene layer aiming at modulating the pressure exerted on the sensors. The response strength of each sensor is proportional to the indentation level and ranges from 0 to 189 femtoFarads (fF). The acquisition frequency of each capacitive pad is 20 Hz.

---

[1] Developed at the Italian Institute of Technology (IIT), Genoa, Italy.

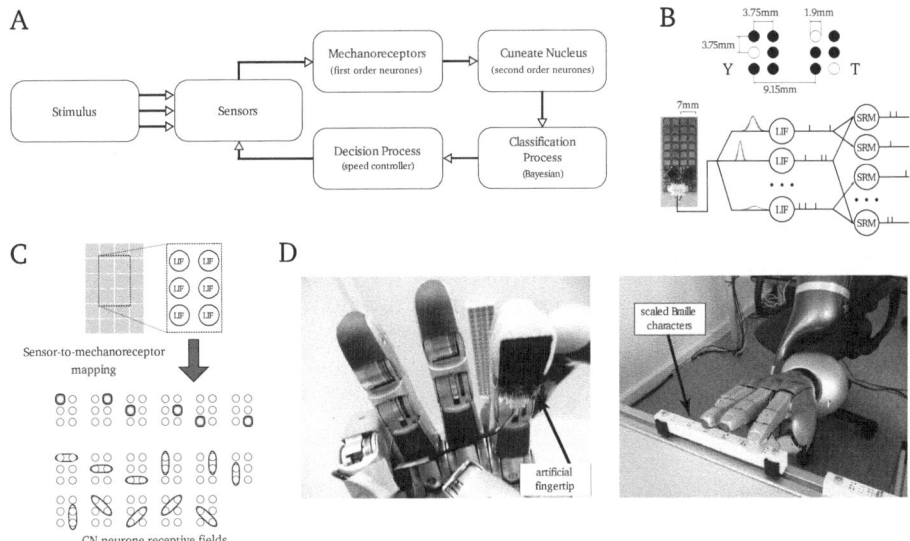

Fig. 1. *Overview of the entire encoding/decoding pathway and robotic setup.* (A) From left to right. We employ Braille characters as tactile stimuli to indent a capacitive artificial touch sensor. A network of Leaky-Integrate-and-Fire (LIF) neurones [5] converts analogue signals from the sensor into spiking activity, mimicking fingertip mechanoreceptors. LIF neurones project onto a network of Spike-Response-Model units [4] implementing second order cuneate nucleus (CN) cells. The outgoing activity is decoded by a Naive Bayesian Classifier whose output allows a speed controller to devise an optimal velocity for the fingertip movement. (B) Top: Examples of scaled Braille characters used as stimuli. Bottom: schematic representation of the encoding process, from artificial sensor signals to CN neurones output activity. (C) Top: LIF neurones modelling mechanoreceptor activity fulfil a topological mapping of fingertip regions. Bottom: CN cell receptive fields are built so as to collect the activity of either a single cell or different possible combinations of two or three adjacent mechanoreceptors. (D) The artificial fingertip mounted on a robotic hand/arm setup (Institute of Robotics and Mechatronics, German Aerospace Center ©).

We developed a simulator reproducing the responses of the artificial fingertip and offering a greater flexibility in data generation and experimental protocols [7]. We modelled the touch sensor responses by means of Gaussian kernels of amplitude 55 fF and standard deviation 1.6 mm. Additionally, we added a white noise to the amplitude and standard deviation of the signals (2.5 fF and 0.1 mm respectively) and we modelled possible position errors due to the experimental setup by adding a Gaussian noise to the location of each stimulus (sd = 0.1 mm).

2.2 Primary Afferent Coding: Analogue-to-Spike Transduction

We implement a network of 12 leaky integrate-and-fire (LIF) neurones [8], [5] to convert analogue touch sensor outputs into spike train patterns (Fig. 1C). We map the capacitance values provided by the touch sensors into current intensities $I(t)$ driving the LIF neurones by applying a multiplicative gain factor of -390 pA/fF (determined by comparing output LIF spike trains against recorded mechanoreceptor responses [9]). The dynamics of the membrane potential $V(t)$ of each LIF neurone is:

$$C \cdot \frac{dV(t)}{dt} = -g \cdot (V(t) - V_{\text{leak}}) - I(t) \tag{1}$$

where $C = 0.5$ nF denotes the membrane capacitance, $g = 25$ nS the passive conductance, $V_{\text{leak}} = -70$ mV the resting membrane potential, and $I(t)$ the total synaptic input of a neurone. The membrane time constant is then $\tau = C/g = 20$ ms. Whenever the membrane potential $V(t)$ reaches the threshold $V_{\text{thr}} = -50$ mV the neurone emits an action potential. Then, its membrane potential is reset to $V_{\text{reset}} = -100$ mV and the dynamics of $V(t)$ is frozen during a refractory period $\Delta t_{\text{ref}} = 2$ ms. We also use a "threshold fatigue" [5] to model the phenomenon of "habituation". It consists in increasing the threshold V_{thr} by a value A_{thr} each time the neurone discharges, making it harder for the neurone to spike again (i.e. preventing it from responding in a highly tonic manner even in the presence of strong inputs). In the absence of spikes, the threshold decreases exponentially back to its resting value V_{restThr}:

$$\frac{dV_{\text{thr}}(t)}{dt} = -\frac{V_{\text{thr}}(t) - V_{\text{restThr}}}{\tau_{\text{thr}}} \tag{2}$$

with $\tau_{\text{thr}} = 100$ ms, $V_{\text{restThr}} = -50$ mV and $A_{\text{thr}} = 50$ mV.

2.3 Second-Order Processing in the Cuneate Network

We model individual cuneate cell responses by means of 49 SRM neurones [4] (see [10], for a previous use of the model) implemented through a simulation environment [11] optimised to reduce execution time. We include a noise model (i.e. escape noise) that follows a stochastic process, thereby providing a linear probabilistic neuronal model.

An input spike arrival at time t induces a membrane potential depolarisation under the form of an EPSP (excitatory postsynaptic potential) $\Delta V(t)$ described by:

$$\Delta V(t) \propto \sqrt{t} \exp(-t/\tau) \tag{3}$$

where the parameter $\tau = 2$ ms determines the decay time constant of the EPSP. If several spikes excite the neurone within a short time window, the EPSPs add up linearly:

$$V(t) = V_r + \sum_{i,j} w_i \, \Delta V(t - \hat{t}_i^j) \tag{4}$$

where i denotes presynaptic neurones, j indexes the spikes emitted by a presynaptic neurone i at times $\hat{t}_i^j$, and $V_r = -70$ mV is the resting potential. The term w_i indicates the synaptic weight of the projection from the presynaptic unit i, defined as:

$$w_i = W \cdot w_i^{0,1} \tag{5}$$

with factor W determining the upper bound of the synaptic efficacy, and $w_i^{0,1}$ being constrained within the range $[0,1]$. We use $W = 0.04$ in our simulations. At each time step, a function $g(t)$ is defined as:

$$g(t) = r_0 \log \left(1 + \exp \left(\frac{V(t) - V_0}{V_f} \right) \right) \tag{6}$$

where the constants $r_0 = 11$ Hz, $V_0 = -65$ mV, $V_f = 0.1$ mV are the instantaneous firing rate, the probabilistic threshold potential, and a gain factor, respectively. A function $A(t)$ determines the refractoriness property of the neurone:

$$A(t) = \frac{(t - \hat{t} - \tau_{abs})^2}{\tau_{rel}^2 + (t - \hat{t} - \tau_{abs})^2} \mathscr{H}(t - \hat{t} - \tau_{abs}) \tag{7}$$

where $\tau_{abs} = 3$ ms and $\tau_{rel} = 9$ ms denote the absolute and relative refractory periods, respectively, $\hat{t}$ the time of the last spike emitted, and $\mathscr{H}$ the Heaviside function. Finally, the functions $g(t)$ and $A(t)$ allow the probability of firing $p(t)$ to be computed:

$$p(t) = 1 - \exp\left(-g(t)A(t)\right) \tag{8}$$

We implement the synaptic connections between mechanoreceptors and cuneate nucleus neurones so as to generate the receptive fields shown in Fig. 1C. Each CN neurone receives non-plastic inputs from either one or a group of two/three adjacent mechanoreceptors depending on the stimulus (see Fig. 1 for details). The dimension and shape of the receptive fields and the synaptic weight distribution of the mechanoreceptor-to-CN projections allow topographical information to be maintained at the level of the second order output space. Also, thanks to the adopted connectivity layout, CN neurones collecting signals from large receptive fields mirror both single primary neurone activation and multiple co-activations, thus enriching the population's spiking dynamics.

2.4 Assessing Neurotransmission Reliability: Metrical Information Analysis

In order to decode neural activities and quantify fine touch discrimination, we apply the recently defined metrical mutual information $I^*(R;S)$ [12]. Unlike Shannon's definition of mutual information [13],[14], this measure takes into account the metrical properties of the spike train space [15],[16],[17] and has been proven to be suitable to decode the responses of human mechanoreceptors obtained via microneurography recordings [10],[12]. The definition of $I^*(R;S)$ relies on a similarity function based on the distances between spike train responses elicited by the entire set of stimuli. The Victor and Purpura Distance was used in the definition of the metrical information [15]. This specific spike metric makes it possible to modulate the importance given to temporal (and rate) coding in the mapping of the spike train space through a cost parameter.

The perfect discrimination condition corresponds to maximum $I^*(R;S)$ and zero conditional entropy $H^*(R|S)$ [12]. It occurs when the size of the largest cluster of responses (to the same stimulus) becomes smaller than the smallest distance between all clusters of responses [12].

2.5 Online Classification of Braille Characters: Naive Bayesian Classifier

To discriminate Braille characters during the reading task, we trained a Naive Bayesian Classifier (NBC) via multinomial distributions. This learning algorithm belongs to the family of probabilistic classifiers relying on Bayes' rule to compute the posterior probability of the sample classes. Despite its simplicity, the NBC has proven to be fast and efficient even when the feature independence hypothesis underlying its application is not fulfilled [18],[19].

Braille characters (i.e. dataset classes) are defined from the spike train activity encoded after the touch sensor signals. With the aim of performing a fast classification, we built the training dataset in two steps. We first swiped the fingertip over the complete set of 26 Braille characters and collected the cuneate neurones' responses. We then binned the 49 CN cells spiking activity with temporal windows of increasing length (see Fig. 3A). The bin size increment was fixed at 10 ms. Each temporal bin is characterised by the spatiotemporal organisation of the firing activity it encloses and is labelled with the corresponding character. The properties of the spiking activity present in each bin (e.g. spike times, spiking neurones), contribute to building a model associating specific Braille characters to the different patterns of activity. In the online task, a character detection occurs whenever the probability distribution's peak exceeds the 90% threshold.

2.6 From Classification to Reading Velocity Control: Kurtosis-Based Assessment of the Likelihood Distribution

Given the procedure adopted to build the activity dataset and the small time bin chosen as temporal increment, characters' probability distributions can be measured frequently while reading. We use such information to compute the excess Kurtosis index which indicates the extent to which a probability distribution is peaked around its values.

At each simulation time step, the Kurtosis index gradient is used to adapt the reading velocity. A positive value indicates that the distribution is narrowing. Such a convergence reflects a decrease in the uncertainty on the character being scanned and triggers a velocity increase proportional to the gradient value. Differently, a distribution's widening induces a deceleration. We investigate whether this velocity modulation decreases the classification time as compared to a constant velocity movement.

3   Results

3.1   Characterisation of Mechanoreceptor Responses

In a previous study, we compared simulated and human mechanoreceptors responses to fingertip skin indentations through Braille-character probes [20]. Modelled primary afferents exhibit receptive fields qualitatively similar to those of human Slow Adaptive I (SA-I) mechanoreceptors, in terms of shape, dimensions and signal-to-noise ratio (see Fig. 2A). SA-I primary afferents show a topological mapping (i.e. their activity correlates with the area of stimulation), demonstrating their role in encoding spatial discontinuities [3]. Fast Adaptive I (FA-I) mechanoreceptors have similar spatial properties but, thus far, no clear experimental evidence has shown whether SA-I or FA-I afferents primarily carry the information needed for Braille character recognition [21].

The first spike jitter distributions of simulated and human SA-I mechanoreceptors, are statistically equivalent in terms of median and shape (Mann-Whitney U test, $P > 0.11$; Kolmogorov-Smirnov test, $P > 0.076$), despite a time lag in the simulated responses of about 2 ms (see Fig. 2B). Thus, modelled mechanoreceptors present the same variability in spike latencies as SA-I afferents, but on a larger time scale.

3.2 Information Content of First and Second Order Tactile Afferent Responses

We investigate neurotransmission reliability at both the first and second order neurones level by decoding mechanoreceptor and CN cell responses to the entire Braille alphabet during scanning. More specifically, we focus on the evolution of the information content over time with the aim of quantifying how rapidly a perfect discrimination of all characters can be achieved after the stimulus onset.

Figure 2C (Top) illustrates the evolution of the metrical information and conditional entropy as spikes flow in while scanning the entire Braille stimuli set at a constant velocity of 30 mm/s. The cost in the Victor and Purpura Distance was chosen so to allow the earliest possible perfect discrimination.

At the mechanoreceptor level, first spikes occur at around 100 ms, and 250 ms later the condition for an errorless stimulus reconstruction is satisfied. In comparison, a small delay is observed at the CN output level, and almost perfect discrimination is possible just as soon. As expected, the metrical information curve exhibits a plateau starting at around 200 ms and lasting approximately 75 ms. This corresponds to the stimulation phase during which the fingertip is already in contact with the first column of Braille dots while the second does not stimulate any sensor yet. The information value at the plateau is about half of the total amount of information transmitted.

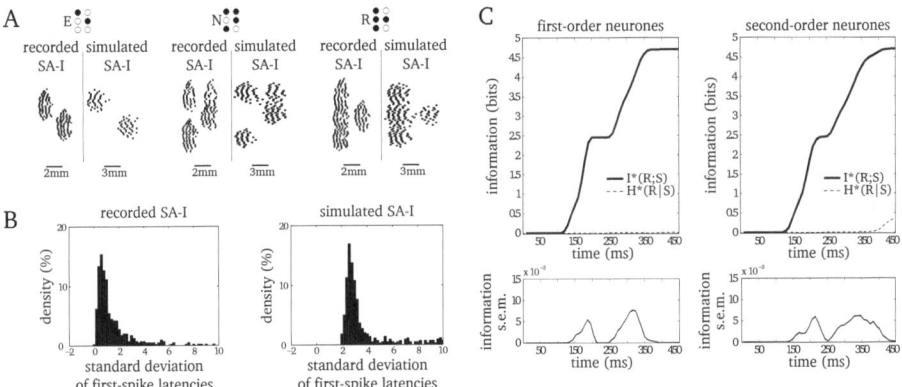

Fig. 2. *Characterisation of mechanoreceptor properties and theoretical information analysis of first and second order afferent responses to Braille stimuli.* (A) Spatial event plots of human SA-I mechanoreceptor responses to scanned Braille characters 'e', 'n' and 'r' and their simulated counterparts (recorded sections adapted from [21]). (B) Distribution of standard deviations (SD) of first-spike latencies for both SA-I (left) and simulated (right) mechanoreceptor responses. (C) Top: Time course of metrical information (full line) and conditional entropy (dashed line) at the output of LIF (left) and CN (right) neurones, as the fingertip scans the Braille characters at 30 mm/s. The 26 Braille characters serve as stimuli, with 20 repetitions per stimulus used. First spikes occur at around 100 ms, and the perfect discrimination condition is reached about 250 ms later. Bottom: Information variability, measured as mean standard error, over time.

3.3   Classification of Braille Characters in a Reading Task

We collect a dataset of character samples and fit a probabilistic classifier to estimate the posterior probability of Braille characters while reading. When the NBC fitting parameters are used for offline classification, a near perfect discrimination is possible as soon as enough spikes are processed by the classifier. In fact, as a result of the binning procedure adopted, data samples built by gathering the spiking activity in small temporal windows (cf. Sect. 2.5) do not carry enough information to allow a correct discrimination; on the contrary, considering longer periods leads to a decrease in the uncertainty (see Fig. 3B). A clear example of the evolution of probability distributions over time is given in Fig. 3C. At the beginning of the scanning movement, the spiking activity of CN neurones responding to letter 'r' (Fig. 3C, left panel) does not allow to distinguish between 'r' and the other Braille characters with a similar dot disposition (i.e. 'l', 'p', 'q', 'v' - see Fig. 1C, left panel inset). But, as time evolves, probability distributions start to peak indicating that the uncertainty decreases till a correct classification is possible (maximal probability reached for letter 'r'). A similar example is shown for letter 'e' (see Fig. 3C, right panel).

In online simulations, when the classifier is tested with a constant scanning velocity of 30 mm/s, 89% of the scanned characters are correctly discriminated (10% false positive, 1% no classification). Differently, if a speed modulation is adopted, the NBC's performance decreases to 78% of correct classifications (18% of false positives, 4% no classification). However, the discrimination of most letters is sensibly the same in both cases. The observed loss in overall performance is mostly due to couples of letters (e.g. 'a' and 'c') whose recognition rate fall considerably (down to approximately 45%) given to the similarity between character dot patterns.

3.4   Online Reading Velocity Modulation

The methodology we adopted for building the spiking-activity based dataset, lends itself to a frequent computation of Braille character probability distributions. We asked whether the probabilities evolution over time could be seen as a possible mechanism underlying the changes in reading speed of blind readers that can be attributed to lexical and sub-lexical processes. Thus, we computed the Kurtosis index gradient on probability distributions at each time step (cf. Sect. 2.6) and, upon multiplication by a constant factor, used it to modulate the reading speed (see Fig. 4A).

Following the velocity modulation approach, global reading performances improve. At the end of the scanning of a single character, the average velocity adopted through modulation is higher than the finger's initial velocity (i.e. when it first encounters the character) which would have been maintained in the absence of dynamic changes to the reading speed (i.e. constant velocity scanning). For a base velocity of 30 mm/s, the speed modulation allowed the average scanning velocity to increase to 35.4 mm/s. Moreover, we observed an average number of 8.6 accelerations per character (see Fig. 4B). This result is coherent with what was observed in human Braille reading experiments where the influence of lexical (i.e. words) and sub-lexical (i.e. characters) content on fingertip velocity intermittencies are investigated [2].

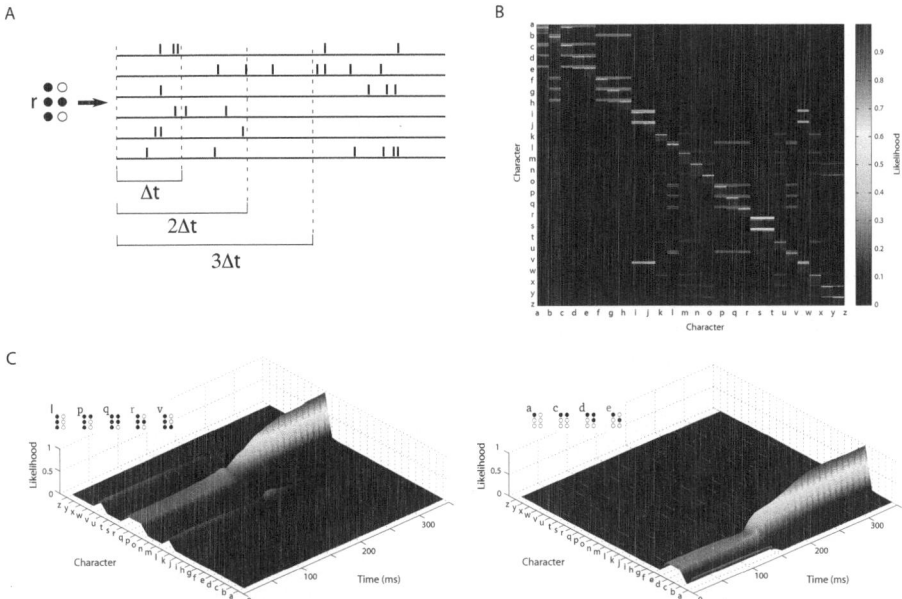

Fig. 3. *Binning procedure and Naive Bayesian classifier output likelihood distributions.* (A) CN spiking activity is binned following time windows of increasing size ($n \times \Delta t$). (B) Confusion matrix of training samples (100 samples per character), indicating classification probabilities as determined by the Bayesian classifier. For each character, samples of increasing duration are considered (matrix rows). Small bins capture little activity and the classifier is not able to distinguish between different patterns (light blue values). Differently, larger bins collect enough information to allow character discrimination (red values). (C) Expanded sections of panel B for letters 'r' (left) and 'e' (right). Initially, characters with similar dot patterns (reported in the insets) are equally probable. As time increases, more activity is collected and the probability distribution peaks on a single value (indicating the character being read).

4   Discussion

Dynamic haptic discrimination in humans involves several processes at different levels of abstraction (e.g. encoding/decoding of afferent signals, sensorimotor control, decision making). At the periphery of the somatosensory pathway, primary afferents precisely encode tactile signals and reliably convey them to downstream structures [9]. At the same time, tactile information is used by cognitive and motor processes to interpret the stimulus and adapt limbic movements to optimise texture exploration.

In Braille reading, the perceptual, cognitive and motor aspects reciprocally and dynamically interact. Notably, it has been demonstrated that blind subjects continuously change reading velocity while surfing a dotted character line, regardless of the semantic of the patterns they encounter and often unconsciously [1]. Nonetheless, besides motor related velocity intermittencies, word frequency and sub-lexical content contribute to the pervasive changes of the reading speed [2].

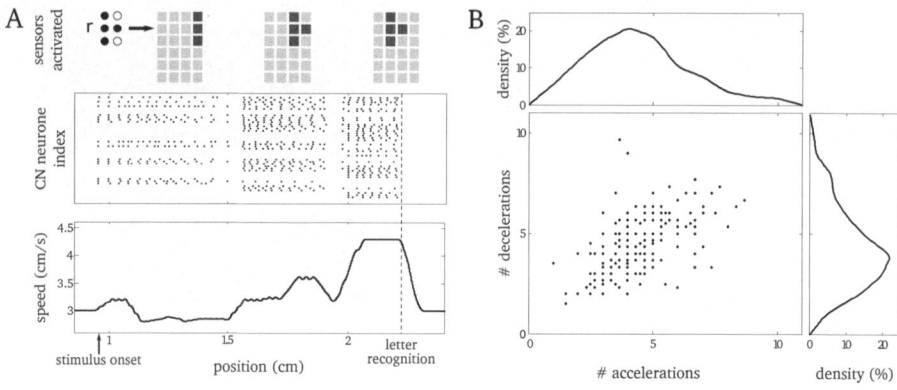

Fig. 4. *CN activity modulation and dynamic adaptation of the reading speed.* (A) Top (left to right): Sensor activation as the stimulus is scanned. Center: The activity of CN neurones is shown as a function of the stimulus position relative to the fingertip. Bottom: Modulation of the reading speed along the Braille line. The fingertip initially moves at a constant velocity. As the character enters the sensor area, the scanning speed starts being modulated on the basis of the computed character probability distributions. Soon after the stimulus has been entirely covered by the finger, a sudden increase in velocity is observed, indicating a peak in the probability distribution. After the classification is performed, the velocity is reset to a constant value (i.e. 30 mm/s). (B) Scatter plot: Number of accelerations and decelerations observed during individual character scanning, for a representative set of samples. Each dot corresponds to a mean value computed over 3 trials of a single character. Depending on the complexity of the pattern being read, a variable number of velocity changes is required. Top and Bottom right: Number of accelerations and decelerations probability distributions.

In this study, we propose a neuro-inspired closed-loop system to investigate haptic discrimination during Braille reading. We integrate tactile information coding at the neuronal level with a probabilistic framework for dynamic stimulus classification and adaptive motor control. We convert the analogue signals from an artificial fingertip into spiking activity and we apply an information theoretical analysis to the first and second afferent output stages. We finally interpret online the output of the second order Cuneate Nucleus neurones through a Naive Bayesian Classifier and use the character probability distribution to determine an efficient reading velocity.

Our results show that signals at the earliest stages of the haptic ascending pathway are conveyed as to allow a complete and fast discrimination by downstream decoders. We demonstrate that a probabilistic approach allows to efficiently recognise all Braille characters in an online reading task and we show that a dynamic adaptation of the finger velocity allows a faster recognition. Finally, we observe a number of accelerations which is coherent with human experiments outcomes and we argue that a probabilistic signal interpretation can account for the velocity intermittencies not induced by motor control operations.

We are currently investigating how the interferences created by different scanning velocities on the activity of CN neurones influence the classification process. We are also integrating a cerebellar model in the closed-loop system to explore in further detail motor and cognitive contributions to velocity corrections. We finally propose to investigate whether the speed modulation can help improve discrimination performance in the case of noisy fingertip movements.

Acknowledgments. Granted by the EC Project SENSOPAC (no. IST-028056-IP) and by the Délégation Générale de l'Armement (DGA). The authors thank the CASPUR Consortium (www.caspur.it) for providing the high performance computing facilities.

References

1. Hughes, B., Van Gemmert, A.W.A., Stelmach, G.E.: Linguistic and perceptual-motor contributions to the kinematic properties of the braille reading finger. Human Movement Science 30(4), 711–730 (2011)
2. Hughes, B.: Movement kinematics of the braille-reading finger. Journal of Visual Impairment & Blindness 105(6), 370–381 (2011)
3. Johansson, R.S., Flanagan, J.R.: Coding and use of tactile signals from the fingertips in object manipulation tasks. Nature Reviews Neuroscience 10(5), 345–359 (2009)
4. Gerstner, W., Kistler, W.: Spiking Neuron Models. Cambridge University Press (2002)
5. Chacron, M.J., Pakdaman, K., Longtin, A.: Interspike interval correlations, memory, adaptation, and refractoriness in a leaky integrate-and-fire model with threshold fatigue. Neural Computation 15(2), 253–278 (2003)
6. Cannata, G., Maggiali, M., Metta, G., Sandini, G.: An embedded artificial skin for humanoid robots. In: Proc. IEEE International Conference on Multisensor Fusion and Integration for Intelligent Systems MFI 2008, pp. 434–438 (2008)
7. Bologna, L.L., Brasselet, R., Maggiali, M., Arleo, A.: Neuromimetic encoding/decoding of spatiotemporal spiking signals from an artificial touch sensor. In: Proceedings of the 2010 International Joint Conference on Neural Networks (IJCNN), vol. 10, pp. 1–6 (2010)
8. Lapicque, L.: Recherches quantitatives sur l'excitation électrique des nerfs traitée comme une polarisation. Journal de Physiologie et Pathologie General 9, 620–635 (1907)
9. Johansson, R.S., Birznieks, I.: First spikes in ensembles of human tactile afferents code complex spatial fingertip events. Nature Neuroscience 7(2), 170–177 (2004)
10. Brasselet, R., Johansson, R., Arleo, A., Bengio, Y., Schuurmans, D., Lafferty, J., Williams, C., Culotta, A.: Optimal context separation of spiking haptic signals by second-order somatosensory neurons. In: Advances in Neural Information Processing Systems 22, pp. 180–188 (2009)
11. Carrillo, R.R., Ros, E., Boucheny, C., Coenen, O.J.D.: A real-time spiking cerebellum model for learning robot control. BioSystems 94(1-2), 18–27 (2008)
12. Brasselet, R., Johansson, R.S., Arleo, A.: Quantifying neurotransmission reliability through metrics-based information analysis. Neural Computation 23(4), 852–881 (2011)
13. Shannon, E.: A mathematical theory of communication. The Bell System Technical Journal 27, 379–423, 623–656 (1948)
14. Rieke, F., Warland, D., de Ruyter van Stevenick, R., Bialek, W. (eds.): Spikes: Exploring the neural code. MIT Press, Cambridge (1997)
15. Victor, J.D., Purpura, K.P.: Nature and precision of temporal coding in visual cortex: a metric-space analysis. Journal of Neurophysiology 76(2), 1310–1326 (1996)
16. Schreiber, S., Fellous, J.M., Whitmer, D., Tiesinga, P., Sejnowski, T.J.: A new correlation-based measure of spike timing reliability. Neurocomputing 52-54, 925–931 (2003)
17. van Rossum, M.C.: A novel spike distance. Neural Computation 13(4), 751–763 (2001)
18. Zhang, H.: The optimality of naive Bayes. In: Proceedings of the FLAIRS Conference, vol. 1(2), pp. 3–9. AAAI Press (2004)

19. Truccolo, W., Friehs, G.M., Donoghue, J.P., Hochberg, L.R.: Primary motor cortex tuning to intended movement kinematics in humans with tetraplegia. The Journal of Neuroscience 28(5), 1163–1178 (2008)
20. Bologna, L.L., Pinoteau, J., Brasselet, R., Maggiali, M., Arleo, A.: Encoding/decoding of first and second order tactile afferents in a neurorobotic application. Journal of Physiology-Paris 105(1-3), 25–35 (2011)
21. Phillips, J., Johansson, R., Johnson, K.: Representation of braille characters in human nerve fibres. Experimental Brain Research 81(3), 589–592 (1990)

Backwards Maneuvering Powered Wheelchairs
with Haptic Guidance

Emmanuel B. Vander Poorten, Eric Demeester, Alexander Hüntemann,
Eli Reekmans, Johan Philips, and Joris De Schutter

KU Leuven, Dept. of Mechanical Engineering,
Celestijnenlaan 300B, 3001 Heverlee, Belgium
{emmanuel.vanderpoorten,eric.demeester,alexander.huntemann,
johan.philips,joris.deschutter}@mech.kulueven.be,
{eli.reekmans}@intermodalics.eu
http://www.mech.kuleuven.be

Abstract. This paper describes a novel haptic guidance scheme that helps powered wheelchair users steer their wheelchair through narrow and complex environments. The proposed scheme encodes the local environment of the wheelchair
as a set of collision-free circular paths. An adaptive impedance controller is constructed upon these circular paths. The controller increases resistance when nearing obstacles and simultaneously helps the user to change motion towards a safer
circular path. To test the algorithm, a commercial powered wheelchair was interfaced and equipped with necessary sensors and an in-house built haptic joystick.
The user was asked to drive backwards into a narrow elevator with and without
navigation assistance. Although there is still room for improvements, the first results are promising. Thanks to the assistance the user can perform this maneuver
successfully in most of the cases without even looking backwards.

Keywords: haptic guidance, navigation assistance, robotic wheelchair, shared
control.

1 Introduction

Accurately steering vehicles backwards, e.g. for parking or maneuvering, remains a
non-trivial, error-prone task. Since 2003 where Toyota sold the first intelligent park
assist system in the Toyota Prius, more and more car manufacturers promote similar
park assist systems. These devices help drivers when executing parallel parking maneuvers. Typically the user keeps the control over the velocity, while steering is managed
by the car. While the need for assistance during backwards maneuvers with powered
wheelchairs is presumably much higher than for cars, as many users have big trouble even looking backwards in their chair, manufacturers of powered wheelchairs are
not that far as their counterparts in the automotive sector. The large difference in market size and available budget explains to a large extent the development delay. Further
explanations can be found in the different vehicle kinematics, the large variability in
environmental conditions compared to the more or less structured car parking tasks,
but also the large variability in level of expertise and capability of the drivers. Powered wheelchair drivers form a very heterogeneous public covering people with physical

P. Isokoski and J. Springare (Eds.): EuroHaptics 2012, Part I, LNCS 7282, pp. 419–431, 2012.
© Springer-Verlag Berlin Heidelberg 2012

and/or cognitive disabilities with varying level of expertise. Navigation assistance systems that help during navigation should account for this large variability and be able to cope with it in a reliable manner. The above elements explain why car technology cannot be simply transferred to wheelchair navigation assistance.

One seemingly appealing manner to overcome the large heterogeneity in the population of wheelchair users would lie in the development of technology to execute maneuvering tasks fully autonomously, thus without the need for user interventions. While this might from a technological point of view become possible, from the point of view of the user this is not always attractive. People in general, and people with disabilities in particular, like to be in control. The knowledge to be capable of executing quite complex tasks can boost moral. This is the case for youngsters who are eager to learn and gain control or for older people that might be afraid to loose control. For people with multiple sclerosis or dementia for example exposure to challenging navigation tasks may help them remain capable and alert, whereas the absence of stimuli might speed up degradation of earlier competences. Also providing too good assistance might have a detrimental effect [1]. The final goal of the EC-funded FP7 project RADHAR (http://www.radhar.eu), which stands for Robotic ADaptation to Humans Adapting to Robots, exists in developing technology that allows automatic adjustment of the level of navigation assistance adapted to each specific user at each instant in time. This assistive technology should enable the user to execute complex navigation tasks in a safe, intuitive and rewarding manner.

This paper describes a novel haptic guidance scheme that was developed within the framework of RADHAR project that helps users maneuver backwards with a powered wheelchair. The assistance simplifies the navigation, while keeping the user in full control of the wheelchair. The layout of the paper is organised as follows: section 2 summarizes briefly the major principles of steering commercial powered wheelchairs; section 3 reviews a number of efforts described in literature that provide navigation assistance to powered wheelchair drivers. The difference between what we like to call 'unilateral' and 'bilateral' assistance schemes is shortly explained. Section 4 describes a novel bilateral navigation assistance scheme that haptically guides the user along collision-free paths. Experimental results of this control scheme are presented and discussed in section 5. Finally, conclusions are drawn and directions for future work are sketched in 6.

2   Powered Wheelchair Navigation Principles

Powered wheelchairs are non-holonomic vehicles that are operated by various types of input devices ranging from proportional manual joysticks, over chin or tongue joysticks towards scanning eye-tracking devices [2] or even exotic brain computer interfaces [3]. This section gives a rough sketch of the basic operating principle of powered wheelchairs through proportional manual joysticks. These two-dimensional joysticks are most commonly used in practice as input devices. The navigation assistance schemes developed further on in this paper rely on a haptic variant of a traditional 2DOF joystick.

2.1 Non-holonomicity of Powered Wheelchair

Similar to cars, commercial powered wheelchairs are non-holonomic vehicles, some exceptions (e.g. [4]) left out of consideration. Dissimilar to cars is the type of non-holonomicity. Powered wheelchairs have two driven wheels, the velocity (ω_1, ω_2) of which is controlled separately. When both wheels rotate at the same speed, assuming equal wheel diameters, the wheelchair moves on a straight line. When both wheels rotate at the same speed but in opposite direction, the wheelchair turns on the spot. This relation is expressed as

$$v_{wch} = r_{wheel}(\omega_1 + \omega_2)/2, \qquad (1)$$

$$\omega_{wch} = r_{wheel}(\omega_1 - \omega_2)/B_{ax}. \qquad (2)$$

Here v_{wch} and ω_{wch} are respectively the translational and rotational velocity of the wheelchair expressed in a local wheelchair coordinate frame Σ_{wch}. The frame Σ_{wch} is rigidly attached to the wheelchair in the center between the two driven wheels (see Fig.1). The distance between both driven wheels along the wheel axis is B_{ax}. r_{wheel}, the wheel radius, for the left and right wheel is assumed equal and constant for simplicity. Depending on the type of wheelchair front-, mid- or back-wheels are actuated. The non-actuated wheels, often refered to as castor wheels, stabilise the wheelchair but also disturb the idealized relations (1) and (2). Wheel slippage, pressure variations in tires and other dynamic effects are other factors that affect the navigation. For reasons of simplicity all these disturbance factors are not accounted for in this work and relations (1) and (2) are employed.

2.2 Mapping of Joystick Deflection to Wheelchair Motion

The most straightforward manner to operate a non-holonomic wheelchair is probably by employing a traditional proportional manual joystick, with a linear relation between joystick deflection and commanded wheelchair speed. This mapping is conceptually depicted in Fig.1. For a joystick deflection (x_j, y_j) expressed in coordinate system frame Σ_j attached to the joystick base, the commanded wheelchair rotational and translational velocity are:

$$v_{wch} = k_v y_j, \qquad (3)$$

$$\omega_{wch} = -k_\omega x_j \qquad (4)$$

with appropriate velocity gains k_v and k_ω. When maintaining this particular joystick position over a longer period of time, the wheelchair will follow a circular trajectory $\mathcal{C}_m$, expressed in a base reference frame Σ_b with radius given by:

$$r_{wch} = |v_{wch}|/|\omega_{wch}|. \qquad (5)$$

When ignoring the dynamic effects of the wheelchair, a general wheelchair trajectory can be approximated as a sequence of instantaneous motions on circular paths $\mathcal{C}_m$ that follow joystick motions (x_j, y_j) over time. The traversal speed along the circular path is proportional to the joystick deflection amplitude $r_j = \sqrt{x_j^2 + y_j^2}$. It is straightforward

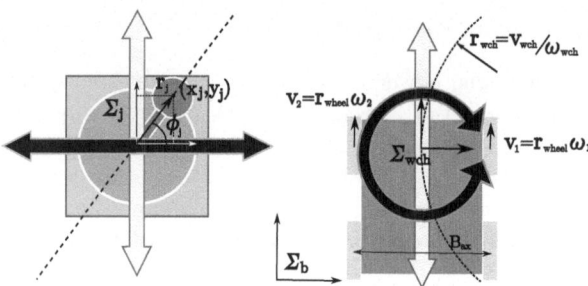

Fig. 1. One to one mapping of joystick position (left) to wheelchair velocity (right). Each position (x_j, y_j) expressed in frame Σ_j corresponds to a linear and rotational velocity (v_{wch}, ω_{wch}) w.r.t. the local wheelchair coordinate frame Σ_{wch}, resulting in an instantaneous circular path with radius $r_{wch} = v_{wch}/\omega_{wch}$ in base coordinate frame Σ_b.

to verify that for all joystick positions forming a constant angle $\phi_j + i\pi$, $i = \{0, 1\}$ with the horizontal X-axis of Σ_j, the wheelchair will move along the same circular path $\mathscr{C}_m$, albeit with different speeds. By extension, all two-dimensional joystick positions can be encoded in such way to a one-dimensional set of circular paths. For computational reasons a discrete number of paths with radius r_{wch}^m is employed:

$$r_{wch}^m = \left| \frac{k_v}{k_\omega} \tan(m\Delta\phi) \right|, \quad \text{for } m = 0, \ldots, N-1 \text{ and } N \in \mathbb{Z}_0^+$$

$$\text{and} \quad \Delta\phi = \frac{2\pi}{N-1} \quad \text{where } N \to \infty. \tag{6}$$

Each circular path segment is represented as a (v, ω, dt) tuple where the path's length equals $v \cdot dt$ and the orientation change $\omega \cdot dt$. $m\Delta\phi = \{0, \pi\}$ corresponds to pure rotational motion, whereas $m\Delta\phi = \{\pi/2, 3\pi/2\}$ corresponds to pure translational motion.

Note that mapping (3-4) requires a user who wants to drive backwards to the right to hold his/her joystick to the left. This is quite counterintuitive and complicates delicate back and forth maneuvering. Some wheelchair manufactures therefore foresee an opposite mapping for backwards motion.

3   Navigation Assistance for Powered Wheelchairs

Maneuvering powered wheelchairs and more in particular maneuvering such devices backwards remains a challenging and potentially dangerous task, even for experienced wheelchair users. Seemingly simple and frequently occuring maneuvers like door-passing or taking an elevator require in reality considerable motor and cognitive skills. The relatively large dimensions of wheelchairs compared to the size of the environment, the limited response speed and large inertia can lead to dangerous situations, possibly inflicting serious harm upon driver, bystanders or damage the environment.

For this reason, so-called 'smart' wheelchairs are being developed since the early eighties. These systems try to simplify navigation tasks and assist in transporting the user safely to a desired location [5]. Some of the more advanced systems such as TAO,

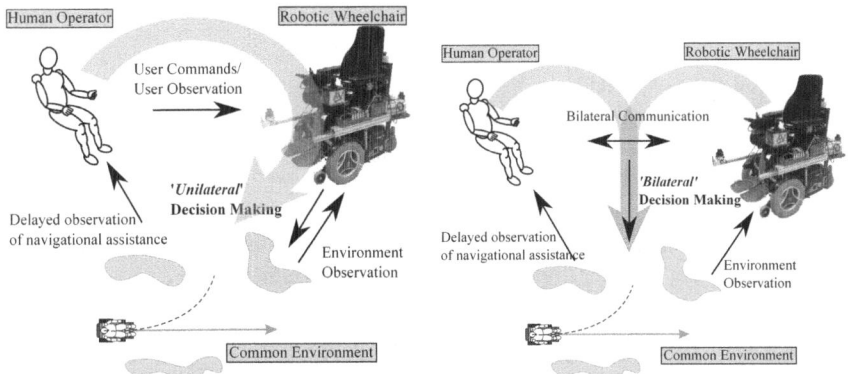

(a) Traditional shared control schemes display asymmetry in the decision making process. Users only understand decisions after actual displacement took place.

(b) Bilateral shared control schemes feature a deeper level of control sharing. The user can negotiate directly with the wheelchair over a haptic communication channel.

Fig. 2. Navigation assistance through unilateral or bilateral shared control

NavChair, Rolland or SmartChair foresee multiple assistance modes that are activated depending on the context. Such systems partially take over the control from the user and adjust the user's inputs by providing assistance for wall-following, collision avoidance or door-passage tasks. Since the user only perceives the decisions by the navigation system *after* wheelchair displacement, such approaches might unexpectedly cause frustration and might actually complicate maneuvering tasks. For instance this is the case when the provided assistance does not correspond to the needed or the expected assistance. The latter is often referred to as *mode confusion* in literature [6, 7]. Shared control approaches that follow this pattern (Fig.2a) exhibit a certain asymmetry in the decision making process and will be referred to as *unilateral* shared control approaches hereafter. With good knowledge of the wheelchair behaviour the user might be able to anticipate the wheelchair's response. Yet, in complex or adaptive assistive schemes this is an error-prone and possibly dangerous process.

To avoid mode confusion problems, researchers started recently to experiment with haptic feedback for wheelchair navigation [8–15] and tele-operated mobile robots [16–18]. By setting up a fast, *bilateral* communication channel between the user and the wheelchair controller, control can be shared more profoundly (Fig.2b). The user can directly negotiate with the controller over this haptic channel and is given the final word, as he/she can *overrule* unwanted wheelchair actions. Next to challenges in designing robust haptic display hardware for this application, a major challenge remains in the design of intuitive bilateral shared control methods that fully exploit the opportunities of the haptic channel and that do not cause additional fatigue of its user. The next section describes a novel haptic navigation assistance strategy that is designed to this end. Experiments, described in section 5 show how this scheme is effectively used to simplify complex maneuvering tasks such as backwards docking into an elevator.

4   Haptic Guidance along Circular Paths

In this section the novel haptic guidance algorithm is introduced. In contrast to earlier works in literature on haptic guidance, the proposed algorithm takes the non-holonomic nature of the wheelchair motion carefully into account. The approach further incorporates the wheelchair's complex (2-dimensional) geometry in order to determine safe passage through confined spaces (subsection 4.1). An adaptive impedance controller is designed (subsection 4.2) to haptically feed the encoded environment information back to the user, assisting him/her to move safely and with confidence.

4.1   Environment Encoding as a Set of Collision-Free Paths

As explained in subsection 2.2, an arbitrary wheelchair trajectory can be seen as a concatenation of multiple circular paths. At each instant of time the user holding the joystick in position (x_j, y_j) commands the wheelchair to move along a path with radius r_m given by (6). The time that the user can hold the joystick in the same position without the wheelchair colliding to an obstacle is a measure of the safety of such a circular path. For example if the distance to an obstacle along this path is large, this path can be considered a relatively *safe* path and the navigation assistance should not interfere. If, on the other hand, the time is short this means that a collision is imminent and haptic guidance away from the collision is required.

Following this principle, the first part of the algorithm encodes the wheelchair's environment as a set of collision-free paths. This takes place as follows:

1. a local map of the environment is built and refined at a fixed update rate. Ranging data from laser sensors together with odometry data is used to build online a map of the environment and to localise the wheelchair within this map. Prior knowledge on the environment can be employed to improve the accuracy of the local map or to speed up the map building process.
2. next, intersections are calculated between the local map and a template of N circular paths departing from the wheelchair. For this, the wheelchair is simulated to follow with a predefined speed (v, ω) each circular path and the earliest time of intersection dt is recorded. These time-values, representative for the collision-free lengths, are stored in an $N \times 1$ vector **dt**.

Figure 3 depicts an encoding of the environment into a set of collision-free paths.

4.2   Adaptive Impedance Controller to Render Collision-Free Paths

Supplied with a vector of path lengths **dt**, a model-free impedance controller ([19, 20] displayed in Fig.4), is programmed next. The target impedance $\mathbf{Z}_d$, written in the Laplace domain is:

$$\mathbf{Z}_d(s) = \frac{\mathbf{k}(\phi)}{s} + \mathbf{b}. \tag{7}$$

Both stiffness **k** and damping **b** contain a radial and a tangential component. The radial component affects the wheelchair speed along a certain trajectory. The tangential

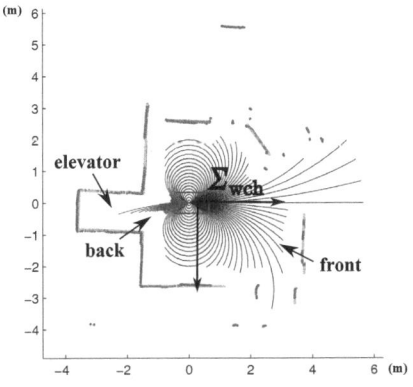

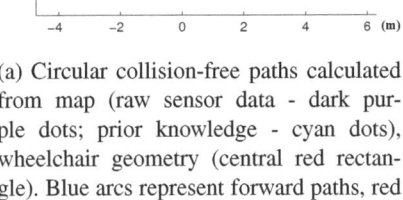

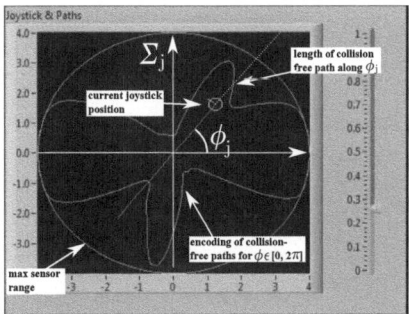

(a) Circular collision-free paths calculated from map (raw sensor data - dark purple dots; prior knowledge - cyan dots), wheelchair geometry (central red rectangle). Blue arcs represent forward paths, red lines are backward paths.

(b) GUI showing encoded environment mapped on joystick frame Σ_j. Small circle represents the joystick position. Lines originating from joystick circle proportional to amplitude of radial and tangential forces.

Fig. 3. Environment encoding as a set of collision-free paths

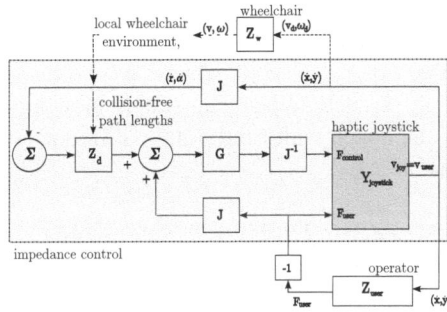

Fig. 4. Impedance controller providing haptic guidance along collision-free paths

component tries to bend the wheelchair motion towards a different circular path. A natural manner to translate collision-free paths towards a stiffness map **k** would exist in attributing higher resistance to shorter collision-free paths and vice versa. In this work a power relation with power p between radial stiffness and path length is employed as described in Table 1. For $p < 1$ the stiffness increases slowly at first and then rises fast when nearing an obstacle. We also performed experiments with a linear relationship between stiffness and collision-free distance. However, in these experiments, we felt that obstacles further away had a relatively large and undesired influence on the stiffness as compared to obstacles nearer by. Given that we do not (yet) take wheelchair velocity into account in the stiffness computation, we found that the polynomial relationship gave better results. Further analysis is deemed necessary to understand this relationship better and see if/how it affects overall system stability.

Table 1. Algorithm 1

Calculating stiffness k_{dt} from path length dt
Input: dt
Output: k_{dt}

if ($dt \geq dt_{max}$)
$\quad k_{dt} = k_{min}$;
else if ($dt \leq dt_{min}$)
$\quad k_{dt} = k_{max}$;
else
$\quad k_{dt} = k_{max} - (k_{max} - k_{min}) \cdot \left(\frac{dt - dt_{min}}{dt_{max} - dt_{min}} \right)^p$;

Under the assumption of perfect masking of joystick dynamics, the radial force displayed to the user is given as:

$$F_r = \begin{cases} -F_{r,0} - k(\phi_j) \cdot r_j - b_r \cdot \dot{r} & \text{for} \quad r \geq r_{nz} \\ -b_r \cdot \dot{r} & \text{for} \quad r < r_{nz} \end{cases} \qquad (8)$$

where the joystick position is given as $r_j = \sqrt{x_j^2 + y_j^2}$ and $\phi_j = \tan(y_j, x_j)$. The stiffness $k(\phi_j) = k_{dt}$ for $dt = dt(m)$. The factor b_r is the radial component of the damping **b**. For stability reasons a neutral zone with radius r_{nz} is programmed around the midpoint in (8). Offset $F_{r,0}$ is such that $F_r|_{(r=r_{nz})}$ is zero, when $\dot{r} = 0$.

In order to pass by an obstacle rather than simply slowing down in front of it, a tangential force F_{ϕ_m} is calculated with central finite differences from the surrounding radial stiffness values as in:

$$F_{\phi_m} = \begin{cases} -F_{\phi_m,0} - \frac{k_{m+1} - k_{m-1}}{\phi_{m+1} - \phi_{m-1}} \cdot r_j - b_\phi \cdot \dot{\phi} & \text{for} \quad r_j \geq r_{nz} \\ -b_\phi \cdot \dot{\phi} & \text{for} \quad r_j < r_{nz} \end{cases} \qquad (9)$$

with b_ϕ the tangential component of the damping **b** and with joystick angle $\phi_m \in [(m - 1/2)\Delta\phi, (m + 1/2)\Delta\phi)$ for $m = 0, \ldots, N - 1$ and $\Delta\phi$ defined in (6). Note that to avoid overly complex notations, the special cases $m = 0, N$ are ignored in (9). The offset $F_{\phi_m,0}$ is a constant such that $F_{\phi_m}|_{(r_j = r_{nz})}$ is zero, when $\dot{\phi} = 0$.

5   Experimental Setup and Validation

5.1   Experimental Setup

To validate the proposed algorithms a test setup was built around a Corpus C500 front-wheel drive wheelchair by Permobil AB. The system is depicted in Fig.5. A metal frame was added to the wheelchair to mount additional hardware. Two Baumer MDFK-10G8124/N64 encoders were integrated on the outgoing axes of the driven wheels. These are used for odometry. Two Hokuyo URG-04LX laser sensors with a field of view of $270°$ were placed at diagonal corners of this frame. In this way a full $360°$

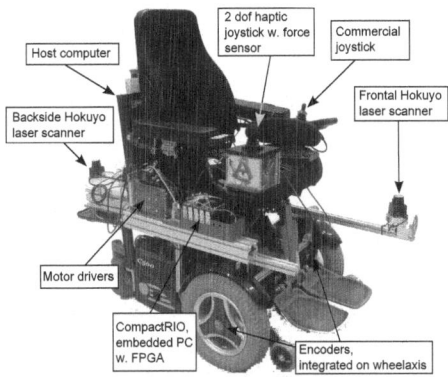

Fig. 5. Permobil Corpus C500 wheelchair interfaced and upgraded for providing haptic navigation assistance

2-dimensional scan of the surroundings of the wheelchair is captured. A host computer mounted at the back of the seating reads out the laser scanners and builds 2D maps of the environment. The host computer communicates over UDP with a NI CompactRIO 9074 central processing unit (cRIO). Labview RT, a real-time OS based upon VxWorks, is installed on this embedded PC. The haptic control loop runs here at a rate of 1kHz. A recent in-house designed 2DOF haptic joystick is mounted in line with the armrest of the wheelchair. This is a compact and powerful joystick showing high continuous output force at the handle (up to 40N, thanks to 2 Maxon RE30 motors, a cable-based reduction mechanism with 1 : 10 reduction ratio and Maxon ADS_E 50/5 PWM current-controlled motor drives). Also, high resolution position measurement (Scancon 2RMHF5000, 20.000 p.p.r. after quadrature encoding) and interaction forces measurements (HBM, 1-LY11-1.5/120 strain gauges glued on the joystick handle) are available.

5.2   Experiment Description

With cardboard boxes an artificial environment was built up to represent an elevator. The user is asked to maneuver the wheelchair backwards inside this elevator starting from a fixed pose. The maneuvering capability with or without navigation assistance is measured during the execution of this task. Parameters that were recorded are time until completion and the number of collisions. At this stage of the research all experiments are conducted by one single able-bodied user (male, 33 years) with limited expertise in conducting powered wheelchairs. Both experiments were conducted 10 times. Three types of experiments were conducted and executed in random order:

type 1: Navigation without Guidance. The user was allowed to look backwards over the shoulder during these experiments. Note that this way of operation is not possible or very tiring for many typical wheelchair users.

type 2: Navigation with Visual Guidance. The user was asked to maneuver the wheelchair while observing the GUI of Fig.3. Guidance forces were calculated and displayed on the GUI (alongside collision-free path lengths) but not applied to the user.

Table 2. Summary of results. (Time in s, average and standard deviation calculated for successful runs only.)

run	1	2	3	4	5	6	7	8	9	10	av.	stdev	coll.
type 1	13.23	13.15	10.11	14.31	11.94	9.45	9.69	10.46	12.07	10.14	11.14	1.47	1
type 2	11.00	12.09	11.47	14.01	9.6	10.34	17.51	11.84	9.82	15.73	11.60	1.48	4
type 3	25.4	10.43	21.86	15.23	19.48	13.37	26.65	13.02	40.83	17.94	18.23	5.89	4

type 3: Navigation under Haptic Feedback. The user was asked to drive 'blindly' inside the elevator solely relying on his sense of touch and the control scheme of section 4. For this, 432 circular paths were used, and the other parameters in Table 1 were: $p = 0.15$, $dt_{min} = 0.3$m and $dt_{max} = 4$m. Some additional precautions were taken. In order to avoid that big jumps in path lengths, due to sensor noise, produce sudden jumps in impedances/and output forces, the circular path data are smoothened by applying a Gaussian convolution mask to vector **dt**. Also, as environment scanning and processing takes place only at a rate of 10Hz, calculated impedance values are upsampled to 1kHz by extrapolating the last two measurements.

5.3 Summary of Results

Table 2 and Fig.6a summarize the results from the different experiments. At this moment, navigating the wheelchair while looking backwards over the shoulder is still superior, but, also here, collisions could not be avoided. Indeed, the task is quite challenging as the elevator is narrow, leaving only about 10cm of space at both sides between wheelchair and door post. With only visual or haptic guidance, the amount of successful executions dropped to $6/10$. This score might seem low, but it must be stressed that the user did not have to look backwards over the shoulder. So such navigation strategy could come in handy to help especially those users that experience problems in looking over their shoulder while steering a wheelchair. Note also that without looking backwards and without guidance successful backwards maneuvering was close to impossible.

At this moment GUI-based navigation is still faster than navigating solely based upon haptic guidance (Table 2). Under haptic guidance the user is somehow 'palpating' the environment to feel where the passage is. The GUI on the other depicts this passage at once, leading to a faster execution. On the other hand, the haptic guidance warns the user when a collision is near. Figure 6b shows an exemplary trajectory where the user turns the wheelchair after such warning and successfully completes the task.

There is still room for improvement. Fig.6c shows that the encoding into circular paths has its limitations. An extension towards other than circular paths will be investigated. Also, at some instants in time, parts of the environment were not observed (Fig.6d). By constructing the wheelchair trajectories from a local map rather than from pure sensory data, paths can be calculated more reliably.

6 Conclusions and Future Work

This article presented a new haptic guidance scheme that is designed to help powered wheelchair users navigate their wheelchair more reliably in their daily environment.

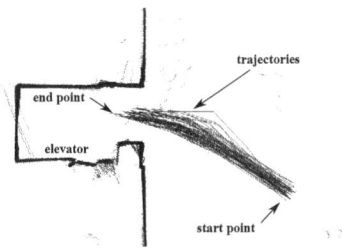

(a) All trajectories of the experiments plotted out. The clock is stopped when the wheelchair's frontal wheels are inside the elevator.

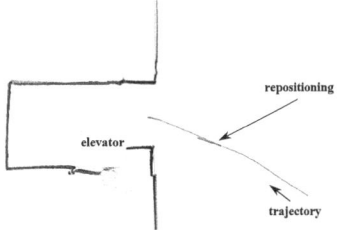

(b) example trajectory under haptic guidance. Through haptic feedback the user turns wheelchair closeby elevator and makes a successful entry

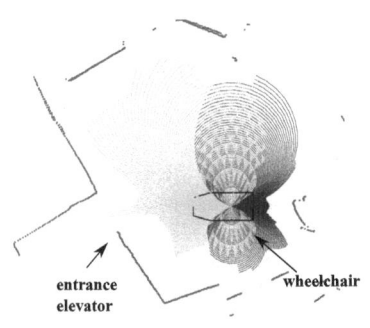

(c) limited guidance when only few circular paths guide towards elevator.

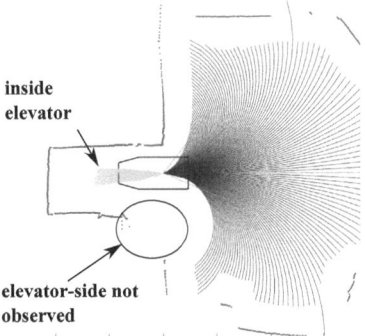

(d) guidance can be wrong when existing obstacles are not observed.

Fig. 6. Navigation trajectories and possible causes of faulty or limited guidance

The guidance scheme contains one component that slows the wheelchair down when approaching an obstacle. Another component bends the wheelchair's motion towards safer paths. The scheme was experimentally validated on a backwards maneuver to drive a wheelchair into an elevator. The results showed that it is feasible to perform such maneuver solely based on haptic guidance, without the user requiring to look backwards over the shoulder or to focus on a GUI screen. Further improvements are still needed however. Improvements could include the use of non-circular paths and the use of the wheelchair's velocity and dynamics in the calculation of the resistance force. Moreover, we would like to test the combination of haptic and visual feedback, and to tune the stiffness-distance relationship further.

Acknowledgments. This research was conducted in the framework of the RADHAR project, funded by the European Commission's 7th Framework Programme FP7/2007-2013 Challenge 2 Cognitive Systems, Interaction, Robotics under grant agreement No. 248873 and under support of a Marie Curie Reintegration Grant, PIRG03-2008-231045. We would also like to thank one anonymous reviewer for the constructive remarks and links to interesting references.

References

1. Powell, D., O'Malley, M.: Efficacy of shared-control guidance paradigms for robot-mediated training. In: World Haptics Conference 2011, Istanbul, pp. 427–432 (June 2011)
2. Matsumoto, Y., Ino, T., Ogusawara, T.: Development of intelligent wheelchair system with face and gazebased interface. In: Proc. of 10th IEEE Int. Workshop on Robot and Human Interactive Communication, pp. 262–267 (2001)
3. Vanacker, G., Millán, J., Lew, E., Ferrez, P., Moles, F., Philips, J., Van Brussel, H., Nuttin, M.: Context-based filtering for assisted brain-actuated wheelchair driving. Computational Intelligence and Neuroscience, Article ID 25 130, 12 pages (2007)
4. Asada, H., Wada, M.: The superchair: A holonomic omnidirectional wheelchair with a variable footprint mechanism. Progress Report, Total Home Automation and Health/Elderly Care Consortium (March 31, 1998)
5. Simpson, R.C.: Smart wheelchairs: A literature review. J. Rehabil. Res. Dev. 42(4), 423–436 (2005)
6. Sheridan, T.: Telerobotics, Automation and Human Supervisory Control. The MIT Press, Cambridge (1992)
7. Lankenau, A.: Avoiding mode confusion in service robots - the bremen autonomous wheelchair as an example. In: Proc. of the 7th Int. Conf. on Rehabilitation Robotics (ICORR 2001), Evry, France, pp. 162–167 (2001)
8. Hong, J.-P., Kwon, O.-S., Lee, E.-H., Kim, B.-S., Hong, S.-H.: Shared-control and force-reflection joystick algorithm for the door passing of mobile robot or powered wheelchair. In: Proc. IEEE Region 10 Conference TENCON 1999, vol. 2, pp. 1577–1580 (1999)
9. Luo, R., Hu, C., Chen, T., Lin, M.: Force reflective feedback control for intelligent wheelchairs. In: Proc. IEEE/RSJ Int. Conf. Intel. Rob. and Syst., pp. 918–923 (1999)
10. Protho, J., LoPresti, E., Brienza, D.: An evaluation of an obstacle avoidance force feedback joystick. In: Conf. of Rehabilitation Engineering and Assistive Technology Soc. of North America (2000)
11. Kitagawa, H., Kobayashi, T., Beppu, T., Terashima, K.: Semi-autonomous obstacle avoidance of omnidirectional wheelchair by joystick impedance control, vol. 4, pp. 2148–2153 (2001)
12. Lee, S., Sukhatme, G., Kim, G., Park, C.-M.: Haptic control of a mobile robot: a user study. Iros 3, 2867–2874 (2002)
13. Fattouh, A., Sahnoun, M., Bourhis, G.: Force feedback joystick control of a powered wheelchair: preliminary study. In: IEEE Int.Conf. on Systems, Man and Cybernetics, vol. 3, pp. 2640–2645 (October 2004)
14. Lee, S., Sukhatme, G.S., Jounghyun, G., mo Park, K.C.: Haptic teleoperation of a mobile robot: A user study. Presence: Teleoperators & Virtual Environments (2005)
15. Bourhis, G., Sahnoun, M.: Assisted control mode for a smart wheelchair. In: IEEE 10th Int. Conf. on Rehabilitation Robotics, ICORR, pp. 158–163 (June 2007)
16. Park, J., Lee, B., Kim, M.: Remote control of a mobile robot using distance-based reflective force. In: IEEE Int'l Conf. on Robotics and Automation, pp. 3415–3420 (September 2003)
17. Diolaiti, N., Melchiorri, C.: Tele-operation of a mobile robot through haptic feedback. In: IEEE Int'l Workshop on Haptic Virtual Environments and Their Applications, pp. 67–72 (2002)

18. Farkhatdinov, I., Ryu, J.-H.: Improving mobile robot bilateral teleoperation by introducing variable force feedback gain. In: IEEE/RSJ Int'l Conf. on Intelligent Robots and Systems, pp. 5812–5817 (October 2010)
19. Tischler, N., Goldenberg, A.: Stiffness control for geared manipulators. In: IEEE International Conference on Robotics and Automation, Proceedings 2001 ICRA, vol. 3, pp. 3042–3046 (2001)
20. Carignan, C., Naylor, M., Roderick, S.: Controlling shoulder impedance in a rehabilitation arm exoskeleton. In: IEEE International Conference on Robotics and Automation, ICRA 2008, pp. 2453–2458 (May 2008)

Direct Touch Haptic Display Using Immersive Illusion with Interactive Virtual Finger

Maisarah Binti Ridzuan[1], Yasutoshi Makino[2], and Kenjiro Takemura[3]

[1] Graduate School of Science and Technology, Keio University, 3-14-1,
Hiyoshi, Kohoku-ku, Yokohama, 223-8526, Japan
maisarahr@gmail.com
[2] Center for Education and Research of Symbiotic, Safe and Secure System Design,
Keio University, 4-1-1, Hiyoshi, Kohoku-ku, Yohokama, 223-8526, Japan
makino@sdm.keio.ac.jp
[3] Department of Mechanical Engineering, Keio University, 3-14-1,
Hiyoshi, Kohoku-ku, Yokohama, 223-8526, Japan
takemura@mech.keio.ac.jp

Abstract. In attempt to improve haptic interaction and manipulation technique in surface computers, we propose an interactive haptic feedback display that utilizes 'direct touch' on a surface computer without using a mediation device. The proposed haptic display lets users interact with the virtual objects inside a pressure-sensitive screen directly using a virtual finger as if the user's finger can penetrate through the screen surface and sink in to the digital world inside the screen, as well as receive haptic feedback direct to their fingers from the display. Using the proposed display, users were able to perceive touch sensation with haptic feedback when they come in contact with virtual object surfaces and discriminate different stiffness using vibratory and pseudo-haptic feedback respectively. This innovation creates a new breakthrough in virtual interaction with more accessible and realistic haptic interaction approach for surface computing.

Keywords: Immersive Display, Direct Touch, Haptic, Virtual Reality, Pseudo-Haptics.

1 Introduction

Surface computers and touch screens have been rather common among users. In surface computing devices, users interact directly with touch-sensitive screens, which could be any intuitive surfaces or everyday objects such as tabletops [1] and bathroom mirrors [2].It is a new way of interacting and working with computers in a more natural way: using their hands and gestures. For more sensible and wider breadth of information, haptic systems are increasingly being integrated into surface computing interactions. Researches have created many different methods to allow haptic feedback to be displayed on surface computers such as using magnets [3], ultrasonic vibrators [4], or using surface acoustic wave [5]. Yet, most approaches in the past still require the use of stylus or gloves which limits the computer interaction to be as natural as the real world. These kinds of interactions lack in haptic information and

P. Isokoski and J. Springare (Eds.): EuroHaptics 2012, Part I, LNCS 7282, pp. 432–444, 2012.

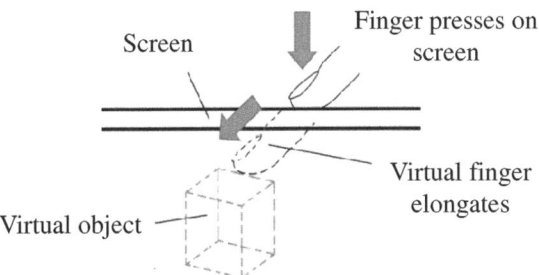

Fig. 1. Proposed display concept

manipulation capabilities. Therefore, more natural and seamless interaction between users and the digital world with realistic haptic feedback technology is in demand for surface computers.

In this paper, we propose a new way to produce a haptic feedback for a surface computing device. Fig. 1 illustrates our proposed concept. Basically, we utilize the combination of haptic and visual information to produce an immersive effect. When user presses the screen surface, a virtual finger is rendered inside the screen and elongates from the touch point in accordance with the fingertip pressing force. If the user's finger and the virtual finger on the screen appear parallel with each other, user can feel as if his or her finger is extending and shrinking inside the screen. The basic idea of this concept is similar to a previous study [6]. However, our concept differs from the previous study [6] as they used a stylus which was capable of producing a haptic feedback with a motor, while in our case, no mediation device was used between the finger and the screen.

We propose two different haptic feedback schemes. One of it utilizes a vibrator attached to the screen. When user touches a virtual object inside the screen with his or her elongated virtual finger, the contact surface vibrates so that the user can feel as though something comes in contact with his or her elongated virtual fingertip. Here, we aim to simulate a contact of object during direct touch with a simple vibrator.

After touching the surface of the virtual object, as the user pushes its surface even further, we utilize a second method called "Pseudo-Haptic" technique which was proposed in [7]. By changing the visual deformation depth of the virtual object in the screen in accordance with the pressing force, user can feel different stiffness sensation of its surface. For example, when pressing a virtual object's surface with the same pressing force, the object with greater surface deformation depth displays a softer texture to users. In contrast, a small depth of deformation with the same pressing force provides the users with a stiffer sensation. Using these two methods, we aim to produce a more realistic haptic sensation during direct touch, thus improving haptic interaction and manipulation technique for surface computing.

This new technique allows users to interact seamlessly while having visual and haptic information spatially coincident with each other. Furthermore, with direct touch technology manipulation technique, immersion for surface computing can be implemented more effectively, thus providing more accurate handling for Computer Aided Design (CAD), game, virtual shopping or virtual aid applications for education.

2   Direct Touch Haptic Display with Virtual Finger

2.1   Hardware and Software Implementation

The hardware implementation of our haptic display system involves a fingertip normal force sensor, an amplifier, an analogue-to-digital (AD) converter and an output touch panel display as shown in Fig. 2. An Iconia tablet computer [Acer] with Windows 7 operating system is used as an output touch panel display for the system. The tablet computer is chosen for its compact design, and compatibility with graphics and programming interface. Meanwhile, the touch recognition function in the tablet computer allows it to receive touch information input directly from the touch screen.

We used strain gauges for measuring the fingertip normal force on the screen. The force sensor system in this research utilizes two pairs of strain gauges on both right and left side of the support plate, attached to two pairs of parallel 0.3 mm thick brass plates. Each pair of brass plates at each sides act as a bending beam with parallel strain to measure the force applied on the display. Half-bridge-gauge system is implemented for each sides of the display as shown in Fig. 3. The strain gauge pairs collectively produce an output voltage V which is in proportional relation with the fingertip normal force F at a fixed position in the middle of the screen. This relation is shown in Fig. 4.

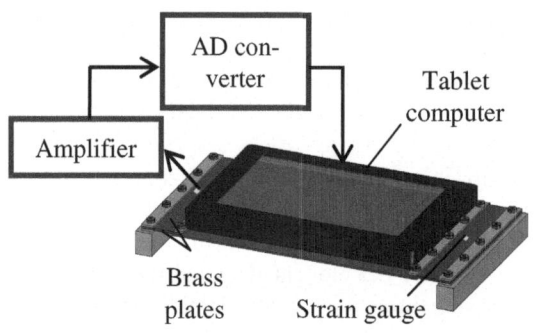

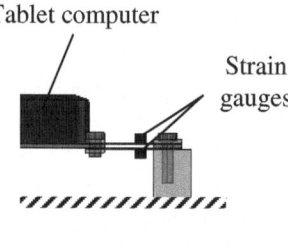

Fig. 2. Display system

Fig. 3. Half-bridge-gauge implementation on the side of display

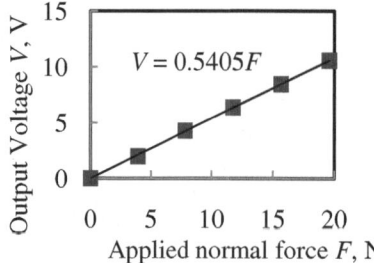

$$V = 0.5405F$$

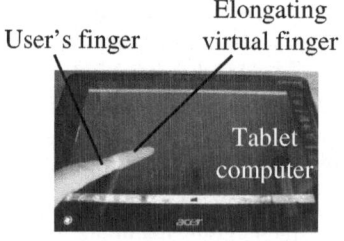

Fig. 4. Linear relation between F and V

Fig. 5. Direct touch using immersive illusion on display

In order to realize the device described in Fig. 1, we need to measure not only fingertip pressure force but also a finger posture to produce a virtual finger which appear extending straight with a real finger from the user's perspective. In this paper, however, we assume that finger contact position, its posture, and user's eyes positions are always fixed to simplify the fabrication of prototype system. We plan to measure the finger posture as one of our future works.

2.2   Graphics and Visual Display

The visual display rendered for this research is generated using OpenGL. The location of a vertex of an object is defined by 3 coordinates; x, y, and z. The origin of the system is positioned at the surface of the screen for z axis, and in the middle on the screen for x and y axis.

In the virtual space for this research, we have chosen a three dimensional perspective view from OpenGL library. According to perspective view, all objects farther away from the screen look smaller to the viewer and ultimately they disappear into a vanishing point. The virtual finger will elongate along the z axis towards the vanishing point in the middle of the screen. Therefore, virtual finger of the display will always appear elongating towards the middle of the screen as shown in Fig. 5.

The virtual finger is a combination of a cylinder with flexible length and two spheres which share the same diameter. The spheres are placed at both ends of the cylinder such that the centre of the sphere coincides with the centre of the circular face of the cylinder at both ends. The origin of the virtual finger is located at one end of the prototype.

The origin of the virtual finger is set to always coincide with the touch point of user's finger on screen, which is acquired from the touch-screen on the tablet computer. Once a touch is registered on the screen, the touch point acts as the initial point for elongation of the virtual finger into the virtual space inside the screen. Meanwhile, the length of the projected virtual finger is determined by the strain gauge's output voltage V yield from the fingertip normal force. The virtual finger is set to not appear during touch manipulation and only appear when 'press' from user's fingertip is registered, which is normal force over 1 N.

3   Direct Touch Sensation Using Vibration Feedback

As users manipulate the screen, users focus their sight to the virtual finger inside the screen rather than their own finger. When the virtual finger come into contact with object's surfaces in virtual space, mechanical haptic feedback can be sent to the user's finger by vibrating the display screen. The vibration is aimed to simulate the same sensation as when humans' fingertip comes into contact with object surfaces in the real world. If the presentation of visual information from the virtual finger coincide with tactile information from their own fingertip, users might feel as though the virtual finger is attached their body. This body awareness study was done using the well-known Rubber Hand Illusion [8]. We plan to create a similar illusion with the rubber hand illusion in this research. Using vibration as a mechanical haptic feedback, we aim to create an illusion that virtual finger is a part of user's finger, such that the vibration comes from the tip of the virtual finger in virtual space, not from the user's fingertip on the display.

3.1   Procedure and Evaluation

Human receptors of the skin are well-known to have their optimal sensitivity ranges; Pacinian corpuscle nearing 250 Hz, and Meissener's corpuscles nearing 50 Hz. For early stage evaluation of the display, we have selected these two frequencies to excite respective receptors and investigate which one is preferable to simulate illusion for virtual finger to be felt as part of user's body. Accordingly, we prepared one and a half period of two different frequency sinusoidal waves (250 Hz for 6 ms and 50 Hz for 30 ms) as vibration. Users were asked to press the screen at a fixed position such that the virtual finger elongates to reach a rectangular rendered surface further inside the virtual space as shown in Fig. 6. As the tip of the virtual finger comes into contact with the virtual surface, vibration was added to the screen with a voice-coil motor, which is secured to the screen. Then the users were asked about the perceived location of the contact; at the end of their finger, at the end of the virtual finger, or other. It is assumed that the outcome frequency of the display at the fingertip position is the same as the input vibration since the voice-coil motor was secured as close as possible to user's fingertip position on the screen. 10 people, involving 7 male and 3 female took part in the evaluation. They were all range from 22 years old to 25 years old.

3.2   Result

The result is shown in Fig. 7. When the frequency was 250 Hz, more users were able to feel the contact sensation at the tip of the virtual finger rather than and the end of their own finger. However, with 50 Hz vibration, more users perceived the contact sensation at the

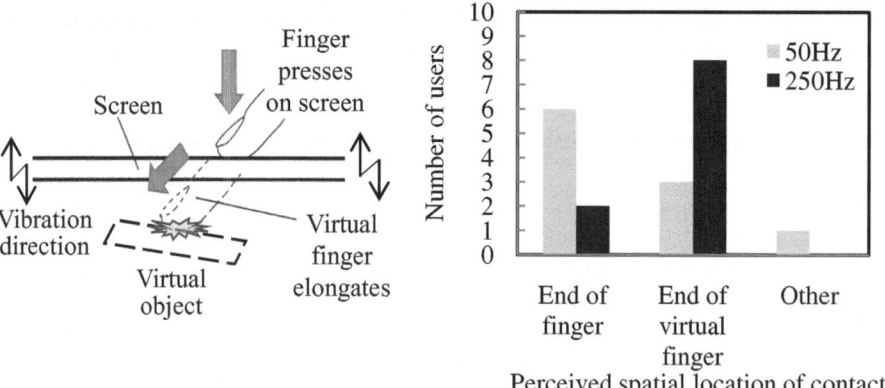

Fig. 6. Vibratory feedback during contact with virtual object

Fig. 7. Perceived contact of touch localization result

tip of their fingers. This suggests that 250 Hz is more suitable frequency vibration to exhibit the sensation of the virtual finger being a part of the users' body during manipulation.

This result is probably because at 250 Hz, Pacinian corpuscle, which is a cutaneous mechanoreceptor with poor localization information, reaches its sensitivity peak at this point. Therefore, during the virtual manipulation, user's ability to localize the contact using tactile sensation becomes poor. Consequently, during the sensory conflict between tactile and visual information, users' perception on the spatial localization of the contact comes from the visual information (which is the virtual finger touching the virtual surface). Therefore, instead of feeling the screen vibrating, users perceived the source of the vibration to come from inside of the screen, which act as a cue when the virtual finger comes into contact with a virtual object. Moreover, similar to rubber hand illusion [8], even though users are aware that the virtual finger is not physically attached to the users, from the result, we can say that users can feel a sense of body ownership and awareness of contact with virtual object surfaces using the virtual finger.

4   Virtual Surface Stiffness Display Using Pseudo-haptics

To attain wider breath of information with less complex structures in the direct touch haptic display, we propose a technique using pseudo-haptic feedback to create illusion of displayed haptic properties to users when the user is pressing on the screen to manipulate the virtual finger.

The pseudo-haptic illusion technique used in this research is proposed to display surface stiffness of an object in the virtual space. The surface stiffness is defined as the value of the surface displacement D along the force direction over the fingertip normal force F when a surface is pressed as shown in Fig. 8. The relation of the fingertip normal force F and the surface displacement can be written as

$$D = \frac{1}{k_s} F \qquad (1)$$

where k_s is the surface stiffness.

When a surface of an object is pressed, the virtual finger's elongation from the surface is equal to the surface displacement, D_v in the virtual space. Meanwhile, the

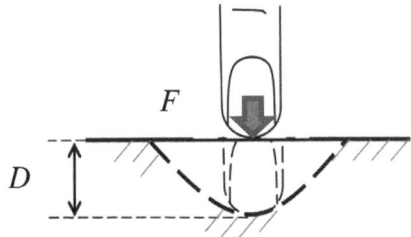

Fig. 8. Surface stiffness is defined by fingertip normal force F and surface displacement D display

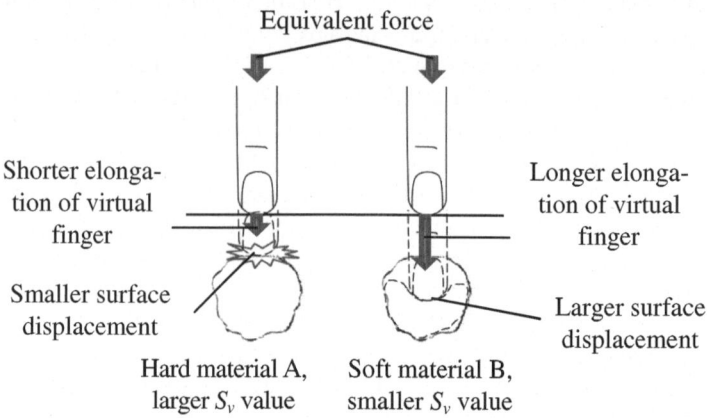

Fig. 9. Example of S_v value modification to display surface stiffness

fingertip normal force is determined by measuring the output voltages V of strain gauges. Therefore, the relation of the surface displacement and fingertip force can be written as

$$D_v = \frac{1}{S_v}V \qquad (2)$$

where S_v is the surface stiffness in the virtual space. In this research, the variations of the S_v values are used to transpose the effect of surface stiffness when pressing on surfaces of virtual objects. Fig. 9 illustrates the modification technique of the S_v value during the simulation of surface stiffness in two different objects. The S_v value is higher when pressing object A in virtual space with the virtual finger compared when pressing object B with the same amount of force. This designates a larger force needed to deform the object A compared to object B. Therefore, a larger S_v value ratio indicates a 'stiffer' surface of a virtual object, and conversely, a smaller S_v value ratio indicates a 'less stiff' surface.

4.1   Surface Stiffness Discrimination between Virtual Samples

The aim of this evaluation is to measure user's ability to discriminate virtual sample with varying surface stiffness value. It is inferred that when user presses on a virtual object, the stiffness perceived subjectively by the user is higher when the surface stiffness setting is increased.

Evaluation Setup. A simple rectangular surface is rendered using non-uniform rational B-spline to simulate a deformation depth change as virtual finger presses on it. In contrast with evaluation in section 3, the rectangular surface is virtually positioned right below the screen as though user can virtually touch the surface by touching the screen. When 'press' from fingertip is detected on the screen, the surface will appropriately deform with the elongation of the virtual finger so that no offset in fingertip force is produced. The virtual samples are colored, textured and positioned the same way such that no difference can be identified when users are visually introduced to the virtual samples. The virtual samples of six are set with different surface stiffness S_v; 0.1, 0.2, 0.5, 1.0, 2.0, and 2.5 V per unit length (u.l.).

Procedure and Evaluation. A total of 11 people, involving 8 male and 3 female took part in this evaluation. They were all range from 22 years old to 25 years old, with no known perception disorders. All of the users used their right hand which is their dominant hand to perform the task. This evaluation is conducted with seven-rank-evaluation scores, ranging from -3 to +3. Score -3 is applied for 'soft' and +3 for 'stiff'. Before the evaluation is conducted, the users had to investigate all the virtual samples randomly without giving any scores. This is conducted to provide the users with a range of available surface stiffness feeling for the evaluation scoring. For each trial, the user first had to place their index finger on a fixed position on the screen such that the virtual finger elongation and deformation of the virtual sample is clearly visible to the user as shown in Fig. 10. Then, users had to push the screen several times to manipulate the virtual finger and deform the virtual sample. The surface stiffness score is given by user after evaluating the surface stiffness of the virtual sample. The same trial is done with six virtual samples for all 11 users.

Results. The score recorded for each user is standardized to fit a normal distribution with an average of 0, and standard deviation of 1. The average of the standardized scores for each virtual sample across 11 users is calculated and shown in Fig. 11. Using analysis of variance, the result for each sample shows statistically significant difference

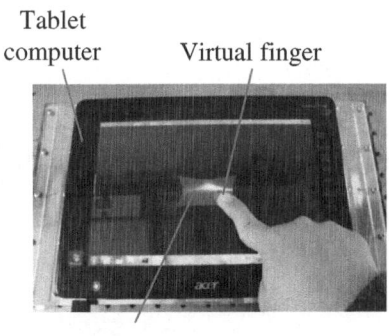

Fig. 10. Virtual sample's surface stiffness evaluation

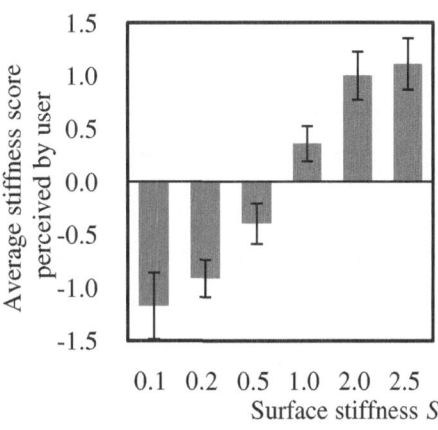

Fig. 11. Result of virtual sample's surface stiffness discrimination

with over 99% confidence level between all pairs of means. The result suggests that stiffness score increases as surface stiffness of virtual sample increase. This indicates that even though the users were pressing on the same screen, the average stiffness score of each virtual sample is higher as the surface stiffness value S_v increase. When asked about the discrimination method, most users explained that when a higher surface stiffness is set for a virtual sample, a larger normal fingertip force is needed to deform of the virtual sample's surface equally during the interaction. Consequently, the users had to push harder on the screen and receive larger reactive force.

With this result, we can clarify that the visual illusion from pseudo-haptic feedback has assisted the user to discriminate the surface stiffness of a virtual sample from the others. Furthermore, it is proven that users can effectively discriminate surface stiffness values between two or more virtual samples using the proposed pseudo-haptic display method.

4.2   Surface Stiffness Discrimination between Virtual and Real Samples

Evaluation in section 4.1 proves that the direct touch haptic display is capable of producing pseudo-haptic property of surface stiffness with discriminable intensity. However, it is still not clear whether the surface stiffness in virtual environment is realistic enough or comparable with in the real world. Therefore, an evaluation is conducted to compare the surface stiffness of virtual samples with that of real samples.

Evaluation Setup. Four real samples were prepared using silicone rubber. Their properties are shown in table 1. For the evaluation, each real sample is used as reference for comparison with 7 other virtual samples. The virtual samples are colored, textured and positioned to appear visually similar. Similar to evaluation in 4.1, the virtual samples are placed directly below the screen so that no force is needed to elongate the virtual finger to reach the sample. Therefore, stiffness perception can be judged only by the force used to deform the sample.

Table 1. Details of real samples

Sample number (real)	1	2	3	4
Surface stiffness k_s, N/mm	9.53	4.50	3.85	3.66
Virtual surface stiffness S_r, V/u.l	21.9	10.3	8.8	6.8

Each real sample is measured for its surface stiffness k_s by adding compression force from 0 to 5 N and measuring their surface deformation depth. The surface stiffness of each real sample k_s is converted into equivalent virtual surface stiffness value. The conversion method involves determining the displacement in the virtual world D_v measured from outside the screen as D, and using the calibration value between output voltage V and fingertip normal force F. The values of the surface stiffness k_s and reference virtual surface stiffness values S_r are shown in table 1. The surface stiffness of virtual samples for comparison S_v varied from the reference surface stiffness S_r by 7 factors f; 0.4, 0.6, 0.8, 1.0, 1.2, 1.4, and 1.6. The relation between the reference surface stiffness S_r and virtual sample's surface stiffness for comparison S_v is given by

$$f = \frac{S_v}{S_r} \tag{3}$$

This means that the surface stiffness of virtual sample for comparison S_v is equal to the reference surface stiffness S_r when stiffness factor f is 1. During the evaluation, the real samples are partially covered such that only the upper surfaces of the real samples are visible to the users as shown. This is so that users will judge the surface stiffness based on the surface of the sample, instead of the overall solid shape of the sample.

Procedure and Evaluation. Eight people, from the age of 22 to 25, took part in this evaluation. There were 4 male and 4 female with no known perception disorders. All of the users used their right hand which is their dominant hand to perform the task. This evaluation is also conducted with seven-rank-evaluation scores, with minimum score -3 for 'different' and maximum score +3 for 'similar'. Before the evaluation is conducted, the users had to press on each of the virtual samples randomly with respective real sample without giving any scores to provide the users with a range of available surface stiffness comparison for the evaluation scoring. For each pair evaluation, users had to evaluate the surface stiffness of a real sample as the reference stimulus before testing on the surface stiffness of a virtual sample as the comparison stimulus as shown in Fig. 12. The similarity score is given by user after investigating the surface stiffness of the both the real and virtual samples. After one trial, the window in the tablet computer is changed and a new virtual sample is introduced randomly to the user. The same procedure is done for all four real samples, which summed up the trials to 28.

Fig. 12. Real and virtual sample's surface stiness discrimination setup

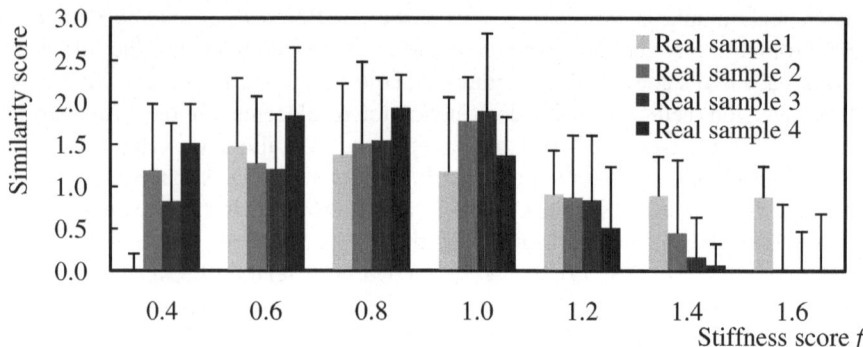

Fig. 13. Real and virtual sample's stiffness sensation comparison

Result. The similarity score ranging from -3 to +3 in this evaluation is standardized with the same method as in evaluation 4.1. The average standardized similarity score for each stiffness factor f is calculated, and subtracted with the minimum average score such that the least similar score is set to 0. The final similarity score for each real sample's stiffness factor is shown in the Fig. 13. Zero in the similarity score axis shows the most dissimilar sensation between real and virtual sample.

The discrimination result for each real sample indicates that the similarity score increases with increment of stiffness factor f until it reaches a peak, before decreasing again. The peak of the similarity score specifies the most similar surface stiffness comparison S_v to real sample's surface stiffness k_s. Since we have set the virtual sample's surface stiffness to be equivalent to real sample's surface stiffness at $f=1$, we expected to see the peak of the similarity score to come just about stiffness factor $f=1$.

From Fig. 13, virtual sample with the most similar stiffness feeling with real samples 2 and 3 resulted at stiffness factor 1 as expected. However, surface stiffness of real sample 1 and 4 were perceived most similar to virtual samples with surface stiffness factor of 0.6 and 0.8 respectively. The result for each real sample statistically shows over 95% confidence level for the difference between all pairs of means of similarity score using the analysis of variance. As aimed, the result proves that using the proposed technique, virtual samples can produce a stiffness sensation similar to that of real objects if set to its most similar surface stiffness S_v value.

Discussions. The difference from expected stiffness factor result for real sample 1 and 4 may result from psychological and measurement factor. In psychological factor, we believe that the most similar stiffness feeling for all the real samples varies among users individually due to subjectiveness of perception. In measurement error, the measurement for surface stiffness k_s may have produced its own error which could have influenced the final comparison result. There is also dissimilarity of deformation shape between real sample and virtual sample for users to discriminate. Moreover, the error in calibration value for converting silicone sample's surface stiffness k_s to virtual sample's surface stiffness S_r is also a possible factor for the error in final evaluation.

Since reference object is always tested first, a global perceptual offset may be produced form the order effect. We plan to treat this influence in a more detailed evaluation in the future.

5   Conclusion

In this paper, an interactive haptic feedback display that utilizes direct touch interaction between finger and the virtual world on a surface computer without using a mediation device is proposed. The display allows users to experience immersive illusion of user's finger penetrating through the screen to manipulate the virtual space inside the screen. As designed, the virtual finger was successfully rendered and elongates from the contact point inside screen according to user's fingertip normal force during user's manipulation.

By adding vibration to the display surface, users can feel the sensation of directly touching the surface of virtual objects using the virtual finger. From the evaluation, it seems that vibration with 250 Hz produces greater illusion of contact sensation than vibration with 50 Hz. We can also conclude that the display is capable to simulate the virtual finger to feel part of user's hand, instead of just an interactive object in the virtual space.

Moreover, with the proposed pseudo-haptic feedback, our evaluation clarifies that when pressing on the virtual surface, users are able to perceive different stiffness feeling of virtual samples with varying surface stiffness setting. With further evaluation, we clarify that the surface stiffness displayed by the pseudo-haptic technique in this research can produce a similar stiffness sensation as perceived from real objects.

Acknowledgments. This work was supported in part by Grant-in-Aid for the Global Center of Excellence Program for "Center for Education and Research of Symbiotic, Safe and Secure System Design" in Japan.

References

1. Microsoft Surface, http://www.microsoft.com/surface
2. Electric Mirror, http://www.electricmirror.com
3. Weiss, M., Wacharamanotham, C., Voelker, S., Borchers, J.: FingerFlux: Near-surface Haptic Feedback on Tabletops. In: Proceedings of ACM Symposium on User Interface Software and Technology 2011, pp. 615–620. ACM Press, New York (2011)
4. Otokawa, K., Maeno, T., Konyo, M.: 4A-5 Tactile Display of Surface Texture by use of Amplitude Modulation of Ultrasonic Vibration. In: Proceeding of the 2006 IEEE Ultrasonics Symposium, pp. 62–65. IEEE Press, Vancouver (2006)
5. Kotani, H., Takasaki, M., Mizuno, T., Nara, T.: Glass Substrate Surface Acoustic Wave Tactile Display With Visual Information. In: Proceedings of the 2006 IEEE International Conference on Mechatronics and Automation 2006, pp. 1–6. IEEE Press, Harbin (2006)

6. Withana, A., Kondo, M., Yasutoshi, M., Kakehi, G., Sugimoto, M., Masahiko, I.: ImpAct: Immersive Haptic Stylus to Enable Direct Touch and manipulation for Surface Computing. In: Computers in Entertainment – Special Issue: Advances in Computer Entertainment Technology, vol. 8(2). Article 9. ACM Press, New York (2010)
7. Lécuyer, A., Burkhardt, J.M., Etienne, L.: Feeling Bumps and Holes Without a Haptic Interface: The Perception of Pseudo-Haptic Textures. In: Proceedings of the ACM Computer Human Interaction International Conference in Human Factor in Computing System 2004, pp. 239–246. ACM Press, Vienna Austria (2004)
8. Botvinick, M., Cohen, J.: Rubber Hands 'Feel' Touch That Eyes See. Nature 391, 756–756 (1998)

Development and Applications
of High-Density Tactile Sensing Glove

Takashi Sagisaka, Yoshiyuki Ohmura, Akihiko Nagakubo*,
Kazuyuki Ozaki**, and Yasuo Kuniyoshi

Department of Mechano-Informatics,
Graduate School of Information Science and Technology,
The University of Tokyo, 7-3-1 Hongo,
Bunkyo-ku, Tokyo 113-0033, Japan
{sagisaka,ohmura,kuniyosh}@isi.imi.i.u-tokyo.ac.jp

Abstract. To understand principles of human hand dexterity, investigation of dynamic contact control by the hand is essential. Here, we developed a novel high-density tactile sensing glove lined with flexible printed circuit boards (FPCs) which were embedded with 1-mm-sized pressure sensitive elements at 1052 points per hand. Those FPCs are arranged in such a way that did not interfere with natural hand movements. In order to demonstrate how the glove captures the hand's sensation and action, we conducted two experiments while human subjects wore the glove. First, we investigated the relationship between pressure distribution and external force exerted on a grasped object. The results showed that the external force correlates to the pressure distribution and that the direction of the external force can be roughly estimated from the pressure distribution. Second, we observed object discrimination by a blindfold subject. As a result, the difference of exploratory patterns arises from difference of objects used in the task was successfully observed from the data of the glove.

Keywords: Flexible PCB, Hand skill analysis, Haptic exploration.

1 Introduction

The human hand is highly dexterous. Even trivial actions such as grasping can be difficult for robots to replicate in the real world. This is due to a lack of understanding of how humans realize object manipulation through dynamic sensory-motor interactions.

To clarify motor control mechanisms used by humans, simultaneous measurement of sensory inputs and motor outputs is necessary. Contact with the environment is a direct cause of tactile sensation and often a direct consequence

---

* Intelligent Systems Research Institute, National Institute of Advanced Industrial Science and Technology, Tsukuba Central 2, 1-1-1 Umezono, Tsukuba, Ibaraki 305-8568 JAPAN. nagakubo.a@aist.go.jp
** Product Development Planning Office, NIPPON MEKTRON, LTD., 1-12-15 Shiba-Daimon, Minato-ku, Tokyo 105-8585 JAPAN. kzozk@mektron.co.jp

P. Isokoski and J. Springare (Eds.): EuroHaptics 2012, Part I, LNCS 7282, pp. 445–456, 2012.
© Springer-Verlag Berlin Heidelberg 2012

of motor actions at the same time. Therefore, tactile sensory information due to contact must be rich in information about control of object manipulation.

Also, contact is often instantaneous, weak and can occur at any point on the surface of the hand. Thus, it must be measured as a seamless distribution over the entire surface of the hand by highly sensitive tactile sensors. However, because skin deforms dynamically according to joint movements, thick and stiff tactile sensors interfere with joint movements when they are distributed all over the hand at a high density. To address all of these concerns, we designed a novel tactile sensing glove lined with dense arrays of tactile sensors which were realized by embedding pressure-sensitive elements in thin FPCs [1].

The objective of this paper is to summarize preliminary results of experiments performed while a human wore the glove. The paper is organized as follows. The next section summarizes existing studies of contact measurement on the human hand. In section 3, key elements of the design of our tactile sensing glove are described. Section 4 gives a detailed explanation of the two experiments we conducted to demonstrate how our glove can capture the action of the human hand. Lastly, section 5 states our conclusions and future goals.

2   Related Works

Johansson et al. [2,3] conducted detailed studies on human tactile sensation using invasive means, in which metal electrodes were directly inserted into the upper arm to receive afferent signals generated by mechanoreceptors distributed under the skin of the hand. Though this approach is highly accurate in capturing true tactile signals issued by the human subject, the use of only this approach limits the measurement to tens of mechanoreceptors, thereby limiting the field of view to a small area such as the fingertips. Moreover, electrode insertion restricts the subject's movement.

On the other hand, many researchers have tried non-invasive measurement techniques by attaching various touch-sensitive device to the surface of the human hand. For example, Nikonovas et al. [4] and Lee et al. [5] attached nearly twenty force-sensing resistors(FSRs), Matsuo et al. [6] attached 160 units of mechanical switches, Sato et al. [7] attached pressure sensitive rubber with stitched electric wires, making up a matrix of 212 pressure sensitive units, and TekScan® developed a sensor glove named the Grip^{TM} system using pressure sensitive matrix of 349 sensor elements embedded in thin films. The latter two examples show that matrix designs are advantageous for realizing higher spatial resolution. However, to assure free joint movements, none of them cover whole surface of the hand seamlessly. For example, in the case of Sato et al. [7], only half of the frontal face of the fingers and thumb are covered, and in the case of Grip^{TM} system, the center of the palm is completely excluded from sensor coverage.

Based on our survey, a detailed and non-invasive means of measuring contact of the human hand are necessary to understand human hand dexterity. However, thus far, tactile sensing gloves whose sensors cover the entire hand without interfering with the wearer's hand movements have yet to be realized. This is

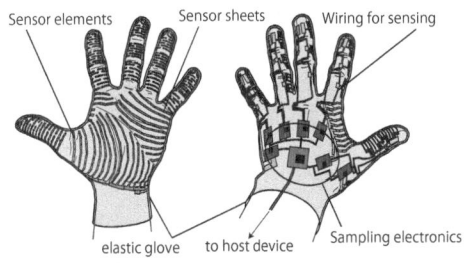

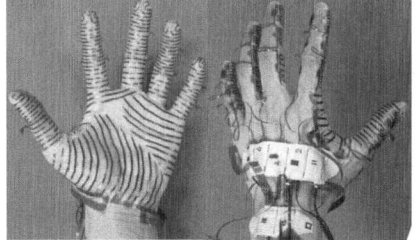

(a) Main components of the glove   (b) Photography of the glove

Fig. 1. Whole image of the developed tactile sensing glove

due to difficulties in covering not only small and curved but also dynamically deformable surfaces of the hand.

3 Development of Tactile Sensing Glove

We designed a new tactile sensor based on our previous study [8] of a tactile sensor module which consists of a signature branch-shaped sensor sheet with adaptability for complex curved surfaces. The newly developed tactile sensor sheet is more compact, tight, and thin, with sensing resolution matching that of the two-point discrimination in the human hand. The sensor sheet is specially designed so as to ensure moveability around joints and creases of the thumb, the fingers, and the palm.

The overall structure of the tactile sensing glove is shown in Figure 1. Sensor sheets are encapsulated in a base glove made of thin layers of elastomer for durability. The back of the hand is not considered as the sensing area because this area is rarely used for manipulation in everyday situations.

In the first stage of the designing, the target distance between adjacent sensor elements is set to 2 mm on the fingertips, to 3-5 mm on other areas of the fingers and on the thumb, and to 6 mm on the palm, considering the distribution of two-point discrimination on the human hand [9]. This results in a total number of approximately 1000 sensor elements on each hand. Each sensor element is aimed to be under 1 mm in size.

The basic structure of the developed sensor element is shown in Figure 2a, and the pressure response of the element is shown in Figure 2b. This sensor element is designed to be practically insensitive to bending so it can properly read out applied pressure even when it is curved dynamically according to skin deformation. Each sensor element occupies only 1.4 mm by 0.6 mm area in the sensor sheet, and the total thickness of the sensor sheet is 120 μm.

The sensor measures contact resistance between the conductive film and the two separate electrodes placed opposite to it. When the pressure reaches threshold, around 0.4 kPa in the case of an element depicted in Figure 2b, the conductive film makes contact with both of the two electrodes, and a kilo-ohm

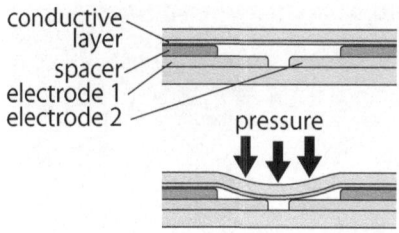

(a) Basic structure of pressure sensitive sensor element

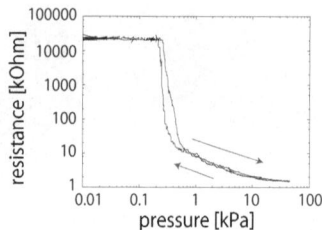

(b) Pressure-resistance profile of the sensor element

Fig. 2. The developed pressure sensitive element for tactile sensing glove

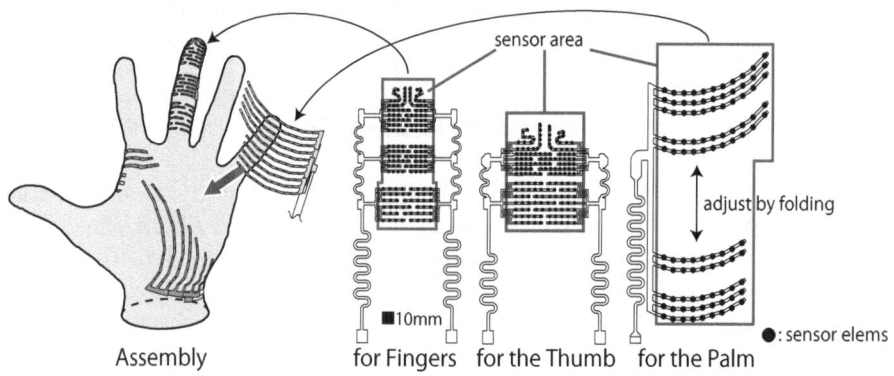

Fig. 3. Shapes of the developed sensor sheets

resistance appears between the electrodes. The resistance is converted into voltage, and then the voltage is read by a 12 bit A/D converter mounted on the sampling board. After initial contact, the resistance value decreases as pressure increases in a nearly linear manner on the log-log scale.

Shapes of the developed sensor sheets are shown in Figure 3. Each branch of the sensor sheet is 1.0 mm-wide on the fingers and the thumb, and 1.5 mm-wide on the palm. Because skin deforms along creases, each branch is placed approximately in parallel to its nearest crease.

4   Experiments

Two experiments were conducted to explore what can be known from the pressure distribution measured by the glove.

Pre-process for Tactile Sensor Data. Resistance values on all sensor elements are sampled at 125 frames-per-second. Although each sensor element has a different pressure-resistance profile, we chose to directly use the resistance value in analysis presented in this paper without converting each sensor element's read-out resistance into units of pressure. This is because the goal of this experiment was to investigate whether differences in resistance patterns alone are sufficient for characterizing manipulative actions. Because resistance value decreases logarithmically with increase in pressure, the following normalization is applied.

$$v_j = \begin{cases} 0 & (r_j > R_0) \\ 1 & (r_j < R_1) \\ \left(\log \frac{r_j}{R_0}\right) / \left(\log \frac{R_1}{R_0}\right) & (\text{else}) \end{cases}, \qquad (1)$$

where r_j is the measured resistance value of a sensor element j, R_0 and R_1 are the upper and the lower limits of resistance, defining the range of interest.

4.1 Force Estimation Experiment

Even a simple action such as grip takes substantial skill. For example, to hold an object such as a rod steady, one must be able to sense forces acting on the object. We hypothesized that pressure patterns measured by the sensor glove will be able to identify external forces acting on a grasped object.

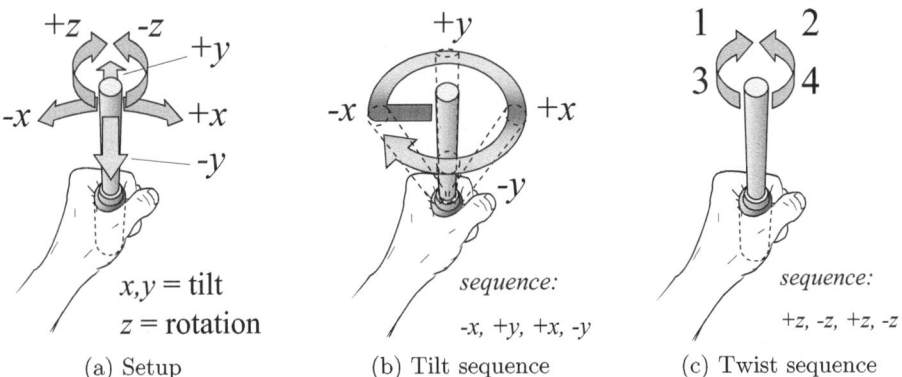

(a) Setup   (b) Tilt sequence   (c) Twist sequence

Fig. 4. Setup and the test conditions for the first experiment

Figure 4a illustrates the experimental setup. We defined components of external forces $\mathbf{f} = (f_x, f_y, f_z)^T$ as two tilt directions x and y, and one twist direction z. During the experiment, a rod was held by the left hand while wearing the tactile sensing glove and external force was exerted on the rod by the right hand instead of contact with environment. First, learning data sets were collected under seven conditions. In one condition, no external force was applied on the rod. In four of the conditions, the subject grasped the rod while the rod was tilted in the -x, +x, -y, and +y directions. In the last two conditions, the rod was twisted

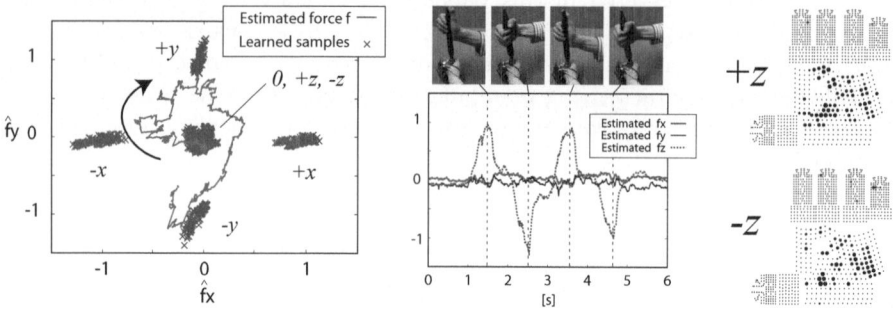

(a) Tilt sequence results are shown. Clusters of learned samples are shown with each class label.

(b) Twist sequence results

(c) Tactile sensor data from twist sequence

Fig. 5. Results of force component estimation

in the -z and +z directions. Each data set contains tactile sensor frames from 1-second recording, and each data frame $\mathbf{v} = (v_1, \ldots, v_q)^T$ consists of q (=1052) variables corresponds to normalized values of each sensor element.

Secondly, the linear regression model that converts tactile sensor vector $\mathbf{v}$ into estimated force vector $\hat{\mathbf{f}}$ was obtained as follows. For each condition i, data set was acquired as $\mathcal{D}_i = \{(\mathbf{f}_i, \mathbf{v}_{1,i}), \ldots, (\mathbf{f}_i, \mathbf{v}_{N_i,i})\}$, where $\mathbf{f}_{\pm \mathbf{x}} = (\pm 1, 0, 0)^T$, $\mathbf{f}_{\pm \mathbf{y}} = (0, \pm 1, 0)^T$, $\mathbf{f}_{\pm \mathbf{z}} = (0, 0, \pm 1)^T$, $\mathbf{f}_0 = (0, 0, 0)^T$, and total data set was defined as $\mathcal{D} = \{\mathcal{D}_{+x}, \mathcal{D}_{-x}, \mathcal{D}_{+y}, \mathcal{D}_{-y}, \mathcal{D}_{+z}, \mathcal{D}_{-z}, \mathcal{D}_0\}$. Canonical correlation analysis (CCA) gives following linear transformation: $\mathbf{s}_i = A^T (\mathbf{f}_i - \bar{\mathbf{f}})$, $\mathbf{t}_i = B^T (\mathbf{v}_i - \bar{\mathbf{v}})$, where A and B are the matrices containing the canonical correlation vectors that maximize the correlation between $\mathbf{s}$ and $\mathbf{t}$. Because $\mathbf{s}$ and $\mathbf{t}$ are the same in terms of linear approximation, we now reach a linear regression equation:

$$\hat{\mathbf{f}} = \left(BA^{-1}\right)^T (\mathbf{v} - \bar{\mathbf{v}}) + \bar{\mathbf{f}}. \tag{2}$$

Following seven training sets, test data $\mathbf{v}$ is collected through the tilt sequence depicted in Figure 4b and the twist sequence shown in Figure 4c. The sequences of force components vector $\hat{\mathbf{f}}$ estimated from each test data are shown in Figure 5a and 5b. From the results of tilt sequence, we can tell that the direction of force vector is successfully estimated. Also, from the twist sequence results, the estimated direction of twist force corresponds to exerted twist. Furthermore, we can say that the change in the pressure distribution by twist is independent to those by tilt in x and y directions because $\hat{f}_x$ and $\hat{f}_y$ stay at around zero during the course of the task sequence.

It came as a surprise to us that we were able to estimate twist information based on our sensor glove, as our tactile sensors were designed to measure force exerted in the normal rather than tangential direction. Usually, twist can be estimated by tangential forces on the surface of the hand. However, in our case, twist caused independent changes in pressure distribution, allowing us to discriminate twist directions. We believe that this can be explained by either the

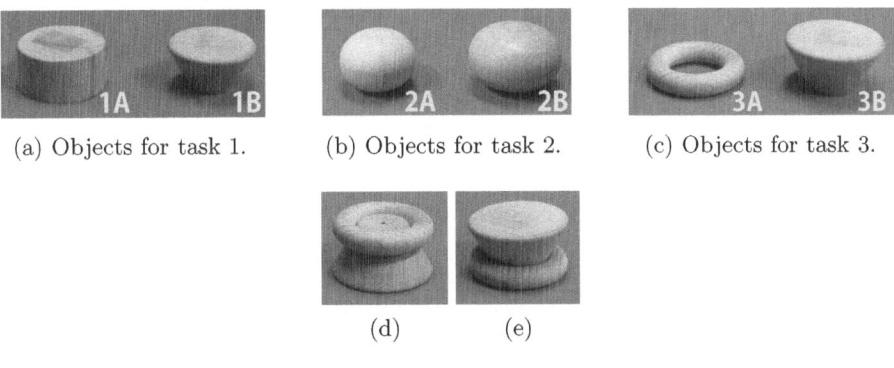

(a) Objects for task 1.   (b) Objects for task 2.   (c) Objects for task 3.

(d)   (e)

Fig. 6. Objects used in the second experiment are shown. In each subfigure (a) to (c), object A is in the left, and object B is in the right. Objects for task 1 are fixed on a flat surface to avoid being recognized by the difference of movement. Objects for task 3 are combined as shown in (d) and (e) to regularize the height. The diameters of the objects are between 50 to 60 mm.

deformation of the soft pads on the surface of the hand or by involuntary modulations of grasp force in order to avoid slippage. The change in tactile sensor data caused by twist are depicted in Figure 5c.

4.2   Haptic Discrimination Experiment

In order to demonstrate how the glove can capture actions in active sensing, the second experiment was focused on haptic discrimination. Lederman[10] showed that humans select optimal exploratory movements according to what they want to know. With this in mind, we compared glove data measured in discrimination tasks between different pairs of objects.

Experimental Setup. We designed tactile-sense-dominant tasks using five different objects shown in Figure 6. In each task, two objects, A and B, are used for two-class discrimination. Task difficulty varied across pairings. Task 3, for example, is thought to be easier than tasks 1 and 2 because the geometry on the top of the objects are comparatively more dissimilar. The three tasks were performed in consecutive blocks, starting with task 1 (20 trials) followed by task 2 (12 trials) and then followed by task 3 (15 trials). In each trial, one object, either A or B, was presented in random order.

The subject was permitted to touch and see those objects associated with labels A and B before each task to make himself familiar with the objects. The object was presented in a fixed location on the table, and was occluded by a board which stood between the table and the subject, and subject wore earplugs and headphones emitting white noise to shut audio keys out. After an object was set on the table, the subject was permitted to touch the object by his gloved hand until he was able to discriminate it as object A or B. The subject was directed

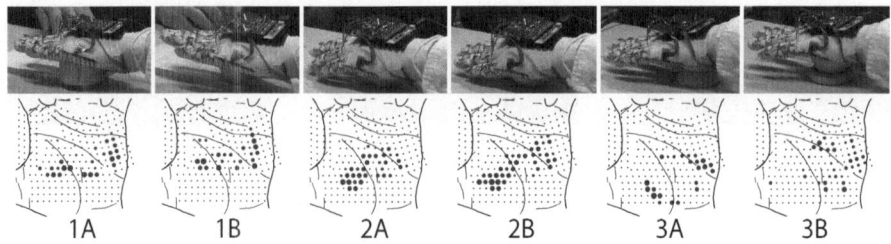

Fig. 7. Examples of tactile sensor data measured in haptic exploration are shown. Each dotted graphics corresponds to the sensor sheets on the palm.

to lift the hand from an object immediately after discrimination. Finally, the subject answered A or B orally, and the next trial began.

The fingers and the thumb were not permitted to be used because humans possibly rely on postural information such as joint angle while manipulating objects with their fingers and thumb in order to recognize them. Moreover, the subject was directed to explore object's surface with minimal displacement of the object. To prevent the fingers from touching the table surface during discrimination, objects were chosen to have heights of around 3-4 cm. There was only one, male, right-handed subject (first author). The glove was worn on his left hand.

Results. Figure 7 shows examples of tactile sensor data and sample frames from a video footage recorded during exploration. The accuracy of subject's answer in tasks 1, 2 and 3 was 100%, 93.3%(=14/15), and 100% respectively.

Due to the physical constraints of the task (no finger use, prohibition of object movement) the subject was forced to gently push against an object in the most accessible way, which was from above. In the case of task 1 and 2, where objects have similar geometry on their top surface, no qualitative difference was noted between tactile sensor pattern observed between objects A and B. However, in the data observed from object A of task 3, we observed a consistent torus-shaped pattern, distinct from that of object B.

Also, based on the video footage, it is found that the subject touched each object several times until discrimination, changing the position of contact on the palm. This action was also evident in the time sequence of mean values of tactile sensor elements: μ_v. Figure 8a is an example time sequence for over the course of several trials for object A in task 1. Multiple peaks observed in μ_v indicate that the subject touched the object multiple times in each trial. Also, from a gradual decrease in the number of peaks with increasing trial count shown in Figure 8b, we can infer that the subject gradually adjusted to the discrimination task along the course of each block.

For the rest of the analysis, we focused on the the last peak in μ_v of each trial, corresponding to the touch right before answering. Assuming that the last touch contains critical information for decision-making, it is expected that variance of pressure pattern is smaller than those observed in earlier touches. The variance

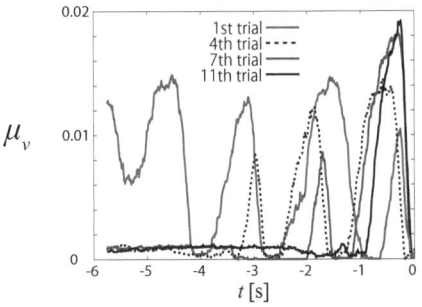

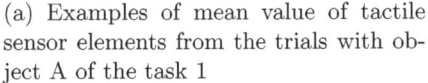

μ_v

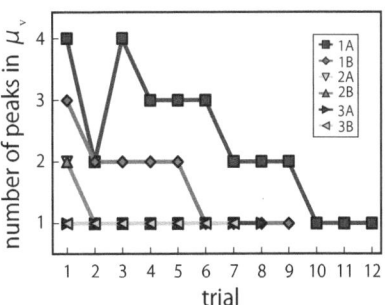

(a) Examples of mean value of tactile sensor elements from the trials with object A of the task 1

(b) Decrease in number of peaks observed in μ_v according to increase in trial count

Fig. 8. Multiple times of touch observed in tactile sensory data measured in haptic discrimination trials

of tactile sensory data during the n-th last touch of each trial with multiple peaks is shown in Figure 9. With respect to task 2 and 3, only the data during the last touch is plotted because multiple peaks were observed in less than two trials, which was too few to calculate variance. Results of task 1, indicated by the blue line and red line in Figure 9, support our speculation that the tactile sensory data during the last touch has the smallest variance, supposedly containing the most critical information for decision-making. Also, we can tell that task 3 gives notably large variance during the last touch compared to the other tasks. This corresponds to the speculation that task 3 is easiest among three tasks.

A variance in tactile sensory data is defined as follows to investigate distribution of contact position. First, assuming from hand actions observed during each touch that the hand changes contact intensity without significantly changing the position of contact, the tactile sensory data giving the maximum value in μ_v is used as the representative sensory data during each touch. To normalize variation in intensity of contact, the data is divided by its L2 norm. Next, a pair of tactile sensory data are compared pixel-wise. Consequently, variance s_n is defined by

$$s_n = \sum_{i \in I_n} |\hat{\mathbf{v}}_{i,n} - \mathbf{m}_n|^2 / \#I_n, \qquad (3)$$

$$\mathbf{m}_n = \sum_{i \in I_n} \hat{\mathbf{v}}_{i,n} / \#I_n, \qquad (4)$$

where i is the trial with a certain object of a certain task, I_n is the collection of trials i where μ_v has n-th last peak, $\hat{\mathbf{v}}_{i,n}$ is the normalized tactile sensor data at n-th last peak, and $\mathbf{m}_n$ is the centroid of $\hat{\mathbf{v}}_{i,n}$.

Finally, in order to visualize to what extent the contact location on the palm converged during the last touch, the center of pressure (COP) of the data from

454   T. Sagisaka et al.

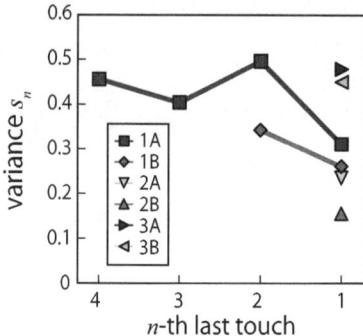

Fig. 9. Variance of tactile sensory data in n-th last touch was plotted to show that the last touch has smaller variance compared to earlier touches

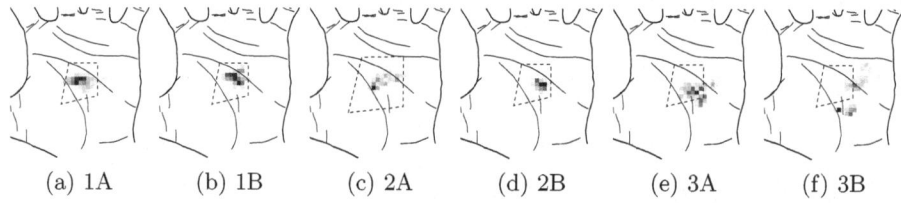

(a) 1A   (b) 1B   (c) 2A   (d) 2B   (e) 3A   (f) 3B

Fig. 10. Histograms of COP point in the last touch of each trial are shown. Bin size of the histogram is 3 mm by 3 mm. The dashed line drawn in the middle of the palm shows the contour of the area where the center of the object can move while outline of the object is still fully covered by the palm. The black cell indicates the maximum value of the histogram, and the white cell indicates zero.

the last touch for each object was plotted (Figure 10). Now, COP is approximated by

$$\text{COP} = \sum_j \mathbf{x}_j \tilde{v}_j / \sum_j \tilde{v}_j, \tag{5}$$

where $\mathbf{x}_j$ is the position of the tactile sensor element j on the palm. We can see from the figure that, in comparison with the area depicted by the dashed contour (where object's center can move within the palm), tasks 1 and 2 yielded more compact COP distributions, whereas task 3 yielded a broader distributions and therefore a higher relative variance (Figure 9).

Discussion. Difference in exploration strategies can be found by comparing tactile sensor data during task 3 versus tasks 1 and 2. Both Figure 9 and 10 indicate that the position of the hand relative to object while making decision is distributed broader in task 3 compared to tasks 1 and 2. This can be explained by the fact that objects in task 3 can be recognized regardless of the position of contact on the palm because their geometry on the top surface are very different.

It is important to note that although the number of peaks in each trial converges to one in every task (Figure 8b), we observed that it was not necessarily the case that towards the end of the experiment, subjects only needed to touch the object once. Based on the video footage, we observed that the subject corrected the palm positions based on extremely slight contacts, without activating any tactile sensor element, even in trials where there was only one recorded peak in μ_v. Therefore, as the subject adapted to the discrimination task, he developed the ability to gain information from contacts that were too weak for the present design of the tactile sensing glove.

5   Conclusions and Future Work

To study operating principles of human hand dexterity, we developed high-density tactile sensing glove to measure detailed pressure distribution over the seamless frontal surface of the human hand, and reported data from two preliminary experiments using the glove.

The results of the first experiment showed that the external force exerted on the grasped object correlates to the pressure distribution measured by the glove and that the direction of the external force can also be estimated. Moreover, the results suggest that not only normal but also the tangential forces correlate to pressure distributions on the surface of the hand while the hand grips an object steadily.

In the second experiment, differences in haptic exploratory patterns for different object was successfully observed from the data measured by the glove. We also found that we need more sensitive tactile sensor to investigate fine sensory-motor control which allow humans to explore haptic information rapidly, even when forced to use the palm where sensitivity is relatively low compared to other parts of the hand such as the finger tips.

As a short-term goal, we will further validate these preliminary findings by recruiting more subjects. We will also improve tactile sensor elements and other electronic components to realize more detailed measurement of weak haptic control. We also plan to analyze manipulative actions involving finger use by integrating motion sensors with the tactile sensing glove.

References

1. Sagisaka, T., Ohmura, Y., Nagakubo, A., Kuniyoshi, Y., Ozaki, K.: High-density conformable tactile sensing glove. In: 11th IEEE-RAS International Conference on Humanoid Robots (Humanoids 2011), pp. 537–542 (2011)
2. Macefield, V.G., Häger-Ross, C., Johansson, R.S.: Control of grip force during restraint of an object held between finger and thumb: responses of cutaneous afferents from the digits. Exp. Brian Res. 108(1), 155–171 (1996)
3. Johansson, R.S., Birznieks, I.: First spikes in ensembles of human tactile afferents code complex spatial fingertip events. Nature Neuroscience 7(2), 170–177 (2004)

4. Nikonovas, A., Harrison, A., Hoult, S., Sammut, D.: The application of force-sensing resistor sensors for measuring forces developed by the human hand. J. Eng. Med. (I. Mech. E. Part H) 218(2), 121–126 (2004)
5. Lee, J.H., Lee, Y.S., Park, S.H., Park, M.C., Yoo, B.K., In, S.M.: A study on the human grip force distribution on the cylindrical handle by intelligent force glove (i-force glove). In: 2008 International Conference on Control, Automation and Systems, pp. 966–969 (2008)
6. Matsuo, K., Murakami, K., Hasegawa, T., Kurazume, R.: A decision method for the placement of tactile sensors for manipulation task recognition. In: 2008 IEEE International Conference on Robotics and Automation, pp. 1641–1646 (2008)
7. Sato, S., Shimojo, M., Seki, Y., Takahashi, A., Shimizu, S.: Measuring system for grasping. In: 5th IEEE International Workshop on Robot and Human Communication, pp. 292–297 (1996)
8. Ohmura, Y., Kuniyoshi, Y., Nagakubo, A.: Conformable and scalable tactile sensor skin for curved surfaces. In: Proc. IEEE Int. Conf. on Robotics and Automation, pp. 1348–1353 (2006)
9. Johansson, R.S., Vallbo, A.B.: Tactile sensory coding in the glabrous skin of the human hand. Trends in Neurosciences 6, 27–32 (1983)
10. Lederman, S.J.: Hand Movements: A Window into Haptic Object Recognition. Cognitive Psychology 19(3), 342–368 (1987)

Presentation of Sudden Temperature Change Using Spatially Divided Warm and Cool Stimuli

Katsunari Sato and Takashi Maeno

Graduate School of System Design and Management, Keio University,
4-1-1 Hiyoshi, Kohoku-ku, Yokohama, Kanagawa 223-8526, Japan
{katsunari.sato,maeno}@sdm.keio.ac.jp

Abstract. We propose a thermal display that can present a sudden temperature change using spatially divided warm and cold stimuli; the display exploits two characteristics of human thermal perception: the spatial resolution of thermal sensation is low, and the thermal threshold depends on the adapting temperature. Experimental results confirmed that users perceived separate individual thermal stimuli as a single stimulus when the thermally stimulated area was small because of the low spatial resolution. The spatially distributed warm and cold stimuli enabled users to perceive the thermal sensation rapidly even if the cold stimulus was suddenly presented after the warm stimulus and vice versa. Furthermore, our thermal display successfully made the skin more sensitive to both warm and cold stimuli simultaneously by using spatially divided warm and cold stimuli, each of which separately adjusts the adapting temperature.

Keywords: thermal display, thermal perception, spatial stimuli, adapting temperature.

1 Introduction

Presentation of the thermal sense, by which we perceive an object's temperature and material, has been a great challenge in the development of haptic displays. It has been revealed that the thermal sense is one of the important properties in human texture perception [1][2]. For example, Shirado and Maeno [2] identified four primary properties of haptic perception: roughness, coldness, moistness, and hardness. On the basis of these studies, many haptic displays that can present not only mechanical skin deformation but also a temperature change in the skin have been developed [3][4].

We also aim to develop a thermal display as part of a haptic display, in particular the glove-type or grip-type, which are generally used in both research and practical applications. To achieve this aim, we consider two main requirements for the thermal display. The first is the device size: the thermal display should be small and light enough for integration with other haptic displays, i.e., tactile and kinesthetic ones. The second is ease of control: the thermal display should present the change in skin temperature that occurs when a human touches an object.

To develop a method that could satisfy these requirements, we focus on the Peltier device, in which a heat flux is created between the two sides of the device by applying

P. Isokoski and J. Springare (Eds.): EuroHaptics 2012, Part I, LNCS 7282, pp. 457–468, 2012.

voltage. It works as either a cooler or a heater depending on the polarity of the and the rate of temperature change can be controlled by adjusting the voltage. Furthermore, because the Peltier device can be implemented in a relatively small size even though it requires equipment for heat radiation, it can be easily mounted in other haptic displays. Because of these advantages, many studies have successfully represented the thermal sense using a Peltier device [3][4][5][6], and we also adopt it.

However, the time response of the Peltier device becomes problematic when it is mounted on glove-type or grip-type haptic displays. When a human touches an object, the temperature of the skin surface changes rapidly at the moment of contact, and the human seems to perceive this change and recognize the object's temperature or material [7]. Therefore, the Peltier device must represent such sudden temperature changes, but this is difficult to realize, especially when the temperature change becomes large, because the human skin is in continuous contact with the surface of the display.

It may be possible to improve the time response by applying high energy to the Peltier device using large radiation equipment or by developing a higher-efficiency Peltier device. Another approach is to use the characteristics of human thermal perception [8]. If the thermal display is effectively designed on the basis of these characteristics, its performance could be improved, even though it consists of conventional Peltier devices.

To present a sudden temperature change in glove-type or grip-type haptic displays using a Peltier device, we also focus on the characteristics of human thermal perception and propose a thermal display using spatially divided warm and cold stimuli (Fig. 1).

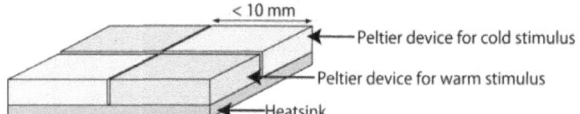

Fig. 1. Configuration of proposed thermal display

2   Proposed Thermal Display

2.1   Spatially Divided Thermal Stimuli

The configuration of the proposed thermal display is shown in Fig. 1. Peltier devices that act as warm or cold stimuli are located alternately on the heat sink. We expect this configuration to present a sudden temperature change efficiently because of two characteristics of human thermal perception: the spatial resolution is low, and the thermal threshold depends on the skin's adapting temperature.

2.2   Spatial Resolution

The spatial resolution of tactile stimulation on the body has been measured in many ways, and the results have described a tactile spatial resolution of high accuracy. However, the resolution of thermal stimulation is low, mainly because of spatial summation: when humans simultaneously touch two objects of different temperature, they perceive a temperature intermediate between those of the two objects [9]. For example, it is reportedly difficult to resolve two low-intensity warm stimuli presented 150 mm apart on the forearm [10]. The ability to localize warm stimuli is better on the dorsal surface of the hand, where the resolution is approximately 19 mm [11].

Because of this characteristic, we consider that users could not perceive the position of the thermal stimuli when their area was smaller than the spatial resolution. Therefore, we expect users to feel that the entire surface of the thermal display is warm (or cold) even when only the two Peltier devices acting as warm (or cold) stimuli are activated.

2.3   Adapting Temperature

The thermal threshold, i.e., the ability to perceive variations in temperature, depends on many factors such as the rate of temperature change [12][13], the stimulated site [14] and area [15] on the body, and the adapting (or baseline) temperature of the skin [13]. We focused on the influence of the skin's adapting temperature. When the skin temperature is around 28°C, the thresholds for detecting warm and cold stimuli are around 1°C and 0.1°C, respectively. As the skin temperature increases, the warm and cold thresholds decrease and increase, becoming 0.2°C and 1.1°C, respectively, when the skin temperature reaches around 40°C [13].

Using this characteristic, Akiyama et al. [8] proposed that a slight increase and decrease in the adapting temperature in neighboring skin regions produced by two adjacent Peltier devices can be applied to lower the thermal threshold and thus make the skin more sensitive to both warm and cold stimuli. We also apply this method in order to present a sudden temperature change using spatially distributed thermal stimuli.

2.4   Presentation of Sudden Temperature Change

Figure 2 shows examples of the representation of temperature changes using the proposed thermal display. First, each Peltier device for providing warm and cold stimuli increases or decreases its surface temperature, respectively, in order to make the skin sensitive to both warm and cold stimuli simultaneously. Because the threshold of the thermal sense depends on the rate of temperature change, the user cannot perceive these thermal stimuli for controlling the adapting temperature if the temperature change is slow enough [16]. Furthermore, the user may perceive the same temperature as the initial one because of the effect of spatial summation.

If a warm stimulus begins (e.g., the user touches a warm object in the virtual world), the Peltier devices acting as warm stimuli are activated, and the user rapidly feels a warm sensation. Then, when the user subsequently touches a cold object in the virtual world, the Peltier elements acting as cold stimuli are activated. Because the temperature of the skin in contact with the cold Peltier device maintains the adapting temperature even though the warm stimulus is presented, the user can quickly perceive the subsequent cold stimulus. After the cold stimulus stops (e.g., the user releases the virtual object), the temperature of the Peltier elements acting as cold stimuli returns to the adapting temperature.

Note that the proposed thermal display does not represent the absolute temperature of the skin surface correctly but presents the comparative temperature change. Therefore, a model for representing thermal sensation using the proposed display should be constructed for practical use.

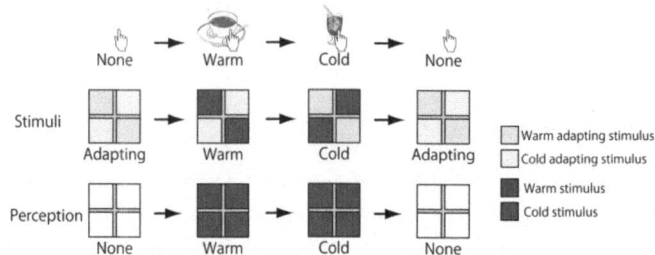

Fig. 2. Examples of presentation of thermal sense

Furthermore, we consider that the proposed method could offer not only improved temporal response, but also a smaller device. A thermal display using Peltier devices requires equipment for radiation, such as a heat sink and fan, on the backside of the display surface. To present greater temperature changes, we must use larger radiation equipment. However, because the Peltier devices for warm and cold stimuli are arranged alternately in the proposed device, we expect that the heat produced by the devices could be mutually canceled in the heat sink. In the future, we will investigate whether this cross-interaction of heat could improve the radiation efficiency of our thermal display.

3   Experiments

In this section, we examined the efficiency of the proposed thermal display. First, we conducted an experiment on position identification of thermal stimuli in order to check whether the user could perceive the position of the stimulus in the proposed thermal display (experiment 1). Then, we examined the thermal response time and showed that the proposed thermal display can present sudden temperature changes (experiment 2).

3.1   Materials and Methods

Participants

Five volunteers (three males and two females, average age 26.6 years) participated in experiment 1, and six volunteers (three males and three females, average age 24.5

years) participated in experiment 2. One participant was the author, and the others were naive to the hypotheses being investigated. Before the experiment, all of them were instructed about the experimental procedures.

The thermal stimuli were presented to a fingertip on the dominant hand because most haptic displays present haptic sensations to the fingertip. All participants were healthy and right-handed: their average finger temperature was around 33°C in a room where the air temperature was maintained at around 24°C.

Equipment

Figure 3a shows the thermal display used in the experiments. Four Peltier devices (8.3 mm × 8.3 mm × 2.4 mm; TEFC1-03112; Nihon Tecmo Co.) were placed on the heat sink (30 mm × 30 mm × 20 mm; made from aluminum), and a cooling fan was used to improve the devices' performance. The Peltier devices were attached to the heat sink using double-coated heat transfer tape.

The controller of the Peltier devices consisted of a laptop PC, two microprocessors (Arduino UNO; Arduino) and four motor servo driver chips (TA7291P; Toshiba Co.), as shown in Fig. 3b. The laptop PC calculated the voltages applied to each Peltier element and sent these values to two microprocessors through a serial connection. The received voltage was output from two digital ‑ analog (DA) converting pulse-width modulation (PWM) output pins, which were connected to driver chips that were in turn connected to the Peltier devices. The power supplies for the motor drivers were two DC adapters (5V2A; Akizuki Denshi Tsusho Co.). The voltages applied to the Peltier devices were updated at 200 Hz.

Using this setup, we first examined the relationship between the voltage applied to a Peltier element (i.e., the output from motor driver circuit) and the temperature changes on the device's surface. We placed a thermistor (103JT-025, Semitec Co.) on the surface of the Peltier element and applied arbitrary voltages to it at an air temperature of 24°C. Figure 3c shows the measurement results. The horizontal and vertical axes represent the applied voltage and the temperature change during the first 1 s, respectively. The strengths of the thermal stimuli in the following experiments were determined using this relationship.

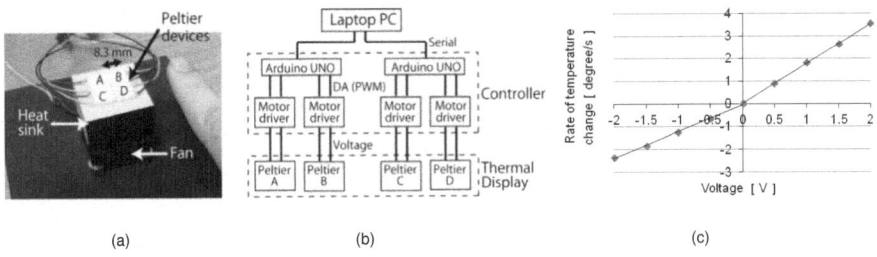

(a)   (b)   (c)

Fig. 3. Experimental equipment. (a) Appearance and (b) configuration of thermal display, and (c) relationship between voltage and rate of temperature change.

Procedures

Experiment 1

This experiment examined the accuracy with which the position of thermal stimuli applied by the four Peltier devices to the finger pad was identified.

Each participant placed the tip of the right index finger so that the finger pad contacted all four Peltier devices, which were kept at around 33°C (the applied voltage was 0.35 V). Participants then pushed the start button for the thermal stimulus using the left hand. The thermal stimulus began 1–2 s (randomly selected) after they pushed the start button. The maximum duration of the thermal stimulus was 3 s. When the participants recognized whether the thermal stimulus was warm or cold, they pushed the stop button for the thermal stimuli using the left hand. They stated the type of stimulus (warm or cold) and its position. We also recorded the duration of the stimulus. After this procedure, the participants rested so that their skin temperature returned to its default state.

Seven spatial patterns of warm and cold stimuli were used (Fig. 4). In one, all the Peltier devices provided stimuli, and in each of the others, two provided stimuli, so the total number of stimulus patterns was $2 \times 7 = 14$. Each pattern was presented to the participants five times. The order in which the patterns were presented was randomly selected.

The strength of each stimulus (i.e., the voltage applied to the Peltier devices) was important in this experiment because the participants might recognize the spatial pattern of the stimuli from its strength, especially when the stimulus came from all the Peltier devices. Greenspan and Kenshalo [15] described the relationship between the thermal threshold and the area of the stimulation: the warm threshold can be halved when the area of the warm stimulation doubles, whereas the cold threshold can be halved with a four-fold increase in the cold stimulation area. We first determined the strength of stimuli from two of the Peltier devices ("Two Peltiers" patterns in Fig. 4) as 2°C/s (1.75V and -1.6V for warm and cold stimuli, respectively). Then, we conducted pilot experiments and determined the strength of stimuli presented from all the Peltier devices ("All Peltiers" pattern in Fig. 4) so that the participants perceived the same strength from each experimental pattern. In this case, the voltages of the warm and cold stimuli for all the Peltier devices were 1.0 V and -0.95 V, respectively.

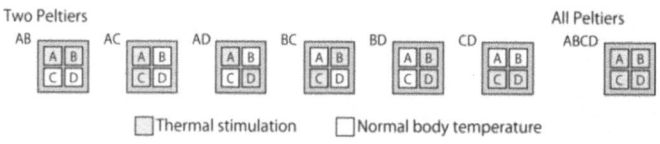

Fig. 4. Spatial patterns of thermal stimuli

Experiment 2

This experiment examined the efficiency of spatially divided thermal stimuli and the control of the adapting temperature for presentation of a sudden temperature change.

We assumed that a user touched warm and cold (or cold and warm) objects consecutively, as shown in Fig. 2, as an example of a sudden temperature change.

Each participant placed the tip of the right index finger on the thermal display and pushed the start button using the left hand. A 5 s pre-stimulus began when they pushed the start button in order to increase or decrease the skin temperature, and then the main stimulus began. When the participants recognized the temperature change caused by the main stimulus, they pushed the stop button using the left hand, and the duration of the main stimulus was recorded. The maximum duration of the main thermal stimulus was 5 s. Then, participants assigned the perceived strength of the stimulus to one of three levels: when they perceived strong simulation, they scored it as 3; when it was weak, they scored it as 1. After this procedure, the participants rested for 20 s so that their skin temperature returned to its default state.

The six thermal stimuli used in this experiment (shown in Fig. 5) were classified into three groups: I. the stimuli from all of the Peltier devices, II. the stimuli from two Peltier devices without adapting stimuli, and III. the stimuli from two Peltier devices with adapting stimuli. Groups I and III represent the conventional and proposed methods, respectively, and group II was used to examine the effect of controlling the adapting temperature. Each group consisted of pre- and main stimuli in order to present warm and cold (or cold and warm) stimuli consecutively. No stimulus was applied in the last 1s of the pre-stimuli except for the adapting temperature for the main stimuli; this represents a situation in which a human disengages a finger from one object and then touches a second object. Each pattern was presented to the participants five times. The order in which they were presented was selected at random.

The strength of each main stimulus was the same as in experiment 1. The voltages for the warm and cold adapting stimuli were 0.65 V and -0.3 V, respectively. In this case, the skin temperature increased or decreased around 1.5°C from the normal skin temperature in 5 s. After the pre-stimulation, the skin temperature changed by around 2.0°C and 4.0°C in 4 s when the stimuli were presented from all the Peltier devices and from two of them, respectively. These temperature changes were measured in the pilot experiment conducted by the authors (finger temperature was 33°C) using the thermistor.

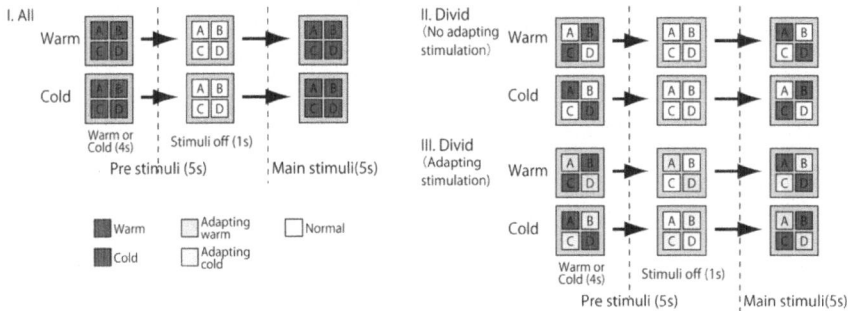

Fig. 5. Pattern of thermal stimulus in experiment 2

3.2   Results

Experiment 1

The graphs in Fig. 6 show the position recognition results for warm and cold stimuli. The horizontal and vertical axes represent the spatial pattern of the stimulus and the average correct answer ratio, respectively. Each dot represents the results for one participant; we adopted the average of five experiments for each stimulus pattern. The bars and error bars represent the average and standard deviation, respectively, of all the participants. Participant K.S. is the author. The correct answer ratio was calculated as the average of the ratio at which the participants correctly recognized the existence of a stimulus. In this case, because the participants indicated only whether a stimulus existed for each Peltier device, the chance level is 0.5. For example, when the participants answered that stimuli came from Peltier devices A and C and the stimuli actually came from Peltier device A, two of the four devices, A, B, and D, were recognized correctly, so the ratio was $(1.0+1.0+0.0+1.0)/4 = 0.75$.

The correct answer ratio for all participants is not good (0.4–0.6; around chance level) regardless of the spatial pattern of the stimuli. Some participants correctly recognized the stimuli positions (0.75–9.0; above chance level) in the case of two adjacent stimuli, such as patterns AC and CD. The average recognition times were 1.99 s, 2.04 s, and 2.28 s for patterns AD, BC, and ABCD, respectively, for both warm stimuli, whereas they were 1.58 s, 1.69 s, and 1.42 s for cold stimuli. Using two-factor (spatial pattern and warm or cold) ANOVA with only one observation (adopting each participant's averaged time for five experiments), we determined that no significant difference was obtained between the various spatial patterns of stimulation [$F(2, 29) = 0.23$, $p = 0.80$], whereas significant differences appeared in the results produced by changes in the type (warm or cold) of stimuli [$F(1, 29) = 45.36$, $p < 0.01$]. All participants correctly recognized the type of stimulation for both warm and cold stimuli regardless of the stimulus pattern.

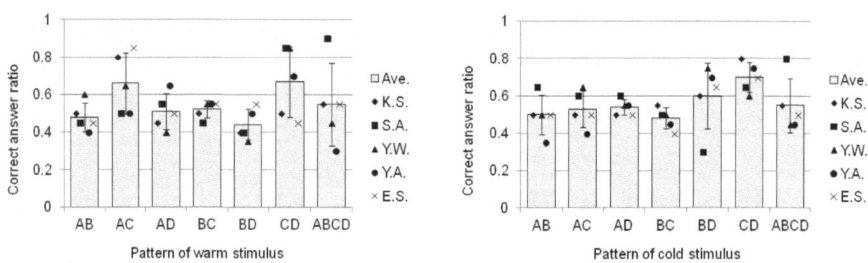

Fig. 6. Correct answer ratios in experiment 1

Experiment 2

The graphs in Figs. 7 and 8 represent the response time and perceived strength, respectively, for warm and cold stimuli. The horizontal and vertical axes represent the stimulus pattern and the average perceived time or strength, respectively. Each dot represents the average results of five experiments for one participant. The bars and error bars represent the average and standard deviation, respectively, of all the participants. Participant K.S. is the author.

The average perceived time for both warm and cold stimuli was shortest for the proposed method (warm, 1.7 s; cold, 1.7 s) and longest for the conventional method (warm, 2.9 s; cold, 2.5 s). Employing two-factor (stimuli patterns and warm or cold) ANOVA with only one observation (adopting each participant's averaged time for five experiments), we determined that significant differences in the results were produced by changes in the patterns of the stimuli [$F(2, 35) = 11.35$, $p < 0.01$]. In particular, the results of multiple comparisons (Ryan's method) demonstrate that significant differences appear between groups I and II [$t(35) = 3.72$, $p < 0.01$] and again between groups I and III [$t(35) = 5.58$, $p < 0.01$] for warm stimuli. For cold stimuli, significant differences appear between the groups I and III [$t(35) = 3.39$, $p < 0.01$].

Furthermore, the average perceived strength was highest for the proposed method for both warm and cold stimuli: 2.7 and 2.2, respectively. Employing two-factor (stimuli patterns and warm or cold) ANOVA with only one observation (adopting each participant's averaged strength for five experiments), we determined that significant differences in the results were produced by changes in the patterns of the stimuli [$F(2, 35) = 4.85$, $p < 0.05$]. In particular, the results of multiple comparisons (Ryan's method) demonstrate that significant differences appear between groups of I and II [$t(35) = 2.82$, $p < 0.05$] and again between groups of I and III [$t(35) = 4.03$, $p < 0.01$] for warm stimuli, whereas no significant differences appeared for cold stimuli.

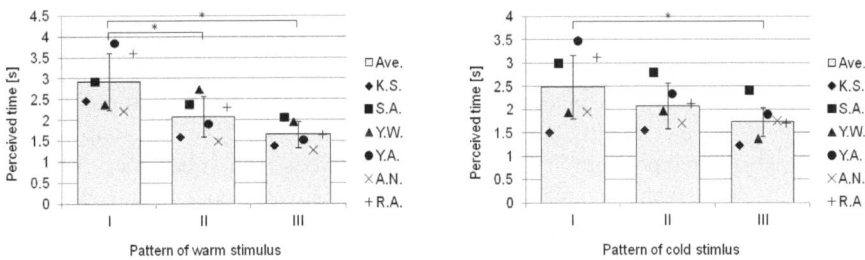

Fig. 7. Perceived time of warm (left) and cold (right) stimuli in experiment 2

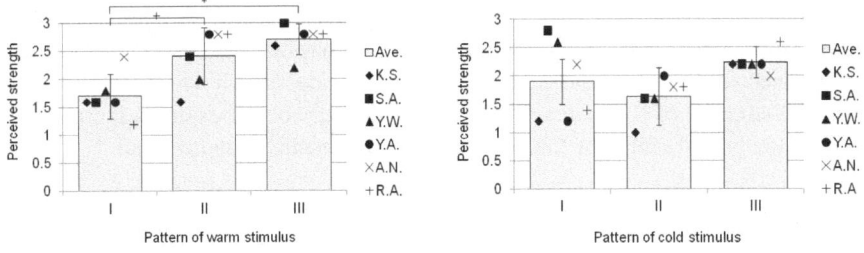

Fig. 8. Perceived strength of warm (left) and cold (right) stimuli in experiment 2

3.3   Discussion

The results of experiment 1 indicated that the participants could not correctly recognize the position of thermal stimuli on the fingertip. Some participants recognized the position correctly, but only when the stimuli were presented from adjacent Peltier devices. Therefore, we consider that users of the proposed thermal display could not recognize the position of each warm and cold stimulus because they are located alternately.

According to the results of experiment 2, the participants' perceptions of the three groups of stimuli were the most rapid and the strongest when the proposed method was used, although the strengths of the stimuli were almost same: the averaged recognition times for each stimulus were almost the same in experiment1. The result that the averaged perceived time of the cold stimulus is shorter than that of the warm stimulus corresponds to the results of conventional studies [12][13][14][15]: humans are more sensitive to cold stimuli. Although no significant differences appear between groups II and III, the averaged perceived time and strength for group III are better than those for group II. Therefore, we consider that controlling the adapting temperature is also important for the proposed thermal stimuli. Furthermore, we expect that a greater change in the adapting skin temperature would improve the perceived time and strength for group III. From these results, we concluded that the proposed thermal display, which consists of spatially divided warm and cold stimuli and controls the adapting temperature, can present a sudden temperature change at the fingertip better than a conventional display. We plan to expand the proposed method beyond the fingertip to the palm or other parts of the body.

Unfortunately, the proposed method did not work effectively for some participants, especially when a cold stimulus was applied. Since humans are more sensitive to cold stimuli, it is a possible that the difference between stimuli patterns became small for a cold stimulus. Furthermore, we attribute this result to differences in the initial skin temperature: if the initial temperature is high, it becomes difficult to decrease the temperature enough to make the skin sensitive to cold stimuli. We expect the proposed method to become more effective when the thermal display measures the skin temperature in real time and controls the adapting temperature more precisely.

Interestingly, the participants perceived the stimuli more strongly with the adapting temperature than without it. Of course, the temperature of the skin that is in contact with the Peltier devices for warm or cold stimuli was the warmest or coldest, respectively, when the stimuli pattern was in group III. However, according to conventional studies of spatial summation, the adapting stimuli appear to weaken the perceived strength of the main stimuli. In the future, we will examine the perceived thermal strength presented by the proposed thermal display in more detail.

4   Conclusion

In this paper, we proposed a thermal display that can present a sudden temperature change using spatially divided warm and cold stimuli for implementation in a glove- or grip-type haptic display. The display is based on two characteristics of human thermal perception: the spatial resolution of thermal sensation is low, and the thermal threshold depends on the adapting temperature. We experimentally confirmed the effectiveness of our thermal display. The participants perceived separate thermal stimuli as a single stimulus when the thermally stimulated area was small. The spatially distributed warm and cold stimuli enabled the participants to perceive the thermal sensation rapidly even when the cold stimulus was presented suddenly after the warm stimulus and vice versa.

This study evaluated mainly the perceived duration of a temperature change at the fingertip, but for practical use, we have to evaluate the perceived strength of the thermal stimulation. It is also desirable to evaluate the proposed display on other parts of the body, such as the palm, so it can be implemented in smart phones or game controllers.

Acknowledgement. This work was partly supported by Grant-in-Aid for JSPS Fellows and Keio University Global COE program (Symbiotic, Safe and Secure System Design).

References

1. Yoshida, M.: Dimensions of Tactual Impressions (1). Japanese Psychological Research 10(3), 123–137 (1968)
2. Shirado, H., Maeno, T.: Modeling of Human Texture Perception for Tactile Displays and Sensors. In: The First Joint Eurohaptics Conference and Symposium on Haptic Interface for Virtual Environment and Teleoperator Systems, pp. 629–630 (2005)
3. Kammermeier, P., Kron, A., Hoogen, J., Schmidt, G.: Display of Holistic Haptic Sensations by Combined Tactile and Kinesthetic Feedback. Presence, MIT Press Journals 13(1), 1–15 (2004)
4. Sato, K., Shinoda, H., Tachi, S.: Design and Implementation of Transmission System of Initial Haptic Impression. In: SICE Annual Conference 2011, pp. 616–621 (2011)
5. Yamamoto, A., Cros, B., Hasgimoto, H., Higuchi, T.: Control of Thermal Tactile Display Based on Prediction of Contact Temperature. In: Proc. IEEE Int'l Conf. Robotics and Automation (ICRA 2004), pp. 1536–1541 (2004)
6. Ho, H.-N., Jones, L.A.: Development and Evaluation of a Thermal Display for Material Identification and Discrimination. ACM Trans. Applied Perception 4, 1–24 (2007)
7. Ho, H.-N., Jones, L.A.: Thermal Model for Hand-Object Interactions. In: Proc. IEEE Symp. Haptic Interfaces for Virtual Environment and Teleoperator Systems (HAPTICS 2003), pp. 461–467 (2003)
8. Akiyama, S., Sato, K., Makino, Y., Maeno, T.: Presentation of Thermal Sensation through Preliminary Adjustment of Adapting Skin Temperature. In: Proc. of Haptics Symposium, pp. 355–358 (2012)
9. Yang, G., Kwon, D., Jones, L.A.: Spatial acuity and summation on the hand: The role of thermal cues in material discrimination. The Psychonomic Society, Inc., Attention, Perception, & Psychophysics 71(1), 156–163 (2009)
10. Vendrik, A.J.H., Eijkman, E.G.: Psychophysical Properties Determined with Internal Noise. In: Kenshalo, D.R. (ed.) The Skin Senses, pp. 178–193. Charles Thomas (1968)
11. Nathan, P.W., Rice, R.C.: The Localization of Warm Stimuli. Neurology 16, 533–540 (1966)

12. Kenshalo, D.R., Homes, C.E., Wood, P.B.: Warm and Cool Thresholds as a Function of Rate of Stimulus Temperature Change. Perception & Psychophysics 3, 81–84 (1968)
13. Kenshalo, D.R.: Correlations of Temperature Sensitivity in Man and Monkey, A First Approximation. In: Zotterman, Y. (ed.) Sensory Functions of the Skin with Special Reference to Man, pp. 305–330. Pergamon Press (1976)
14. Stevens, J.C., Choo, K.C.: Temperature Sensitivity of the Body Surface over the Life Span. Somatosensory and Motor Research 15, 13–28 (1998)
15. Greenspan, J.D., Kenshalo, D.R.: The Primate as a Model for the Human Temperature-Sensing System: 2. Area of Skin Receiving Thermal Stimulation. Somatosensory Research 2, 315–324 (1985)
16. Stevens, J.C.: Thermal Sensibility. In: Heller, M.A., Schiff, W. (eds.) The Psychology of Touch, pp. 61–90. Lawrence Erlbaum (1991)

Perceptually Robust Traffic Control in Distributed Haptic Virtual Environments

Clemens Schuwerk, Rahul Chaudhari, and Eckehard Steinbach

Technische Universität München
Institute for Media Technology
{clemens.schuwerk,rahul.chaudhari,eckehard.steinbach}@tum.de

Abstract. In this paper we present a traffic control scheme for server to client communication in distributed haptic virtual environments (VE). We adopt a client-server architecture where the server manages the state consistency of the distributed VE, while haptic feedback is computed locally at each client. The update rate of network traffic from the server to the client is dynamically adapted by exploiting characteristics and limitations of human haptic perception. With this, an excellent trade-off between network communication efficiency and perceptually robust rendering of haptic feedback is achieved. Subjective tests with two users collaboratively manipulating a common object show a packet rate reduction of up to 99% from the server to the clients without deteriorating haptic feedback quality.

Keywords: Weber's law, perceptual coding, distributed haptics.

1 Introduction

Incorporating the haptic modality into human-machine or human-human interaction mediated by technical systems is a growing research field. It has been shown repeatedly that haptic feedback improves both task performance and the sense of immersion into virtual or remote environments [1]. In shared applications, where several users can collaboratively work together, the addition of the haptic modality to the audio-visual modalities also improves the sense of togetherness and leads to better task performance [2]. Existing communication infrastructures like the Internet make haptic exploration of distant real or virtual environments feasible.

Shared haptic virtual environments (SHVEs) can be used for virtual prototyping, assembly simulations, teaching/training applications as well as networked multiplayer games with haptic feedback. All these applications struggle with strong demands on the communication network due to haptic transparency and stability reasons, like standard telepresence and teleaction (TPTA) systems do [3]. The high temporal resolution of the human haptic perception system requires a high stimulus refresh rate of $1kHz$. Thus 1000 packets/second are transmitted between the server and clients in both directions. To address these issues in TPTA systems, perceptual haptic data reduction schemes have been

P. Isokoski and J. Springare (Eds.): EuroHaptics 2012, Part I, LNCS 7282, pp. 469–480, 2012.
© Springer-Verlag Berlin Heidelberg 2012

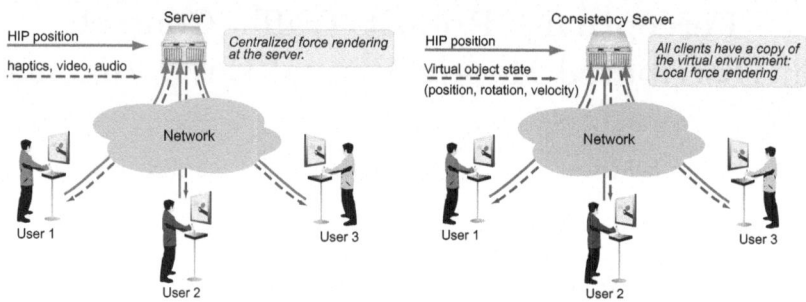

Fig. 1. Architectures for SHVEs: standard force architecture with centralized force rendering (left-hand side), distributed haptic virtual environment with consistency server (right-hand side) (note: haptic interface point is represented as HIP)

introduced [3]. They successfully achieve up to 95% reduction in packet rates, while keeping the distortion introduced into the haptic feedback below human perception thresholds.

A number of network architectures can be conceived for building SHVEs. The goal of such architectures is to support stable and perceptually transparent visual-haptic collaboration between a large number of users, while consuming as less network resources as possible. The most straightforward one is the client-server architecture, where each user sends the current position of his haptic device to the server, as depicted in Figure 1 left-hand side. Based upon these incoming position data, reaction forces are calculated at the server individually for each client and sent back. The haptic feedback is accompanied by video and audio feedback to facilitate a convincing presentation of the shared scene at the client side. Thus, this architecture extends the common TPTA scenario to a multiuser application. The primary drawback of this architecture is that the server becomes a bottleneck as the number of users logging in to the shared virtual environment increases. Moreover, the communication network closes a global control loop between the humans and the remote virtual entity being controlled. The stability of this control loop necessitates low-delay communication of haptic data packets, when operating over packet-switched networks like the Internet. In the remainder, we name this architecture as the *Standard Force Architecture (SFA)*.

A useful modification of this approach is to have local copies of the entire virtual environment (VE) at each client, so that forces can be rendered locally based on the current VE state. Such an architecture constitutes a so-called *Distributed Haptic Virtual Environment* (DHVE). Here, the state of the VE needs to be kept consistent across all clients in the system for smooth and natural client-to-client interactions. This task is handled by a consistency server (see Figure 1 right-hand side), where a centralized physics engine updates the movement of objects. Now, instead of sending *forces* from the server to the client, the *state* of the virtual objects in the VE is sent. The advantage of this approach, compared to the SFA, is that the haptic control loop is now closed locally and force feedback can be calculated directly at the client side. This benefit comes at the cost of increased effort for maintaining a consistent state of the VE across all clients. In [4] Lee and Kim show that such an architecture is stable, regardless of the delay in the communication channel, but not transparent. Transparency depends on the round trip delay and physical properties of the virtual environment, like object mass, stiffness and friction coefficients.

Adopting such a distributed architecture for our work, we will study the reduction of the packet rate from the consistency server to the client while maintaining haptic transparency for the user. A reduced packet update rate from the server to the clients leads, however, to deviations of the state of the VE among the clients. These deviations may lead to perceivable artifacts in the force feedback displayed to the user, when the VE state is updated suddenly. In order to ascertain that the distortion introduced in the haptic feedback remains below the human haptic perceptual thresholds, we propose a method to adaptively downsample the update rate based on the limitations of human haptic perception. We neglect communication delay in this work, as delay leads to unavoidable inconsistencies in the VE states across the network and this is the topic of other related work, which will be reviewed in the next section. We also adopt the above explained *Standard Force Architecture* as a reference, to which the DHVE architecture will be compared.

The remainder of this paper is structured as follows. In Section 2 we review related work in the area of distributed haptic environments. In Section 3 we shortly explain Weber's Law and its use for perceptual packet rate reduction in teleoperation systems. In the following Section 4 we explain our proposed packet rate reduction scheme. Performance evaluation of the proposed architecture and corresponding subjective tests are described in detail in Section 5 and the results are presented in Section 6. Finally, in Section 7 we summarize our results and conclude our work with a discussion of the limitations and how they can be addressed in future work.

2   Related Work

One of the first experiments exploring the role of touch in shared virtual environments was conducted by Basdogan et al. in 1998. Their subjective experiment involved two humans collaborating in different applications, e.g., in the ring on a wire game. Performance was measured objectively and the user was asked several questions after each experiment in order to measure their subjective sense of togetherness. The presented results in [2] and [5] clearly show that haptic feedback, in extent to visual feedback, improves task performance and increases the sense of togetherness between the collaborators.

These results motivated more research on haptic feedback in shared and distributed virtual environments. Hikichi et al. presented a new distributed client-server architecture for shared virtual environments in 2001 [6], where force rendering was moved from the server to the client. Here, the server-to-client packet rate is controlled according to the network state to prevent congestion.

Thus, the packet rate might drop below the usual haptic update rate. At the clients, prediction and interpolation techniques are used to supplement missing VE state update information. Additionally, *Differential Pulse Code Modulation* (DPCM) is used to decrease the bitrate on the communication channel. The evaluation of the system showed improved performance and higher subjective ratings in presence of network delay compared to the traditional client-server architecture, where position information is transmitted on the forward and force samples on the reverse channel. This gain can be explained with the shift of the haptic rendering from the server to the client and the opening of the formerly closed global control loop between the user and the remote VE. Within such a framework, delay leads to inconsistency in the VE state instead of destabilizing the overall system. Contrary to this approach, we are adapting the transmission rate of the VE state updates on the reverse channel according to the current activity of the human and its perception of force.

The forward channel is studied by Ishibashi et al. in [7] where they control the packet transmission rate from the clients to the server according to the network load, by simply adapting the rate to the current average network delay. Please note that we leave the packet rate on the forward channel untouched and are only studying the return path from the server to the clients in this paper. Of course, we plan to take both channels into account in future work.

Another network adaptive transport scheme for DHVEs is presented by Lee et al. in [8]. Each haptic media unit (MU) is categorized into three different priorities, whereby the lowest priority MU can be skipped and the highest priority one has to be transmitted reliably, by sending redundant packets. At the server and the client, a dead reckoning scheme is used for predicting the movement of objects. Based on the quality of the object position prediction, the sender sets the priority of each MU. Additionally, haptic MUs are aggregated into single packets according to the current network state and the priority level. Every client has its own physics engine which updates the object's position locally. In contrast, in our case, only the server runs a physics engine, assuring global state consistency between the VEs. The system shows improved performance and reduced transmission rate over networks with transmission delay and jitter. The prioritization of packets in [8] is based only on position, whereas our introduced scheme takes resulting forces into account. As the user will feel force changes rather than position changes, we claim that our approach is more general, as position changes as well as physical properties of the virtual environment are subsumed into the force computation and rendering process.

3   Haptic Data Reduction

Transmitting haptic signals over packet-switched networks is characterized by strict delay constraints as well as high packet rates. Previous work has shown that these challenges can be successfully addressed by the so-called perceptual deadband approach [3]. It allows us to significantly reduce the amount of haptic data as well as the network packet rate without impairing the system's quality

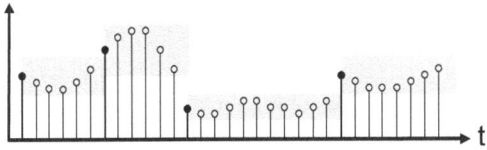

Fig. 2. Principle of perceptual deadband-based data reduction (adopted from [3])

as shown by human user studies [3]. These data reduction algorithms are based on Weber's law of Just Noticeable Differences (JNDs):

$$\frac{\Delta I}{I} = k \quad \text{or} \quad \Delta I = k \cdot I, \tag{1}$$

where I is the initial stimulus and ΔI is the so-called Difference Threshold (or the JND). It describes the smallest amount of change of stimulus I which can be detected as often as it cannot be. The constant k (called the deadband parameter k from now on) describes the linear relationship between ΔI and the initial stimulus I.

Based on Weber's Law, perceptually insignificant changes in the force feedback data stream are deemed undetectable and corresponding haptic samples can be dropped. Thus, only if the difference between the most recently sent sample and the current value exceeds human perception thresholds (grey areas in Figure 2), defined by the deadband parameter k, a new signal update is triggered (black dots in Figure 2). This new sample also redefines the deadband width. At the receiver, a simple zero-order hold strategy or a linear extrapolator allows for upsampling the irregularly received signal updates to a constant and high sampling rate required for the local control loops.

4   Traffic Control for Server-to-Client Updates

In this section, we propose a perceptually motivated traffic control scheme for the server-to-client communication in a DHVE. Each client sends the current position of his haptic device (also called haptic interface point (HIP)) to the server. As we only focus on the return channel, we keep the rate on the forward channel at 1 kHz which corresponds to the rate of the local haptic loops at the clients. The server receives the current position of all clients and calculates the resultant updates of the virtual environment based on its physics engine. This update information is then sent back to the clients whenever it is deemed relevant for the client's perception of the VE. More specifically, position, rotation and velocity of each object, as well as the current HIP position of every client is transmitted to all clients. Every client in turn updates its own local copy of the virtual environment. In between the updates, the clients extrapolate the objects' movements linearly by using the received velocity. This also means that an object at the client is not movable if its current velocity is zero and the client does not receive any updates from the server, although the user feels a resistive force when pushing against the object due to local force rendering.

Nevertheless, the object's position and rotation might change drastically due to a received status update. This change can be felt as an abrupt force change. The magnitude of this change is influenced by several factors. Firstly, the most important factor is the discrepancy between the current client's object state and the newly received state update. This difference is influenced by the time between two update packets and the object's velocity. In [8], this position change at the client side is the basis for deciding if a packet should be sent or not. However, besides position/rotation change, other factors such as stiffness of the object, penetration depth into the object, friction, etc. also influence force computation and hence force perception. Hence, we base our decision - whether to transmit a packet or not - on the perceptibility of the resultant force change:

$$if: \quad |f_{estimated} - f_{actual}| \geq k \cdot |f_{actual}|, \quad then: send\ update \qquad (2)$$

The perceptibility of the force change is determined based on the Weber's law and the deadband approach explained in Section 3. This is done at the server by calculating the current virtual object status at the clients by performing the same extrapolation mechanism as each client does, based on the last sent update packet. Taking the current HIP position of the client, the current globally valid object's state at the server, the estimated object state at the client and mechanical properties of the VE into account, the force that the user is currently feeling is estimated ($f_{estimated}$). The actual force the user should feel (f_{actual}) and $f_{estimated}$ are compared using the perceptual deadband approach and a new VE update is triggered when the inequality in (2) holds. Thus, the packet rate from the server to the client is adapted to the user's action, as well as to all parameters that influence force rendering and the human force perception. This distinguishes our approach from the ones adopting uniform downsampling or rate adaptation based solely on position differences.

5   Psychophysical Evaluation

To evaluate the performance of the data reduction scheme described in Section 4, psychophysical experiments were conducted.

5.1   Task Description

We design our evaluation methodology around a *pursuit tracking* task, wherein two users collaboratively push/pull a virtual cube tracking the movements of a reference cube as closely as possible. The mass of the cube is set such that a single user cannot perform the task alone, and collaboration is necessary. This choice of task is motivated by the bidirectional nature of haptics, which may lead to widely varying haptic interactions across subjects, and thus to widely varying subjective experiences and average packet rates. With two humans involved, this issue is

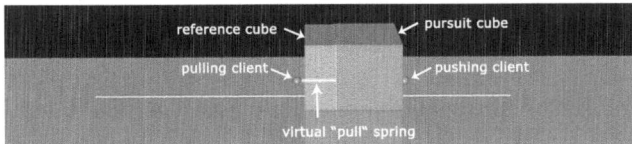

Fig. 3. The dynamic virtual environment used for the pursuit tracking task

compounded further. The specified reference signal, namely the position of the reference cube as a function of time, helps to reduce the variability between different experiment runs and establish a common basis for evaluation across subjects [9]. Additionally, one of the two humans involved in the task is the same throughout the whole experiment.

The pursuit tracking task consists of a 1-DoF dynamic virtual environment (refer to Figure 3). In this VE, the "reference" cube is to be pursued as closely as possible with the "pursuit" cube. The reference cube serves only as a visual reference, and cannot be haptically touched. The subjects can, on the other hand, haptically interact with the pursuit cube and push or pull it. The pulling user is attached to the pursuit cube with a virtual "pull" spring. Contact between the user and the cube is lost if the spring force exceeds 2 Newtons. This is done to ensure that a single user alone is not capable of executing the task successfully. The reference cube moves from right to left and back to the start position twice during a single run and its velocity is controlled to have a sinusoidal profile.

5.2   Subjects

In total, 14 subjects participated in the evaluation, 1 of them female and 13 male. The participants' age was in the range of 25-43 with a mean of 29 years. Two of them were left-handed, the others right-handed. Four of them were researchers working in the area of haptic data compression and daily users of desktop haptic devices. The remaining subjects have frequently participated in subjective tests for evaluation of haptic compression schemes in the past and were therefore familiar with haptic devices. Nevertheless, before the actual experiment, they were trained to ensure they were at par with the more experienced participants for the given task.

5.3   Experimental Setup

The experiment consists of two sub-experiments:

1. Sub-experiment 1: Standard Force Architecture as explained in Section 1. The physics engine and force rendering is centralized, force values as well as a video representation of the VE are sent to the clients. To reduce the haptic packet transmission on the return channel, deadband coding in combination with a first-order linear predictor [3] is applied for each client.

2. Sub-experiment 2: Distributed haptic virtual environments as explained in Section 1. On the reverse channel from the server to the clients, the traffic control scheme introduced in Section 4 is applied and the virtual objects' state is transmitted instead of force values whenever the criterion in Eq. (2) is fulfilled. Thus, forces are rendered locally but the physics engine remains centralized to ensure a globally consistent VE state.

We consider only the haptic transmission rates in these experiments, neglecting the rate needed for transmitting the video stream in sub-experiment 1. In both sub-experiments there is no packet rate reduction scheme applied on the forward channel, thus current position signals are transmitted to the server at the rate of $1\ kHz$.

Each sub-experiment had 10 runs, within which the deadband parameter k is varied between 0% and 50% to find a setting where the packet rate is maximally reduced while maintaining satisfactory force feedback to the user. The order of sub-experiments changed randomly from one subject to another, along with the order of the parameter values within each sub-experiment. This was done to prevent prediction of trends by the subjects and to avoid any systematic manifestation of fatigue in the results of the evaluation. The seating posture and hand-device configuration was standardized across subjects. In order to prevent distractions due to noise emanating from the haptic device, subjects wore headphones playing music. In addition, distractions due to visual observation of the haptic device were avoided by blocking the view to the haptic device with a cardboard. The PHANTOM Omni haptic device (Sensable Technologies Corp., Woburn, MA) was used for the experiments.

5.4   Experiment

To eliminate individual differences related to experience with the experimental setup, or differences in understanding the instructions, a training session was conducted at the beginning of the experiment. Herein, the task to be performed was explained to the subjects and they had the chance to get familiar with the task execution, the experimental setting and the parameter variations. We used a within-subject experimental design, with all participants executing the same haptic task under the same parameter settings. The range of the deadband parameter k for both sub-experiments was determined in pilot tests.

At the beginning of every sub-experiment, the psychological scale of the subjects was calibrated according to Table 1. Two reference stimuli for the specific sub-experiment, corresponding to the ratings of 100 and 0 were displayed to the user. The stimulus corresponding to the rating of 100 was the best (most "natural" haptic interaction, with no distortion artifacts due to data reduction) that the system could offer (no packet rate reduction). The one corresponding to the rating of 0 was the worst (deviating most from the "natural" haptic interaction, high downsampling of update packets), while still allowing successful task completion. After the completion of every run, subjects were asked to rate the haptic perception in comparison to the reference stimuli on the scale shown in Table 1. Note that intermediate ratings between the ones indicated where also allowed.

Table 1. Used rating scheme

Description	Rating
no difference	100
perceptible difference, but not disturbing	75
slightly disturbing difference	50
disturbing difference	25
strongly disturbing difference	0

6   Results

The results of the subjective evaluation in both sub-experiments for the whole deadband parameter range and the resulting mean packet rate from the server to the clients are shown in Figure 4. According to the rating scale in Table 1, the threshold denoting the quality degradation that is not disturbing is illustrated with the horizontal line at a rating of 75. For the *SFA* with deadband-based data reduction (sub-experiment 1), we can find the corresponding deadband parameter k to be around 8%, leading to a mean overall packet rate of 70 packets per second from the server to the two clients (left-hand side of Fig. 4). For the proposed approach (sub-experiment 2), these numbers change to $k = 14\%$ and 10 packets per second (right-hand side of Fig. 4). This corresponds to a packet rate reduction by a factor of 14 for sub-experiment 1 and 100 for sub-experiment 2, respectively, when compared to the packet rate ($1\ kHz$) for the case of no compression. The distributed architecture has the advantage that a single globally valid state of the VE is transmitted to all clients, rather than individual force values for every client, calculated centrally at the server. Thus, multicasting can be used for updating the connected clients' VE states, reducing load at the server's network interface. On the contrary, in the Standard Force

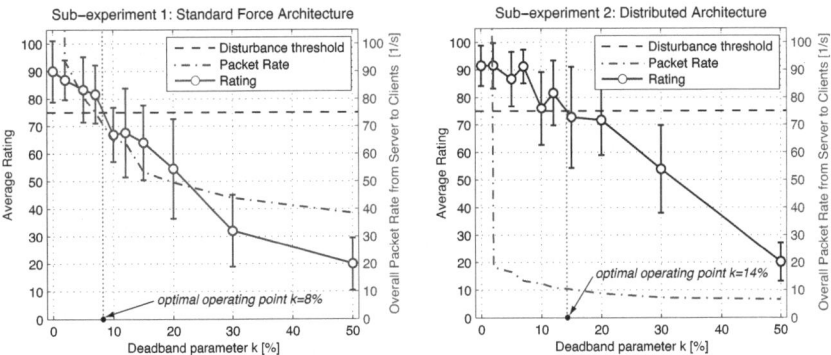

Fig. 4. Average ratings as a function of the deadband parameter k, their standard deviations plotted as error bars and the corresponding packet rates.

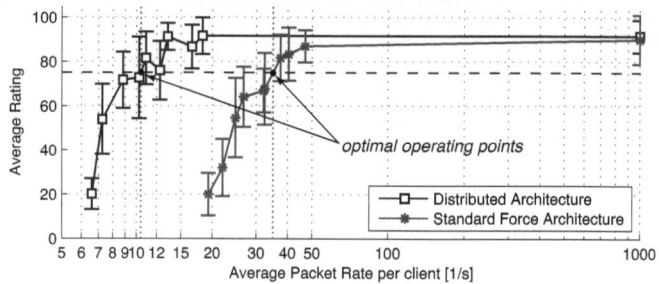

Fig. 5. Resulting packet rates per client and user ratings in the two sub-experiments

Architecture an individual packet is sent to each client, leading to an approximately linearly growing packet rate with the number of connected clients.

Statistical analysis was conducted to investigate the influence of the communication architecture (sub-experiment 1 or sub-experiment 2) on the user's ratings. Due to the small sample size, a normal distribution could not be assumed [10]. Thus, Friedman's Test was used, which shows a statistically significant difference between the ratings for the two architectures ($\chi^2(1) = 9.85, p = .0017$).

In Figure 5, the resulting *average packet rate to a single client* for the different deadband parameters k (0 to 50%) and the corresponding subjective ratings are shown on a semi-logarithmic scale. The desired operating point should lie as far up (high subjective rating) and as much towards the left as possible (low average packet rate). Please note that, contrary to Figure 4, the average packet rate per client is shown. In this case the comparison of the two approaches is fairer, as the packet rate will not increase with the number of connected clients in the distributed architecture due to multicasting. The onset of disturbing degradation is felt at an average packet rate of around 10 packets per second for the distributed architecture, while it was felt at an average rate of 35 packets for the standard approach (half of the overall packet rate in Figure 4). This gain can be explained as follows. When a user gets in contact with a rigid object, high frequently transient forces lead to a large number of triggered force updates in the Standard Force Architecture [11]. On the other hand, since the object's position/velocity is not changing significantly at first due to small penetration depth during the contact event, thus no update packets have to be sent in the distributed architecture.

A further illustration of the adaptation to mechanical properties of the virtual environment and the users' action is shown in Figure 6. Rigid objects in virtual environments are commonly modeled with high stiffness values to ensure realistic force feedback to humans, when touching the object [12]. With lower stiffness values the user can penetrate deeper into the object, leading to the impression of a soft rather than a rigid object. But for example when modeling a virtual spring, as we do in our experiment, lower stiffness values are an explicitly desired specification. The stiffness property is an important factor when rendering forces and thus also heavily influences the user's perception. In Figure 6, the packet

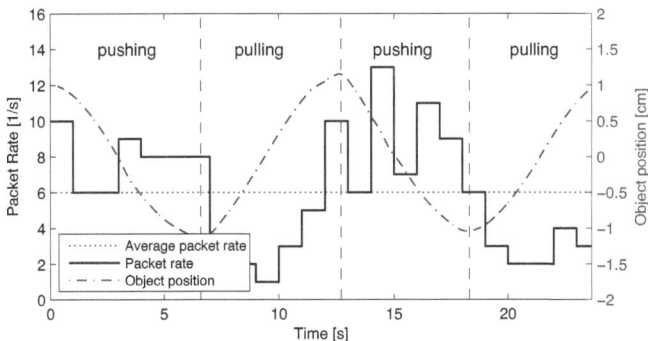

Fig. 6. Packet rates during pushing and pulling within one single run. Different stiffness values are implemented for the object stiffness (pushing) and the virtual "pull" spring (pulling) and the packet rate is adapted accordingly.

rate during a single run with only one user involved for a deadband parameter $k = 15\%$ in the distributed architecture is shown. It can be clearly seen how the packet rate is adapted in each of the four phases to the virtual environment's specifications and the current user's action. While pulling the cube, the packet rate is smaller as a virtual spring with a lower stiffness value, compared to the object's stiffness, is modeled and thus larger position errors of the object's state at the client are allowed by the server's traffic control scheme, because they lead to smaller force changes.

7   Conclusions and Future Work

This work introduces a perceptually robust traffic control scheme for server to client communication in distributed haptic virtual environments. The proposed approach shows good packet rate reduction performance. For the collaborative push/pull task, an average of only around 10 packets per second have to be sent to the clients to achieve satisfactory haptic feedback. In comparison, for a standard client server architecture with centralized force rendering, an average of 35 packets per second per client are needed. Packet rate control is based on end-user force perception, which subsumes changes in all factors that affect force computation. This guarantees that perceptual requirements (reflected by subjective ratings) will never be violated, unlike state-of-the-art approaches.

In the future, we plan to investigate the influence of communication delay on the transparency of the overall system as well as the integration of packet rate reduction schemes on the forward channel to the server. They are important, as in both architectures investigated in this paper, the server becomes a bottleneck as more clients simultaneously collaborate in the virtual environment and everyone sends his current HIP position at a very high packet rate. Furthermore, we will conceive a more general multiple DoF task, where multiple users can freely interact with each other.

Acknowledgement. This work has been supported, in part, by the European Research Council under the European Union's Seventh Framework Programme (FP7/2007-2013) / ERC Grant agreement no. 258941 and, in part, by the German Research Foundation (DFG) under the project STE 1093/4-1.

References

1. Dennerlein, J.T., Martin, D.B., Hasser, C.: Force-feedback improves performance for steering and combined steering-targeting tasks. In: Proceedings of the SIGCHI Conference on Human Factors in Computing Systems, pp. 423–429. ACM, New York (2000)
2. Basdogan, C., Ho, C.H., Slater, M., Srinivasan, M.: The role of haptic communication in shared virtual environments. In: Proceedings of the Third PHANTOM Users Group Workshop, Artificial Intelligence Laboratory and the Research Laboratory of Electronics at the Massachusetts Institute of Technology, pp. 85–89 (1998)
3. Hinterseer, P., Hirche, S., Chaudhuri, S., Steinbach, E., Buss, M.: Perception-based data reduction and transmission of haptic data in telepresence and teleaction systems. IEEE Trans. on Signal Processing 56(2), 588–597 (2008)
4. Lee, S., Kim, J.: Dynamic network adaptation scheme employing haptic event priority for collaborative virtual environments. In: Proceedings of the First International Conference on Immersive Telecommunications, ImmersCom 2007, pp. 12:1–12:6 (2007)
5. Basdogan, C., Ho, C.H., Srinivasan, M.A., Slater, M.: An experimental study on the role of touch in shared virtual environments. ACM Transactions on Computer-Human Interaction (TOCHI) - Special Issue on Human-Computer Interaction and Collaborative Virtual Environments 7, 443–460 (2000)
6. Hikichi, K., Morino, H., Fukuda, I., Matsumoto, S., Yasuda, Y., Arimoto, I., Iijima, M., Sezaki, K.: Architecture of haptics communication system for adaptation to network environments. In: IEEE International Conference on Multimedia and Expo, ICME 2001, pp. 563–566 (August 2001)
7. Ishibashi, Y., Kanbara, T., Hasegawa, T., Tasaka, S.: Traffic control of haptic media in networked virtual environments. In: Proceedings of IEEE Workshop on Knowledge Media Networking, pp. 11–16 (2002)
8. Lee, S., Moon, S., Kim, J.: A network-adaptive transport scheme for haptic-based collaborative virtual environments. In: Proceedings of 5th ACM SIGCOMM Workshop on Network and System Support for Games, NetGames 2006 (2006)
9. Chaudhari, R., Steinbach, E., Hirche, S.: Towards an objective quality evaluation framework for haptic data reduction. In: Proceedings of the IEEE World Haptics Conference, Istanbul, Turkey (June 2011)
10. Field, A.: Discovering Statistics Using SPSS, 3rd edn. Sage Publications Ltd. (April 2009)
11. Kuchenbecker, K., Fiene, J., Niemeyer, G.: Improving contact realism through event-based haptic feedback. IEEE Transactions on Visualization and Computer Graphics 12(2), 219–230 (2006)
12. Basdogan, C., Srinivasan, M.A.: Haptic Rendering in Virtual Environments. In: Handbook of Virtual Environments: Design, implementation, and applications. Lawrence Erlbaum Associates Publishers (2002)

An Ungrounded Pulling Force Feedback Device Using Periodical Vibration-Impact

Takuya Shima[1] and Kenjiro Takemura[2]

[1] Graduate School of Science and Technology, Keio University, Yokohama, Japan
ta.ku.ya@z8.keio.jp
[2] Department of Mechanical Engineering, Keio University, Yokohama, Japan
takemura@mech.keio.ac.jp

Abstract. This paper presents a novel ungrounded force feedback device which is capable of making the user confuse as if a pulling force is displayed by utilizing non-linear characteristics of human perception. The force feedback device is mainly composed of a spring-mass system and a voice coil motor. By hitting a weight against a wall periodically, the device generates a vibration impact to the user. The device is 140 x 41.5 x 41.5 mm^3 with total mass of 230 g. A sensitivity of the pulling force perception are evaluated and presented in this paper.

Keywords: Ungrounded feedback device, Pulling force, Impulsive force, Spring-mass system, Sympathetic vibration.

1 Introduction

A force feedback device, which is a next generation human-machine interface capable of presenting a reaction force between human and virtual object, is expected to be applied to a variety of systems. The typical applications for the force feedback device are tele-operation and navigation systems. In case of tele-operation, by using a force feedback device, kinesthetic sensations can be displayed to an operator. Presenting deep sensory information together with audio-visual information enables an operator to get a natural feeling as if he/she is actually working on site. Particularly, in extreme conditions such as space, deep ocean level, nuclear power plant etc., tele-operation is rather necessary. For this reason, practical force feedback devices are required in order to improve a work efficiency.

In a navigation system on the other hand, a force feedback device capable of displaying deep sensory information enables a user to intuitively identify the directions to go. As smart phones get huge interests around the world, many people become possible to find the way he/she should take to get a destination even in foreign countries. This is a kind of personal navigation system. Traditionally, a navigation system provides users with guidance information through auditory and visual information. However, a navigation using only these two sensory channels poses two problems. First, since users have to focus their hearing and visual attention to the navigation, they usually have to stop walking in

P. Isokoski and J. Springare (Eds.): EuroHaptics 2012, Part I, LNCS 7282, pp. 481–492, 2012.

order to follow the guidance provided. Secondly, using these two information can lead a misinterpretation of information because the system is not intuitive. In contrast, navigation using deep sensory information can provide a safe and intuitive guidance, since deep sensory information is displayed directly to the body without distracting other sensory channels. Therefore, users can focus their sight and hearing to their surroundings. In particular, if the deep sensory information is displayed as a pulling force, it is possible to offer more natural navigation to a user, because the force reminds the user of the direction to destination.

By using force feedback devices in the applications mentioned above, new value can be added to the traditional interface. However, most of force feedback displays such as PHANToM [1] and SPIDER [2] are all fixed to table/floor, i.e., grounded. Therefore, their application areas are limited, because they are bounded to their operating space ranges. For this reason, it has been impossible to apply the grounded force feedback devices to, for example, navigation system. Although some ungrounded force feedback devices which utilize the gyro effect [3] and the variation of angular momentum [4] have been proposed for mobile use, these devices can display only a short-time torque. Hence, development of an ungrounded force feedback device which can continuously display pulling force is required in order to expand the application area of force feedback devices.

Ungrounded devices, which are able to display the pulling force by utilizing non-linear characteristics of human perception, have been developed [5-6]. One is a device which uses a slider-crank mechanism to create asymmetric acceleration of a weight [5]. The other is a device which uses a rotating link mechanism to generate periodical impulsive force by hitting a weight against a wall [6]. However, these devices generate unnecessary fluctuations of displayed force in the direction perpendicular to the direction of pulling force displayed, and they produce an adverse effect on perception of the force. In addition, it is difficult to reduce the size and weight of the devices, because they use two symmetric devices in parallel in order to compensate the fluctuations.

Hence, in this study, we propose a novel ungrounded pulling force feedback device using periodical vibration-impact. The device may not generate any fluctuations of displayed force, which are observed in the previously developed devices.

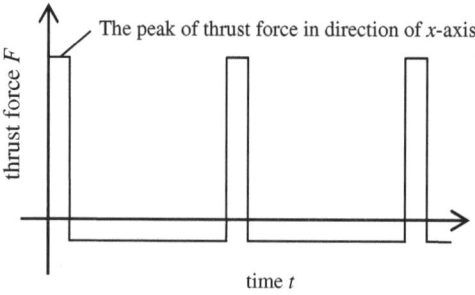

Fig. 1. Conceptual diagram of thrust force wave

Furthermore, the sensitivity of the pulling force perception with the developed device is evaluated.

2   Proposed Method

It is possible to make humans confuse as if the pulling force is displayed by utilizing non-linear characteristics of human perception [7-9]. Namely, as shown in Fig. 1, if periodically steep inertial forces are displayed to a user in a single direction, it is possible for the user to feel like being pulled in the direction. In order to generate such a periodical force, we propose a pulling force feedback device as shown in Fig. 2. The force feedback device is mainly composed of a voice coil motor (VCM) and a spring-mass system. Natural vibration of the spring-mass system is generated by the VCM. Then the weight is hit against a wall with equal intervals in order to display a precipitous thrust force to the user. The reason why we use the natural vibration is to efficiently obtain motional energy.

Since the motion of the system is in a single dimension, the device may not generate any fluctuations in other direction. Therefore, it is expected that this device is able to display more accurate pulling force than the pulling force feedback devices previously developed [5-6]. In addition, it may be possible to make the device smaller and lighter than the previous devices because the proposed device has much simpler configuration than the others.

3   Development of Pulling Force Feedback Device

3.1   Spring-Mass System

According to the data reported by Amemiya et. al., the human perception sensitivity against the periodical pulling force becomes highest when frequency of the force is 10 Hz [7]. Therefore, we decide to make the spring-mass vibrate with 10 Hz using the VCM (Akrbis, AVM20-10-C6.3) which generates 2 N per ampere. The proposed spring-mass system consists of a weight, a spring, a linear guide rail, and the VCM.

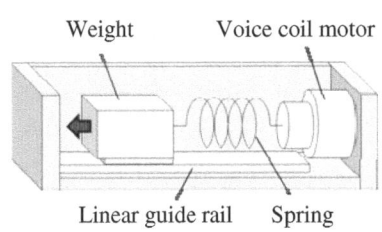

Fig. 2. Conceptual diagram of proposed device

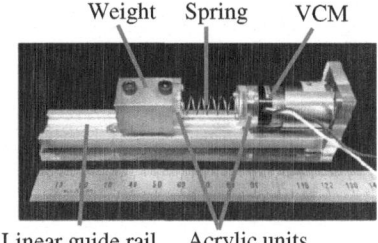

Fig. 3. Spring-mass system

Fig. 3 shows the developed spring-mass system. The weight is made of brass, because the density is relatively large compared with other metal material. Moreover, it has a square section in order to have a larger surface area compared with a circular column, resulting in making the net force larger. Considering the input current to the VCM and the total weight of the device, the volume is determined to be 22 x 16 x 16 mm³. Then the mass of the weight is 49.2 g. In order to limit the motion of the weight in a single direction, it is set on the linear guide rail (DryLin, NS-01-17) which is composed of a rail with 15 g and a carriage with 1.7 g. In addition, there is an acrylic unit which is in contact with the weight in order to connect the weight and the spring. The carriage, the acrylic unit and the weight move together, and their mass m is 51.5 g in total.

With regard to the weight, we determine the spring constant k of the spring to generate natural vibration of the spring-mass system with 10 Hz. Fig. 4 shows the model of the system. As shown in the figure, the system has two-degree-of-freedom of motion, because the movable part of VCM is not mechanically fixed. Therefore, the motion of the system is represented by the following equation of motion.

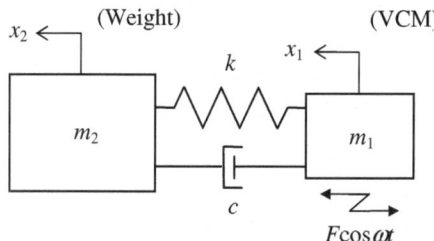

Fig. 4. Model of the system

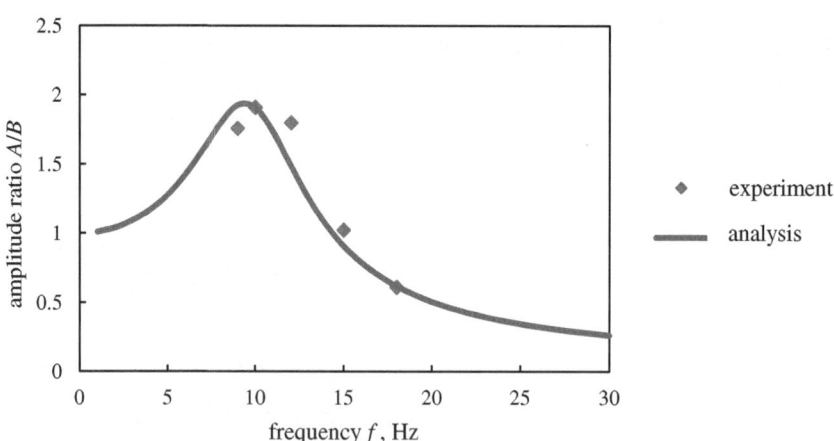

Fig. 5. Comparison of collision experiment with numerical analysis (c =2.05 Ns/m)

$$\begin{cases} m_1\ddot{x}_1 = -c(\dot{x}_1 - \dot{x}_2) - k(x_1 - x_2) + F\cos\omega t \\ m_2\ddot{x}_2 = -c(\dot{x}_2 - \dot{x}_1) - k(x_2 - x_1) \end{cases} \tag{1}$$

By solving this equation, we calculate amplitude of the weight A and that of the movable part of VCM B, and optimize k in order to maximize an amplitude ratio A/B with 10 Hz force input by the VCM.

First, in order to estimate the damping constant c of the system, we produced a spring-mass system by using a spring of $k = 216$ N/m, which is determined by a simple vibration theory without damping, and measured the amplitudes of the weight and the movable part of VCM. Then, we compared the measured result with a numerical analysis result, and estimated a value of c. As shown in Fig. 5, they become close when the value of c is 2.05 Ns/m. Finally, we calculated an optimum value of k by using the estimated damping constant c in order to maximize the amplitude ratio with 10 Hz. As shown in Fig. 6, A/B reaches its peak at $k = 268$ N/m. Therefore, we decide to use the spring with spring constant k of 265 N/m.

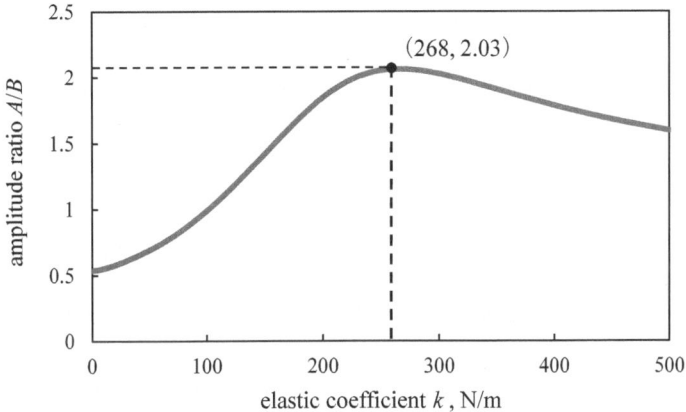

Fig. 6. Relation elastic coefficient and fraction of the amplitud（c = 2.05 Ns/m

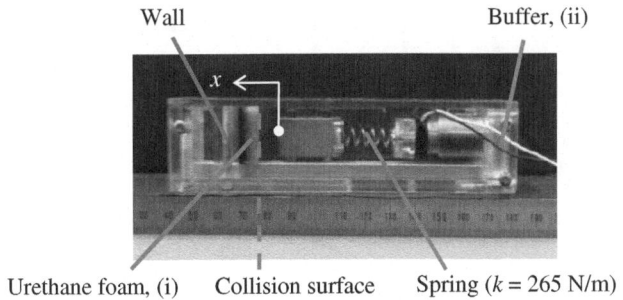

Fig. 7. Force feedback device

3.2   Pulling Force Feedback Device

Fig. 7 shows the pulling force feedback device we developed. As can be seen, the device is composed of the spring-mass system designed in Section 3.2, an acrylic outer frame, an acrylic wall, an urethane foam, and a high shock absorption urethane buffer. The direction in which the pulling force is displayed is the positive direction of the x-axis. We define the maximum thrust force generated in the positive x-direction as F_a, the maximum thrust force generated in the negative x-direction as F_b, and the thrust force rate as F_a/F_b. Moreover, we define the collision time of the weight against the wall as Δt. The developed device could adjust F_a/F_b and Δt by changing the thickness of the urethane foams.

The acrylic wall is able to move and could be fixed in any position along the x-axis. Namely, it is possible to adjust the position of the collision surface in order to make F_a/F_b maximum. Moreover, the urethane foam is set on the surface of the acrylic wall, indicated in Fig. 7 as (i), and its thickness d could be adjusted to change the collision time Δt. In this study, we prepare five urethane foams whose thicknesses are $d = 2$ mm, 4 mm, 6 mm, 8 mm, and 10 mm. The collision time Δt may increase when the urethane foam become thicker. In addition, we could also increase the thrust force ratio F_a/F_b by setting the high shock absorption urethane buffer between the VCM and the outer frame, indicated in Fig. 7 as (ii). The device is 140 x 41.5 x 41.5 mm^3, and the total mass without the urethane foam and the buffer is 230 g.

3.3   Collision Experiment

The thrust forces generated by the device were measured using a load cell. Fig. 8 shows the example of time history of the thrust forces when the collision surface is 3 mm away from the coordinate origin shown in Fig. 7. Sine-wave current with amplitude and frequency of 1 A and 10 Hz is input into the VCM, and the thickness

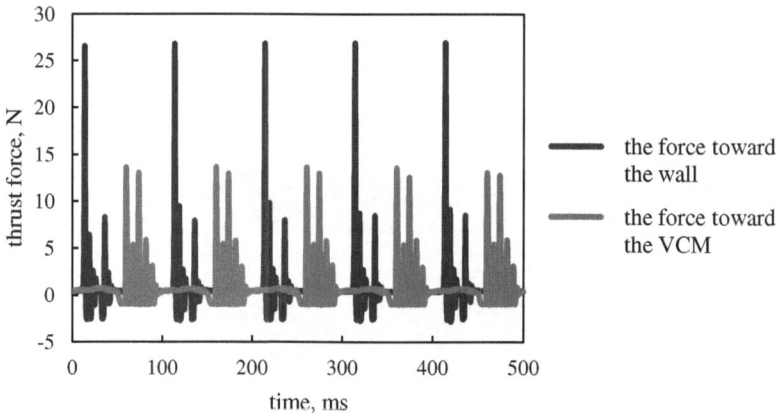

Fig. 8. Waveform of thrust force

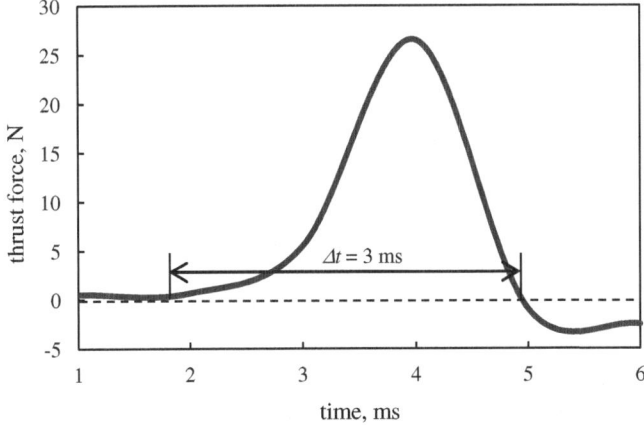

Fig. 9. Collision time

Table 1. Thickness of urethane foams and measured result

d [mm]	F_a [N]	F_b [N]	F_a/F_b	Δt [ms]
2	28.62	3.70	7.73	3
4	21.49	3.51	6.12	4
6	12.44	3.50	3.56	7
8	10.09	3.45	2.92	10
10	4.35	2.30	1.89	20

Table 2. Presence and absence of the buffer and measured result

	F_a [N]	F_b [N]	F_a/F_b	Δt [ms]
With the buffer	28.62	3.70	7.73	3
Without the buffer	26.05	13.10	1.99	3

of urethane foam is 2 mm. Since the forces toward the wall and the VCM are measured separately, the two waveforms are not synchronized. Furthermore, preloads are previously applied to keep the device in contact with the load cell. Then the minus value of the thrust forces may be measured in the experiment. We measured the thrust forces for 10 second to get 100 peak thrusts which emerged every 0.1 second. Then, the maximum value of positive and negative x-direction force, F_a and F_b, are calculated as an average of the 100 peak values. Furthermore, The collision time Δt is defined as in Fig. 9, which is an enlargement of Fig. 8, i.e., the collision time Δt is defined as the time period during the thrust force is measured positive.

First, we measured F_a, F_b, F_a/F_b, and Δt with different urethane foam thickness d. Note that, the 6 mm thick buffer is set. Table 1 shows the result. As can be seen, F_a/F_b decreases when the urethane thickness increases. On the other hand, Δt increases when the urethane thickness increases. Secondly, we also measured F_a, F_b, F_a/F_b, and Δt with/without the buffer when the urethane thickness d is 2 mm. Table 2 shows the result. As can be seen, F_a/F_b increases by setting the buffer, and Δt is kept constant with/without the buffer.

4   Sensory Evaluation

4.1   Sensitivity of Pulling Force Perception

The sensitivity of the pulling force perception must depend on the thrust force ratio F_a/F_b, because human perception has non-linear characteristics [10]. Furthermore, it must depend on Δt as well, because human cannot perceive the thrust force if the collision time Δt is too short [6]. Therefore, we validate the produced device with regard to the sensitivity of the force perception by the pair comparison method [11].

First, we prepare five experimental conditions (A_1, A_2, A_3, A_4, A_5). Each condition has a different thickness of the urethane foam. Other than the urethane thickness, experimental conditions are the same, i.e., the buffer is set, sine-wave current with amplitude and frequency of 1 A and 10 Hz is input into the VCM. Table 3 shows d, F_a, F_b, F_a/F_b, and Δt with each experimental condition. Test subjects grip the device as shown in Fig. 10, and compare each combination of conditions, (A_1, A_2),

Table 3. Measured result of $A_1 \sim A_5$

	d [mm]	F_a [N]	F_b [N]	F_a/F_b	Δt [ms]
A_1	2	28.62	3.70	7.73	3
A_2	4	21.49	3.51	6.12	4
A_3	6	12.44	3.50	3.56	7
A_4	8	10.09	3.45	2.92	10
A_5	10	4.35	2.30	1.89	20

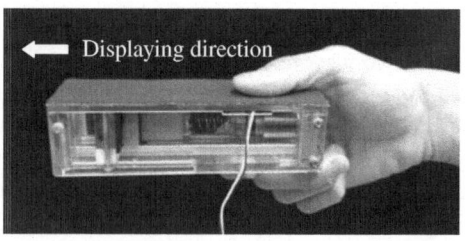

Fig. 10. Way to grip the device

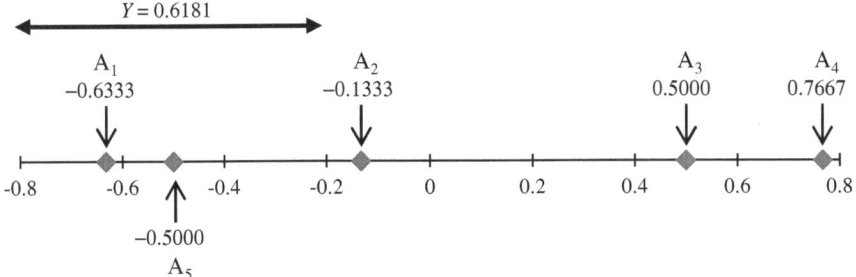

Fig. 11. Evaluation result $(A_1 \sim A_5)$

Table 4. Measured result of $B_1 \sim B_6$

	d [mm]	F_a [N]	F_b [N]	F_a/F_b	Δt [ms]
B_1	4	10.01	3.53	2.83	5
B_2	4	9.87	6.38	1.55	5
B_3	8	10.09	3.45	2.92	10
B_4	8	10.01	6.71	1.49	10
B_5	10	10.30	4.05	2.54	15
B_6	10	10.20	6.87	1.49	15

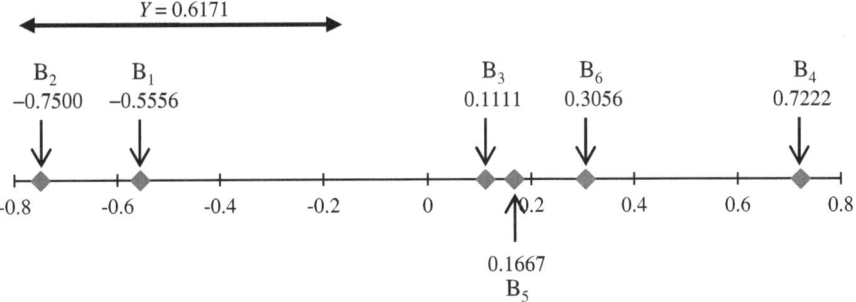

Fig. 12. Evaluation result $(B_1 \sim B_6)$

(A_1, A_3), …, (A_4, A_5), to evaluate the difference of the displayed pulling forces. The evaluations are conducted without sight, and subjects score the conditions on a 5-point scale of -2, -1, 0, 1, 2 (2: detected high, -2: detected low). Number of subjects is 6 male/female in their 20s.

Fig. 11 shows the evaluation result. In the figure, values on the line mean each condition's average preference-score. If the difference of the value between two experimental conditions is larger than a yardstick Y shown in the figure, there is a significant difference between them at the level of 5%. Hence, there is a significant difference between (A_3, A_4) and (A_1, A_2, A_5), and an order of the sensitivity is $A_4 \sim A_3$ > $A_2 \sim A_5 \sim A_1$. Since A_1 and A_2 have a lower score than A_3 or A_4 despite A_1 and A_2 have large F_a/F_b as shown in Table 4.1, it could be found difficult to perceive the pulling force when the collision Δt is too short.

On the other hand, it is difficult to say that the low evaluation score of A_5 is due to the collision time Δt because F_a and F_a/F_b of A_5 are smaller than those of A_3 and A_4. Hence in the next experiment, we newly prepare six conditions (B_1, B_2, B_3, B_4, B_5, B_6). Table 4 shows d, F_a, F_b, F_a/F_b, and Δt of each condition. With all conditions, F_a is equalized to be about 10 N by adjusting the amplitude of the input current and the urethane foam thickness; B_1 and B_2 with 4 mm, B_3 and B_4 with 8 mm, B_5 and B_6 with 10 mm, respectively. Moreover, the buffer is removed with B_2, B_4, and B_6 to make F_a/F_b smaller. The experiment is again conducted without sight, and test subjects score the conditions on a 5-point scale of -2, -1, 0, 1, 2. Number of subjects is 6 male/female in their 20s.

Fig. 12 shows the evaluation result. As can be seen, there is a significant difference between (B_1, B_2) and (B_3, B_4, B_5, B_6), and the order of the sensitivity is $B_4 \sim B_6 \sim B_5 \sim B_3$ > $B_1 \sim B_2$. Therefore, it is found difficult to perceive the pulling force when the collision time Δt is too short, even if F_a is equalized. Moreover, the sensitivity of the pulling force perception strongly depends on Δt, not on the thrust force ratio F_a/F_b, because there is no significant difference between B_1/B_2, B_3/B_4, and B_5/B_6.

The reason why there is no significant difference between B_3, B_4, B_5 and B_6 is that the evaluation score of B_1 and B_2 are extremely low. Then, we removed B_1 and B_2 from the analysis and analyze the other four conditions again. Fig. 13 shows the result. As can be seen, there is a significant difference between B_4 and (B_3, B_5, B_6), and the order of the sensitivity is B_4 > $B_6 \sim B_3 \sim B_5$. B_4 has higher evaluation score compared with B_3 because total time of negative x force generation is shorter. This shorter negative x force generation causes difficulty in thrust force perception, resulting in making the subject feel the positive x force stronger. Note that, the total times of negative x force generation are 5 ms with B_3 and 2 ms with B_4. Therefore, making the total time of negative x force shorter is required for the proposed pulling force feedback device. Furthermore, since B_4 has a higher score than B_6, the sensitivity of the pulling force perception is the highest if Δt is 10 ms.

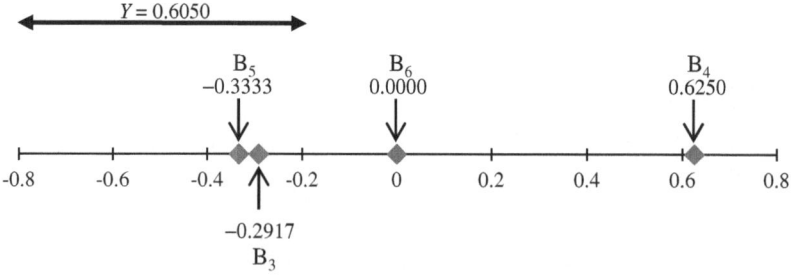

Fig. 13. Evaluation result （$B_3 \sim B_6$）

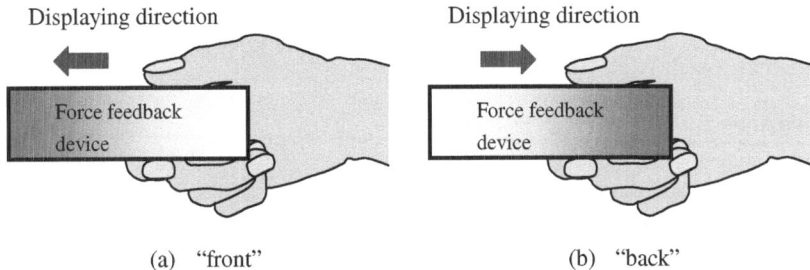

Fig. 14. Evaluation of directional sense

4.2 Evaluation of Directional Perception

In this section, to evaluate a directional perception with the experimental condition B_4 which has the highest score in the previous evaluations, we make test subjects grip the device in two different ways. One is the way in which they perceive the pulling force in "front" (elbow-to-hand direction) as shown in Fig. 14 (a). The other is the way in which they perceive the pulling force in "back" (hand-to-elbow direction) as shown in Fig. 14 (b).

First, we make the subject grip the device in one of the two ways without sight. Then, we drive the VCM, and make them answer which direction they perceive the force, "front" or "back". Following the same steps, we changed the way to grip the device randomly. Each subject evaluates the force displayed direction totally ten times, 5 in the ways of "front" and 5 in the ways of "back".

As a result, the correct answer rate is 95 % in "front" and 100 % in "back". Therefore, we confirmed that there is no dependence of directional perception on the way to grip the device.

5   Conclusions

In this paper, we developed the ungrounded force feedback device which makes users feel as if the pulling force is displayed by using asymmetric periodical impact. Since the motion of the system is in a single direction driven by the VCM, the device does not generate any fluctuations of displaying force in the unwanted direction. We evaluated the sensitivity of the pulling force perception using the developed device. Then, we make sure that it is possible to display the pulling force to the user with the developed device. In addition, the sensitivity of the pulling force perception strongly depends on the collision time Δt, and it becomes the highest when the collision time Δt is 10 ms. And the total time of negative x force generation should be shorter in order to display stronger pulling force. Furthermore, as the result of the evaluation of direction perception, the correct answer rate is higher than 95 % in the two different ways of grasping the device. Our future work focus on making the device smaller and lighter aiming for portable use, because it is not required to be grounded.

References

1. Massie, T.H., Salisbury, J.K.: The PHANToM Haptic Interface: A Device for Probing Virtual Objects. In: ASMEWAM, Symposium on Haptic Interfaces for Virtual Environment and Teleoperator Systems, vol. 55-1, pp. 295–300 (1994)
2. Sato, M.: SPIDER and virtual Reality. In: World Automation Congress, IFMIP 043, pp. 1–7 (2002)
3. Yano, H., Yoshie, M., Iwata, H.: Development of a Non-Grounded Haptic Interface Using the Gyro Effect. In: Proc. of HAPTICS 2003, pp. 32–39 (2003)
4. Tanaka, Y., Masataka, S., Yuka, K., Fukui, Y., Yamashita, J., Nakamura, N.: Mobile torque display and haptic characteristics of human palm. In: Proc. the 2001 International Conference on Augmented Teleexistence, pp. 115–120 (2001)
5. Amemiya, T., Ando, H., Maeda, T.: Virtual force display: Direction guidance using asymmetric acceleration via periodic translational motion. In: Proc. World Haptics Conference, pp. 619–622. IEEE Computer Society (2005)
6. Hamaguchi, H., Amamiya, T., Maeda, T., Ando, H.: Design of repetitive knocking force display for being-pulled illusion. In: 19th IEEE International Symposium on Robot and Human Interactive Communication, pp. 33–37 (2010)
7. Amemiya, T., Ando, H., Maeda, T.: Phantom-DRAWN: Direction Guidance using Rapid and Asymmetric Acceleration Weighted by Nonlinearity of Perception. In: Proc. of ICAT 2005, pp. 201–208 (2005)
8. Amemiya, T., Ando, H., Maeda, T.: Directed Force Perception When Holding a Nongrounding Force Display in the Air. In: Proc. of Eurohaptics 2006, pp. 317–324 (2006)
9. Amemiya, T., Kawabuchi, I., Ando, H., Maeda, T.: Double-layer Slider-crank Mechanism to Generate Pulling or Pushing Sensation without an External Ground. In: 2007 IEEE/RSJ International Conference on Intelligent Robots and Systems, pp. 2101–2106 (2007)
10. Annett, J., Morris, P., Holloway, C., Roth, I.: Human information processing Part1, pp. 57–69. The Open University Press (1974)
11. Amasaka, K., Nagasawa, S.: Basics and applications of sensory evaluation Part 1, pp. 193–203. Japanese Standards Association (2000) (in Japan)

A Feasibility Study of Levels-of-Detail in Point-Based Haptic Rendering

Wen Shi and Shahram Payandeh

Experimental Robotics and Haptics Laboratory
School of Engineering Science
Simon Fraser University
Burnaby, British Columbia
Canada, V5A 1S6

Abstract. This paper presents a study for defining the levels-of-detail (LOD) in point-based computational mechanics in haptic rendering of objects. The approach uses the description of object as a set of sampled points. In comparison with the finite element method (FEM), point-based approach does not rely on any predefined mesh representation and depends on the point representation of the object. Different from solving the governing equations of motion representing the entire object based on pre-defined mesh representation which is used in FEM, in point-based modeling approach, the number of points involved in the computation of displacement/deformation can be adaptively re-defined during the solution cycle. This frame work can offer an implementation for the notion of LOD techniques for which can be used to tune the haptic rendering environment for increased realism and computational efficiency. This paper presents some initial experimental studies in implementing LOD in such environment.

1 Introduction

Interactive virtual training environment has been gaining popularity over the past decades. One of the main challenges of such development is the modeling of deformable objects such as soft tissues, organs, tendon, and skin. The two main requirements for modeling such environment is the availability and knowledge of material and biological properties of objects and the need for the efficient/fast computational framework for achieving a desired interactive haptic rate.

One of the proposed approaches for modeling is based on the mass-spring methodology which represents the object as discretized mass points with spring connectivity. For example, (LeDuc, Payandeh, Dill) [12] and (Shi,Payandeh) [4] modeled a 1D suture object, and (Payandeh, Zhang, Cha)[8], (Mollemans, Schutyser) [22] and (Shi, Payandeh) [5] modeled a deformable cloth-like object representing soft tissue. Another approach for modeling deformable objects is based on finite element method (FEM) and its various extensions. For example, (Berkley, Weghorts, Gladstone, ..)[9] presented a simplified computational set-up for real-time approach for 3D soft tissue simulation, (Berkley,Turkiyyah,Berg,

P. Isokoski and J. Springare (Eds.): EuroHaptics 2012, Part I, LNCS 7282, pp. 493–504, 2012.

..)[10] and (Marlatt,Payandeh)[7] presented a fast linear solution for real-time user interaction. In general, due to the large sizes of system matrices and the definition of shape functions, most of the solution approaches require an increased in computational load which in general not suitable for interactive applications.

An approach for modeling deformable objects is referred to as the meshless method (Belytschko,Krongauz,Organ, ..)[13] which uses descretized element nodes (or points) to construct the object and uses the governing equations of motions through usage of radial basis weighting functions. (Fries,Matthies)[14] presented one of the earliest meshless methods in physics-based animation of objects. (Pauly,Keiser,Adams, ..)[15] presented a meshless elastic and plastic deformation framework. (Guo,Qin) [16] proposed a meshless paradigm for modeling the point-sampled deformable object in real time. (Adams,Wicke,Ovsjanikov, ..)[17] proposed a similar meshless framework for modeling multiple deformable objects for user interaction purposes. One of the extensions of meshless method is referred to as the point-based approach. (Muller,Keiser,Nealen, ..) [18] proposed a point-based modeling approach for graphic simulation of elastic and plastic objects. (Muller,Charypar,Gross) [19] presented a point-based approach for animating elasto-plastic solids, where a novel computational scheme based on displacement gradient is introduced using neighborhood optimization. Some experimental comparisons for haptic rendering of object using FEM and point-based method was presented in (Shi,Payandeh) [21]. In this analysis it was shown that the point-based approach can offer a comparable result to that of FEM and hence can be used as an accurate representation and solution for haptic interaction. In this paper, we demonstrate the implementation of a similar approach which can be suitable for haptic rendering. Comparing with mass-spring model, the point-based approach incorporate implicitly the material properties of object which has a significant advantage on obtaining an increased degree of accuracy of the solution. Comparing with FEM, the proposed modeling frame-work can be potentially applied to real time rendering applications by adaptively including sample points in the computational model at the interactive rate.

The paper is organized as follow: Section (2) presents an overview of the point-based computational mechanics; Section (3) presents a definition for LOD which was used in this paper; Section (4) presents some experimental studies of our proposed LOD and Section (5) presents some concluding remarks.

2   An Overview of the Point-Based Mechanics

Point-based mechanics approach discretizes the control volume of an object into distributed points with no predefined connectivity or meshes (Figure (1)). Instead of using the polynomial interpolation functions used in FEM, functions such as Radial Basis Functions (RBF) are selected for interpolation of a continuous function based on a set of sample points [6].

Initially, through a weighting function, each point is defined to have a given mass in relation to its neighbor points. This given mass is computed by dividing the mass of the object by the number of points which need to be maintained

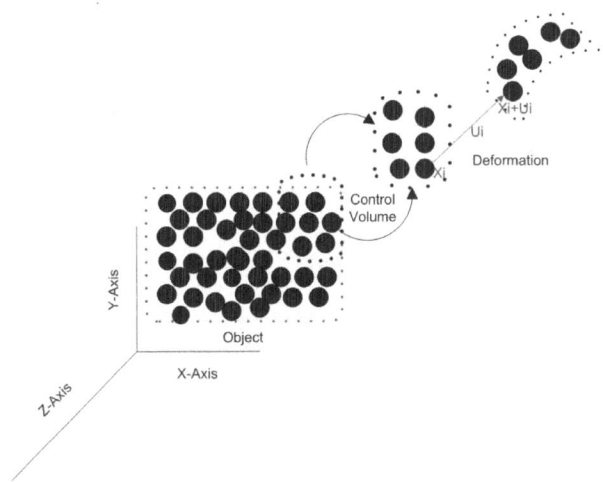

Fig. 1. Illustration of the point-based model. The volume of the object is discretized into points. The actual displacement/deformation of the object is achieved by the displacement of each point inside the control volume. X_i is the position vector of the ith point, $X_i + U_i$ is its new position after being deformed through the displacement vector U_i.

during each solution cycle. For example, the governing equation of motion of an object can be written as:

$$\rho \ddot{X} = -\nabla_U \mathcal{E} + f_{ext}, \tag{1}$$

where $\nabla_U \mathcal{E}$ is defined as the spatial gradient of strain energy function with respect to displacement vector. The following equation defines the computation of strain which is used in the definition of $\mathcal{E}$ (strain energy function).

$$\epsilon = 0.5(\nabla U + [\nabla U]^T + [\nabla U]^T \nabla U), \tag{2}$$

where U is the displacement vector of a point. One of the main computational challenges for the solution to above equation is the computation of the gradient (i.e. ∇U). The following presents an overview of such computational approach.

In RBF approaches, Smoothed Particle Hydrodynamics (SPH) ([11]) is commonly used for the numerical computation of the gradient function. Moving least square (MLS) is also a computational tool belonging to the RBF methodologies which reconstructs continuous functions from a set of unorganized point samples (discrete points). A weighted least squares measure can further be computed which biases the result towards the region around a point at which the reconstructed value is being evaluated[2],[3].

The selection of weighting functions used in the computational model plays a very important role. In modeling deformable objects, there are several criterion which can be used for selecting a suitable weighting function. First the value of the weighting function should always be positive since for example, mass and density of points are always positive; Secondly, the weighting function should be decreased in its value when the distance between the reference point and its neighbor points increase (the influence is local and vanishes as the distance increases) . In general, several weighting functions can be defined[19], which are modified versions of composite quadratic tensor-product [16]:

$$\omega(r) = \begin{cases} \frac{315}{64\pi h^9}(h^2 - ||r||^2)^3 & ||r|| < h, \\ 0 & ||r|| \geq h. \end{cases} \tag{3}$$

$$\omega(r) = \begin{cases} \frac{15}{\pi h^6}(h - ||r||)^3 & ||r|| < h, \\ 0 & ||r|| \geq h. \end{cases} \tag{4}$$

$$\omega(r) = \begin{cases} \frac{15}{2\pi h^3}(-\frac{r^3}{2h^3} + \frac{r^2}{h^2} + \frac{h}{2r} - 1) & ||r|| < h, \\ 0 & ||r|| \geq h. \end{cases} \tag{5}$$

$$\omega(r) = \begin{cases} 1 - 2\frac{r^2}{h} & 0 < ||r|| < 0.5h, \\ 2(1 - \frac{r}{h})^2 & 0.5h < ||r|| < h, \\ 0 & ||r|| \geq h. \end{cases} \tag{6}$$

where r is the distance vector between any two points and h is a cut off distance. The cut off distance, h, in the weighting function has two main properties. First, when h is decreasing, it localizes the weights to points which are within the reduced neighborhood of the current reference point. When h increases, it can offer a uniform and smooth contribution of the neighbor points. Weighting function defined in equation (4) and equation (5) are only used for fluid modeling. Equation (4) is used for solving the particle clustering problem and equation (5) is used for viscosity computation. Equation (6) is also for modeling elastic object.

For example, the density of each point can be computed as:

$$\rho_i = \sum_j m_j \omega(r_{ij}), \tag{7}$$

where r_{ij} is the distance between point i and j. The volume of each point v_i is defined as:

$$v_i = \frac{m_i}{\rho_i}. \tag{8}$$

The gradient function can be written as:

$$\nabla U = \begin{bmatrix} u_{xx} & u_{xy} & u_{xz} \\ u_{yx} & u_{yy} & u_{yz} \\ u_{zx} & u_{zy} & u_{zz} \end{bmatrix} = \begin{bmatrix} \nabla u_x \\ \nabla u_y \\ \nabla u_z \end{bmatrix}, \tag{9}$$

where $\nabla u_x = [u_{xx}, u_{xy}, u_{xz}]$, $\nabla u_y = [u_{yx}, u_{yy}, u_{yz}]$, and $\nabla u_z = [u_{zx}, u_{zy}, u_{zz}]$ denote the derivatives with respect to the three coordinate axes (x,y and z).

Using MLS, the spatial derivative ∇U evaluated at the ith point can be written as:

$$\nabla u_{x_i} = M_i^{-1}(\sum_j (u_{x_j} - u_{x_i}) r_{ij} \omega(r_{ij}))$$

$$\nabla u_{y_i} = M_i^{-1}(\sum_j (u_{y_j} - u_{y_i}) r_{ij} \omega(r_{ij})) \tag{10}$$

$$\nabla u_{z_i} = M_i^{-1}(\sum_j (u_{z_j} - u_{z_i}) r_{ij} \omega(r_{ij})),$$

where for example u_{x_i} denotes the x-component of the displacement vector at the ith point and the matrix M_i is computed at point i is:

$$M_i = \sum_j r_{ij} r_{ij}^T \omega(r_{ij}). \tag{11}$$

The value for ∇U is also used to evaluate the strain defined in equation (2). The strain energy at point i can be calculated as $\mathcal{E}_i = v_i \frac{1}{2}(\sigma \cdot \epsilon)$ where $\mathcal{E}_i$ is the stored strain energy and v_i is the volume of the point. The body force at the jth point can be defined as:

$$f_{body(j)} = -\nabla_{U_j} \mathcal{E}_i = -v_i \sigma_i \nabla_{U_j} \epsilon_i, \tag{12}$$

where $f_{body(j)}$ is the body force evaluated at point j, $\mathcal{E}_i$ is the strain energy at point i, v_i is the volume of point i, σ_i is magnitude of stress, ϵ_i is the associated strain. Based on the Newton's Third Law, the body forces acting on point i is computed as the negative sum of all the body force acting at its neighbor points,

$$f_{body(i)} = -\sum_j f_{body(j)}. \tag{13}$$

Using the expression for ϵ and ∇U, we can determine the expression of $\nabla_{U_j} \epsilon_i$ (details are omitted for brevity). The expressions for the internal body forces are defined as:

$$f_{body(i)} = -2v_i \mathbf{J} \sigma_i d_i, f_{body(j)} = -2v_i \mathbf{J} \sigma_i d_j \tag{14}$$

where $\mathbf{J}$ is the Jacobian matrix, $d_i = M_i^{-1}(-\sum_j r_{ij}\omega_{ij})$, and $d_j = M_i^{-1}(r_{ij}\omega_{ij})$, v_i and v_j are the volume of points i and j.

3   A Definition for Levels-of-Detail

The proposed framework offers an approach where global deformation of an object can be computed using small number of points in the local deformation solution. In order to establish relationships between the global and local deformation of object as a function of the sampled points, we study how the resolution of the base model of the object (i.e. object defined by set of foundation points) can affect the haptic frame rate by increasing the resolution of the base point. As pointed-out before, the major computational challenge in our approach is the computation of the displacement gradient tensor (∇U) evaluated at each point. Intuitively the basic approach is that we assume the local deformation in only affected by points in the proximity of the contact region. In this case, the contact force propagate through-out the body where the resultant internal forces have less propagation beyond certain threshold away from the contact area. A localization or levels-of-detail (LOD) can then be implemented to capture the above hypothesis where one should be able to increase the realism of the haptic interaction while maintaining a high haptic rate. Here we propose an approach by defining influence ranges as a function of the distance of a point from the contact area. For example, in Figure (2) and at the instant of contact a dense sample of points of object inside a volume located at the tip of a probe can be considered as points located at the proximity of the contact area.

Fig. 2. Schematic diagram for Levels-of-detail - proximity between the tip of the probe and the points within the half sphere object is detected by a sphere-sphere collision detection algorithm. Neighbor points that fall within the influence range will be selected and incorporated into the computational model

The local and internal deformation can propagate from the contact region to points defining the boundary of the object. Our initial algorithm associates the larger radius of influence propagation to smaller number of sample points from the foundation points. We proposed a pseudo-random point selection algorithm which selects points within pre-set ranges/radii under pre-defined weighting factors defined through a random number generator. For example, the selection of weighting factor for the localization range (i.e. range within the contact area) is set to be large since we require a high density of points in this range. On the other hand, the selection weighting factor for the influence range is set to be small since away from the contact area since we only need a limited number of points to propagate the deformation from the contact region to the boundary points. This approach is analogous to the LOD scheme used in the finite element model where the mesh density varies from fine mesh in the region of interest to the coarse mesh in the boundary region. Figure (3) illustrates the conceptual schematic based on the localization distance threshold.

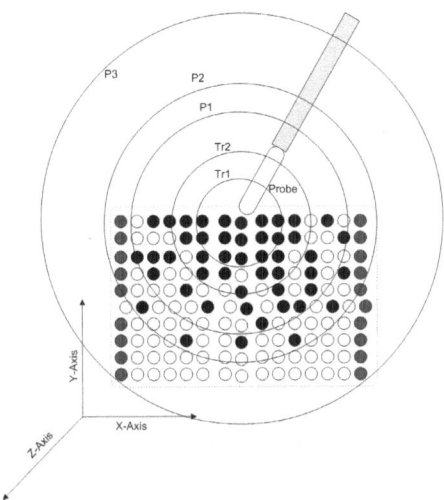

Fig. 3. Schematic of the proposed LOD algorithm. A probe contacted an object represented by the blue point at the contact point. Two localization distances T_{r1}, T_{r2} and two propagation distances P_1, P_2 are initialized. P_3 is the last propagation distance which can include all the boundary points. Red points are the boundary points; Points selected by the algorithm are shaded in black.

In Figure (3) we define localization distance T_r to be a linear function of the cut-off distance h: $T_r = kh$, and we use two different constants ($k_1 = 1$ and $k_2 = 1.5$) to define the threshold of the localization distances ($T_{r1} = h, T_{r2} = 1.5h$) as the selection ranges. Once being defined, T_{r1} and T_{r2} will remain constant through-out the computation. Similarly, we define propagation distance, P_i, where i indicates the ith level of propagation. We also define P_i to be a linear function of h. By changing the constants (k_1, k_2), the threshold distance of each propagation level can vary.

Once the contact is detected, the dynamic selection algorithm initiated to select the points in terms of the contact region and the propagation of deformation. We incorporate selected points into the computation model and render the deformation effect. The selection function first generates a random number Ran between 0 and 10 and determines the selection weighting factor. Then, it determines the distances between the current positions of neighbor points and the contact point. After that, it selects the points by comparing the distances with the threshold distance based on the selection weighting factor. If the distance is less than T_{r1}, the selection weighting factor is 1, and the point is directly selected. If the distance is between T_{r1} and T_{r2}, the weighting factor of selection is 0.8 (the probability of $Ran > 2$ is 4/5). The points within the propagation levels are also selected based on the weighting factor of the propagation level (i). K_i is a monotonically increasing integer in terms of propagation level i, which controls the weighting factor of the propagation points being selected. The weighting factor of selection for level i is $\frac{10-K_i}{10}$. The upper bound of K_i is set to be 10.

4   Experimental Studies

We conduct experimental studies to compare and analyze how the number of the propagation levels and the selection factors can affect the computational model used in haptic rendering. The experiment is carried on a PC with a dual core AMD Athlon 3.0GHz CPU, 3G RAM and nVidia 8600 graphics card with 1G RAM of memory. Figure (4) shows the initial foundation points which were used to represent the experimental object. Figures (5) and (6) show some examples of our experimental results comparing the different propagation levels and their selection weighting factors along with the system update rate. In the experiment, we apply a constant pulling force at the 111-th point in the model, and plot the magnitudes of the displacements of all the points in the model at the nine seconds in the simulation. The localization distances Tr_1 and Tr_2 remain unchanged and the selected neighbor points are shaded in brown and blue.

Fig. 4. Foundation points representing the object used in the proposed LOD studies

We analyze how different cut-off distances (T_r) of the localization can affect the frame rate in the LOD model. Table (1) compares the run time of the different localization as a function of cut off distances on different foundation points. As we increase the distance thresholds T_{r1} and T_{r2}, the rendering time per frame increases since more points are incorporated into the computational model.

5   Conclusion

In this paper, we investigated a 3D point-based object modeling approach for haptic rendering. It was shown previously that this approach can offer a comparable results to that of the FEM [21]. The proposed framework allows interaction

Cases	(a)
LOD Levels	Two localization levels L_1, L_2; One propagation level, $P_1 = 5h$, $K_1 = 6$
Update Frame rate	16.7
Displacement plot	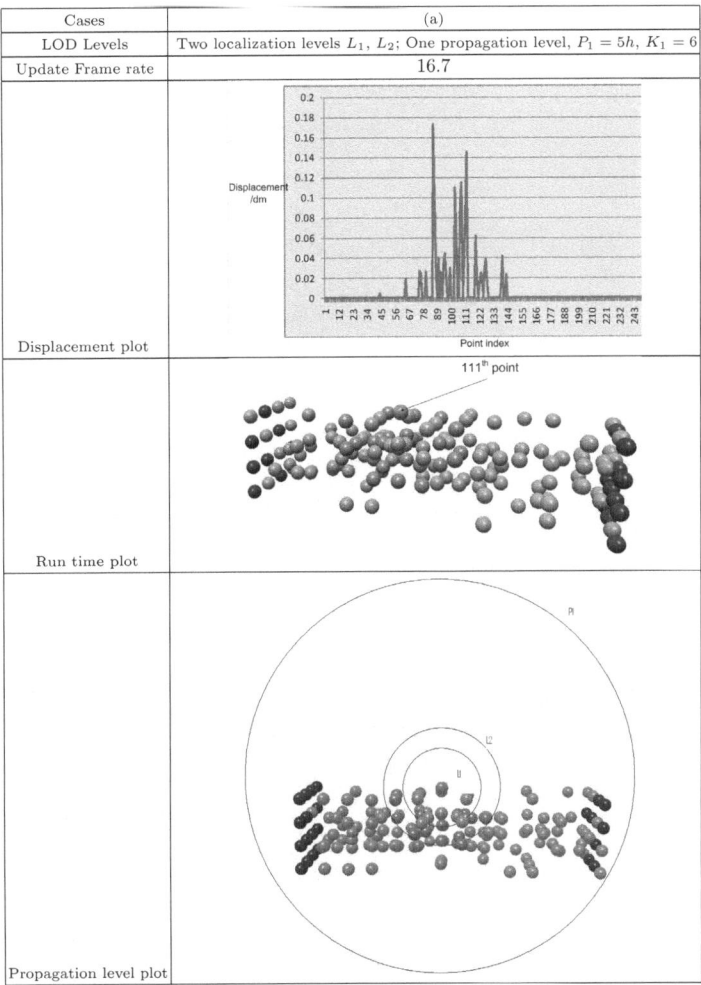
Run time plot	
Propagation level plot	

Fig. 5. Comparison of different propagation levels. First row is the definition of the propagation level; Second is the associated frame rate; Third row is the displacement plot of the points; Fourth row is the visualization of the displacement; Fifth row is the schematics of the propagation level. In this experiment, we have define one propagation level with low selection weighting factor (0.4). The local points are colored in brown and dark blue, the propagation points are colored in green.

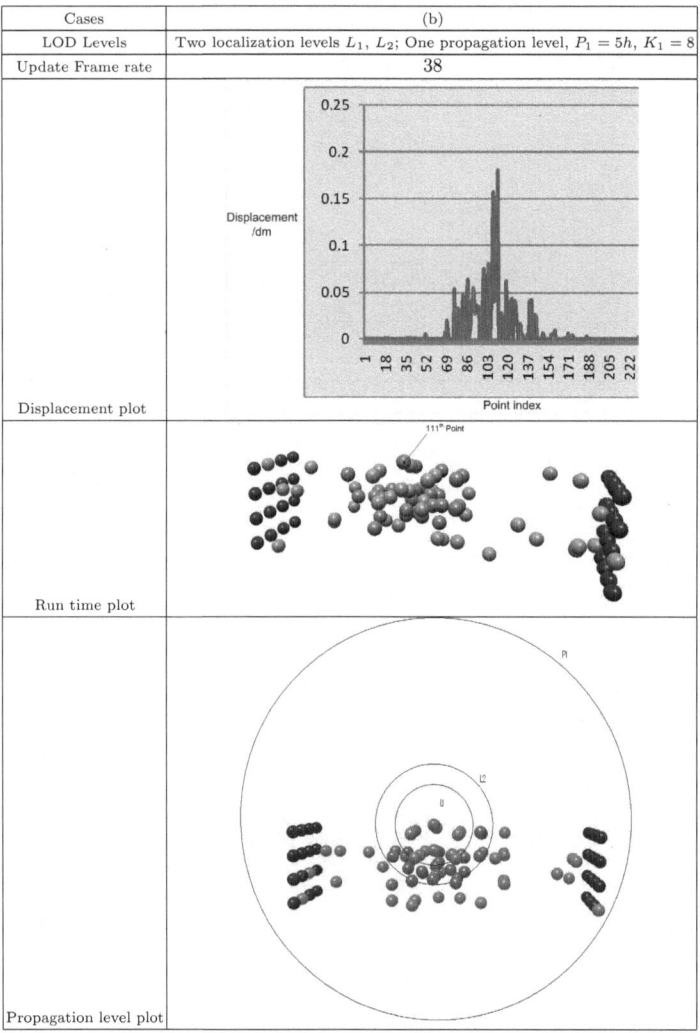

Cases	(b)
LOD Levels	Two localization levels L_1, L_2; One propagation level, $P_1 = 5h$, $K_1 = 8$
Update Frame rate	38
Displacement plot	
Run time plot	
Propagation level plot	

Fig. 6. Comparison of different propagation levels. First row is the definition of the propagation level; Second is the associated frame rate; Third row is the displacement plot of the points; Fourth row is the visualization of the displacement; Fifth row is the schematics of the propagation level. In this experiment, we have define one propagation level with low selection weighting factor (0.2). Brown and dark blue points are local points and green points are propagation points.

Table 1. Analysis of the proposed point-based LOD in haptic rendering

	resolution	Frame rate(fps) for $T_{r1} = h, T_{r2} = 1.5h$	Frame rate(fps) for $T_{r1} = 1.5h, T_{r2} = 2h$	Frame rate(fps) $T_{r1} = 2h, T_{r2} = 2.5h$
(a)	224 points	83	56	46
(b)	500 points	50	39	24
(c)	1200 points	40	33	20

where user can manipulate the point-based object model using a haptic device. We present a notion of levels-of-detail (LOD) where foundation points representing the object can be grouped into different levels. Regions can be populated with various densities of points. We have presented an approach were based on the distances of points from the contact area, they can be populated with different densities in various geometrical regions away from the contact point while always including regions defined by the boundary points. Feasibility of the proposed LOD approach was studied to investigate the practicality of the proposed approach in changing the resolution of the points during the haptic interaction. This framework allows both inclusion of the material properties of the object and also direct mapping to various medical imaging modalities where scanned point-clouds can be mapped into the haptic interaction framework.

References

1. Salisbury, K., Conti, F., Barbagli, F.: Haptic Rendering: Introductory Concepts. In: IEEE Computer Graphics and Applications, pp. 24–32 (2004)
2. Nealen, A.: An As-Short-As-Possible Introduction to the Least Squares, Weighted Least Squares and Moving Least Squares Methods for Scattered Data Approximation and Interpolation. Internal Report, TU Darmstadt (1990)
3. Levin, D.: The approximation power of moving least-sqaures. Mathematics of Computation 67(224), 1517–1531 (1998)
4. Shi, H.F., Payandeh, S.: Real-Time Knotting and Unknotting. In: Proceedings of IEEE International Conference on Robotic Automation, pp. 2570–2575 (2007)
5. Shi, F., Payandeh, S.: On suturing Simulation with Haptic Feedback. In: Ferre, M. (ed.) EuroHaptics 2008. LNCS, vol. 5024, pp. 599–608. Springer, Heidelberg (2008)
6. Hardy, R.L.: Multiquadric Equations of Topography and Other Irregular Surfaces. Journal of Geophysical Research 76(8), 1905–1915 (1971)
7. Marlatt, S., Payandeh, S.: Modelling the Effect of Rayleigh Damping on the Stability of Real-time Finite Element Analysis. In: Proceedings of World Haptics (2005)
8. Payandeh, S., Zhang, H., Cha, J.: Toward Interactive Haptic Simulation of Cutting. International Journal of Virtual Technology and Multimedia 1(2), 172–186 (2010)
9. Berkley, J., Weghorst, S., Gladstone, H., Raugi, G., Berg, D., Ganter, M.: Banded Matrix Approach to Finite Element Modelling for Soft Tissue Simulation. Virtual Reality 4(2), 203–212 (1999)
10. Berkley, J., Turkiyyah, G., Berg, D., Ganter, M., Weghorst, S.: Real-Time Finite Element Modeling for Surgery Simulation: An Application to Virtual Suturing. IEEE Transactions on Visualization and Computer Graphics 10(3), 314–325 (2004)

11. Monaghan, J.J.: Smoothed Particle Hydrodynamics. Annu. Rev. Astron. 30, 543–574 (1992)
12. LeDuc, M., Payandeh, S., Dill, J.: Toward Modeling of a Suture Task. In: Proceedings of Graphics Interface (GI), pp. 273–279 (2003)
13. Belytschko, T., Krongauz, Y., Organ, D., Fleming, M., Krysl, P.: Meshless methods: An overview and Recent Developments. Computer Methods in Applied Mechanics and Engineering 139(1-4), 3–47 (1996)
14. Fries, T., Matthies, H.: Classification and Overview of Meshfree Methods. Technical Report, TU Brunswick, Germany Nr. (2003)
15. Pauly, M., Keiser, R., Adams, B., Dutr, P., Gross, M., Guibas, L.J.: Meshless animation of fracturing solids. In: International Conference on Computer Graphics and Interactive Techniques, pp. 957–964 (2005)
16. Guo, X., Qin, H.: Meshless methods for physics-based modeling and simulation of deformable models. Science in China Series F: Information Sciences 52(3), 401–417 (2009)
17. Adams, B., Wicke, M., Ovsjanikov, M., Wand, M., Seidel, H.-P., Guibas, L.J.: Meshless Shape and Motion Design for Multiple Deformable Objects. Computer Graphics Forum 29(1), 43–59 (2010)
18. Muller, M., Keiser, R., Nealen, A., Pauly, M., Gross, M., Alexa, M.: Point Based Animation of Elastic, Plastic and Melting Objects. In: Proceedings of Eurographics/ACM SIGGRAPH Symposium on Computer Animation (2004)
19. Muller, M., Charypar, D., Gross, M.: Particle-Based Fluid Simulation for Interactive Applications. In: Proceedings of Eurographics/SIGGRAPH Symposium on Computer Animation (2003)
20. Gerszewski, D., Bhattacharya, H., Bargteil, A.W.: A Point-based Method for Animating Elastoplastic Solids. In: Proceedings of Eurographics/ ACM SIGGRAPH Symposium on Computer Animation (2009)
21. Shi, W., Payandeh, S.: Towards Point-Based Haptic Interactions With Deformable Objects. In: ASME 2010 World Conference on Innovative Virtual Reality (WINVR), pp. 259–265 (2010)
22. Mollemans, W., Schutyser, F., Cleynenbreugel, J.V., Suetens, P.: Tetrahedral Mass Spring Model for Fast Soft Tissue Deformation. In: Proceedings of International Conference on Suegery Simulation and Soft Tissue Modeling, pp. 145–154 (2003)

Haptic Rendering of Cultural Heritage Objects at Different Scales

K.G. Sreeni, K. Priyadarshini, A.K. Praseedha, and Subhasis Chaudhuri*

Vision and Image Processing Laboratory, Department of Electrical Engineering,
Indian Institute of Technology Bombay, Powai, Mumbai-400076
{sreenikg,pkumari,praseedha,sc}@ee.iitb.ac.in

Abstract. In this work, we address the issue of virtual representation of objects of cultural heritage for haptic interaction. Our main focus is to provide a haptic access of artistic objects of any physical scale to the differently abled people. This is a low-cost system and, in conjunction with a stereoscopic visual display, gives a better immersive experience even to the sighted persons. To achieve this, we propose a simple multilevel, proxy-based hapto-visual rendering technique for point cloud data which includes the much desired scalability feature which enables the users to change the scale of the objects adaptively during the haptic interaction. For the proposed haptic rendering technique the proxy updation loop runs at a rate 100 times faster than the required haptic updation frequency of 1KHz. We observe that this functionality augments very well to the realism of the experience.

Keywords: Haptic rendering, HIP, proxy-based rendering, voxel based rendering, image pyramid, virtual museum, stereoscopic display.

1 Introduction

In the recent years digital technology is paving a way into safeguarding cultural heritages, and it also offers a great promise for enhancing access to them. A user's experience of accessing such cultural objects can be made more realistic and immersive by incorporating the recently evolving haptic technologies. Museum of Pure Form [2], a virtual reality system placed inside several museums and art galleries around Europe is an attempt to use of haptic technologies in cultural heritage applications. The incorporation of haptics in cultural heritage applications also helps in letting visually impaired people feel the exhibits that are behind glass enclosures, making even very fragile objects available to the scholars and allowing museums to show off a range of artefacts that are currently in storage due to lack of space. Further a joint hapto-visual rendering improves the immersivenes of the kinesthetic interaction. Some existing systems also allow users to interact with museum exhibition pieces via the internet [16]. It is required that such a system should enable the users to hapto-visually explore ancient monuments and heritage sites like Taj Mahal. However, currently available haptic systems are unable to handle objects at different scales.

---

* This work was supported in part by a DST grant on Indian Digital Heritage and another by MCIT on perception engineering.

P. Isokoski and J. Springare (Eds.): EuroHaptics 2012, Part I, LNCS 7282, pp. 505–516, 2012.

As a part of our exercise in preserving our cultural heritage we propose a simple multilevel hapto-visual rendering technique with depth data of cultural heritage objects. With mesh models of objects there are effective rendering techniques in haptics like god object rendering algorithm as proposed in [17]. However this algorithm fails in the case of point cloud based models. Further cultural objects appear at various different scales, and the user needs to experience the object at different levels of details. A mesh-based haptic technique is not amenable to scale changes as it requires the mesh to be pre-computed at all scales which is not feasible. In this paper, we propose a fast, proxy based rendering technique capable of working with point cloud based $3 - D$ models. Additionally the proposed method is amenable to haptic rendering at various scales. In order to render the model at different levels of resolution, we generate depth at each point of the model by reading the contents of depthbuffer in OpenGL and create a Monge surface from it. We show that the user's experience can be improved by allowing the user to interact with the object at multiple resolutions. This feature allows the user to feel the object more precisely at a closer level when needed and zoom out when context is desired. We have also developed a graphical user interface to make accessibility easier. Moreover, the easy availability of $3 - D$ models makes it a cost-effective system to savour the experience of various cultural heritage sites. The key contribution in this paper include how to render a Monge surface represented by a non-uniform point cloud data and how to handle scale change for zooming in and out during haptic interaction.

2   Literature Review

In the haptic rendering literature there are mainly two different approaches: Polygon (geometry) based rendering and Voxel based rendering. A good introduction to the basic haptic rendering technique is given by [7], [14]. Traditional haptic rendering method is based on a geometric surface representation which consists of mainly triangular or polygonal meshes. In polygon based rendering, each time the haptic interface point (HIP) penetrates the object, the haptic rendering algorithm calculates the closest surface point on the polygonal mesh and the corresponding penetration depth. If $\mathbf{d}$ is the vector representing the depth of penetration in the model, the reaction force can be calculated as $\mathbf{F} = -k\mathbf{d}$, where k is the stiffness constant, a physical property of the associated surface. The above method has problems while determining the appropriate direction of the force while rendering thin objects. Zilles and Salisbury [17], and Ruspini et al. [13] independently introduced the concept of god-object and proxy algorithm, respectively, which can solve the problems associated with thin objects.

In the God-Object rendering method [17], the authors use a second point in addition to the HIP called "god-object", sometimes called the ideal haptic interface point (IHIP). While moving in free space the god-object and the HIP are collocated. However, as the HIP penetrates the virtual object, the god-object is constrained to lie on the surface of the virtual object. The position of the god-object can be determined by minimizing the energy of the spring between the god-object and the HIP taking into account constraints represented by the faces of the virtual object [7]. If (x, y, z) are the coordinates of the proxy lying on the virtual object and (x_h, y_h, z_h) represents the coordinates of the HIP, the spring energy is given by

$$L = \frac{(x - x_h)^2}{2} + \frac{(y - y_h)^2}{2} + \frac{(z - z_h)^2}{2} + \sum_{i=1}^{3} l_i(A_i x + B_i y + C_i z - D_i) \qquad (1)$$

where L is the cost function to be minimized, l_1, l_2, l_3 are Lagrange multipliers and (A_i, B_i, C_i, D_i) are the homogeneous coefficients for the constraint plane equations on which the proxy lies. The 'force shading' technique (haptic equivalent of Phong shading) introduced by Morgenbesser and Srinivasan refined the above algorithm while rendering smooth objects [11]. Mesh based haptic rendering is not amenable to object scaling as the constraint equation for the planes (A_i, B_i, C_i, D_i) must be recomputed.

Volume haptic rendering technique is another alternative rendering technique used in haptics. The most basic representation for a volume is the classic voxel array in which each discrete spatial location has a one-bit label indicating the presence or absence of material. Avila et al. have used additional physical properties like stiffness, color and density during the voxel representation [1]. The voxmap-point shell algorithm uses the voxel map for stationary objects and point shell for dynamic objects [10], [12]. Point shell has been defined as a set of point samples and associated inward facing normals. However, these normals are not available and one needs to compute the normal at every location. The external surface ∂O of a solid object O can be described by the implicit equation as [6]

$$\partial O = \{(x, y, z) \in R^3 \mid \phi(x, y, z) = 0\},$$

where ϕ is the implicit function (also called the potential function) and (x, y, z) is the coordinate of a point in $3 - D$. In other words, the set of points for which the potential is zero defines the implicit surface. This has found applications in haptic rendering. This technique also suffers from the thin object problem. Lee et al. have proposed a rendering technique with point cloud data which computes the distance from HIP to the closest point on the moving least square (MLS) surface defined by the given point set[8]. Here the same problem occurs as in the distance field based rendering technique, since we do not keep track of HIP penetration and therefore is not good in rendering thin objects. El-Far et al. used axis aligned bounding boxes to fill the voids in the point cloud and then rendered with a god object rendering technique[4]. Leeper et al. described a constraint based approach of rendering point cloud based data where the points are replaced by spheres or surface patches of approximate size [9]. Another proxy based technique of rendering dense 3D point cloud was proposed in [15], where the surface normal is estimated locally from the point cloud.

3   Proposed Method

The rendering technique we propose is a proxy based method which does not use polygonal meshes for the reasons mentioned earlier. In practice, most of the cultural objects are carefully preserved and a dense $3 - D$ scan is performed on these objects to create virtual $3 - D$ model in the form of .obj, .ply, .3ds file, etc. Instead we directly use the point cloud data defining the models. As mentioned in the introduction, we get the depth data in the form of $z_i = f(x_i, y_i)$ where x_i and y_i are discrete values and z_i is the height of each sampled point from a reference plane $(z = 0)$ and i and j can take values depending on the size of the model. We haptically render the sampled surface of the object approximated by the depth values. In order to haptically render the object, we need to find the collision of HIP with the bounding surface and hence the penetration depth of HIP into the surface. The key factors in haptic rendering algorithm are-

1. The magnitude of the haptic force should be proportional to the penetration depth of HIP from the surface.
2. The direction of force should be normal to the surface at the point of contact of the proxy.

By taking these factors into consideration, the proposed algorithm tries to move the proxy over the object surface in short steps during the interaction so that each time it finds the most appropriate proxy position, the new position minimizes the distance between HIP and proxy and at the same time applies the reaction force normal to the surface at the point of contact. Initially, let us assume that the surface is known or the values of z is known for all values of x and y in continuum.

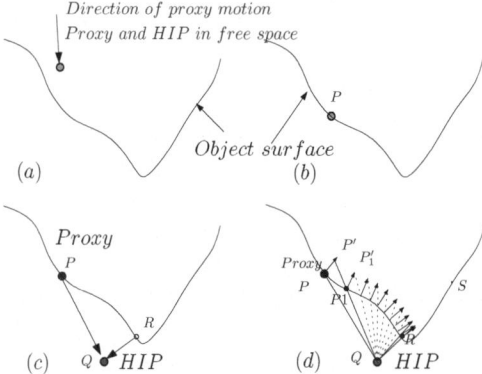

Fig. 1. Illustration of the proposed method to find the penetration depth of HIP into the surface

To understand our procedure let us look at the situation in Fig. 1. The bounding surface of the object is shown with the curve. In free space, proxy and HIP are collocated and is shown with green circle above the surface. Let the HIP and the proxy be in free space at a time $t = t_0$ as shown in Fig. 1(a). The HIP and hence the proxy are together moving towards the surface in a direction as shown by the arrow. At $t = t_1$, let HIP and proxy touch the surface at point P as in Fig. 1(b). Up to this point the proxy moves with the HIP. If the HIP is moved further in the direction it penetrates the surface and let Q as shown in Fig. 1(c) be the HIP position at time $t = t_2$. Now the proposed algorithm finds the most appropriate position of proxy at R where the distance QR is minimum, and the penetration depth QR is calculated in the direction exactly opposite to the surface normal at R. To find the point R from the starting point P we use the successive approximation method and move the proxy P to a distance $\delta\mathbf{n}$ along the normal to the surface at point P' and draw a line $P'Q$ to the current HIP as shown in

Fig. 1(d). The point P_1 on the surface at which the line intersects is found and is updated as the current proxy position. Again we move the proxy point P_1 along the normal at the current proxy position to P_1' and the process is repeated until the final proxy position R is attained. This is a greedy method but works well for smooth surfaces. If the surface has a fine texture, the line $P'Q$ may intersect the surface at multiple points and the process may converge to a poor, local minima, yielding a jerky haptic interaction. In case of multiple intersections, the one closest to P' is selected. The length of the vector $\delta\mathbf{n}$ determines the rate of convergence of the process. A large value of $\delta\mathbf{n}$ may lead to some spurious interaction when PP' may intersect the surface along the segment RS instead of PR. Hence a smaller value of $\delta\mathbf{n}$ (0.1mm is used in our algorithm) is preferred for rendering purposes. Quite naturally, it is required that the proxy position updation is performed within 1 ms of time, so that the user's interaction with the object through the haptic device is unhindered and is carried at 1 KHz. Till now, we have assumed

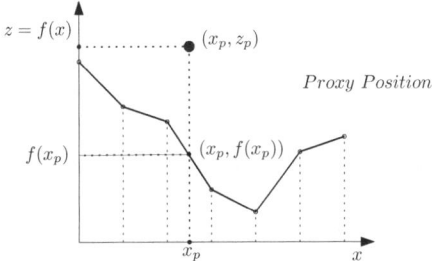

Fig. 2. Surface approximation from depth values

the surface to be known. Now, we try to approximate the surface from the given depth values. In case of $2-D$ depth data, we project the proxy onto the X-Y plane and the corresponding depth value is obtained by interpolating the neighbourhood depth values to form a continuous function $z = f(x,y)$. For better understanding, we consider a one-dimensional function $z = f(x)$ as shown in Fig. 2. As we have the function defined only at sampled points we interpolate the function at x_p to find the value of $f(x_p)$. Since the available points are sampled quite densely, bilinear interpolation is sufficient to find the bounding surface as shown in Fig. 2. In order to check the collision of HIP with the function we perform the following. At a given proxy position (x_p, z_p) we check for the function value $f(x_p)$. If $f(x_p) > z_p$ proxy has touched the surface, otherwise it is free to move towards the HIP. Extending the concept to $2-D$ depth data, let $\mathbf{X}_h = [x_h, y_h, z_h]^T$ denotes the HIP point and $\mathbf{X}_p = [x_p, y_p, z_p]^T$ denotes the proxy point. Collision can be easily checked here by comparing z_p with the depth interpolated at the projected point $z = f(x,y)$. The proxy movement during the rendering is managed by equation 2.

$$
\begin{aligned}
\mathbf{X}_p^{(k+1)} &= \mathbf{X}_p^{(k)} + \delta\mathbf{n} \qquad if \qquad z_p < f(x_p, y_p) \\
&= \mathbf{X}_p^{(k)} + |\delta\mathbf{n}| \frac{(\mathbf{X}_h - \mathbf{X}_p)^{(k)}}{|\mathbf{X}_h - \mathbf{X}_p|^{(k)}} \qquad otherwise
\end{aligned} \tag{2}
$$

This allows a smooth interaction when the force is withdrawn out of the object. The updated proxy then slowly moves towards the new HIP position.

4 Rendering

Rendering part of our work concerns with both haptic rendering and graphic rendering. Haptic rendering involves generating software controlled forces and feeding it to the users to provide them the sensation of touch. Any haptic rendering technique must include two steps:

1. detection of collision of the HIP with the object.
2. force computation if a collision is detected.

If $z_p < f(x_p, y_p)$ then the proxy has touched the object and a force needs to be fed back by the haptic device. Subsequently, the reaction force is computed as $\mathbf{F} = -k\mathbf{d}$ where k is the Hooke's constant, and $\mathbf{d}$ is the penetration depth given by $\mathbf{d} = |\mathbf{X}_h - \mathbf{X}_p|$, where $\mathbf{X}_h$ is the HIP position and $\mathbf{X}_p$ is the proxy position. Here we assume the stiffness to be constant everywhere on the surface of the object, but it can also be a function of position, provided the material property of the object is well documented. Fig. 3a shows proxy and HIP positions while rendering an arbitrary surface. For illustration we have selected only a small part of the depth map around the active region. The blue ball is the computed proxy touching the surface while the HIP is penetrated inside the surface. The HIP is shown with a red ball in the scene and the line from HIP to proxy is normal to the surface point at the proxy position. In order to show the surface of the object

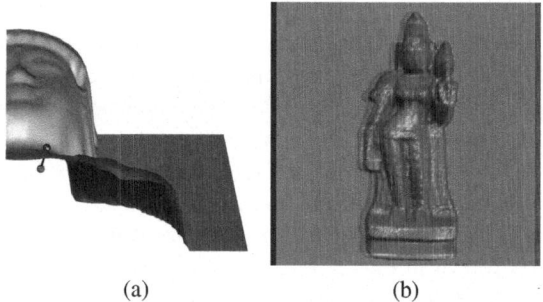

(a)   (b)

Fig. 3. (a) Illustration of proxy and HIP positions for an arbitrary surface. (b) Stereoscopic view of an Indian heritage object. (Data Courtesy: *www.archibaseplanet.com*)

graphically for simultaneous visual immersion, we display the image as a simple quad mesh out of the depth values. The normal is computed at each vertex. Although, we use point cloud data for haptic rendering, using the same for graphic rendering would result in gaps in the visually rendered object. Hence we have opted for the mesh-based graphical display in order to give a better perception to the viewer. We have used the

stereoscopic display technique for creating the effect of depth in the image by presenting two offset images in different colours separately to the left and right eye of the
viewer. A human observer combines these $2 - D$ offset images to recreate the $3 - D$
perception. Anaglyphic glasses can be used to filter offset images from a single source,
separated to each eye to give the perception of a $3 - D$ view to the users. Fig. 3b shows
the $3 - D$ view of an Indian heritage object as displayed on the screen.

4.1   Rendering at Different Scales

As mentioned earlier, heritage objects come at various physical scales- a few cm^2 for
coins and bas-reliefs to a several km^2 for ancient ruins like Hampi. In a virtual museum,
one should be able to experience objects of all sizes at different scales to get a sense of
overall structure to a finer details from the same data set. Hence, we have implemented
adaptive scaling in both graphic and haptic domains. In order to scale the surface we
resize depth data of resolution $N \times N$ depending on the level we want, with $N \times N$ as
the lowest level. If we load the level $N \times N$ into the haptic space the full object can be
rendered visually as well as haptically. Users can select the level as well as the region of
interest at run time either using buttons in the haptic device or using keyboard functions.
Additionally, we have developed a graphical user interface for easy acessibility. The
pink window in Fig. 4 represents the selected region to be zoomed in.

Depending on the scale selected by the user, only the corresponding depth data is
dynamically loaded into the active haptic space and an appropriate haptic force is rendered. As only a limited subset of data is loaded, the rendering is very fast. In general,
at higher levels of resolution, the user should be able to view higher depth value at each
point and also more finer details. The haptic force also vary accordingly. Hence in order
to incorporate realistic haptic and graphic perception, we need to appropriately scale
the depth values at each level of depth map. Further, trying to map a large physical dimension over a small haptic work space (typically about 4 inch cube of active space)
leads to a lot of unwanted vibrations (something similar in concept to aliasing) during
rendering. Hence the depth values need to be smoothed before being downsampled and
mapped into the haptic work space. The next section explains the generation of different
levels of depth data.

Fig. 4. Illustration of selection of a window for graphic and haptic rendering. (Data Courtesy:*www.cc.gatech.edu/projects/large_models*)

5 Multi-scale Data Generation

The aim of this work is to allow users to have access to the cultural heritage at different levels of details. To obtain the depth data at different levels of details, we perform Gaussian low-pass filtering followed by down sampling with a factor M where M can be any integer. We can also use fractional values of M, but it requires rational function approximation methods. In our work, we illustrate with $M = 2$. For that the data pyramid offers a flexible and convenient multiresolution format that mirrors the different levels of details [5]. It consists of the available highest resolution depth data and a series of successively lower resolution data. Low-pass filtering before sub-sampling is done to prevent aliasing of data. Consequently, instability in the haptic domain is also prevented as smoothing removes higher frequency components responsible for micro textures on the surface. The presence of micro textures would have made sensing more realistic, but this makes the haptic rendering process miss to some extent a full understanding of the object at hand. The fine texture is experienced when the object is rendered at a finer scale by zooming into the object. The base, or zero level of the pyramid is equal to the original depth map (g_0). Level 1 of the pyramid corresponds to depth map g_1 which is reduced or low-pass filtered version of g_0. Each value in level 1 is computed as a weighted average of values in level 0 within a 5×5 window. Each value in level 2 (g_2) is then obtained from values of level 1 by applying the same pattern of weights. Fig. 5 shows the Gaussian pyramid of depth map. The depth value at each point at the level l is given by the following equation:

$$g_l(i,j) = \sum_{m=-2}^{2} \sum_{n=-2}^{2} w(m,n)g_{l-1}(2i+m,2j+n), \qquad (3)$$

For levels $1 < l < N+1$, and nodes (i,j), $0 < i < C_l$, $0 < j < R_l$, the upper level of the pyramid can be represented in the above given form. Here N refers to the number of levels in the pyramid, while C_l and R_l are the dimensions at level l. The weighting pattern $w(m,n)$ is the Gaussian kernel [3].

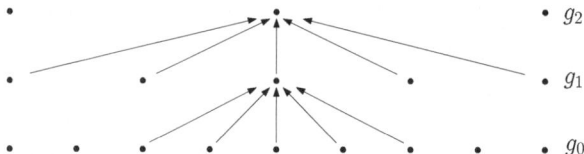

Fig. 5. A one-dimensional graphical representation of the process which generates a Gaussian pyramid. Each row of dots represents nodes within a level of the pyramid. The value of each node in the zero level is the magnitude of the corresponding depth map. The value of each node at a higher level is the decimated and weighted average of node values in the preceding level.

6 Results

The proposed method was implemented in visual C++ in a Windows XP platform with a CORE 2QUAD CPU @ 2.66 GHZ with 2 GB RAM. We have experimented with various models of cultural heritage objects and a few of them are displayed below. The Fig. 6 shows the model of Ganesh, visually rendered in OpenGL. For haptics rendering we use HAPI library. The blue ball represents the position of the proxy constrained to be on the surface. The discrete position in the model is displayed in a fixed 200×200 haptic space. The size and spatial resolution of the model depend on two factors: the active space of the haptic device used to render the model, and the resolution at which the model should be displayed. We use a 3-DOF haptic device from NOVINT with a 4 inch cube of active space. While interacting with the object haptically, the average proxy updation time is 0.0056 ms which is much less than the required upper bound of 1 ms, and hence the user has very smooth haptic experience. The average time required for dynamic data generation and loading it into the haptic space depends on the resolution of input depth data and it was observed to be around 6.5 s and 2.0 s respectively for depth data with resolution of 800×800. As explained in the previous section, Fig. 6a corresponds to the lowest level of details. We also carry out the rendering at finer levels of details by successively zooming into the heritage object. These are shown in Fig. 6b and Fig. 6c.

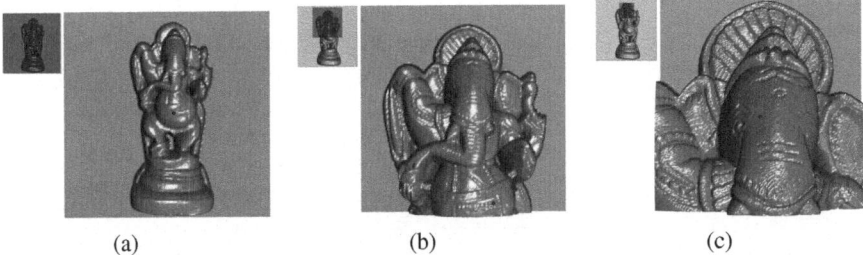

Fig. 6. Model of Ganesh, at (a) least level of details (b) at double the resolution and (c) at the finest resolution. (Data Courtesy: *www.archibaseplanet.com*)

In above cases, each figure consists of two parts where the left part is the reference for the users to select the part of the object they wish to explore haptically. The right part of the figure corresponds to the selected region at the appropriate resolution for haptic rendering. Fig. 6c shows the scaled up version of Fig. 6b. It is quite clear from Fig. 6c that the users are able to feel even minute details of the sculpture and have visual perception of closeness in depth. Hence they can have a more realistic experience. The object rendered in Fig. 7 is a special case that illustrates how one can handle holes in the model. For any haptic rendering, holes in the model are difficult to accommodate as the proxy would sink through the hole and the user will perceive a wrong depth in the region around the holes. We avoid this by defining a base plane on which the object lies. Wherever, there is a hole, the depth at that point is replaced by $z(x,y) = z_{MAX}$ where z_{MAX} is the maximum depth. This object has several holes, but the users reported a very good experience even in presence of such holes. Fig. 7a allows rendering at a coarser level while Fig. 7b and Fig. 7c allow rendering at a much finer scale.

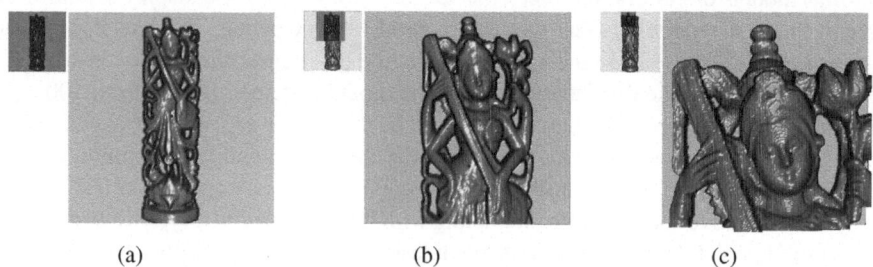

Fig. 7. Model of Goddess Saraswati at (a) level 1 (b) level 2 (c) level 3. (Data Courtesy: *www.archibaseplanet.com*)

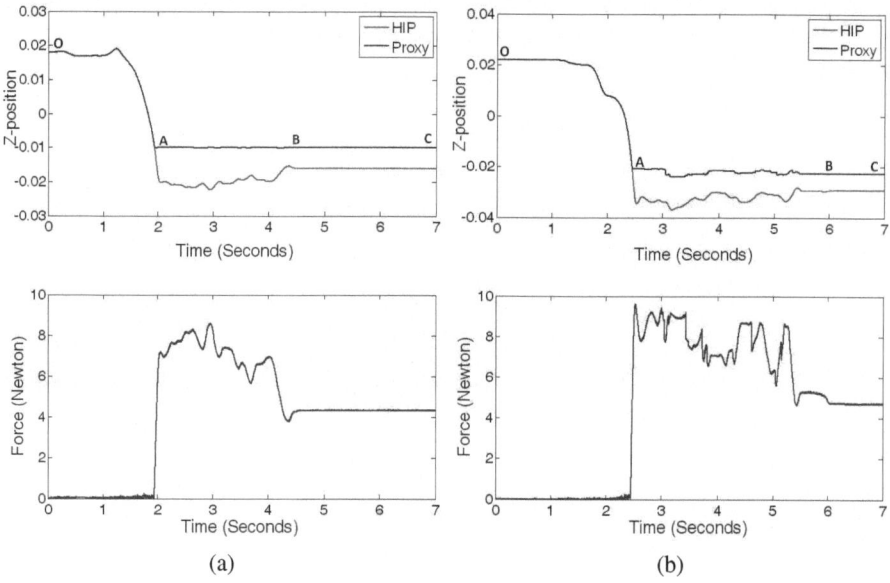

Fig. 8. Force vs. time graph for a particular interaction with the depth data a) on a flat region b) on a curved region (The top figure correspond to changes in z-coordinate only)

Validation of result is often a difficult task during haptic rendering, we demonstrate this using Fig. 8 that shows the reaction force versus time relation while haptically interacting with the depth data. The red line and the blue line in the figure shows the z-component of the HIP and the proxy point, respectively, during the interaction. The reaction force on the haptic device is also shown during the same time interval. In free space the HIP and proxy positions are almost the same as shown in part OA of the HIP position and hence the reaction force on the haptic device is zero. As the HIP penetrates the object the proxy stays on the surface according to the iteration method discussed in section 3. The proxy point moves continuously during interaction, whenever there is a change in HIP position. This is shown with the part AB in the curve. After the point B the HIP position is kept constant inside the objects. As soon as the HIP is kept constant, the proxy quickly attains a stable position as shown in the Section BC in the curve. Fig. 8a shows the plots corresponding to the interaction on a flat region and Fig. 8b the same on a curved region, when a larger variation in force is observed.

Fig. 9. Illustration of hapto-visual immersion of a subject for a virtual cultural heritage object. On the right, the user wearing anaglyphic glasses is holding the FALCON haptic device while interacting with the cultural heritage model displayed on the screen.

In Fig. 9, we show the actual set up of our virtual haptic museum. A user wearing the anaglyphic glasses watches the stereoscopic visual rendering of the artefact and at the same time haptically interacts with the object with his hand. This provides an excellent hapto-visual immersion of the subject into the virtual object. However, for the visually impaired users, the selection of scale and the location for rendering cannot be based on the small navigation window on the screen. For such subjects, we use the buttons available on the haptic device for the user to explore the object at different scales and locations.

7 Conclusions

In this work we have proposed a new technique of rendering cultural heritage objects represented as depth map data. Our primary goal is to provide access of cultural heritage objects and sites to the visually impaired people. Additionally, our method gives a better immersive experience to the sighted persons. We include scalability and stereoscopic display of $3 - D$ models as additional features to enhance the realism in experience. We conducted experiments with several $3 - D$ models of cultural significance. We also tested the rendering technique with some subjects and observed that hapto-visual rendering of virtual $3 - D$ models using the proposed method greatly augmented the user's experience.

References

1. Avila, R.S., Sobierajski, L.M.: A haptic interaction method for volume visualization. In: IEEE Visualization Conference, vol. 0, p. 197 (1996)
2. Bergamasco, M., Frisoli, A., Barbagli, F.: Haptics technologies and cultural heritage applications. In: IEEE Proceedings of the Computer Animation, CA 2002, p. 25. IEEE Computer Society, Washington, DC (2002)
3. Burt, P., Adelson, E.: The laplacian pyramid as a compact image code. IEEE Transactions on Communications 31(4), 532–540 (1983)
4. El Far, N.R., Georganas, N.D., El Saddik, A.: An algorithm for haptically rendering objects described by point clouds. In: Proceedings of the 21th Canadian Conference on Electrical and Computer Engineering, Niagara, ON, Canada (2008)
5. Gluckman, J.: Scale variant image pyramids. In: IEEE Computer Society Conference on Computer Vision and Pattern Recognition, 2006, vol. 1, pp. 1069–1075 (June 2006)
6. Kim, L., Kyrikou, A., Desbrun, M., Sukhatme, G.: An implicit-based haptic rendering technique. In: Proceeedings of the IEEE/RSJ International Conference on Intelligent Robots (2002)
7. Laycock, S.D., Day, A.M.: A survey of haptic rendering techniques. Computer Graphics Forum 26(1), 50–65 (2007)
8. Lee, J.-K., Kim, Y.J.: Haptic rendering of point set surfaces. In: World Haptics Conference, vol. 0, pp. 513–518 (2007)
9. Leeper, A., Chan, S., Salisbury, K.: Constraint based 3-dof haptic rendering of arbitrary point cloud data. In: RSS Workshop on RGB-D Cameras, University of Southern California (June 2011)
10. Mcneely, W.A., Puterbaugh, K.D., Troy, J.J.: Six degree-of-freedom haptic rendering using voxel sampling. In: Proc. of ACM SIGGRAPH, pp. 401–408 (1999)
11. Srinivasn, M.A., Morgenbesser, H.B.: Force shading for haptic shade perception. In: Proceedings of the ASME Dynamic Systems and Control Division, vol. 58, pp. 407–412 (1996)
12. Renz, M., Preusche, C., Ptke, M., Hirzinger, G.: Stable haptic interaction with virtual environments using an adapted voxmap-pointshell algorithm. In: Proc. Eurohaptics, pp. 149–154 (2001)
13. Ruspini, D.C., Kolarov, K., Khatib, O.: The haptic display of complex graphical environments. In: Proc. of ACM SIGGRAPH, pp. 345–352 (1997)
14. Salisbury, K., Conti, F., Barbagli, F.: Haptic rendering: Introductory concepts. IEEE Computer Graphics and Applications 24(2), 24–32 (2004)
15. Sreeni, K.G., Chaudhuri, S.: Haptic Rendering of Dense 3D Point Cloud Data. In: IEEE Haptic Symposium, Vancouver, BC, Canada, March 4-7 (2012)
16. Zhang, C., Li, J.: Interactive browsing of 3d environment over the internet. In: Proc. SPIE Visual Communications and Image Processing, VCIP 2001, pp. 509–520 (2001)
17. Zilles, C.B., Salisbury, J.K.: A constraint-based god-object method for haptic display. In: IEEE/RSJ International Conference on Intelligent Robots and Systems, vol. 3, p. 3146 (1995)

Haptic Communication Tools for Collaborative Deformation of Molecules

Jean Simard and Mehdi Ammi

Universit de Paris-Sud,
CNRS/LIMSI, Btiment 508,
F-91403 Orsay Cedex, France
jean.simard@limsi.fr

Abstract. Several previous studies have investigated collaborative approaches for processing complex environments. Beyond the improvement of performance and working efficiency, these studies highlighted two important constraints, which limit the efficiency of these approaches. First, social loafing which is linked to the redundancy of roles in the same group. Second, coordination conflicts which are linked to the limits of communication in standard collaborative environments. This paper addresses these issues by providing an efficient group structure to overcome the social loafing, which is then coupled with haptic metaphors to improve communication between partners. The experimental study, conducted in the context of molecular docking, shows an improvement for group efficiency as well as communication between partners.

Keywords: collaboration, coordination, haptic, molecular docking.

1 Introduction

In the field of molecular modeling, docking is a search process which aims to bind two or more molecules to predict the best complex of molecules. This study of molecular conformations and interactions allows biologists to understand the functions of the manipulated molecules. During the docking process, the biologists analyze and identify the best structural and chemical complementarity between two molecules [8] in order to find the best assembly solution. Moreover, they must consider the flexibility of the molecule [13] by deforming the geometric structure at different scales (intermolecular, intramolecular level, atomic level).

Today, several solutions based on Virtual Environments (VE) are proposed to process these complex problems. The objective is to introduce the experience and skills of biologists during the different steps of the docking process. However, docking relevant molecules of large size is beyond the capability of a single biologist working alone. Collaborative Virtual Environments (CVE) provide new approaches to deal with these complex problems [15,16]. Hutchins [6] showed that collaboration between several users improved global efficiency for the realization of a given task. The experimental study conducted on the collaborative control of an airplane cockpit showed that group efficiency is more

P. Isokoski and J. Springare (Eds.): EuroHaptics 2012, Part I, LNCS 7282, pp. 517–527, 2012.
© Springer-Verlag Berlin Heidelberg 2012

important than the sum of individual work. This phenomenon, called "workload distribution", was defined by Hollan [4] as follows:

> *Unlike traditional theories, however, [the theory of distributed cognition] extends the reach of what is considered cognitive beyond the individual to encompass interactions between people and with resources and materials in the environment.*

Based on a "workload distribution" approach, Simard [16] performed several experiments to study the contribution of CVE, in particular closely coupled collaboration, for the processes of complex docking. The results highlighted two important constraints: (1) social loafing and (2) coordination conflicts.

(1) **Social loafing** is defined by Schermerhorn [14] as

> *The tendency of group members to do less that they are capable of as individuals.*

Social loafing has negative consequences on the group efficiency. In fact, the inaction of some members of the group induces a misbalanced workload. Kraut [9] proposed an efficient solution for this issue. It consists of assigning different roles for each member. Each user is in charge of a part of the task process which acts as an incentive for better group performance.

(2) **Coordination conflicts** are due to imprecise or incomplete communication. This leads to poor coordination of actions during closely coupled collaborations (e.g. manipulation of the same structure, selection of the same artefact) [7,16].

Several solutions were investigated to improve communication during closely coupled manipulations. The use of haptic feedback to support different levels of communication was widely investigated. Basdogan [1] studied the role of this channel for implicit communication (i.e. haptic feedthrough) during collaborative manipulation tasks. The experimental results showed that collaboration through the haptic channel significantly improved group efficiency and sense of togetherness in CVE. Based on this study, Oakley [11] proposed to improve haptic communication for tasks of 2D UML diagram creation. The components of the proposed approach were, to: (1) push, (2) pull, (3) attract, (4) be attracted by, or (5) reduce the speed of the partners cursor (damping), when the two cursors approach each other. These haptic functions significantly improved group efficiency. Moreover, users found these tools useful even if they caused some fatigue and frustration. Moll [10] proposed similar tools for 3D environments. The experimental results showed that the haptic communication tool significantly improved the communication of objectives and spatial information.

In this paper, we propose to improve the collaborative manipulation of molecules during a docking process. The proposed approach combines an efficient group structure to overcome social loafing, with haptic metaphors to improve communication between partners. The paper is structured as follows. In Section 2, we present the employed group structure and the proposed haptic communication metaphors. Section 3 describes the experimental protocol for the evaluation. Section 4 presents the results. Finally, we will conclude and present some perspectives in Section 5.

2   Proposed Approach

To identify the best group structure and the required communication tools, it is necessary to know the different tasks involved. Biologists identify two main tasks during the docking process [2]:

- The relative movement and rotation of the overall molecules: involves moving and rotating one molecule with regard to the second molecule to enable the assembly of the two structures.
- The deformation of the geometric structures: involves manipulating the molecule at the level of atoms and residues (i.e. group of atoms) to change the general shape of the structure. The objective is to find the most suitable configuration in order to assemble the manipulated molecule with the second structure.

To enable a relevant assembly of the two molecules, these different tasks must be managed simultaneously.

2.1   Proposed Collaborative Configuration of Work

Previous studies [15] show that collaborations, where partners have similar roles lead to an increase in social loafing. To deal with this situation, we propose a new configuration of collaborative work where each member of the group has an identified role. Based on the analysis of the activity of the docking process [16], we propose the two following roles: the *coordinator* and the *operator*.

- The *coordinator* is the global leader of the docking process. The role consists of analysing and exploring the overall structures of the molecules, and then testing the relevant assembly solutions by moving and orienting one molecule relative to the second structure. If the two conformations (i.e. overall shape) are not adapted for the assembly, the coordinator will first identify the required deformations and corresponding points of manipulation, and then entrust the corresponding tasks to the *operator* by designating the residues to manipulate and the corresponding spatial targets. Based on these two tasks, we provide two tools to the *coordinator*. The first tool enables the control of the overall position of one molecule. It links the molecule to a haptic arm (i.e. Desktop PHANToM) through a spring-damper force model. The second tool enables the designation of the targets to manipulate. Given the numerous structures that are to be simultaneously manipulated, we propose to use two *operators* for monomanual configuration of work. In fact, the bimanual mode shows some limits for the manipulation of complex structures Guiard [3].
- The *operator* manipulates and deforms the molecule by grabbing designated atoms or residues[1]. We provide the *operator* with a tool that enables the selection and grasping of the designated targets [15].

---

[1] Internal structures of the molecule composed by 10 to 50 atoms.

2.2   Proposed Haptic Communication Metaphors

Based on the proposed collaborative working configuration (*coordinator* and *operators*), we propose a designation metaphor to improve communication between the *operators* and the *coordinator*. This metaphor, inspired by the works of Moll [10], enables the indication of a region of interest (ROI) on the 3D structure of the molecule. It includes two components: (1) a visual component and (2) a haptic component. The visual component highlights, through the involved target, the 3D structure of the molecule. The haptic component enables active notification of new designated targets. Moreover, it provides an active guidance tool to facilitate reaching the designated targets on the 3D structure.

We summarize the working of the designation metaphor as follows:

1. the *coordinator* $\mathcal{A}$ identifies the required target.
2. the *coordinator* $\mathcal{A}$ designates the target.
3. a visual feedback highlights the target with a neutral color (gray atoms ar residues). All users ($\mathcal{A}$, $\mathcal{B}$ and $\mathcal{C}$) are notified about the new designation through a haptic vibration.
4. the *operator* $\mathcal{B}$ or $\mathcal{C}$ (for the example, the *operator* $\mathcal{B}$) accepts the target. The target is now highlighted with the same color as the cursor of the *operator* $\mathcal{B}$. The vibrations are stopped.
5. the *operator* $\mathcal{B}$ is attracted to the target through the following spring-damper force model:

$$F(\mathbf{x}) = \begin{cases} k\,(t - t_0)\,(\mathbf{x} - \mathbf{x_t}) - b\dfrac{\partial \mathbf{x}}{\partial t} & \text{if } t \geq t_0 \\ 0 & \text{if } t < t_0 \end{cases}$$

where $\mathbf{x}$ is the cursor's position, $\mathbf{x_t}$ is the target's position, t_0 the time of acceptation of the target and k and b respectively the spring and damping constants. The force is saturated when over 4 N.
6. the process ends when user $\mathcal{B}$ selects the target.

3   Experiment

This section presents the experimental evaluation of the proposed metaphors in the context of molecular docking.

Hypothesis. Based on the identified constraints of collaborative work and corresponding performance factors [15], we propose to investigate the following hypothesis.

$\mathcal{H}_1$ **Better efficiency** The proposed working configuration and the haptic communication metaphor will improve the global efficiency of the group.
$\mathcal{H}_2$ **Better coordination** The haptic communication metaphor will improve communication between users and therefore, improve coordination.

Hardware Setup. Experiments were conducted on a collaborative platform, coupling standard desktop workstations with a large screen display, for a public and global view fig 1. This platform is integrating biologists solutions with virtual reality softwares: VMD [5] is used for the molecular visualization, NAMD [12] for the molecular simulation and IMD [17] to create an interactive molecular simulation. One haptic interface is plugged on each desktop workstation with a VRPN server [18] which communicates with NAMD through VMD and IMD: 2 Omni PHANToM for deformation tools, 1 Omni PHANToM for designation tool and 1 Desktop PHANToM for the molecule's manipulation tool.

All participants are sitting in front of the large screen and can verbally communicate without restriction. The *coordinator* is placed in the center and the two *operators* are placed on each side of the *coordinator*.

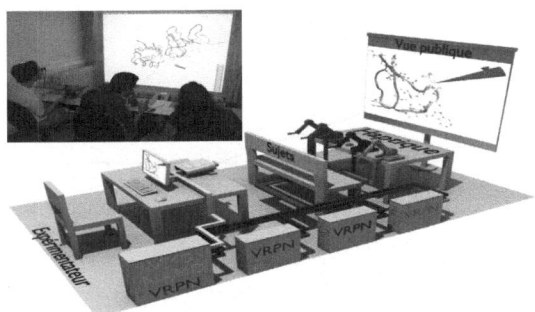

Fig. 1. Collaborative platform for molecular deformations

Molecular Deformation Task. The proposed experiment consists of presenting a molecule with an *initial conformation* (i.e. a given shape and position) fig. 2. Participants are then asked to move and deform the molecule to reach a *target conformation* fig. 2. The *target conformation* provides the best geometrical complementarity with the second molecule. The participants evaluate the similarity between the *current conformation* and the *target conformation* with the RMSD score. This score, used by biologists, is defined by the following formula:

$$\text{RMSD}\left(\mathbf{d}, \mathbf{g}\right) = \sqrt{\frac{1}{N} \sum_{i=1}^{N} \|d_i - g_i\|^2} \tag{1}$$

where N is the total number of atoms and c_i, g_i are respectively atoms i from the *current conformation* $\mathbf{d}$ and the *target conformation* $\mathbf{g}$. The RMSD score is displayed on the right of the screen in a blue bar. The smallest reached RMSD score is displayed with an orange bar inside the blue one.

As developed above, the *coordinator* is in charge of the control of the overall position of the molecule as well as the designation of residues and atoms to

manipulate. The *operators* select the designated residues (*current residue* on
fig. 2) and pull them to a given position (*target residue* on fig. 2). Finally, some
parts of the molecule are fixed (*fixed residue* on fig. 2) to avoid the displacement
of the molecule outside of the working space.

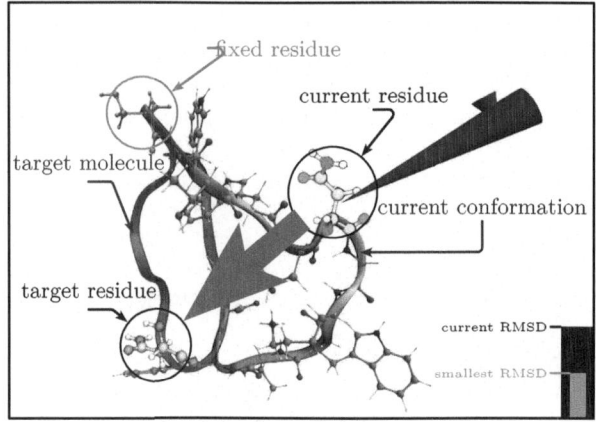

Fig. 2. Visual display to carry out the deformation process

Procedure. The experiment presents three successive steps:

1. The experimenter presents the objectives of the experiment, the tasks, the
   roles (one *coordinator* and two *operators*)to the group. The group then de-
   liberates and assigns each member a role.
2. In the second step we successively present the various tools to the participants through elementary experiments without effective evaluation. First, we
   present the platform and the manipulation tools (without the haptic communication metaphor) to the *operators*. A second training scenario is proposed
   to introduce the haptic metaphor (to the *operator* and *coordinator*). Finally,
   the tool to move the molecule is presented to the *coordinator* during a third
   training scenario.
3. The third step consists of presenting the effective experiential scenario. Participants are evaluated for both molecules (Ubiquitin then NusE:NusG) with
   and without haptic communication metaphor. The conditions are counterbalanced across the groups (see below the description of this condition).
   This step begins with a short period of exploration (1 mn) during which
   the groups elaborate an overall strategy. Then the participants begin the
   manipulation of the molecules in order to reach the smallest RMSD score
   (8 mn).

Subjects. 1 woman and 23 men ($\mu = 27.4, \sigma = 3.8$) participated in the experiment. They were all students, researchers or researcher assistants in bioinformatics, linguistic, virtual reality or acoustic. They were all French speakers and had

no visual or audio deficiency. No remuneration was given to the participants. To reduce learning effects during the experiment, we chose participants who already had experience with molecular deformation in virtual reality platforms.

Experimented Conditions. Two main factors were investigated in this experiment: the presence of communication metaphor and the complexity of the experimented molecules.

$[\mathcal{V}_{i1}]$ **Haptic communication metaphor** , This within subject counterbalanced variable has 2 modalities: "without metaphor" or "with metaphor". The "without metaphor" condition provides the designation tools with visual feedback only; the "with metaphor" condition provides the designation tools with visuo-haptic feedback.

$[\mathcal{V}_{i2}]$ **Complexity of the molecules** , This within subject variable has 2 modalities: "Ubiquitin" and "NusE:NusG". The complexity of both molecules is defined by the size of the molecules (number of atoms and of residues) and the nature of the task. The molecule Ubiquitin presents 1231 atoms (76 residues); the corresponding task concerns the deformation of internal structures of the molecule. The complex of molecules NusE:NusG (a set of two molecules) is composed of NusE with 1294 atoms (80 residues) and NusG with 929 atoms (59 residues); the corresponding task concerns the deformation and the movement of the molecule NusG (the backbone[2] of the NusG is entirely fixed in the virtual environment).

Objective Measurements. The analysis is based on the following objective measures:

$[\mathcal{V}_{d1}]$ **Smallest RMSD score** , Smallest RMSD score reached during the task fulfilment.

$[\mathcal{V}_{d2}]$ **Time of the smallest RMSD score** , Completion time to reach the smallest RMSD score during the realization of the task.

$[\mathcal{V}_{d3}]$ **Frequency of selections** , Number of selections realized by the *operators* during the deformation divided by the total duration of the task.

$[\mathcal{V}_{d4}]$ **Mean time of acceptation of targets** , Duration between a new designation (by the *coordinator*) and the acceptation of this designation (by an *operator*).

$[\mathcal{V}_{d5}]$ **Mean time to reach targets** , Duration between the acceptation (by the *operator*) and the selection of the target (by the *operator*).

$[\mathcal{V}_{d6}]$ **Number of accepted selections** , Number of fulfilled designations done by the *coordinator* that have been accepted by an *operator*.

$[\mathcal{V}_{d7}]$ **Mean speed of the *coordinator*** , Mean speed of the *coordinator*'s endeffector during the whole task.

4   Results and Discussion

All the results were analyzed using an analysis of variance with the Wilcoxon signed-rank test.

---

[2] Main internal structure of a molecule mainly composed of carbon atoms.

4.1   Improvement of Efficiency

The fig. 3(a) shows that there is not a significant effect of the haptic communication metaphor $[\mathcal{V}_{i1}]$ on the smallest RMSD score $[\mathcal{V}_{d1}]$ ($W = 87$, $p = 0.348$). However, the fig. 3(b) shows a significant effect of the haptic communication metaphor $[\mathcal{V}_{i1}]$ on the completion time to reach the smallest RMSD score $[\mathcal{V}_{d1}]$ for the complex NusE:NusG ($W = 36$, $p = 0.008$) with a decrease of -48.3 %. However, there is no significant effect of the haptic communication metaphor $[\mathcal{V}_{i1}]$ for the molecule Ubiquitin ($W = 13$, $p = 0.547$). The complex NusE:NusG presents the most difficult scenario due to the important number of residues to deform. On a simple scenario (Ubiquitin), there is no gain and no loss of working efficiency with the haptic communication metaphor. In fact, simple tasks involve less designations which limits the effect of the metaphor on the performance results of the overall process.

We observe on fig. 3(c) that *operators* significantly decreased the frequency of selection $[\mathcal{V}_{d3}]$ by -12.8 % with the haptic communication metaphor $[\mathcal{V}_{i1}]$ ($W = 401$, $p = 0.009$). The completion time performance of the groups was better (-48.3 % for the complex NusE:NusG) or at least the same (for the molecule Ubiquitin)

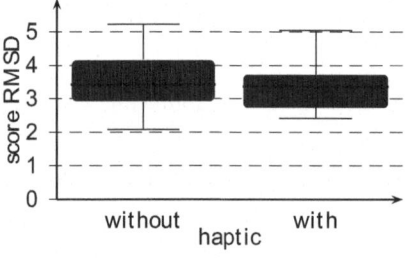

(a) Smallest RMSD scorereached $[\mathcal{V}_{d1}]$ during the realization of the task according to the two investigated conditions in haptic $[\mathcal{V}_{i1}]$

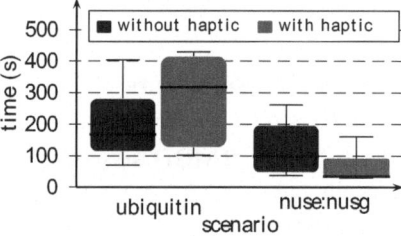

(b) Completion time to reach the smallest RMSD score $[\mathcal{V}_{d2}]$ according to the two investigated molecules $[\mathcal{V}_{i2}]$

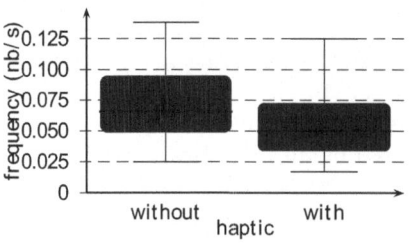

(c) Frequency of selections $[\mathcal{V}_{d3}]$for the *operators* according to the two investigated conditions in haptic $[\mathcal{V}_{i1}]$

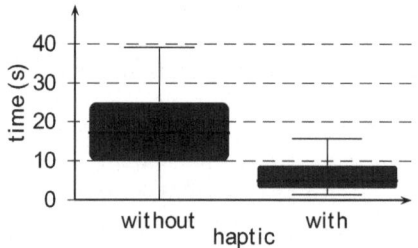

(d) Time between the acceptation and the selection for the *operators* $[\mathcal{V}_{d5}]$ according to the two investigated conditions in haptic $[\mathcal{V}_{i1}]$

Fig. 3. Results related to the working efficiency

with haptic, even if the frequency of selection iwas reduced: the efficiency of the groups was always better.

Finally, fig. 3(d) shows that the haptic communication metaphor $[\mathcal{V}_{i1}]$ presented a significant improvement in the time between the acceptation and the selection steps $[\mathcal{V}_{d5}]$ ($W = 473$, $p \ll 0.05$). The time was decreased by -64.3 %. Based on these results, the $\mathcal{H}_1$ hypothesis is validated.

4.2 Improvement of Coordination

Fig. 4(a) shows that the haptic communication metaphor $[\mathcal{V}_{i1}]$ introduced a significant decrease (-51.5 %) on the mean time of targets acceptation $[\mathcal{V}_{d4}]$ ($W = 404$, $p = 0.008$). Moreover, fig. 4(b) shows that the haptic communication metaphor $[\mathcal{V}_{i1}]$ significantly reduced the rate of acceptation $[\mathcal{V}_{d6}]$ by 25.7 % ($W = 93.5$, $p = 0.004$). Finally, the haptic communication metaphor $[\mathcal{V}_{i1}]$ had a significant effect of 25.7 % on the mean speed of the *coordinator* $[\mathcal{V}_{d7}]$ as shown on the fig. 4(c) ($W = 15$, $p = 0.004$).

The mean time to reach the target $[\mathcal{V}_{d5}]$ was directly linked tothe communication between the members of the group. During the deformation process,

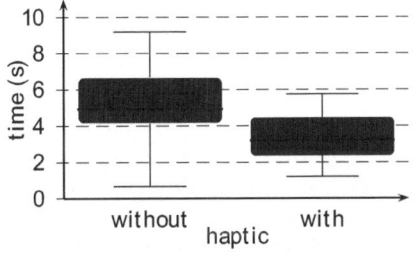

(a) Time between a designation by the *coordinator* and the acceptation by an *operator* $[\mathcal{V}_{d4}]$ according to the two investigated conditions in haptic$[\mathcal{V}_{i1}]$

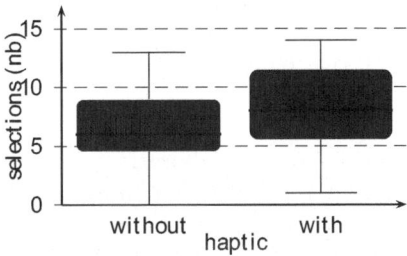

(b) Number of designations accepted by the *operators* $[\mathcal{V}_{d6}]$ according to the two investigated conditions in haptic $[\mathcal{V}_{i1}]$

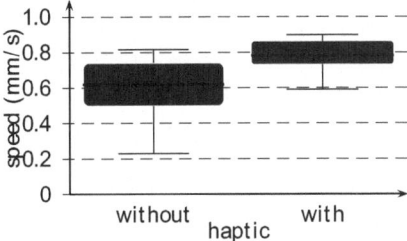

(c) Mean speed of the *coordinator*'s end-effector $[\mathcal{V}_{d7}]$ according to the two investigated conditions in haptic $[\mathcal{V}_{i1}]$

Fig. 4. Results related to the improvement of the coordination

the *operator* needs to be aware of the new designations in order to perform the corresponding manipulations. Two strategies of communication can be adopted. First, the *operator* detects the new designations based on the visual feedback. However, this strategy is constrained by the complexity of the molecule. In fact, the designation targets may be hidden by the structures of the molecule. In the second strategy, the *coordinator* verbally indicates the new designations to the *operators*. However, verbal communication is not precise enough to indicate 3D positions in the 3D virtual space. The haptic communication metaphor addresses these two constraints. The haptic tool provides an active notification of all new designations through a vibration feedback even if the *operators* are working on other regions. Moreover, the haptic metaphor enables the active guidance of the *operator* to efficiently reach the target on the molecular structure.

The fig. 4(b) shows that the rate of unaccepted designations was significantly reduced. Moreover, 4(c) shows that the *coordinator* worked significantly faster. Indeed, the *coordinator* must wait until the *operators* accept the new designations. Since the *operators* are more effective at identifying and selecting the designated targets (i.e. active notification and gestural guidance), the *coordinator* can designate more targets with better acceptation rate. These results show that the communication with the haptic metaphor is faster and provides congruent spatial information. Based on these results, the H_2 hypothesis is validated.

5   Conclusion

This paper presents a new strategy to address the two important constraints of closely coupled collaboration: social loafing and coordination conflicts. The proposed approach is based on a suitable group structure presenting two identified roles, and an efficient communication metaphor for the active notification and designation of targets in 3D complex environments. The experimental results, obtained in the context of complex docking process, show that the haptic metaphor significantly improves the performance and efficiency of the group working on complex tasks. In fact, simple tasks involve fewer designations which limit the contribution of the metaphor. Moreover, the results show that the metaphor improves communication through an active notification procedure coupled with an efficient gestural guidance strategy to reach effectively the designated targets. Based on these encouraging results, we will, in future works, investigate other steps of the collaborative docking process, for instance, the simultaneous deformation of the same molecular structure which requires a strong coordination of actions. Furthermore, to support some spatial information, we propose to study and integrate an audio component into the collaborative metaphor.

References

1. Başdoğan, Ç., Ho, C.H., Srinivasan, M.A., Slater, M.: An experimental study on the role of touch in shared virtual environments. ACM Transaction on Computer-Human Interaction 7(4), 443–460 (2000)

2. Férey, N., Nelson, J., Martin, C., Picinali, L., Bouyer, G., Tek, A., Bourdot, P., Burkhardt, J.M., Katz, B., Ammi, M., Etchebest, C., Autin, L.: Multisensory VR interaction for protein-docking in the CoRSAIRe project. Virtual Reality 13(4), 273–293 (2009)
3. Guiard, Y.: Asymmetric division of labor in human skilled bimanual action: the kinematic chain as a model. Journal of Motor Behavior 19(4), 486–517 (1987)
4. Hollan, J., Hutchins, E.L., Kirsh, D.: Distributed cognition: toward a new foundation for human-computer interaction research. ACM Transaction on Computer-Human Interaction 7(2), 174–196 (2000)
5. Humphrey, W.F., Dalke, A., Schulten, K.: VMD: Visual Molecular Dynamics. Journal of Molecular Graphics 14(1), 33–38 (1996)
6. Hutchins, E.L.: How a cockpit remembers its speeds. Cognitive Science 19(3), 265–288 (1995)
7. Kankanhalli, A., Tan, B., Wei, K.K.: Conflict and performance in global virtual teams. Journal of Management Information Systems 23(3), 237–274 (2007)
8. Kessler, N., Perl-Treves, D., Addadi, L., Eisenstein, M.: Structural and chemical complementarity between antibodies and the crystal surfaces they recognize. Proteins: Structure, Function, and Bioinformatics 34(3), 383–394 (1999)
9. Kraut, R.E.: Applying social psychological theory to the problems of group work, ch. 12, pp. 325–356. Morgan Kaufmann, San Francisco (2003)
10. Moll, J., Sallnäs, E.L.: Communicative Functions of Haptic Feedback. In: Altinsoy, M.E., Jekosch, U., Brewster, S. (eds.) HAID 2009. LNCS, vol. 5763, pp. 1–10. Springer, Heidelberg (2009)
11. Oakley, I., Brewster, S.A., Gray, P.D.: Can you feel the force? An investigation of haptic collaboration in shared editors. In: Proceedings of EuroHaptics, pp. 54–59 (January 2001)
12. Phillips, J.C., Braun, R., Wang, W., Gumbart, J., Tajkhorshid, E., Villa, E., Chipot, C., Skeel, R.D., Kalé, L., Schulten, K.: Scalable molecular dynamics with NAMD. Journal of Computational Chemistry 26(16), 1781–1802 (2005)
13. Qi, S., Krogsgaard, M., Davis, M.M., Chakraborty, A.K.: Molecular flexibility can influence the stimulatory ability of receptor–ligand interactions at cell–cell junctions. Proceedings of the National Academy of Sciences of the United States of America 103(12), 4416–4421 (2006)
14. Schermerhorn, J., Hunt, J.G., Osborn, R.N., Uhl-Bien, M.: Organizational behavior, 11th edn. John Wiley & Sons, Inc. (January 2009)
15. Simard, J., Ammi, M., Auvray, M.: Study of synchronous and colocated collaboration for search tasks. In: Fellner, D. (ed.) Proceedings of Joint Virtual Reality Conference (JVRC), pp. 51–54. Eurographics Association (September 2010)
16. Simard, J., Ammi, M., Mayeur, A.: How to improve group performances on collocated synchronous manipulation tasks? In: Proceedings of Joint Virtual Reality Conference. JVRC – EuroVR-EGVE. Eurographics Association, Nottingham (2011)
17. Stadler, J., Mikulla, R., Trebin, H.R.: IMD: a software package for molecular dynamics studies on parallel computers. International Journal of Modern Physics 8(5), 1131–1140 (1997)
18. Taylor II, R.M., Hudson, T.C., Seeger, A., Weber, H., Juliano, J., Helser, A.T.: VRPN: a device-independent, network-transparent VR peripheral system. In: Proceedings of the ACM Symposium on Virtual Reality Software and Technology. Virtual Reality Software and Technology, pp. 55–61. ACM, New York (2001)

Towards a Standard on Evaluation of Tactile/Haptic Interactions

Ian Sinclair[1], Jim Carter[2], Sebastian Kassner[3], Jan van Erp[4],
Gerhard Weber[5], Linda Elliott[6], and Ian Andrew[7]

[1] MPB Technologies Inc., Montréal, Canada
ian.sinclair@mpbc.ca
[2] University of Saskatchewan, Saskatoon, Canada
carter@cs.usask.ca
[3] Technische Universität Darmstadt, Darmstadt, Germany
s.kassner@emk.tu-darmstadt.de
[4] TNO Human Factors, Soesterberg, Netherlands
jan.vanerp@tno.nl
[5] Technische Universität Dresden, Dresden, Germany
gerhard.weber@tu-dresden.de
[6] U.S. Army Research Lab, Fort Benning, USA
linda.r.elliott@us.army.mil
[7] HF Engineer, United Kingdom
andyand@talktalk.net

Abstract. Tactile and haptic interaction is becoming increasingly important; ergonomic standards can ensure that systems are designed with sufficient concern for ergonomics and interoperability. ISO (through working group TC159/SC4/WG9) is developing international standards in this subject area, dual-tracked as both ISO and CEN standards. A framework and guidelines for tactile/haptic interactions have recently been published as ISO 9241-910 and ISO 9241-920 respectively. We describe the main concepts and definitions in support of a new standard that describes how to evaluate tactile/haptic interactions and how to link this evaluation to previous standards. The new standard addresses three major aspects of the evaluation of a tactile/haptic system− the validation of system requirements, the verification that the system meets the requirements, and the overall usability of the system. Several measurement and analysis techniques are discussed, such as the calculation of scores for the determination of effectiveness. Tactile/haptic measurements have to be repeatable, and as an example we discuss how an appropriate model of the interaction with a virtual wall can be formed and used in evaluating a device.

Keywords: Guidelines, haptics, human computer interaction, standards, evaluation.

1   Introduction

Ergonomic standards go beyond providing consistency and interoperability. They help enhance usability in a number of ways, including improving effectiveness and

P. Isokoski and J. Springare (Eds.): EuroHaptics 2012, Part I, LNCS 7282, pp. 528–539, 2012.
© Springer-Verlag Berlin Heidelberg 2012

avoiding errors, improving performance, and enhancing the comfort and well-being of users. Ergonomic standards provide a basis for analysis, design, evaluation, procurement, and even for arbitrating issues of international trade. Therefore, an ISO expert group has been working on standards documents for haptic interaction since 2005. ISO TC159/SC4/WG9 has reported on its progress at several conferences [11, 9, 10, 2] and published its first standard ISO 9241-920 Guidance on tactile and haptic interactions [6] in 2009; this was followed by a second standard, ISO 9241-910 Framework for tactile and haptic interaction [7] published in 2011.

As of 2012, the following countries are actively participating in WG9: Canada, USA, UK, The Netherlands, Sweden, Germany, South Korea, and Japan. Drafts produced by WG9 undergo a thorough review process, including rounds of commenting and voting on the drafts by National Technical Advisory Groups.

ISO TC159/SC4/WG9 is currently focusing on the evaluation of tactile/haptic interactions, to be published as ISO 9241-940. This paper presents a preliminary view into this future standard and invites participation in its evolution.

2   Framework for Evaluation

ISO TC159/SC4/WG9 recognizes that there are three major aspects of evaluation: validation, verification and usability, which relate the user, the requirements and the system under consideration.

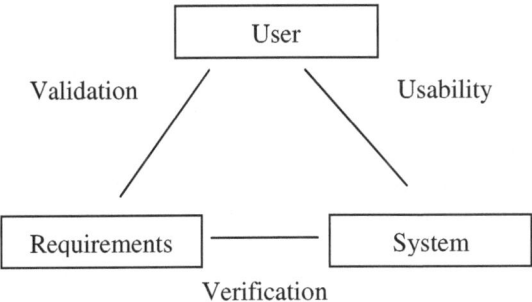

Fig. 1. User, requirements, system triangle

2.1   Validation

Validation evaluates the accuracy and completeness of the requirements' specifications. Boehm [1] states that validation involves answering the question: "Are we building the right system?" According to ISO/IEC-15288 [5], "This process performs a comparative assessment and confirms that the stakeholders' requirements are correctly defined."

Where criteria can be established from existing sources (e.g. ISO 9241-910), they should be specified in the requirements. Where a criterion cannot be established, usability testing (Section 2.3) becomes more important than validation.

Validation during development is generally performed by means of requirement reviews. Validation can also be performed on the set of user requirements that are used as input to an acquisition process (e.g. in a request for proposal). The quality of a validation is highly dependent on the quality of the communications between the user and the developer, especially on the clarity of the requirements documentation. Additional sources of pre-validated haptic requirements can come from International Standards such as ISO 9241-920 [6].

2.2   Verification

Verification tests the accuracy and completeness of the system and its operations. Boehm [1] states that verification involves answering the question: "Are we building the system right?" According to ISO/IEC 15288 [5], "The purpose of the Verification Process is to confirm that the specified design requirements are fulfilled by the system."

Verification requires specific criteria which can be measured by an appropriate technique. If specific criteria are not available, usability testing is an essential alternative to verification.

Verification during the development process is generally performed by testing that the system meets the requirements. Verification during acquisition is often limited to comparing user requirements to published specifications of a system, trusting that the published specifications have been properly verified. The quality of the verification is dependent on identifying the appropriate measures and measurement techniques. ISO 9241-940 will be suggesting measures and measurement techniques appropriate for haptic interactions.

2.3   Usability

Usability tests how well a user can operate or use a system. ISO 9241-11 [3] defines usability as "the extent to which a product can be used by specified users to achieve specified goals with effectiveness, efficiency and satisfaction in a specified context of use". It defines effectiveness as "the accuracy and completeness with which users achieve specified goals"; efficiency as "the resources expended in relation to the accuracy and completeness with which users achieve goals"; and satisfaction as "positive attitudes to the use of the product and freedom from discomfort in using it". It also defines context of use as, "users, tasks, equipment (hardware, software and materials), and the physical and social environments in which a product is used".

Usability testing answers the question, "Is the system right for the users and their tasks within the context of use?" Usability tests can be carried out during both development and system acquisition.

3   Evaluating Tactile/Haptic Interactions

The possible configurations for tactile/haptic interactions are large in number, and the possible tests that could be devised to evaluate such interactions are correspondingly numerous. To help make sense of the field, we list several types of interaction that can be evaluated in Table 1. Possible users range from experienced practitioners to the general public. The lists are not exhaustive, but form a representative set from which examples can be drawn and for which test procedures may be defined.

Table 1. Examples of interactions

Type	Auxiliary Display	Example	Test
Reality Simulation	Visual Display	Driving simulation	Lane change test
Virtual Display	Visual Display	Displaying earth map on a simulated globe	Tracing mountain ranges and valleys
Visual Control Panel	Visual Display	Selection button icons with haptic effects	Make selections in rapid sequence
Haptic Control Panel	Audible tones	Selection button icons with haptic effects but no visual counterpart	Navigate space to understand and make selections
Haptic Control Space	None	Controls for radio, air conditioning, seat positions while driving	Navigate control space to understand and make selections
Real Space Sensor	None	Cane with proximity sensors and haptic indications in the grip	Ease of navigating in a maze, especially by visually impaired users
Tactile Active Scanning	None	Braille reader with user-directed line pacing	Repeat verbally out loud a selection of Braille text
Tactile Passive Warning	None	Cell phone with silent ringer "tones" for selected callers	Recognize individual callers while distracted on other tasks

4   Validation

The validation of requirements for a tactile/haptic interaction begins with a clear statement of the goal of the interaction. Each requirement is then held up against this goal statement in order to judge its relevance to the goal.

For example, the goal could be to allow an artist to paint a scene in a virtual world. Detailed requirements can be laid out – camel hair brush to be held in either hand, the pressure of the brush on the virtual canvas to be reflected back to the user, simulating both water color and oil painting techniques. These should be translated to technical specifications – the required dynamic range and resolution of the brush pressure, for example.

The process can be aided by the construction of one or more operating scenarios. Such a scenario could reveal additional requirements – the speed of the brush stroke, the selection of colors from an on-screen pallet, for example. If the initial set of requirements had omitted a requirement such as brush stroke speed, then this can be inserted during the validation process.

If technical specifications are missing from requirements, then additional tests may be called for. For example, the range of brush pressures could be ascertained by having the artist paint a real canvas that is lying on a weigh scale. Several artists may be called upon to repeat the test, thereby accessing a range of opinions and thus extending the versatility of the planned system.

During the validation process, it will likely be necessary to prioritize the requirements. Limits can be encountered in both budget and available technology, leaving some desirable options to a future system. These may provide useful goals in the further development of tactile/haptic systems.

5   Verification

Tactile/haptic devices are generally constructed on electromechanical principles, independently of the tactile/haptic scenario software used in the interaction. Tactile devices are primarily directed at skin stimulation. They may use mechanical, thermal, chemical or electrical stimulation, although most rely on mechanical stimulation. Haptic devices are primarily directed at the kinesthetic senses, cues in the human body related to the sensing of joint angles, limb position and muscle tension. ISO 9241-910 presents general requirements for such devices, while Annexes A and B to that standard give examples of tactile and haptic devices.

5.1   Requirements for Devices

Whether a device is developed in a university laboratory or offered for sale, it should have a set of specifications attached to its performance. This allows comparison of devices. Meaningful comparison is possible only if the same measurement technique has been used to measure the parameter of interest.

As a principle of system engineering, each requirement should be measurable and testable. There is no point in setting a requirement that cannot be tested.

5.2   What Should Be Measured?

Lab measurements can be exacting and tedious, depending on the quantity to be measured. If a developer wants to characterize the device completely, then every clause in the specification should be either noted or measured. Table 2 shows examples of tactile/haptic system characteristics.

Table 2. Typical tactile/haptic system characteristics

Ergonomic	Temporal & General	Force & Torque	Environmental
Device-body interface	Bandwidth	Peak force & torque	Mobility
Degrees of freedom	System latency	Max continuous force & torque	Size
Motion range	Device latency	Min displayable force & torque	Weight
Working position	Maximum stiffness	Resolution of force & torque	Ease of installation
Limb support	Reliability	Dynamic range of force & torque	Ease of maintenance
	Fidelity	Peak acceleration	Thermal safety
	Modifiable to a task	Static friction	Electrical safety
	Adaptable to a task	Free space motion resistance	Mechanical safety
		Inertia	Acoustic noise

5.3 Measurement Resolution

The measurement of many attributes of haptic devices can be subtle. As a rule, the equipment used to measure an attribute should be about ten times higher in resolution than the resolution of the measurement that a researcher or user expects to make. At the same time, measurements of high resolution are not always required to determine if a device is suitable. It may suffice to measure the workspace to the nearest centimeter, rather than to sub-millimeter precision.

5.4 Context of Measurement

In order to allow for the widest possible scope of device evaluation, the draft standard presents verification techniques in two sections – one for the examiner who does not have measurement equipment at his or her disposal, and another for the laboratory equipped to make a variety of measurements.

6 Verification Example

6.1 Background

To illustrate the principle of verification, consider the example of a motorized haptic system. As a rule of thumb, when the maximum stiffness of the system is exceeded, unintended vibration will occur. Rather than making a passive presentation of touch, the system introduces energy into the interaction that reveals it to be a poor simulation of what it is trying to represent.

A convenient test for maximum stiffness is to present the device with a virtual wall with adjustable spring constant and no damping. The maximum stiffness is the highest spring constant of the wall that can be explored by the device without generating vibration.

The handle must be gripped by the user in the same manner that it would be gripped during normal operation. If held tightly, the measured stiffness will be slightly higher than its actual value in normal operation; if held loosely, the measured stiffness will be slightly lower than it would be in normal operation.

The following implicitly assumes a kinesthetic device.

6.2   Equipment

Consider a haptic simulation consisting of a virtual wall with a variable amount of springiness - the surface of the wall is modeled by a spring governed by Hooke's law. The haptic device is connected to a small virtual sphere, so that the sphere moves synchronously with the movement of the haptic device. The wall has a return force F proportional to the depth of penetration d by the virtual probe into the wall, where k is the constant of proportionality (also known as the spring constant).

$$F = -k \cdot d \tag{1}$$

The computer software should be set up so as to

1. give the user control over a point virtual probe by means of the haptic device under test;
2. locate a virtual wall, typically the local x, y or z Cartesian reference plane;
3. allow the user to place the virtual probe onto the surface of the wall, but to pull it back at will;
4. allow the user to increase or decrease the spring constant of the wall.

6.3   Test Procedure

The maximum spring stiffness that does not cause vibration is found by a bracketing technique

1. Set the wall spring stiffness to any value near the expected maximum stiffness.
2. If vibration occurs, reduce the spring constant.
3. If vibration does not occur, increase the spring constant
4. Continue increasing and decreasing the spring constant, bracketing ever closer the maximum stiffness where vibration does not occur.
5. When the smallest iteration crosses the line between vibration and no vibration, the maximum stiffness where vibration does not occur has been located.

A convenient way to vary the stiffness is to use a slide bar. When the mouse is used to select the slide bar, the roller on top of the mouse will move the slide bar, thus changing the stiffness. The user can vary the stiffness with one hand, sight unseen, while focusing attention on the device while it is controlled with the other hand.

7   Evaluating Usability

7.1   General

Data should be quantified so that a decision can be made as to the required qualities of the interaction. At the top level, the three components of usability are tested – effectiveness, efficiency and user satisfaction.

The purpose of validation in this context is to determine if a tactile/haptic interaction is usable for the purpose for which it was designed.

In order to validate the interaction, the examiner should ensure that goals have been specified to define the intention of the interaction. In many cases, the goals are set most concretely by positing a specific situation in which the interaction is to be used. For example, a hand-held touch screen has a number pad that gives a specific frequency of vibration when each number is pressed. A specific situation may be to dial a ten-digit phone number using the touch pad.

7.2   Test

A repeatable test procedure is then constructed from the situation. The procedure should include some means of measuring at least three components of usability set by ISO 9241-11 [3]:

Effectiveness measures can include:

- the success of each attempt to reach the goal
- percentage of goals achieved
- percentage of users successfully completing the task
- average accuracy of completed tasks

Efficiency measures can include:

- time to complete a task
- time taken on first attempt
- time spent on correcting errors

Satisfaction measures can include:

- frequency of complaints
- rate of voluntary use
- user rated ease of use

During a typical evaluation of usability, a representative set of users will place themselves in the test situation. They will follow the test procedure in an attempt to reach the goal by means of the tactile/haptic interaction.

7.3   Data Collection

The analyst collects the data on the three components of usability, plus optional auxiliary data on specific aspects of the interaction. The data is analyzed using common statistical procedures and the results are reported. A useful means of reporting is the common industry format for usability test reports, ISO 25062 [8].

For instance, in a dialing test with the hand-held touch screen, the success of dialing phone numbers and the time taken is measured by a monitoring computer. The experience of the user is assessed by a questionnaire, in which the user rates his impressions on a number of bipolar scales. In practice, higher quality data can be collected by comparing two or more means of interaction. Typically, a small number of variables will be altered between trials, so that the effect of the variables can be compared and a conclusion drawn as to which one is the best (or better) way of achieving the goal.

Additionally, trials may be run with and without vibrations in the touch pad of the previous examples. A variation may be tested with just a haptic 'click' under the character "5" in the centre of the number cluster, to see if the user is better able to centre the fingers while dialing the phone number.

7.4   Effectiveness

Some effectiveness measures include:

- reading speed
- speed of identifying an icon
- targeting speed
- moving speed
- reaction time

Tests will arrange for a score of effectiveness. A score of success p may be derived directly from the achieved result. If the goal of each run is yes/no success, then the number of successes n in m tries may be assessed. The score will then be $p = n/m$.

A number of users should test the tactile/haptic interaction, thereby reducing the possibility of individual bias. The exact number depends on the desired degree of certainty required for the test. Users should be selected as randomly as possible from the typical user group.

7.5   Efficiency

In a typical tactile/haptic interaction, the time taken to achieve the goal is measurable. Efficiency may then be conveniently calculated from the ratio of the individual scores used to determine effectiveness divided by the time taken to achieve each score.

The basic efficiency would be the mean of these scores. More useful is the decision as to which of two test runs is more efficient, and by how much.

7.6   Satisfaction

User satisfaction with an interaction may be assessed by any one of twelve methods presented in ISO 16982 [4], Usability methods for human-centered design. Methods vary from observation of users and questionnaires to expert opinion without direct user involvement. As set out in that standard, the choice depends on many factors – whether one is acquiring, designing or operating a system; whether the task to be performed is simple or complex, and whether the task is well known to the general population or relatively obscure.

In the present considerations of evaluation, we shall assume the use of a questionnaire. The questions thus presented can serve for a user-filled questionnaire, but also as the basis of an interview or a check sheet for the opinion of a subject expert.

We shall also assume the use of ordinal scales to rate user satisfaction, with the neutral opinion in the middle. The possible answers to a question such as "This interaction was easy to use" may range, for example, from "Strongly agree" to "strongly disagree".

8   Example of Evaluating Usability

8.1   Background

We consider the evaluation of the usability of a simulation of a surgical procedure. Reality simulation such as this may involve several interaction modalities – 6-DOF haptic feedback, stereo vision and high quality audio. The interaction may be intended for exploration, training or entertainment. In the case of surgery, the interaction would be one of precision, so fidelity and convincing immersion are important features.

We suppose that the user has experience with the actual scenario, and that he would judge the simulation against his experience with that scenario. As a case example, we may further suppose that the scenario is a surgical procedure involving the removal of a tumor from a brain. A skull section has been removed, and the surgeon is using a cauterizer, a bipolar coagulator and a suction tool.

The goal of the simulation is a realistic rendering of the visual scene and the feel of a surgical instrument as the surgeon wields it during a surgical procedure.

8.2   Test

The test will consist of parallel scans across the surface of the meninges (the membrane that covers the brain). The tool is a pen-like cauterizer which we shall refer to as a stylus for convenience and generality. The surgeon holds it as one would hold a pencil.

Fifty traces are made, each of approximately the same length. The stylus rides the surface of the meninges as the tool is moved by the surgeon. A very light force is maintained on the membrane. Success for each scan is measured by the ability of the surgeon to maintain a constant and correct force on the stylus while executing a scan of constant curvature. The centre of the position curve will typically be at the elbow of the surgeon, as he executes adduct-abduct motion of the elbow while keeping the wrist locked.

The raw data measured would be the downward force and the position of the tip of the stylus. The derived data could be any of a number of possibilities:

- the standard deviation of the force
- the distance along the arc during which the force exceeds a maximum or a minimum value
- the difference between the average value of the force and the target value
- the standard deviation of the radial position of the trace
- the distance along the arc during which the radial position deviated from the ideal radius by more than a certain value
- the distance along the arc during which the radial position deviated from the ideal parallel distance from the preceding trace

Considering the goal of the test (to make parallel lines at a certain pressure), we shall choose success to be a complete trace made within force limits and within radial distance limits of the preceding trace.

8.3   Analysis

Effectiveness will be measured by the percentage of the fifty traces that are successful. Efficiency will be measured by the success rate divided by the time taken to complete the fifty traces.

Satisfaction will be measured by the bipolar response to the statement, "Is this a realistic simulation of the actual operating scenario?"

The questionnaire could be expanded to include evaluating statements that solicit more fine-grained responses (*strongly agree, agree, neutral, disagree*, or *strongly disagree*).

"Does the stylus feel correct in my hand?"
"Does the stylus have the correct inertia?"
"Were there erroneous forces that I had to resist as I drew the traces?"
"Has the visual rendition been like an actual surgical site?"
"Has the feel been exactly matched to the visual rendition of the operating scene?"

8.4   Ramifications

The evaluation could be used

- to assess if a surgical simulator is satisfactory for training surgeons.
- to compare different brands of surgical simulator.

- to compare different haptic devices for use in a surgical simulator.
- to compare different orientations of the same haptic device.
- to compare different haptic and visual models of the brain surface.

9   Getting Involved

TC159/SC4/WG9 is continuously working to ensure that all guidelines are technically correct and feasible. You can get involved as an expert member of TC159/SC4/WG9, actively developing drafts of the planned work items. Even as a casual advisor, the members of WG9 are very interested in your opinions on tactile/ haptic-related terms and definitions, and hearing about your experience with measures for haptic devices or human performance.

References

1. Boehm, B.W.: Software Engineering Economics. Prentice-Hall, Inc., Upper Saddle River (1981)
2. Carter, J., van Erp, J.: Ergonomics of Tactile and Haptic Interactions. In: Proceedings of the Human Factors and Ergonomics Society 2006, pp. 674–688 (2006)
3. ISO: ISO 9241-11, Ergonomic requirements for office work with visual display terminals (VDTs) Guidance on usability. International Organization for Standardization (1998)
4. ISO: ISO/TR 16982: 2002 Ergonomics of human-system interaction – Usability methods supporting human-centered design. International Organization for Standardization (2005)
5. ISO: ISO/IEC 15288, Systems and software engineering – System life cycle processes. International Organization for Standardization (2008)
6. ISO: 9241-920 Ergonomics of human-system interaction – Part 920 Guidance on tactile and haptic interactions. International Organization for Standardization (2009)
7. ISO: 9241-910 Ergonomics of human-system interaction – Part 910 Framework on tactile and haptic interaction. International Organization for Standardization (2011)
8. ISO: ISO/IEC 25062: 2006 Software engineering – Software product Quality Requirements and Evaluation (SQuaRE) – Common Industry Format (CIF) for usability test reports. International Organization for Standardization (2011)
9. van Erp, J., Carter, J., Andrew, I.: ISO's work on tactile and haptic interaction guidelines. In: Proceedings of Eurohaptics 2006, pp. 467–470 (2006)
10. van Erp, J.B.F., Kern, T.A.: ISO's Work on Guidance for Haptic and Tactile Interactions. In: Ferre, M. (ed.) EuroHaptics 2008. LNCS, vol. 5024, pp. 936–940. Springer, Heidelberg (2008)
11. van Erp, J.B.F., Kyung, K.-U., Kassner, S., Carter, J., Brewster, S., Weber, G., Andrew, I.: Setting the Standards for Haptic and Tactile Interactions: ISO's Work. In: Kappers, A.M.L., van Erp, J.B.F., Bergmann Tiest, W.M., van der Helm, F.C.T. (eds.) EuroHaptics 2010, Part II. LNCS, vol. 6192, pp. 353–358. Springer, Heidelberg (2010)

Dynamics Modeling of an Encountered Haptic Interface for Ball Catching and Impact Tasks Simulation

M. Solazzi, D. Pellegrinetti, P. Tripicchio, A. Frisoli, and M. Bergamasco

Scuola Superiore Sant'Anna, TeCIP Institute, PERCRO Lab, Pisa, Italy
m.solazzi@sssup.it

Abstract. This paper deals with a model-based control strategy implemented on an encountered haptic interface developed for the simulation of ball catching tasks. A dynamical model of a reference device has been developed and validated by experimental results. This model was applied to increase the control performance and to simulate realistic impacts. The control strategy to generate the haptic interface trajectories consistent with the simulation of ballistic motion of virtual objects has been defined. At the impact instant the perceptively correct kinetic energy is transferred from the device end-effector to the user hand adopting a velocity scaling rule. Experimental results confirm control accuracy in fast dynamics trajectory tracking.

Keywords: Dynamics Modeling, Encountered Haptics, Impacts Simulation.

1 Introduction

Haptic interfaces are typically used for simulating slow dynamic human-robot interaction, mainly in applications simulating virtual objects manipulation or shape exploration. Encountered type haptic interfaces have been designed to display the user a direct contact with the end effector when a collision occurs between his avatar and an object in the virtual environment [3,4]. Moreover, encountered haptic interfaces allow to interact with objects with fast dynamics in large virtual environments. The task of simulating catching and throwing balls, such as in juggling [2,8], is an example of application requiring high dynamic performance of encountered haptic devices. While the mentioned works mainly focus on the psychophysical investigation about the most effective haptic rendering for impact simulation, in this paper we present a control strategy to achieve high dynamic performances of the device and a trajectory planning to simulate realistic impacts with virtual objects following a generic motion law.

Catching tasks represent also a relevant scenario for neurophysiologic studies related to motor learning and adaptation. In particular, in the study of human adaptation to altered gravity conditions, such as in the case of astronauts in long lasting space missions, the paradigm of launching and catching balls has been efficiently exploited to investigate motor learning [6,7]. The study of the human reaction and adaptation to altered gravity conditions is important for the feasibility analysis of long term spatial missions, where the capabilities of the crew to operate in unfamiliar conditions and to readapt to standard environment after landing are fundamentals.

P. Isokoski and J. Springare (Eds.): EuroHaptics 2012, Part I, LNCS 7282, pp. 540–551, 2012.

The objective of this work was to devise a general control framework for the simulation of high dynamics ball launching and catching tasks in conditions of altered gravity. To this purpose, due to the fast dynamics of the involved phenomena and the need of simulating quickly varying impact forces, a specific dynamic model-based strategy control for an encountered haptic interface was devised and tested.

Such a model requires a precise knowledge of the dynamic performance of the employed haptic device, both to improve the performance of the device in trajectory tracking and to realize more realistic simulation of impact forces. This was achieved through a feed forward compensation of the dynamics to reduce the settling time and to exploit the full dynamic bandwidth of the device.

In this paper we report the design and experimental evaluation of the proposed control for encounter haptic devices.

The paper is organized as follows. The second section describes the experimental system that has been used; the third section presents the adopted control strategy, while trajectories planning and implementation are presented in section four. Finally, experimental results of the control performance are presented in section five.

2   System Description

The experimental apparatus consists of the GRAB haptic interface, a 6 DoF magnetic hand tracker, a host PC for the motion tracker data acquisition and the simulation of the virtual environment, and a visualization system (Fig. 1).

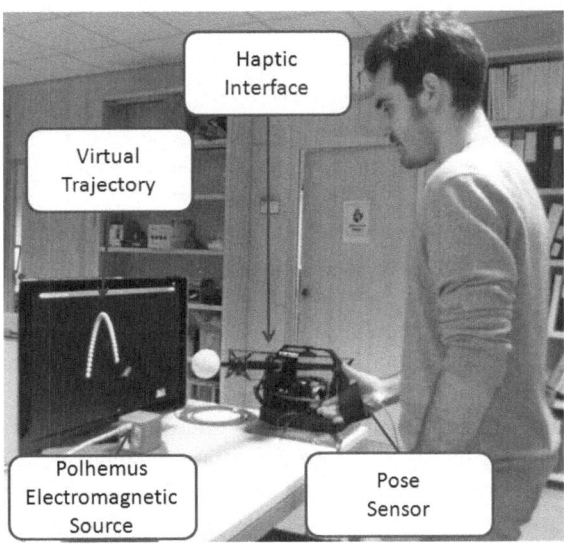

Fig. 1. Overall view of the experimental system

The GRAB is a 3-dof mechanism [5] composed of three robotic links with an equivalent R-R-P kinematics. A standard tennis ball is linked to the end-effector of the mechanism, to simulate the hand grasping of a real ball during the impact phase. The workspace, the kinematics and the inertial reference frame (X_0, Y_0, Z_0) of the device are represented in Fig. 2. X_{ee} is the end-effector position referred to the inertial frame.

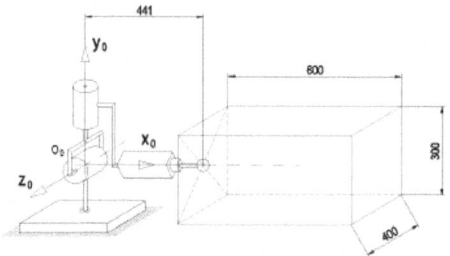

Fig. 2. GRAB interface kinematics and workspace

The virtual environment has been developed by means of the eXtreme Virtual Reality (XVR) [1] framework and runs on the Simulation host (see Fig. 3). The position and the orientation of the hand of the user are measured by a Polhemus Liberty motion tracking system.

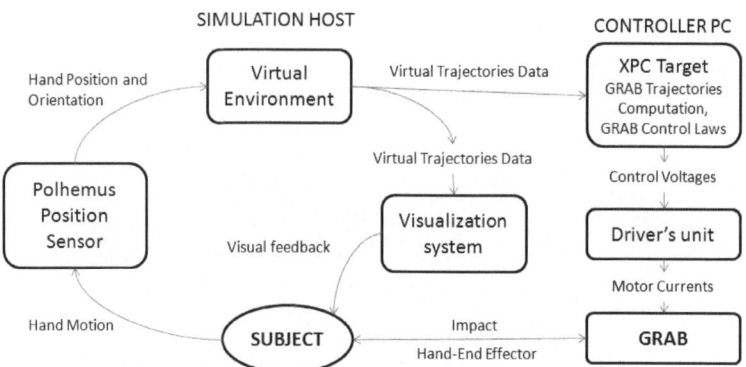

Fig. 3. Schematic view of the system functioning

The simulation host is responsible for the generation of virtual ball trajectories. Impact virtual velocity vectors and the required GRAB flight times to impact are also computed, and sent together with the impact points as inputs to the controller PC. It computes and implements adequate trajectories for the end-effector by means of a trajectory planner and a dynamical model based position control. A visualization system allows the subject to observe the virtual trajectories and its respective parameters. The data flows through the different units are represented in Fig. 3.

3   System Dynamic Modeling

According to the kinematic scheme of Fig. 2 a dynamical model of the haptic interface was defined. Denavit-Hartenberg convention was used to solve the direct, inverse and differential kinematics. Mechanism dynamics equations were derived as:

$$B_m \cdot K^{(-1)} \cdot \ddot{q}_m + C_m \cdot K^{(-1)} \cdot \dot{q}_m + g_m = K^T \cdot \tau_m \tag{1}$$

where q_m represents the column vector of the motor angular positions, $B_m(q_m)$ the inertia matrix of the mechanism, $C_m(q_m, \dot{q}_m)$ the Coriolis matrix, $g_m(q_m)$ the column vector of the gravity torques and τ_m the column vector of the actuated motor torques. The transmission matrix K defines the relation $q_m = K \cdot q$ between joint angles q and motor angles q_m.

The geometrical and the inertial properties of the haptic interface were estimated by the CAD model. The estimation of the inertia matrix B_{est} was computed in terms of joint variables q. Similarly the Coriolis matrix $C_{est}(q, \dot{q})$, gravity joint torques $g_{est}(q)$ and transmission matrix K_{est}, were estimated from the CAD model.

3.1   Dynamics Model Based Position Control

The following control law was assumed:

$$\tau_m = \tau_m^{gc} + \tau_m^{dc} + \tau_m^{PD} \tag{2}$$

where τ_m^{gc} term is a standard gravity compensation terms. The transmission matrix estimation is exploited to express the functional dependence of g_{est}, with respect to q_m^{meas}:

$$\tau_m^{gc} = g_{est}(K_{est}^{(-1)} \cdot q_m^{meas}) \tag{3}$$

The functional dependence expressed in equation (3) holds also for B_{est}, C_{est} and for the other kinematics variables. From now on, it will be taken as a tacit assumption.

The τ_m^{dc} term is a feed-forward contribution which compensates the dynamical behavior of the GRAB. We used the estimation of the inertia and Coriolis matrices, and the desired motor angular positions, indicated with q_m^{des}. These ones are chosen in order to exploit GRAB own dynamics within the limits of the maximum of its performances. Due to the inertial characteristic of the device and the actuators torque limits the maximum acceleration at the center of the workspace is 25 [m/s²].

Inverse kinematics was used to compute the desired motor angular positions from the desired end-effector trajectories, indicated with X_{ee}^{des}, which are naturally defined on the workspace. We have:

$$\tau_m^{dc} = K_{est}^{(-T)} \left(B_{est} \cdot K_{est}^{(-1)} \cdot \ddot{q}_m^{des} + C_{est} \cdot K_{est}^{(-1)} \cdot \dot{q}_m^{des} \right) \tag{4}$$

The final contribution is the action of a standard PD controller:

$$\tau_m^{PD} = K_P \cdot (q_m^{des} - q_m^{meas}) - K_D \cdot \dot{q}_m^{meas} \qquad (5)$$

where the $\dot{q}_m^{meas}$ is computed by a first order filter.

A scheme of this model based control strategy is represented in the Fig. 4:

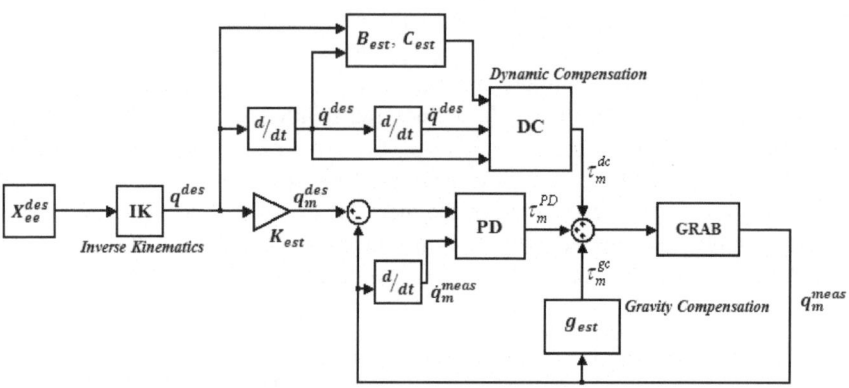

Fig. 4. Control algorithm scheme

3.2   Comparison of the Proposed Method with a Standard PD Position Control: Experimental Results

The effect and the accuracy of the system dynamics can be analyzed by comparing the performance of the model based position control and a standard PD solution optimized in fast response. The performance of the system under the two controls is reported in Fig. 5.

The plots show the response of the system to a generic input signal with demanding performance: the device was required to cover a distance of 0.23 [m] in a time of 0.25 [s] with a null final velocity. In the upper three plots the performance of the presented control law is depicted and the output closely matches the input. In the lower plots the response of the PD control feedback without the τ_m^{dc} compensation is shown for comparison. The proportional and derivative gains are set to the same values in both cases. Adding forward dynamics compensation resulted both in a faster response at the beginning of the motion and a reduced overshoot at the end.

Fig. 6 shows the actual motor torques exerted by the PD controller against the motor torques estimated by the dynamics simulation. The imposed motion at the GRAB end-effector is a circle of radius 0.1[m] in the (y_0, z_0) plane at a frequency of 1.43 [Hz] that does not demand particularly high performances. The estimates are good approximations of the real actuation torques.

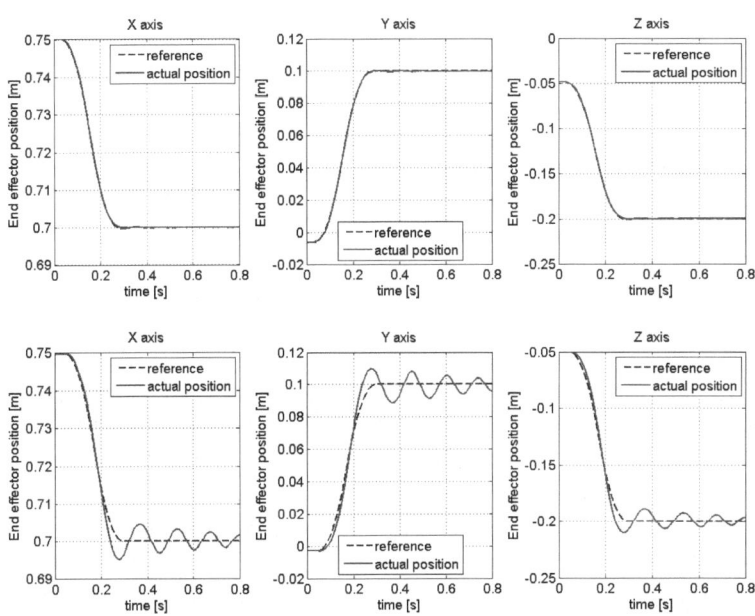

Fig. 5. Control law responses with (upper) and without (lower) dynamic compensation

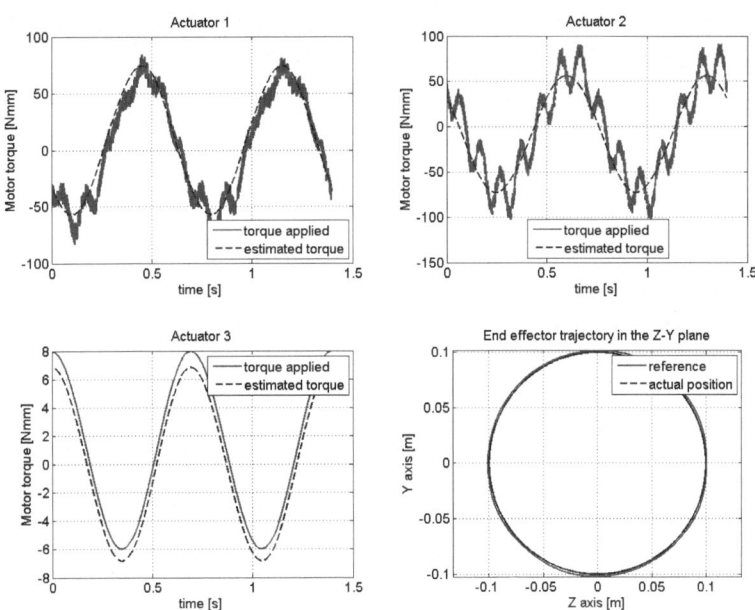

Fig. 6. Actual VS estimated motor torques and imposed trajectory

4   Haptic Rendering of Impacts

4.1   General Framework

Virtual trajectories were defined with respect to the virtual environment reference frame (X_o, Y_o, Z_o). The imposed kinematics motion laws follows ballistic motions with a defined initial velocity in a constant gravity field which is chosen to be consistent with earth's gravity (-9.81 [m/s^2] along the Y_o axis). The gravity field parameter can be changed in order to simulate impact and catching tasks in altered gravity conditions.

The Polhemus motion tracker provides the position and orientation measures of an electromagnetic sensor. The sensor is attached to the hand of the subject at the center of the palm. We defined a target plane passing through the sensor position and oriented as the palm of the user. Given the virtual trajectories and the time-varying target planes, the impact points were computed in real time as their intersection. Once the impact point X_f was obtained, the impact time T_i were retrieved according to the motion law, as well as the impact velocity v.

A condition for the activation of the GRAB control was then introduced. When the distance between the impact point and the palm center was below a given threshold (0.05 [m]) and the time to impact (given by the difference between the impact time T_i and the current time) was equal to the given value chosen to be the real GRAB flight time T_f, the activation command was sent to the control algorithm. Once the impact had occurred, the control algorithm drove the end-effector to a standard rest position with null velocity.

The input variables sent to the trajectory planner implemented on the control algorithm are: the activation commands, the impact points $X_{ee}(T_f) = X_f$, the virtual velocities at the impact points v and the real GRAB flight times T_f.

The GRAB trajectories were then computed in real time by the trajectory planner, and have to satisfy an important condition on the impact velocity. This condition is discussed in the next subsection.

4.2   Energy Modeling

The chosen rule to simulate a realistic impact perceived at the hand of the subject was based on a kinetic energy balance. In detail the kinetic energy of the equivalent inertia of the haptic device reduced to the mechanism end-effector must be equal to the kinetic energy of the simulated virtual tennis ball with its velocity at the impact point. This assumption was supported by psychophysical evidences from experiments previously conducted in [2,8] by the same authors . The inertia properties reduced to the end-effector are functions of the impact points, and are computed from the inertia matrix and the Jacobian position matrix estimations. The differential kinematics solution provides the end-effector Jacobian position matrix, indicated with J^P, which allows us to write:

$$V_{ee} = J^P(q) \cdot \dot{q} \tag{6}$$

where the left term of the equation represents the end-effector velocity evaluated with respect to the (X_o, Y_o, Z_o) frame.

From the geometrical properties we can write the end-effector Jacobian position matrix estimation with respect to the desired joint positions, indicated with J^P_{est}.

The equivalent end-effector Inertia, indicated with M, is computed:

$$M = J^{P\,(-T)}_{est} \cdot B_{est} \cdot J^{P\,(-1)}_{est} \tag{7}$$

As we have done in section 3.1, we can transform the end-effector position variables of the desired trajectories, defined in the workspace, into the joint positions by means of the inverse kinematics solution. Then, given the impact points X_f, the equivalent end-effector inertia is immediately computed and the following energy balance at impact is required:

$$\frac{1}{2} v^T \cdot m \cdot v = \frac{1}{2} V_{ee}^T \cdot J^{P\,(-T)}_{est} \cdot B_{est} \cdot J^{P\,(-1)}_{est} \cdot V_{ee} \tag{8}$$

where m is the theoretical mass value of a standard tennis-ball.

Assuming to preserve the direction between the virtual impact velocity and the real one of the end-effector we can write:

$$V_{ee}(T_f) = c \cdot v \tag{9}$$

where c is a positive real number. From the kinetic energy balance we find c to be:

$$c = \sqrt{\frac{v^T \cdot m \cdot v}{V_{ee}^T \cdot J^{P\,(-T)}_{est} \cdot B_{est} \cdot J^{P\,(-1)}_{est} \cdot V_{ee}}} \tag{10}$$

4.3   Trajectory Planning

The GRAB trajectories were computed in real time by the trajectory planner depending on the input data described in section 4.1 and the impact velocity vector V_{ee} computed in section 4.2. The following conditions were imposed to simulate the impact with the virtual object in a realistic way:

- The starting point X_0.
- The device flight time to impact (T_f).
- The impact velocity vector $V_f = V_{ee}(T_f)$.
- The impact point (X_f).

Among the possible solutions satisfying these conditions, the parabola was chosen as the lowest order trajectory. We defined the shape of the parabola satisfying the first and the last requirements and allowing the end-effector to reach the computed impact velocity direction at the final time T_f. The motion law was calculated to make the end effector cover the overall trajectory in the time T_f with final speed $|V_f|$. The shape of the parabola was computed with respect to the frame (x,y,z) with origin at X_0 and defined as follow:

$$\hat{x} = \frac{X_f - X_0}{|X_f - X_0|}; \quad \hat{y} = -\hat{x} \wedge \hat{V}_f; \quad \hat{z} = \hat{x} \wedge \hat{y} \tag{11}$$

where $\hat{V}_f = V_f / |V_f|$ is the versor of the final velocity.

In this frame the parabola was contained in the (x,z) plane and its equation is:

$$z = Ax^2 + Bx \tag{12}$$

The required conditions on the impact point and the impact velocity direction allow to compute the two parameters A and B:

$$B = -\text{sgn}(\hat{V}_f \cdot \hat{z}) \cdot \tan[\text{acos}(\hat{V}_f \cdot \hat{x})] ; \quad A = \frac{-B}{|X_f - X_0|} \tag{13}$$

When $V_f = 0$ and V_f is parallel to the x-axis defined in equation (11), we can choose an arbitrary y-axis perpendicular to the x-axis. Further we chose the degenerate solution of equation (12) with A=0 and B=0 (the straight line solution).

The total arc length S_f of the parabola from the starting point to the impact point can be compute as follows:

$$S_f = \int_0^{|X_f - X_0|} \sqrt{(dz / dx)^2 + 1} \cdot dx \tag{14}$$

By imposing the following arc length motion law:

$$\begin{cases} \dot{s}(\tau) = A_s \cdot \tau & \text{for } 0 \le \tau \le \tau^* \\ \dot{s}(\tau) = -A_s \cdot \tau + \dot{s}(\tau^*) & \text{for } \tau^* \le \tau \le T_f \end{cases} \tag{15}$$

The parameters A_s and τ^* are computed imposing the condition on the impact velocity norm and the condition of the total arc length:

$$\begin{cases} \dot{s}(T_f) = |V_f| \\ s(T_f) = S_f \end{cases} \tag{16}$$

The mathematical solution results:

$$\tau^* = T_f - \frac{S_f}{|V_f|} + \frac{\sqrt{S_f^2 + \frac{1}{2}T_f^2 \cdot S_f^2 - S_f \cdot T_f \cdot |V_f|}}{|V_f|}; \quad A_s = \frac{|V_f|}{2\tau^* - T_f} \quad (17)$$

under the following condition:

$$S_f \geq \frac{1}{2}T_f \cdot |V_f| \quad (18)$$

The actuation torques and the inertial properties of the device limit the maximum value for the acceleration A_s. Given the shape of the parabola and the impact parameters $|V_f|$ and S_f, the minimum value for T_f can be computed according to equation (17) in order to assure that the device performs correctly the trajectory.

Fig. 7 shows the complete haptic algorithm for the impact rendering.

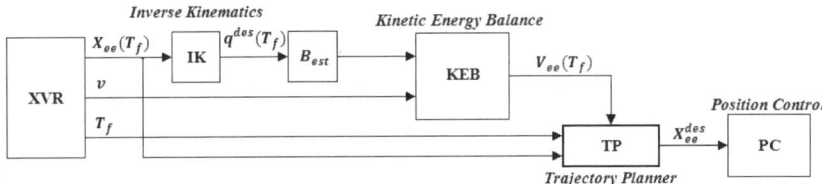

Fig. 7. Haptic rendering scheme

As shown in the bottom right picture of Fig. 8 the virtual tennis-ball and the real end-effector trajectories do not lie on the same plane in the most general situation, but with the proposed method we can ensure that the two trajectories become tangent in a neighborhood of the impact point.

4.4   Experimental Results

Position control performance is plotted in Fig. 8 and Fig. 9 for one exemplary case of simulation of virtual catching of a flying ball. In the first three plots of Fig. 8 the desired position coordinates of the end-effector (blue dotted lines) are compared with the references generated by the control (red solid lines). The fourth plot shows how the control assures the matching and tangency between the end-effector trajectory (red solid line) and the virtual ball trajectory (black solid line) at the impact point.

The virtual ball reached the impact point with a velocity of $(0,-2.34,-1.2)$ [m/s] which, according to the kinetic energy balance performed at the impact point, corresponds to a device velocity V_f equal to $(0,-1.02,-0.52)$ [m/s]. The device flight time (T_f) was set to 0.3 [s] and the value of the maximum end effector acceleration required to perform the planned trajectory was 22 [m/s^2], very close to the maximum capability of the device (25 [m/s^2]).

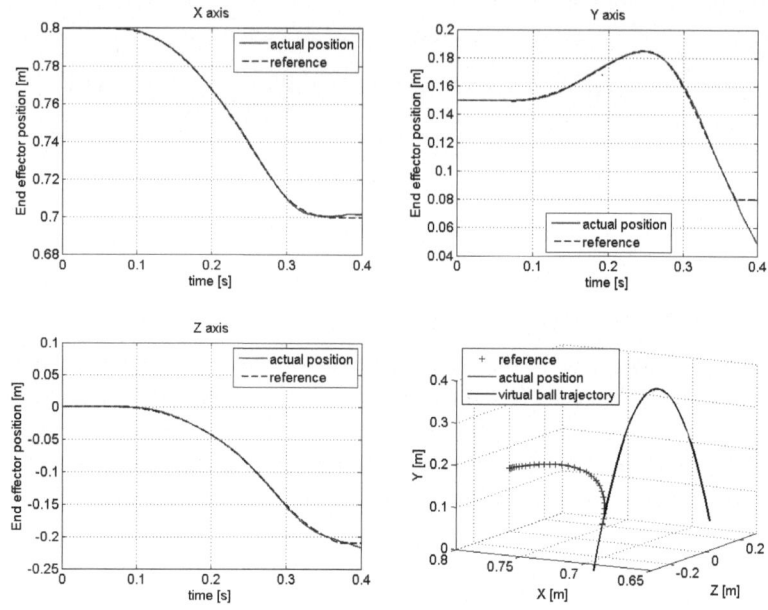

Fig. 8. GRAB position tracking in the Cartesian space and comparison with the virtual ball trajectory

Fig. 9 shows the control performances on velocity tracking. The actual values (red solid lines) are close to the desired values (blue dotted lines) during the flight and to the final target velocity V_f (black solid line) at the impact point. As shown in the two figures, when the end-effector reaches the hand of the subject, the desired positions assumes a constant value and the velocities drop to zero, but the position control is shut down so that the GRAB continues its motion with its own dynamics.

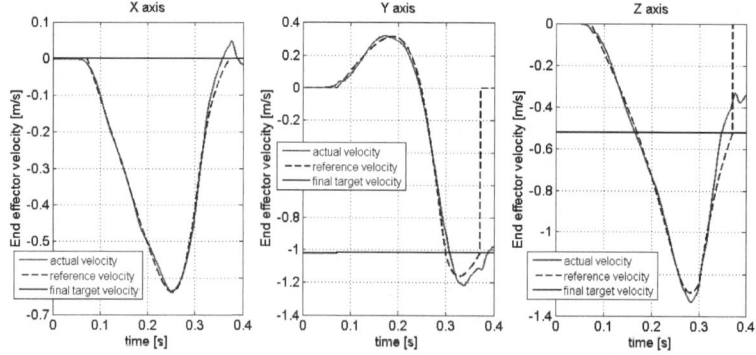

Fig. 9. GRAB velocity tracking in the Cartesian space

5   Conclusions

In the present work a dynamics model-based position control strategy for encountered haptics was proposed and experimentally tested to demonstrate the feasibility of the approach. Given a ballistic motion of a virtual object, a general framework for the generation of control state-space trajectories exploiting the full dynamic range of the haptic device has been implemented. A catching task has been performed and a realistic stimulus at the impact with the hand of the subject is provided under the assumption of a kinetic energy balance. The dynamics model has been applied to estimate the feasible trajectory parameters and to compute the inertial properties of the device at the impact point. Results from experimental tests proved the accuracy of the dynamics model and the effectiveness of the developed control in the tracking of fast dynamics trajectories. Future works foresee the application of the implemented framework for conducting neurophysiological and behavioral studies on motor learning in altered gravity conditions.

Acknowledgements. This work was supported by VERE, an IP project funded under the European Seventh Framework Program, Future and Emerging Technologies (FET), Grant Agreement Number 257695, and by the Italian Space Agency (ASI) funded project CRUSOE "CRUising in Space with Out-of-body Experiences".

References

1. Carrozzino, M., Tecchia, F., Bacinelli, S., Cappelletti, C., Bergamasco, M.: Lowering the development time of multimodal interactive application: the real-life experience of the XVR project. In: Proc. Advances in Computer Entertainment (2005)
2. Tripicchio, P., Ruffaldi, E., Avizzano, C.A., Bergamasco, M.: Control strategies and perception effects in co-located and large workspace dynamical encountered haptics. In: Proceedings of the World Haptics, WHC 2009 (2009)
3. McNeely, W.: Robotic graphics: a new approach to force feedback for virtual reality. In: Virtual Reality Annual International Symposium, pp. 336–341 (1993)
4. Tachi, S., Maeda, T., Hirata, R., Hoshino, H.: A construction method of virtual haptic space. In: Proceedings of the 4th International Conference on Artificial Reality and Tele-Existence, ICAT 1994 (1994)
5. Bergamasco, M., Avizzano, C.A., Frisoli, A., Ruffaldi, E., Marcheschi, S.: Design and validation of a complete haptic system for manipulative tasks. In: Advanced Robotics, vol. 20(3), pp. 367–389 (2006)
6. d'Avella, A., Cesqui, B., Portone, A., Lacquaniti, F.: A new ball launching system with controlled flight parameters for catching experiments. J. Neurosci. Methods 196(2), 264–275 (2011)
7. Senot, P., Zago, M., Lacquaniti, F., McIntyre, J.: Anticipating the effects of gravity when intercepting moving objects: differentiating up and down based on nonvisual cues. J. Neurophysiol. 94(6), 4471–4480 (2005)
8. Ruffaldi, E., Tripicchio, P., Avizzano, C.A., Bergamasco, M.: Haptic Rendering of Juggling with Encountered Haptic Interfaces. Presence: Teleoperators and Virtual Environments 20(5), 480–501 (2011)

Novel Interactive Techniques for Bimanual Manipulation of 3D Objects with Two 3DoF Haptic Interfaces

Anthony Talvas, Maud Marchal, Clément Nicolas, Gabriel Cirio,
Mathieu Emily, and Anatole Lécuyer

IRISA/INRIA Rennes, France
{anthony.talvas,maud.marchal,anatole.lecuyer}@inria.fr

Abstract. This paper presents a set of novel interactive techniques adapted to two-handed manipulation of objects with dual 3DoF single-point haptic devices. We first propose the *double bubble* for bimanual haptic exploration of virtual environments through hybrid position/rate controls, and a bimanual viewport adaptation method that keeps both proxies on screen in large environments. We also present two bimanual haptic manipulation techniques that facilitate pick-and-place tasks: the *joint control*, which forces common control modes and control/display ratios for two interfaces grabbing an object, and the *magnetic pinch*, which simulates a magnet-like attraction between both hands to prevent unwanted drops of that object. An experiment was conducted to assess the efficiency of these techniques for pick-and-place tasks, by comparing the *double bubble* with viewport adaptation to the *clutching* technique for extending the workspaces, and by measuring the benefits of the *joint control* and *magnetic pinch*.

Keywords: Haptic interfaces, Bimanual manipulation, Virtual manipulation, Interaction techniques.

1 Introduction

In the field of haptics and virtual reality, two-handed interaction with virtual environments (VEs) is a domain that is slowly emerging while bearing very promising applications. Examples of these are surgery training [1], rehabilitation [2], industrial prototyping [3], and 3D graphics [4]. More generally, the use of two hands in haptics allows to realize tasks in a more natural way, as most tasks done in real life are bimanual in a way or another: from simple cases such as using scissors to more complex ones such as playing the guitar.

Numerous haptic devices have been proven to be suitable for bimanual interaction, which can be either single-point interfaces, in which case they will be represented by proxies in the VE, or multi-fingered interfaces, with which it is much easier to mimic the behaviors of an actual hand. Most of these interfaces, however, have small workspaces, which is a strong limitation when a user wants to carry out tasks as simple as pick-and-placing an object in a large VE. Several

P. Isokoski and J. Springare (Eds.): EuroHaptics 2012, Part I, LNCS 7282, pp. 552–563, 2012.
© Springer-Verlag Berlin Heidelberg 2012

hardware and software solutions were proposed to address this issue, notably the *Bubble* technique [18] which proved to be suitable for simultaneous grasping of objects and exploration in a VE with a bimanual whole-hand haptic interface [9]. However, while multi-fingered interfaces make it easier to firmly hold an object with two hands through the use of the fingers, it is still currently very difficult to grab and carry an object with proxies controlled by two single-point haptic interfaces, especially in large environments. Notably, a grasped object tends to slip from virtual hands if the contacts between them are not strongly maintained, and current bimanual navigation techniques tend to add to the difficulty of keeping those contacts over time.

In this paper, we introduce novel metaphors and interaction techniques to improve bimanual interaction with dual single-point haptic interfaces. Our major contributions are a *double bubble* technique with viewport adaptation for bimanual haptic exploration of large VEs, as well as the *magnetic pinch* and *joint control* techniques for facilitating the grasping and carrying of virtual objects with two virtual proxies. The paper is structured as follows: section 2 covers the related work on bimanual interaction with large VEs, section 3 details the proposed techniques, section 4 presents the experiment conducted to evaluate the techniques and section 5 discusses the results of the evaluation. Finally, conclusion and perspectives are presented in section 6.

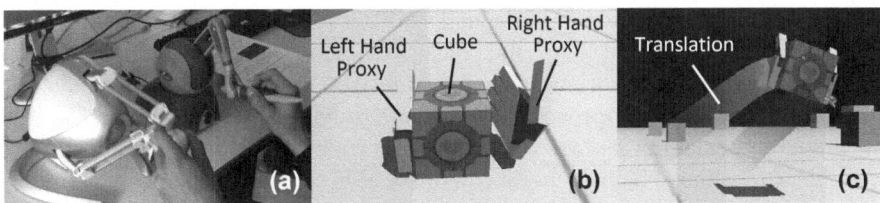

Fig. 1. Example of a bimanual pick-and-place task in a large VE as addressed in this paper. (a) Bimanual haptic setup made of two single-point devices. (b) Grasping a virtual cube with two proxies. (c) Carrying and displacing the cube using our novel interactive techniques.

2   Related Work

This section presents the current state of the art in bimanual haptic interaction in large VEs, by first giving an overview of the existing bimanual haptic devices, then exposing the previously proposed hardware and software solutions for extending their workspaces, notably the *bubble* technique, and finally evoking the grasping of objects with two single-point haptic interfaces.

Several haptic devices allow bimanual interaction with VEs, whether being devices that were specifically designed for such use, or generic devices that were either adapted to this context or used as is. Some are single-point interfaces, such as the SPIDAR G&G [5], the DLR bimanual haptic interface [3] or the more widespread PHANToM series. Others are multi-fingered interfaces, such as the MasterFinger-2 [6], SPIDAR-8 [7] and Bimanual HIRO [8], which enable interaction through 4, 8 and 10 fingertips respectively. Within this category of multi-fingered devices, the Haptic Workstation is a special case, not only providing interaction through the fingertips but also through the palm of both hands [9]. Multi-fingered interfaces are especially suited for grasping virtual objects thanks to their numerous interaction points, although it was shown that single-point devices could also allow grasping with a full control of all degrees of freedom (DoF) of an object through a soft-fingers method [10]. A disadvantage of the aforementioned devices is that their workspaces are limited, and as such are not well suited for working in large VEs, hence bringing the need for techniques that increase the available workspace.

Existing solutions for increasing the workspace of bimanual interfaces can be divided into hardware-based and software-based approaches. A straightforward hardware approach consists in increasing the workspace provided by each haptic device to fit that of the VE, either through a bigger frame or a redundant DoF. However the best result obtained for bimanual devices was the reach of human arms, in the cases of the DLR interface [3] and the Haptic Workstation [9]. Another solution for handling large VEs is the use of mobile haptic interfaces, i.e. haptic devices fixed on a mobile robot. Bimanual examples of these are the Mobile Haptic Grasper [12] and VISHARD7-based mobile interface [13]. However, while potentially providing an infinite planar workspace [11], these devices are still limited in vertical reach. Other hardware approaches solve the workspace issue by providing additional DoF to the user to handle navigation in the VE. For instance, a 3DoF foot pedal was used for controlling the motion of a two-armed robot in a remote environment [14].

While the hardware approaches do manage to solve the workspace issue to a certain extent, such devices are not necessarily widespread. Software approaches have the advantage of being generic and applicable to any haptic device available to the user with no further requirements, although the majority of them are not bimanual-specific. A first technique consists in applying a scaling factor to match the real workspace provided by the haptic devices with a virtual volume defined in the VE [15], although reducing the accuracy of motions in the virtual space. Another approach is the *clutching* technique, which consists in holding down a button to temporarily interrupt the coupling between the device and the proxy while the user recenters the device. The *Dual Shell* method is an extension of this technique, that automatically handles the clutching when predefined boundaries are reached, without requiring the potentially counterintuitive manipulation of a button [16]. The use of rate control was also proposed to control the velocity of the virtual proxy through the position of the haptic device [17]. This technique infinitely increases the workspace in all directions, however it is far from being intuitive and appears to be an acquired skill.

This leads to the *Bubble* technique, which uses position control inside predefined spherical boundaries of the device, and rate control when the device leaves those boundaries [18]. This technique was used more recently for bimanual interaction with complex VEs through the Haptic Workstation, by allowing users to translate and rotate the camera by moving both hands outside the bubble in the same direction [9]. This technique showed to be efficient for simultaneous navigation and manipulation with the Haptic Workstation, since it allows interaction with both palms and fingers. However, it remains more difficult to use with single-point interfaces. In this case, picked objects are frequently dropped during the translations of the virtual workspace through rate control, especially when using two different interfaces with physical workspaces of different size and shape.

The problem of grasping virtual objects with two single-point haptic interfaces has not yet been specifically addressed, as few bimanual techniques focus on two-handed haptic manipulation. In this context, several issues arise, although they have received little attention. Virtual springs between multiple contact points [6] were used for the haptic rendering of grasping, while repulsion forces [3,8] were computed for the prevention of collisions between a user's hands and the haptic devices. Previous work from the area of augmented reality could be used for the modulation of the stiffness of objects [19] during a manipulation with haptic devices exhibiting different gains. Therefore, the question of how to facilitate the carrying of objects with dual single-point interfaces remains fully open, and we provide a first answer in this paper.

3   Novel Techniques for Improving Bimanual Interaction with Dual Single-Point Haptic Interfaces

We propose a set of new interaction techniques for improving the exploration of large VEs with two haptic interfaces, and the manipulation of objects with two 3DoF haptic devices represented by simple proxies that enter in contact with virtual objects through single contact points. We first present two haptic exploration techniques: the *double bubble*, which allows free motion with both hands in a VE, and a viewport adaptation method that maintains both virtual proxies on screen at all times. Then, we present two haptic manipulation techniques: the *magnetic pinch*, which uses a simulated spring to keep the virtual proxies from dropping a picked object, and the *joint control*, which solves issues related to different control modes between the two hands.

3.1   Double Bubble

In the *double bubble* technique, the workspace of each haptic device is defined by two areas, each associated to a control mode. An inner area controls the proxy directly in position, and an outer area, starting at the boundaries of the inner area and extending up to the physical limits of the device, controls the virtual workspace in speed within the VE. Besides using two interfaces instead of one, two major differences separate the *double bubble* from the previously mentioned *bubble* technique. The first difference is the use of a rectangular parallelepiped for the boundaries of the *bubbles* instead of a sphere, to better fit the physical workspaces of the devices. We can notably think of PHANToM devices which have a higher width than their height or depth. The second difference is the presence of a visual feedback added to the haptic feedback when the devices leave the boundaries, in the form of a trail behind the rate-controlled proxies. The technique is illustrated in Figure 2.

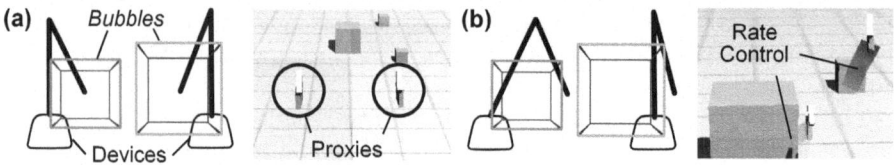

Fig. 2. Control modes of the *double bubble*. (a) Devices inside the *bubbles* : position control. (b) Devices outside the *bubbles* : rate control.

3.2   Viewport Adaptation

Since each device is attached to a *bubble* independent from the other, a method is required to keep both proxies on the screen, as these can move infinitely in completely opposite directions. Thus, we developed a method to ensure both virtual workspaces stay in the screen. This is accomplished by setting the distance of the camera to the center of the scene to a value proportional to the distance between the leftmost border of the left workspace and the rightmost border of the right workspace, plus an arbitrary margin (Figure 3). Given the left virtual workspace of center $\mathbf{l} = (l_x, l_y, l_z)$ and width w_l, and the right workspace of center $\mathbf{r} = (r_x, r_y, r_z)$ and width w_r, the position of the camera is computed following Equations (1-3).

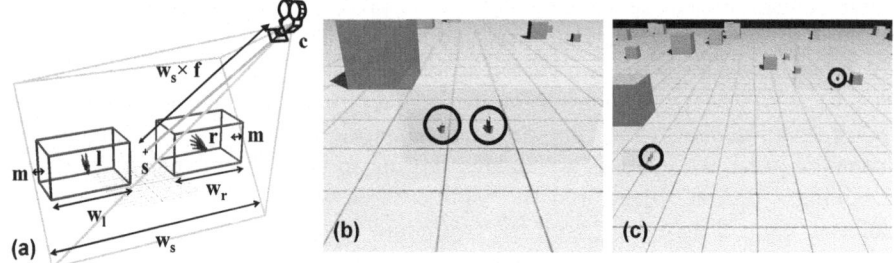

Fig. 3. Viewport adaptation. (a) Computation of the camera position. (b-c) Automatic viewport adaptation from different relative positioning of the proxies (circled).

The center of scene $\mathbf{s}$ is first computed from both workspace centers following:

$$\mathbf{s} = \left(\frac{l_x + r_x}{2}, \frac{l_y + r_y}{2}, max(l_z, r_z) \right) . \tag{1}$$

The width of the displayed scene w_s is then computed from the widths of both workspaces w_l and w_r, as well as an arbitrary margin m that ensures that the virtual workspace boundaries do not leave the borders of the screen:

$$w_s = \sqrt{(r_x - l_x)^2 + (r_y - l_y)^2} + w_l/2 + w_r/2 + 2m . \tag{2}$$

Finally, the position of the camera **c** is computed following:

$$\mathbf{c} = \mathbf{s} + w_s \times d \times \mathbf{a} . \tag{3}$$

where d is a scalar that depends on the camera field of view, and **a** is an arbitrary vector that determines the angle from which the scene is displayed.

3.3   Magnetic Pinch

In order to facilitate the picking of virtual objects with two single-point interfaces, we propose two haptic manipulation techniques triggered whenever a grasping situation is detected. Three conditions are considered to determine whether both hands are grasping an object or not, according to the contact normals, the contact forces, and the relative position of both hands (Figure 4):

1. The angle between the contact normals must be under a certain threshold.
2. Both contact forces must exceed a threshold in order to discriminate simple contacts with an object from a true intent of grasping the object.
3. Two cylinders projected from both proxies following the contact normal and whose radii match the sizes of the proxies must intersect.

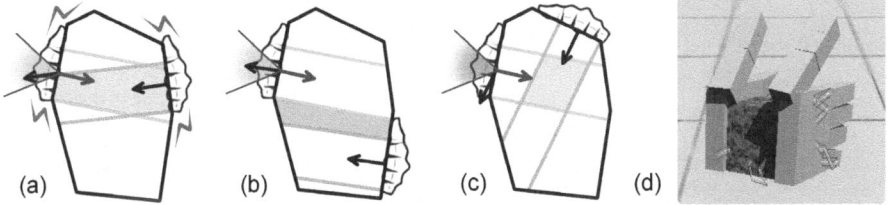

Fig. 4. Different cases of dual contact with a virtual object: (a) Normals nearly colinear and hands face-to-face, (b) Hands not in front of each other, (c) Normals far from colinearity. (d) Visual feedback of the *magnetic pinch*, symbolized by red bolts.

Once the grasping is initiated, the *magnetic pinch* takes effect, simulating a spring pulling both hands towards the picked object to prevent unintentional drops. For each haptic device, a force $\mathbf{F}_h$ is generated following:

$$\mathbf{F}_h = -k_h \times \left(1 - \frac{g_s}{\|\mathbf{o} - \mathbf{p}\|}\right) \times (\mathbf{o} - \mathbf{p}) . \tag{4}$$

where **p** is the position of the first interface, **o** is the position of the second interface, g_s is the size of the grasped object (the distance between the two contact points when the grasping is initiated), and k_h is the stiffness of the spring. The spring is removed as soon as the user gives enough force to end the contact of the hands with the object, hence dropping it.

Additionally, the position of the grasped object $\mathbf{g}_p$ can be constrained to the central point between the positions of the two virtual proxies **l** and **r**, further reducing the risk of unwanted drops. For this, we use another spring of stiffness k_o, with a force $\mathbf{F}_o$, following:

$$\mathbf{F}_o = -k_o \times \left(\frac{1 + \mathbf{r}}{2} - \mathbf{g}_p \right) . \tag{5}$$

The spring feels as if the hands were "magnetized" to the object, and small red bolts are visually displayed to highlight this effect.

3.4   Joint Control

The *double bubble* metaphor may introduce a difference in control modes and/or scaling factors when activated. In order to reduce the impact of these differences when pick-and-placing a virtual object, we introduce the notion of *joint control*. During a grasping situation, both devices use a common control/display ratio (average of both) and common bubble size (minimal dimensions), and enter rate control simultaneously when at least one device leaves its *bubble*. This technique allows easier exploration of a VE when holding an object between virtual hands controlled by two different haptic interfaces.

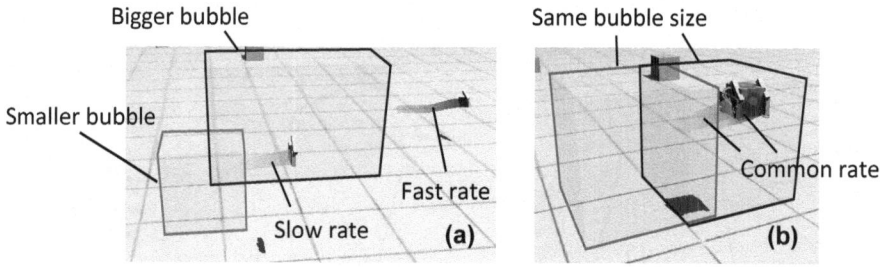

Fig. 5. Illustration of *joint control*. (a) Difference in *bubble* size and workspace translation speed without *joint control*. (b) Carrying an object with *joint control*.

4   Evaluation

To assess the efficiency of the proposed techniques, we conducted an experiment involving a simple pick-and-place task, where users had to pick a cube and place it at a given position. To evaluate the *double bubble* technique, we compared it to the *clutching* technique for workspace extension, and the benefits of the *magnetic pinch* and *joint control* were also measured for grasping facilitation.

4.1   Method

Population. Thirteen participants (2 females and 11 males) aged from 20 to 26 (mean = 22.8, sd = 1.7) performed the experiment. None of the participants had any known perception disorder. All participants were naïve with respect to the proposed techniques, as well as to the experimental setup and the purpose of the experiment.

Experimental Apparatus. The participants were seated at 1m in front of a 24 inch widescreen monitor. The experiment was conducted using two different haptic interfaces. The participants manipulated a Falcon (Novint Technologies Inc., Albuquerque, New Mexico, USA) in their left hand, and a PHANToM Omni (Sensable Technologies, Wilmington, Massachusetts, USA) in their right hand, both placed in front of the screen as shown in Figure 6. Visual feedback was rendered at a refresh rate of 50 Hz, while the haptic rendering rate was 1,000 Hz. Physical simulation was performed using Nvidia PhysX at a rate of 1,000 Hz to match the update frequency of the haptic loop. A virtual coupling mechanism was used between the haptic interfaces and the virtual proxies by simulating a spring-damper system between each haptic device and its corresponding proxy.

Virtual Environment. The VE was composed of a 100m-wide ground plane with four potential target planes, of 1m of width, placed at the corners of a 6m-wide square around the center of the VE. The target plane of each trial were colored in red, and the other planes were colored in white. The cube to be manipulated had a width of 30cm and a mass of 3g, and was placed at the center of the VE. The proxies controlled by each haptic device were physically represented by cubes of 20cm of width, and were positioned 2m away from each other and 5m away from the central cube at the start of each trial. The cube was thus lying beyond the limits of the workspaces. The proxy controlled by the left device was visually represented by a blue left hand, and the right proxy was represented by a green right hand. Figure 6 shows the scene as displayed at the beginning of a trial.

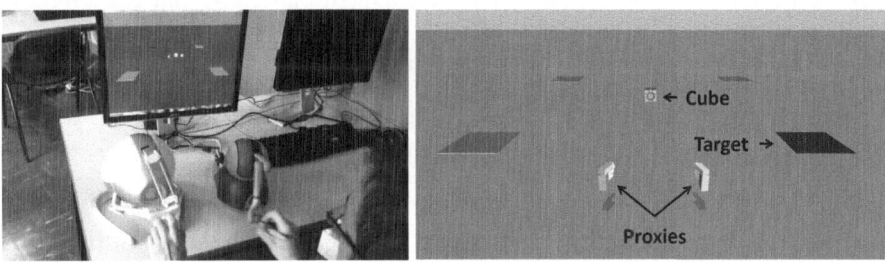

Fig. 6. Apparatus and virtual environment used in the experiment

Procedure. At the start of each trial, both haptic devices and proxies were set to their starting positions. The subject had to pick the cube from both sides, carry it towards the red target and make the cube contact with the target, thus ending the trial. A black screen warned the subject about the beginning of the next trial.

Experimental Conditions. We used a within-subject design to evaluate the four different conditions. In the control condition *Ctrl*, the participants were able to use the *clutching* technique when they reached the limits of the workspaces.

The three other conditions corresponded to: (1) *DB* (*double bubble*), (2) *MP* (*clutching* with *magnetic pinch/joint control*) and (3) *DB+MP* (a combination of *double bubble* and *magnetic pinch/joint control*). All the conditions were tested 44 times (11 times per target). The order between the different conditions was counterbalanced across participants, and for each condition, the order between the targets was randomized. The experiment lasted around 1 hour.

Collected Data. For each trial and each participant, the completion time and number of drops were recorded. The completion time is the time elapsed between the moment the proxies leave their starting positions and the moment the cube touches its target plane. The number of drops is the number of hits recorded between the cube and any part of the ground plane that is not the target plane. At the end of the experiment, participants had to complete a subjective questionnaire in which they had to grade the different techniques according to different criteria. The participants could rate the criteria from 1 (very bad) to 7 (very good). The different criteria were: (1) Global appreciation, (2) Efficiency, (3) Learning, (4) Usability, (5) Fatigue, and (6) Realism.

4.2   Experiment Results

Completion Time. We conducted a statistical analysis from the completion time data collected during the experiment. For each participant, statistics (mean M, standard deviation SD) were computed on the 44 trials in each condition. A Friedman test on the completion time (in seconds) revealed a significant effect of the technique ($\chi^2 = 27.66$, $p < 0.001$). Follow-up post-hoc analysis revealed that completion time in both the *MP* ($M = 14.16$, $SD = 7.14$) and *DB/MP* ($M = 8.43$, $SD = 2.91$) conditions were significantly shorter that in the control ($M = 21.41$, $SD = 13.19$) and *DB* ($M = 20.06$, $SD = 14.63$) conditions ($p < 0.001$ in all cases), and that the *DB+MP* condition led to significantly shorter times than the *MP* condition as well ($p < 0.001$).

Number of drops. Similarly, a statistical analysis was conducted on the number of drops for all trials of each participant. A Friedman test showed a significant effect of the technique ($\chi^2 = 25.52$, $p < 0.001$). Post-hoc analysis showed that the *MP* ($M = 4.22$, $SD = 9.45$) and *DB/MP* ($M = 2.36$, $SD = 2.33$) conditions led to significatively less drops than the control ($M = 7.88$, $SD = 6.37$) and *DB* ($M = 8.79$, $SD = 6.77$) conditions ($p < 0.001$ in all cases).

4.3   Subjective Questionnaire

We perfomed a Friedman test to analyse the answers of the participants to the subjective questionnaire. The reported p-values were adjusted for multiple comparisons (alpha-level p=0.05). We found a significant effect for 5 criteria: Global appreciation ($\chi^2 = 4.62$, $p < 0.001$), Efficiency ($\chi^2 = 4.92$, $p < 0.001$),

Learning easiness ($\chi^2 = 4.50$, $p < 0.001$), Use easiness ($\chi^2 = 4.80$, $p < 0.001$) and Fatigue ($\chi^2 = 4.46$, $p < 0.001$).

Post-hoc analysis showed that the *DB+MP* condition was preferred to both the control and *DB* for all criteria: Global appreciation ($p < 0.001$ and $p < 0.001$ respectively), Efficiency ($p < 0.001$ and $p < 0.001$), Learning ($p < 0.001$ and $p < 0.001$), Usability ($p < 0.001$ and $p < 0.001$) and Fatigue ($p < 0.001$ and $p < 0.001$). The *MP* condition was also preferred over the control and *DB* for 3 criteria: Global appreciation ($p = 0.029$ and $p = 0.028$), Learning ($p = 0.032$ and $p = 0.009$) and Usability ($p = 0.027$ and $p = 0.008$), plus a fourth criteria for the *DB*: Efficiency ($p = 0.020$).

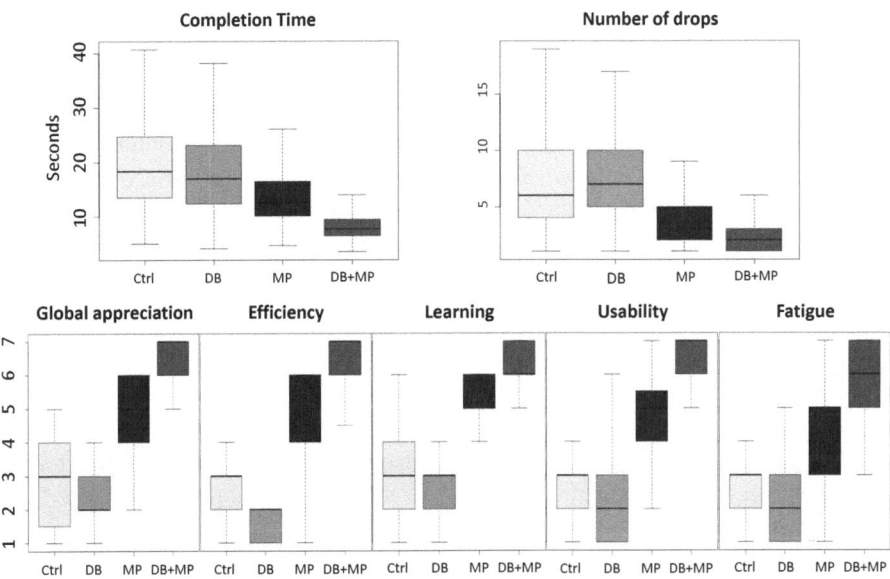

Fig. 7. Box plots of the completion times, number of drops and subjective ratings for the significative criteria, for all conditions. They are delimited by the quartile (25% quantile and 75% quantile) of the distribution of the condition over the individuals. The median is represented for each trial.

5   Discussion

We proposed the *double bubble* technique with viewport adaptation for bimanual haptic navigation in a large VE as well as two haptic manipulation techniques, the *magnetic pinch* and *joint control*. The conducted experiment showed that the manipulation techniques improved performance and subjective appreciation for a pick-and-place task over the *double bubble* and *clutching* navigation techniques, while the combination of all of the proposed techniques led to the best results.

The *double bubble*, used alone, performed as good as the *clutching* technique without outperforming it, in terms of completion time, drop rate, and subjective appreciation. The technique allows to translate the workspace in a VE in a smoother way than the *clutching* technique, by removing the need to move the devices back and forth several times. We are planning to allow rotations of the viewport with the technique in future work.

The experiment showed that the *magnetic pinch* and *joint control* significantly reduced completion times and dropping rates compared to the conditions that did not use them. In addition, the subjective appreciation also favored the conditions which used these techniques over those that did not, globally and more particularly for learning and usability. These results strongly indicate that the *magnetic pinch* and *joint control* techniques, by stabilizing the grasping of a virtual object with virtual proxies, are efficient for facilitating pick-and-place tasks. Additionally, while the *magnetic pinch* inherently adds an unrealistic behaviour through the magnetic attraction, it does not seem to hinder the global realism of the scene, as no significant difference in the participants perception of realism was reported for the different conditions.

The best results were obtained with the combination of all of the proposed techniques. The *double bubble* showed its full potential when used jointly with the *magnetic pinch* and *joint control*, outperforming the combination of the latter techniques with *clutching*. The *double bubble* allows users to perform the task in a simpler and faster way than the *clutching* technique, which imposes frequent stops of both proxies to recenter the two haptic devices.

6   Conclusion and Perspectives

In this paper, we presented novel interaction techniques for bimanual haptic manipulation of virtual objects with two single-point haptic interfaces. The *double bubble* allows to move the workspaces of both virtual proxies inside the VE through hybrid position/rate controls, and a viewport adaptation technique keeps both proxies on screen at all times. The *magnetic pinch* prevents dropping of picked objects using a virtual spring between proxies. Finally, the *joint control* allows better handling of picked objects when moving around.

An experiment with a pick-and-place task showed that the *magnetic pinch* and *joint control* could lead to faster completion of the task with less unwanted drops of the object and overall better user appreciation compared to conditions that did not use them. They are thus efficient for simplifying the picking and carrying of an object. The *double bubble*, when used jointly with the aforementioned techniques, reduced even further the time needed to complete the task, outperforming the *clutching* technique. Overall, the combination of all of these techniques was shown to be very efficient for extending the workspaces of different haptic interfaces and allowing bimanual manipulation of objects with single-point interfaces in large VEs.

Future work will focus on supporting 6DoF devices and more complex proxies that generate multiple contact points with objects. We also plan to increase the interaction possibilities of the techniques, and apply them to more complex tasks and applications like industrial prototyping or medical training.

References

1. Sun, L.-W., Van Meer, F., Bailly, Y., Yeung, C.K.: Design and Development of a Da Vinci Surgical System Simulator. In: IEEE ICMA, pp. 1050–1055 (2007)
2. Li, S., Frisoli, A., Avizzano, C., Ruffaldi, E., Lugo-Villeda, L., Bergamasco, M.: Bimanual Haptic-desktop platform for upper-limb post-stroke rehabilitation: Practical trials. In: Proc. of IEEE ROBIO, pp. 480–485 (2009)
3. Hulin, T., Sagardia, M., Artigas, J., Schaetzle, S., Kremer, P., Preusche, C.: Human-Scale Bimanual Haptic Interface. In: Proc. of the 5th Int. Conf. on Enactive Interfaces, pp. 28–33 (2008)
4. Faeth, A., Oren, M., Sheller, J., Godinez, S., Harding, C.: Cutting, Deforming and Painting of 3D meshes in a Two Handed Viso-haptic VR System. In: IEEE VR, pp. 213–216 (2008)
5. Murayama, J., Bouguila, L., Luo, Y., Akahane, K., Hasegawa, S., Hirsbrunner, B., Sato, M.: SPIDAR G&G: A Two-Handed Haptic Interface for Bimanual VR Interaction. In: Proc. of EuroHaptics, pp. 138–146 (2004)
6. Garcia-Robledo, P., Ortego, J., Barrio, J., Galiana, I., Ferre, M., Aracil, R.: Multifinger Haptic Interface for Bimanual Manipulation of Virtual Objects. In: Proc. of IEEE HAVE, pp. 30–35 (2009)
7. Walairacht, S., Koike, Y., Sato, M.: String-based Haptic Interface Device for Multifingers. In: Proc. of IEEE VR, p. 293 (2000)
8. Endo, T., Yoshikawa, T., Kawasaki, H.: Collision Avoidance Control for a Multi-fingered Bimanual Haptic Interface. In: Kappers, A.M.L., van Erp, J.B.F., Bergmann Tiest, W.M., van der Helm, F.C.T. (eds.) EuroHaptics 2010, Part II. LNCS, vol. 6192, pp. 251–256. Springer, Heidelberg (2010)
9. Ott, R., De Perrot, V., Thalmann, D., Vexo, F.: MHaptic: a Haptic Manipulation Library for Generic Virtual Environments. In: Proc. of Cyberworlds, pp. 338–345 (2007)
10. Barbagli, F., Salisbry Jr., K., Devengenzo, R.: Enabling multi-finger, multi-hand virtualized grasping. In: Proc. of IEEE ICRA, pp. 809–815 (2003)
11. Formaglio, A., Prattichizzo, D., Barbagli, F., Giannitrapani, A.: Dynamic Performance of Mobile Haptic Interfaces. IEEE Trans. on Robotics 24, 559–575 (2008)
12. de Pascale, M., Formaglio, A., Prattichizzo, D.: A mobile platform for haptic grasping in large environments. Virtual Reality 10, 11–23 (2006)
13. Peer, A., Buss, M.: A New Admittance-Type Haptic Interface for Bimanual Manipulations. IEEE/ASME Transactions on Mechatronics 13, 416–428 (2008)
14. Peer, A., Unterhinninghofen, U., Buss, M.: Tele-assembly in Wide Remote Environments. In: Proc. of IEEE/RSJ IROS (2006)
15. Fischer, A., Vance, J.M.: PHANToM haptic device implemented in a projection screen virtual environment. In: Proc. of Workshop on Virtual Environments, pp. 225–229 (2003)
16. Isshiki, M., Sezaki, T., Akahane, K., Hashimoto, N.: A Proposal of a Clutch Mechanism for 6DoF Haptic Devices. In: Proc. of ICAT, pp. 57–63 (2008)
17. Zhai, S.: User performance in relation to 3D input device design. Computer Graphics 32, 50–54 (1998)
18. Dominjon, L., Lecuyer, A., Burkhardt, J.M., Andrade-Barroso, G., Richir, S.: The "Bubble" Technique: Interacting with Large Virtual Environments Using Haptic Devices with Limited Workspace. In: Proc. of First Joint Eurohaptics Conference and Symposium on Haptic Interfaces for Virtual Environment and Teleoperator Systems, pp. 639–640 (2005)
19. Jeon, S., Harders, M.: Extending Haptic Augmented Reality: Modulating Stiffness during Two-Point Squeezing. In: Proc. of IEEE Haptics Symposium (2012)

Shaking a Box to Estimate the Property of Content

Yasuhiro Tanaka[1] and Koichi Hirota[2]

[1] Faculty of Engineering, The University of Tokyo
yas1010jp@gmail.com
[2] Graduate School of Frontier Sciences, The University of Tokyo
k-hirota@k.u-tokyo.ac.jp

Abstract. The fact that we can guess the properties of the contents in a box or a bottle by shaking it is interesting. If this experience can be realized virtually, it would be possible to use it as a means to transmit information from a portable information appliance to a user via haptic interaction. This paper describes an approach for implementing such a device; the development of a haptic device and control system as well as the modeling and simulation of a virtual box and its contents are presented in this paper. The prototype system was evaluated, and different model parameters were experimentally tested.

Keywords: shaking, estimation of properties, localization of action point, haptic interaction.

1 Introduction

People can guess the contents of a box or a bottle by shaking it. For example, by shaking a box containing a steel ball, one can estimate the size and weight of the ball. If it becomes possible to simulate the interaction of shaking a box with various contents and people can identify the virtual contents, the interaction can be a useful means for transmitting information through haptic sensation. A straightforward approach to realize this virtual interaction is to implement a device that generates force in response to motion in a manner similar to that of an actual box and its contents.

This paper describes the implementation and evaluation of a prototype system. A device that is capable of generating inertial force while measuring motion was designed and developed, a control system that reflects the behavior of the model of the device was implemented, and experiments to evaluate the ability of a user to determine the parameters of the virtual object model were performed.

2 Related Research

The generation of haptic sensation has been investigated by various approaches. Most early studies dealt with devices that were grounded to the earth; these include well-known devices such as the PHANToM [1], SPIDAR [2], and HapticMaster [3]. A grounded device is advantageous in that it can measure absolute position and obtain

P. Isokoski and J. Springare (Eds.): EuroHaptics 2012, Part I, LNCS 7282, pp. 564–576, 2012.

feedback of the reactive force from the environment. However, the range of operation of these devices is limited by the scale of the device.

The spread of portable information devices has led researchers to investigate non-grounded haptic devices. Devices of this type can be categorized into wearing-type and non-wearing-type. A typical approach to the wearing-type device is adding a force-feedback mechanism to a glove device; CyberGrasp [4] is a well-known example. There are some studies that investigate wearing-type devices; research on the Pseudo-Force-Feedback Display was aimed at producing the sensation of pressure by pinching the fingertip [5]; the Wearable Haptic Display was intended to generate shear force as well as pressure by controlling the tension of a belt around the finger [6]. It is interesting that in spite of the side effect of reaction force on the back of the fingertip, these devices are successful in providing the sensation of touch. A disadvantage of devices of this type is that it is annoying to wear such a device in daily life.

The non-wearing-type devices are held by the user. The ability to generate a force from a non-grounded device is limited; inertial force, electrostatic force, magnetic force, and the drag and lift force of air are common approaches. Among them, inertial force has been investigated in many studies; the GyroCube [7] and the Wearable Force Display using brakes [8] are based on the idea of generating torque by changing the angular momentum of a flywheel. The GyroMaster [9] and the Haptic Direction Indicator [10] also uses a gyroscopic effect. The authors also have been investigating the inertial force display [11], which is a direct background of this research.

There have been studies on devices that approximate the action of shaking an object; Linjama and Kaaresoja demonstrated feasibility of vibration feedback for simplified gesture-input interaction [12]. The Ubiquitous Haptic Device [13] was devised to simulate the impact of collision while shaking a box with solid contents. Shoogle [14] produced shaking interactions by generating vibration and sound. The Virtual Rolling Stone [15] has demonstrated an approach to produce the sensation of an object rolling inside a tube; also, in this study, vibration is used as a parameter for estimating the status of an object. In comparison with these studies, ours is more interested in generating interaction forces in addition to vibration.

One interesting idea of the inertial force display is embodied by the Virtual Force Display [16]. This device is capable of producing the sensation of unidirectional force using the reciprocal motion of a weight. By generating different forces in response to forward and backward motions, the user feels a unidirectional force because of the non-linearity of the force sensation; also, in the subsequent researches, improvement in smoothness of the induced force has been investigated [17, 18]. In addition, a previous study investigated the sensation of touch using a non-wearing device. The Ungrounded Pen-shaped Kinesthetic Display [19] provides sensation contact force to the fingers by causing a relative displacement between the tip and grip section of a pen-type device. Although this device cannot constrain the pen in space, the idea of presenting contact force by a kind of illusion should be useful.

The intensity and bandwidth of an interaction force in a shaking operation varies widely depending on the properties of the contents and the box. However, the sensation that is received by the operator is restricted by the properties of perception. Based on the knowledge of the characteristics of skin mechanoreceptors, the maximum frequency of response is approximately 1000 Hz [20]; hence, a device that emulates shaking should be expected to present vibrations over this frequency.

3   Emulation of Shaking

When a user shakes a box with contents, a force between the box and its contents is generated, for example, by friction and collision, and that force causes a reaction by the box that is perceived by the user. If the action of a user on a device is measured, the phenomenon inside the box can be simulated based on the measured action, and the reaction of the box can be fed back to the user by the device; the user would then be able to have a similar experience to that of shaking an actual box.

3.1   Getting Information by Shaking

Most information through haptic sensation is obtained by actively touching and operating objects; for example, texture is perceived more precisely by stroking, and stiffness is felt through a deforming operation. Similarly, information on the contents of a box can be acquired by shaking the box. Estimating the various properties of the contents just by shaking is a common experience; the approximate amount, size, density, viscosity, etc., can be recognized.

From the viewpoint of engineering, estimation of the properties of the contents is the process of creating a kind of mental model based on the relationship between the motion and reacting forces (see Figure 1). If the operator can create a mental model reflecting the dynamic properties of the actual object, then information on the object can be successfully transferred to the operator. By using a device in place of an actual object, it becomes possible to change properties to enable the user create a different mental model.

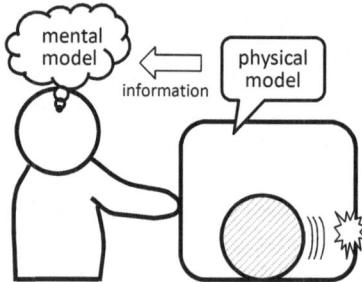

Fig. 1. Transmission of Information via Haptic Interaction

3.2   Motion of Center of Gravity

The device is a substantiation of a virtual object model; hence, it is assumed that the mass of the device is equal to the mass of the model, and the position of the frame of the device is equal to the position of the box in the model. In addition, the external force on the model is equal to the force applied to the device; hence, the total momentum of the device is equal to that of the model.

$$m_f + m_m = M_f + M_m \ , \tag{1}$$

$$m_f \dot{x}_f + m_m \dot{x}_m = M_f \dot{X}_f + M_m \dot{X}_m \ , \tag{2}$$

$$x_f = X_f \ . \tag{3}$$

where m_f, m_m, x_f, and x_m are the mass and position of the box and content of the model, respectively, and M_f, M_m, X_f, and X_m the respective mass and position of the frame and weight of the device.

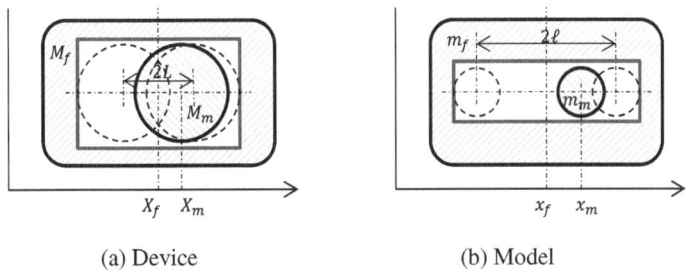

(a) Device   (b) Model

Fig. 2. Device and Model

By temporally integrating Eq. 2, assuming initial positions to be 0,

$$m_f x_f + m_m x_m = M_f X_f + M_m X_m \tag{4}$$

The following relational expression is obtained by dividing both sides by the mass:

$$\frac{m_f x_f + m_m x_m}{m_f + m_m} = \frac{M_f X_f + M_m X_m}{M_f + M_m} \tag{5}$$

This means that the center of gravity of the device must be identical to that of the model. Also, eliminating M_f and m_f from equation (5), yields

$$\frac{x_m - x_f}{X_m - X_f} = \frac{M_m}{m_m} \tag{6}$$

Note that $X_m - X_f$ and $x_m - x_f$ are the position of the weight and contents relative to the frame and the box, respectively. Hence, control of the device is realized by moving the weight in the device proportionally to the motion of the content.

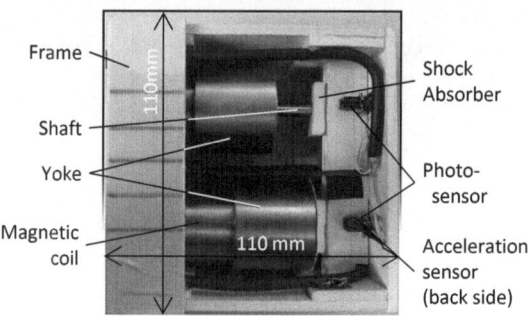

Fig. 3. Implementation of Prototype Device

4   Device and Control System

4.1   Implementation of the Device

Implementation of the device is shown in Figure 3. A pair of voice coil motors (VCMs) (AVM40-20, Akribis Systems) was installed in the frame; the masses of the yoke and coil of each VCM were 205 g and 63.3 g, respectively, and the maximum output force was 59.1 N. Since the mass of the yoke was relatively large, the yoke was regarded as the weight of the device. The yoke was supported by a linear bearing. Each VCM was connected to a motor driver (OCA-5/100, Elmo); the current-loop bandwidth of the driver was 4 kHz.

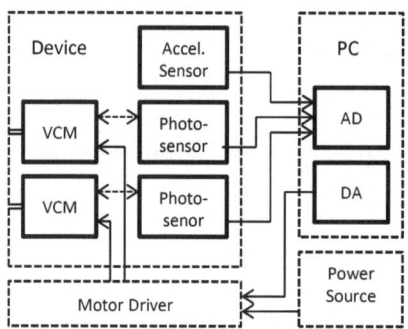

Fig. 4. System Configuration

The structure of the system is shown in Figure 4. The position of each yoke was measured by a photosensor (RPR-220, ROHM); the output voltage of the sensor does not change linearly with distance; hence, the function of the voltage-distance relationship was identified beforehand and it was converted for the purpose of controlling the device. Acceleration of the frame of the device was measured by an accelerometer (MM-2860, Sunhayato) whose range of measurement was set to ±14.7 m/s².

The computation of the control was performed by a PC (Intel Core i7-950, 3.06 GHz), and an AD/DA board (PCI-360116, Interface) was used to input and output analog signals from sensors and to the motor drivers.

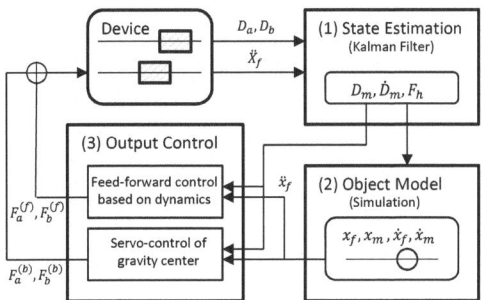

Fig. 5. Structure of Control System

4.2   Control System

The control system estimates force from the user, simulates the behavior of a virtual box and its contents, and computes the actuation force that makes the device behave similarly to the model. The structure of the control system is shown in Figure 6.

(1) State Estimation

As stated earlier, it is assumed that the device moves in one axis or one degree of freedom. The device has two weights, and in the following description, values related to them are denoted by subscripts a and b. The equation of motion for the device is formulated as follows:

$$M_f\ddot{X}_f = F_h - (F_a + F_b), \quad M_m\ddot{X}_a = F_a, \quad M_m\ddot{X}_b = F_b, \tag{7}$$

where F_h is the external force on the device and M_m is the mass of the weight; note that the masses of the two weights are same. From these equations, theoretically, the external force can be computed by the following equation:

$$F_h = M_f\ddot{X}_f + F_a + F_b. \tag{8}$$

The value of the first term is computed using the acceleration measured by the sensor, and the second and the third terms are computed using the values of the current monitor output of the two motor drivers. The noise level of the acceleration sensor is relatively high; hence, it is difficult to compute the force using this equation. Our implementation employed a Kalman filter algorithm to estimate the state of the device including external forces. Because, in this estimation process, there is no need to deal with the two weights separately,

$$D_m = (X_a - X_f) + (X_b - X_f), \quad F_m = F_a + F_b, \tag{9}$$

so, the equation of motion could be simplified as

$$M_f\ddot{X}_f = F_h - F_m, \; M_m(\ddot{X}_f + \ddot{D}_f) = F_m. \tag{10}$$

In our implementation of the filter, the state variable and the observation and control values were defined, respectively, as follows:

$$\mathbb{x} = [D_m, \dot{D}_m, F_h], \; \mathbb{y} = [D_m, \ddot{X}_f], \; \mathbb{u} = [F_m]. \tag{11}$$

The state transition equation and observation equation of the system were defined based of the equations of motion (Eq. 10). Also, the covariance matrices were determined based on the noise level of the sensors and the frequency of the variation of values.

(2) Object Model

The model deals with the situation where a solid object in the box moves in one dimension; the motion causes friction, and the object bounces back with certain attenuation if it collides with the box. The simulation is performed in discrete time steps of Δt.

The equations of motion of the box and its contents are defined as follows:

$$m_f\ddot{x}_f = F_h - f_r, \; m_m\ddot{x}_m = f_r, \tag{12}$$

where

$$f_r = \mu m_m g \frac{\dot{x}_f - \dot{x}_m}{|\dot{x}_f - \dot{x}_m|}. \tag{13}$$

Also, μ and g are the friction coefficient and gravitational acceleration, respectively. The equations of motion are solved numerically by Euler's method to obtain the position and velocity of the box and model after each.

Collision between the box and contents is considered to occur in the case of

$$|x_f - x_m| > l \; . \tag{14}$$

When a collision is detected, first, the interval within the time step before the collision is computed. Since, in our application, the change of velocity within a time step Δt is relatively small, this interval was computed simply by the ratio of the distance of motion and the distance before the collision.

Then, the box and its contents are brought back to the position of the collision, and the velocities of the box and its contents after collision are computed. The relationship of the velocities before and after the collision is determined by the law of conservation of momentum

$$m_f\dot{x}_f^{(a)} + m_m\dot{x}_m^{(a)} = m_f\dot{x}_f^{(b)} + m_m\dot{x}_m^{(b)}, \tag{15}$$

$$\frac{\dot{x}_f^{(a)} - \dot{x}_m^{(a)}}{\dot{x}_f^{(b)} - \dot{x}_m^{(b)}} = e, \tag{16}$$

where superscripts (b) and (a) indicate the values before and after the collision respectively, and e is the coefficient of restitution.

Finally, the position and velocity of the box and its contents at the end of the time step are computed assuming they continued frictional motion for the remainder of the time step, and the acceleration of the box $(\ddot{x}_m = \ddot{X}_m)$ is obtained from the change of velocity during the time step.

(3) Output Control

This process computes the output force that gives the device the acceleration that was computed in the simulation of the model. From the equation of motion (Eq. 10), the force that must be generated by the actuator is

$$F_m = F_h - M_f \ddot{X}_f .$$
(17)

It should be noted that the intensity of the collision force is relatively high, and it can exceed the maximum output force of the actuator. Our system employed a framework that outputs the impulse of the collision in multiple steps; the maximum force is limited by F_{max}; hence, the maximum impulse that can be output in a single time step is limited to $F_{max}\Delta t$. The rest of the impulse is carried over to subsequent steps.

As stated above, the device has two actuators, which are intended to virtually move the action point of force (i.e., the line on which the virtual content moves), on the plane containing axes of two actuators, in the direction perpendicular to the axes (denoted as the Y axis). The perception of the action point is thought to be caused by the relationship of force and torque; hence, the method for controlling torque was investigated. The torque that is generated by the two actuators can be changed, while keeping the total force unchanged, by changing the ratio of force that is distributed to these actuators.

We suppose there is a virtual device that is acting force F_m on Y_m. Also, there is an actual device that has two actuators on Y_a and Y_b and acting forces F_a and F_b, respectively. If the force and torque on the virtual and actual device are same,

$$F_m = F_a + F_b,$$
(18)

$$Y_m F_m = Y_a F_a + Y_b F_b,$$
(19)

this leads to the following relationship:

$$F_a^{(f)} = w_a F_m, \quad F_b^{(f)} = w_b F_m$$
(20)

where

$$w_a = \frac{Y_b - Y_m}{Y_b - Y_a}, \quad w_b = \frac{Y_a - Y_m}{Y_a - Y_b},$$
(21)

and the superscript (f) was added to indicate that these forces are considered as a feed-forward component of the control.

However, as stated in Section 3, the center of gravity of the device should coincide with that of the model. The feed-forward control does not necessarily conform to this constraint because of noise, friction, and other disturbances from the environment. Hence, a feedback control that imposes the constraint was integrated.

The target position and velocity of the weight in the device are computed using Eq. 6 as follows:

$$D_m^* = \frac{m_m}{M_m}(x_m - x_f), \quad \dot{D}_m^* = \frac{m_m}{M_m}(\dot{x}_m - \dot{x}_f) . \tag{22}$$

In our implementation, the target values for each VCM were determined by distributing D_m^* and $\dot{D}_m^*$ in the same proportion to the force:

$$D_a^* = w_a D_m^*, \; D_b^* = w_b D_m^*, \; \dot{D}_a^* = w_a \dot{D}_m^*, \; \dot{D}_b^* = w_b \dot{D}_m^* , \tag{23}$$

where D_a and D_b are the positions of the weight relative to the frame, or $D_a = X_a - X_f$ and $D_b = X_b - X_f$. The superscript $*$ indicates that it is a target value. Using these values, the feedback forces are computed as follows:

$$F_a^{(b)} = K(D_a^* - D_a) + C(\dot{D}_a^* - \dot{D}_a), \; F_b^{(b)} = K(D_b^* - D_b) + C(\dot{D}_b^* - \dot{D}_b), \tag{24}$$

where K and C are coefficients that define the feedback gain, and they were determined empirically to maintain stability in the system. The superscript (b) was added to indicate that these forces are considered as feed-forward components of the control.

4.3   Operation of the Control System

The behavior of the control system was confirmed by examining the target and actual values in the control process. Figure 6 shows the position of the center of gravity in the frame and actuation forces of the two actuators. The time step of computation and control was set to $\Delta t = 0.25$ ms. The model parameters were set to $m_f + m_m = 610$ g, $l = 10$ mm, $y = 10$ mm, $m_m = 200$ g, $e = 0.4$, and $\mu = 0$. The plot on the right hand side shows magnification in time at around 200 ms. Note that the baseline of the force drifted relatively slowly because of the feedback control, and the feed-forward component was added to the baseline; the ratios of $F_a^{(f)}$ and $F_b^{(f)}$ were approximately equal to $w_a = 0.25$ and $w_b = 0.75$, respectively.

5   Evaluation

Preliminary experiments were carried out to evaluate the ability of the prototype system to transmit model parameters via shaking interaction.

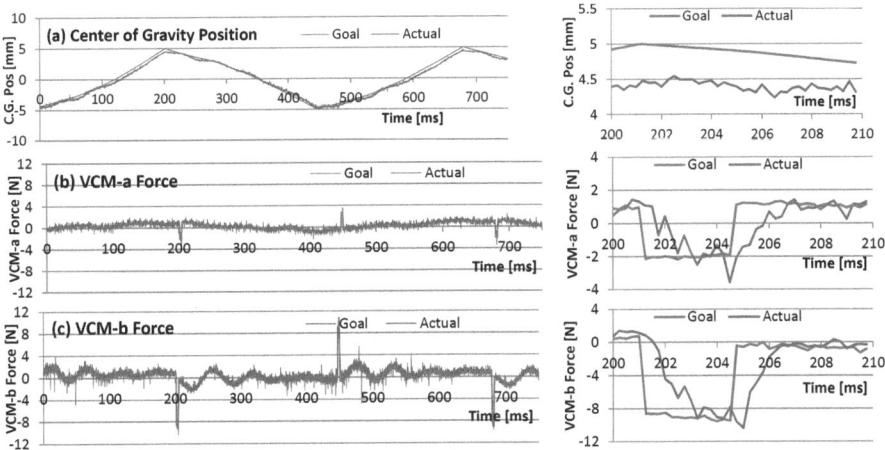

Fig. 6. Behavior of Device

5.1 Method

The experiments employed the same object model as described in Section 4. The model has four parameters: (a) the position of the action y, (b) the mass of the contents m_m, (c) the reflection coefficient e, and (d) the friction coefficient μ. In each experiment, one of the four parameters was changed while the other parameters were kept at the default values. Although, in implementation using actual box and content, change of one parameter can have effect on other parameters, such cross-effects were ignored in this experiment. The default values and the variations of each parameter are listed in Table 1. In all experiments, the total mass of the model was $m_f + m_m = 610$ g, and the range of motion of the content was kept to $l=10$ mm all through the experiment.

Table 1. Variation of parameters in experiment

Parameter	Default value	Variation	Unit
Position	0	-20, -10, 0, 10, 20	mm
Mass	50	50, 100, 150, 200, 250	g
Restitution	0.4	0.0, 0.1, 0.2, 0.3, 0.4	-
Friction	0.0	0.00, 0.01, 0.02, 0.03, 0.04	-

In the experiment on position, test subjects were asked to answer the perceived position by choosing from five options. The number of trials was 25 per person, with 5 trials for each position of action. The order of the trials was determined randomly.

Experiments on other parameters were performed by pairwise comparison; the subjects were asked to shake the device presenting two different models in series and to answer regarding the differences in the parameters, which were larger or the same. The number of trials was 25 per person, which covers all combinations of parameter variations for the former and latter parameters. The order of the trials was also determined randomly.

Each shaking time was a few seconds, although no restriction on shaking time was set. The manner of holding and shaking the device was standardized to avoid variation among the subjects; each subject was asked to hold the device using both hands, keep the device horizontal, and shake it right and left. The number of subjects was 10; they were all males aged 22 to 24 years old.

5.2   Results

The results of the experiment are shown in Figure 7. In plate (a) the ratio of the answered position was plotted with respect to the presented position. In plates (b) to (d), the ratio of selected options was plotted depending on the differences between the former and latter parameters. The results of the experiment on position suggest that a difference of 20 mm significantly reduces failure in discrimination. It was interesting that many subjects remarked that, in some conditions, they felt the action point outside the range of variation. The results of the experiments on the other parameters also suggested that the subjects could, to some extent, discriminate between models that had different parameters. If the discrimination threshold is defined as the crossing point of "sameness" in the other plots, the thresholds of mass, restitution, and friction were approximately 32 g, 0.17, and 0.012, respectively.

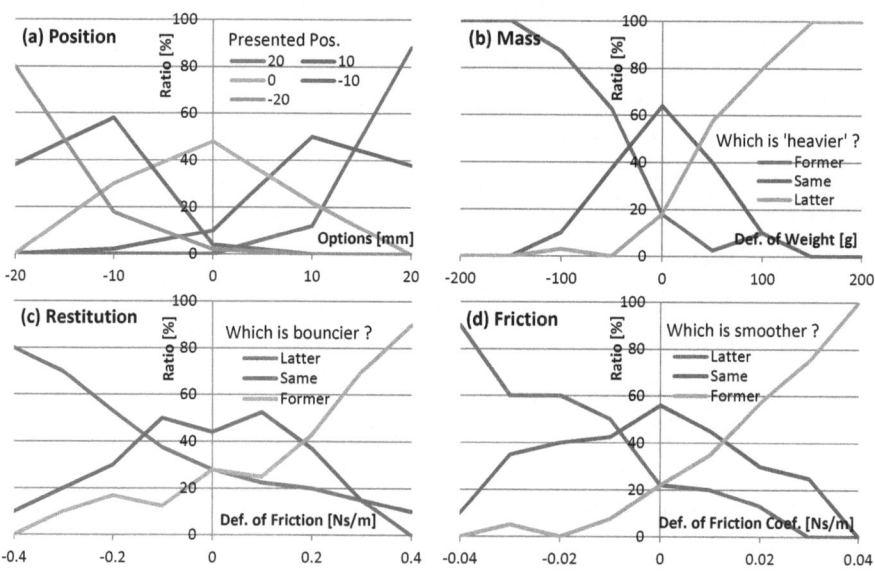

Fig. 7. Result of Experiment

5.3   Discussion

Based on the discrimination threshold, the amount of information that is presented by the shaking interaction can be computed. The approximate number of steps for each parameter that the user can discriminate is obtained by dividing the range of parameter variation in the experiment by twice the threshold; 3.13 (1.65 bit), 1.17 (0.22 bit), and 1.67 (0.74 bit) for mass, restitution, and friction, respectively. Also, the difference in position can be discriminated approximately at 3 steps (1.58 bit). It should be noted that these thresholds are just roughly estimated values and need further inspection from both sensorial characteristic and statistical processing.

6   Conclusion

This paper described the implementation of a prototype system that virtually realizes the interaction of shaking a box. An inertial force display that generates force and torque for a one-dimensional shaking motion was implemented and control methods for the device were discussed. In addition, the model of a box with solid contents was integrated for interaction. Through experiments, it was suggested that the parameters of the model such as the position of action of the force, the mass of the contents, restitution, and friction are transmitted to the user by the interaction.

In our future work, we are going to carry out experiments that fully examine characteristics of discriminating model parameters, including comparison with real model. Also, we are interested in increasing the degrees of freedom of the device. The device reported in the paper was assumed to be shaken in one direction, and this restricted the interaction with the device. Devices that are capable of generating two- or three-dimensional forces are needed.

References

1. Sensable HP, http://www.sensable.com/industries-application-development.htm
2. Sato, K., Igarashi, E., Kimura, M.: Development of non-constrained master arm with tactile feedback device. In: Proc. ICAR 1991, vol. 1, pp. 334–338 (1991)
3. Iwata, H.: Artificial reality with force-feedback: development of desktop virtual space with compact master manipulator. In: Proc. SIGGRAPH 1990, pp. 165–170 (1990)
4. CyberGlove Systems, http://www.cyberglovesystems.com/products/cybergrasp/overview
5. Inaba, G., Fujita, K.: Pseudo-force-feedback display by fingertip tightening. Trans. VRSJ 12(1), 95–102 (2007) (Japanese)
6. Minamizawa, K., Kajimoto, H., Kawakami, N., Tachi, S.: Wearable haptic display to present gravity sensation - preliminary observations and device design. In: Proc. World Haptics 2007, pp. 133–138 (2007)
7. Sakai, M., Fukui, Y., Nakamura, N.: Effective output patterns for torque display "GyroCube". In: Proc. ICAT 2003, pp. 160–165 (2003)

8. Ando, H., Obana, K., Sugimoto, M., Maeda, T.: A wearable force display based on brake change in angular momentum. In: Proc. ICAT 2002, pp. 16–21 (2002)
9. Yoshie, M., Yano, H., Iwata, H.: Development of non-grounded force display using gyro moment effect. Trans. VRSJ 7(3), 329–337 (2002)
10. Antolini, M., Bordegoni, M., Cugini, U.: A Haptic Direction Indicator Using the Gyro Effect. In: Proc. IEEE World Haptics 2011, pp. 251–256 (2011)
11. Hirota, K., Sekiguchi, Y.: Inertial force display - concept and implementation. In: Proc. ISUC 2008, pp. 281–284 (2008)
12. Linjama, J., Kaaresoja, T.: Novel, minimalist haptic gesture interaction for mobile devices. In: Nordi CHI 2004, pp. 457–458 (2004)
13. Sekiguchi, Y., Hirota, K., Hirose, M.: The design and implementation of ubiquitous haptic device. In: Proc. World Haptics 2005, pp. 527–528 (2005)
14. Williamson, J., Smith, R.M., Hughes, S.: Excitatory multimodal interaction on mobile devices. In: Proc. CHI 2007, pp. 121–124 (2007)
15. Yao, H.Y., Hayward, V.: An experiment on length perception with a virtual rolling stone. In: Proc. Euro Haptics 2006, pp. 325–330 (2006)
16. Amemiya, T., Ando, H., Maeda, T.: Directed force perception when holding a non-grounding force display in the air. In: Proc. Euro Haptics 2006, pp. 317–324 (2006)
17. Amemiya, T., Maeda, T.: NOBUNAGA: Multicylinder-Like Pulse Generator for Kinesthetic Illusion of Being Pulled Smoothly. In: Ferre, M. (ed.) EuroHaptics 2008. LNCS, vol. 5024, pp. 580–585. Springer, Heidelberg (2008)
18. Amemiya, T., Maeda, T.: Impact of Pulse Width and Pulse Oscillation Interval on Perception of Pseudo-Attraction Force. In: Proc. IEEE SMC 2009, pp. 1724–1729 (2009)
19. Kamuro, S., Minamizawa, K., Tachi, S.: Ungrounded pen-shaped kinesthetic display for sketch-based modeling. Trans. VRSJ 16(3), 459–468 (2011)
20. Shimoga, K.B.: A survey of perceptual feedback issues in dexterous telemanipulation. II. Finger touch feedback. In: Proc. VRAIS 1993, pp. 271–279 (1993)

Inside the Boundaries of the Physical World: Audio-Haptic Feedback as Support for the Navigation in Virtual Environments

Luca Turchet, Niels Nilsson, and Stefania Serafin

Aalborg University Copenhagen, Media Technology Department,
Lautrupvang 15, 2750, Ballerup, Denmark
{tur,ncn,sts}@create.aau.dk

Abstract. One of the main issues in creating virtual environments in a laboratory setting where users can navigate is the fact that laboratories have a limited physical space. One way of compensate this issue has been to redirect users to perform specific paths, keeping for example the illusion of walking straight while indeed subjects were walking in circles. In this paper we investigate whether audio-haptic feedback and haptic feedback alone help in directing users to walk away from the boundaries of a physical space while experimenting with a simulated virtual environment. Specifically, haptic feedback was provided at feet level by using a pair of shoes enhanced with actuators, and auditory feedback of different footsteps was also provided interactively. Results show that it is possible to use auditory and haptic feedback to provide users with navigational cues in virtual environments.

Keywords: walking, audio-haptic feedback, virtual environments.

1 Introduction

One of the main issues encountered when navigating in virtual environments is the fact that the physical limitations of a laboratory prevent users from freely navigate in the virtual world as if they were in the physical world. To cope with such limitations, several approaches have been proposed. As an example, the redirected walking technique proposed in [5] creates a visual illusion where subjects feel as if they were walking straight while instead they are walking in circles. Several variations of such illusion have been exploited, such as redirected walking in place [6], which combines the walking in place technique [15] with redirected walking. The use of haptic feedback at feet level to facilitate navigation is not widely explored in the research community. An exception is the work presented in [2], where an alternative navigation system based on haptic feedback is proposed. Similar work was also presented in [10], where a system that changes the physical texture perceived at the ground is proposed.

In previous research, we described a system able to simulate the auditory and haptic sensation of walking on different materials and presented the results of a preliminary surface recognition experiment [9]. This experiment was conducted

P. Isokoski and J. Springare (Eds.): EuroHaptics 2012, Part I, LNCS 7282, pp. 577–588, 2012.
© Springer-Verlag Berlin Heidelberg 2012

under three different conditions: auditory feedback, haptic feedback, and both. Stimuli were presented to the participants while walking in the laboratory, in such a way to maintain the tight sensorimotor coupling that is natural during walking and foot interaction. This is true for the auditory channel, but even more so for the haptic channel. However, limitations given by the physical space of the laboratory prevented subjects from freely walking as if they were navigating in the real world.

In this paper, we propose a solution to allow subjects to almost freely walk in the real world by providing interactive auditory and haptic feedback. Our hypothesis is that the provided feedback redirects the subjects and drives them inside the boundaries of the physical space. Our system presents several applications related to navigation techniques based on non visual feedback. As an example, the system can allow visual impaired users to be properly directed when navigating in a space with obstacles. This is the case not only for a laboratory setting, but also for navigation in a real world by using a portable version of the system.

2   The Interactive Multimodal System

We recently developed a multimodal interactive architecture with the goal of creating audio-haptic-visual simulations of walking-based interactions [8]. The architecture used during the proposed experiments consists of a motion capture system (MoCap), an head-mounted display (HMD), two soundcards, twelve loudspeakers, an Arduino Diecimila board, two amplifiers, two haptic shoes and two computers (see Figure 1). Users are required to wear a pair of shoes enhanced with sensors and actuators able to provide haptic feedback during the act of walking. In addition, markers are placed on the top of each shoe in correspondence of both heel and toe in order to track the user's locomotion by means of the MoCap. Furthermore, markers are also placed on top of the HMD, in order to track orientation and position of the users' head. Concerning the surround sound system, the configuration of the twelve loudspeakers is illustrated in Figure 2. Eight loudspeakers are placed on the ground at the vertices of a regular octagon, while four loudspeakers are placed in correspondence to the vertices of the the rectangular floor at the height of 1.40 m. All the loudspeakers are used for delivering soundscapes. The eight loudspeakers at floor level are used to deliver footstep sounds. Such configuration was chosen according to the results of preliminary studies on footstep sounds delivery methods.

A multimodal synthesis engine able to reproduce visual, auditory and haptic feedback was also developed on the basis of the architecture described above.

The auditory feedback was obtained by the combination of a footstep and a soundscape sound synthesis engine. The footstep sounds synthesizer used was the one proposed in [14], which allows the real-time simulation of footstep sounds on several different materials. In the proposed experiments, the footstep sounds synthesis was driven interactively during the locomotion of the subject wearing the above mentioned shoes according to the algorithms described in [12].

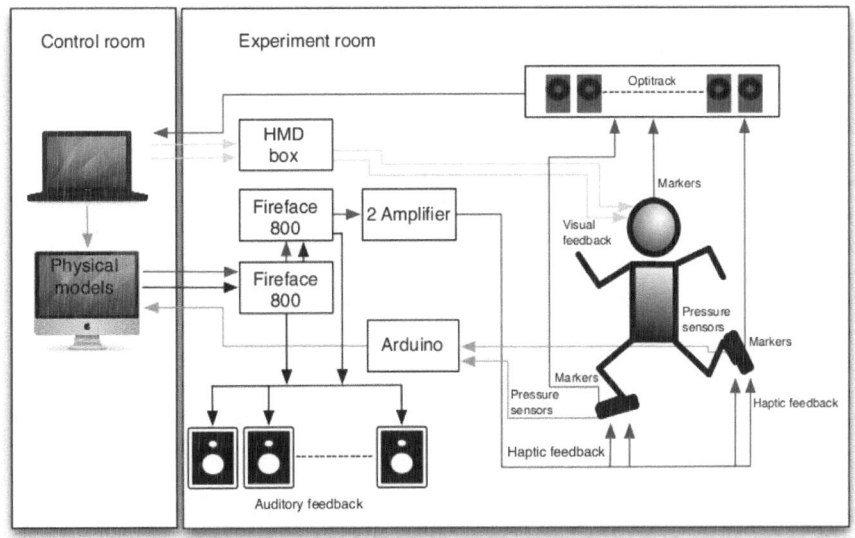

Fig. 1. Schematic representation of the overall architecture developed

During the proposed experiments the footstep sounds were diffused according to a delivery method consisting of the combination of static and dynamic diffusion. For static diffusion we intend that the footstep sounds are diffused simultaneously to the eight loudspeakers using the same amplitude for each loudspeaker. Conversely, during the dynamic diffusion the user position tracked by the MoCap is used to diffuse the footstep sounds according to the vector base amplitude panning algorithm [4] which allows to place under the user's feet the virtual sound source containing the footstep sounds. In this way the sound follows the user trajectories during his/her locomotion, and therefore the eight loudspeakers deliver the footstep sounds with different amplitudes. The approach we followed combines the two methods: the static diffusion was used when the user walked in the central zone of the walking area (see the grey circle in Figure 2), while as soon as the user stepped outside such zone the dynamic diffusion was provided.

As concerns the soundscapes diffusion, the environmental sounds were delivered dynamically using a sound diffusion algorithm based on ambisonics [7].

Regarding the haptic feedback, it was provided by means of the haptic shoes previously described. The haptic synthesis is driven by the same engine used for the synthesis of footstep sounds, and is able to simulate the haptic sensation of walking on different surfaces, as illustrated in [12].

Concerning the visual feedback provided by the HMD, three outdoor scenarios were developed in order to provide a visual representation of the physically simulated surfaces rendered in the audio-haptic engine. As an example, during the experiments a forest, a snowy landscape and a beach were visually rendered to match the physically simulated forest underbrush, snow and sand.

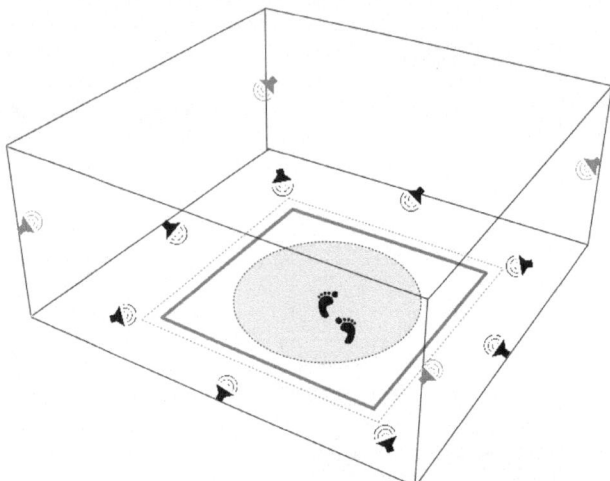

Fig. 2. Loudspeakers' configuration (the 8 loudspeakers used for the footstep sounds diffusion are indicated in black). The area available for the walk is indicated by the rectangular dotted line on the floor. The inner grey rectangle indicates the region inside which the audio-haptic feedback was not changed. The light grey circle indicates the area in which footstep sounds were delivered according to the static diffusion.

3   Method

During the experiments the area of the laboratory available for the users to walk was divided in two zones: an inner zone in which the audio-haptic feedback was not modified ("walking area") and the zone in which the feedback was changed in order to redirect the user inside the walking area. The two zones consisted of two rectangles 2.30x2.40 m and 2.50x2.60 m disposed as indicated in Figure 2 by the grey and the dotted line respectively. The perimeters of both the rectangles were indicated on the floor by means of scotch tape strips. No indications about the boundaries of the walking area were provided at visual level, so the users had to base their walks only on the feedback received sonically and haptically.

Two different techniques were chosen among all the possible methodologies available combining in various ways the auditory and the haptic feedback in order to redirect the walk inside the walking area once one or both the feet had exceeded it:

Method 1: the surface type is changed both at auditory and haptic level
Method 2: the auditory feedback is stopped while the haptic feedback (on the same surface) is kept

In both cases the change of the feedback is provided only in correspondence of the foot which exceeds the boundaries of the walking area. This means that if one foot goes outside the boundary while the other remains inside, the feedback generated by the latter is not affected.

The two approaches were chosen with the aim of being on one hand enough informative and useful for the purpose of warning the walker, and on the other hand intuitive. These choices were made also taking into account the results of previous studies conducted using the haptic shoes and the audio-haptic footsteps synthesizer [13,9]. Such studies revealed on one hand the role of dominance of the auditory feedback on the haptic one, and on the other hand the difficulty in distinguishing at haptic level the surface materials. Therefore the choice of modifying only the haptic feedback would have been not enough informative for the purpose in question. However we did not want to use the auditory feedback alone because the addition of the haptic feedback allows to increase the realism of the interaction [11]. In addition one of the specification requirements was that the change in the feedback had to be as less disruptive as possible for the sensation of the user of being in the virtual environment. As regards the first technique, the audio-haptic change of the surface material was designed in order to avoid the passage from an aggregate surface to a solid one and vice versa (e.g. snow-wood). This in addition would have been perceived by the users as highly incoherent with the surface type visually provided (e.g. a snowy environment). Therefore only materials belonging to the same typology (aggregate-aggregate or solid-solid) were used. Concerning the second technique, the complete stop of the auditory feedback while continuing the haptic one, was chosen because the haptic shoes through the activation of the actuators generate not only the haptic sensation but also a sound having the same nature (but not the same audio quality) of that provided at auditory level. The choice of this design was also supported by the fact that stopping both the auditory and the haptic feedback would have resulted in a too drastic change.

All these considerations led to two hypotheses:

Hypothesis 1: both techniques are informative enough for the task of correctly warning the walker, but the first technique is more helpful compared to the second one

Hypothesis 2: the application of the second technique results as less negatively affecting the sensation of feeling present in the virtual environment compared to the first one

4   Experiment Design

We conducted an experiment whose goal was to assess the limits of the proposed techniques and which of the two was preferred by the users. Each condition of the experiment tested both the two techniques in a different virtual environment: a forest, in condition 1, a snowy landscape in condition 2, and a beach in condition 3. Such scenarios were built in order to fit coherently with the synthesized audio-haptic footsteps on forest underbrush, deep snow and sand respectively.

The reason for choosing three surface materials and three virtual environments was to assess whether the stimulus type affected the quality of the results. Concerning the first technique, the change in the audio-haptic feedback consisted of passing from forest underbrush to gravel, from snow to sand, and from sand to forest underbrush, in the three environments respectively. Conversely in the second technique, the auditory feedback was stopped while the haptic feedback was kept on the same surface.

An experimenter was placed in the room where the experiment was performed by participants, in order to prevent them to fall down because of eventual balance losses as well as in order to take them inside the walking area in case they were not able to follow the audio-haptic cues. The scotch tape lines placed on the floor indicating the two rectangles illustrated in Figure 2 were used by the experimenter to check the participants' behavior in proximity of the boundaries of the walking area.

4.1   Task

During the three conditions participants were asked to wear both the HMD and the haptic shoes previously mentioned and to walk twice in the simulated virtual environment. In each walk one of the two techniques for warning the walker was provided. The order of the techniques was randomized across participants. For each walk the time available to the users lasted two minutes.

Before performing the experiment participants tried both the techniques and were instructed on the task they had to perform. The task consisted of freely navigating the landscape visually provided, modifying the direction of the walk according to the feedback change in order to remain inside the boundaries of the walking area.

Immediately after each walk participants were asked to evaluate (by voice) the following statements on a 9 points Likert scale (1=strongly disagree, 9=strongly agree):

– "The change in feedback made it clear to me that I was leaving the area which I was supposed to stay within."
– "The change in feedback, occurring when I reached the edge of the area which I was supposed to stay within, disrupted the sensation of being there in the virtual environment."

At the end of the two walks participants were asked to answer to the following questions in order to compare the two techniques:

– "Which of the two feedback types made it the easiest to determine that you had reached the edge of the area which you were supposed to stay within?"
– "Which of the two feedback types was the least disruptive for your sensation of being there in the virtual environment?"

Finally at the end of the experiment subjects were asked to evaluate the system they interacted with, by means of a questionnaire built on the Virtual Experience Test (VET) proposed in [1]. Such survey aims to measure holistic virtual environment experiences based upon five dimensions of experiential design: sensory, cognitive, affective, active, and relational. Nevertheless only the questions relative to the sensory and cognitive dimensions were utilised since our main goal was not to measure the presence level experienced by participants. Rather we were interested on one hand in their evaluations of the sensory hardware and content quality, as well as of the consistency of the sensory information, and on the other hand in the ratings of their perceived ability to complete the proposed task and to understand the environments rules.

The utilized questions are illustrated in section 5.1, and were evaluated on a 9 points Likert scale (1=strongly disagree, 9=strongly agree). The order of the questions was randomized using a 5x5 Latin square to reduce questionnaire order bias. Finally, participants were also given the opportunity to leave an open comment on their experience interacting with the system.

4.2   Participants

Thirty participants were divided in three groups (n = 10) to perform the experiment. The three groups were composed respectively of 8 men and 2 women, aged between 24 and 33 (mean=26.08, standard deviation=3.08), 8 men and 2 women, aged between 23 and 35 (mean=27.8, standard deviation=4.56), and 8 men and 2 women, aged between 20 and 30 (mean=23.3, standard deviation=2.83). Participants took on average about 6 minutes to complete the walking experiment, and about 5 minutes to complete the questionnaire. All participants reported normal hearing conditions and no locomotion problems.

5   Results

The collected answers were analyzed and compared between the three conditions. Results are illustrated in Table 1. The first noticeable element emerging from these data is that the first technique led to a higher number of successful completion of the task (i.e. participants correctly understood the warning and therefore moved back inside the walking area) and a lower number of unsuccesses compared to the second technique. An in-depth analysis using chi-square test, shows significant difference between the two conditions ($\chi^2(1) = 25.05$, p < 0.001).

Considering the total number of successes and unsuccesses it is possible to notice that participants reached the boundaries of the walking area an amount of times greater during the use of the first technique compared to when the second technique was presented.

Furthermore the total number of successes and unsuccesses of both the techniques, was higher for the forest environment rather than the other two, as well as for beach respect to the snowy landscape. The statistical analysis conducted by means of a t-test revealed significant differences for forest compared to snowy landscape ($t(38) = 3.628$, p < 0.001) and for sand compared to snowy landscape ($t(38) = 3.106$, p $= 0.003$). As regards the evaluation of the clearness of the feedback in providing the information on which direction taking in order to stay within the walking area, the average scores are always higher for the first technique. The statistical analysis conducted by means of a t-test revealed that such differences are significant (t(58) = 2.418, p = 0.019). As concerns the evaluations of the level of presence disruption, higher evaluations were given for the second technique in two cases out of three, with the exception of the beach environment (overall no significant difference was measured).

Concerning the preferences expressed by participants after having tried both the techniques, it is possible to notice that the first technique was clearly preferred to the second one. Indeed the first technique was found more useful and intuitive than the second one, as well as less disruptive for the presence level (with the exception of the beach environment, for which the percentages are identical). In addition an exact binomial test revealed that the preferences for the technique 1 are significant (p = 0.001 in the first case and p = 0.042 in the second one).

Table 1. Results of the experiment

	Forest		Snow		Beach	
	Technique 1	Technique 2	Technique 1	Technique 2	Technique 1	Technique 2
Number of successes	70	42	48	32	59	29
Number of unsuccesses	17	28	16	25	20	37
Total number of successes/unsucesses	87	70	64	57	79	66
Clearness	7.9 ± 0.87	6.1 ± 2.55	7.5 ± 1.77	7.2 ± 1.61	6.9 ± 1.8	5.3 ± 2.45
Presence disruption	3.8 ± 2.48	4.2 ± 2.29	5.7 ± 2	6.1 ± 2.33	5 ± 2.26	4.3 ± 2.35
Preference easiest	90%	10%	70%	30%	80%	20%
Preference least disruptive	80%	20%	80%	20%	50%	50%

5.1   System Evaluation

As mentioned in section 4.1, at the end of the experiment participants were provided with a questionnaire in order to evaluate the system they interacted with. Results are shown in Table 2; questions from Q1 to Q7 are relative to the sensory experimental design dimension, while questions from Q8 to Q11 concern the cognitive one.

As regards the sensory input (visual, aural, and haptic), as well as the perception of those stimuli, it is possible to notice that there are not very high evaluations for the investigated parameters. This can be seen as an indication of a lack in the utilized technology for both the sensory hardware and software that create the sensations. Even if the evaluations are never low, the questionnaire results points toward an improvement of the system technology. Negative correlation was found between the evaluations of the questions Q1, Q3, and Q5 relative to the hardware and the questions Q2, Q4, and Q6 relative to the provided content.

As concerns the cognitive dimension, the evaluations about how well the environment supported task engagement through the clarity of task explanations, the perceived task interest and the explanation of environment rules, are mostly around average. However these data are relative to the tasks conducted with both the techniques, therefore the interpretation of such result should take into account the fact that while the first technique was well understood, for the second one the participants' performances were less successful.

Overall, almost always the average scores for both the sensory and the cognitive dimensions, are higher for the snowy landscape compared to the other two, revealing that the best interaction occurred with that virtual environment.

Table 2. Results of the questionnaire on the system

Questions	Forest	Snow	Beach
Q1: I found the visual display hardware to be of high quality	4.4 ± 2.1	5.8 ± 1.3	5.4 ± 2.1
Q2: I found the visual content of the environment to be of high quality	4.6 ±1.9	6.1 ± 1.4	5.1 ± 1.8
Q3: I found the audio hardware to be of high quality	6.1 ± 1.6	6.8 ± 1.1	4.6 ± 1.9
Q4: I found the audio content of the environment to be of high quality	6 ± 1	7.1 ± 0.8	5.3 ± 1.8
Q5: I found the haptic hardware to be of high quality	4.3 ± 1.8	6.4 ± 2.3	4.1 ± 1.5
Q6: I found the haptic content of the environment to be of high quality	4.9 ± 2	5.6 ± 2	3.6 ± 1.6
Q7: I found that the sensory information of the virtual environment was consistent	5.5 ± 2.1	6.1 ± 2.4	5.4 ± 2.5
Q8: I found that the content in the virtual environment was helpful in informing me of my current task	4.8 ± 2	4.2 ± 2.8	4 ± 2.3
Q9: I found the user interface to be helpful in informing me of my current task	4 ± 2.2	5.4 ± 2.6	5 ± 1.5
Q10: I thought that the virtual environment made it clear what I was and was not allowed to do	4 ± 1.8	4.4 ± 3.2	3.8 ± 1.9
Q11: I thought that the tasks I was able to do in the virtual environment were interesting	3.4 ± 1.9	4.2 ± 2.5	5.1 ± 2.5

6   Discussion

The results of the experiment are clear: the participants' performances were better with the first technique, and participants preferred it to the second one. Our initial hypotheses were only partially confirmed. Indeed on one hand only the first technique led to a correct warning of the walker in most of the cases, while the second technique presented a too high number of unsuccesses compared to the successful completion of the task. On the other hand, the second technique was also rated as the more negatively affecting the sensation of feeling present in the virtual environment compared to the first one.

A deeper observation of the participants' performances revealed that also the second technique proved to work well. Indeed most of the successful completion happened towards the end of the task, indicating that participants needed a learning phase longer than the one they tried at the beginning of the experiment. Moreover, this aspect was reported in the comments of some participants. However one of the participants also reported that in presence of the second technique he was satisfied of the proposed feedback since the continuous presence of the haptic allowed to mask the sensation of walking on a laboratory surface therefore increasing the realism of his interaction.

Concerning the performances at the moment of reaching the boundaries, for the second technique it was observed a delay in the choice of which direction to take in order to come back inside the walking area: participants took longer time to realize where they had to place the next step compared to the first technique. Furthermore, it was noticed that while for the first technique rarely participants placed both the feet outside the walking area before correctly coming back, this instead happened more frequently for the second technique. As regards the negative effect of the techniques on sense of presence, it is not possible to conclude that the first technique was significantly perceived as the less disruptive. However, results seem to point toward this direction since in two cases out of three the first technique was preferred and rated as the best one from this point of view. Nevertheless, evaluations of the presence disruption parameter of Table 1 are not low neither for the first technique, revealing that the impact of the use of the proposed techniques on the presence level is still too high.

An unexpected result of this research was that the the total number of successes and unsuccesses of both the techniques, was quite different between the three virtual environments. Indeed it was higher for the forest environment rather than the other two, as well as for beach respect to the snowy landscape. This is an indication that the average quantity of motion was different in the three simulations. In addition, a difference in the gaits was also observed: participants tended to walk more and faster in the forest environment rather than the other two, while the lowest average velocity was held during the navigation of the snowy landscape where in addition some of the participants tended to remain stopped, exploring visually the environment rather than by walking. On one hand this result seems to be related to the fact that participants felt present in the virtual environment they were exploring; on the other hand the noticed differences in the gaits could also be related to the different degrees of compliance of the three surfaces simulated at auditory and haptic level.

As concerns the system evaluation, it emerges the necessity of improvements of the system technology both at software and hardware level. In particular for a better experience while physically navigating the simulated virtual environment it would be needed a totally wireless system; indeed the wires of the HMD as well as those coming out from the haptic shoes limit a lot the freedom of the navigation and in some cases the participants became aware of this aspect during the interaction with the environment. Improving these aspects would allow also to increase the presence level experienced by users, since the quality of sensory hardware, as well as the sensory content, has a widespread positive effect on reported presence, as shown in a review done by [3].

7 Conclusion

In this study two techniques for warning walkers were proposed and evaluated. The first technique consisted of the change of the surface type both at auditory and haptic level; the second one consisted in the stop of the auditory feedback only, while the haptic feedback continued on the same surface.

Results show that by means of the proposed techniques it is possible to navigate a walker inside the walking area once one or both the feet had exceeded it, even if the first technique produced better performances and was preferred by participants. However, results also show that the use of these techniques produced a not negligible disruption of the sense of presence.

An unexpected results of the proposed study was that the quantity of motion produced during participants' navigations was different in the three simulated environments. This aspect could be related to the type of the simulated surface and deserves to be investigated in a deeper way.

References

1. Chertoff, D.B., Goldiez, B., LaViola, J.J.: Virtual experience test: A virtual environment evaluation questionnaire. In: Proceedings IEEE VR 2010, pp. 103–110 (2010)
2. Frey, M.: Cabboots: shoes with integrated guidance system. In: Proceedings of the 1st International Conference on Tangible and Embedded Interaction, pp. 245–246. ACM (2007)
3. Lee, K.M.: Why presence occurs: evolutionary psychology, media equation, and presence. Presence: Teleoperators and Virtual Environments 13, 494–505 (2004)
4. Pulkki, V.: Compensating displacement of amplitude-panned virtual sources. In: Proceedings of the AES 22nd International Conference, pp. 186–195 (2002)
5. Razzaque, S., Kohn, Z., Whitton, M.C.: Redirected walking. In: Proceedings of EUROGRAPHICS, pp. 289–294. Citeseer (2001)
6. Razzaque, S., Swapp, D., Slater, M., Whitton, M.C., Steed, A.: Redirected walking in place. In: Proceedings of the Workshop on Virtual Environments, EGVE 2002, Eurographics Association, Aire-la-Ville, Switzerland, pp. 123–130 (2002)
7. Schacher, J.C., Neukom, M.: Ambisonics Spatialization Tools for Max/MSP. In: Proceedings of the International Computer Music Conference (2006)
8. Serafin, S., Turchet, L., Nilsson, N.C., Nordahl, R.: A multimodal architecture for simulating natural interactive walking in virtual environments. PsychNology Journal 9(3), 245–268 (2012)
9. Serafin, S., Turchet, L., Nordahl, R., Dimitrov, S., Berrezag, A., Hayward, V.: Identification of virtual grounds using virtual reality haptic shoes and sound synthesis. In: Proceedings of Eurohaptics Symposium on Haptic and Audio-Visual Stimuli: Enhancing Experiences and Interaction, pp. 61–70 (2010)
10. Takeuchi, Y.: Gilded gait: reshaping the urban experience with augmented footsteps. In: Proceedings of the 23nd Annual ACM Symposium on User Interface Software and Technology, pp. 185–188. ACM (2010)
11. Turchet, L., Burelli, P., Serafin, S.: Haptic feedback for enhancing realism of walking simulations. In: IEEE Transactions on Haptics (submitted, 2012)

12. Turchet, L., Nordahl, R., Berrezag, A., Dimitrov, S., Hayward, V., Serafin, S.: Audio-haptic physically based simulation of walking on different grounds. In: Proceedings of IEEE International Workshop on Multimedia Signal Processing, pp. 269–273. IEEE Press (2010)
13. Turchet, L., Serafin, S., Dimitrov, S., Nordahl, R.: Conflicting Audio-Haptic Feedback in Physically Based Simulation of Walking Sounds. In: Nordahl, R., Serafin, S., Fontana, F., Brewster, S. (eds.) HAID 2010. LNCS, vol. 6306, pp. 97–106. Springer, Heidelberg (2010)
14. Turchet, L., Serafin, S., Dimitrov, S., Nordahl, R.: Physically based sound synthesis and control of footsteps sounds. In: Proceedings of Digital Audio Effects Conference, pp. 161–168 (2010)
15. Usoh, M., Arthur, K., Whitton, M.C., Bastos, R., Steed, A., Slater, M., Brooks Jr., F.P.: Walking > walking-in-place > flying, in virtual environments. In: Proceedings of the 26th Annual Conference on Computer Graphics and Interactive Techniques, pp. 359–364. ACM Press/Addison-Wesley Publishing Co. (1999)

Tactile Sensibility through Tactile Display: Effect of the Array Density and Clinical Use

M. Valente[1], F. Cannella[1], L. Scalise[2], M. Memeo[2], P. Liberini[3], and D.G. Caldwell[1]

[1] Istituto Italiano di Tecnologia (Genova) at Department of Advanced Robotics
massimiliano.valente@iit.it, ferdinando.cannella@iit.it,
darwin.caldwell@iit.it
[2] Università Politecnica delle Marche (Ancona) at the Dipartimento di Ingegneria Industriale e
Scienze Matematiche
l.scalise@univpm.it, mc.memeo@gmail.com
[3] Università degli Studi di Brescia, at the Clinica Neurologica
paolo.liberini@spedalicivili.it

Abstract. Sensing in humans is carried out in different modalities, among them, the touch is essential in every-day life allowing most of all the manual activities. Less known is the fundamental role in the screening of some neurological diseases. In fact it is possible, trough the assessment of the level of tactile perception, to evaluate during clinical/medical diagnosis the nervous system health state. It is therefore fundamental to determine valid measurement procedures aiming to this task. The analysis has started with the investigation of tactile frequency sensibility, on healthy subjects, who examined tactile stimuli with devices called arrays. Arrays are formed by passive pins, which follow the shape of the stimuli. There were different arrays depending on the number of pins on them. The only task of subjects was to recognize the higher between couples of stimuli (one grating) with different frequency. The aim of these tests is that of creating a diagnostic scale for tactile neurodegenerative syndromes and diseases. Results show a significant role of pins density with a performance that falls when pins distance is higher than 1 mm. Preliminary clinical tests on 3 patients with tunnel carpal syndrome suggest positive development for use as a diagnostic tool.

Keywords: Tactile perception, psychophysics, tactile device design and evaluation, clinical diagnosis.

1   Introduction

Most of the physical interactions between the human being and the environment are generated trough the tactile sense. First scientific studies on such issue are from the middle of 19th century; and an important step has been the definition of the JND (Just Noticeable Displacement) [1]. Later a straightforward relationship between stimulus, receptor, afferent nerve fiber and subjective sensation, with regard to specificity has been highlighted [2-7]. In these studies, several tactile stimulators were also proposed

P. Isokoski and J. Springare (Eds.): EuroHaptics 2012, Part I, LNCS 7282, pp. 589–600, 2012.

in order to investigate the role of the tactile stimulus. Up to now, the evaluation of somatic sensation measurement devices are based on Von Frey filaments or Semmes-Weinstein monofilaments. Even if it is quite simple device, used in screening exam they are well designed to give an input as a light touch or a pain or a vibration or a warmth&cold to the human skin [8-12].

Even if, these devices are widely used with positive results [13-17], the weak aspect is that the variety of nervous system disorders or injuries cause impaired tactile sensibility. So, for instance, the crude touch (von Frey) hairs or two-point calipers may often fail to confirm this disorder [18]. Several works focus on the reliability of these techniques of investigation; in particular, they are not accurate in repeatability because of several environmental factors such as: temperature and humidity [19-22] and longevity and recovery stress [23-24]. It is therefore evident how well-defined procedures must be accurately followed by experts in order to keep such modifying and interfering parameters under control [25-26].

The aforementioned aspects were taken in account by the authors and their proposed solution is a high level of reliability obtainable by a well-defined procedure and a new test rig. Stimuli have been designed to exciting all the mechanoreceptors, in particular the RA [27]. Moreover the sensory system examination was focused on the fingertip because it is one of the most sensitive parts of the human body The core of this work is based on the assumption that the finest grating spacing that humans can resolve (spatial acuity limit) corresponds closely to the distance between sensory receptor units at the fingertips [19].

This high resolution of cutaneous sensation permits to monitoring every small difference between the measurements. The concept is based on the well known experimental tests widely used in haptic: the fingertip moving on gratings (they permit to measure spatial acuity of fingertip skin) so it receives a known input; the subject undergoing test to report an answer [28-29]. Gratings are built to simulate different haptic sensitivity [30-33] and to correlate them with peripheral neuropathy [15-18]; in fact arrays are formed by passive pins which move following shape of stimuli [34-36]. For our tests, we have used 5 arrays with the following pins distribution: 7x7, 9x9, 11x11, 13x13, 17x17 (respectively numbered as array number 1, 2, 3, 4, 5) and the bare finger which corresponds to the array number 6 (table 1).

Table 1. Arrays resolution

Number	Resolution	Pin distance [mm]
1	7x7	2,00
2	9x9	1,50
3	11x11	1,20
4	13x13	1,00
5	17x13	0,75
6	Bare finger	No pins

A sample of the tactile arrays realized for the reported tests is depicted in Figure 1.

Fig. 1. (Top) Tested arrays with different spatial resolution; (Down) one tactile displays on a grating

2   The Test-Rig

The proposed test-rig has been designed in order to assure subjects' comfort during experimental tests, in this way they are able to focus on the single grating. The proposed device is reported in Figure 2.

Testers' right arms will rest on a metallic support where is placed also a non-slip surfaced, the only movement enabled is that necessary to explore gratings. This support let to efficiently investigate gratings because of the friction reduction between the base and the driven sliders.

The gratings, that's to say tactile cues, are 17; each of them has two elements separated by a level part. Each elements, a stimulus is formed by ridge series (height of 1mm) with different wavelengths and always there is the reference one. Wavelengths chosen are 3.77, 4.10, 4.43, 4.76, 5.09 (the zero reference), 5.42, 5.75 6.08, 6.41 expressed in mm . These data can be translated in spatial frequency, that is 0.26, 0.24, 0.22, 0.21, 0.20, 0.18, 0.17, 0.16, 0.15 respectively, expressed in 1/mm.

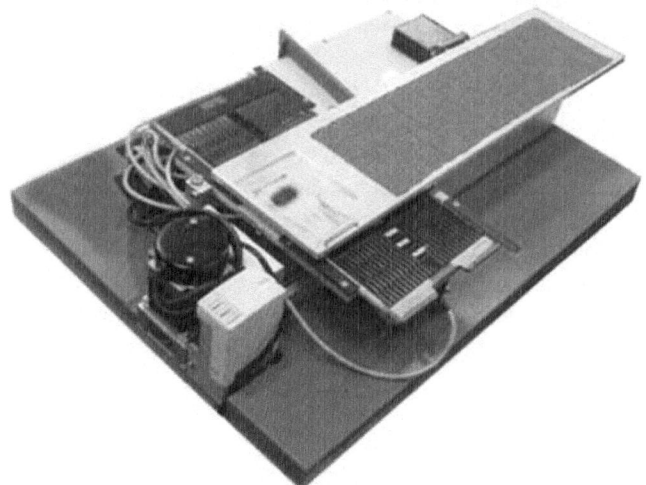

Fig. 2. Test-rig

Table 2. Gratings

Number	1st	2nd	Difference
1	5,09	6,41	-1,30
2	5,09	5,75	-0,66
3	5,09	5,42	-0,33
4	5,09	5,25	-0,16
5	4,93	5,09	-0,16
6	4,76	5,09	-0,33
7	4,43	5,09	-0,66
8	3,77	5,09	-1,32
9	5,09	5,09	0
10	5,09	3,77	1,32
11	5,09	4,43	0,66
12	5,09	4,76	0,33
13	5,09	4,93	0,16
14	5,25	5,09	0,16
15	5,42	5,09	0,33
16	5,75	5,09	0,66
17	6,41	5,09	1,32

3   Experiment

3.1   Method

Tested Subjects

Subjects without any previous experience in tactile psychophysical experiments had been chosen among volunteers. Tests had been conducted with seven persons 4 males (aged: 22-40) and, 3 females (aged: 24-32). All subjects were right handed and used their right index finger to execute the experiments; they were not allowed to see the gratings. Most of them were university students while the others are not workers heavily using hands.

Before starting tests, subjects had to fill the "Demographic Quest Frequency Sensibility", in which each subject indicated the volunteer ID, gender, age, occupation, fingerprint, how hard is the use of hands in his/her work or in free time, if he/she have drunk or have taken drug for medical reasons before tests.

Every subject has executed six tests, each with a different array (1-6); the array sequence was randomly determined for each subject. In to data from 42 tests have been used for data processing and analysis.

During experimental tests, subjects' fingers were covered with talcum to prevent friction in the interface grating/finger o array/finger. Total test time was of about 25 minutes.

Test Procedure

The patient's task is to report the number position of stimulus with higher spatial frequency and this information is recorded in the GUI. To improve subjects' attention on the tactile sensation, they wore darken glasses and earphones.

Every subject undergoes tests on rig using different arrays and bare finger. The initial subject's finger position is in the center of the grating; then it starts the exploration on first stimulus followed by the second one (Fig. 3). These two explorations constitute a trial, 185 trials constitute a complete test. The rig is controlled by a program written in MATLAB. It provides a GUI (Graphical User Interface) for setting the subject parameters, controlling the test rig. In order to guarantee a useful homogeneity of test is a specific *shuffle* function which enables to make the grating sequence in random order, shown in the GUI windows and updated thanks to a MATLAB function called *global*.

The patient's task is to report the number position of stimulus with higher spatial frequency and this information is recorded in the GUI. This data is introduced in the GUI which does automatically also an answer and starts next trial. In this way the procedure is automated and the experimental test duration is approximately 35 minutes.

Fig. 3. Grating exploration

Results and Discussion

Analysis has been focused on subjects answers on each gratings with same stimuli (we considered gratings with same comparison, filled like first stimulus or like second stimulus, like same grating) to calculate the psychometric curve; from these it had been found the threshold (at 0.8) and JND for each subject. Moreover it has been calculated the delta between first and second stimuli of each grating to analyze the probability of corrects answers for each delta. Results were analyzed using a single factor analysis of variance (one-way ANOVA) which determinates if there is a significant difference in the means. ANOVA analysis generates a p-value, were the lower the p-value the higher is the probability that there is a significant difference in the means. When ANOVA analysis had found a significant difference in the means, a post hoc Fisher test was conducted to compare all possible pair of means are significantly different. In this study ANOVA and Fisher test uses a level of significance of $\alpha=0.05$.

Fig. 4 shows the psychometric curve for each array, ANOVA analysis indicates there is a significant effect of the tactile array on the threshold of the perceived a frequency differences (F=4.99, p=0.001). Moreover fisher test shows that the significant difference occurs between arrays 4, 5, 6 (6 is bare finger) and 1, 2, 3. Fig. 5 (Left) shows the differences among bare finger threshold and array 1, 2 and 3.

The same results in JND analysis, ANOVA indicates significant effect (F=4.87, p=0.001). Also for JND results, Fisher test shows that the significant differences are between arrays 4, 5, 6 and 1, 2, 3. We do not found significant differences among bare

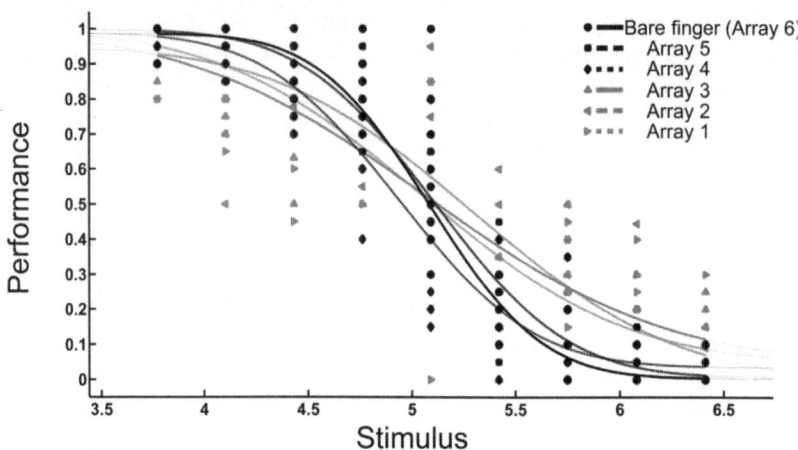

Fig. 4. Experimental psychometric curves as function of the six tested arrays

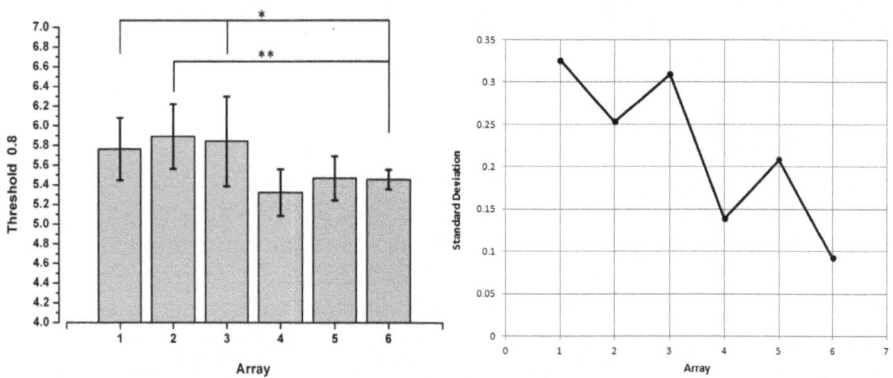

Fig. 5. (Left) Threshold 0.8, (Right) Standard Deviation

finger and arrays 5 and 4. It is important to note that the standard deviation of JND (Fig. 5 Right) reaches the minimum always for the array 6 (bare finger).

Analysis on correct answers for different arrays confirms a major accuracy of bare finger respect to all the other arrays: ANOVA of rights answers indicates that there is a significant effect of the tactile array on the recognition (F=10.36, p<0.001), in particular the Fisher test shows that bare finger result is significant with all the arrays except the array 5 (t=0.93, Prob=0.349); moreover the array 5 is significant with all the arrays except the array 4 (t=1.81, Prob=0.071) and with bare finger obviously (Fig. 6).

The delta analysis expected output chart of correct answer was a line V shaped: the volunteers could recognize the larger differences of frequency on 100% of times, and

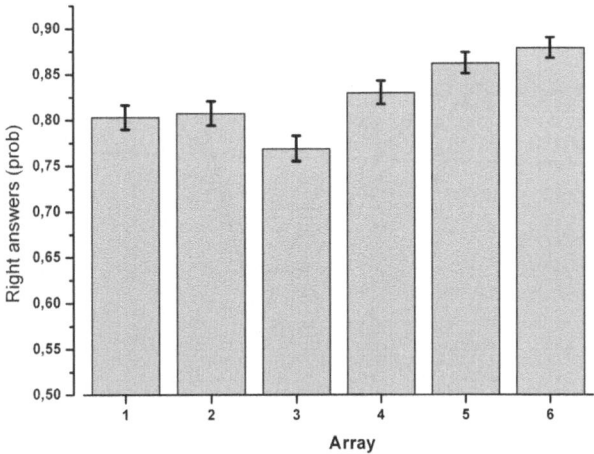

Fig. 6. Corrects answers for each array

the smaller differences was recognized less precision and the difference zero (grating 5.09-5.09) was about 50% of answers yes-no (because nothing answers was correct in this case). Graphs in Fig. 7 show that this idea in partially wrong: if the negative delta frequencies falling uniformly, positive delta shows a great step after delta zero. After analyzing with ANOVA the correct answers distribution on delta it has been found significant differences between same (in modulus) deltas (Table. 3).

These results indicate that the tactile performance is not damaged from a signal loss if the "resolution" is inside 1 pin for mm2. The little and not significant difference among bare finger and array 5 and 4 could attributed to the fact that the pin arrays don't stimulate all groups of mechanoreceptors, in particular Ruffini corpuscle, more sensible to skin stretch.

Table 3. Delta

|Delta|	t value	probability	Significant
1.32	1.301	1	No
0.99	3.255	0.041	Yes
0.66	7.067	<0.001	Yes
0.33	12.449	<0.001	Yes

We found very interesting results from Delta analysis: the difference between negative delta and positive delta indicates a possible saturation phenomenon for fingertip mechanoreceptors when the subjects fill for first a stimulus with frequency higher of second stimulus. This fact can explain the better performance when subjects filled a lower frequency for first.

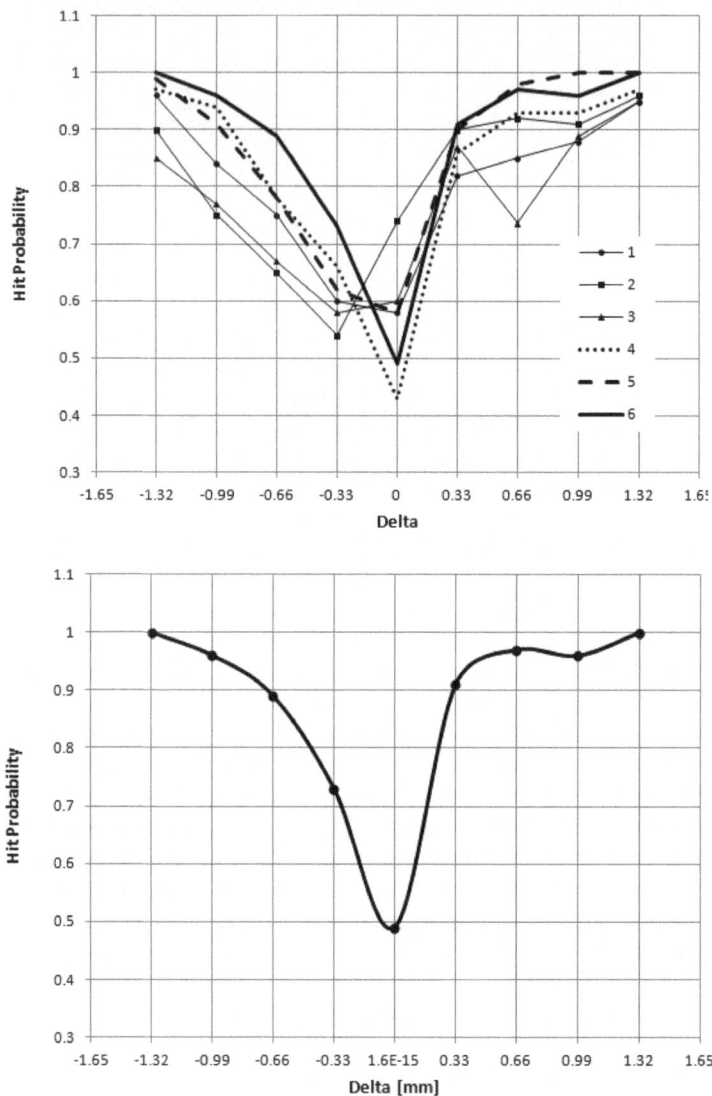

Fig. 7. Corrects answers: (Top) each of the six arrays, (Down) all the arrays

4 Clinical Testing: Patients Affected by Carpal Tunnel Syndrome

We have selected 3 subjects (all of female gender) and with an average age value of 58 years, the third subject is left-handed and she complains of a problem on left hand, but the test was mad with right hand. Classical neurological non-invasive test haven't found any tactile problem on patients.

The experimental tests have been performed in the same way of previous ones; the only difference is that in this case number of trials has been decreased because of the presence of the pathology of carpal tunnel syndrome and because we didn't want to force their attention.

The first intent was to perform tests on the right and left hand, so as to check whether there were differences in tactile sensitivity, but after about 50 trials subjects were already tired thus only the right hand was tested.

Patients' results are summarized in Table 4.

Table 4. Patient results

	Threshold 0.8	JND	Prob. Correct
Patient 1	6.379	0.829	0.66
Patient 2	6.686	11.896	0.48
Patient 3	4.4579	0.036	0.82

Our test shows tactile deficit on tactile frequency recognition for patient 1 and 2. This deficit seems alike to performances with arrays 2 or 3 for patient 1 and the deficit seems worse than the worst arrays for patient 2.

Patient 3 showed very high precision but she call the second stimulus like higher for all gratings with same stimulus (5.09-5.09).

This tests series with patients is only a pilot experience because patient's profiles don't matching with our experimental subjects, but we think that our device could be helpful for tactile neurodegenerative disease diagnostic.

5   Conclusions and Future Works

The experimental test results on healthy subjects show that the tactile resolution has not several damages if the resolution of tactile display is higher of one pin for mm^2. The comparison, between results of healthy subjects with pins arrays and the three patients with tunnel carpal syndrome, exceeded our expectations and two patients on three had results same or worse of those obtained with the low resolution arrays.

This result will be verified by a new series of experimental session on patients with tactile deficit already checked with traditional diagnostic systems.

With the aim to make the test appropriate to patient attention, we would modify the testing procedure for making this quicker then now but still accurate.

Moreover we want refining the discrimination task, for healthy subjects, we will add new gratings with smaller difference between two stimuli for a good check of the variance among the arrays.

Moreover for taking in account the patient fingertip size [38] and stiffness [39], that influence the tactile acuity, the test-rig will be equipped with two devices: one for measuring the finger size and another one for skin stiffness, in order to evaluate the reduction of the stimulus input due to the structure of the finger and not due to some peripheral neuropathy.

The slight difference between the bare finger and the array 5, even is statically not evident, should show the influence of the strain in the tactile perception. The array pins transmit only the vertical displacement given by stimuli shape without the horizontal forces given by the friction, so, most probably, the SAII were involved less than in the bare finger test; moreover the absence of stick-slip, given by the skin sliding on the stimuli surfaces, reduces even the vibrations, so it seems that the Pacinians are less involved too [40-43]. That happens only with this array, because the high number of the pins reproduce quite accurately the stimuli surface.

The next step will deal with the aim of this work: establish a tactile sensitivity scale. For doing that, 50 people (to have a statistically significant data records) will be tested with arrays and bare finger. Then, the tactile acuity (subject answers) with the parameters (subject's fingertip size, skin stiffness, force and displacement) will be cross-correlated and their influence will be highlighted: only in this way the tactile sensitivity levels can be established correctly.

Those results will permit more accurately to evaluate the peripheral neuropathies than at the state of art.

It is important to highlight that the novel system here proposed rig is not aiming to substitute the current sensitivity tools and methodologies, but test rig has been conceived to support the current peripheral neuropathy screening tools, providing a quantitative and repeatable measurement procedure. Actually it takes 35 minutes, but the goal is to reduce it up to 15 minutes in order to use the rig to test neurodegenerative patients.

Acknowledgments. The authors wish to acknowledge all the students of Università Politecnica delle Marche who spent their time in experimental tests that permitted to accomplish this work.

References

[1] Weber, E.H.: The sense of touch (Ross, H.E. (ed. & trans.)). Academic Press, London (1978) (Original work published, 1834)

[2] Blix, M.: Experimentela bidrag till lösning af frågan om hudnervernas specifika energi. Uppsala LäkFör. Förh. 18, 427–440 (1883)

[3] Donaldson, H.H.: On the temperature sense. Mind 10, 399–416 (1885)

[4] Anonymous. Berlin: Prof. Eulenburg's report to the Physiological Society in Berlin (December 12, 1884); Nature 31, 259 –260 (1885)

[5] Von Frey, M.: Treatise on the sensory functions of the human skin. In: Hand-werker, H.O., Brune, K. (eds.) Classical German Contributions to Pain Research, IASP 5th World Congress, Hamburg, pp. 69–131 (1987); (original article printed, 1896)

[6] Norrsell, U., Finger, S., Lajonchere, C.: Cutaneous sensory spots and the "law of specific nerve energies": history and development of ideas. Brain Research Bulletin 48(5), 457–465 (1999)

[7] Weinstein, S.: Intensive and extensive aspects of tactile sensitivity as a function of body part, sex, and laterality The skin senses, DR Kenshalo, 195–222 (1968)

[8] https://www.stoeltingco.com/stoelting/productlist13c.aspx?home=Physio&searchtext=domes&catid=3129

[9] http://www.bioseb.com/bioseb/anglais/default/item_id=577_cat_id=_Von%20Frey%20Filaments.php

[10] http://www.komkare.com/diagnostics/reflx_sens/bsln_disc.html
[11] http://www.resto-medical.com/Neurological-hammer-c16.html
[12] http://www.aecc-spinecentre.co.uk/products/instruments-11/tuning-fork-c1256 -3492.aspx
[13] Brodal, A.: Neurological Anatomy in Relation to Clinical Medicine, 2nd edn. Oxford University Press, New York (1969); 3 edn. (January 15, 1981)
[14] Bell-Krotoski, J., Tomancik, E.: The repeatability of testing with Semmes-Weinstein filaments. The Journal of Hand Surgery 12A, 155–161 (1987)
[15] Sangyeoup, L., Hyeunho, K., Sanghan, C., Yongsoon, P., Yunjin, K., Byeungman, C.: Clini-cal Usefulness of the Two-site Semmes-Weinstein Monofilament Test for Detecting Dia-betic Peripheral Neuropathy. J. Korean Med. Sci. 18, 103–107 (2003)
[16] Keizer, D., Fael, D., Wierda, J.M., Van Wijhe, M.: Quantitative Sensory Testing With Von Frey Monofilaments in Patients With Allodynia: What Are We Quantifying? Clini-cal Journal of Pain 24(5), 463–466 (2008)
[17] Pestronk, A., Florence, J., Levine, T., Al-Lozi, M.T., Lopate, G., Miller, T., Ramneantu, I., Waheed, W., Stambuk, M.: Sensory exam with a quantitative tuning fork. Rapid, sensi-tive and predictive of SNAP amplitude. Neurology 62(3), 461–464 (2004)
[18] Keizer, D., van Wijhe, M., Post, W.J., Wierda, J.M.: Quantifying allodynia in patients suffering from unilateral neuropathic pain using von frey monofilaments. Clin. J. Pain. 23(1), 85–90 (2007)
[19] Komiyama, O., Gracely, R.H., Kawara, M., Laat, A.D.: Intraoral measurement of tactile and filament-prick pain threshold using shortened Semmes-Weinstein monofilaments. Clin. J. Pain. 24(1), 16–21 (2008)
[20] Johnson, K.O., Van Boven, R.W., Phillips, J.R.: JVP DOMES. Operation Manual (January 1997)
[21] Werner, M.U., Rotbøll-Nielsen, P., Ellehuus-Hilmersson, C.: Humidity affects the per-formance of von Frey monofilaments. Acta Anaesthesiol. Scand. 55, 577–582 (2011)
[22] Andrews, K.: The effect of changes in temperature and humidity on the accuracy of von Frey hairs. J. Neurosci. Methods 50(1), 91–93 (1993)
[23] Booth, J., Young, M.J.: Differences in the performance of commercially available 10-g monofilaments. Diabetes Care 23(7), 984–988 (2000)
[24] McGill, M., Molyneaux, L., Spencer, R., Heng, L.F., Yue, D.K.: Possible sources of dis-crepancies in the use of the Semmes-Weinstein monofilament. Impact on Prevalence of Insensate Foot and Workload Requirements. Diabetes Care 22(4), 598–602 (1999)
[25] Berquin, A.D., Lijesevic, V., Blond, S., Plaghki, L.: An adaptive procedure for routine meas-urement of light-touch sensitivity threshold. Muscle Nerve 42(3), 328–338 (2010)
[26] Smith, K.D., Emerzian, G.J., Petrov, O.: A comparison of calibrated and non-calibrated 5.07
[27] Johansson, R.S.: Tactile sensibility in the human hand: receptive field characteristics of mechanoreceptive units in the glabrous skin area. The Journal of Physiology 281, 101–125 (1978)
[28] Vega-Bermudez, F., Johnson, K.O.: Spatial Structure of Primary Afferent Re-ceptive Fields in the Somatosensory System. Society Neuroscience Abstracts 17, 840 (1991)
[29] Meenes, M., Zigler, M.J.: An experimental study of the perception of roughness and smoothness. Am. Journal of Psychology 34, 542–549 (1923)
[30] Morley, J.W., Goodwin, A.W., Darian-Smith, I.: Tactile discrimination of gratings. Expe-rimental Brain Research 49(2), 291–299 (1983)

[31] Goodwin, A.W., Morley, J.W.: Sinusoidal movement of a grating across the monkey's fingerpad: Effect of contact angle and force of the grating on afferent fiber responses. J. Neurosci. 7, 2192–2202 (1987)

[32] Drewing, K., Lezkan, A., Ludwig, S.: Texture Discrimination in Active Touch: Effects of the Extension of the Exploration and their Exploitation. In: IEEE World Haptics Conference 2011, Istanbul, Turkey, June 21-24, pp. 21–24 (2011)

[33] Wexler, M.: Vincent Hayward Weak Spatial Constancy in Touch. In: IEEE World Haptics Conference 2011, Istanbul, Turkey, June 21-24, pp. 21–24 (2011)

[34] Garcia-Hernandez, N.V., Tsagarakis, N.G., Caldwell, D.G.: Effect of the Tactile Array Density on the Discrimination of Edge Patterns: Implications for Tactile Systems Design. In: International Conference on Advanced Robotics, ICAR 2009 (2009)

[35] Garcia-Hernandez, N.V., Tsagarakis, N.G., Caldwell, D.G.: Human Tactile Ability to Discriminate Variations in Small Ridge Patterns thorugh a Portable-Wearable Tactile Display. In: ACHI 2010, pp. 38–43 (2010)

[36] Garcia-Hernandez, N.V., Tsagarakis, N.G., Caldwell, D.G.: Feeling through Tactile Displays: A Study on the Effect of the Array Density and Size on the Discrimination of Tactile Patterns. IEEE T. Haptics 4(2), 100–110 (2011)

[37] Gleeson, B., Horschel, S., Provancher, W.: Design of a fingertipmounted tactile display with tangential skin displacement feedback. IEEE Transactions on Haptics 3(4), 297–301 (2010)

[38] Peters, R.M., Hackeman, E., Goldreich, D.: Diminutive Digits Discern Deli-cate Details: Fingertip Size and the Sex Difference in Tactile Spatial Acuity. The Journal of Neuroscience 29(50), 15756–15761 (2009)

[39] Thomas, C.K., Westling, G.: Tactile unit properties after human cervical spinal cord injury. Brain 118(6), 1547–1556 (1995)

[40] Johnson, K.O.: The Roles and Funcions of Cutaneous Mechanoreceptors. Current Opinion in Neurobiology 11(4), 455-461

[41] Wiertlewski, M., Hudin, C., Hayward, V.: On the 1/f Noise and Non-Integer Harmonic Decay of the Interaction of a Finger Sliding on Flat and Sinusoidal Surfaces. In: IEEE World Haptics Conference 2011, Istanbul, Turkey, June 21-24 (2011)

[42] Solazzi, M., Provancher, W.R., Frisoli, A., Bergamasco, M.: Design of a SMA Actuated 2-DoF Tactile Device for Displaying Tangential Skin Dis-placement. In: IEEE World Haptics Conference 2011, Istanbul, Turkey, June 21-24 (2011)

[43] Kandel et al.: Principles of Neural Science. McGraw Hill (2000)

Contact Force and Finger Angles Estimation for Touch Panel by Detecting Transmitted Light on Fingernail

Yoichi Watanabe[1], Yasutoshi Makino[2], Katsunari Sato[1], and Takashi Maeno[1]

[1] Graduate School of System Design and Management, Keio University,
4-1-1, Hiyoshi, Kohoku-ku, Yokohama, 223-8526, Japan
[2] Center for Education and Research of Symbiotic, Safe and Secure System Design,
Keio University, 4-1-1, Hiyoshi, Kohoku-ku, Yokohama, 223-8526, Japan
chronicle@z3.keio.jp,
{makino,katsunari.sato,maeno}@sdm.keio.ac.jp

Abstract. In this paper, we propose a new method that can estimate a normal force and finger posture for touch panel devices by measuring transmitted light on fingernail. When a user touches a light source with their finger, the light (mainly red component) can be seen at a fingernail since the red light can go through the finger tissue. Based on this characteristic, we can estimate the manipulatory force by detecting the light intensity at the fingernail because the intensity of the transmitted light increases according to the applied force. We can also estimate the relative angles of the finger to a touch panel device. Even though the transmission through the finger tissue is diffusive, we can know the movement of the light source under the finger with a camera attached onto the fingernail. Therefore, a transient trajectory pattern of the light source is useful for estimating the relative direction between the finger and the device. When the posture of the finger is estimated, we can use "rotate" motion for manipulation. Our proposed method can be used as a new input device.

Keywords: Man-Machine Interface, Tactile Sensor, Force Measurement.

1 Introduction

In these days, we can use many small touch panel devices including smart phones, tablet PCs, portable games and so on. These devices can arrange many different types of buttons on the screen depending on applications. On the other hand, there is one drawback referred as "fat finger problem [1]." When a user tries to touch a button on the screen, they cannot see it at the moment they push. Their finger interrupts seeing the button and may push a wrong one. The "fat finger problem" spoils usability of touch panel devices.

In this paper, we show a new interface that uses the "fat finger problem" positively. When a finger touches a screen, the device can change the images under the finger since nobody can see there. The modification of the image under the finger will not affect usability. Spatiotemporal light patterns under the finger can go through the

P. Isokoski and J. Springare (Eds.): EuroHaptics 2012, Part I, LNCS 7282, pp. 601–612, 2012.

fingertip. The light pattern can be detected on the fingernail with a camera attached on it. The intensity of transmitted light and its position depend on contact conditions. We experimentally revealed following two facts: 1) the intensity of the transmitted light increases when a user strongly pushes a screen and 2) the light position on the camera on the fingernail moves when a user changes a contact angles. In this paper, we show that we can estimate 1 directional force along z-axis (shown in Fig. 1) and 3 rotational contact angles independently. One of the advantages of this method is that we can use any conventional touch panel devices by attaching camera onto a fingernail. Resulting touch condition can be used for producing tactile feedback for improving usability [2]. In order to give appropriate tactile feedback, we should know a given force and touch angles.

There are some previous researches which tried to detect touch conditions with the device attached on a fingernail. Mascaro et al. proposed vision-based method which detected the change of colors of the fingernail and estimated applied forces [3]. Sun et al. proposed an external camera method for measuring fingertip forces by imaging the fingernail and surrounding skin [4]. Makino et al. proposed audio-based method [5]. A piezoelectric device on a fingernail was used to estimate touching objects. They estimated the objects based on a conventional voice recognition scheme. Tanaka et al. used microphones to record texture related sounds at the fingertip [6]. Nakatani et al. detected finger deformation to estimate a pressure force by mounting a small deformation sensitive device on user's fingernail [7].

These previous studies measured touch information passively. In contrast, our method changes a light patter under the finger actively. This may gives richer information for estimating touch conditions though the proposed method can be applied only for a touch panel device.

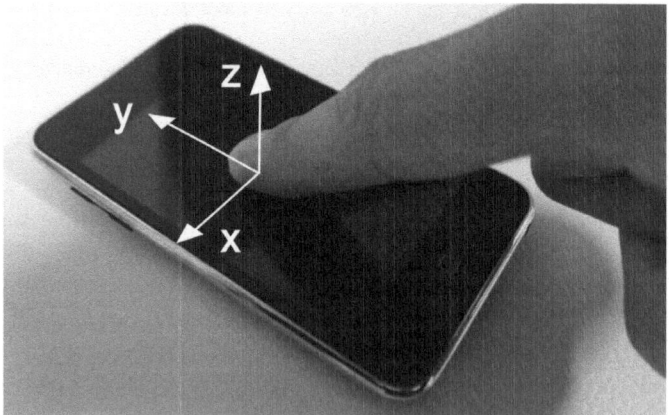

Fig. 1. Coordinate system of touch point

There are studies which utilized input pressure or rotational condition for touch pad devices. Rekimoto et al. proposed a pressure sensitive touch pad [8]. They showed examples of applications which enable users to zoom in/out a map and to scroll a list easily. Ramos et al proposed a pressure sensitive stylus system for improving an input capability [9]. Stewart et al. investigated pressure-based input for mobile devices [10]. Wang et al. proposed a method to estimate a finger orientation by detecting a contact shape of finger [11]. Rotational information was used for an orientation-sensitive widget like a dial. These methods can add new functions to a touch panel device.

2   Principles

2.1   Physical Values for Measurement

Figure 2 shows the basic settings of our proposed method. We can detect following two physical values with this configuration.

1. Light intensity transmitted through the finger tissue and the fingernail.
2. A position of the light source on the light sensor

In this research, we used a camera as the light sensor. Based on the light intensity, z-directional force (i.e. normal force to the screen) can be estimated. In contrast, x, y and z rotational angles can be estimated by the position of the transmitted light.

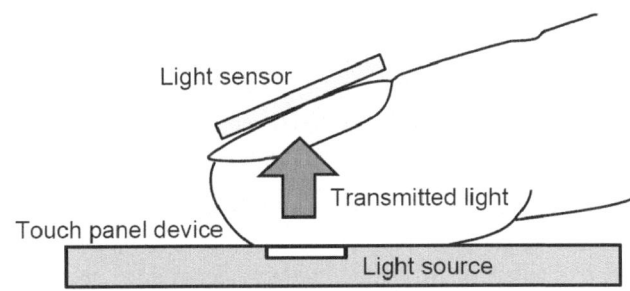

Fig. 2. Proposed method detecting transmitted light on the fingernail

2.2   Estimation of z-Directional Force

Here we show a simple model which explains how to estimate z-directional force by using light intensity. We assume that light decay in the finger tissue is low. In this case, the total amount of light intensity detected at the fingernail is proportional to the total amount of incident light from the screen. When the light intensity distribution is uniform incident light from the screen is proportional to the contact area between the finger and touch panel.

To understand a relationship between the applied z-directional force and the size of the contact area, we modeled finger as an elastic sphere. When we assume Hertz contact, a radius of contact area a is represented in equation (1)

$$a = P^{1/3} \left[\frac{3}{4} R \frac{1-v}{E} \right]^{1/3}$$

(1)

Here the R represents the radius of fingertip, the E means the Young's modulus and the v means the Poisson's ratio and the P represents applied z-directional force. Therefore, the contact area S is proportional to the 2/3th power of z-directional force.

$$S \propto a^2 \propto P^{2/3}$$

(2)

As we assumed before, the light intensity at the fingernail is thought to be proportional to the contact area. As a result total amount of the light intensity I is represented as follows

$$I \propto P^{2/3}$$

(3)

The equation (3) means that the total amount of the light intensity I increases monotonically according to the applied force. Thus, the system can estimate z-directional force uniquely.

2.3   Estimation of Rotation Angles around the x and y-axis

Then we explain how to estimate rotational angles around the x and y axes. In this case, the light position at the fingernail changes depending on the angles. In order to make it simple, we assume a fingertip as a half-circle (Fig.3). We can discuss both angles with the same model. We used following assumptions.

- The finger touches the screen with a point contact.
- Closest point from the contact point is the brightest position on the camera.

When the contact angle α is 0, the brightest position is center of the camera. When the finger rotates with its angle α the brightest point moves. The distance from the center is given as $r\sin\alpha$. Contact angle can be uniquely determined because of monotonic increase from -90 to 90 degrees. It is also clear that the distance between the fingernail and the contact point, which can be written $r\cos\alpha$, becomes short according to the increase of the touch angle. Though we assumed that the light decay in the finger tissue is low, the intensity of light depends on a propagating distance. This may affect the force estimation.

2.4   Rotating Angle around z-axis

In order to detect rotating angle of finger around the z-axis we propose the method which uses spatiotemporal light source motion. We assume that the light source under the finger constantly moves clockwise with its period T. Observation of the center of the transmitted light on the fingernail for T seconds makes it possible to estimate a rotational angle around the z-axis. Figure 4 shows how it works. The x and y axes in the figure represent the coordinate systems of the attached camera. The x-axis means the longitudinal direction and the y-axis means the transversal direction of user's finger. When the finger touches the screen as shown in Fig. 4-(a), the x directional position of the brightest point on the fingernail becomes largest at time 0. In contrast, the y position becomes largest at the time 0 in the case of Fig. 4-(b). Consequently, the initial phase of the light trajectory on the camera represents the contact direction (rotation angle around the z-axis).

We are now planning to use this method as a time sharing scheme. The system provides a static light source for estimating z-directional force and x- y-axes rotations; then it moves clockwise for estimating z-axis rotation.

In the following experiments, we show that 1) applied force can be estimated by using light intensity and 2) touch angles can be measured by light position. Finally we show that motion of the light source can be detected at the fingernail which enables estimation of z-axis rotation angle.

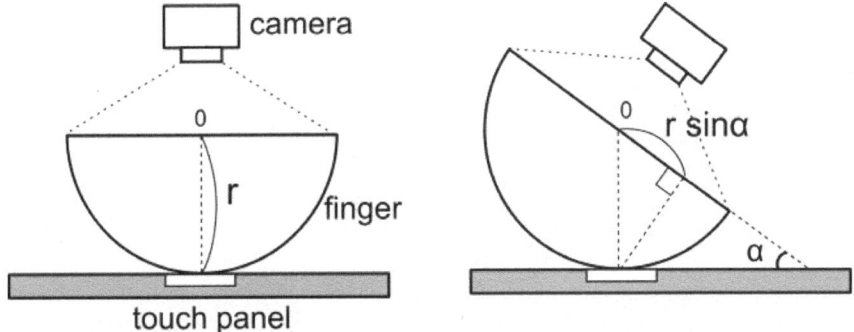

Fig. 3. Rotation angle is converted into the motion of brightest point

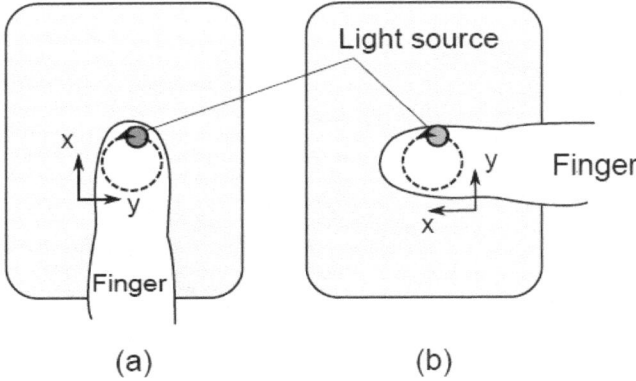

Fig. 4. Rotation angle around the z-axis is detected with moving light source

3   Experiments

3.1   Experimental Setting

Figure 5 shows our experimental setting. We used iPod touch as the touch panel device and developed its application. The application detects the contact point of a finger and renders a white shine circle at the detected point while the rest area is black.

A commercially available camera which has 640×480 image pixels was used as the light sensor. The camera was fixed on the fingernail of user's right index finger with a double headed tape. To eliminate the optical external noises, we used a light-blocking tape at the border between the camera and the fingernail.

3.2   Experiment 1 : Estimation of z-Directional Force

At first, we examined the relation between z-directional normal force and the transmitted light intensity. We measured both applied force to the device and the light intensity of the camera. As the source, we used a circle which had 5.5 millimeters (35 pixels) in diameter. The light intensity upon the circle in a dark room was about 36.6 Lx while it was 11.1 Lx without drawing any circles.

Transmitted light through the finger was mainly red component. To decrease noises from the other components (Green and Blue), we used only the red component for estimation in all the experiments in this paper. A gray scale picture with 256 gradations was generated from the red light. In this experiment, the mean value of brightness in the whole area was calculated from the picture. A user pushed the touch panel in a quasistatic manner with keeping his finger angle constant.

Figure 6 shows the result. The horizontal axis represents the applied force and the vertical axis represents the brightness. Brightness shows detectable light intensity, which was expressed with 256 gradations. The fitted curve in the graph was derived from a relation given in the equation (3) using the least-squares method. The graph shows that the averaged light intensity monotonously increases in accordance with the increase of the applied force. Therefore, it is clear that the applied z-directional force can be uniquely estimated by using the regression curve.

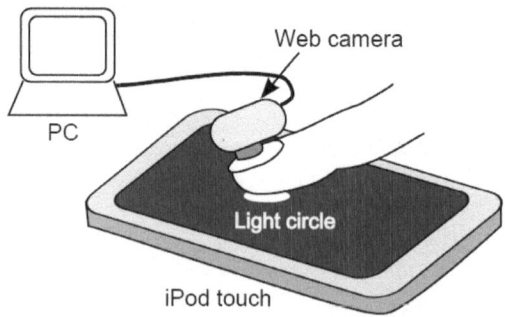

Fig. 5. Schematic diagram of experimental setting

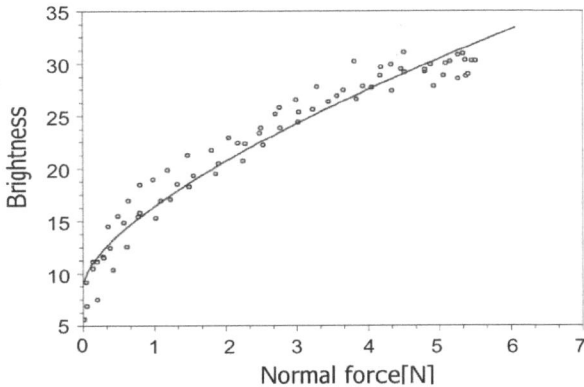

Fig. 6. Relation between z-directional normal force and the transmitted light intensity

3.3 Experiment 2 : Estimation of Rotation Angle around the x-axis

We examined whether a finger tilt angle can be coded as a light position of the camera image. Center of the light intensity was measured by using red-color-based gray scale picture as well as the experiment 1. The picture was converted into binary image by using a particular threshold level, which is experimentally decided. A center of the bright area was computed from the binary image.

Six subjects (3 female and 3male aged from 20 to 50) were participated in this experiment so that we tried to know individual differences. They were asked to tilt their finger freely as shown in Fig. 7. In this experiment we only consider the rotation around the x-axis so as to make it simple. The actual contact angle was visually calculated with the marker put on the iPod touch and the camera on the fingernail.

Figure 8 shows the experimental results. The horizontal axis shows the tilt angle and the vertical axis shows the position of light source detected on the fingernail. It is safe to say that the position of light source moves monotonously when the angle is 20 to 40 degrees for every subjects. The tilt angle can be uniquely estimated by using the individually calibrated regression curves. Also we briefly examined that the rotation angle around the y-axis can be estimated by using the change of the y directional position.

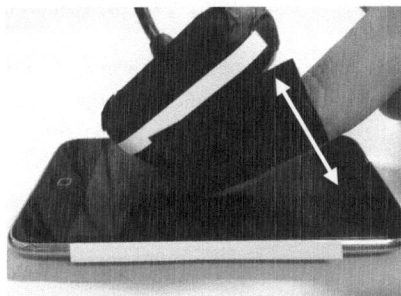

Fig. 7. Measuring method of Experiment 2

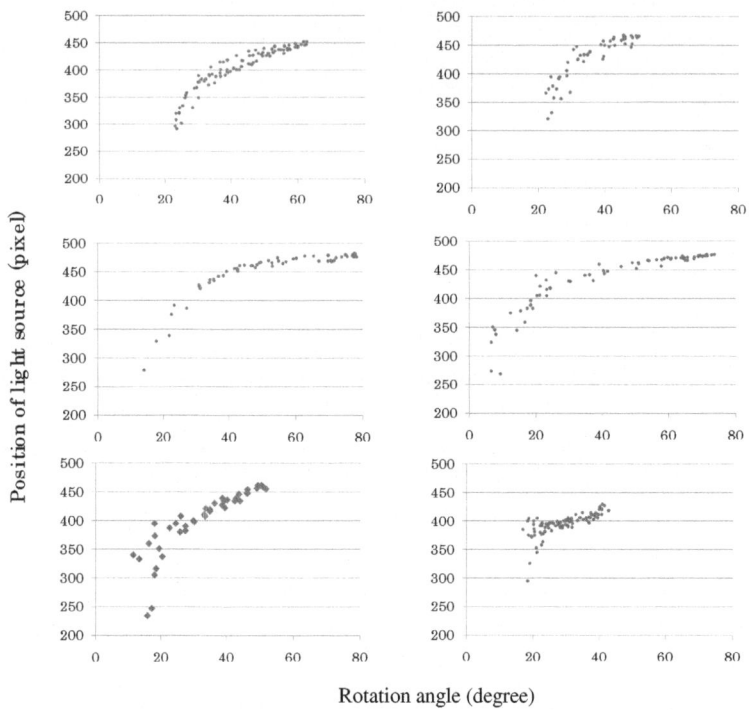

Position of light source (pixel)

Rotation angle (degree)

Fig. 8. Relation between the rotation angle around x-axis and the position of light source

3.4   Experiment 3: Interferences between the Normal Force and the Tilt Angle

The tilt angle of the finger affects the transmitted light intensity, which is used for estimating pressing force, since the distance from the contact point changes as is shown in Fig.3. We evaluated the interference between the normal force and the tilt angle.

We used a circle which had 7.8 millimeters (55 pixels) in diameter. One user pushed the touch panel changing the tilt angle of his finger. The angles were 20, 40, 60 degree respectively.

Figure 9 shows the result. The horizontal axis represents the applied force and the vertical axis represents the averaged light intensity. Each color represents the angle: the green is 20 degree, the red is 40 degree and the blue is 60 degree. The graph shows that the averaged light intensity monotonously increases when the applied force increases in any tilt angle. In order to estimate the applied z-directional force, the system estimate the tilt angle based on the light position on the camera, at first. Then the system uses the regression curve corresponds to the tilt angle. Finally, the system can know the z-directional applied force.

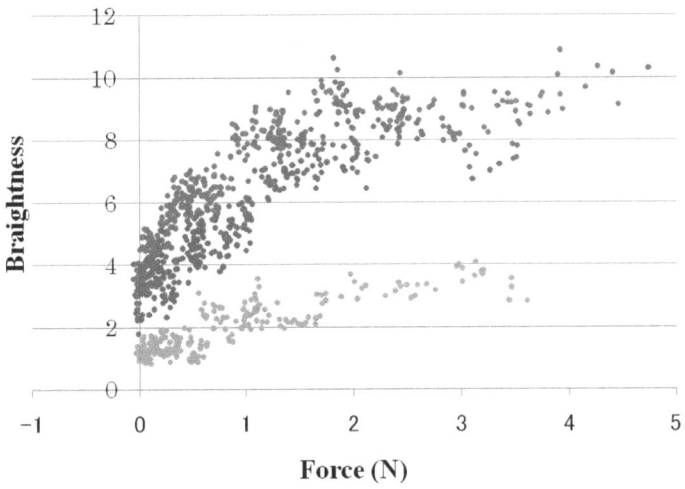

Fig. 9. Relation between z-directional force and the transmitted light intensity which depends on the finger angle. Green: 20 degree, Red: 40 degree and Blue: 60 degree.

3.5   Experiment 4 : Estimation of Rotation Angle around the z-axis

Finally, we used moving light source and observed the motion of it on the fingernail with a camera. A circle which had 5.5 millimeters (35 pixels) in diameter was rendered when the finger touches the screen. The circle moved clockwise at a rate of approximately 2π radian per second whose trajectory radius was 3.1 millimeters. The camera on the fingernail measured a center of the light intensity.

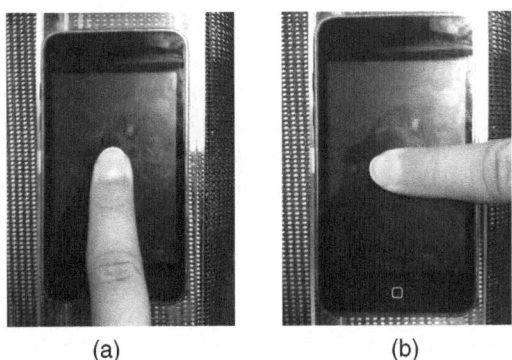

(a)   (b)

Fig. 10. Rotation angle measured in the Experiment 4

Figure 10(a) shows two contact conditions of a finger. The circle started moving from the upper side of the device. That is to say, the circle started moving from the tip of the finger for Figure 10(a), while in Figure 10(b), the circle started moving from the right side of the finger.

Figure 11 shows experimental results. Averaged values of 6 trials are plotted. The right side of the graphs shows the tip of the finger. The both axes show the position on the camera in pixel. The center of the light distribution started moving from the tip of the finger in Fig. 11 (a). In contrast, it started moving from the right side of the finger in Fig. 11 (b). In this way the camera on the fingernail can detect motion of the light source under the finger. Therefore the finger's rotational angle around the z-axis can be detected with this method.

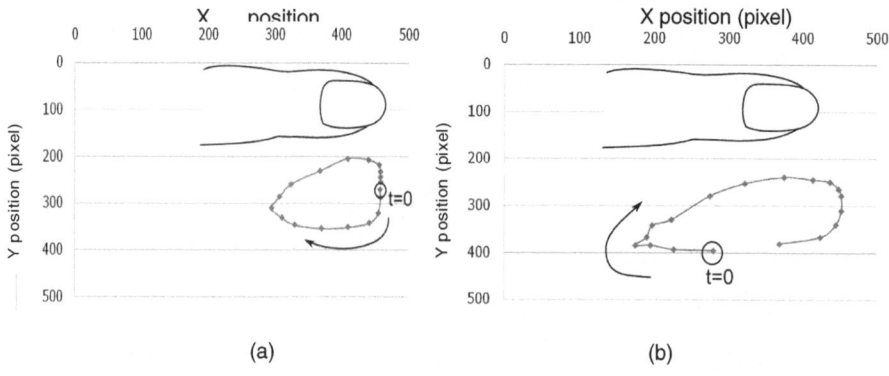

(a)   (b)

Fig. 11. Detected motion of the light source on the fingernail

4   Applications

Our proposed method allows users to input information in many ways. Usually, a touch panel device can detect two-dimensional position on the screen. Users input data only based on their finger position. That means, user need to change their finger position such like a swipe action for input. In contrast, our method enables users to input data by changing a finger contact angles, i.e. user does not need to move their finger. User can input many commands only by changing a touch angles at the same position on the screen.

We prototyped applications for achieving two possible usages on an iPod touch. One enables user to input data by rotating action around the z-axis as shown in Fig. 12. The visual knob on the screen can be turned by rotating the finger. The other application enables user to control an UFO object by changing the finger contact angles at arbitrary positions on the screen as shown in Fig. 13. These types of operability cannot be achieved with a conventional touch panel device.

We asked 7 subjects about usability of our prototypes. Their positive responses were: "I can input data without moving my finger," "I can choose many kinds of input ways," "This enables me to input data like a joystick." There were also negative responses. "Accuracy of the angle estimation is not sufficient for precise control," "The size of the device is too large," "It is annoying to put the heavy device on my fingernail." Some of these disadvantages may be solved if we use a smaller camera instead of the commercially available webcam. Both reducing the size of the device and making its estimation accuracy higher are our future works.

Fig. 12. Data input by rotating finger to turn the visual knob on the screen

 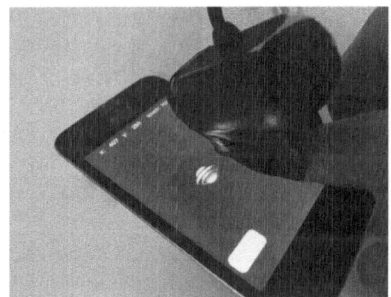

Fig. 13. Data input by changing contact angles of the finger at arbitrary positions on the screen

5   Conclusion

In this paper, we proposed a new method which estimates finger contact conditions on a touch panel device. A camera on a fingernail measured transmitted light coming from a screen through fingertip.

We showed four results. First, z-directional normal force was non-linearly proportional to a transmitted light intensity on the fingernail. Second, when a light source under the finger moved, its center position seen on the camera also moved, which determines tilt angles between the finger and the device. Third, we evaluated the interferences between an applied force and tilt angles. Finally, the rotation angle around the z-axis was estimated by using a moving light source.

This method can be used for any conventional touch panel device and is useful for achieving many different kinds of input methods.

Acknowledgement. This work was supported in part by Grant-in-Aid for the Global Center of Excellence Program for "Center for Education and Research of Symbiotic, Safe and Secure System Design" in Japan.

References

1. Voida, S., Tobiasz, M., Stromer, J., Isenberg, P., Carpendale, S.: Getting Practical with Interactive Tabletop Displays: Designing for Dense Data, "Fat Fingers," DiverseInteractions, and Face-to-Face Collaboration. In: Proc. of ITS 2009, pp. 23–25 (2009)
2. Tashiro, K., Shiokawa, Y., Aono, T., Maeno, T.: Realization of Button Click Feeling by Use of Ultrasonic Vibration and Force Feedback. In: Proc. of WorldHaptics 2009, pp. 1–6 (2009)
3. Mascaro, S., Asada, H.: Measurement of Finger Posture and Three-Axis Fingertip Touch Force Using Fingernail Sensors. IEEE Transactions on Robotics and Automation 20(1), 26–35 (2004)
4. Sun, Y., Hollerbach, J.M., Mascaro, S.A.: Measuring Fingertip Forces by Imaging the Fingernail. In: International Symposium on Haptic Interfaces for Virtual Environment and Teleoperator Systems (HAPTICS 2006), pp. 125–131 (2006)
5. Makino, Y., Murao, M., Maeno, T.: Life Log System Based on Tactile Sound. In: Kappers, A.M.L., van Erp, J.B.F., Bergmann Tiest, W.M., van der Helm, F.C.T. (eds.) EuroHaptics 2010, Part I. LNCS, vol. 6191, pp. 292–297. Springer, Heidelberg (2010)
6. Tanaka, Y., Horita, Y., Sano, A., Fujimoto, H.: Tactile sensing utilizing human tactile perception. In: World Haptics 2011, pp. 621–626 (2011)
7. Nakatani, M., Kawasoe, T., Shiojima, K., Koketsu, K., Kinoshita, S., Wada, J.: Wearable contact force sensor system based on fingerpad deformation. In: World Haptics 2011, pp. 323–328 (2011)
8. Rekimoto, J., Schwesig, C.: PreSenseII: Bi-directional Touch and Pressure Sensing Interactions with Tactile Feedback. In: CHI 2006 Extended Abstracts on Human Factors in Computing Systems (2006)
9. Ramos, G., Boulos, M., Balakrishnan, R.: Pressure widgets. In: Proceedings of the SIGCHI Conference on Human Factors in Computing Systems (CHI 2004), pp. 487–494 (2004)
10. Stewart, C., Rohs, M., Kratz, S., Essl, G.: Characteristics of Pressure-Based Input for Mobile Devices. In: 28th International Conference on Human Factors in Computing Systems (CHI 2010), pp. 801–810 (2010)
11. Wang, F., Cao, X., Ren, X., Irani, P.: Detecting and leveraging finger orientation for interaction with direct-touch surfaces. In: Proceedings of the 22nd Annual ACM Symposium on User Interface Software and Technology (UIST 2009), pp. 23–32 (2009)

Electrostatic Modulated Friction as Tactile Feedback: Intensity Perception

Dinesh Wijekoon[1], Marta E. Cecchinato[2], Eve Hoggan[1], and Jukka Linjama[3]

[1] Helsinki Institute for Information Technology (HIIT)
Department of Computer Science, University of Helsinki, Finland
firstname.lastname@helsinki.fi
[2] Department of General Psychology, Università degli Studi di Padova, Italy
marta.cecchinato.1@studenti.unipd.it
[3] Senseg Ltd Helsinki, Finland
jukka.linjama@senseg.com

Abstract. We describe the preliminary results from an experiment investigating the perceived intensity of modulated friction created by electrostatic force, or electrovibration. A prototype experimental system was created to evaluate user perception of sinusoidal electrovibration stimuli on a flat surface emulating a touch screen interface. We introduce a fixed 6-point Effect Strength Subjective Index (ESSI) as a measure of generic sensation intensity, and compare it with an open magnitude scale. The results of the experiment indicate that there are significant correlations between intensity perception and signal amplitude, and the highest sensitivity was found at a frequency of 80 Hz. The subjective results show that the users perceived the electrovibration stimuli as pleasant and a useful means of feedback for touchscreens.

Keywords: electrovibration, tactile, intensity, perception.

1 Introduction

Touch provides a rich channel through which to communicate and has been the focus of much research [1] [2] [3] [4] [5]. There are several different sub-modalities within touch. Tactile feedback (commonly vibrotactile mechanical stimulation of the skin) is the best understood and most used in HCI. Electrovibration is part of the same sensory system but has been relatively underutilized in interactive systems due to the lack of available hardware. However, recent advancements in touchscreen electrovibration technology mean that mainstream usage is now becoming possible. Electrovibration is created using electrostatic friction between a surface and a user's skin [5]. Passing an electrical charge into the insulated electrode creates a small attractive force when a user's skin comes into contact with it. By modulating this attractive force a variety of sensations can be generated [6].

The properties of electrovibration are quite different to those of standard vibrotactile feedback. In order to use electrovibration feedback effectively in user interfaces, it is important to identify the available design parameters. Tacton and Haptic Icon

P. Isokoski and J. Springare (Eds.): EuroHaptics 2012, Part I, LNCS 7282, pp. 613–624, 2012.
© Springer-Verlag Berlin Heidelberg 2012

[7] [8] designers use parameters such as intensity, waveform, rhythm and spatial locations, but the electrovibration domain has not been investigated with such an approach as yet. Given the importance of intensity as a parameter in vibrotactile and other haptic research [8], this paper investigates the perceived intensity levels of electrovibration feedback and identifies the most distinguishable levels of intensity.

We report the results of a psychophysical experiment and subjective evaluation of electrovibration intensity with a focus on amplitude and frequency manipulations, and the effects of gender and handedness on perception. The results of this research will aid designers in the creation of electrovibration feedback for touchscreen interaction with perceptually discriminable intensity levels.

2   Related Work

2.1   The History of Electrovibration

In 1950, Mallinckrodt [9] identified that electric flow in brass light sockets generates a different feeling when a current is flowing. Later, through experiments conducted on this phenomenon, he observed that such sensations were created through induced electricity.

Electrocutaneous stimulation has also been studied as a psychophysical phenomenon. But in electrocutaneous stimulation, a real electric current is passed through the body as a nerve-stimulating agent. In Higashiyama's work [6] they discuss the various combinations of electrode placements to create pleasant feedback (the authors defined the term "pleasant feedback" as a less painful or less uncomfortable feeling).

Following on from this, a micro fabricated electrostatic display was tested by Tang *et al.* [10]. An electrode array with different resolutions was created where each electrode could be controlled separately. The study showed that more than 7.6 mm distance is needed to feel each electrode separately. The same concept was used by Fukushima *et al.* in their "Palm Touch Panel" [11] [12]. The Panel stimulated the palm of a hand using electrodes with controlled electric shocks instead of electrovibration.

The main focus of electrovibration research is in the usage of haptic feedback for telepresentation interfaces [1] and for visually impaired people, with studies targeted towards finding the thresholds in sensation using different materials [13] [14].

2.2   Current Electrovibration Systems

Parallel to academic research of galvanic electrotactile stimulation, and dedicated isolated electrode arrays for electrovibration, Senseg[1] started creating a commercially applicable system solution for utilizing electrovibration in 2007. With advancements in grounding arrangements, display material and structure, and stimulus signal characteristics, Senseg was able to bypass lot of the challenges mentioned by [1, 11, 14, 15]. The key emphasis has been in integrating electrovibration feedback with mobile

---

[1] http://www.senseg.com

devices using active surfaces called *tixels*, the tactile pixel [15]. The experiment reported here makes use of Senseg's tixel.

In 2009, Disney Research announced "Tesla Touch", which uses similar kind of mechanism to produce haptic feedback. The key difference is that Tesla Touch implemented on an experimental desktop-based multi-touch system while the Senseg implementation works on mobile and handheld devices. Bau *et al.* [5] have studied the noticeable differences in frequency and amplitude changes in different frequencies. 80Hz, 120Hz, 180Hz are reported to have more than 25% of Just Noticeable Difference (JND) and for 270Hz and 400Hz, the JND is less than 15%.

Amberg *et al.* [4] have presented a tactile input device called STIMTAC which works by changing the friction similar to Senseg and TeslaTouch. But the only difference is that it reduces the friction by creating an air bearing between the finger and surface in addition to the friction. STIMTAC adopted the friction reduction technique from T-PaD designed by Winfield *et al.*, and later introduced a larger area version of T-PaD (LATPaD) in 2010 [16] [17]. One of the main drawbacks of those systems is the lack of transparency; therefore it is not possible to integrate it into any current touchscreen displays. But in 2011 Levesque *et al.* [2] presented a group of frictional widgets using the same mechanism. There they used a transparent surface on top of a graphic display. Although the implementation focuses on reducing the friction, they concluded that programmable friction could increase the awareness of the system and improve the appreciation of an interface.

Although there are several existing implementations of electrovibration for touchscreen interaction, there have been very few studies into user perception of electrovibration stimuli. In studies with vibrotactile stimulation using Eccentric Vibration Motors (ERM) and Linear Resonator Actuators (LRA), it has been shown that 250Hz is the optimum frequency where people feel the vibration with a very low level of displacement [18]. The research described here attempts to analyse electrovibration feedback in a similar manner to establish the optimum frequency and amplitude levels that can be used to design electrovibration feedback with different intensity levels for touchscreen interaction.

2.3   Electrovibration Intensity

The perceived intensity of electrovibration depends on many factors. The basic principle behind electrovibration is the attractive electric Coulomb force between the isolated electrode and the touching finger [9]. It can be conveniently expressed as Maxwell pressure:

$$P = \frac{1}{2} \varepsilon E^2 \tag{1}$$

Where ε is dielectric constant, E is electric field $E = V/d$, V = voltage difference between electrode and conductive tissue in finger, d = effective distance between charges. Distance is comprised of the electrical insulator thickness, and (effective) finger skin thickness. The basic mechanism for electrovibration is thus quadratic with respect to the stimulus voltage. One consequence of this quadratic behavior is that it doubles the frequency of sinusoidal drive signals.

When the attraction force is modulated with a stimulus signal, it vibrates the finger skin and can be felt as a normal vibration. For a finger sliding on the tixel surface, the normal vibration is coupled with the lateral force of kinetic friction between the skin and surface. The resulting modulation of friction force creates a large lateral vibration of the skin. The factors that affect the modulated friction include the friction coefficient (nonlinear) and finger biomechanics, modulation signal shape, and the user's potential grounding (effective voltage).

These factors are carefully taken into account in the experimental setup design. We used a plastic film that has relatively stable friction properties, and has an even thickness profile.

3   Experiment: Evaluating Perceived Intensity

The aim of this experiment was to investigate what levels of frequency and amplitude in electrovibration increase the perceived intensity of the stimuli. This was studied with a magnitude estimation approach, and using a given index scale, ESSI (outlined below). A qualitative survey was also performed to assess the participants' overall perception of the stimuli.

3.1   Magnitude Estimation

Psychophysical scaling has a history as long as psychophysics. In 1850, Fechner came up with the concept of arithmetical representation of sensations [19]. Each psychophysical scaling task can be categorized in to one of the following:

(a) Subjects rate the sensation in an ordinal scale;
(b) Subjects adjust the stimuli to create equal sensory intervals;
(c) Subjects assign a number to represent the magnitude of the sensation (Magnitude Estimation).

In 1988, Gescheider [19] pointed out that magnitude estimation is the most used scaling mechanism in psychophysical evaluations. According to this method, the user starts rating the sensation by giving an arbitrary value at the starting point of the test. But later the user builds up his/her own scale around the given previous ratings.

3.2   ESSI Rating Scale

Alongside the magnitude estimations, we propose a practical perception strength index scale, Effect Strength Subjective Index (ESSI) for evaluating electrovibration perception. The design of the ESSI rating is based on previous studies with vibration alert perception, where a 4-value strength scale was used [20]. This numerical index makes use of Likert scales, which are widely applied in psychophysics, and in in acoustic and visual research [21].

The ESSI scale is a 6-value discrete scale where 0 represents the non-perceived signals and 5 represents signals that are considered to be 'too strong'. The value 3 was defined as "Medium" and it is considered to be the optimum value for stimuli.

3.3   Test Hardware

The test system uses a PC as the signal source, an audio amplifier and transformer, and the tixel test surface with conductive and insulating layers. The insulating layer was made from a screen protector film (iPad Screen protector with a thickness of 125 um). Around the test surface, a ground reference surface was arranged (electrostatic matt) that the user could touch with their non-dominant hand. This provided a stable grounding of the user potential, leaving the drive voltage between the dominant hand and the test surface. The drive signal was floating, and there was a 5 Mohm serial resistor in series with the output. The system was capable of driving voltages up to 2.4 kV peak to peak in the frequency range of 40 to 600 Hz. The system schematics are shown in Figure 1.

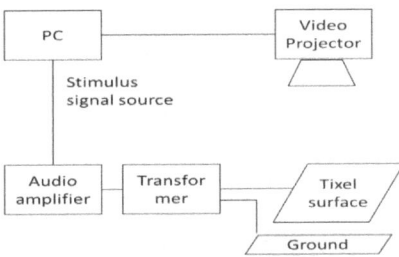

Fig. 1. Test system hardware. The user slides their finger on the tixel surface, where the visual gesture hint is projected. The non-dominant hand is placed on the ground surface.

3.4   Stimuli

All the stimuli used in the experiment were pure sine waves. There were 13 different signal patterns used in this test. In the original pilot studies we conducted to identify the frequency scale, we tested 40Hz-200Hz with 40Hz intervals. In that study we found that 80Hz was the most sensitive vibration level and using a maximum of 200Hz did not cover a large enough scale. Hence we increased the range to 320Hz. In the pilot study, the intensity ratings of 120Hz stimuli were not significantly high despite that fact that 125Hz should be the most sensitive level of vibration for the skin. Therefore, in the study reported here, we examined a range of frequencies using the following scale: 40Hz, 80Hz, 160Hz, 240Hz and 320Hz. 0dB, -6dB and -12dB were used as amplitudes. -6dB and -12dB were selected since they resemble the signals closer to the half and the quarter of the original signal. 0dB corresponds to a 1.2 kV peak. All combinations of frequency and amplitude were used in the test except 240Hz/-12dB and 320Hz/-12dB. These stimuli were excluded from the experiment after pilot studies revealed them to be unperceivable.

3.5   Subjects

We recruited 26 volunteer participants (mean age: 25,5; s.d.: 4,4), mostly students or members of the University of Helsinki, who gave their informed consent to participate in the study and received two movie tickets as compensation. There were an equal number of males and females, left-handed and right-handed participants. Handedness was determined by the Edinburgh Handedness Inventory [22], which showed that three participants were actually ambidextrous, even though they were not aware of it.

3.6   Methodology

Before the actual test began, all participants took part in short tutorial at the beginning of the session to acquaint themselves with the Tixel hardware and electrovibration feedback. The participants' task was to make a continuous circle gesture with the index finger of their dominant hand, tracing an image projected from above onto the test bed (see Figure 2). We instructed participants to keep the other hand on the grounding surface while doing the gesture.

There were two phases in the experiment: magnitude estimation and ESSI ratings. Firstly, a magnitude estimation method was used where participants assigned an arbitrary number of their own choice to the intensity of a baseline stimulus presented. Participants were then presented with a series of further stimuli and asked to assign an intensity number reflecting its perceived intensity relative to the baseline. Overall there were 26 tasks in the experiment using a random baseline stimulus (chosen from the 13 frequency/amplitude combinations). For example, a participant could be presented with the 80Hz 0dB stimulus as the baseline and asked to assign a weight value to the 40Hz -12dB stimulus in comparison to the baseline (see Fig. 2b).

The second phase of the experiment focused on ESSI ratings. Users felt the stimuli in a random order, and rated them one by one using the ESSI scale.

Before the test there was a short questionnaire, and it was repeated after the test with additional qualitative questions of user perceptions. The experiment took approximately 30 minutes to complete. Due to the repetitiveness of the task, participants could interrupt their performance at any time, and the subjects lifted their index finger from the system between each task.

Fig. 2. (a) left: Test setup (b) right: Magnitude Estimation Test

3.7   Magnitude Estimation Results

Figure 3 shows the average magnitude estimation for each amplitude and frequency combination. Given that participants could assign any intensity baseline rating they wished, all data were normalized prior to analysis. A partial correlation (taking frequency into account) showed a significant correlation of 0.79 (p = 0.01) between amplitude and perceived intensity. This correlation suggests that higher amplitude levels results in the perception of increasing intensity. Although there is no significant correlation between frequency and perceived intensity, the graph indicates that 80Hz stimuli are rated higher than all other frequencies.

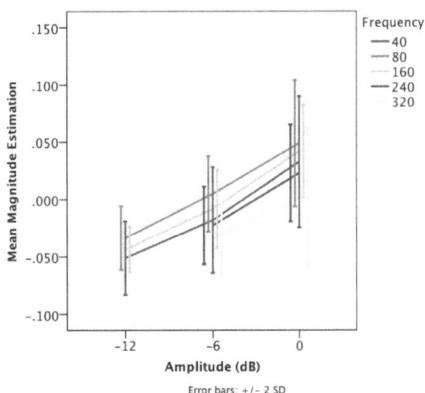

Fig. 3. The mean magnitude estimation for each amplitude level grouped by frequency showing a positive correlation between intensity and amplitude

Analysis of the intensity estimates using a MANOVA showed a significant main effect for amplitude (F=367.2, df = 2, p = 0.001) and also for frequency (F = 35, df = 4, p = 0.001). There was no significant interaction between the two factors. *Post hoc* Tukey's HSD tests showed many significant differences. The most notable results being stimuli using 80Hz and 0dB or 160Hz and 0dB are perceived to be significantly more intense than all other stimuli.

Gender

The participants were evenly balanced in terms of gender. The average intensity estimations for each gender are shown in Figure 4(a). An ANOVA showed no significant differences in the magnitude estimations for gender type (F = 0.004, df = 1, p = .949).

Handedness

Half of the participants (13) stated that they were right-handed while the other half considered themselves to be left-handed. The average intensity estimations for each hand are shown in figure 4(b). An ANOVA showed no significant differences in intensity estimations for each hand (F = .832, df = 1, p= .362).

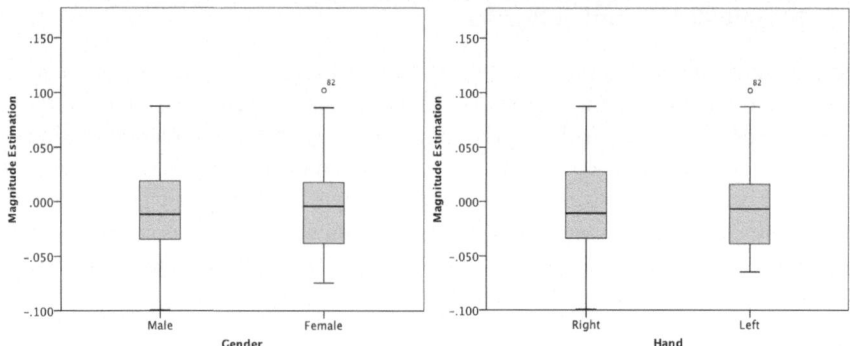

Fig. 4. (a) left: Average Intensity Estimations by Gender; (b) right: Average Intensity Estimations by Dominant Hand

ESSI Results

Figure 5 shows the average ESSI rating for each amplitude/frequency combination. A partial correlation (taking frequency into account) showed a significant correlation of 0.801 (p = 0.01) between amplitude and ESSI rating. This correlation suggests that higher amplitude levels results in the perception of increasing intensity. Although there is no significant correlation between frequency and ESSI, the graph indicates that 80Hz stimuli are rated higher than all other frequencies. This corresponds with the magnitude estimation results.

Analysis of the intensity estimates for each stimulus using a MANOVA showed a significant main effect for amplitude (F=388, df = 2, p = 0.000) and also for frequency (F = 30.8, df = 4, p = 0.000). There was no significant interaction between the two factors. *Post hoc* Tukey tests show similar results to those shown in Table 1. Once again, stimuli using 80Hz and 0dB were perceived to be significantly more intense than all other stimuli except 160Hz and 0dB.

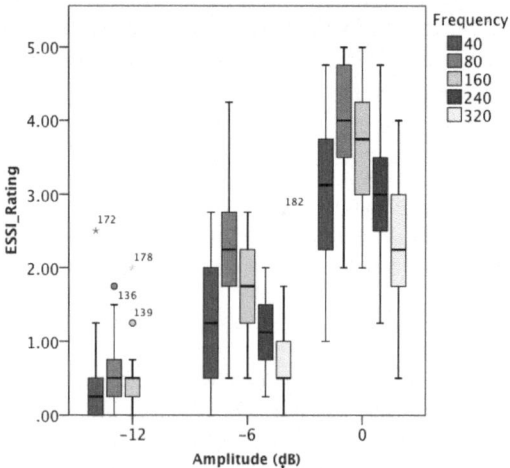

Fig. 5. Average ESSI ratings (max 5)

Gender

Once again, an ANOVA showed no significant difference between ESSI ratings and gender as shown in Figure 6a (F=0.001, df = 1, p=.977).

Handedness

An ANOVA showed a significant difference in ESSI ratings for left and right handed users (F=4.5, df = 1, p=0.033). The results indicate that right-handed users rate the stimuli as 18% more intense than left-handed users (Figure 6b).

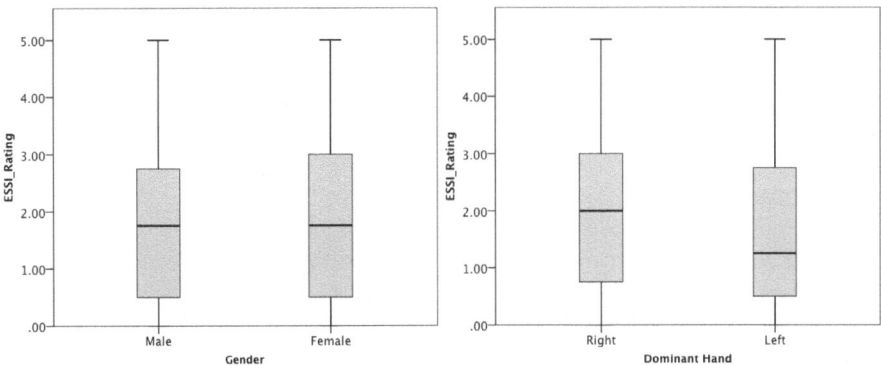

Fig. 6. (a) left: Average ESSI Rating by Gender. (b) right: Average ESSI Rating by Dominant Hand

Questionnaires

The post-study questionnaire explored the participants' general reaction to the system and stimuli. There were 8 statements using a 5-point Likert scale, where 1 corresponded to "strongly disagree" and 5 corresponded to "strongly agree". Examples of statements included are *"I can easily feel this kind of feedback"*, *"I think this kind of feedback is ticklish"*, *"I can barely feel this kind of feedback"*, *"I like the feedback"* and *"I think this kind of feedback hurts physically"*.

A qualitative analysis shows that overall 88% of the users stated a neutral or positive attitude towards the system. Of these, men and left-handed individuals, in particular, perceived the system positively, rating it with an average value of 4.23 on the Likert scale to the questions regarding how much they liked the system. On the other hand, women and right-handed people seem to perceive the stimuli 4,95% more intense than men. This information though is not statistically significant.

4   Discussion

The results of the magnitude estimation and ESSI ratings show that electrovibration stimuli using 80Hz sine waves produced the highest ratings. An input signal of 80Hz actually creates a 160Hz Coulomb force vibration on user's finger. Therefore, 160Hz provides the most intense sensations.

In terms of amplitude, we observed a linear change of intensity in a logarithmic scale. Hence we can suggest that the perceived intensity is logarithmically proportional to the supplied signal.

We selected a set of test participants by carefully maintaining a balance between their gender and the handedness. The analyzed results of magnitude estimation did not show any significance with both factors. In the ESSI test, gender did not show any significant difference but the right-handed participants rated the stimuli as slightly more intense than left-handed participants. Serrien *et al.* [23] found that the left hemisphere, which is strongly linked with the right hand, is dominant when controlling complex skills of either hand, especially in visuomotor conflict situations. In both cases (our study and the one presented by Serrien *et al.*) participants had to perform continuous drawing movements. This may explain the possible differences in electrovibration intensity perception between left and right-handed users.

5   Conclusions

This paper provides a brief overview of electrovibration as tactile feedback, and describes an experimental setup for a preliminary investigation of the perceived intensity of modulated friction created by electrovibration. We introduced and used a fixed 6-point Effect Strength Subjective Index (ESSI) as a measure of generic sensation intensity, and an open magnitude scale. The early results of the experiment indicate that, for continuous sinusoidal signals, there are significant differences in intensity perception for different amplitude/frequency combinations. The highest intensity was found at 80 Hz, which corresponds to 160 Hz as friction modulation. These results are evident in both the magnitude estimation data and the ESSI ratings.

In follow-up studies it will be necessary to recruit participants from a larger age range. The study reported here was conducted with 26 participants and all were between ages of 20 to 36. Hence there was not enough variation in age to analyze the result with respect to aging. In terms of touchscreen feedback, in order for electrovibration intensity to be used as a design parameter effectively, it will be necessary to combine it with other parameters to see if there are any negative effects when used in combination. Furthermore, future studies will investigate electrovibration stimuli with other combinations of modalities such as audio and visual.

Acknowledgements. We would like to thank Ville Mäkinen for useful comments, and Zohaib Gulzar, and Juha Virtanen for their contribution to the actual setup implementation

References

1. Yamamoto, A., Nagasawa, S., Yamamoto, H., Higuchi, T.: Electrostatic tactile display with thin film slider and its application to tactile telepresentation systems. IEEE Transactions on Visualization and Computer Graphics 12(2), 168–177 (2006)

2. Levesque, V., Oram, L., MacLean, K., Cockburn, A., Marchuk, N., Johnson, D., Colgate, J.E., Peshkin, M.: Frictional widgets: enhancing touch interfaces with programmable friction. In: Proceedings of the 2011 Annual Conference Extended Abstracts on Human Factors in Computing Systems, Vancouver, BC, Canada (2011)

3. Brewster, S., Chohan, F., Brown, L.: Tactile feedback for mobile interactions. In: Proceedings of the SIGCHI Conference on Human Factors in Computing Systems, San Jose, California, USA (2007)

4. Amberg, M., Giraud, F., Semail, B., Olivo, P., Casiez, G., Roussel, N.: STIMTAC: a tactile input device with programmable friction. In: Proceedings of the 24th Annual ACM Symposium Adjunct on User Interface Software and Technology, Santa Barbara, California, USA (2011)

5. Bau, O., Poupyrev, I., Israr, A., Harrison, C.: TeslaTouch: electrovibration for touch surfaces. In: Proceedings of the 23nd Annual ACM Symposium on User Interface Software and Technology, New York, New York, USA (2010)

6. Higashiyama, A., Rollman, G.: Perceived locus and intensity of electrocutaneous stimulation. IEEE Transactions on Biomedical Engineering 38(7), 679–686 (1991)

7. Brown, L.M., Brewster, S.A.: Multidimensional Tactons for Non-Visual Information Display in Mobile Devices. In: MobileHCI 2006 (2006)

8. Maclean, K., Enriquez, M.: Perceptual design of haptic icons. In: Proceedings of Eurohaptics (2003)

9. Mallinckrodt, E., Hughes, A.L., William Sleator, J.: Perception by the Skin of Electrically Induced Vibrations. Science 118(3062), 277–278 (1953)

10. Tang, H., Beebe, D.: A microfabricated electrostatic haptic display for persons with visual impairments. IEEE Transactions on Rehabilitation Engineering 6(3), 241–248 (1998)

11. Fukushima, S., Kajimoto, H.: Palm touch panel: providing touch sensation through the device. In: Proceedings of the ACM International Conference on Interactive Tabletops and Surfaces, Kobe, Japan (2011)

12. Hiroyuki, K., Yonezo, K., Susumu, T.: Forehead electrotactile display. In: Proceedings of the Virtual Reality Society of Japan Annual Conference, vol. 11 (2006)

13. Agarwal, A., Nammi, K., Kaczmarek, K., Tyler, M., Beebe, D.: A hybrid natural/artificial electrostatic actuator for tactile stimulation. In: 2nd Annual International IEEE-EMB Special Topic Conference on Microtechnologies in Medicine & Biology (2002)

14. Strong, R., Troxel, D.: An Electrotactile Display. IEEE Transactions on Man-Machine Systems 11(1), 72–79 (1970)

15. Linjama, J., Mäkinen, V.: E-Sense screen: Novel haptic display with Capacitive Electrosensory Interface. In: HAID 2009, 4th Workshop for Haptic and Audio Interaction Design, Dresden, Germany (2009)

16. Winfield, L., Glassmire, J., Colgate, J.E., Peshkin, M.: T-PaD: Tactile Pattern Display through Variable Friction Reduction. In: Second Joint EuroHaptics Conference, 2007 and Symposium on Haptic Interfaces for Virtual Environment and Teleoperator Systems, World Haptics 2007, Tsukaba (2007)

17. Marchuk, N., Colgate, J., Peshkin, M.: Friction measurements on a Large Area TPaD. In: IEEE Haptics Symposium, Waltham, MA, USA (2010)

18. Verrillo, R.T.: Psychophysics of vibrotactile stimulation. The Journal of the Acoustical Society of America 77(1), 225–232 (1985)

19. Gescheider, G.A.: Psychophysical Scaling. Annual Review of Psychology 39, 169–200 (1988)

20. Linjama, J., Puhakka, M., Kaaresoja, T.: User Studies on Tactile Perception of Vibrating Alert. In: HCI International, Crete, Greece (2003)

21. Keelan, B.W.: Handbook of Image Quality, Characterization and Prediction. CRC Press (2002)
22. Oldfield, R.C.: The assessment and analysis of handedness: The Edinburgh inventory. Neuropsychologia 9(1), 97–113 (1971)
23. Serrien, D.J., Spapé, M.M.: The role of hand dominance and sensorimotor congruence in voluntary movement. Exp. Brain Res. 199, 195–200 (2009)

Stability of Model-Mediated Teleoperation: Discussion and Experiments

Bert Willaert[1], Hendrik Van Brussel[1], and Günter Niemeyer[2]

[1] Dept. of Mechanical Engineering, KULeuven, Belgium
bert.willaert@mech.kuleuven.be
[2] Willow Garage Inc., Menlo Park, CA, USA
gunter.niemeyer@stanford.edu

Abstract. The design of a bilateral teleoperation system remains challenging in cases with high-impedance slave robots or substantial communication delays. Especially for these scenarios, model-mediated teleoperation offers a promising new approach. In this paper, we present a first stability discussion. We examine the continuous behavior using general control principles and discuss how the model structure and its predictive power affects system lag and stability. We also recognize the unavoidability of discrete model jumps and discuss measures to isolate events and prevent limit cycles. The discussions are illustrated in a single degree of freedom case and supported by single degree of freedom experiments.

Keywords: Teleoperation, Model-mediation, Stability.

1 Introduction

Bilateral teleoperation systems allow the human operator to act on a remote environment via a robotic slave device while providing force feedback through a master device. Typically, the user's desire to feel directly connected to the environment is opposed to the need for stable interactions under all operating conditions. This well-known trade-off between transparency and stability is most challenging for systems characterized by either (1) limited response times, or (2) substantial communication time-delays.

The most conservative approaches, e.g. based on wave variables, can achieve stability even with large response times and long communication delays, but fail to provide transparency [1]. At the other extreme, more transparent approaches, e.g. the traditional position-force architecture, can hide the slave dynamics from the user's perception, but are prone to instability and often need high damping or low scaling factors [2,3]. Other methods vary the transparency versus stability trade-off in realtime monitoring for instabilities [4,5,6], most often by temporarily adding damping to provide stability at the cost of loosing transparency.

In the above works, controllers communicate position, force, or combined sensory information between master and slave. A fundamentally different approach involves transmission of models for the environment [7,8]. This has been called *impedance reflecting*, *virtual-reality based* or *model-mediated* teleoperation and

P. Isokoski and J. Springare (Eds.): EuroHaptics 2012, Part I, LNCS 7282, pp. 625–636, 2012.

proposes to achieve transparency with an indirect connection. Models of various complexity have been used, including a mass-spring-damper model [9], a spring-damper model [10], a pure spring model [11,12] or merely a rigid wall with variable location [8]. By choosing the model structure consistent with the environment, most previous works have been able to focus on transparency with limited stability considerations.

In this work we concentrate on the closed-loop stability of the model-mediated framework proposed in [13], which defines a bilateral controller communicating abstracted model and task information: the model captures the environment while the task encodes the user intent. At the slave side, the *model estimation* uses sensory data to continuously estimate a model of the environment. At the master side, this model is used by the *model rendering* to generate the haptic feedback. The human response to the haptic rendering is monitored by the *task estimation* to generate a task description. Back at the slave side, the *task execution* regulates the slave to accomplish the incoming task as well as possible using the current model of the environment.

As the model should estimate the actual environment, many works implicitly assume the existence of a single correct or best model fit. For an unchanging environment and with sufficient excitation, the model estimator converges and stability is achieved by assumption. However, any model is a simplification and the value of teleoperation is highest for operations in unstructured environments which are at best difficult to model. Therefore, model convergence can not be expected. Instead the model should be treated as a time-varying signal. It should capture the currently most salient aspect of the environment and continuously adjust as the user explores and operates.

We examine closed-loop stability, **allowing environments to differ from the assumed model structure**. This encompasses two aspects: First, model errors lead to continuous *model adjustments* which close the loop with appropriate stability needs. We use classic loop gain and phase principles to review the behavior and discuss the implications of model structure. Second, discrete *model jumps* need to occur and have the potential to disturb the user or trigger limit cycles. We implement appropriate mitigation strategies. We guide and illustrate the discussion via a single degree of freedom case study and use the same framework to confirm the conclusions in experiments.

2   Model and Task Signals

To examine system stability, we must view the model and task descriptions as time-varying, continuous signals. To further illustrate and understand this perspective, let us consider a case study in a single degree of freedom. We first select a model structure and then present the corresponding controller elements, following the high-level model-task loop.

2.1 Model Structure

The underlying motivation for this study is the use of teleoperation in daily life scenario's. When considering daily life scenario's, many environments consist

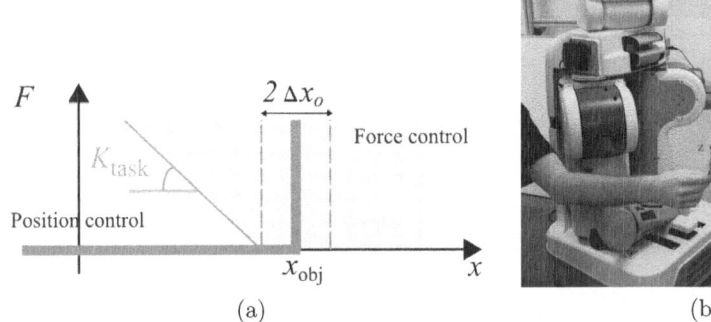

Fig. 1. (a) The model: the simple, rigid contact model is represented by the thick gray line: forces are zero for $x < x_{obj}$ while they are positive if $x = x_{obj}$. (b) The experimental setup: the slave is the left 7-d.o.f. arm of the PR2 robot and the master is a 3-d.o.f. haptic device designed at the University of Leuven.

of moderately hard objects and tasks usually require moving, manipulating, or assembling these objects. Hence we select a model that contains the location of a hard object. The model is very simple while capturing the most salient environment feature, being the combination of and transition between free-space and object contact. We already see that as objects in the real world change or are moved, this model will need to adjust and becomes time-varying.

The model, as shown in Fig. 1(a), allows free space motion along a single degree of freedom up to a rigid, immovable object located at x_{obj}. The model is quasi-static, relating force F only to position x: the force is zero if the position falls in front of the object ($F = 0$; $x < x_{obj}$), while it is positive if the position matches the object location ($F > 0$; $x = x_{obj}$).

2.2 Model Estimation

An instance of the model is maintained at the slave side to estimate the current environment. To update its parameter x^s_{obj}, the model estimation first has to decide whether the slave is in contact or in free space, corresponding to the vertical and horizontal branch of the model as shown in Fig. 1(a). To do so, the model estimation uses the following basic contact detector:

$$\text{contact state} \begin{cases} \text{'in contact' if } F_e \geq F_{\text{threshold}} \\ \text{'free space' if } F_e < F_{\text{threshold}} \end{cases} \tag{1}$$

where F_e stands for the measured or estimated interaction force with the environment and $F_{\text{threshold}}$ has to be chosen depending on the noise level of the measured or estimated interaction force F_e. While in contact, the object location is updated and estimated using a first order low-pass filter of bandwidth λ:

$$\dot{x}^s_{obj} = \begin{cases} \lambda(x_s - x^s_{obj}) \text{ if 'in contact'} \\ 0 \qquad\qquad \text{if 'free space'} \end{cases} \tag{2}$$

2.3 Model Rendering

A separate model instance is maintained at the master side for rendering the haptic feedback, with an object location x_{obj}^m. The master model is updated with the most recent incoming slave model

$$x_{\mathrm{obj}}^m(t) = x_{\mathrm{obj}}^s(t - T_d), \tag{3}$$

where T_d is a possible communication delay. The object is rendered as a one-sided pure stiffness, which means that the master device only exerts forces on the human operator if he/she has virtually penetrated the object. To define this behavior, a haptic proxy is used:

$$F_m = K_p^m(x_{\mathrm{proxy}} - x_m) \quad \text{with } x_{\mathrm{proxy}} = \begin{cases} x_m & \text{if } x_m \leq x_{\mathrm{obj}}^m \\ x_{\mathrm{obj}}^m & \text{if } x_m > x_{\mathrm{obj}}^m \end{cases} \tag{4}$$

Because the model represents a infinitely stiff object, the gain K_p^m should be set as high as practically possible.

2.4 Task Estimation

We employ a simple task representation to encode the user's intent, namely a pair of position and force values $(x, F)_{\mathrm{task}}$. The force value F_{task} is set to the human force acting against the model, while the position value x_{task} is set to the master proxy location. The proxy location maps the user's position onto the model, i.e. it creates a position that is consistent with the displayed model. As such, the task is feasible against the model, independent of any practical limitations on the gain K_p^m.

2.5 Task Execution

The task $(x, F)_{\mathrm{task}}^m$ estimated by the master is communicated to the slave

$$(x, F)_{\mathrm{task}}^s(t) = (x, F)_{\mathrm{task}}^m(t - T_d), \tag{5}$$

and compared against the most recent environment model. If the task lies in the grey zone in Fig. 1(a), i.e. if $F_{\mathrm{task}}^s > K_{\mathrm{task}}(x_{\mathrm{obj}}^s - x_{\mathrm{task}}^s - \Delta x_0)$, the user is most likely expecting contact and the slave is placed under force control using a natural-admittance type controller [14]. Otherwise, the slave is operated in position control mode.

3   Continuous Closed-Loop Stability

Model mediated systems are nonlinear and preclude simple analyses. Nevertheless we can gain substantial insight by examining the high-level loop between master and slave. We see why model mediation can provide real improvements and how the model selection and its predictive power affects the stability.

3.1   Classic Stability: Gain Stabilization

Consider the classic stability problem of a telerobot with direct force feedback contacting a stiff environment [2,3]. The user's movements are measured and commanded to the slave, executed, causing the environment forces, which communicated back to and displayed to the user. From a linear system's perspective, an overall transfer function maps user motion to force feedback. Its gain is the environment stiffness. Its phase lag stems from two sources: (a) communication delay and (b) slave controller response lag. So if the slave tracking or communication are slow *and* the environment contact is stiff, the transfer function shows both large gain and phase and predicts instability. Commonly, to stabilize this system, the force feedback gain is lowered. Equivalently, the transfer function gain is reduced, equivalent to gain stabilization.

3.2   Model-Mediated Stability: Phase Stabilization

Model mediation can be seen as phase stabilization. Phase stabilization requires phase lead, which provides signals earlier in time, and hence is achieved by prediction. Model mediation effectively predicts and displays the environment interaction force that will happen when the task is executed, by assuming knowledge of the environment. Stability is thus determined by the model's predictive power, i.e. over which time horizon and how accurately predictions of environment interactions can be made.

Consider the model-mediated loop: user tasks are sent to the slave and eventually lead to feedback of model adjustments. The loop gain describes the size of model adjustments that result from a given task. Following the case study, when the user advances, an object may slide or deflect. Thus a model adjustment in the form of a shifted object location is necessary. The loop gain describes the size of the location shift for the size of the user's movement.

More generally, model adjustments occur when the predicted motions and/or forces differ from the observed values. Larger prediction errors require larger adjustments. Thus the size of prediction errors correlate to the loop gain. A model with greater predictive power lowers the loop gain and improves stability.

Meanwhile the loop phase describes how quickly the adjustments are received. This stems from three sources: (a) the communication delay, (b) how quickly the user commands are executed (slave controller response lag), and (c) the model estimator. Faster model estimation lowers the loop lag and improves stability, subject to a lower bound on lag given by the other two sources. In our case study, the single model parameter can be *estimated* very quickly.

The selection of an appropriate model structure and complexity must balance these loop gain and phase needs. Choosing a more appropriate structure will help predictive power. Choosing a simpler model will lower the number of parameters and increase the possible estimation speed. Of course, if the other sources of lag dominate the estimation time constants, the balance will shift to favor greater predictive power and devalue simplicity.

4 Discrete Model Jumps and Stability

Under normal operating conditions, model adjustments happen continuously and slowly. However, sudden and drastic updates to the model may be necessary when new environment features are discovered. In our case study, imagine detecting a previously unknown object. The model should immediately jump to the newly discovered location. But, due to any lag source, the master will be leading the slave when the contact is detected. Immediately displaying this object to the user requires the master to jump backwards and triggers a potentially dangerous discontinuity in force levels.

In general, the information encoded in such temporal model discontinuities or jumps is important. It indicates a radical change in the expected environment and should not be filtered or removed from the user's perception. However it poses the danger of introducing a discontinuity into the task, which could trigger a jump in the slave movement or force, and ultimately excite unmodelled dynamic elements in the environment, robot, or controllers. In turn, with an imperfect model structure, the estimation may react with another drastic or discrete jump, escalating into a limit cycle or other stability problems. Hence, model jumps should remain isolated and contained from propagation. Note that longer lags between master and slave will enlarge the size of necessary model jumps and increase the need for explicit handling and containment.

Explicit handling will appear in nearly all subsystems. Model estimation must determine when it is necessary to initiate jumps. Model rendering must make sure the users are informed without confusion and without triggering impulsive reactions in the task. And the task execution must decide how to interpret a task that was based on an obsolete model without exciting further jumps. In the following, we demonstrate this in our case study.

4.1 Triggering Model Jumps

The model estimation can determine two distinct events that require a model jump, i.e. the detection of an unknown object and the disappearance of an expected object. These situations are detected and acted upon as follows:

Detection: if $(x_s < x^s_{obj} - \Delta x_0$ and 'in contact') reset $^s_{obj} = x_s$,
Removal: if $(x_s > x^s_{obj} + \Delta x_0$ and 'free space') reset $^s_{obj} = +\infty$.

4.2 Rendering Model Jumps

A previous user study has compared several methods of rendering model jumps [15]. In our implementation, we select a gradual, constant-time introduction or removal of the object if the user is or would be in contact with the object:

Detection: if $(x_m > x^s_{obj})$ introduce x^m_{obj} at x_m and move to x^s_{obj}
over a period of time t_{move},
Removal: if $(x_m > x^m_{obj})$ move x^m_{obj} from x^m_{obj} to x_m
over a period of time t_{fade} before removal.

Table 1. Gains and parameters used for the experiments

K_p^s: 2000 N/m	G_v:	20 Ns/m	Δx_o:	0.004 m
K_v^s: 10 Ns/m	m_d:	7 kg	t_{move}:	500 msec
K_{task}: 250 N/m	K_p^m:	4000 N/m	t_{fade}:	300 msec

4.3 Task execution under Model Jumps

Task execution depends on the current environment model. Explicit handling is thus necessary to avoid discontinuities in the controller when the model jumps.

Just before an object removal the controller is most likely in force control mode, as the user is pressing against the expected object. At the moment of the removal, the slave position x_s is leading the object location x_{obj}^s by Δx_0 (see 4.1). The removal switches the controller into position control mode, while the task position x_{task}^s is still the *old* object location. To ensure smoothness, a position offset is added to the current task and then gradually removed over a period of time t_{fade}.

At the moment of an object detection the interaction force is equal to the threshold force for the contact detector $F_{threshold}$ while the task force F_{task}^s is still at zero. If the threshold is small, no transition measure is necessary. Otherwise, a force offset may be introduced and gradually removed.

5 Experiments

Following the single degree of freedom case study, we substantiate our discussions in experiments performed in three stages. First, we confirm the limitations of two classical control architectures. Next, we demonstrate that the model-mediated approach performs well independently of lags if the model adequately captures the physical environment. Finally, we show how performance degrades when lag appears and the environment is inconsistent with the assumed model. In all tests, the system begins out of contact. The user moves to and interacts with the unknown environment. The figures 2(a) - 3(d) show the data, including the master (blue) and slave (green) position and force signals plotted against time as well as each other. The position graphs further show the object locations in dashed lines, as estimated by the slave and rendered at the master.

5.1 Experimental Setup

The experimental setup is shown in Fig. 1(b). The slave is the left 7-d.o.f. arm of the PR2 robot. The master is a haptic device designed at the KU Leuven. The Cartesian y-axes of both devices are linked to create a 1-d.o.f. system, using the transpose of the Jacobians to transform all control forces to motor torques. The master device is low impedance (mass $\approx$ 0.4 kg, $F_{friction} \approx$ 1 N) and the applied motor torques provide a good estimate of the user forces. The PR2 arm has a higher impedance (mass$\approx$ 7 kg and $F_{friction} \approx$ 3 N) and uses a 6 axis force/torque sensor integrated in the wrist. The threshold for the contact detector, $F_{\text{threshold}}$, was set to 1 N. Tuning of the position and force controller

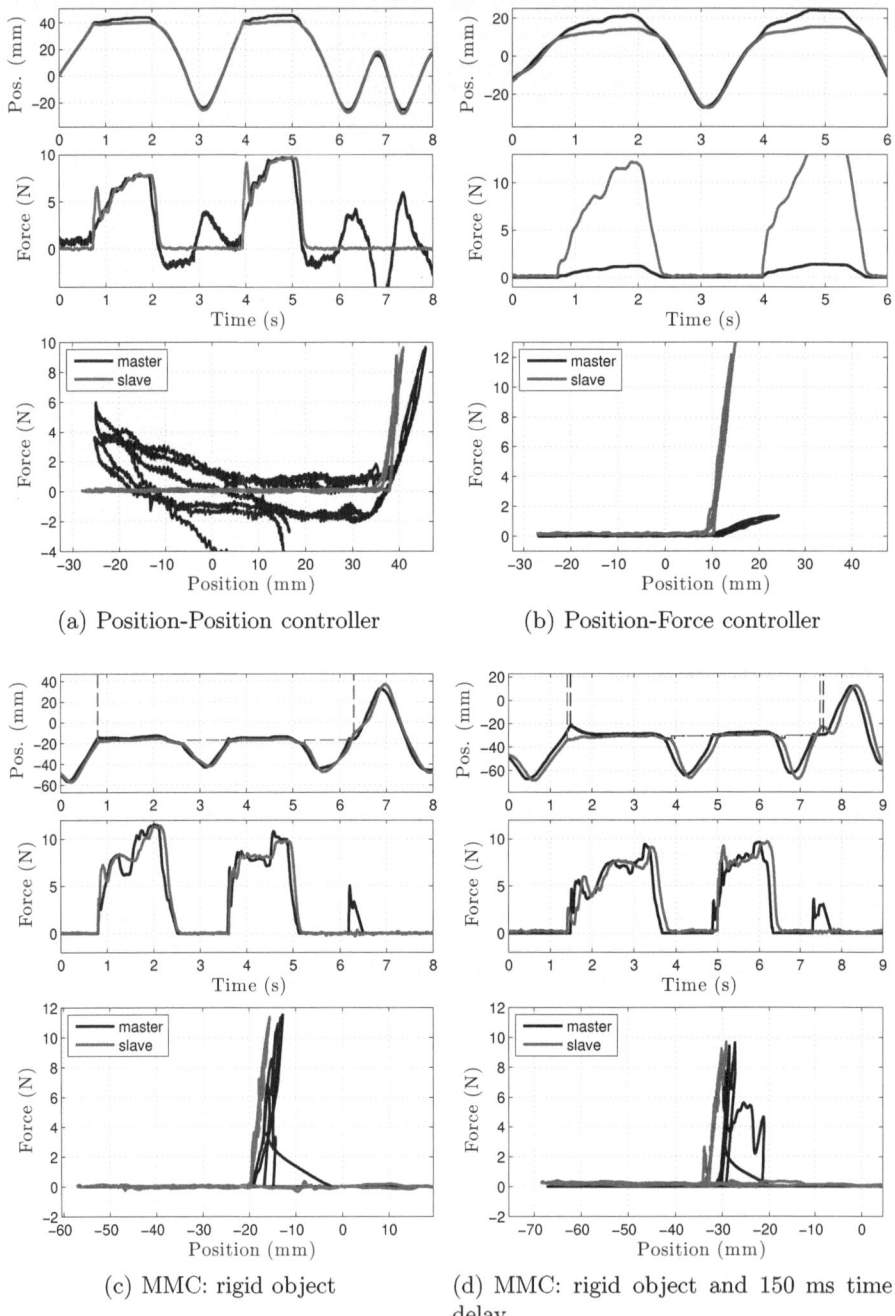

(a) Position-Position controller

(b) Position-Force controller

(c) MMC: rigid object

(d) MMC: rigid object and 150 ms time delay.

Fig. 2. (a-b) Poor performance with classical control schemes. (c-d) Model-mediated controller (MMC) and a rigid environment: good performance achieved by a well-matched model independent of the overall lag.

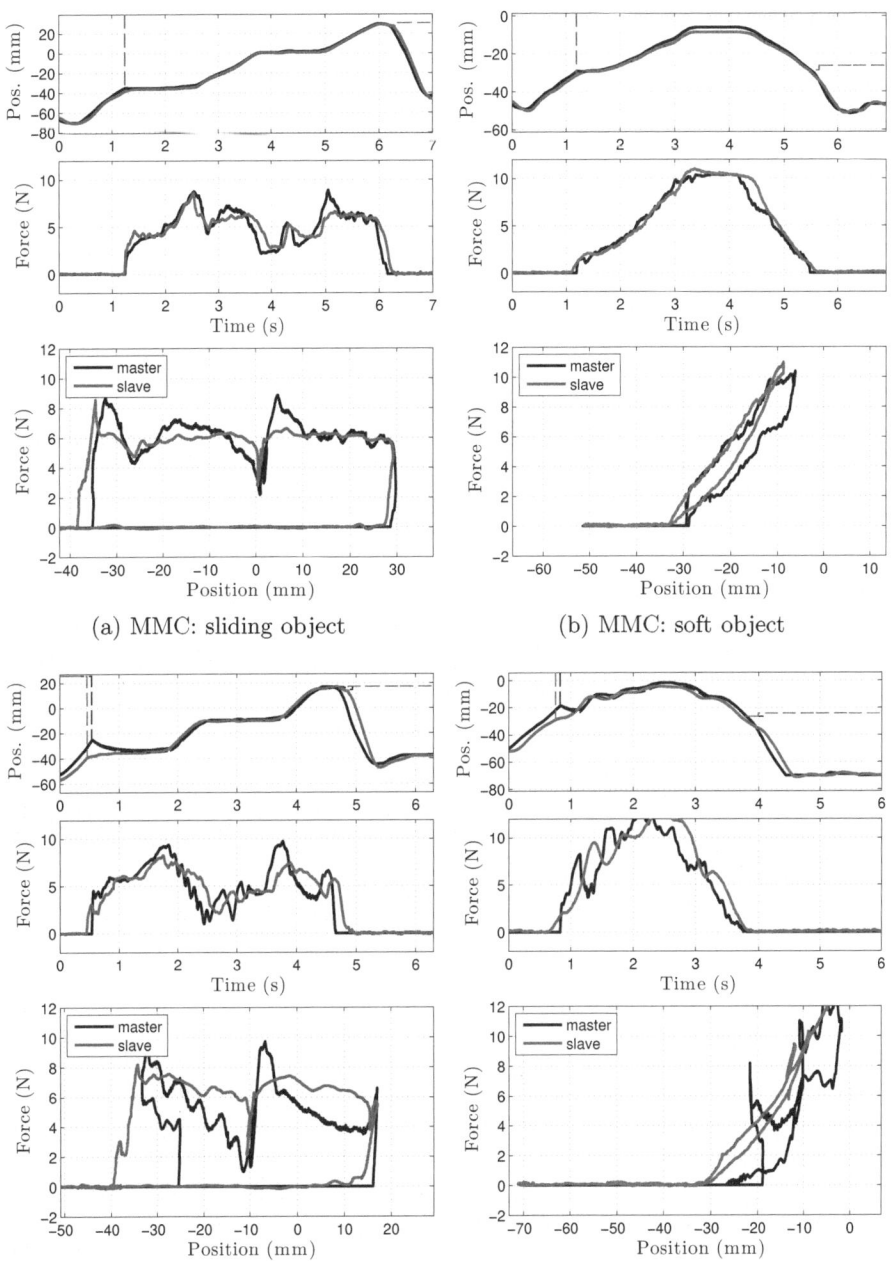

(a) MMC: sliding object

(b) MMC: soft object

(c) MMC: sliding object and 150 ms delay. (d) MMC: soft object and 150 ms time delay.

Fig. 3. Model-mediated controller (MMC) and a sliding and a soft object: (a-b) good performance despite the poorly matched model and (c-d) poor performance due to the extra lag

results in a 2.7 Hz ($\sqrt{\frac{K_p^s}{M_s}}$) and 0.5 Hz bandwidth ($\frac{G_v}{M_s}$) respectively. The latter results in poor force tracking which is clearly visible in the experimental results. Flexibility and backlash in the PR2 arm lead to an uncertainty in its gripper position and thereby the model estimation of $\Delta x_0 = 4$mm. The model update filter is set to a bandwidth of 10 Hz. Table 1 summarizes all properties.

5.2 Experimental Results

Classical Architectures: Model-Free controllers. As mentioned earlier, stable and transparent interaction with very different environment types is really challenging in case of a slave robots with a limited response time, even in the absence of significant communication time delay. The performance of the two extreme classical controllers is a good illustration of this.

Fig. 2(a) shows the system under position-position control, where the master and slave track each other's position. This stable architecture is able to achieve a 1:1 force scale, but the user feels the slave inertial and residual friction forces, overlaid on the environment forces. For a high-impedance slave robot, especially in free space, this controller results in poor transparency.

Fig. 2(b) shows the system under position-force control, when the feedback force is directly applied to the master. While this completely hides the slave's inertia and friction, a 10:1 force reduction is required in order to obtain stable interaction. This reduction is necessitated by the high mass and slow response times of the PR2-arm [2,3] and results in poor transparency during contact.

Model-Mediation: Consistent Environment. In the second set of experiments, the system uses model-mediated control to interact with one fixed hard object. Fig. 2(c) and 2(d) show the behavior without and with and artificial round-trip communication delay of 150 ms. Note that with a 2.7Hz bandwidth, the slave has a settling time greater than 150 ms and adding the communication delay is no worse than doubling the overall lag.

Overall, Fig. 2(c) and 2(d) show excellent position and force tracking in both contact and free space. These experiments show that in case the model adequately captures the physical environment, a model-mediated controller can give the free-space feeling of the position-force architecture while achieving the stability and 1:1 force scaling of the position-position architecture.

As the model adequately captures the physical environment, there are no significant continuous model adjustments. However, discrete model jumps are needed when the object is first detected (at 0.8s/1.5s in Fig. 2(c)/ 2(d)). At these times the slave lags the master (by 5 mm/15 mm) with the additional time delay creating more lag. As described in 4.2, the master fades in the model leading to the observed gradual re-convergence of master and slave positions. Especially in Fig. 2(d), this effect is visible in the position-force graph as a non-physical artifact which directly alerts the user of the object detection without triggering subsequent effects.

Discrete jumps are also needed after the object is secretly removed (at 6s/7s). Unaware of the removal, the system allows the user to apply forces against the rendered model. When the slave attempts to track these forces, it moves and detects the object disappearance (see 4.1). The master renders this removal by fading out the model and fading the forces to zero (see 4.2). Again, the non-physical effect, clearly visible in the position-force graph, alerts the user of the discontinuity without triggering additional effects.

Model-Mediation: Inconsistent Environments. In the last set of experiments we challenge the model-mediated controller with inconsistent environments that are not fixed (sliding object) or not hard (soft object), again without and with a round-trip delay of 150 ms. Fig. 3(a) shows the model-mediated controller in case of pushing a hard object over a surface ($F_{fr,\mathrm{obj}} \approx 6$ N). Upon initial contact, the user's force rises while the object remains stationary. Once the user's force exceeds the static friction, the object slides. The model is continuously updated allowing the user to feel the object motion. Motion ceases when the applied force drops. Fig. 3(b) shows the same controller in case of compressing a fixed soft object($K_{\mathrm{obj}} \approx 600$ N/m). Again the model is continuously updated so that both position and forces are tracked and the user feels a soft object. The experiments show that, despite the significant loop lag due to the limited response time of the slave, the predictive power of the model is sufficiently high for these contact scenario's.

For the experiments with the additional time-delay, this is no longer the case. Fig. 3(c) and 3(d) show the delayed model-mediation interacting with the sliding object and the soft object. In comparison to Fig. 3(a) and 3(b), we see a clear distortion of the position-force graphs. This performance degradation can be explained by the combination of the limited predictive power of the model, i.e. the need for significant model updates (high loop gain), in combination with a big loop lag. A model with better predictive power is necessary to retain performance under the delay here. For example, a spring model [10,11,12] may be better suited to handle soft objects in this case.

6   Conclusions

In this work, we further formalized the model-mediated framework described in [13]. We discussed stability and introduced the notion of continuous *model adjustments* versus discrete *model jumps*.

We discussed qualitatively how the model choice impacts the overall system stability during continuous model adjustments: the selection of a model structure has to balance the loop gain and phase. Better predictive power reduces loop gain while faster model estimation lowers lag. Systems with longer inherent lag will favor models with better predictive power.

The experiments demonstrate this in two ways: (1) as long as the model captures the environment adequately, i.e. the model has an excellent predictive power (small loop gain), extra lag in the closed-loop does not result in performance degradation. (2) if the model is, however, not consistent, i.e. the model has a limited predictive power (higher loop gain), the performance does degrade if the overall loop lag increases. For scenarios when the environment can have very difference characteristics and large lags are unavoidable, the use of multiple models in one controller should be explored in future work.

This is one of the reasons why we also recognized the need for discrete model jumps and discussed how and why these discrete events should remain isolated, i.e. they should not trigger subsequent effects.

Being a first systematic stability discussion of model-mediated teleoperation, we hope this research not only demonstrates the value of model-mediation but opens a tantalizing avenue to further explore the relationship of a model's predictive power with other system parameters.

References

1. Niemeyer, G., Slotine, J.-J.E.: Stable adaptive teleoperation. IEEE Journal of Oceanic Engineering 16, 152–162 (1991)
2. Daniel, R.W., McAree, P.R.: Fundamental limits of performance for force reflecting teleoperation. The Int. J. of Robotics Research 17(8), 811–830 (1998)
3. Willaert, B., Corteville, B., Reynaerts, D., Van Brussel, H., Vander Poorten, E.B.: A mechatronic analysis of the classical position-force controller based on bounded environment passivity. The Int. J. of Robotics Research 30(4), 444–463 (2011)
4. Hannaford, B., Ryu, J.: Time domain passivity control of haptic interfaces. In: Proc. of the Int. Conf. on Robotics and Automation, pp. 1863–1869 (2001)
5. Love, L.J., Book, W.J.: Force reflecting teleoperation with adaptive impedance control. Trans. on Systems, Man and Cybernetics (Part B) 34(1), 159–165 (2004)
6. Ryu, D., Song, J.-B., Choi, J., Kang, S., Kim, M.: Frequency domain stability observer and active damping control for stable haptic interaction. In: Proc. of the Int. Conf. on Robotics and Automation, pp. 105–110 (2007)
7. Hannaford, B.: A design framework for teleoperators with kinesthetic feedback. IEEE Transactions on Robotics and Automation 5(4), 426–434 (1989)
8. Mitra, P., Niemeyer, G.: Model-mediated telemanipulation. The International Journal of Robotics Research 27(2), 253–262 (2008)
9. Hashtrudi-Zaad, K., Salcudean, S.E.: Adaptive transparent impedance reflecting teleoperation. In: Proc. of the Int. Conf. on Robotics and Automation, Minneapolis, Minnesota, pp. 1369–1374 (April 1996)
10. Achhammer, A., Weber, C., Peer, A., Buss, M.: Improvement of model-mediated teleoperation using a new hybrid environment estimation technique. In: Proc. of the Int. Conf. on Robotics and Automation, pp. 5358–5363 (2010)
11. Tzafestas, C., Velanas, S., Fakiridis, G.: Adaptive impedance control in haptic teleoperation to improve transparency under time-delay. In: Proc. of the Int. Conf. on Robotics and Automation, pp. 212–219 (2008)
12. Willaert, B., Goethals, P., Reynaerts, D., Van Brussel, H., Vander Poorten, E.: Transparent and Shaped Stiffness Reflection for Telesurgery. In: Advances in Haptics, ch. 13, pp. 259–282. IN-TECH (2010)
13. Park, J.: Improving Teleoperation by using models and tasks. PhD thesis, University of Stanford (2009)
14. Newman, W.S.: Stability and performance limits of interactions controllers. Journal of Dynamic Systems, Measurement, and Control 114(4), 563–570 (1992)
15. Mitra, P., Gentry, D., Niemeyer, G.: User preferences and performance in model mediated telemanipulation. In: World Haptics Conf., pp. 268–273 (2007)

Novel Thin Electromagnetic System for Creating Pushbutton Feedback in Automotive Applications

Ingo Zoller, Peter Lotz, and Thorsten A. Kern

Continental Automotive GmbH, VDO-Str. 1, 64832 Babenhausen, Germany
{ingo.zoller,peter.lotz,
thorsten.alexander.kern}@continental-corporation.com

Abstract. The integration of artificial haptics on flat surfaces is a usability requirement resulting from modern control devices based on capacitive touchscreens. In automotive applications, aspects like high signal to noise ratio of the feedback, mechanical grounded situation of the devices and general robustness in thermal and climatic conditions have to be considered. This paper presents an electromagnetic actuator module for active tactile feedback for automotive applications, allowing a homogenous and intense feedback on a surface of at least 100x60 mm². The package of the actuator is thin, guaranteeing easy product integration even in complex multimedia control units. The resulting challenges on parallel guidance, sensing capabilities and resulting performance are sketched and summarized into an outlook for the future application in a series production context.

Keywords: haptic, actuator, electromagnetic, automotive.

1 Context

Products with active tactile feedback inside a car cover all elements which are in direct contact with its occupants. The more classic applications are active steering wheels with tactile and/or kinesthetic support [1] and break- or throttle-pedals [2]. Research additionally focuses on tactile signals to create situative awareness or give assistive clues [3]. In addition to these feedback loops directly related to the task of driving, comfort functions like multimedia- and climate-control elements are also subject to active tactile feedback. Numerous solutions were presented for tactile feedback on automotive touchscreens, but lately faceplates (fig. 1a) and remote control touch devices like touchpads (fig. 1b) are also equipped with active haptics. The kind of feedback generated by these devices is usually designed to mimic the impression of real pushbuttons.

The intention to create a situative and configurable tactile feedback to confirm the activation of a function is not at all new. Different technologies are in application to create such feedback. The type of technology in use is always strongly influenced by the volume available. In mobile application vibrotactile solutions range from standard rotary motors with eccentric discs [4] over high current versions of such motors for the creation of shorter and sharper pulses (TouchSense© by Immersion) to voice-coil

P. Isokoski and J. Springare (Eds.): EuroHaptics 2012, Part I, LNCS 7282, pp. 637–645, 2012.
© Springer-Verlag Berlin Heidelberg 2012

systems which allow a broader frequency response. Lately other systems can be seen on the market reducing building volume and costs to create vibrotactile systems with broad perceptional bandwidth by an intelligent way to mix just few oscillating frequencies [5]. Beside these classic actuators based on rotary or linear movement, lately mobile devices can be noted based on piezoelectric actuators, or electroactive polymers. Ultrasonic [6] and electrostatic [7] devices gain increased attention too. Originally they were not invented for confirmation of a function, but for tactile texture simulation. However they show some fascinating performance in that point.

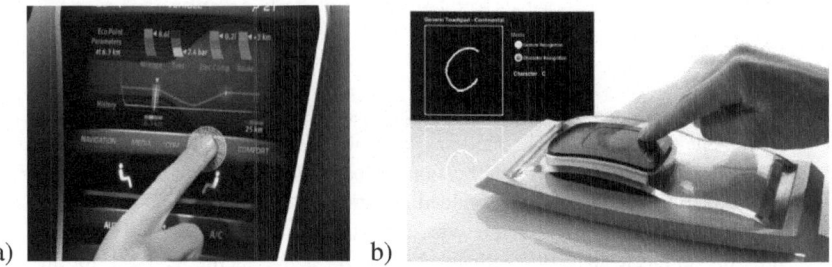

Fig. 1. Faceplate with active haptic feedback (a) ; Stand-alone touchpad with handwriting recognition and haptic feedback (b)

Although this bunch of technology exists and some of those already have reached a series level in mobile devices, only few solutions can be applied to automotive applications too. This fact is due to market requirements and technical reasons: Automotive industry has a certain delay in the application of new technologies. Typical design sequences for automotive products last 3 years, which makes them less interesting for young emerging companies which are required to make profitable business within a short period of time. In addition quality requirements are high, making an entrance into this market almost impossible for anyone but an established Tier 1 or Tier 2 supplier. Besides these general challenges, automotive operating conditions differ in two significant aspects from mobile devices: Their operation is truly one-handed, which cancels any approach to create the haptic feedback in the holding hand. In addition the feedback has to be very precise and strong to create a good signal to noise ratio in the vibrating surrounding of a moving car. This makes actuators for haptic feedback which are still suitable for automotive to become very special.

2   Requirement Considerations

The design requirements for an actuator to create precise and strong haptic feedback are not at all obvious. Analysis of the performance of real pushbuttons indicate, that the actuator has to create extremely sharp oscillations (>3G) which are short in time (damped, fig. 2 left). In addition a mechanical switch always operates at a certain pretension in force (e.g. 2. to 5 N) and after a clearly defined path of travel (e.g. 0.25 to 0.5 mm) (fig. 2 right). Accordingly the functional elements for an actuator to create haptic feedback on a decorative surface can be identified as follows:

- Force-source to create an acceleration of a mass weighting approx. 100 grams
- Parallel guidance to match the force-displacement curve independent from the area touched
- Force-sensing mechanism to guarantee a reproducible switching point

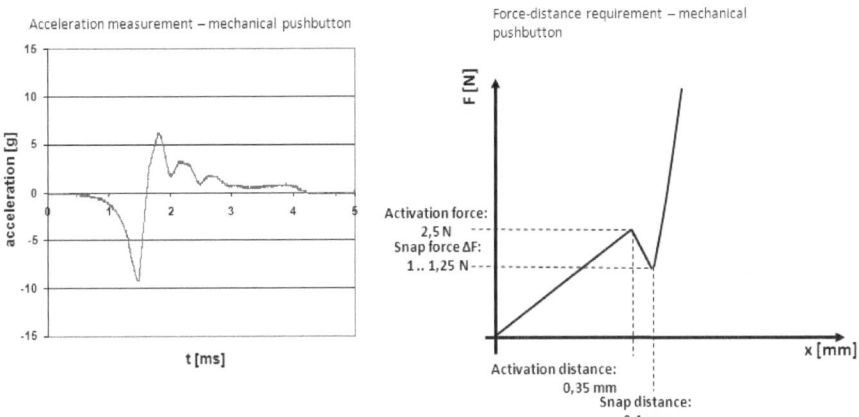

Fig. 2. Acceleration - measurement and force-distance-requirements of a mechanical pushbutton

For all components packaging requirements are formulated to limit thickness of the device to below 6 mm.

For the actuator subject in this paper an actuation system compromising of the following components had been designed:

- Electromagnetic actuator based on punch-bended pole- and anchor with PCB based coils
- Parallel guidance mechanism based on flexible metallic hinges
- Optical SMD displacement sensors

3   Actuation Principle

3.1   Electromagnetic System

The magnetic system consists basically of two steel plates and one PCB located between the steel plates. The PCB contains the electromagnetic coils, the coil cores are supplied by one of the steel plates. Thus a simple yet effective electromagnet is formed (fig. 3). This system offers significant advantages compared with a conventional electromagnet. The inductance of the coils is small, thus allowing for very short raise and fall times and very short magnetic pulses. The PCB-based coils are capable of carrying effective current densities   in the range of 10 A/mm^2 over the whole PCB thickness, while the direct connection between PCB and supportive metal parts ensures an effective heat transfer and thus prevents overheating of the coil system.

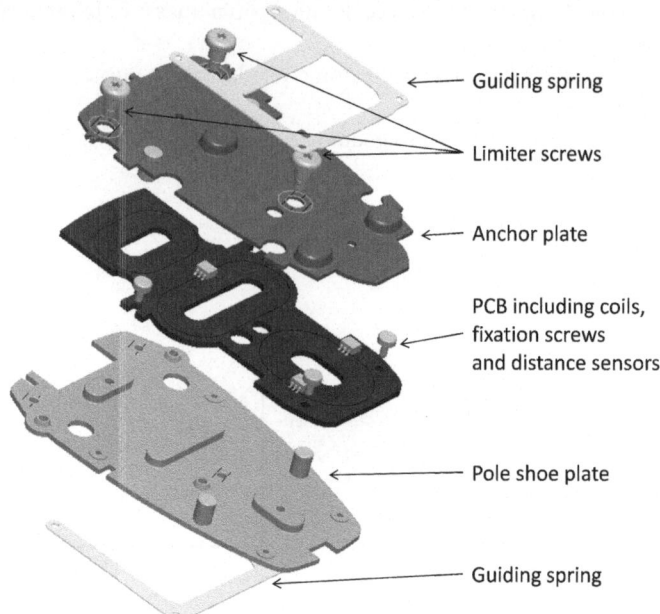

Fig. 3. Basic actor design – PCB between two steel plates. The springs represents the elastic component of the module, at the same time they provide the parallel guiding system.

Distance / Force measurement

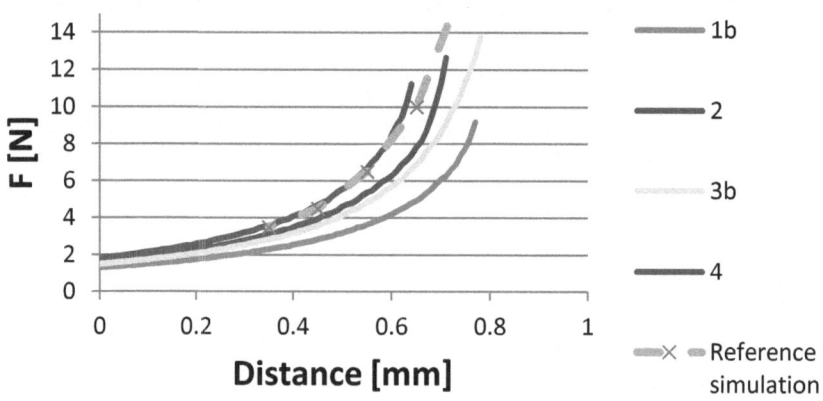

Fig. 4. Distance / magnetic force measurement using pole shoe plates out of different manufacturing processes – sample 4 is closest to CAD target. Spring is excluded from measurement, thus only the magnetic snap force ΔF (see Fig. 2 b) depending on distance is measured. Distance is measured from mechanical rest position.

Electromagnetic simulation was used to calculate expected force displacement curves. Measurements showed a good agreement between these calculated curves and real measurements (fig. 4). The influence of mechanical tolerances and magnetic imperfections were included into the simulation of the system, and after several optimization loops the magnetic system was finalized to be manufactured by well-established, cost-efficient processes capable technologies suitable for automotive use. The current density / force displacement curves allow for significant dynamic ranges in feedback both due to force profile and acoustic feedback options.

3.2   Parallel Guidance

Customers explicitly ask for a homogeneous movement of the surface, with a preferred movement range of less than 0.3 mm. The intention behind this is to visually inhibit actual movement of the input device. For a high-quality feel of the context-sensitive feedback it is required to activate the haptic pulse at the same user input force anywhere inside the active area. Both aspects are achieved by a parallel guiding system free from play (fig. 3 and 5b). Two parallel springs guide the anchor plate with respect to the pole shoe plate. At the same time they provide the return force for the actuator module. The transfer rod between the two spring arms transfers a share of the force to the other side of the guiding system to prevent a rotation and tilting of anchor the anchor plate and stabilize the parallel guiding system. As a result homogeneous force-displacement curves are reached almost independent of the location on the actuator surface (fig. 5). The system leads to an almost vertical movement of the anchor plate, and a homogeneous force at the switching point over the active area.

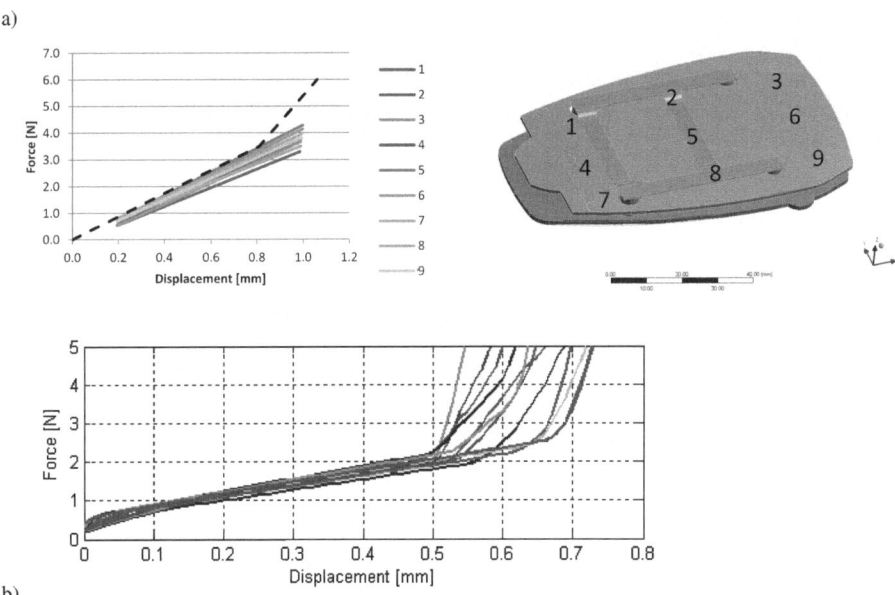

Fig. 5. FEM-results of simulated force displacement curves at several locations on actuator surface (a) compared with real measurement results (b). Only mechanical spring force is shown, snap force ΔF (see Fig. 2b) is excluded.

3.3   Sensors

Sensing the user input is a crucial issue within the whole system. Additionally, the automotive context requires a high robustness against electromagnetic interference, operation in a broad temperature range and reliability during a lifetime which is much higher than of any consumer device. In contrast, for a valuable haptic impression sensors with a high sensitivity and precision are necessary. Table 1 summarizes these requirements. An user gets a constant and valuable impression at repeated inputs if the travelling of the surface varies less than 10% around the predefined switching point. In combination with the small distance the surface can travel at all (see sec. 2) and the resolution of the ADC of the used µC the sensitivity of the sensor has to be better than 700 mV/mm. To detect fast movements the sensor has to provide a cutoff frequency above 500 Hz.

Table 1. Sensor requirements

Parameter	Value
tolerance switching point	± 10%
Sensitivity	700 mV/mm
cutoff frequency	> 500 Hz

To fulfill all of these different and opposed requirements as much as possible an integrated reflective interrupter is used in this haptic input device. The most important advantages are: The optical functional principle of the distance measurement does not interfere with any electromagnetic fields; the sensitivity can be adjusted by two resistors R_F and R_L (Fig. 6a) and the design of the reflective surface (Fig. 6b). The elevation of the reflective surfaces from the anchor plate level controls the operating point of the sensor. Hence, the output of the sensor is almost linear for the whole movement of the input surface (Fig. 7).

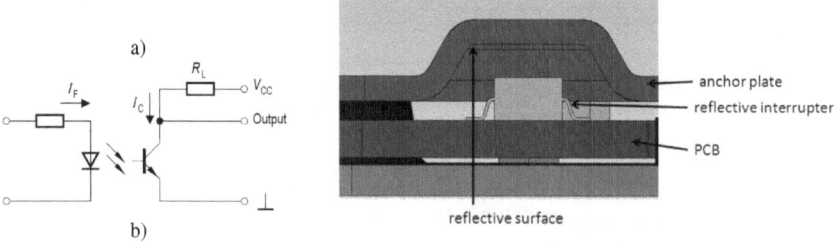

Fig. 6. Equivalent circuit of the reflective interrupter (a); Assembly of sensor and reflective surface (b)

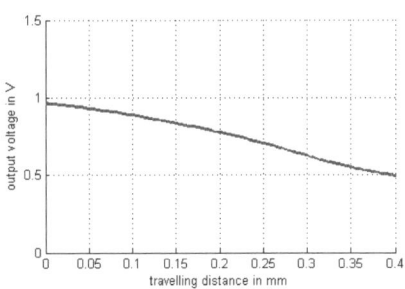

Fig. 7. Sensor output characteristics

4   Performance

4.1   Acceleration Amplitude Range

Simulation results lead to the expectation that the maximum acceleration achievable is mostly determined by the air gap remaining at the switching point. Accelerations of more than 10 G are achievable. Smaller accelerations can be controlled by limiting the available current. Initial customer evaluations [unpublished, confidential] showed that this performance exceeds the actual needs. A nominal acceleration of 3 G corresponds to user expectations (fig. **8**) in the intended context.

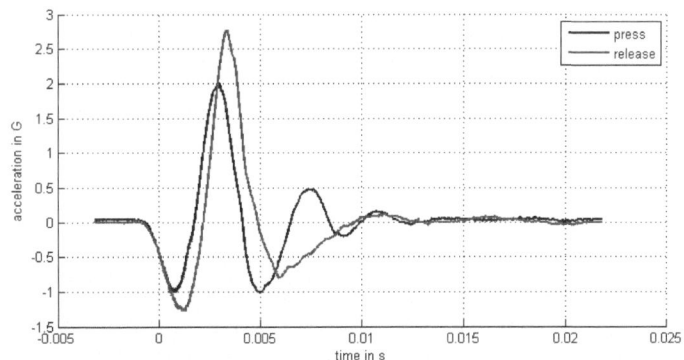

Fig. 8. Examples of haptic output characteristic with a real finger as a load

4.2   Impulse Response Times

To correlate the haptic response with the user input action it is required to match input action and response to a time window of less than 10 msec [8] As a result of the low inductance design current raise times of 1 to 2 msec are achievable (fig. 9), the correlation between input action and haptic feedback thus is determined by processing speed of the controlling hard- and software.

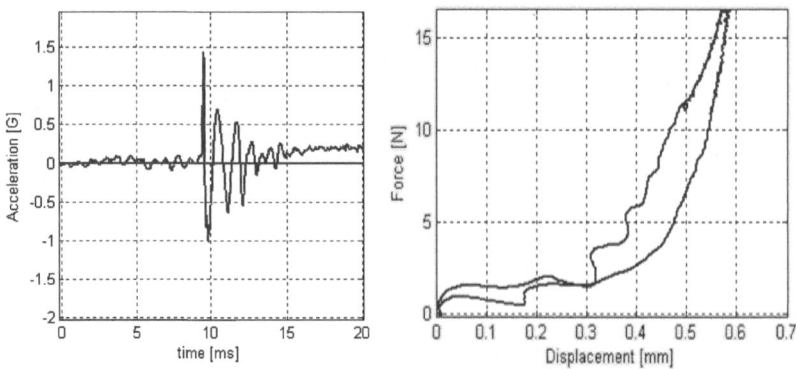

Fig. 9. Example of an acceleration measurement (left), and dynamic force-displacement curve (right) measured at 15 mm/sec in an artificial measurement setup with stiff contact element

5   Application

This actuator is subject to be used in high volume series-production touchpad-like device, which offers input options similar to a mechanical keypad as well as working

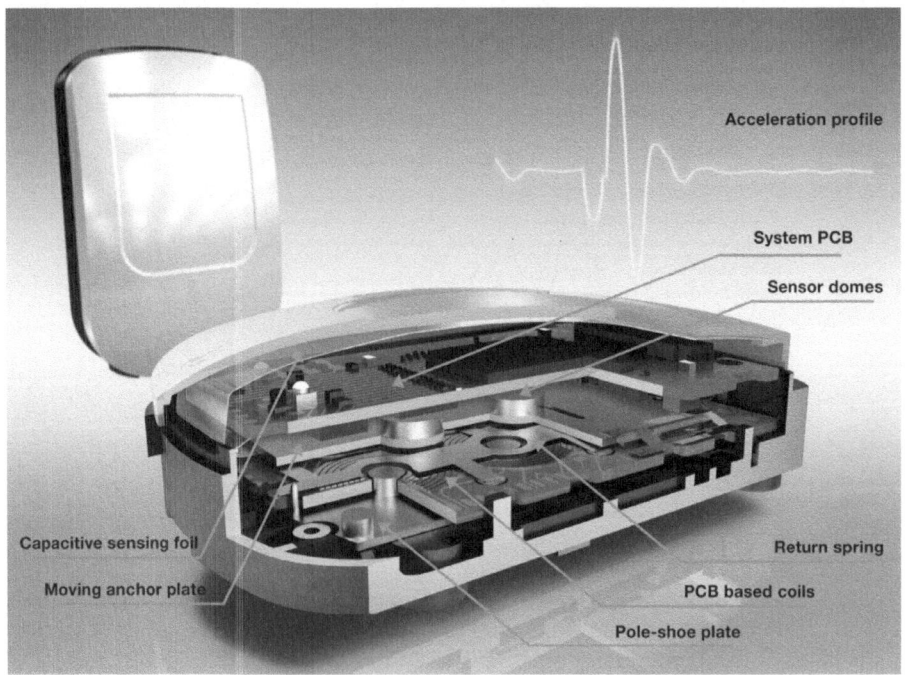

Fig. 10. Sample application for our actuator module - early design stage

like a conventional computer touchpad. The advantages compared with purely passive solutions are a clear tactile feedback for key-like inputs where a tactile feedback would be expected, while at the same time avoiding tactile feedback for gesture-like or invalid inputs. Thanks to the robust and cost-effective design it is especially suited for cost-sensitive automotive applications.

6   Summary

A novel actuation unit for high fidelity haptic feedback in an automotive context had been presented. The actuator itself is based on a classic electromagnetic system being designed to a thin package. The general technical framework for these kinds of system with requirements on parallel guidance and sensor technology were sketched. The overall performance of the device was presented based on acceleration measurements in conjunction with power requirements. An outlook was formulated showing a design study illustrating the intention of the design in a series-product context.

References

1. Morioka, M., Griffin, M.J.: Frequency dependence of perceived intensity of steering wheel vibration: effect of grip force. In: Second Joint EuroHaptics Conference 2007 and Symposium on Haptic Interfaces for Virtual Environment and Teleoperator Systems, World Haptics 2007, March 22-24, pp. 50–55 (2007)
2. Mulder, M., Abbink, D.A., van Paassen, M.M., Mulder, M.: Design of a Haptic Gas Pedal for Active Car-Following Support. IEEE Transactions on Intelligent Transportation Systems 12(1), 268–279 (2011)
3. Morrell, J., Wasilewski, K.: Design and evaluation of a vibrotactile seat to improve spatial awareness while driving. In: 2010 IEEE Haptics Symposium, March 25-26 (2010)
4. Kern, T.A. (ed.): Engineering Haptic Devices, p. 472. Springer (2009) ISBN 978-3-540-88247-3
5. Makino, Y., Maeno, T.: Dual Vibratory Stimulation for Mobile Devices. In: 2011 IEEE World Haptics Conference (WHC), June 21-24 (2011)
6. Bau, O., Poupyrev, I., Israr, A., Harrison, C.: TeslaTouch: electrovibration for touch surfaces. In: Proceedings of the 23nd Annual ACM Symposium on User Interface Software and Technology (UIST 2010), pp. 283–292. ACM, New York (2010)
7. Marchuk, N.D., Colgate, J.E., Peshkin, M.A.: Friction measurements on a Large Area TPaD. In: 2010 IEEE Haptics Symposium, March 25-26, pp. 317–320 (2010)
8. Kaaresoja, T., Anttila, E., Hoggan, E.: The effect of tactile feedback latency in touchscreen interaction. In: 2011 IEEE World Haptics Conference (WHC), June 21-24, pp. 65–70 (2011)

Erratum: Development of Intuitive Tactile Navigational Patterns

Christos Giachritsis[1,*], Gary Randall[1], and Samuel Roselier[2]

[1] BMT Group Ltd, Goodrich House, 1 Waldegrave Road, Teddington,
Middlesex, TW11 8LZ, UK
{cgiachritsis,grandall}@bmtmail.com
[2] CEA, LIST, Sensory & Ambient Interfaces Lab.
18 Route du Panorama, 92265 Fontenay-aux-Roses CEDEX, France
samuel.roselier@gmail.com

P. Isokoski and J. Springare (Eds.): EuroHaptics 2012, Part I, LNCS 7282, pp. 136–147, 2012.
© Springer-Verlag Berlin Heidelberg 2012

DOI 10.1007/978-3-642-31401-8_57

The Acknowledgments section is missing from the paper starting on page 136 of this publication. This should read as follows:

Acknowledgments. This work was supported by the EC as part of the HaptiMap FP7 project (ICT-2007.7.2/224675).

The original online version for this chapter can be found at
http://dx.doi.org/10.1007/978-3-642-31401-8_13

Author Index